physica status solidi (b)

ISSN 0370-1972 · SUBJECT AND AUTHOR INDEX (b) VOLUMES 85 TO 90 · 1978

ISSN 0031-8965 · SUBJECT AND AUTHOR INDEX (a) VOLUMES 45 TO 50 · 1978

physica status solidi (a)

KADEMIE-VERLAG · BERLIN

Classification Scheme

1. Structure of Crystalline Solids
   - 1.1 Perfectly Periodic Structure
   - 1.2 Solid-State Phase Transformations
   - 1.3 Alloys. Metallurgy
   - 1.4 Microstructure (Magnetic Domains See 18; Ferroelectric Domains See 14.4.1)
   - 1.5 Films
   - 1.6 Surfaces
2. Non-Crystalline State
3. Crystal Growth
4. Bonding Properties
5. Mössbauer Spectroscopy
6. Lattice Dynamics. Phonons
7. Acoustic Properties
8. Thermal Properties
9. Diffusion
10. Defect Properties (Irradiation Defects See 11)
    - 10.1 Metals
    - 10.2 Non-Metals
11. Irradiation Effects (X-Ray Diffraction Investigations See 1 and 10)
12. Mechanical Properties (Plastic Deformations See 10)
    - 12.1 Metals
    - 12.2 Non-Metals
13. Electron States
    - 13.1 Band Structure
    - 13.2 Fermi Surfaces
    - 13.3 Surface and Interface States
    - 13.4 Impurity and Defect States
    - 13.5 Elementary Excitations (Phonons See 6)
      - 13.5.1 Excitons
      - 13.5.2 Plasmons
      - 13.5.3 Polarons
      - 13.5.4 Magnons
14. Electrical Properties. Transport Phenomena
    - 14.1 Metals. Semi-Metals
    - 14.2 Superconductivity. Superconducting Materials and Devices
    - 14.3 Semiconductors
      - 14.3.1 Films
      - 14.3.2 Surfaces and Interfaces
      - 14.3.3 Devices. Junctions (Contact Problems See 14.3.4)
      - 14.3.4 High-Field Phenomena, Space-Charge Effects, Inhomogeneities, Injected Carriers (Electroluminescence See 20.3; Junctions See 14.3.3)
    - 14.4 Dielectrics
      - 14.4.1 Ferroelectrics
15. Thermoelectric and Thermomagnetic Properties
16. Photoconductivity. Photovoltaic Effects
17. Emission of Electrons and Ions
    - 17.1 Field Emission Microscope Investigations
18. Magnetic Properties
    - 18.1 Paramagnetic Properties
    - 18.2 Ferromagnetic Properties
      - 18.2.1 Ferromagnetic Films
    - 18.3 Ferrimagnetic Properties
    - 18.4 Antiferromagnetic Properties

(Continued on cover three)

physica status solidi (b)

basic research

physica status solidi (a)

applied research

SUBJECT AND AUTHOR INDEX

Volumes 85 to 90, 1978 Volumes 45 to 50, 1978

ISSN 0370-1972/0031-8965

Edited by

E. BORCHARDT

AKADEMIE-VERLAG · BERLIN

Subscriptions and orders for this Subject and Author Index
or single copies of physica status solidi (b) and physica status solidi (a) should be directed
in the DDR:

to the Postzeitungsvertrieb,
to a book-shop, or
to the Akademie-Verlag, DDR-108 Berlin, Leipziger Straße 3 – 4;

in the other socialist countries:
to a book-shop for foreign language literature or
to the competent news-distributing agency;

in the BRD and BERLIN (WEST):
to a book-shop or
to the wholesale distributing agency Kunst und Wissen, Erich Bieber, D-7000 Stuttgart 1, Wilhelmstr. 4 – 6;

in ÖSTERREICH:
to the Globus-Buchvertrieb, A-1201 Wien, Höchstädtplatz 3;

in the other Western European countries:
to Kunst und Wissen Erich Bieber GmbH, CH-8008 Zürich, Dufourstr. 51;

in USA and CANADA
to Verlag Chemie International, Inc., 175 Fifth Avenue, New York, N.Y. 10010, USA;

in other countries:
to the international book- and journal-selling trade,
to Buchexport, Volkseigener Außenhandelsbetrieb der Deutschen Demokratischen Republik,

DDR-710 Leipzig, Postfach 160, or
to the Akademie-Verlag, DDR-108 Berlin, Leipziger Straße 3 – 4.

Herausgeber und verantwortlich für den Inhalt des Registers:
Dipl.-Phys. E. Borchardt, DDR-102 Berlin, Neue Schönhauser Straße 20.

Verlag:
Akademie-Verlag, DDR-108 Berlin, Leipziger Straße 3 – 4; Fernruf 2 23 62 21 und 2 23 62 29; Telex-Nr. 114 420;
Bank: Staatsbank der DDR, Berlin, Kto.-Nr. 6836-26-20712.

Veröffentlicht unter der Lizenznummer 1310 und 1620 des Presseamtes beim Vorsitzenden des Ministerrates der Deutschen Demokratischen Republik.

Gesamtherstellung:
VEB Druckerei „Thomas Müntzer", DDR-582 Bad Langensalza.

Bestellnummer des Registers 1068/85 R 11.

Printed in the German Democratic Republic.
AN(EDV) 20 831/20 735

Preface

The eleventh index volume contains volumes 85 to 90 of "physica status solidi (b) -- basic research" as well as volumes 45 to 50 of "physica status solidi (a) — applied research", i.e. all volumes having appeared in 1978 containing 1579 papers.

The general scheme of the last index volume was maintained. The basis for quoting is the subject classification given for every paper in the respective issue. Nevertheless, before taking over these subject classifications into the index volume they were once more carefully revised and in some cases improved and completed in order to get a most uniform and clear classification for all papers. The average number of quotations per paper in this index volume is nearly the same as in the last one, namely about 3.0.

Again I have to express my thanks to the readers for their interest, and I will always readily take up any critical remarks and proposals for improvements. I am especially obliged to Prof. Dr. P. Görlich, Prof. Dr. E. Gutsche, Dr. H.-J. Hänsch, Dr. H. Lange, Dr. S. Oberländer, and Dipl. Phys. K. Wannewitz for giving me advice and unselfish help. Furthermore I have to thank Mrs. E. Schwan for her exact and reliable secretary's work at the manuscript.

Berlin, January 1979

Eva Borchardt

Classification Scheme

1. Structure of Crystalline Solids
   1.1 Perfectly Periodic Structure
   1.2 Solid-State Phase Transformations
   1.3 Alloys. Metallurgy
   1.4 Microstructure (Magnetic Domains See 18; Ferroelectric Domains See 14.4.1)
   1.5 Films
   1.6 Surfaces
2. Non-Crystalline State
3. Crystal Growth
4. Bonding Properties
5. Mössbauer Spectroscopy
6. Lattice Dynamics. Phonons
7. Acoustic Properties
8. Thermal Properties
9. Diffusion
10. Defect Properties (Irradiation Defects See 11)
    10.1 Metals
    10.2 Non-Metals
11. Irradiation Effects (X-Ray Diffraction Investigations See 1 and 10)
12. Mechanical Properties (Plastic Deformations See 10)
    12.1 Metals
    12.2 Non-Metals
13. Electron States
    13.1 Band Structure
    13.2 Fermi Surfaces
    13.3 Surface and Interface States
    13.4 Impurity and Defect States
    13.5 Elementary Excitations (Phonons See 6)
        13.5.1 Excitons
        13.5.2 Plasmons
        13.5.3 Polarons
        13.5.4 Magnons
14. Electrical Properties. Transport Phenomena
    14.1 Metals. Semi-Metals
    14.2 Superconductivity. Superconducting Materials and Devices
    14.3 Semiconductors
        14.3.1 Films
        14.3.2 Surfaces and Interfaces
        14.3.3 Devices. Junctions (Contact Problems See 14.3.4)
        14.3.4 High-Field Phenomena, Space-Charge Effects, Inhomogeneities, Injected Carriers (Electroluminescence See 20.3; Junctions See 14.3.3)
    14.4 Dielectrics
        14.4.1 Ferroelectrics
15. Thermoelectric and Thermomagnetic Properties
16. Photoconductivity. Photovoltaic Effects
17. Emission of Electrons and Ions
    17.1 Field Emission Microscope Investigations

18. Magnetic Properties
    18.1 Paramagnetic Properties
    18.2 Ferromagnetic Properties
        18.2.1 Ferromagnetic Films
    18.3 Ferrimagnetic Properties
    18.4 Antiferromagnetic Properties

19. Magnetic Resonance

20. Optical Properties
    20.1 Spectra. Optical Constants (X-Ray Spectra See 20)
    20.2 Lasers
    20.3 Luminescence (X-Ray Spectra See 20)

21. Metals. Alloys
    21.1 Transition Metals (Group Sc to Zn) and Their Alloys
        21.1.1 Fe and Fe Alloys
    21.2 Alkali Metals
    21.3 Alkali-Earth Metals
    21.4 Rare-Earth Metals
    21.5 Actinides
    21.6 Noble Metals
    21.7 Semi-Metals (Semiconducting Alloys See 22.7)

22. Semiconductors and Ionic Crystals
    22.1 Elements
        22.1.1 Germanium
        22.1.2 Silicon
        22.1.3 Group VI Elements
    22.2 II–V Compounds
        22.2.1 Arsenides
        22.2.2 Phosphides
        22.2.3 Antimonides
        22.2.4 Mixed Crystals
    22.3 IV–IV, II–V, and III–VI Compounds
        22.3.1 Mixed Crystals
    22.4 II–VI Compounds
        22.4.1 Sulphides
        22.4.2 Selenides
        22.4.3 Tellurides
        22.4.4 Mixed Crystals
    22.5 Halides
        22.5.1 Silver Halides
        22.5.2 Alkali Halides
        22.5.3 Alkali-Earth Halides
        22.5.4 Mixed Crystals
    22.6 Simple Oxides (CdO, ZnO See 22.4)
        22.6.1 Mixed Crystals
    22.7 Semiconducting Intermetallic Compounds
    22.8 Tertiary and Higher Compounds
        22.8.1 Salts of Oxy-Acids (Carbonates, Phosphates, Silicates Including Mica etc.)
        22.8.2 Spinels, Garnets, Ferrites
    22.9 Organic Semiconductors

23. Solidified Gases

For comments on the use of the classification scheme see p. 6.

Some Comments on the Use of the Subject and Author Index

The classification system[1]) comprises 20 broad sections each devoted to a major field of solid state physics and 3 broad sections devoted to the main groups of substances. A section may contain a number of subsections dealing with aspects of this field or special substance, respectively, which tend to receive particular attention. A paper dealing with this field will either be classified under one or more of the subsections or, if none of these are relevant, it will just appear under the general section heading (i.e. sections and subsections are equivalent in this sense). For example, a paper dealing with problems of paramagnetism would appear under 18.1 but not 18, whereas a paper on diamagnetism would be quoted under 18; a paper dealing with GaP would appear under 22.2.2, not under 22 or 22.2, whereas a paper on AlN would be quoted under 22.2, and a paper on a II–IV or a IV–VI compound would be quoted under 22.

All papers will be quoted under each relevant section. There is made a distinction between one or two principal entries (refering to the main subject matter of the paper) and secondary entries (refering to subsidiary subject matter and substance, respectively). In the principal entries the title of the paper is given in its original language and in English. In addition, every principal entry is marked by * in front of the title of the paper. In the secondary entries the title is only given in English.

Each entry gives the volume, the number of the issue, the initial and final page numbers, and the year in which the paper appeared. A paper of physica status solidi (b) or physica status solidi (a) is marked by (b) or (a) in front of the line, respectively. The letter "K" before the page numbers indicates a short note published in the offset part of the journal. Each entry is followed by references to the other entries. These references give the subject classifications and the pages in the index where the other entries appear. The references to the principal entries are given in italics.

Review articles appear at the head of the list of entries under a given subject and are given the subtitle "Review Articles". These articles are listed in the chronological order of their publication, irrespective of being a paper of physica status solidi (b) or physica status solidi (a). The subject classification for a review article consisting of several parts is based on the article as a whole rather than on the individual parts.

The entries for original papers, short notes, and errata are made together in a single list for each subject and this follows the entries for review articles. These are again ordered according to data of publication, irrespective of being a paper of physica status solidi (b) or physica status solidi (a). Original papers are sometimes published in several parts under the same principal title (in general with special subtitles). The various parts of papers of this type are treated separately and each part is classified according to its particular contents.

In quoting errata the following rule is obeyed. If the erratum refers to a paper quoted in the preceding index, it is listed in the chronological order of its publication, otherwise the erratum is quoted together with the paper to which it refers.

---

[1]) See cover two and three, and pages 4 and 5. For the case of losing the cover by bookbinding we repeat the classification scheme on pages 4 and 5.

In using the subject index attention should be given to the following details:

subject	quoted in	remarks
representations of space groups	1 and 13.1	
Alfven waves, helicon waves	13 and 20	
hot electrons	13 and 14.3.4	
transport phenomena	14 or subsection of 14	if not clearly to be assigned to 8 or 15
galvanomagnetic effects	14 or subsection of 14	not in 18
electro-optical effects	14 and 20 (or subsection of 14 and 20)	
piezoelectricity, electrostriction	14 or subsection of 14	not in 12 or subsection of 12; different electromechanical effects are quoted in 12 or subsection of 12
ionic conduction	14.4	not in 14.3
magneto-optical effects	18 and 20 (or subsection of 18 and 20)	possibly in additional sections, e. g. Faraday effect in 13
ferromagnetic films	18.2.1	not additionally in 1.5
X-ray spectra	20	but not X-ray studies of structure
MIS and MOS structures	the section assigned to the semiconductor	not in the section assigned to the oxide or the metal
MIM structures	the section assigned to the insulator	generally not in the section assigned to the metal
epitaxial layers	the section assigned to the layer	not in the section assigned to the substrate

The author index gives the names of the authors in alphabetical order. Each name is followed by the bibliographic details of the paper and the numbers of the pages on which the paper appears in the subject index. A paper of physica status solidi (b) or physica status solidi (a) is again marked by (b) or (a) in front of the number of the volume, respectively. The principal entry is again given in italics. When several papers have been published by the same author they are arranged according to the dates of publication, irrespective of being a paper of physica status solidi (b) or physica status solidi (a).

Subject Index

1. Structure of Crystalline Solids

Original Papers and Short Notes

M. V. Eremin and Yu. V. Rakitin
Channel Model in Isotropic Exchange Theory (III)
(b) Vol. **85**, No. 2, p. 783–788 (1978).
See also *13*, p. *143*; 22.5.4, p. 489.

R. Ciszewski, A. Jabłonka, A. Sobaszek, and D. Sikorska
*Temperature Shift of the Maximum of the Neutron Critical Scattering in an Iron Monocrystal
(b) Vol. **85**, No. 2, p. K73–K75 (1978).
See also 18.2, p. 282; 21.1.1, p. 385.

R. Ciszewski, A. Jabłonka, and A. Sobaszek
*Experimental Anisotropy of the Neutron Critical Scattering from a Ferromagnet Monocrystal
(b) Vol. **85**, No. 2, p. K77–K79 (1978).
See also 18.2, p. 282; 21.1.1, p. 385.

M. W. Muller
*Site Ordering and Magnetostriction in Garnet (II): Iron
(b) Vol. **86**, No. 1, p. 345–348 (1978).
See also *18.3*, p. *295*; 22.8.2, p. 518.

R. Lenk
*Multiple Scattering in Condensed Materials (I): The Embedding Field and the Effective Medium
(b) Vol. **86**, No. 2, p. 603–613 (1978).
See also 13, p. 144.

V. V. Gagulin, N. V. Fadeeva, A. G. Belous, A. V. Titov, K. P. Mitrofanov, M. V. Plotnikova, S. P. Soloviev, and Yu. N. Venevtsev
*New Seignette-Magnets with Hexagonal Barium Titanate Structure
(a) Vol. **48**, No. 1, p. 183–189 (1978).
See also 5, p. 61; *14.4.1*, p. *255*; 18.2, p. 289; 18.4, p. 301; 19, p. 310; 22.8.1, p. 514.

H. Kiess and R. Clarke
*Pyroelectric Effect on PTS
(a) Vol. **49**, No. 1, p. 133–136 (1978).
See also *14.4*, p. *250*; 15, p. 261; 22.9, p. 527.

M. V. EREMIN, V. N. KALINENKOV, and YU. V. RAKITIN
Two-Centre Exchange Interactions between Orbitally Degenerate Ions (I): One Ion in Orbitally Degenerate State
(b) Vol. **89**, No. 2, p. 503–512 (1978).
See also *13*, p. *147*.

H. BERTAGNOLLI and H. G. HERTZ
*Preservation and Loss of Structural Features of Solid Acetic Acid during the Melting Process
(a) Vol. **49**, No. 2, p. 463–472 (1978).
See also *2*, p. *50*; 22.9, p. 527.

M. V. EREMIN, V. N. KALINENKOV, and YU. V. RAKITIN
Two-Centre Exchange Interactions between Orbitally Degenerate Ions (II): Two Ions in Orbitally Degenerate States
(b) Vol. **90**, No. 1, p. 123–134 (1978).
See also *13*, p. *148*.

P. M. VAN DEN BROEK and J. F. CORNWELL
*Clebsch-Gordan Coefficients of Symmetry Groups
(b) Vol. **90**, No. 1, p. 211–224 (1978).
See also *13*, p. *148*.

K. LOCHNER, TH. HINRICHSEN, W. HOFBERGER, and H. BÄSSLER
*Polymerization of a Polydiacetylene Single Crystal under Hydrostatic Pressure
(a) Vol. **50**, No. 1, p. 95–100 (1978).
See also 22.9, p. 528.

R. E. VANDENBERGHE
*Comment on Neutron Diffraction Studies of the Spinel Oxide $CuMn_2O_4$
(a) Vol. **50**, No. 1, p. K85–K88 (1978).
See also 22.8.2, p. 521.

A. M. ALTSHULER, YU. KH. VEKILOV, and G. R. UMAROV
Characteristic Features of the Electronic Spectrum and the Stability of Ternary Diamond-Like Semiconductors
(b) Vol. **90**, No. 2, p. 733–740 (1978).
See also *13.1*, p. *161*; 22, p. 421.

S. TSUNEKAWA and H. TAKEI
*Twinning Structure of Ferroelastic $LaNbO_4$ and $NdNbO_4$ Crystals
(a) Vol. **50**, No. 2, p. 695–702 (1978).
See also 10.2, p. 122; 14.4.1, p. 257; 22.8.1, p. 516.

1.1 Perfectly Periodic Structure

Original Papers and Short Notes

V. K. Garg and M. P. Verma
*Erratum to: Evaluation of Some Important Lattice Sums on NaCl and CsCl Structures
(b) Vol. **85**, No. 1, p. 417 (1978).
See also 12.2, p. 138; 22.5.2, p. 478.

F. Fujimoto
*Periodicity of Crystal Structure Images in Electron Microscopy with Crystal Thickness
(a) Vol. **45**, No. 1, p. 99 – 106 (1978).

Z. Janáček and J. Kuběna
*Integrated Intensity of the Laue Diffraction from an Elastically Bent Crystal
(a) Vol. **45**, No. 1, p. K55 – K57 (1978).
See also 22.1.2, p. 429.

H. D. Carstanjen, J. Dünstl, G. Löbl, and R. Sizmann
*Lattice Location and Determination of Thermal Amplitudes of Deuterium in α-$PdD_{0.007}$ by Channeling
(a) Vol. **45**, No. 2, p. 529 – 536 (1978).
See also 11, p. 124; 21, p. 354.

C. Boulesteix, G. Nihoul-Boutang, and B. Yangui
*Influence of Upper Laue Areas on Fine Structure of Bend Contours near a High Symmetry Axis
(a) Vol. **45**, No. 2, p. 639 – 645 (1978).
See also 1.4, p. 29; 22.6, p. 491.

D. Gignoux and H. Asmat
*Magnetic Properties and Structures of TbAlGa and HoAlGa
(a) Vol. **45**, No. 2, p. K149 – K153 (1978).
See also *18.4*, p. *298*; 21, p. 355; 21.4, p. 396.

M. V. Kruglov, V. N. Shchemelev, and G. G. Kareva
*Detection of Atomic Plane Displacements in Near-Surface Layers of Crystals Using the Angular Dependence of Photoemission during Bragg Diffraction of X-Rays
(a) Vol. **46**, No. 1, p. 343 – 350 (1978).
See also 1.6, p. 40; 17, p. 269; 22.1.2, p. 431.

G. J. Ashwell
*Organic Metals: Chain Stabilisation in Bipyridinium-TCNQ Salts
(b) Vol. **86**, No. 2, p. 705 – 715 (1978).
See also 14, p. 203; 22.9, p. 524.

V. K. Bashenov, A. M. Mutal, and V. V. Timofeenko
*Interaction Parameters in III–V Semiconductor Solid Solutions
(b) Vol. **86**, No. 2, p. K139 – K142 (1978).
See also 1.2, p. 17; 22.2.4, p. 456.

J. van Landuyt, G. A. Wiegers, and S. Amelinckx
*A New Type of Deformation Modulated Superstructure in 1T-VSe_2 and Its Relation with Other Superstructures in Transition Metal Dichalcogenides
(a) Vol. **46**, No. 2, p. 479–492 (1978).
See also 1.2, p. 17; 22, p. 412.

M. Polcarová
*X-Ray Diffraction on a Crystal Containing a Misfit Boundary (I): General Solution
(a) Vol. **46**, No. 2, p. 567–575 (1978).

J. Beauvillain
*Diffusion des électrons par un cristal. Introduction d'une notion de cohérence. Application aux diagrammes de lignes
(Electron Scattering in Crystals. Introduction of the Notion of Coherence. Application of Line Diagrams)
(a) Vol. **46**, No. 2, p. 659–670 (1978).

S. Siek, A. Szytuła, and J. Leciejewicz
*The Crystal and Magnetic Structure of $CeMn_2Si_2$
(a) Vol. **46**, No. 2, p. K101–K105 (1978).
See also *18.4*, p. *299*; 22.8, p. 504.

O. Brümmer, H.-R. Höche, and J. Nieber
*Some Remarks on Extremely Asymmetric X-Ray Diffraction in the Laue Case of Grazing Incidence
(a) Vol. **46**, No. 2, p. K131–K133 (1978).
See also 22.1.2, p. 432.

I. P. Jones
*Systematic Electron Diffraction by H.C.P. Metals (I): Theory and Main Beam Transmission
(a) Vol. **47**, No. 1, p. 163–178 (1978).
See also 21, p. 358.

M. Polcarová
*X-Ray Diffraction on a Crystal Containing a Misfit Boundary (II): Special Cases
(a) Vol. **47**, No. 1, p. 179–186 (1978).

B. Yangui, M. Rekik, G. Nihoul-Boutang, C. Jouanin, C. Boulesteix, and D. Renard
*Theoretical and Experimental Study of the Transmitted Intensity as a Function of Thickness when the Incident Beam Is Parallel to a Crystallographic Axis
(a) Vol. **47**, No. 1, p. 311–320 (1978).
See also 1.4, p. 30; 22.6, p. 493.

B. Borie
*On the Observation of Forbidden Bragg Maxima for White Tin
(a) Vol. **47**, No. 1, p. K39–K42 (1978).
See also 6, p. 71; 21.7, p. 406.

D. W. Field
*Comment on "On the Observation of Forbidden Bragg Maxima for White Tin"
(a) Vol. **47**, No. 1, p. K43 (1978).
See also 6, p. 71; 21.7, p. 406.

M. David, G. van Tendeloo, J. van Landuyt, and R. Gevers
*The Critical Voltage Effect in Transmission Electron Microscopy (X): Experimental Observations in the Presence of Non–Systematic Reflections
(b) Vol. 87, No. 2, p. 419—432 (1978).
See also 22.1.1, p. 425; 22.6, p. 494.

I. P. Jones
*Systematic Electron Diffraction by H.C.P. Metals (II): The Critical Voltage Effect
(a) Vol. 47, No. 2, p. 385—395 (1978).
See also 21, p. 359.

S.-L. Chang
*On the 'Solution' Dispersion Surfaces of Two-Beam X-Ray Dynamical Bragg Reflection for Infinitely Thick Crystals
(a) Vol. 47, No. 2, p. 717—722 (1978).
See also 22.1.1, p. 426.

V. G. Baryshevskii, A. O. Grubich, and I. Ya. Dubrovskaya
*Diffraction of Radiation from Channelled Charged Particles
(b) Vol. 88, No. 1, p. 351—358 (1978).
See also *11*, p. *127*.

C. S. G. Cousins, L. Gerward, and J. Staun Olsen
*Multiple Diffraction in Crystals Studied by an X-Ray Energy-Dispersive Method
(a) Vol. 48, No. 1, p. 113—119 (1978).
See also 22.1.2, p. 434.

F. Studer et B. Raveau
*Etude par microscopie électronique des phénomènes d'intercroissance dans les siliconiobate de baryum (I):Les composés $(Ba_3Nb_6Si_4O_{26})_n$. $Ba_3Nb_4Ti_4O_{21}$ avec n compris entre $^3/_2$ et $^{29}/_2$
(Electron Microscopic Study of Intergrowth Phenomena in Barium Siliconiobates. (I): The Compounds $(Ba_3Nb_6Si_4O_{26})_n$. $Ba_3Nb_4Ti_4O_{21}$ with n between $^3/_2$ and $^{29}/_2$)
(a) Vol. 48, No. 2, p. 301—312 (1978).
See also *1.4*, p. *32*; 22.8, p. 506.

C. Meyer, Y. Gros, B. Bochu, A. Collomb, J. Chenavas, J. C. Joubert, and M. Marezio
*Synthesis, Crystal Structure, and Mössbauer Study of a Series of Perovskite-Like Compounds $[ACu_3](M, Fe)_4O_{12}$
(a) Vol. 48, No. 2, p. 581—586 (1978).
See also *5*, p. *62*; 22.8.1, p. 515.

F. Studer et B. Raveau
*Etude par microscopie électronique des phénomènes d'intercroissance dans les siliconiobates de baryum (II): Intercroissances multiples et microdomaines
(Electron Microsocopic Study of Intergrowth Phenomena in Barium Siliconiobates (II): Multiple Intergrowths and Microdomains)
(a) Vol. 49, No. 1, p. 189—205 (1978).
See also *1.4*, p. *33*; 22.8, p. 507.

G. van Tendeloo and S. Amelinckx
*A High Resolution Study of Ordering in Au_4Mn (I): The Microstructure of Fully Ordered Au_4Mn
(a) Vol. 49, No. 1, p. 337—346 (1978).
See also *1.4*, p. *33*; 21.1, p. 377; 21.6, p. 404.

A. Siems
*Quantitative Untersuchungen an Kikuchi-Strukturen (I): Energieanalyse von Intensitätsprofilen
(Quantitative Study of Kikuchi Structures (I): Energy Analysis of Intensity Profiles)
(a) Vol. **49**, No. 2, p. 621–628 (1978).
See also **6**, p. **77**; **13.5.2**, p. **199**; **22.1.2**, p. **437**.

G. van Tendeloo, R. de Ridder, and S. Amelinckx
A High Resolution Study of Ordering in Au_4Mn (II): The Evolution of Ordering in Au_4Mn
(a) Vol. **49**, No. 2, p. 655–666 (1978).
See also *1.4*, p. *33*; 21.1, p. 379; 21.6, p. 405.

Z. Janáček, J. Kuběna, and V. Holý
*X-Ray Laue Diffraction from a Bent Crystal. Integrated Intensity
(a) Vol. **50**, No. 1, p. 285–291 (1978).
See also 10.2, p. 121; 22.1.2, p. 438.

R. de Ridder, D. van Dyck, and S. Amelinckx
*Diffuse Intensity Contours Observed in Electron Diffraction Patterns of Transition Metal Tri- and Di-Chalcogenides: A Structural Interpretation
(a) Vol. **50**, No. 1, p. 349–362 (1978).
See also *1.4*, p. *34*; 22, p. 420.

A. Bombik, B. Leśniewska, A. Oleś
*Atomic and Magnetic Structure of $HoVO_3$
(a) Vol. **50**, No. 1, p. K17–K20 (1978).
See also *18.3*, p. *296*; 22.8.2, p. 520.

Z. Jirák, S. Vratislav, and P. Novák
*Study of Cubic and Tetragonal Structures in the System $Mn_xCr_{3-x}O_4$
(a) Vol. **50**, No. 1, p. K21–K24 (1978).
See also 22.8.2, p. 520.

T. Tuomi, M. Tilli, V. Kelhä, and J. D. Stephenson
*Pendellösung Fringes in Synchrotron X-Ray Topographs of a Wedge-Shaped Silicon Crystal
(a) Vol. **50**, No. 2, p. 427–431 (1978).
See also 22.1.2, p. 439.

Th. Karakostas, G. Nouet, G. L. Bleris, S. Hagege, and P. Delavignette
*Grain Boundary Analysis in TEM (I): Practical Determination of Bicrystal Orientations
(a) Vol. **50**, No. 2, p. 703–709 (1978).
See also *1.4*, p. *35*.

C. Riekel, M. Thomas, and R. Schöllhorn
*Structure Refinement on Stoichiometric $TiTe_2$ by Neutron Diffraction
(a) Vol. **50**, No. 2, p. K231–K234 (1978).
See also 22, p. 421.

1.2 Solid-State Phase Transformations

Review Article

V. S. POSTNIKOV, V. V. POSTNIKOV, and V. M. FEDOROV
Metastability and Superconductivity in Thin Films
(b) Vol. **87**, No. 1, p. 11–23 (1978).
See also *1.6*, p. *40*; *14.2*, p. *213*; 21, p. 352; 22, p. 408.

Original Papers and Short Notes

G. NEUMANN and H. VOGT
*Rayleigh Wing Scattering in Disordered Sodium Nitrate
(b) Vol. **85**, No. 1, p. 179–184 (1978).
See also 2, p. 42; *6*, p. *64*; 20.1, p. 319; 22.8.1, p. 510.

TH. NATTERMANN
Static and Dynamic Critical Behaviour of Uniaxial Ferroelectrics and the Phase Transition in TGS
(b) Vol. **85**, No. 1, p. 291–300 (1978).
See also *14.4.1*, p. *253*; 22.9, p. 522.

YU. I. KHARKATS
*Multiple Phase Transitions in Disordering Systems
(b) Vol. **85**, No. 1, p. 371–379 (1978).

K. S. ALEKSANDROV, I. P. ALEKSANDROVA, V. F. SHABANOV, V. I. YUZVAK, Y. Z. NOZIK, and L. I. FIKIN
Mechanism of Ferroelectric Phase Transition in $NaNH_4SeO_4 \cdot 2H_2O$
(a) Vol. **45**, No. 1, p. 53–62 (1978).
See also *14.4.1*, p. *253*; 19, p. 303; 20, p. 315; 22.8.1, p. 510.

E. WIESER, W. BRÜCKNER, B. THUSS, and U. GERLACH
*Mössbauer Effect Studies on Modified VO_2. $V_{1-x}Fe_xO_{2-y}$ and $V_{1-x-z}Fe_xMo_zO_2$
(a) Vol. **45**, No. 1, p. 123–131 (1978).
See also *5*, p. *60*; 18.1, p. 277; 22.6.1, p. 499.

H. DEXPERT et P. CARO
*Processus réactionnels en phase solide pour certains composés des terres rares
(Reaction Processes in the Solid Phase of Some Rare Earth Compounds)
(a) Vol. **45**, No. 1, p. 297–304 (1978).
See also 21.4, p. 396.

I. M. BELII, F. F. KOMAROV, V. S. TISHKOV, and V. M. YANKOVSKII
*Formation of Chemical Compounds by Ion Bombardment of Thin Transition Metal Films
(a) Vol. **45**, No. 1, p. 343–352 (1978).
See also *1.5*, p. *36*; 11, p. 123; 21.1, p. 369.

Č. Koňák, L. Budík, and L. Novák
*Electro-Optical Properties of NH_4Cl Single Crystals near Their Order–Disorder Phase Transition
(a) Vol. **45**, No. 1, p. K63—K66 (1978).
See also 14, p. 203; 20, p. 315; 22.5, p. 474.

F. W. Boswell, A. Prodan, W. R. Vaughan, and J. M. Corbett
*On the Ordering of Fe Atoms in Fe_xNbS_2
(a) Vol. **45**, No. 2, p. 469—481 (1978).
See also *1.3*, p. *25*; 22.8, p. 503.

T. Gotô, T. Fujimura, and K. Kamiyoshi
*Ultrasonic Investigation of the Phase Transitions in $NH_4Cl_{1-x}Br_x$ Mixed Crystals
(a) Vol. **45**, No. 2, p. 513—519 (1978).
See also *7*, p. *80*; 14.4, p. 248; 22.5.4, p. 489.

M. van Sande, G. van Tendeloo, J. van Landuyt, and S. Amelinckx
*A Study by Transmission Electron Microscopy and Diffraction of the Ternary Alloy System Cu_2AuPd
(a) Vol. **45**, No. 2, p. 553—558 (1978).
See also *1.4*, p. *29*; 21.1, p. 370.

M. Krupski and J. Stankowski
*New High-Pressure Phase in $[Ni(NH_3)_6]$ $(ClO_4)_2$ Detected by EPR
(a) Vol. **45**, No. 2, p. 585—589 (1978).
See also 19, p. 304; 22.8, p. 503.

P. Konsin
*Structural Phase Transitions of Antiferroelectric and Displacive Modulated Types Caused by Electron–Phonon Interaction
(b) Vol. **86**, No. 1, p. 57—66 (1978).
See also 6, p. 67; 14.4.1, p. 254; 22, p. 410.

F. Sacchetti, P. Bosi, F. Dupré, G. Frollani, F. Menzinger, and M. C. Spinelli
*Observation of Precursory Phonons of an F.C.C.–H.C.P. Transition
(b) Vol. **86**, No. 1, p. 77—81 (1978).
See also *6*, p. *67*; 21.1, p. 371.

J. C. Castro and M. de Souza
*The OCN^-Vibrational Spectrum and the Phase Transitions in KCN_xKCl_{1-x} Single Crystals
(b) Vol. **86**, No. 1, p. 137—144 (1978).
See also *6*, p. *67*; 12.2, p. 138; 20.1, p. 321; 22.5, p. 474; 22.5.4, p. 489.

M. Schwab
*Phase Transition in the Mixed Crystals $Tb_pY_{1-p}AsO_4$
(b) Vol. **86**, No. 1, p. 195—203 (1978).
See also 13.4, p. 172; 18.4, p. 298; 20.1, p. 322; 22.8.1, p. 511.

J. Sznajd
*Cross-Over from Discontinuous to Continuous Phase Transition under an External Magnetic Field
(b) Vol. **86**, No. 1, p. K11—K14 (1978).
See also 18.1, p. 277; 18.4, p. 298; 22.6, p. 492.

B. K. Chakrabarti, B. P. Barua, and S. K. Sinha
*Coupling of the Martensitic Transformation with the Superconducting Transition
(b) Vol. **86**, No. 1, p. K55—K58 (1978).
See also *14.2*, p. *214*; 21, p. 356.

V. T. Bublik and V. N. Leikin
*Calculation of the Pseudobinary Alloy Semiconductor Phase Diagrams
(a) Vol. **46**, No. 1, p. 365—372 (1978).
See also 22.2.1, p. 446; 22.2.2, p. 450; 22.2.3, p. 452.

A. T. Macrander and R. K. Crawford
*Erratum to: Density of Solid Argon at Melting
(a) Vol. **46**, No. 1, p. 397 (1978).
See also 23, p. 530.

N. N. Koren, V. V. Krasnova, and E. E. Matyas
*Investigation of Alloys of the System GeS–GeSe
(a) Vol. **46**, No. 1, p. K1—K3 (1978).
See also *1.3*, p. *25*; 22.7, p. 501.

G. C. Vezzoli
*Optical and Electrical Effects during the Solidification of Mercury
(a) Vol. **46**, No. 1, p. K49—K53 (1978).
See also 14.1, p. 206; 20.1, p. 324; 21, p. 356.

H. Vogt and G. Neumann
Hyper-Rayleigh Scattering in Disordered Sodium Nitrate
(b) Vol. **86**, No. 2, p. 615—620 (1978).
See also 6, p. 69; *20.1*, p. *324*; 22.8.1, p. 512.

F. Milia, M. Voudouris, A. Kandušer, A. Levstik, and R. Blinc
*^{75}As Nuclear Quadrupole Resonance Study of the Phase Transition in Na_2HAsO_4
(b) Vol. **86**, No. 2, p. K119—K121 (1978).
See also 14.4.1, p. 254; *19*, p. *306*; 22.8.1, p. 512.

V. K. Bashenov, A. M. Mutal, and V. V. Timofeenko
Interaction Parameters in III–V Semiconductor Solid Solutions
(b) Vol. **86**, No. 2, p. K139—K142 (1978).
See also *1.1*, p. *11*; 22.2.4, p. 456.

J. van Landuyt, G. A. Wiegers, and S. Amelinckx
A New Type of Deformation Modulated Superstructure in 1T-VSe_2 and Its Relation with Other Superstructures in Transition Metal Dichalcogenides
(a) Vol. **46**, No. 2, p. 479—492 (1978).
See also *1.1*, p. *12*; 22, p. 412.

S. Grande, H.-D. Mecke, L. A. Shuvalov
*Deuterium Resonance of $KD_3(SeO_3)_2$ Single Crystals above and below Phase Transition
(a) Vol. **46**, No. 2, p. 547—555 (1978).
See also *4*, p. *57*; 19, p. 306; 22.8, p. 504.

A. I. Otko, N. M. Nesterenko, and L. V. Povstyanyi
*Phenomenological Approach to Structural Phase Transitions in Trigonal Double Molybdates and Tungstates
(a) Vol. **46**, No. 2, p. 577—587 (1978).
See also 22.8.1, p. 512.

A. Delunas and F. Aymerich
*On Electrolytic Amorphous Antimony
(a) Vol. **46**, No. 2, p. 703—706 (1978).
See also *2*, p. *44*; 21.7, p. 406.

B. Pura and J. Przedmojski
*X-Ray Critical Scattering in $NaNO_2$
(a) Vol. **46**, No. 2, p. K115—K117 (1978).
See also 22.8.1, p. 512.

U. Roy, K. Petrov, I. Tsolovski, and P. Peshev
*Erratum to: On the Structural Phase Transition and the Thermal Expansion of Spinel Lithium Titanate
(a) Vol. **46**, No. 2, p. K171 (1978).
See also 8, p. 85; 22.8.1, p. 513.

I.O. Bashkin, E. G. Ponyatovskii, and M. E. Kost
*Phase Transformations in Hydrides of Rare-Earth Metals under Pressure (II): The Sc–H and Yb–H Systems under High Hydrogen Pressure
(b) Vol. **87**, No. 1, p. 369—372 (1978).
See also 12.1, p. 135; 21.4, p. 397.

A. Ya. Gordon
Soft Mode Contribution to the Ferroelectric Heat Capacity
(b) Vol. **87**, No. 1, p. K57—K60 (1978).
See also 6, p. 71; 8, p. 85; *14.4.1*, p. *254*.

A. Migdał-Mikuli, E. Mikuli, M. Rachwalska, and S. Hodorowicz
*Investigation of the Phase Situation in Polycrystalline $[Mg(NH_3)_6](ClO_4)_2$ by Adiabatic Calorimetry and X-Ray Diffraction Methods
(a) Vol. **47**, No. 1, p. 57—64 (1978).
See also 8, p. 85; 22.8, p. 505.

R. Kind, S. Plesko, and J. Roos
*Structural Phase Transitions in the Perovskite-Type Layer Compound $NH_3(CH_2)_3NH_3MnCl_4$
(a) Vol. **47**, No. 1, p. 233—240 (1978).
See also 19, p. 307; 22.9, p. 525.

R. Wolf, G. van Tendeloo, J. van Landuyt, and S. Amelinckx
*Evidence for the Occurrence of the AABB Structure in the Gold–Manganese System
(a) Vol. **47**, No. 1, p. 241—244 (1978).
See also 1.3, p. 25; 1.4, p. 30; 21.1, p. 373; 21.6, p. 402.

Ya. A. Kraftmakher and T. Yu. Pinegina
*Anomalies of Resistivity and Thermopower of Samarium at High Temperatures
(a) Vol. **47**, No. 1, p. K81—K83 (1978).
See also *14.1*, p. *208*; *15*, p. *260*; 21.4, p. 397.

B. I. Swanson
*Displacive Phase Transformations in K_2SnBr_6
(a) Vol. **47**, No. 1, p. K95—K98 (1978).
See also 6, p. 71; 20.1, p. 327; 22.8, p. 505.

I. G. Ratishvili
*Stepped Ordering in Binary Solid Solutions
(b) Vol. **87**, No. 2, p. 461—471 (1978).
See also 1.3, p. 26; 21, p. 359.

E. Wiszt
*A Contribution to the Phenomenological Theory of the Anomalies of the Elastic Constants near a Structural Phase Transition
(b) Vol. **87**, No. 2, p. 675—688 (1978).
See also 12.2, p. 139; 22.5, p. 475.

V. K. Bashenov, Y. K. Markolenko, and V. V. Timofeenko
*Semiconductor–Metal Transition of Diamond-Type Crystals
(b) Vol. **87**, No. 2, p. K73—K75 (1978).
See also *13.1*, p. *155*; 22.1, p. 422.

R. Blinc and F. C. de Sa Barreto
*Electric Field Induced Lifshits Point in Smectic Liquid Crystals
(b) Vol. **87**, No. 2, p. K105—K109 (1978).
See also *2*, p. *45*; 14.4, p. 249.

H. J. Bauer, M. Becker, H. Pretsch, and M. Zwick
*Magnetic Study of the Activation Energy of the Phase Transition Kinetics between Nickel and Nickel Hydride
(a) Vol. **47**, No. 2, p. 445—450 (1978).
See also 18.2, p. 287; 21.1, p. 374.

D. Bartzokas, C. Manolikas, and J. Spyridelis
*Electron Microscopic Study of the Destabilization of Stabilized γ-Phase of Indium Sesquisulphide
(a) Vol. **47**, No. 2, p. 459—467 (1978).
See also 1.4, p. 31; 22.3, p. 458.

G. van Tendeloo, R. Wolf, J. van Landuyt, and S. Amelinckx
*Electron Diffraction and High Resolution Electron Microscopic Study of Ordering in the Gold–Manganese System (I): The Orthorhombic and Monoclinic I Structure in $Au_{11}Mn_4$
(a) Vol. **47**, No. 2, p. 539—554 (1978).
See also 1.3, p. 26; 1.4, p. 31; 21.1, p. 374; 21.6, p. 403.

H. Metzger, H. Jo, S. C. Moss, and D. G. Westlake
Single Crystal X-Ray Study of the Superstructure Modulation and Long-Range Order in V_2D
(a) Vol. **47**, No. 2, p. 631—638 (1978).
See also 1.3, p. 26; 21.1, p. 374.

M. Godlewska and M. Rachwalska
*An Adiabatic Calorimetric Study of the Phase Situation in 2,2-Dinitropropane
(a) Vol. 47, No. 2, p. 661—665 (1978).
See also 8, p. 86; 22.9, p. 525.

P. Airo
*On the Superstructure of the Alloy Au–18 at%Mg
(a) Vol. 47, No. 2, p. K107—K110 (1978).
See also *1.3*, p. *26*; 21.3, p. 394; 21.6, p. 403.

Y. Kimura and M. Umeda
*Temperature Dependence of the Electrical Resistivity of Laves Phase Compounds V_2 (Hf, Zr)
(a) Vol. 47, No. 2, p. K111—K113 (1978).
See also *14.1*, p. *208*; 21, p. 359.

W. Brückner, W. Moldenhauer, and H. Oppermann
*Conductivity Transitions in (Ga + Mo)-Doped VO_2
(a) Vol. 47, No. 2, p. K129—K132 (1978).
See also *14.3*, p. *221*; 22.6, p. 494.

C. R. Whitehouse and A. A. Balchin
*Non-Stoichiometry in ZrS_2 and $ZrSe_2$
(a) Vol. 47, No. 2, p. K173—K176 (1978).
See also 22, p. 414.

V. M. Ryzhkovskii
On Structural and Magnetic Properties of the Mn_2As–Mn_2Sb Quasibinary System
(a) Vol. 47, No. 2, p. K177—K180 (1978).
See also *18.3*, p. *296*; *18.4*, p. *300*; 22.8, p. 505.

R. R. Levitskii, N. A. Korynevskii, and I. V. Stasyuk
*Distribution Functions and Thermodynamical Properties of KD_2PO_4 and $ND_4D_2PO_4$ Type Crystals
(b) Vol. 88, No. 1, p. 51—63 (1978).
See also *14.4.1*, p. *255*; 22.8.1, p. 514.

E. Heinze and S. Grande
Nuclear Spin Relaxation in the Dipolar System in Solid p-Alkoxybenzylidene-p-n-Alkylanilines
(b) Vol. 88, No. 1, p. K23—K25 (1978).
See also 6, p. 74; *19*, p. *309*; 22.9, p. 526.

R. Wolf, G. van Tendeloo, J. van Landuyt, and S. Amelinckx
Electron Diffraction and High Resolution Electron Microscopic Study of Ordering in the Gold–Manganese System (II): The Monoclinic II Structure in $Au_{11}Mn_4$ and Its Relation to Other LPAPB Phases
(a) Vol. 48, No. 1, p. 39—52 (1978).
See also 1.3, p. 27; 1.4, p. 31; 21.1, p. 375; 21.6, p. 403.

M. A. Arriandiaga, M. J. Tello, J. Fernandez, H. Arend, and J. Roos
*Calorimetric Study of Order–Disorder Phase Transitions in Long-Chain $(C_nH_{2n+1}NH_3)_2\ CuCl_4$ Compounds
(a) Vol. 48, No. 1, p. 53—57 (1978).
See also 22.8, p. 506.

M. Chabin, F. Gilletta, and C. Ridou
*Thermal Properties of $TlCdF_3$ and $RbCaF_3$ near Their Phase Transitions
(a) Vol. 48, No. 1, p. 67—70 (1978).
See also *8*, p. *87*; 22.8.1, p. 514.

J. Rużyłło
*Model of Structural Transformations on the Chemically Etched Silicon Surface during Thermal Oxidation
(a) Vol. 48, No. 1, p. 199—202 (1978).
See also 1.6, p. 41; *3*, p. *55*; 22.1.2, p. 434; 22.6, p. 494.

R. Florian, J. Pelzl, M. Rosenberg, H. Vargas, and R. Wernhardt
*Photoacoustic Detection of Phase Transitions
(a) Vol. 48, No. 1, p. K35—K38 (1978).
See also *7*, p. *81*; 22.8, p. 506.

J. Barton, E. Wieser, and M. Müller
*Investigation of the α–γ Phase Transformation by Annealing of a Cold-Worked Fe–Mn Base Alloy
(a) Vol. 48, No. 1, p. K51—K53 (1978).
See also 5, p. 62; 21.1.1, p. 388.

O. Simmich and H. Löffler
*A Simple Model to Describe Isothermic Phase Transformations in Binary Alloys
(a) Vol. 48, No. 1, p. K67—K70 (1978).
See also 10.1, p. 104; 21, p. 361.

H. Konwent and J. Lorenc
A Thermodynamic Theory of the Improper Antiferroelectric Phase Transition in ADP
(b) Vol. 88, No. 2, p. 747—755 (1978).
See also *12.2*, p. *140*; *14.4.1*, p. *256*; 22.8.1, p. 515.

M. Rachwalska and T. Stanek
* An Adiabatic Calorimetric Study of the Phase Situation in $[Ni(H_2O)_6](ClO_4)_2$
(a) Vol. 48, No. 2, p. 297—300 (1978).
See also 8, p. 87; 22.8, p. 506.

E. I. Terukov, W. Reichelt, M. Wolf, H. Hemschik, and H. Oppermann
*On the Influence of the Substitution of O^{16} by O^{18} in a Few Vanadium Oxides on Their Semiconductor–Metal Transition
(a) Vol. 48, No. 2, p. 377—381 (1978).
See also 6, p. 75; 14.3, p. 223; 22.6, p. 495.

Ya. L. Shaiovich and R. A. Zvinchuk
*Conditions of Magnetite Precipitation during Decomposition of Wüstite $Fe_{1-x}O$
(a) Vol. 48, No. 2, p. 543—550 (1978).
See also 10.2, p. 116; 22.8.2, p. 519.

A. T. MACRANDER
*Density of Solid Krypton at Melting and Isochoric Equation of State of Solid Krypton and Solid Argon
(a) Vol. **48**, No. 2, p. 571—579 (1978).
See also 8, p. 87; 23, p. 530.

A. V. GAISANYUK, K. I. KUGEL, V. A. PETROV, and V. V. SOKOLOV
*Resistivity of the ZrC–C System: An Example of Percolation Behaviour
(a) Vol. **48**, No. 2, p. K131—K133 (1978).
See also *14*, p. *204*; 22.3, p. 458.

A. M. KAPITONOV, E. M. SMOKOTIN, A. P. KORENKO, and G. N. STEPANOV
*Ultrasonic-Wave Velocities in Cr_2O_3 near the Magnetic Phase Transition
(a) Vol. **48**, No. 2, p. K203—K206 (1978).
See also 7, p. 82; *18*, p. *275*; 22.6, p. 496.

I. FÖLDVÁRI, R. VOSZKA, and Z. MORLIN
*The Properties of Ni Ions in NaCl Single Crystals (I): Vacuum Ultraviolet, Ionic Conductivity, and X-Ray Diffraction Studies
(b) Vol. **89** No. 1, p. 235—240 (1978).
See also *14.4*, p. *250*; *20.1*, p. *332*; 22.5.2, p. 484; 22.5.4, p. 489.

A. YA. GORDON
*On Phase Transitions in Two-Dimensional Ferroelectrics
(b) Vol. **89**, No. 1, p. K69—K71 (1978).
See also *14.4.1*, p. *256*.

H. E. MÜSER, W. KUHN, and J. ALBERS
*Optical Investigation of the Cubic–Tetragonal Phase Transition in $BaTiO_3$ Single Crystals
(a) Vol. **49**, No. 1, p. 51—58 (1978).
See also *14.4.1*, p. *256*; 20, p. 317; 22.8.1, p. 515.

H.-J. ULLRICH, U. REINHOLD, S. DÄBRITZ, P. PAUFLER, K. KLEINSTÜCK, and B. PIETRASS
*On the Lattice Transformation in Plastically Deformed V_3Si Single Crystals Studied by the Kossel Technique
(a) Vol. **49**, No. 1, p. 323—330 (1978).
See also *10.1*, p. *106*; 14.2, p. 215; 21.1, p. 377.

L. LEBIODA, S. HODOROWICZ, and K. LEWINSKI
*Low-Temperature Phase-Transition in Urea
(a) Vol. **49**, No. 1, p. K27—K30 (1978).
See also 22.9, p. 527.

P. MÜLLER
Preparation of Superconducting Nb_3Au Diffusion Layers
(a) Vol. **49**, No. 1, p. K75—K79 (1978).
See also *14.2*, p. *215*.

W. A. THEINER and H. P. GESERICH
*Anderson Transition in $FeSi_2$ Films
(b) Vol. **89**, No. 2, p. 441—447 (1978).
See also *1.5*, p. *39*; 2, p. 50; 14.1, p. 210; 20.1, p. 334; 21.1.1, p. 389.

Y. L. KHAIT
*Some Observable Consequences of Short-Term Local Phase and Polymorphic Transformations and Changes of Inner States in Solids
(b) Vol. **89**, No. 2, p. K111—K114 (1978).

M. RAZEGHI and B. HOULIER
*EPR Study of Gd^{3+} around the Ferroelastic Transition Point of $Pb_3(PO_4)_2$
(b) Vol. **89**, No. 2, p. K135—K137 (1978).
See also *19*, p. *313*; 22.8.1, p. 516.

M. WAUTELET
*The Landau Theory of Phase Transitions: An Example of the Catastrophe Theory
(b) Vol. **89**, No. 2, p. K199—K200 (1978).

W. AUCH, W. STEUDLE, J. U. VON SCHÜTZ, and H. C. WOLF
The Phase Transition in the 1:1 CT-Complex Anthracene/TCNB at 206 K. A NMR Study
(a) Vol. **49**, No. 2, p. 563—567 (1978).
See also 6, p. 77; *19*, p. *313*; 22.9, p. 528.

M. MAČEK, V. KRAŠEVEC, and V. MARINKOVIĆ
*On the Relief Transformation Strains in γ-Phase Cr_5Al_8
(a) Vol. **49**, No. 2, p. 767—771 (1978).
See also *1.4*, p. *34*; 21.1, p. 379.

M. I. ZAKHAROVA, S. A. KIROV, and A. G. KHUNDZHUA
*Formation of Metastable and Equilibrium Phases in the Decomposition of β Solid Solution in Zr Alloys
(a) Vol. **49**, No. 2, p. 803—810 (1978).
See also 1.3, p. 28; 21.1, p. 379.

T. I. CHEKMASOVA and I. P. ALEKSANDROVA
*NMR Investigation of the High-Pressure Phase in $LiNH_4SO_4$
(a) Vol. **49**, No. 2, p. K185—K188 (1978).
See also *14.4.1*, p. *256*; 19, p. 313; 22.8.1, p. 516.

K. NARAYANDAS, M. RADHAKRISHNAN, and C. BALASUBRAMANIAN
Annealing Effect of Silver–Tellurium Films
(a) Vol. **49**, No. 2, p. K213—K215 (1978).
See also 2, p. 51; *14.1*, p. *211*; 21.6, p. 405; 22.1.3, p. 442.

W. REHWALD and A. VONLANTHEN
*Ultrasonic Studies of the Structural Phase Transition in Squaric Acid
(b) Vol. **90**, No. 1, p. 61—66 (1978).
See also 7, p. 82; 12.2, p. 141; 22.9, p. 528.

V. I. LITVINOV, V. M. BAGINSKII, and A. A. MITROFANOV
*The Phase Diagram of a System with Two Order Parameters. Mott Transition and Ferromagnetism
(b) Vol. **90**, No. 1, p. 423—427 (1978).
See also *13*, p. *148*; 14, p. 204; 18, p. 275.

B. Mrygoń
*Scaling Hypothesis for Non-Ideal System
(b) Vol. **90**, No. 1, p. K99—K101 (1978).
See also *8*, p. *89*; 21.1, p. 380.

H. Möhwald and A. Thaer
*Raman Study of the Phase Transition in the CT-Crystal Anthracene–TCNB
(a) Vol. **50**, No. 1, p. 131—138 (1978).
See also 6, p. 78; 20.1, p. 336; 22.9, p. 528.

I. A. Oxton and O. Knop
*Infrared Spectra of $SiHD_3$ in Crystalline Germane
(a) Vol. **50**, No. 1, p. K55—K58 (1978).
See also *20.1*, p. *337*; 23, p. 531.

J. Maetz, M. Müllner, and H. Jex
*The Cubic to Tetragonal Phase Transition in $RbCaF_3$ Studied with Neutron Diffraction
(a) Vol. **50**, No. 1, p. K117—K121 (1978).
See also 22.8, p. 508.

E. G. Batyev and A. V. Efanov
*Polaron State of an Electron Near a Structural Instability
(b) Vol. **90**, No. 2, p. 449—455 (1978).
See also *13.5.3*, p. *200*.

T. Soma
Equation of State under Pressure for Metallic β-Sn
(b) Vol. **90**, No. 2, p. K129—K131 (1978).
See also *4*, p. *59*; *12.1*, p. *136*; 21, p. 366.

F. L. Edelman, V. Z. Latuta, B. N. Zaitsev, and A. A. Khoromenko
*The Structure of Silicon Nitride Films (I): Stoichiometric Silicon Nitride
(a) Vol. **50**, No. 2, p. 573—578 (1978).
See also *1.5*, p. *39*; 22, p. 421.

G. L. Bleris, Th. Karakostas, N. A. Economou, and R. de Ridder
*The Ordered State of In_2Te_3 and Its Relation to the Transition State
(a) Vol. **50**, No. 2, p. 579—586 (1978).
See also *1.3*, p. *28*; 10.2, p. 122; 22.3, p. 459.

A. S. Bhalla, L. E. Cross, and D. A. Payne
*Thermal, Optical, and Elastic Properties of Ferroelastic Isopropyl Ammonium Chloroplatinate (IPACP) and Chlorostannate (IPACS)
(a) Vol. **50**, No. 2, p. 661—667 (1978).
See also 8, p. 90; 12.2, p. 141; 20.1, p. 338; 22.9, p. 528.

V. I. Kolomytsev, V. A. Lobodyuk, and L. G. Khandros
*Structural Changes at Heating of the β-Phase in NiTi Alloy
(a) Vol. **50**, No. 2, p. K127—K129 (1978).
See also 21.1, p. 383.

1.3 Alloys. Metallurgy

Original Papers and Short Notes

P. Turchi, Y. Calvayrac, and F. Plicque
*Effects of Configurational Order in a Ni_3Fe Single Crystal on Debye Temperature Deduced from Elastic Constants and X-Ray Measurements
(a) Vol. **45**, No. 1, p. 229—238 (1978).
See also 6, p. 65; 12.1, p. 134; 21.1, p. 369; 21.1.1, p. 384.

F. W. Boswell, A. Prodan, W. R. Vaughan, and J. M. Corbett
*On the Ordering of Fe Atoms in Fe_xNbS_2
(a) Vol. **45**, No. 2, p. 469—481 (1978).
See also *1.2*, p. *16*; 22.8, p. 503.

M. Wendt and A. Schmidt
*Improved Reproducibility of Energy-Dispersive X-Ray Microanalysis by Normalization to the Background
(a) Vol. **46**, No. 1, p. 179—183 (1978).
See also 22.8.2, p. 518.

J. Takahashi, K. Ozawa, S. Yamaguchi, Y. Fujino, O. Yoshinari, and M. Hirabayashi
*Temperature Dependence of the Thermal Vibrations of Deuterium in a $Pd_{0.8}Au_{0.2}$ Alloy Studied by Ion Channelling
(a) Vol. **46**, No. 1, p. 217—223 (1978).
See also 6, p. 68; *11*, p. *125*; 21.6, p. 401.

G. van Tendeloo and S. Amelinckx
*Erratum to: Long-Period Antiphase Boundary Structures in the Ordered Alloy Au_4Zn(I)
(a) Vol. **46**, No. 1, p. 399 (1978).
See also *1.4*, p. *30*; 21.1, p. 371; 21.6, p. 401.

N. N. Koren, V. V. Krasnova, and E. E. Matyas
*Investigation of Alloys of the System GeS–GeSe
(a) Vol. **46**, No. 1, p. K1—K3 (1978).
See also *1.2*, p. *17*; 22.7, p. 501.

E. Seitz, W. Schmatz, G. S. Bauer, and W. Just
A Trigonal Distortion Field for Substitutionally Dissolved Bi Atoms in Pb
(a) Vol. **46**, No. 2, p. 557—565 (1978).
See also *10.1*, p. *102*; 21, p. 357.

G. van Tendeloo, R. Wolf, D. van Dyck, and S. Amelinckx
*The Study of Ordering in the Au_5Mn_2 Alloy by Means of Structure Images
(a) Vol. **47**, No. 1, p. 105—122 (1978).
See also *1.4*, p. *30*; 21.1, p. 373; 21.6, p. 402.

R. Wolf, G. van Tendeloo, J. van Landuyt, and S. Amelinckx
Evidence for the Occurrence of the AABB Structure in the Gold–Manganese System
(a) Vol. **47**, No. 1, p. 241—244 (1978).
See also *1.2*, p. *18*; 1.4, p. 30; 21.1, p. 373; 21.6, p. 402.

A. J. MORTON
*Faulted Defects in γ-Brass-Type Alloys
(a) Vol. **47**, No. 1, p. 293—304 (1978).
See also *1.4*, p. *30*; 10.1, p. 103; 21.1, p. 373.

O. PAVLOV and Z. BONTCHEVA-MLADENOVA
*Measurements of the Ag_2Te–Ag_2S System at 300 K
(a) Vol. **47**, No. 1, p. K59—K61 (1978).
See also 8, p. 86; *14.3*, p. *221*; 15, p. 260; 22, p. 413.

I. M. CHAPNIK
*Atomic Volume and Semiconducting Properties in Sulfides of Non-Magnetic Transition Metals
(a) Vol. **47**, No. 1, p. K71—K75 (1978).
See also 14.1, p. 207; *14.3*, p. *221*; 22, p. 413.

I. G. RATISHVILI
Stepped Ordering in Binary Solid Solutions
(b) Vol. **87**, No. 2, p. 461—471 (1978).
See also *1.2*, p. *19*; 21, p. 359.

S. V. SEMENOVSKAYA
*Effect of Correlation on the X-Ray Diffuse Scattering
(b) Vol. **87**, No. 2, p. 733—737 (1978).
See also 21.1.1, p. 387.

G. VAN TENDELOO, R. WOLF, J. VAN LANDUYT, and S. AMELINCKX
Electron Diffraction and High Resolution Electron Microscopic Study of Ordering in the Gold–Manganese System (I): The Orthorhombic and Monoclinic I Structure in $Au_{11}Mn_4$
(a) Vol. **47**, No. 2, p. 539—554 (1978).
See also *1.2*, p. *19*; 1.4, p. 31; 21.1, p. 374; 21.6, p. 403.

G. VAN TENDELOO and S. AMELINCKX
*Direct Observation by High Resolution Electron Microscopy of Coincidence Sites in Domain Boundaries in Ordered Gold–Manganese Alloys
(a) Vol. **47**, No. 2, p. 555—564 (1978).
See also *1.4*, p. *31*; 21.1, p. 374; 21.6, p. 403.

H. METZGER, H. JO, S. C. MOSS, and D. G. WESTLAKE
Single Crystal X-Ray Study of the Superstructure Modulation and Long-Range Order in V_2D
(a) Vol. **47**, No. 2, p. 631—638 (1978).
See also *1.2*, p. *19*; 21.1, p. 374.

P. AIRO
*On the Superstructure of the Alloy Au–18 at%Mg
(a) Vol. **47**, No. 2, p. K107—K110 (1978).
See also *1.2*, p. *20*; 21.3, p. 394; 21.6, p. 403.

R. WOLF, G. VAN TENDELOO, J. VAN LANDUYT, and S. AMELINCKX
*Electron Diffraction and High Resolution Electron Microscopic Study of Ordering in the Gold–Manganese System (II): The Monoclinic II Structure in $Au_{11}Mn_4$ and Its Relation to Other LPAPB Phases
(a) Vol. **48**, No. 1, p. 39—52 (1978).
See also *1.2*, p. *20*; 1.4, p. 31; 21.1, p. 375; 21.6, p. 403.

S. PANCHANADEESWARAN, R. W. URE, JR., and J. G. BYRNE
*Positron Trapping at Precipitates in Aluminum–4 wt% Copper Single Crystals
(a) Vol. **48**, No. 1, p. K83—K87 (1978).
See also *10.1*, p. *105*; 21, p. 361.

FARID A. KHWAJA, A. A. KATSNELSON, and V. M. SILONOV
*Many-Electron Effects on the Characteristics of Short-Range Order in the Pseudopotential Approximation
(b) Vol. **88**, No. 2, p. 477—483 (1978).
See also *13*, p. *146*; 21.1, p. 375.

C. KOENIG
*On the Positron Localization in Ordered and Disordered Metallic Alloys
(b) Vol. **88**, No. 2, p. 569—579 (1978).
See also 10.1, p. 105; 21, p. 361.

H. BOSSAC, H.-G. FABIAN, and H. LÖFFLER
*A Quantitative Interpretation of the Decomposition Kinetics of Some Al–Zn–Mg Alloys
(a) Vol. **48**, No. 2, p. 369—375 (1978).
See also *10.1*, p. *105*; 14.1, p. 209; 21, p. 361; 21.1, p. 376; 21.3, p. 395.

M. VAN SANDE, R. DE RIDDER, G. VAN TENDELOO, J. VAN LANDUYT, and S. AMELINCKX
*High Resolution Study of One-Dimensional Long Period Superstructures in Cu_3Sn with Additions of Zinc and Nickel
(a) Vol. **48**, No. 2, p. 383—394 (1978).
See also 1.4, p. 32; 21, p. 361; 21.1, p. 376.

W. H. M. ALSEM, P. M. BRONSVELD, and E. W. VAN ROYEN
*Superlattice Dislocations in Cu_2NiZn
(a) Vol. **48**, No. 2, p. 497—502 (1978).
See also *10.1*, p. *105*; 21.1, p. 376.

M. L. JOHNSON, S. SATERLIE, D. BOICE, and J. G. BYRNE
*The Removal of Defects from Solids as Observed with Positron Annihilation
(a) Vol. **48**, No. 2, p. 551—554 (1978).
See also *10.1*, p. *105*; 21, p. 362; 21.1, p. 376.

W. EIGERMANN, G. MÜLLER-VOGT, and W. WENDL
Solubility Curves in High-Temperature Melts for the Growth of Single Crystals of Rare Earth Vanadates and Phosphates
(a) Vol. **49**, No. 1, p. 145—148 (1978).
See also *3*, p. *55*; 22.8.1, p. 515.

T. J. Magee and J. Peng
*Investigation of Solid–Solid Reactions of Au Films on Silicon
(a) Vol. **49**, No. 1, p. 313–322 (1978).
See also 1.5, p. 38; *14.3.2*, p. *231*; 21.6, p. 404; 22.1.2, p. 437.

P. Raj, S. K. Kulshreshtha, and J. V. Yakhmi
*Mössbauer and X-Ray Studies of $Fe_{1.67}Ge$
(a) Vol. **49**, No. 1, p. K91–K95 (1978).
See also *5*, p. *62*; 18.2, p. 290; 21.1.1, p. 389.

M. I. Zakharova, S. A. Kirov, and A. G. Khundzhua
Formation of Metastable and Equilibrium Phases in the Decomposition of β Solid Solution in Zr Alloys
(a) Vol. **49**, No. 2, p. 803–810 (1978).
See also *1.2*, p. *23*; 21.1, p. 379.

P. J. Flatau
*Kinetics of the Intraband Absorption and Magnetoabsorption Coefficients in Mixed Semiconductors with Composition Fluctuations
(b) Vol. **90**, No. 1, p. 251–260 (1978).
See also *20.1*, p. *335*; 22.4.4, p. 473.

J. M. Roland, X. Quillard, A. Moreau et R. Gay
*Analyse de clichés de diffraction électronique d'alliages Cu–Al par la fonction de Patterson (Analysis of Electron Diffraction Micrographs of Cu–Al Alloys by the Patterson Function)
(a) Vol. **50**, No. 1, p. 41–52 (1978).
See also 14.1, p. 212; 21.1, p. 380.

G. van Tendeloo and S. Amelinckx
*Long-Period Antiphase Boundary Structures in the Ordered Alloy Au_4Zn (II): Growth Mechanism of the LPAPB Structure
(a) Vol. **50**, No. 1, p. 53–66 (1978).
See also *1.4*, p. *34*; 21.1, p. 380; 21.6, p. 405.

E. Japa and K. Krop
Mössbauer Studies of Ordering in Fe–Si Alloys
(a) Vol. **50**, No. 1, p. K89–K91 (1978).
See also *5*, p. *62*; 21.1.1, p. 390.

S. S. Hsu, J. H. Chen, and P. C. Clapp
Acoustic Absorption and Emission Associated with the Martensitic Transformation in an FeNi Alloy
(a) Vol. **50**, No. 2, p. 393–398 (1978).
See also *7*, p. *82*; 21.1, p. 382; 21.1.1, p. 390.

G. L. Bleris, Th. Karakostas, N. A. Economou, and R. de Ridder
*The Ordered State of In_2Te_3 and Its Relation to the Transition State
(a) Vol. **50**, No. 2, p. 579–586 (1978).
See also *1.2*, p. *24*; 10.2, p. 122; 22.3, p. 459.

M. van Sande, R. de Ridder, J. van Landuyt, and S. Amelinckx
*A Study by Means of Electron Microscopy and Electron Diffraction of Vacancy Ordering in Ternary Alloys of the System AlCuNi
(a) Vol. **50**, No. 2, p. 587–599 (1978).
See also *1.4*, p. *34*; 10.1, p. 108; 21, p. 366; 21.1, p. 382.

1.4. Microstructure (Magnetic Domains See 18; Ferroelectric Domains See 14.4.1)

Original Papers and Short Notes

A. P. SUTTON and R. C. POND
*Electron Microscope Images of Coincidence Boundaries
(a) Vol. **45**, No. 1, p. 149–155 (1978).
See also 21, p. 353.

S. S. SHEININ and J. M. CORBETT
*Electron Microscope Images of Twin Boundaries (II): Contrast Mechanisms
(a) Vol. **45**, No. 1, p. 289–295 (1978).
See also 21.1, p. 369.

A. N. ALESHIN, V. YU. ARISTOV, B. S. BOKSHTEIN, and L. S. SHVINDLERMAN
*Kinetic Properties of ⟨111⟩ Tilt Boundaries in Aluminium
(a) Vol. **45**, No. 1, p. 359–366 (1978).
See also 9, p. 91; 21, p. 353.

V. V. KALININ, N. N. GERASIMENKO, and S. I. STENIN
The Effect of Implantation Temperature on the Mechanism of Misfit Dislocation Formation
(a) Vol. **45**, No. 2, p. 377–385 (1978).
See also *10.2*, p. *110*; *11*, p. *124*; 22.1.2, p. 429.

M. VAN SANDE, G. VAN TENDELOO, J. VAN LANDUYT, and S. AMELINCKX
*A Study by Transmission Electron Microscopy and Diffraction of the Ternary Alloy System Cu_2AuPd
(a) Vol. **45**, No. 2, p. 553–558 (1978).
See also *1.2*, p. *16*; 21.1, p. 370.

R. A. VARIN, J. W. WYRZYKOWSKI, W. ŁOJKOWSKI, and M. W. GRABSKI
*Spreading of Extrinsic Grain Boundary Dislocations in Plastically Deformed Aluminium
(a) Vol. **45**, No. 2, p. 565–569 (1978).
See also 10.1, p. 100; 21, p. 354.

C. BOULESTEIX, G. NIHOUL-BOUTANG, and B. YANGUI
Influence of Upper Laue Areas on Fine Structure of Bend Contours near a High Symmetry Axis
(a) Vol. **45**, No. 2, p. 639–645 (1978).
See also *1.1*, p. *11*; 22.6, p. 491.

V. A. ANTONOV, P. A. ARSENEV, and D. S. PETROVA
*Analysis of Spectral and Structural Regularities of Neodymium Ions in Garnet Monocrystals
(a) Vol. **46**, No. 1, p. 55–58 (1978).
See also 20, p. 316; 22.8.2, p. 518.

J. NARAYAN and S. M. OHR
*Residual Image Contrast of Dislocation Loops
(a) Vol. **46**, No. 1, p. 281–284 (1978).
See also *10.1*, p. *101*; 22.6, p. 492.

G. van Tendeloo and S. Amelinckx
*Erratum to: Long-Period Antiphase Boundary Structures in the Ordered Alloy Au_4Zn (I)
(a) Vol. **46**, No. 1, p. 399 (1978).
See also *1.3*, p. *25*; 21.1, p. 371; 21.6, p. 401.

G. A. Andreev and V. A. Klimov
*Flotation Study of Ca^{2+} and Sr^{2+} Precipitation in KCl Crystals
(a) Vol. **46**, No. 2, p. 421–426 (1978).
See also *10.2*, p. *112*; 22.5.2, p. 480.

B. Hughes and G. H. Narayanan
*On Annealing-Induced Prismatic Dislocation Loops and Electrical Changes in Heavily Te-Doped GaAs
(a) Vol. **46**, No. 2, p. 627–637 (1978).
See also *10.2*, p. *112*; 14.3, p. 220; 22.2.1, p. 446.

L. Benguigui and D. Cabib
Determination of the Tilt Angle in the Smectic C Phase of Bis-(Heptyloxy)-Azoxybenzene by Means of Dielectric Measurements
(a) Vol. **47**, No. 1, p. 71–78 (1978).
See also *14.4*, p. *249*; 22.9; p. 525.

G. van Tendeloo, R. Wolf, D. van Dyck, and S. Amelinckx
*The Study of Ordering in the Au_5Mn_2 Alloy by Means of Structure Images
(a) Vol. **47**, No. 1, p. 105–122 (1978).
See also *1.3*, p. *25*; 21.1, p. 373; 21.6, p. 402.

R. Wolf, G. van Tendeloo, J. van Landuyt, and S. Amelinckx
Evidence for the Occurrence of the AABB Structure in the Gold–Manganese System
(a) Vol. **47**, No. 1, p. 241–244 (1978).
See also *1.2*, p. *18*; 1.3, p. 25; 21.1, p. 373; 21.6, p. 402.

A. J. Morton
*Faulted Defects in γ-Brass-Type Alloys
(a) Vol. **47**, No. 1, p. 293–304 (1978).
See also *1.3*, p. *26*; 10.1, p. 103; 21.1, p. 373.

B. Yangui, M. Rekik, G. Nihoul-Boutang, C. Jouanin, C. Boulesteix, and D. Renard
Theoretical and Experimental Study of the Transmitted Intensity as a Function of Thickness when the Incident Beam Is Parallel to a Crystallographic Axis
(a) Vol. **47**, No. 1, p. 311–320 (1978).
See also *1.1*, p. *12*; 22.6, p. 493.

P. Erlenwein
*On a Relationship between the Symmetries of Faceted Cones on Ion Bombarded Single-Crystal Faces and Ejection Patterns
(a) Vol. **47**, No. 1, p. K9–K10 (1978).
See also 11, p. 126; 21, p. 358; 21.6, p. 402.

J. A. Ashby, H. M. Flower, and R. D. Rawlings
*Correlation of Electrical Resistivity with Microstructure in an Fe–Co–2% V Alloy
(a) Vol. 47, No. 2, p. 407—414 (1978).
See also *14.1*, p. *208*; 21.1, p. 374; 21.1.1, p. 387.

D. Bartzokas, C. Manolikas, and J. Spyridelis
Electron Microscopic Study of the Destabilization of Stabilized γ-Phase of Indium Sesquisulphide
(a) Vol. 47, No. 2, p. 459—467 (1978).
See also *1.2*, p. *19*; 22.3, p. 458.

P. Rapps, K.-H. Katerbau, H. Mughrabi, K. Urban, and M. Wilkens
*Electron Diffraction in Crystalline Specimens Containing a High Density of Lattice Defects (II): Numerical Evaluation for the Case of Narrow Dislocation Dipoles — Comparison with Experimental Data
(a) Vol. 47, No. 2, p. 479—488 (1978).
See also *10*, p. *96*.

G. van Tendeloo, R. Wolf, J. van Landuyt, and S. Amelinckx
Electron Diffraction and High Resolution Electron Microscopic Study of Ordering in the Gold–Manganese System (I): The Orthorhombic and Monoclinic I Structure in $Au_{11}Mn_4$
(a) Vol. 47, No. 2, p. 539—554 (1978).
See also *1.2*, p. *19*; 1.3, p. 26; 21.1, p. 374; 21.6, p. 403.

G. van Tendeloo and S. Amelinckx
*Direct Observation by High Resolution Electron Microscopy of Coincidence Sites in Domain Boundaries in Ordered Gold–Manganese Alloys
(a) Vol. 47, No. 2, p. 555—564 (1978).
See also *1.3*, p. *26*; 21.1, p. 374; 21.6, p. 403.

P. L. Rossiter and M. E. Houghton
*Magnetic Properties and Microstructure of an Fe–27.5 Cr–17.5 Co–0.5 Al Alloy
(a) Vol. 47, No. 2, p. 597—608 (1978).
See also *18.2*, p. *287*; 21.1, p. 374; 21.1.1, p. 387.

R. Wolf, G. van Tendeloo, J. van Landuyt, and S. Amelinckx
Electron Diffraction and High Resolution Electron Microscopic Study of Ordering in the Gold–Manganese System (II): The Monoclinic II Structure in $Au_{11}Mn_4$ and Its Relation to Other LPAPB Phases
(a) Vol. 48, No. 1, p. 39—52 (1978).
See also *1.2*, p. *20*; 1.3, p. 27; 21.1, p. 375; 21.6, p. 403.

M. E. Houghton and P. L. Rossiter
*Induced Anisotropy in an Fe–27.5 Cr–17.5Co–0.5 Al Alloy
(a) Vol. 48, No. 1, p. 71—77 (1978).
See also *18.2*, p. *289*; 21.1, p. 375; 21.1.1, p. 388.

T. Cornelissens, G. van Tendeloo, J. van Landuyt, and S. Amelinckx
*Peierls Distortion, Chain Polytypism, and Dislocation Coupling in NbS_3
(a) Vol. 48, No. 1, p. K5—K9 (1978).
Erratum
(a) Vol. 49, No. 2, p. K217 (1978).
See also 10.2, p. 115; 22, p. 415.

V. A. Streltsov
*Shear Stresses Induced by Hydrostatic Pressure near the Surface of an Inclusion
(a) Vol. 48, No. 1, p. K23—K25 (1978).
See also *12.1*, p. *135*; 21, p. 360.

F. Studer et B. Raveau
*Etude par microscopie électronique des phénomènes d'intercroissance dans les siliconiobates de baryum (I): Les composés $(Ba_3Nb_6Si_4O_{26})_n \cdot Ba_3Nb_4Ti_4O_{21}$ avec n compris entre $^3/_2$ et $^{29}/_2$ (Electron Microscopic Study of Intergrowth Phenomena in Barium Siliconiobates (I): The Compounds $(Ba_3Nb_6Si_4O_{26})_n \cdot Ba_3Nb_4Ti_4O_{21}$ with n between $^3/_2$ and $^{29}/_2$)
(a) Vol. 48, No. 2, p. 301—312 (1978).
See also *1.1*, p. *13*; 22.8, p. 506.

M. van Sande, R. de Ridder, G. van Tendeloo, J. van Landuyt, and S. Amelinckx
High Resolution Study of One-Dimensional Long Period Superstructures in Cu_3Sn with Additions of Zinc and Nickel
(a) Vol. 48, No. 2, p. 383—394 (1978).
See also *1.3*, p. *27*; 21, p. 361; 21.1, p. 376.

H. Menniger, G. Voigt, H. Raidt, H. Fabig, and J. Maege
*Scanning Electron Microscopic Investigations of GaP
(a) Vol. 48, No. 2, p. 407—415 (1978).
See also 14.3.2, p. 230; 20.3, p. 346; 22.2.2, p. 451.

M. Treilleux, P. Thévenard, G. Chassagne, and L. W. Hobbs
*Observation of Implanted Potassium Aggregates in MgO Single Crystals
(a) Vol. 48, No. 2, p. 425—430 (1978).
See also 11, p. 128; 22.6, p. 495.

T. Hirata
*Structural Changes during Annealing Vacuum-Evaporated Thin Aluminium Films at Temperatures near the Thermodynamic Melting Point
(a) Vol. 48, No. 2, p. 451—458 (1978).
See also *1.5*, p. *37*; 2, p. 48; 21, p. 361.

M. Lefeld-Sosnowska
*X-Ray Diffraction Contrast of a 60° Dislocation Investigated by Means of Section Topography
(a) Vol. 48, No. 2, p. 565—569 (1978).
See also *10.2*, p. *116*; 22.1.2, p. 435.

G. Weidner
*Application of Hillock Etching of Microdefects to the Investigation of Czochralski Silicon by Transmission Electron Microscopy
(a) Vol. 48, No. 2, p. K105—K107 (1978).
See also 22.1.2, p. 435.

S. Y. Chuang, P. K. Tseng, G. J. Jan, and H. S. Chen
*An Experimental Study of the Angular Correlation of Positron Annihilation Radiations in Electron Irradiated Pd–Ni–Si Metallic Glass
(a) Vol. 48, No. 2, p. K181—K183 (1978).
See also *2*, p. *48*; 21.1, p. 377; 21.6, p. 404.

F. Studer et B. Raveau
*Etude par microscopie électronique des phénomènes d'intercroissance dans les siliconiobates de baryum (II): Intercroissances multiples et microdomaines
(Electron Microscopic Study of Intergrowth Phenomena in Barium Siliconiobates (II): Multiple Intergrowths and Microdomains)
(a) Vol. **49**, No. 1, p. 189—205 (1978).
See also *1.1*, p. *13*; 22.8, p. 507.

K. Abdelmoula, B. Pardo, and D. Renard
*Study by Electron Microscopy of Some Twinned Gold Crystals with Variations in Thickness
(a) Vol. **49**, No. 1, p. 267—278 (1978).
See also 1.5, p. 38; 21.6, p. 404.

G. van Tendeloo and S. Amelinckx
*A High Resolution Study of Ordering in Au_4Mn (I): The Microstructure of Fully Ordered Au_4Mn
(a) Vol. **49**, No. 1, p. 337—346 (1978).
See also *1.1*, p. *13*; 21.1, p. 377; 21.6, p. 404.

V. Dutta, P. Nath, V. D. Vankar, and K. L. Chopra
· Structural Properties of Polycrystalline Ge–Metal Films
(a) Vol. **49**, No. 1, p. 379—385 (1978).
See also *1.5*, p. *38*; 2, p. 49; 22.1.1, p. 427.

V. M. Ievlev, A. V. Bugakov, an V. A. Ammer
*Interphase Coincidence Boundaries in Au–Si and Au–CaF_2 Systems
(a) Vol. **49**, No. 1, p. 413—420 (1978).
See also *1.5*, p. *38*; 3, p. 56; 21.6, p. 404.

J. Gastaldi and C. Jourdan
New Possibilities for Recrystallization Study by X-Ray Synchrotron Radiation Topography
(a) Vol. **49**, No. 2, p. 529—537 (1978).
See also *3*, p. *56*; 21, p. 363.

M. Wilkens and H. Föll
*The Black–White Vector $\boldsymbol{l}$ of Small Dislocation Loops on Transmission Electron Microscope Images
(a) Vol. **49**, No. 2, p. 555—561 (1978).
See also *10*, p. *97*.

G. van Tendeloo, R. de Ridder, and S. Amelinckx
*A High Resolution Study of Ordering in Au_4Mn (II): The Evolution of Ordering in Au_4Mn
(a) Vol. **49**, No. 2, p. 655—666 (1978).
See also *1.1*, p. *14*; 21.1, p. 379; 21.6, p. 405.

L. Trépied and J.-C. Doukhan
*Dissociated $\boldsymbol{a}$-Dislocations in Quartz. Influence on Plastic Deformation
(a) Vol. **49**, No. 2, p. 713—724 (1978).
See also *10.2*, p. *119*; 22.6, p. 496.

M. J. Marcinkowski
*Further Generalizations of the Unified Theory of Grain Boundaries
(a) Vol. **49**, No. 2, p. 725—735 (1978).
See also 10, p. 97.

M. Maček, V. Kraševec, and V. Marinkovič
*On the Relief Transformation Strains in γ-Phase Cr_5Al_8
(a) Vol. **49**, No. 2, p. 767—771 (1978).
See also *1.2*, p. *23*; 21.1, p. 379.

G. van Tendeloo and S. Amelinckx
*Long-Period Antiphase Boundary Structures in the Ordered Alloy Au_4Zn (II): Growth Mechanism of the LPAPB Structure
(a) Vol. **50**, No. 1, p. 53—66 (1978).
See also *1.3*, p. *28*; 21.1, p. 380; 21.6, p. 405.

P. Duval et A. Hauet
*Interprétation des lignes de diffusion observées en diffraction électronique dans l'alliage Cu_3Ga
(Interpretation of Diffuse Intensity Lines Observed in Electron Diffraction in Cu_3Ga Alloy)
(a) Vol. **50**, No. 1, p. 67—72 (1978).
See also 21, p. 365; 21.1, p. 380.

J. Fidler and P. Skalicky
*TEM Investigation of Grown-in Dislocations and Stacking Faults in $SmCo_5$ Single Crystals
(a) Vol. **50**, No. 1, p. 73—81 (1978).
See also 10.1, p. 107; 21.1, p. 380; 21.4, p. 398.

R. de Ridder, D. van Dyck, and S. Amelinckx
*Diffuse Intensity Contours Observed in Electron Diffraction Patterns of Transition Metal Tri- and Di-Chalcogenides: A Structural Interpretation
(a) Vol. **50**, No. 1, p. 349—362 (1978).
See also *1.1*, p. *14*; 22, p. 420.

L. Trepied and J.-C. Doukhan
*Some Precisions on the Brazil Twins in Quartz
(a) Vol. **50**, No. 1, p. K37—K41 (1978).
See also 22.6, p. 497.

R. I. Kuznetsova and N. N. Zhukov
Structural Peculiarities of the Superplastic Behaviour of the Alloy Al–4 wt% Ge Due to Porosity
(a) Vol. **50**, No. 2, p. 385—391 (1978).
See also *10.1*, p. *108*; 21, p. 366.

M. van Sande, R. de Ridder. J. van Landuyt. and S. Amelinckx
*A Study by Means of Electron Microscopy and Electron Diffraction of Vacaney Ordering in Ternary Alloys of the System AlCuNi
(a) Vol. **50**, No. 2, p. 587—599 (1978).
See also *1.3*, p. *28*; 10.1, p. 108; 21, p. 366; 21.1, p. 382.

M. J. Marcinkowski and K. Jagannadham
*The Fine Structure and Energy of Grain Boundaries
(a) Vol. **50**, No. 2, p. 601–610 (1978).
See also *10*, p. *98*.

Th. Karakostas, G. Nouet, G. L. Bleris, S. Hagege, and P. Delavignette
*Grain Boundary Analysis in TEM (I): Practical Determination of Bicrystal Orientations
(a) Vol. **50**, No. 2, p. 703–709 (1978).
See also *1.1*, p. *14*.

L. G. Aravin, R. A. Salukvadze, V. P. Shantarovich, and A. A. Khomenko
*Positron Annihilation in Carbon Structures
(a) Vol. **50**, No. 2, p. K213–K217 (1978).
See also *10.2*, p. *122*; 22.1, p. 423.

1.5 Films

Original Papers and Short Notes

C. J. Rossouw and M. J. Whelan
*Thickness Determination of Thin Graphite and Aluminium Nitride Foils by Transmission Electron Microscopy Using Diffraction and Energy Loss Analysis
(a) Vol. **45**, No. 1, p. 277–288 (1978).
See also 13.5.2, p. 198; 22, p. 409; 22.1, p. 422.

I. M. Belii, F. F. Komarov, V. S. Tishkov, and V. M. Yankovskii
*Formation of Chemical Compounds by Ion Bombardment of Thin Transition Metal Films
(a) Vol. **45**, No. 1, p. 343–352 (1978).
See also *1.2*, p. *15*; 11, p. 123; 21.1, p. 369.

V. D. Vankar, S. R. Das, Prem Nath, and K. L. Chopra
*Structure of Vacuum Evaporated $Cd_xZn_{1-x}S$ Thin Films
(a) Vol. **45**, No. 2, p. 665–669 (1978).
See also 22.4.4, p. 472.

A. Bohg
*Measurement of Stresses in Thin Films on Single Crystalline Substrates
(a) Vol. **46**, No. 2, p. 445–450 (1978).
See also 10.2, p. 112; 22.1.2, p. 432.

J. F. Graczyk
*Defect Structures in Amorphous Germanium Films
(a) Vol. **46**, No. 2, p. 611–617 (1978).
See also *2*, p. *44*; 22.1.1, p. 425.

P. K. Madden and J. E. Harbottle
*The Imaging of Slip Bands in Nickel
(a) Vol. **46**, No. 2, p. 639–652 (1978).
See also 11, p. 125; 21.1, p. 372.

J. H. Basson and C. A. B. Ball
*On the Density of Misfit Dislocations in Double- and Triple-Layer Heterojunction Devices
(a) Vol. **46**, No. 2, p. 707–715 (1978).
See also *10.2*, p. *112*.

A. Banerjee, Prem Nath, V. D. Vankar, and K. L. Chopra
*Properties of $Zn_xCd_{1-x}S$ Films Prepar ed by Solution Spray Technique
(a) Vol. **46**, No. 2, p. 723–728 (1978).
See also *14.3.1*, p. *226*; 16, p. 263; 22.4.4, p. 472.

J. F. Graczyk and P. Chaudhari
*Microdiffraction Patterns from the Finney Model
(b) Vol. **87**, No. 1, p. K5–K8 (1978).
See also *2*, p. *44*.

C. L. Bauer and A. G. Jordan
*Study of Interfacial Reactions in Bimetallic Thin-Film Couples by Contact Resistance Measurements
(a) Vol. **47**, No. 1, p. 321—328 (1978).
See also 14.1, p. 207; 21.1, p. 373.

A. R. Patel, K. V. Rao, and G. K. Shivakumar
*Initial Stages of Bismuth Deposition
(a) Vol. **47**, No. 1, p. K1—K3 (1978).
See also 21.7, p. 406.

N. Y. Chernysheva, G. A. Kachurin, and V. A. Bogatyriov
Amorphisation and Recrystallisation of InSb Ion Implanted Layers
(a) Vol. **47**, No. 1, K5—K8 (1978).
See also *11*, p. *126*; 22.2.3, p. 453.

B. Knappe, H.-R. Müller, and B. Springmann
*Homogeneity of Amorphous Gd–Co Films Prepared by Bias Sputtering
(a) Vol. **47**, No. 2, p. 523—532 (1978).
See also *2*, p. *45*; 18.3, p. 296; 21.1, p. 374; 21.4, p. 398.

V. Dutta, P. Nath, and K. L. Chopra
*Structural and Electrical Properties of Polycrystalline Ge Films
(a) Vol. **48**, No. 1, p. 257—262 (1978).
See also *14.3.1*, p. *227*; 22.1.1, p. 427.

U. Köster
*Crystallization of Amorphous Silicon Films
(a) Vol. **48**, No. 2, p. 313—321 (1978).
See also 2, p. 48; *3*, p. *55*; 22.1.2, p. 435.

K. H. Yang and G. H. Schwuttke
*Analysis of Vapor-Deposited Silicon Carbide Films on Silicon Ribbon Surfaces
(a) Vol. **48**, No. 2, p. 335—343 (1978).
See also 3, p. 55; 22.3, p. 458.

P. Nath, J.-E. Sundgren, and H. T. G. Nilsson
*Structural and Mechanical Properties of Composite Films of Iron Particles in an Oxide Matrix
(a) Vol. **48**, No. 2, p. 345—355 (1978).
See also 10.2, p. 116; 22.6, p. 495.

T. Hirata
*Structural Changes during Annealing Vacuum-Evaporated Thin Aluminium Films at Temperatures near the Thermodynamic Melting Point
(a) Vol. **48**, No. 2, p. 451—458 (1978).
See also *1.4*, p. *32*; 2, p. 48; 21, p. 361.

F. COSANDEY, Y. KOMEM, and C. L. BAUER
*Characterization of [001] Tilt Boundaries Produced by Epitaxial Growth on Bicrystalline Substrates of NaCl
(a) Vol. **48**, No. 2, p. 555–563 (1978).
See also 21.6, p. 403.

K. I. KIROV, S. S. GEORGIEV, E. V. GEROVA, and S. P. ALEKSANDROVA
*Investigation of SiO_2 Layers Deposited by Plasma Decomposition of Tetra-Ethoxy Silane in a Planar Reactor
(a) Vol. **48**, No. 2, p. 609–614 (1978).
See also *14.3.1*, p. *228*; 22.6, p. 495.

M. JEDLIČKA and V. VALVODA
*The Macroscopic Structure of PbO Layers
(a) Vol. **48**, No. 2, p. K101–K104 (1978).
See also 22.6, p. 495.

H.-J. NOWAK, H. LEITZ und W. BUCKEL
*Zur Struktur amorpher metallischer Schichten des Systems Germanium–Kupfer
(On the Structure of Amorphous Metallic Layers of the System Germanium–Copper)
(a) Vol. **49**, No. 1, p. 73–82 (1978).
See also *2*, p. *49*; 21.1, p. 377; 22.1.1, p. 427.

K. DETTMER, H. MÜLLER, and F. R. KESSLER
*Surface Influences of Glass Substrates on Lattice Parameters and Phase Transition Temperature of Evaporated Ge Thin Films
(a) Vol. **49**, No. 1, p. 255–260 (1978).
See also *2*, p. *49*; 22.1.1, p. 427.

K. ABDELMOULA, B. PARDO, and D. RENARD
Study by Electron Microscopy of Some Twinned Gold Crystals with Variations in Thickness
(a) Vol. **49**, No. 1, p. 267–278 (1978).
See also *1.4*, p. *33*; 21.6, p. 404.

T. J. MAGEE and J. PENG
Investigation of Solid–Solid Reactions of Au Films on Silicon
(a) Vol. **49**, No. 1, p. 313–322 (1978).
See also *1.3*, p. *28*; *14.3.2*, p. *231*; 21.6, p. 404; 22.1.2, p. 437.

U. SAXENA and O. N. SRIVASTAVA
*Effect of Nonstoichiometry on the Electrical Conductivity of Erbium Oxide Films
(a) Vol. **49**, No. 1, p. 371–378 (1978).
See also 2, p. 49; 13.1, p. 158; *14.3.1*, p. *228*; 22.6, p. 496.

V. DUTTA, P. NATH, V. D. VANKAR, and K. L. CHOPRA
*Structural Properties of Polycrystalline Ge–Metal Films
(a) Vol. **49**, No. 1, p. 379–385 (1978).
See also 1.4, p. 33; 2, p. 49; 22.1.1, p. 427.

V. M. IEVLEV, A. V. BUGAKOV, and V. A. AMMER
*Interphase Coincidence Boundaries in Au–Si and Au–CaF_2 Systems
(a) Vol. **49**, No. 1, p. 413–420 (1978).
See also *1.4*, p. *33*; 3, p. 56; 21.6, p. 404.

V. Kobzarenko, F. Donika, and S. Radautsan
*Thin Films of the One-Packed Polytype of $ZnIn_2S_4$(I)
(a) Vol. **49**, No. 1, p. K5—K7 (1978).
See also 22.8, p. 507.

H. Nelkowski and K. Petermann
Preparation and Investigation of ZnS–$CuGaS_2$ Heterojunctions
(a) Vol. **49**, No. 1, p. K13—K16 (1978).
See also 3, p. 56; *14.3.3*, p. *236*; *20.3*, p. *348*; 22.4.1, p. 465; 22.8, p. 507.

W. A. Theiner and H. P. Geserich
*Anderson Transition in $FeSi_2$ Films
(b) Vol. **89**, No. 2, p. 441—447 (1978).
See also *1.2*, p. *22*; 2, p. 50; 14.1, p. 210; 20.1, p. 334; 21.1.1, p. 389.

B. Baranov, L. Däweritz, V. B. Gutan, G. Jungk, H. Neumann, and H. Raidt
*Growth and Properties of $Al_xGa_{1-x}N$ Epitaxial Layers
(a) Vol. **49**, No. 2, p. 629—636 (1978).
See also 13.4, p. 182; 14.3.1, p. 228; 20.1, p. 335; 20.3, p. 348; 22.2.4, p. 456.

K. I. Kirov, E. D. Atanasova, and N. A. Ivanov
*Investigation of Al_2O_3 Layers Obtained by DC Planar-Magnetron Sputtering
(a) Vol. **50**, No. 1, p. 363—368 (1978).
See also *14.3.3*, p. *238*; 14.4, p. 252; 22.6, p. 497.

I. P. Grebennik and S. P. Yurchenko
*The Use of Two-Dimensional Electron Diffraction Effects for the Initial Stage Research of γ-Phase Growth in the Metal–Melt Contact Zone
(a) Vol. **50**, No. 1, p. K1—K3 (1978).
See also 21, p. 365; 21.6, p. 405.

F. L. Edelman, V. Z. Latuta, B. N. Zaitsev, and A. A. Khoromenko
*The Structure of Silicon Nitride Films (1): Stoichiometric Silicon Nitride
(a) Vol. **50**, No. 2, p. 573—578 (1978).
See also *1.2*, p. *24*; 22, p. 421.

A. A. Konova and M. G. Michailov
*Crystalline Anodic Silicon Dioxide on Silicon
(a) Vol. **50**, No. 2, p. 627—633 (1978).
See also 22.6, p. 497.

1.6 Surfaces

Review Article

V. S. Postnikov, V. V. Postnikov, and V. M. Fedorov
*Metastability and Superconductivity in Thin Films
(b) Vol. **87**, No. 1, p. 11–23 (1978).
See also 1.2, p. 15; *14.2*, p. *213*; 21, p. 352; 22, p. 408.

Original Papers and Short Notes

B. Michel
*Dilatation Energy of Extended Defects with Arbitrary Shape and Size near Solid Surfaces
(b) Vol. **85**, No. 1, p. K49–K52 (1978).
See also *10*, p. *96*.

J.-M. Derochette
*Adsorption of Nitrogen on Iridium. Influence of Steps on the Surface Potential Studied b Field Emission
(a) Vol. **45**, No. 1, p. 163–169 (1978).
See also *17.1*, p. *272*; 21, p. 353.

M. Höhne
EPR of Lithium in Mechanically Affected Silicon
(b) Vol. **85**, No. 2, p. 525–534 (1978).
See also 13.4, p. 170; *19*, p. *304*; 22.1.2, p. 429.

R. Feder and J. Kirschner
*Surface Structure Determination via LEED Rotation Diagrams and the Relaxation of W (001)
(a) Vol. **45**, No. 2, p. K117–K119 (1978).
See also 21, p. 355.

M. Höhne
Nonrandom Strain in "Crushed" Silicon. EPR of Thermally Excited Lithium Donors
(b) Vol. **86**, No. 1, p. 119–126 (1978).
See also 13.4, p. 172; *19*, p. *305*; 22.1.2, p. 430.

H. Wolf and H. Solbrig
*Model Calculations on Surface-Resonance Fine Structures in Low-Energy Electron Diffraction
(b) Vol. **86**, No. 1, p. K15–K19 (1978).
See also 21.1, p. 371.

M. V. Kruglov, V. N. Shchemelev, and G. G. Kareva
Detection of Atomic Plane Displacements in Near-Surface Layers of Crystals Using the Angular Dependence of Photoemission during Bragg Diffraction of X-Rays
(a) Vol. **46**, No. 1, p. 343–350 (1978).
See also *1.1*, p. *11*; 17, p. 269; 22.1.2, p. 431.

Diep-The-Hung
Erratum to: Effect of Induced Surface Magnetic Moment on Spin-Wave Spectrum
(b) Vol. **86**, No. 2, p. K181 (1978).
See also *13.5.4*, p. *201*; 18.4, p. 299.

G. J. Russell and J. Woods
*An Electron Diffraction Study of Phases in the Cu_xS System
(a) Vol. **46**, No. 2, p. 433—444 (1978).
See also 22, p. 412; 22.4.1, p. 464.

R. Erlandsson, U. Birkholz, and S.-E. Karlsson
*Study of Si_2Te_3 Surface Reactions with Auger Electron Spectroscopy
(a) Vol. **47**, No. 1, p. 85—90 (1978).
See also 22, p. 413.

J. Rużyłło
Model of Structural Transformations on the Chemically Etched Silicon Surface during Thermal Oxidation
(a) Vol. **48**, No. 1, p. 199—202 (1978).
See also *1.2*, p. *21*; *3*, p. *55*; 22.1.2, p. 434; 22.6, p. 494.

L. E. Cross and A. Bhalla
*Domain "Decoration" in $Gd_2(MoO_4)_3$
(a) Vol. **48**, No. 2, p. 431—437 (1978).
See also *14.4.1*, p. *256*; 22.8.1, p. 515.

K. Müller and K. Heinz
*Reconstructed Surfaces
(a) Vol. **49**, No. 1, p. 39—49 (1978).
See also 21, p. 362; 21.6, p. 404; 22.1.2, p. 436.

H. Jagodziński
*Systematic Absences in Low Energy Electron Diffraction
(a) Vol. **49**, No. 1, p. 119—126 (1978).

B. Michel
*A Note on the Interface-Interaction of Point Defects and Volume Defects
(b) Vol. **89**, No. 2, p. K179—K181 (1978).
See also *10*, p. *97*.

R. de Ridder, D. van Dyck, and S. Amelinckx
*A Simple Geometrical Interpretation of the Superstructures $\sqrt{p} \times \sqrt{p}$ ($p = 3, 7, 19, 31, 49, \ldots$) Observed on Surfaces with Hexagonal Symmetry by Means of LEED and RHEED
(a) Vol. **50**, No. 1, p. K67—K71 (1978).
See also 22, p. 420.

A. Kalz, F. Storbeck, and G. Blasek
*Adsorption of H_2O on α-Fe(111) under the Influence of 2.5 keV Electrons
(a) Vol. **50**, No. 2, p. K247—K250 (1978).
See also 11, p. 132; 21.1.1, p. 391.

2. Non-Crystalline State

Original Papers and Short Notes

G. NEUMANN and H. VOGT
Rayleigh Wing Scattering in Disordered Sodium Nitrate
(b) Vol. **85**, No. 1, p. 179—184 (1978).
See also *1.2*, p. *15*; *6*, p. *64*; 20.1, p. 319; 22.8.1, p. 510.

J. RICHTER and S. KOBE
*Density of States and Itinerant Ferromagnetism in Amorphous Systems with Dense Random Packed Hard Sphere Structure
(b) Vol. **85**, No. 1, p. 233—238 (1978).
See also 13, p. 142; *18.2*, p. *281*.

A. V. SHENDRIK and D. M. YUDIN
A New Concept of the Model for the Paramagnetic E'-Centre in the SiO_{2-x} System
(b) Vol. **85**, No. 1, p. 343—349 (1978).
See also *13.4*, p. *169*; *19*, p. *303*; 22.8.1, p. 510.

C. L. BRIANT and J. J. BURTON
*Icosahedral Microclusters. A Possible Structural Unit in Amorphous Metals
(b) Vol. **85**, No. 1, p. 393—402 (1978).
See also 21, p. 352.

M. I. DARBY and G. R. EVANS
*Local Fields in Amorphous Elements
(b) Vol. **85**, No. 1, p. K63—K66 (1978).

M. KITAO, H. IKEDA, H. HASEGAWA, and S. YAMADA
Photoconductivity of Amorphous As_2Se_3
(a) Vol. **45**, No. 1, p. 107—111 (1978).
See also *16*, p. *262*; 22, p. 409.

J. GAZSÓ and J. HAJTÓ
Self-Controlled Laser Beam Chopping Effect in $GeSe_2$ Thick Films
(a) Vol. **45**, No. 1, p. 181—186 (1978).
See also 11, p. 123; *20.1*, p. *320*; 22.7, p. 501.

R. YAMAMOTO, H. MATSUOKA, and M. DOYAMA
*Structural Relaxation of the Dense Random Packing Model for Amorphous Iron
(a) Vol. **45**, No. 1, p. 305—314 (1978).
See also 21.1.1, p. 384.

J. SCHNEIDER, A. HANDSTEIN, and R. HESSKE
*Mictomagnetic to Ferromagnetic Transition in Amorphous Ni–Fe Based Alloys
(a) Vol. **45**, No. 1, p. K47—K50 (1978).
See also *18.2*, p. *282*; 21.1, p. 369; 21.1.1, p. 384.

R. M. MEHRA, A. K. KATHURIA, RADHEY SHYAM, A. L. DAWAR, and P. C. MATHUR
*Determination of the Density of Localized States in a-Se Films
(a) Vol. **45**, No. 1, p. K59—K62 (1978).
See also *13.4*, p. *170*; 14.3.1, p. 226; 22.1.3, p. 441.

G. P. SRIVASTAVA
Electronic and Optical Properties of Crystalline and Amorphous Si
(b) Vol. **85**, No. 2, p. K121—K126 (1978).
See also *13.1*, p. *152*; *20.1*, p. *321*; 22.1.2, p. 429.

J. KOČKA and J. KRIŠTOFIK
*AC Conductivity of $CdAs_2$-Based Glasses
(a) Vol. **45**, No. 2, p. 559—564 (1978).
See also *14.3*, p. *218*; 22.3, p. 457.

R. C. O'HANDLEY and D. S. BOUDREAUX
*Magnetic Properties of Transition Metal–Metalloid Glasses. A Charge Transfer Model
(a) Vol. **45**, No. 2, p. 607—615 (1978).
See also *18.2*, p. *283*; 21.1, p. 370.

A. N. TRUKHIN
Investigation of the Photoelectric and Photoluminescent Properties of Crystalline Quartz and Vitreous Silica in the Fundamental Absorption Region. A Model for Electronic Structure and Migration of Energy in SiO_2
(b) Vol. **86**, No. 1, p. 67—75 (1978).
See also *13.1*, p. *152*; 16, p. 263; 20.3, p. 342; 22.6, p. 491.

E. MAJERNÍKOVÁ and Š. BARTA
Magnetic Freeze-Out Effect in Noncrystalline Semiconductors
(b) Vol. **86**, No. 1, p. 183—194 (1978).
See also *13.1*, p. *152*; *13.5.1*, p. *190*; 18, p. 273; 22, p. 410.

R. GRÜNDLER, K. BREUER, and W. TEWS
Optical Properties of Amorphous and Polycrystalline BeO Thin Films from Electron Energy Loss Measurements
(b) Vol. **86**, No. 1, p. 329—338 (1978).
See also 13.1, p. 153; *14.4*, p. *248*; *20.1*, p. *322*; 22.6, p. 492.

J. DUMAS
*Analysis of the Intensity of the ESR Line in Amorphous Si–Ge Films
(b) Vol. **86**, No. 1, p. K75—K77 (1978).
See also *19*, p. *305*; 22.1.1, p. 424; 22.1.2, p. 431.

P. THOMAS and D. WUERTZ
*Low Temperature Polaronic Hopping Conduction
(b) Vol. **86**, No. 2, p. 541—548 (1978).
See also *14.3*, p. *219*; 15, p. 260; 22.1.1, p. 425; 22.1.2, p. 432.

L. VON HEIMENDAHL
*Improvements of the Equation of Motion Technique for Disordered Systems
(b) Vol. **86**, No. 2, p. 549—556 (1978).
See also *6*, p. *68*.

R. S. Sorbello
*Electromigration in Liquid Metal Alloys
(b) Vol. **86**, No. 2, p. 671—678 (1978).
See also *14.1*, p. *207*; 21.2, p. 392.

W. C. Kok
*Resistivity of Spin Glasses below the Ordering Temperature
(b) Vol. **86**, No. 2, p. K113—K117 (1978).
See also *14*, p. *203*.

V. K. Saxena and R. Kishore
*Disorder Induced Behaviour of Magnetisation at Low Temperatures
(b) Vol. **86**, No. 2, p. K163—K166 (1978).
See also *18.2*, p. *285*.

K. Hübner and A. Lehmann
Chemical Bond and Related Properties of SiO_2(IV): Structure of Amorphous SiO_2
(a) Vol. **46**, No. 2, p. 451—457 (1978).
See also *4*, p. *57*; 22.6, p. 492.

J. F. Graczyk
*Defect Structures in Amorphous Germanium Films
(a) Vol. **46**, No. 2, p. 611—617 (1978).
See also *1.5*, p. *36*; 22.1.1, p. 425.

R. Yamamoto, T. Hoshiai, H. Matsuoka, K. Taira, and M. Doyama
*The Calculations of the Heat of Crystallization and Bulk Modulus of Glassy Iron
(a) Vol. **46**, No. 2, p. 671—674 (1978).
See also *12.1*, p. *135*; 21.1.1, p. 386.

A. Delunas and F. Aymerich
*On Electrolytic Amorphous Antimony
(a) Vol. **46**, No. 2, p. 703—706 (1978).
See also *1.2*, p. *18*; 21.7, p. 406.

H.-R. Müller and R. Perthel
*Model Calculations for the Anisotropy Formation by Resputtering Processes during the Growth of Amorphous Gd–Co Films
(b) Vol. **87**, No. 1, p. 203—211 (1978).
See also 3, p. 54; *18.2.1*, p. *293*; 21.1, p. 372; 21.4, p. 397.

L. Vescan, H. Stötzel, T. Stoica, and A. Kottwitz
Photoconductivity in a-GeSe
(b) Vol. **87**, No. 1, p. 239—246 (1978).
See also *16*, p. *264*; 22.7, p. 501.

J. F. Graczyk and P. Chaudhari
*Microdiffraction Patterns from the Finney Model
(b) Vol. **87**, No. 1, p. K5—K8 (1978).
See also *1.5*, p. *36*.

B. Rauschenbach and W. Hinz
*Sputtering of Flat Glass Surfaces with Noble Gas Ions
(a) Vol. **47**, No. 1, p. 79—84 (1978).
See also *11*, p. *126*.

V. Cháb, L. Štourač, and B. Velický
*The Composition Dependence of Activation Energy in the As_x–Te_{100-x} and $As_2Te_3 + Tl_x$ Glassy Systems
(a) Vol. **47**, No. 1, p. 271—276 (1978).
See also *14.3*, p. *221*; 15, p. 260; 22, p. 413.

K. O. Lee
*The Meyer-Neldel Rule and the DC Conductivity in Amorphous Ge Films
(a) Vol. **47**, No. 1, p. K47—K49 (1978).
See also *14.3.1*, p. *227*; 22.1.1, p. 425.

C. Razzetti and P. P. Lottici
*Resonant Raman Scattering in As_2S_3 Amorphous Thin Films
(b) Vol. **87**, No. 2, p. 479—483 (1978).
See also *20.1*, p. *327*; 22, p. 413.

Y. Waseda and H. S. Chen
*Calculation of Electrical Resistivity and Its Temperature Coefficient of Amorphous Cu_{60}–Zr_{40} Alloy Using *t*-Matrix
(b) Vol. **87**, No. 2, p. 777—782 (1978).
See also *14.1*, p. *208*; 21.1, p. 373.

J. Richter
*Electronic Structure of Amorphous Alloys
(b) Vol. **87**, No. 2, p. K89—K92 (1978).
See also *13*, p. *145*; 18.2, p. 287.

R. Bling and F. C. de Sa Barreto
*Electric Field Induced Lifshits Point in Smectic Liquid Crystals
(b) Vol. **87**, No. 2, p. K105—K109 (1978).
See also *1.2*, p. *19*; 14.4, p. 249.

U. Kh. Kopvillem and V. M. Choodnovskii
*Phonon Avalanche in Glasses
(b) Vol. **87**, No. 2, p. K145—K149 (1978).
See also *6*, p. *72*; 7, p. 81.

B. Knappe, H.-R. Müller, and B. Springmann
*Homogeneity of Amorphous Gd–Co Films Prepared by Bias Sputtering
(a) Vol. **47**, No. 2, p. 523—532 (1978).
See also *1.5*, p. *37*; 18.3, p. 296; 21.1, p. 374; 21.4, p. 398.

N. Terao
*Experimental Study of Crack Propagation in Brittle Amorphous Solids. Bifurcation of a Crack in Glass
(a) Vol. **47**, No. 2, p. 565—575 (1978).
See also *12.2*, p. *140*; 22.8, p. 505.

Z. Vučić, D. Subašić, and Z. Ogorelec
*The Determination of the Crystal–Liquid Interfacial Tension in Germanium
(a) Vol. **47**, No. 2, p. 703—710 (1978).
See also 22.1.1, p. 426.

S. Neov, I. Gerassimova, K. Krezhov, B. Sydzhimov, and V. Kozhukharov
*Atomic Arrangement in Tellurite Glasses Studied by Neutron Diffraction
(a) Vol. **47**, No. 2, p. 743—750 (1978).
See also 4, p. 58; 22.6, p. 494.

Y. Shimada and H. Kojima
*Thermal Stability of Amorphous Ferromagnetic Thin Films Produced by Sputtering
(a) Vol. **47**, No. 2, p. K119—K121 (1978).
See also *18.2.1*, p. *293*; 21.1.1, p. 387.

B. Cvikl
*On Coherent Neutron Diffraction in Nematic Liquid Crystals
(b) Vol. **88**, No. 1, p. 25—36 (1978).
See also *6*, p. *73*.

Gh. Ilonca
*Electrical Resistivity and Magnetoresistivity of Amorphous $Cr_xPd_{82-x}Ge_{18}$ Alloys
(b) Vol. **88**, No. 1, p. 37—41 (1978).
See also *14.3*, p. *221*; 22, p. 414.

I. P. Zvyagin
*Hopping Conductivity in the Presence of an Electromagnetic Wave
(b) Vol. **88**, No. 1, p. 149—154 (1978).
See also 6, p. 73; *14.3*, p. *222*.

S. Ueda and T. Shimizu
*Nuclear Magnetic Resonance of ^{77}Se in Amorphous $Cu_x(As_{0.4}Se_{0.6})_{1-x}$
(b) Vol. **88**, No. 1, p. K1—K3 (1978).
See also *19*, p. *309*; 22, p. 414.

H. A. Brown
*The Bethe-Peierls-Weiss Approximation for Amorphous Magnets
(b) Vol. **88**, No. 1, p. K31—K34 (1978).
See also *18.2*, p. *288*.

D. Wuertz and P. Thomas
*The Pre-Exponential in DC-Hopping Conduction
(b) Vol. **88**, No. 1, p. K73—K77 (1978).
See also *14.3*, p. *222*; 22.1.2, p. 434.

V. K. Saxena
*Constant Coupling Approximation for the Curie Temperature of Ferromagnets with Random Anisotropy
(b) Vol. **88**, No. 1, p. K87—K91 (1978).
See also *18.2*, p. *288*.

V. A. Gritsenko, E. E. Meerson, and S. P. Sinitsa
*Unsteady Silicon Nitride Conductivity in High Electric Fields
(a) Vol. 48, No. 1, p. 31—37 (1978).
See also *14.3.2*, p. *230*; *14.3.4*, p. *243*; 22, p. 415.

K. Hübner
Chemical Bond and Related Properties of SiO_2(V): Pseudopotential Study of Bonding and Structure
(a) Vol. 48, No. 1, p. 147—154 (1978).
See also *4*, p. *59*; *13.1*, p. *157*; 14.4, p. 249; 22.6, p. 494.

M. Kikuchi, K. Fukamichi, T. Masumoto, T. Jagielinski, K. I. Arai, and N. Tsuya
Giant ΔE Effect and Elinvar Characteristics in Amorphous Fe–B Binary Alloys
(a) Vol. 48, No. 1, p. 175—181 (1978).
See also *18.2*, p. *289*; 21.1.1, p. 388.

J. Kočka
*Influence of Contacts on the AC Conductivity of Glasses
(a) Vol. 48, No. 1, p. K59—K62 (1978).
See also *14.3*, p. *222*; 14.3.4, p. 243; 22, p. 415.

G. C. Vezzoli
*Peaked Electrical Conductivity in Liquid Bismuth and Associated Optical Scattering
(a) Vol. 48, No. 1, p. K75—K77 (1978).
See also *14.1*, p. *209*; 20.1, p. 330; 21.7, p. 407.

P. Doubrava, V. L. Averyanov, and J. Škácha
*Spectral Characteristics of Photodarkening of a-$As_{60}Se_{40}$ Films
(a) Vol. 48, No. 1, p. K93—K96 (1978).
See also 16, p. 265; *20.1*, p. *330*; 22, p. 415.

H. Mell, B. Movaghar, and L. Schweitzer
*Magnetic Field Dependence of the Photoconductivity in Amorphous Silicon
(b) Vol. 88, No. 2, p. 531—535 (1978).
See also *16*, p. *265*; 22.1.2, p. 435.

Nguyen van Dong and Tran Quoc Hai
*Electronic Transport in Doped a-Ge and a-Si Prepared by DC Cathodic Sputtering
(b) Vol. 88, No. 2, p. 555—561 (1978).
See also *14.3.1*, p. *227*; 22.1.1, p. 427; 22.1.2, p. 435.

H. Szymczak
*On Magnetostriction in Amorphous Thin Films
(b) Vol. 88, No. 2, p. K97—K101 (1978).
See also *18*, p. *275*; 21, p. 361.

J. D. Patterson, G. R. Gruzalski, and D. J. Sellmyer
*Finite Temperature Calculations for the Random Anisotropy Model for Amorphous Magnets
(b) Vol. 88, No. 2, p. K127—K130 (1978).
See also *18.2*, p. *289*.

U. Köster
Crystallization of Amorphous Silicon Films
(a) Vol. **48**, No. 2, p. 313–321 (1978).
See also *1.5*, p. *37*; *3*, p. *55*; 22.1.2, p. 435.

M. H. El-Fouly and J. T. Edmond
*Electrical Conductivity of Some Chalcogenide Glasses as a Function of Electric Field and Temperature
(a) Vol. **48**, No. 2, p. 395–405 (1978).
See also *14.3.4*, p. *244*; 22, p. 416.

T. Hirata
Structural Changes during Annealing Vacuum-Evaporated Thin Aluminium Films at Temperatures near the Thermodynamic Melting Point
(a) Vol. **48**, No. 2, p. 451–458 (1978).
See also *1.4*, p. *32*; *1.5*, p. *37*; 21, p. 361.

A. Buroff, E. Nebauer, and P. Süptitz
*Influence of an Electric DC Field on the Migration of Silver in Amorphous As_2S_3
(a) Vol. **48**, No. 2, p. K109–K112 (1978).
See also *9*, p. *93*; 14.3.4, p. 244; 22, p. 416.

G. Juška, S. Vengris, and K. Arlauskas
*Electron Lifetime in Amorphous Selenium
(a) Vol. **48**, No. 2, p. K149–K151 (1978).
See also *14.3*, p. *223*; 22.1.3, p. 441.

O. K. Alekseeva, V. I. Mikhailov, and V. P. Shantarovich
*Positron Annihilation in Point Defects of the Glassy As–Se System
(a) Vol. **48**, No. 2, p. K169–K173 (1978).
See also *10.2*, p. *117*; 22, p. 416.

S. Y. Chuang, P. K. Tseng, G. J. Jan, and H. S. Chen
*An Experimental Study of the Angular Correlation of Positron Annihilation Radiations in Electron Irradiated Pd–Ni–Si Metallic Glass
(a) Vol. **48**, No. 2, p. K181–K183 (1978).
See also *1.4*, p. *32*; 21.1, p. 377; 21.6, p. 404.

I. I. Fishchuk and V. N. Rudko
*Theory of the Phononless High Frequency Hall Effect in Disordered Systems
(b) Vol. **89**, No. 1, p. 61–68 (1978).
See also *14*, p. *204*.

J. Richter and W. Schiller
*Residual Resistivity of Disordered Systems
(b) Vol. **89**, No. 1, p. 143–150 (1978).
See also *14.1*, p. *210*; 21, p. 362.

Th. Eifrig
*On the Electronic Properties of Disordered Quasi-One-Dimensional Systems with Electron Correlation
(b) Vol. **89**, No. 1, p. 171–175 (1978).
See also *14.1*, p. *210*; 18.1, p. 279; 22.9, p. 526.

P. Blaudeck and R. Lenk
*Multiple Scattering in Condensed Materials (II): Calculation of the Density of States for Non-Crystalline Copper and Iron
(b) Vol. **89**, No. 1, p. 199—202 (1978).
See also 13, p. 147; 21.1, p. 377; 21.1.1, p. 389.

H.-J. Nowak, H. Leitz und W. Buckel
*Zur Struktur amorpher metallischer Schichten des Systems Germanium–Kupfer (On the Structure of Amorphous Metallic Layers of the System Germanium–Copper)
(a) Vol. **49**, No. 1, p. 73—82 (1978).
See also *1.5*, p. *38*; 21.1, p. 377; 22.1.1, p. 427.

D. Möckel, W. Rössler, and F. Baumann
Electrical Resistivity of Amorphous Sb Films and Sb–Au Films
(a) Vol. **49**, No. 1, p. 103—110 (1978).
See also *14.1*, p. *210*; *14.3.1*, p. *228*; 21.6, p. 404; 21.7, p. 407.

A. V. Moskalonov and J. R. Zakis
X-Ray Induced Paramagnetic States and Luminescence in As_2S_3 and As_2Se_3
(a) Vol. **49**, No. 1, p. 231—235 (1978).
See also *13.4*, p. *181*; 19, p. 312; *20.3*, p. *347*; 22, p. 417.

N. Terao
Experimental Study of Crack Propagation in Brittle Amorphous Solids. Refraction of Propagating Crack by Stress Waves
(a) Vol. **49**, No. 1, p. 237—248 (1978).
See also *10.2*, p. *118*; *12.2*, p. *140*.

K. Dettmer, H. Müller, and F. R. Kessler
*Surface Influences of Glass Substrates on Lattice Parameters and Phase Transition Temperature or Evaporated Ge Thin Films
(a) Vol. **49**, No. 1, p. 255—260 (1978).
See also *1.5*, p. *38*; 22.1.1, p. 427.

U. Saxena and O. N. Srivastava
Effect of Nonstoichiometry on the Electrical Conductivity of Erbium Oxide Films
(a) Vol. **49**, No. 1, p. 371—378 (1978).
See also *1.5*, p. *38*; 13.1, p. 158; *14.3.1*, p. *228*; 22.6, p. 496.

V. Dutta, P. Nath, V. D. Vankar, and K. L. Chopra
Structural Properties of Polycrystalline Ge–Metal Films
(a) Vol. **49**, No. 1, p. 379—385 (1978).
See also 1.4, p. 33; *1.5*, p. *38*; 22.1.1, p. 427.

Y. Waseda and H. S. Chen
*A Structural Study of Metallic Glasses Containing Boron (Fe–B, Co–B, and Ni–B)
(a) Vol. **49**, No. 1, p. 387—392 (1978).
See also 21.1, p. 378.

Y. Waseda and W. A. Miller
*Structural Relaxation in Metallic Glasses
(a) Vol. **49**, No. 1, p. K31—K36 (1978).
See also 21.6, p. 405.

H.-R. MÜLLER
*Influence of Anisotropic Short-Range Order on Magnetic Properties of the Amorphous Ferrimagnet Gd–Co
(a) Vol. **49**, No. 1, p. K63—K66 (1978).
See also *18.3*, p. *296*; 21.1, p. 378; 21.4, p. 398.

R. KOSICIK, P. DUDA, W. KEILIG, and R. MATTHEIS
*The Influence of Ion Bombardment on the Magnetic Anisotropy in Evaporated Amorphous Gd–Co Films
(a) Vol. **49**, No. 1, p. K67—K69 (1978).
See also 11, p. 129; *18.3*, p. *296*; 21.1, p. 378; 21.4, p. 398.

J. ČERMÁK, V. TVAROŽEK, R. HARMAN, and J. KACZÉR
*Contribution to the Technology of Gd–Co Films
(a) Vol. **49**, No. 1, p. K81—K83 (1978).
See also *18.2.1*, p. *294*; 21.1, p. 378; 21.4, p. 398.

W. A. THEINER and H. P. GESERICH
Anderson Transition in $FeSi_2$ Films
(b) Vol. **89**, No. 2, p. 441—447 (1978).
See also *1.2*, p. *22*; *1.5*, p. *39*; 14.1, p. 210; 20.1, p. 334; 21.1.1, p. 389.

W. FUHS, M. MILLEVILLE, and J. STUKE
Drift Mobility and Photoconductivity in Amorphous Silicon
(b) Vol. **89**, No. 2, p. 495—502 (1978).
See also *14.3.1*, p. *228*; *16*, p. *266*; 22.1.2, p. 437.

E. WIMMER, P. WEINBERGER, A. KOSAKOV, and G. LEONHARDT
On the Electronic Structure of As_2O_3, Sb_2O_3, and As_2S_3
(b) Vol. **89**, No. 2, p. 619—624 (1978).
See also *13*, p. *148*; *20*, p. *317*; 22.6, p. 496.

R. ENDERLEIN
On the Theory of Light Propagation through Disordered Solids. Scattering from Configurational Fluctuations
(b) Vol. **89**, No. 2, p. 625—632 (1978).
See also *6*, p. *76*; *20.1*, p. *334*; 22, p. 418.

B. ESSER
On the Theory of Resonance Raman Scattering in Disordered Semiconductors
(b) Vol. **89**, No. 2, p. 633—643 (1978).
See also *6*, p. *76*; 13.1, p. 159; *20.1*, p. *334*; 22, p. 418.

H. BERTAGNOLLI and H. G. HERTZ
*Preservation and Loss of Structural Features of Solid Acetic Acid during the Melting Process
(a) Vol. **49**, No. 2, p. 463—472 (1978).
See also *1*, p. *10*; 22.9, p. 527.

K. MAEDA and S. TAKEUCHI
*Computer Simulation of Deformation in Two-Dimensional Amorphous Structures
(a) Vol. **49**, No. 2, p. 685—696 (1978).
See also 10, p. 97; 12, p. 133.

V. P. ARORA and V. K. AGARWAL
*Dielectric Behaviour of Nematic Phase OHMBBA at Radio Frequencies
(a) Vol. **49**, No. 2, p. 757—760 (1978).
See also *14.4*, p. *251*; 22.9, p. 528.

N. TERAO
*Experimental Study of Crack Propagation in Brittle Amorphous Solids: Cracks Formed in Glassy Carbon
(a) Vol. **49**, No. 2, p. K101—K102 (1978).
See also *12.2*, p. *141*; 22.1, p. 422.

R. DURNÝ
*EPR of Mn^{2+} in Ge–S Glasses
(a) Vol. **49**, No. 2, p. K113—K116 (1978).
See also *19*, p. *313*; 22, p. 419.

F. SCHAUER, V. ŠMÍD, and V. ŽALUDEK
*On the Contact Capacitance and the Field Effect in Chalcogenide Glasses
(a) Vol. **49**, No. 2, p. K157—K161 (1978).
See also 13.3, p. 167; *14.3.4*, p. *245*.

M. MITERA, M. NAKA, T. MASUMOTO, N. KAZAMA, and K. WATANABE
*Effects of Metalloids on the Magnetic Properties of Iron Based Amorphous Alloys
(a) Vol. **49**, No. 2, p. K163—K166 (1978).
See also *18.2*, p. *291*; 21.1.1, p. 390.

K. NARAYANDAS, M. RADHAKRISHNAN, and C. BALASUBRAMANIAN
Annealing Effect of Silver–Tellurium Films
(a) Vol. **49**, No. 2, p. K213—K215 (1978).
See also 1.2, p. 23; *14.1*, p. *211*; 21.6, p. 405; 22.1.3, p. 442.

B. B. SAHAY and P. L. SRIVASTAVA
Isothermal Compressibility for Various Liquid Metals by Harmonic Model Potential
(b) Vol. **90**, No. 1, p. 93—97 (1978).
See also *12.1*, p. *136*; 21, p. 364; 21.1, p. 379; 21.2, p. 393.

N. N. SINHA and P. L. SRIVASTAVA
Resistivity and Thermo-Electric Power of Monovalent, Divalent, and Trivalent Liquid Metals
(b) Vol. **90**, No. 1, p. 369—374 (1978).
See also *14.1*, p. *211*; *15*, p. *261*; 21, p. 364; 21.1, p. 380; 21.2, p. 393.

J. D. PATTERSON
*The Low Temperature Lorenz Number of Spin Glasses
(b) Vol. **90**, No. 1, p. K85—K88 (1978).
See also *8*, p. *89*; 14.1, p. 212; 21.1.1, p. 390.

G. HAUSCH and E. TÖRÖK
*Elastic, Magnetoelastic, and Thermal Properties of Some Ferromagnetic Metallic Glasses
(a) Vol. **50**, No. 1, p. 159—164 (1978).
See also 8, p. 89; 12.1, p. 136; *18.2*, p. *291*; 21.1, p. 381; 21.1.1 p. 390.

R. Yamamoto, T. Mihara, and M. Doyama
*Comments on the Glass Forming Region of the Metal–Metalloid Systems Predicted by the Polk Model
(a) Vol. **50**, No. 1, p. 165–169 (1978).
See also 21, p. 365.

D. I. Tsiulyanu, A. M. Andriesh, and E. P. Kolomeyko
*Arsenic Trisulphide in Isotype Amorphous–Crystalline Heterojunctions
(a) Vol. **50**, No. 1, p. 195–202 (1978).
See also *14.3.3*, p. *237*; 14.3.4, p. 246; 22, p. 420.

T. Jagieliński, K. I. Arai, N. Tsuya, and K. Fukamichi
*Magnetoelastic Effects in Invar-Type Amorphous Fe–B Alloys
(a) Vol. **50**, No. 1, p. K25–K28 (1978).
See also *18.2*, p. *292*; 21.1.1, p. 390.

M. Urner-Wille, T. S. te Velde, and P. G. van Engen
Kerr-Effect Enhancement in Amorphous GdFe Films
(a) Vol. **50**, No. 1, p. K29–K31 (1978).
See also *18.3*, p. *297*; *20*, p. *318*; 21.1.1, p. 390; 21.4, p. 398.

P. Macko and J. Doupovec
Thermal and Electrical Properties of Neutron Irradiated GeS_{11} Glass
(a) Vol. **50**, No. 1, p. K51–K54 (1978).
See also *8*, p. *89*; 11, p. 130; *14.3*, p. *225*; 22, p. 420.

S. D. Shutov and M. S. Iovu
*Dispersive Charge Transport in Vitreous 0.55 $As_2S_3 \cdot 0.45\ Sb_2S_3$
(a) Vol. **50**, No. 1, p. K59–K61 (1978).
See also *14.3.1*, p. *229*; 22, p. 420.

H. A. Brown
*On the Classical-Spin Model of Amorphous Ferromagnets
(b) Vol. **90**, No. 2, p. K165–K168 (1978).
See also *18.2*, p. *292*.

P. T. Landsberg, D. J. Robbins, and E. Schöll
Threshold Switching as a Generation–Recombination Induced Non-Equilibrium Phase Transition
(a) Vol. **50**, No. 2, p. 423–426 (1978).
See also 13.4, p. 185; *14.3.4*, p. *247*; 22.1, p. 423.

G. Schroeder, R. Schäfer, and H. Kronmüller
Magneto-Optical Investigation of the Domain Structure of Amorphous $Fe_{80}B_{20}$ Alloys
(a) Vol. **50**, No. 2, p. 475–481 (1978).
See also *18.2*, p. *292*; *20*, p. *318*; 21.1.1, p. 390.

A. A. Bukharaev and N. R. Yafaev
Laser and Thermal Bleaching of Colour Centres in Sodium Borate Glasses
(a) Vol. **50**, No. 2, p. 711–716 (1978).
See also *13.4*, p. *186*; *20.1*, p. *338*; 22.8.1, p. 516.

L. N. Skuja, A. R. Silin, and J. Mareš
*Decay Time and Polarization Properties of Luminescence Centers in Vitreous Silica
(a) Vol. **50**, No. 2, p. K149–K152 (1978).
See also *20.3*, p. *350*; 22.6, p. 498.

H. Kresse, D. Demus, and A. Wiegeleben
*Dielectric Relaxation of Liquid Crystals with Nematic, Smectic A, and Smectic C Phases
(a) Vol. **50**, No. 2, p. K181–K183 (1978).
See also *14.4*, p. *252*; 22.9, p. 529.

3. Crystal Growth

Review Article

V. V. TITOV
*Ion Implantation — Problems and Perspectives
(a) Vol. **48**, No. 1, p. 13—22 (1978).
See also *11*, p. *123*; 22.1.2, p. 429.

Original Papers and Short Notes

D. BOULANGER and M.-S. MARTIN
Crystal Growth and EPR Study of Cr^{+}-Doped MgTe Single Crystals
(b) Vol. **85**, No. 2, p. 597—601 (1978).
See also 13.4, p. 170; *19*, p. *304*; 22.4.3, p. 470.

M. HIROSE, A. FISCHER, and K. PLOOG
*Growth of Al_2O_3 Layer on MBE GaAs
(a) Vol. **45**, No. 2, p. K175—K177 (1978).
See also *14.3.1*, p. *226*; 22.6, p. 491.

R. W. A. HILLHOUSE and J. WOODS
*Plasma Growth of Rutile Crystals and Their Photoelectronic Properties
(a) Vol. **46**, No. 1, p. 163—172 (1978).
See also 13.4, p. 173; 16, p. 263; 22.6, p. 492.

R. HERGT and P. GÖRNERT
*Growth Rate Anisotropy and Kinetic Coefficients of Vicinal Faces at LPE Garnet Films (I): Growth Rate Anisotropy
(a) Vol. **46**, No. 2, p. 427—432 (1978).
See also 22.8.2, p. 518.

B. ETLINGER, R. TROJKO, and U. V. DESNICA
Erratum to: Thermal Decomposition of $Al_2In_{12}S_{21}$
(a) Vol. **46**, No. 2, p. K173 (1978).
See also 22.3.1, p. 460.

H.-R. MÜLLER and R. PERTHEL
Model Calculations for the Anisotropy Formation by Resputtering Processes during the Growth of Amorphous Gd–Co Films
(b) Vol. **87**, No. 1, p. 203—211 (1978).
See also *2*, p. *44*; *18.2.1*, p. *293*; 21.1, p. 372; 21.4, p. 397.

P. GÖRNERT and R. HERGT
*Growth Rate Anisotropy and Kinetic Coefficients of Vicinal Faces at LPE Garnet Films (II): Kinetic Coefficients
(a) Vol. **47**, No. 1, p. 99—104 (1978).
See also 22.8.2, p. 519.

E. Kuroda, M. Matsuda, and M. Maki
*Growth and Characterization of Silicon Ribbon Crystals Grown with Wetting and Non-Wetting Dies
(a) Vol. **48**, No. 1, p. 105—111 (1978).
See also 14.3, p. 222; 22.1.2, p. 434.

J. Rużyłło
*Model of Structural Transformations on the Chemically Etched Silicon Surface during Thermal Oxidation
(a) Vol. **48**, No. 1, p. 199—202 (1978).
See also *1.2*, p. *21*; 1.6, p. 41; 22.1.2, p. 434; 22.6, p. 494.

J. Borkowicz, J. Korec, and E. Nossarzewska-Orłowska
*Optimum Growth Conditions in Silicon Vapour Epitaxy
(a) Vol. **48**, No. 1, p. 225—228 (1978).
See also 22.1.2, p. 434.

O. A. Troitskii
*Fractionation of Stable Isotopes of Metals by Combined Electric Migration and Zone Melting
(a) Vol. **48**, No. 1, p. 229—234 (1978).
See also 9, p. 93; 21, p. 360.

H. E. Peltner and Ch. Herzig
*New Application of the Knudsen Method for a Direct Determination of Thermodynamic Activities
(a) Vol. **48**, No. 1, p. K79—K81 (1978).
See also 21.6, p. 403.

U. Köster
*Crystallization of Amorphous Silicon Films
(a) Vol. **48**, No. 2, p. 313—321 (1978).
See also *1.5*, p. *37*; 2, p. 48; 22.1.2, p. 435.

K. H. Yang and G. H. Schwuttke
Analysis of Vapor-Deposited Silicon Carbide Films on Silicon Ribbon Surfaces
(a) Vol. **48**, No. 2, p. 335—343 (1978).
See also *1.5*, p. *37*; 22.3, p. 458.

E. V. Shvedov and V. M. Ievlev
*Nucleus Saturation Density and Epitaxy in the Condensation of Metals on Alkali-Halide Crystals (II): Oriented Growth of Islands
(a) Vol. **48**, No. 2, p. 603—608 (1978).
See also *10.2*, p. *116*; 22.5.2, p. 483.

W. Eigermann, G. Müller-Vogt, and W. Wendl
*Solubility Curves in High-Temperature Melts for the Growth of Single Crystals of Rare Earth Vanadates and Phosphates
(a) Vol. **49**, No. 1, p. 145—148 (1978).
See also 1.3, p. 27; 22.8.1, p. 515.

A. A. Kaminskii, V. V. Osiko, S. E. Sarkisov, M. T. Timoshechkin, E. V. Zharikov, J. Bohm
P. Reiche, and D. Schultze
Growth, Spectroscopic Investigations, and Some New Stimulated Emission Data of $Gd_3Ga_5O_{12}$: Nd^{3+} Single Crystals
(a) Vol. **49**, No. 1, p. 305—311 (1978).
See also 20.1, p. **333**; *20.2*, p. *339*; *20.3*, p. *347*; 22.8.2, p. 520.

V. M. Ievlev, A. V. Bugakov, and V. A. Ammer
Interphase Coincidence Boundaries in Au–Si and Au–CaF_2 Systems
(a) Vol. **49**, No. 1, p. 413—420 (1978).
See also *1.4*, p. *33*; *1.5*, p. *38*; 21.6, p. 404.

H. Nelkowski and K. Petermann
Preparation and Investigation of ZnS–$CuGaS_2$ Heterojunctions
(a) Vol. **49**, No. 1, p. K13—K16 (1978).
See also 1.5, p. 39; *14.3.3*, p. *236*; *20.3*, p. *348*; 22.4.1, p. 465; 22.8, p. 507.

J. Gastaldi and C. Jourdan
*New Possibilities for Recrystallization Study by X-Ray Synchrotron Radiation Topography
(a) Vol. **49**, No. 2, p. 529—537 (1978).
See also 1.4, p. 33; 21, p. 363.

Y. Andersson, S. Rundqvist, O. Beckman, L. Lundgren, and P. Nordblad
*Properties of Fe_2P Crystals Prepared from a Liquid Copper Medium
(a) Vol. **49**, No. 2, p. K153—K156 (1978).
See also 18.2, p. 291; 21.1, p. 379; 21.1.1, p. 389.

N. A. Zakharov, V. S. Krikorov, E. F. Kustov, and S. Yu. Stefanovich
New Non-Linear Crystals in the $A_2B_2O_7$ Series
(a) Vol. **50**, No. 1, p. K13—K16 (1978).
See also *14.4.1*, p. *257*; *20*, p. *318*; 22.8, p. 508.

4. Bonding Properties

Original Papers and Short Notes

H. Gottschalk, G. Patzer, and H. Alexander
Stacking Fault Energy and Ionicity of Cubic III–V Compounds
(a) Vol. **45**, No. 1, p. 207–217 (1978).
See also *10.2*, p. *109*; 22.2.1, p. 445; 22.2.2, p. 450; 22.2.3, p. 452.

S. Grande, H.-D. Mecke, and St. Idziak
*High-Resolution Proton Magnetic Resonance in Single Crystals of Potassium Trihydrogen Selenite
(a) Vol. **45**, No. 1, p. K29–K32 (1978).
See also *19*, p. *303*; 22.8, p. 503.

N. I. Kulikov, V. N. Borzunov, and A. D. Zvonkov
The Electronic Band Structure and Interatomic Bond in Nickel and Titanium Hydrides
(b) Vol. **86**, No. 1, p. 83–91 (1978).
See also *13.1*, p. *152*; 21.1, p. 371.

T. Soma
*Local Heine-Abarenkov Model Potential for Covalent Crystals
(b) Vol. **86**, No. 1, p. 263–268 (1978).
See also 13.1, p. 153; 22.1, p. 422; 22.1.1, p. 424; 22.1.2, p. 430.

V. K. Bashenov, D. I. Marvakov, and A. M. Mutal
*Internal Strain Parameter in Diamond-Type Crystals
(b) Vol. **86**, No. 1, p. K7–K10 (1978).
See also 12.2, p. 138; 22.1, p. 422; 22.1.1, p. 424; 22.1.2, p. 431.

O. P. Sharma
A Pseudopotential Study of the Electronic Properties of Noble Metals
(b) Vol. **86**, No. 2, p. 483–492 (1978).
See also *6*, p. *68*; 14.1, p. 207; 21.1, p. 371; 21.6, p. 401.

D. D. Richardson
*Use of Finite Atomic Radius in Dispersion Force Calculations
(b) Vol. **86**, No. 2, p. K97–K100 (1978).
See also *13*, p. *144*.

K. Hübner and A. Lehmann
*Chemical Bond and Related Properties of SiO_2(IV): Structure of Amorphous SiO_2
(a) Vol. **46**, No. 2, p. 451–457 (1978).
See also 2, p. 44; 22.6, p. 492.

S. Grande, H.-D. Mecke, and L. A. Shuvalov
*Deuterium Resonance of $KD_3(SeO_3)_2$ Single Crystals above and below Phase Transition
(a) Vol. **46**, No. 2, p. 547–555 (1978).
See also *1.2*, p. *17*; 19, p. 306; 22.8, p. 504.

L. P. Mokhracheva, V. A. Tskhai, and P. V. Geld
*Charge Distribution and Ionicity of the Chemical Bonds in TiC, VC, and TiN
(b) Vol. 87, No. 1, p. 49—52 (1978).
See also *13*, p. *145*; 22.3, p. 457.

A. Milchev and K. Unger
*Third-Order Contribution to the Effective Ion–Ion Interaction and Formation of Angular Forces in Crystalline Solids (I): General Formulation of the Expression for the Total Energy
(b) Vol. 87, No. 1, p. 227—237 (1978).
See also 13, p. 145; 22.2, p. 443.

A. Milchev and K. Unger
*Third-Order Contribution to the Effective Ion–Ion Interaction and Formation of Angular Forces in Crystalline Solids (II): Application to Tetrahedral Solids
(b) Vol. 87, No. 2, p. 473—478 (1978).
See also 13, p. 145; 22.2, p. 443.

S. K. Sarkar and S. Sengupta
*Lundqvist Three-Body Interaction and Lattice Dynamics of Ionic Crystals
(b) Vol. 87, No. 2, p. 517—525 (1978).
See also 12.2, p. 139; 22.5.2, p. 481.

R. Prakash, L. K. Bansal, and S. C. Goyal
*Photoelastic Constants of the Alkaline Earth Oxides
(b) Vol. 87, No. 2, p. K127—K131 (1978).
See also *12.2*, p. *139*; 22.6, p. 494.

S. Neov, I. Gerassimova, K. Krezhov, B. Sydzhimov, and V. Kozhukharov
*Atomic Arrangement in Tellurite Glasses Studied by Neutron Diffraction
(a) Vol. 47, No. 2, p. 743—750 (1978).
See also *2*, p. *46*; 22.6, p. 494.

P. R. Sarode
*Correlation between the Chemical Shifts in X-Ray Absorption Spectra of Ionic Solids and Contribution to the Chemical Bond
(b) Vol. 88, No. 1, p. K35—K38 (1978).
See also *20*, p. *317*; 22.5.2, p. 482.

T. Soma
*Equation of State under Pressure for Si and Ge
(b) Vol. 88, No. 1, p. K69—K72 (1978).
See also 22.1.1, p. 426; 22.1.2, p. 434.

J. Cornelis, J. Nihoul, F. Brosens, and J. T. Devreese
*Interatomic Potentials from Experimental Phonon Spectra. Application to Au, Cu, K, Ni, Pd, Pt, and Rb
(b) Vol. 88, No. 1, p. K93—K96 (1978).
See also 6, p. 74; 21, p. 360.

K. Hübner
*Chemical Bond and Related Properties of SiO_2 (V): Pseudopotential Study of Bonding and Structure
(a) Vol. **48**, No. 1, p. 147–154 (1978).
See also 2, p. 47; *13.1*, p. *157*; 14.4, p. 249; 22.6, p. 494.

P. Kadura and L. Kühne
From Small to Infinite F.C.C. Clusters
(b) Vol. **88**, No. 2, p. 537–544 (1978).
See also *13*, p. *146*.

V. C. Jain and J. Shanker
The Exchange Charge Shell Model Analysis of Strain Derivatives of Static Polarizability and Effective Charge Parameter of Alkali Halides
(b) Vol. **89**, No. 1, p. 213–219 (1978).
See also 6, p. 75; 14.4, p. 250; 22.5.2, p. 483.

V. K. Bashenov, D. I. Marvakov, and A. G. Petukhov
*Average Structure Factors and Stability of Binary Crystals
(b) Vol. **89**, No. 1, p. K5–K8 (1978).
See also *13.1*, p. *158*; 22, p. 417.

R. N. Singh and P. L. Srivastava
*Model Potential Calculation of Binding Energy and Compressibility of Alkali Metals
(b) Vol. **89**, No. 1, p. K13–K16 (1978).
See also *13*, p. *147*; 21.2, p. 393.

Debjani Bose and A. N. Basu
*Second-Neighbour Interaction Potential in Alkali Halide Crystals
(b) Vol. **89**, No. 1, p. K35–K38 (1978).
See also *12.2*, p. *140*; 22.5.2, p. 484.

P. S. Grover
*Computation of Van Der Waals Forces across Metal Surfaces
(b) Vol. **89**, No. 2, p. K163–K166 (1978).
See also 21, p. 363.

W. F. W. M. van Heugten and J. Th. M. de Hosson
*Interatomic Potentials for Alkali Metals. A Comparative Study
(b) Vol. **90**, No. 1, p. 225–232 (1978).
See also 6, p. 77; 10.1, p. 107; 21.2, p. 393.

P. A. Varotsos
An Estimate of the Pressure Dependence of the Dielectric Constant in Alkali Halides
(b) Vol. **90**, No. 1, p. 339–343 (1978).
See also 6, p. 77; 12.2, p. 141; *14.4*, p. *251*; 22.5.1, p. 477; 22.5.2, p. 485.

V. K. Bashenov, D. I. Marvakov, and A. G. Petukhov
*Bond-Orbital Model of Layer Gallium Selenide
(b) Vol. **90**, No. 1, p. K5–K8 (1978).
See also *13.1*, p. *160*; 22.3, p. 459.

T. Soma
*Equation of State under Pressure for Metallic β-Sn
(b) Vol. **90**, No. 2, p. K129–K131 (1978).
See also 1.2, p. 24; *12.1*, p. *136*; 21, p. 366.

5. Mössbauer Spectroscopy

Review Article

P. P. Seregin, F. S. Nasredinov, and L. N. Vasilev
*A Study of Radiation Defects in Solids by Means of Mössbauer Spectroscopy
(a) Vol. **45**, No. 1, p. 11–45 (1978).
See also *11*, p. *123*; 22.5, p. 474; 22.8, p. 503; 22.9, p. 522.

Original Papers and Short Notes

H. Spiering, S. Hösl, and H. Vogel
*Polarization Effects in Mössbauer Absorption of a $FeCl_2 \cdot 4\,H_2O$ Single Crystal in an External Magnetic Field
(b) Vol. **85**, No. 1, p. 87–93 (1978).
See also 22.5, p. 474.

V. K. Yarmarkin, S. P. Teslenko, and A. I. Knyazev
*Mössbauer Study of Sb and Fe Impurities in Semiconducting and Insulating TiO_2
(a) Vol. **45**, No. 1, p. 63–69 (1978).
See also 9, p. 91; 10.2, p. 109; 22.6, p. 491.

E. Wieser, W. Brückner, B. Thuss, and U. Gerlach
*Mössbauer Effect Studies on Modified VO_2. $V_{1-x}Fe_xO_{2-y}$ and $V_{1-x-y}Fe_xMo_yO_2$
(a) Vol. **45**, No. 1, p. 123–131 (1978).
See also *1.2*, p. *15*; 18.1, p. 277; 22.6.1, p. 499.

J. J. Bara and B. F. Bogacz
Addendum to: Investigations of Thickness Dependences of Mössbauer Absorption and Scattering Lines
(a) Vol. **45**, No. 1, p. K81 (1978).
See also 22.8.1, p. 511.

J. Barton and E. Wieser
*Investigation of Magnetic Preferred Directions in an Fe–Mn Base Alloy
(a) Vol. **45**, No. 2, p. K105–K108 (1978).
See also *18.2*, p. *283*; 21.1, p. 370; 21.1.1, p. 385.

R. G. Pillay and P. N. Tandon
*Hyperfine Magnetic Field at Sn in Heusler Alloys Rh_2MnSn and Rh_2NiSn
(a) Vol. **45**, No. 2, p. K109–K112 (1978).
See also *18.2*, p. *283*; 21, p. 355.

J. V. de Menezes and F. de S. Barros
*On the Mössbauer Spectrum of $FeC_2O_4 \cdot 2H_2O$
(a) Vol. **45**, No. 2, p. K139–K143 (1978).
See also 22.8.2, p. 518.

G. N. Belozerskii
*Superparamagnetism—Determination of the Diffusion Coefficient of the Vector of Magnetization, of Anisotropy Energy, and of Volume Distribution of Particles by Mössbauer Effect
(a) Vol. **46**, No. 1, p. 131—140 (1978).
See also *18.2*, p. *284*.

J. A. Sawicki and B. D. Sawicka
*Correlation of ^{57}Fe Isomer Shift Quadrupole Coupling, and Interatomic Distance in Group-IV Elements
(b) Vol. **86**, No. 2, p. K159—K161 (1978).
See also 22.1.3, p. 441.

M. Kopcewicz
*Mössbauer Study of the Fast Magnetization Reversal Forced in Permalloy and Invar by an External RF Magnetic Field
(a) Vol. **46**, No. 2, p. 675—685 (1978).
See also *18.2*, p. *285*; 21.1, p. 372; 21.1.1, p. 386.

E. Burzo and L. Stănescu
*On the Charge Compensation of Iron Ions in the V_2O_5 Lattice
(a) Vol. **46**, No. 2, p. K163—K166 (1978).
See also 22.6, p. 493.

Ch. S. Kuo and D. Schroeer
*Mössbauer Study of Dy^{3+} in CaF_2
(b) Vol. **87**, No. 1, p. 325—331 (1978).
See also 10.2, p. 113; 22.5.3, p. 487.

S. V. Koptsik, V. A. Povitskii, M. Yu. Messinev, Yu. V. Baldokhin, and N. F. Shchukin
*Mössbauer Study on Shocked Hematite
(a) Vol. **47**, No. 1, p. 39—43 (1978).
See also 18.4, p. 300; 22.6, p. 493.

J. Sitek, J. Kružliak, M. Tomašich, J. Cirák, and M. Prejsa
Phase Analysis of Superconducting Nb—Sn Materials by Mössbauer Spectroscopy
(a) Vol. **47**, No. 1, p. 287—291 (1978).
See also *14.2*, p. *214*.

V.A. Povitskii, A. N. Salugin, Yu. V. Baldokhin, E. F. Makarov, and N. V. Elistratov
*Spin Reorientation in Hematite (α-Fe_2O_3) with Al Impurity
(b) Vol. **87**, No. 2, p. K101—K104 (1978).
See also 18.3, p. 295; 22.8.2, p. 519.

V.V.Gagulin, N. V. Fadeeva, A. G. Belous, A. V. Titov, K. P. Mitrofanov, M. Y. Plotnikova, S. P. Soloviev, and Yu. N. Venevtsev
New Seignette-Magnets with Hexagonal Barium Titanate Structure
(a) Vol. **48**, No. 1, p. 183—189 (1978).
See also *1*, p. *9*; *14.4.1*, p. *255*; 18.2, p. 289; 18.4, p. 301; 19, p. 310; 22.8.1, p. 514.

J. Barton, E. Wieser, and M. Müller
Investigation of the $\alpha-\gamma$ Phase Transformation by Annealing of a Cold-Worked Fe–Mn Base Alloy
(a) Vol. **48**, No. 1, p. K51–K53 (19 8).
See also *1.2*, p. *21*; 21.1.1, p. 388.

R. Chandra, R. P. Tripathi, and S. Lokanathan
*Search for Anisotropy of Mössbauer Absorption in Biotite
(b) Vol. **88**, No. 2, p. 633–637 (1978).
See also 22.8, p. 506.

C. Meyer, Y. Gros, B. Bochu, A. Collomb, J. Chenavas, J. C. Joubert, and M. Marezio
*Synthesis, Crystal Structure, and Mössbauer Study of a Series of Perovskite-Like Compounds [ACu_3] (M, $Fe)_4O_{12}$
(a) Vol. **48**, No. 2, p 581–586 (1978).
See also *1.1*, p. *13*; 22.8.1, p. 515.

G.Langouche, I. Dezsi, M. Van Rossum, J. De Bruyn, and R. Coussement
*On the Existence of a Quadrupole Interaction at ^{57}Fe Implanted in Si and Ge
(b) Vol. **89**, No. 1, p. K17–K19 (1978).
See also 22.1.1, p. 427; 22.1.2, p. 436.

R. Nagarajan and J. K. Srivastava
*Anomalous Mössbauer Lineshapes in a Magnetically Ordered Ga_2O_3–Fe_2O_3 System in Presence of an External Electric Field Applied during the Solid Solution Formation
(b) Vol. **89**, No. 1, p. K79–K83 (1978).
See also *18.1*, p. *279*; 22.6.1, p. 499.

P. Raj, S. K. Kulshreshtha, and J. V. Yakhmi
*Mössbauer and X-Ray Studies of $Fe_{1.67}Ge$
(a) Vol. **49**, No. 1, p. K91–K95 (1978).
See also *1.3*, p. *28*; 18.2, p. 290; 21.1.1, p. 389.

W. Bergholz and W. Schröter
*Precipitation of Cobalt in Silicon Studied by Mössbauer Spectroscopy
(a) Vol. **49**, No. 2, p. 489–498 (1978).
See also 9, p. 94; *10.2*, p. *119*; 22.1.2, p. 437.

J. K. Srivastava and B. W. Dale
Cross Relaxation Studies in the Cr^{3+}-Doped Magnetically Dilute Paramagnetic Al_2O_3–Fe_2O_3 System (I): Effect of Cr^{3+}Concentration
(b) Vol. **90**, No. 1, p. 391–399 (1978).
See also 13.4, p. 184; 22.6.1, p. 499.

E. Japa and K. Krop
*Mössbauer Studies of Ordering in Fe–Si Alloys
(a) Vol. **50**, No. 1, p. K89–K91 (1978).
See also 1.3, p. 28; 21.1.1, p. 390.

J. K. Srivastava and B. W. Dale
*Cross Relaxation Studies in the Cr^{3+}-Doped Magnetically Dilute Paramagnetic Al_2O_3–Fe_2O_3 System (II): Effect of Temperature and External Magnetic Field
(b) Vol. **90**, No. 2, p. 571–574 (1978).
See also 13.4, p. 185; 22.6.1, p. 499.

J. K. Srivastava and S. Muralidhara Rao
*Search for Reclustering of Fe^{3+} Ions in the Incompletely Annealed Cr_2O_3–Fe_2O_3 Solid Solutions
(b) Vol. **90**, No. 2, p. K175—K177 (1978).
See also 22.6.1, p. 500.

Yu. M. Vysotskii, E. A. Vasilev, A. E. Gelyasin, and A. V. Shablovskii
Nuclear Zeeman Effect in the $Mn_{1+x}Sb_{1-x}Sn_x$ System
(a) Vol. **50**, No. 2, p. K205—K208 (1978).
See also *18.2*, p. *292*; *18.3*, p. *297*; 21, p. 367.

6. Lattice Dynamics. Phonons

Review Articles

J. A. Wilson
Modelling the Contrasting Semimetallic Characters of TiS_2 and $TiSe_2$
(b) Vol. **86**, No. 1, p. 11—36 (1978).
See also 13.1, p. 150; *14.1*, p. *205*; 22, p. 408.

V. I. Ivanov-Omskii, L. I. Korovin, and E. M. Shereghii
*Phonon-Assisted Cyclotron Resonance in Semiconductors
(b) Vol. **90**, No. 1, p. 11—32 (1978).
See also 13, p. 142; *19*, p. *303*; 22, p. 408; 22.2.3, p. 452; 22.4.1, p. 463.

P. C. Mathur, R. Shyam, and Sushil Jain
Transport Phenomena in III–V Compound Semiconductors
(a) Vol. **50**, No. 1, p. 11—40 (1978).
See also *13.1*, p. *150*; 13.4, p. 169; *14.3*, p. *217*; 20.1, p. 319; 22.2, p. 443.

Original Papers and Short Notes

V. V. Hizhnyakov and A. V. Sherman
*On the Theory of Resonant Secondary Radiation of Excitons Weakly Interacting with Phonons
(b) Vol. **85**, No. 1, p. 51—61 (1978).
See also *13.5.1*, p. *189*; 20.1, p. 319; 20.3, p. 341; 22, p. 408.

V. G. Vaks, E. V. Zarochentsev, S. P. Kravchuk, V. P. Safronov, and A. V. Trefilov
*Lattice Heat Capacity and Thermal Vibrations in Alkali Metals
(b) Vol. **85**, No. 1, p. 63—74 (1978).
See also *8*, p. *83*; 21.2, p. 392.

E. L. Ivchenko, I. G. Lang, and S. T. Pavlov
*Resonant Secondary Radiation in Polar Semiconductors
(b) Vol. **85**, No. 1, p. 81—85 (1978).
See also *20.3*, p. *341*; 22, p. 408.

G. Neumann and H. Vogt
*Rayleigh Wing Scattering in Disordered Sodium Nitrate
(b) Vol. **85**, No. 1, p. 179—184 (1978).
See also *1.2*, p. *15*; 2, p. 42; 20.1, p. 319; 22.8.1, p. 510.

H. Michel and M. Wagner
*On the Use of Generalized Langevin Equations in Phonon Transport Problems
(b) Vol. **85**, No. 1, p. 195—201 (1978).
See also *13.4*, p. *169*.

P. Kleinert and F. Bechstedt
*Influence of Mass Defect Disorder on Resonance Raman Scattering
(b) Vol. **85**, No. 1, p. 253—260 (1978).
See also *20.1*, p. *319*; 22, p. 409.

V. Wagner, G. Dolling, B. M. Powell, and G. Landwehr
*Lattice Vibrations of Bi_2Te_3
(b) Vol. **85**, No. 1, p. 311–317 (1978).
See also *20.1*, p. *320*; 22.7, p. 501.

E. L. Bokhenkov, E. M. Rodina, E. F. Sheka, and I. Natkaniec
*Inelastic Incoherent Neutron Scattering Spectra at Different Temperatures and Computer Experiment for External Phonon Modes of Naphthalene Crystals
(b) Vol. **85**, No. 1, p. 331–342 (1978).
See also 22.9, p. 522.

K. A. Aristova, C. Trallero Giner, I. G. Lang, and S. T. Pavlov
*Scattering and Decay of Hot Excitons Interacting with Longitudinal Optical Phonons
(b) Vol. **85**, No. 1, p. 351–357 (1978).
See also *13.5.1*, p. *189*; 20.1, p. 320.

F. E. Faradev, N. M. Gasanly, B. N. Mavrin, and N. N. Melnik
*Raman Scattering in Some III–VI Layer Single Crystals
(b) Vol. **85**, No. 1, p. 381–386 (1978).
See also *20.1*, p. *320*; 22.3, p. 457.

P. Turchi, Y. Calvayrac, and F. Plicque
Effects of Configurational Order in a Ni_3Fe Single Crystal on Debye Temperature Deduced from Elastic Constants and X-Ray Measurements
(a) Vol. **45**, No. 1, p. 229–238 (1978).
See also *1.3*, p. *25*; 12.1, p. 134; 21.1, p. 369; 21.1.1, p. 384.

V. Valvoda and J. Ječný
*X-Ray Diffration Study of Anharmonic Thermal Vibrations in CuCl
(a) Vol. **45**, No. 1, p. 269–275 (1978).
See also 22.5, p. 474.

B. G. Helme and P. J. King
The Phonon Viscosity Tensor of Si, Ge, GaAs, and InSb
(a) Vol. **45**, No. 1, p. K33–K37 (1978).
See also *7*, p. *80*; 22.1.1, p. 424; 22.1.2, p. 429; 22.2.1, p. 445; 22.2.3, p. 452.

V. S. Sardaryan and M. M. Arakelyan
*Bound States of Electrons and Optical Phonons under Conditions of One-Dimensional Quantizing
(b) Vol. **85**, No. 2, p. 459–463 (1978).
See also *13*, p. *142*.

S. C. Goyal, R. Prakash, and S. P. Tripathi
*Many-Body Lattice Theory of Elastic Dielectric and Its Application to Solids with NaCl Structure
(b) Vol. **85**, No. 2, p. 477–486 (1978).
Erratum
(b) Vol. **89**, No. 1, p. 335 (1978).
See also 12.2, p. 138; 22.5.2, p. 478.

Yu. V. Gulyaev, S. N. Ivanov, and A. G. Kozorezov
On the Electronic Absorption and Amplification of Sound in Semiconductors at "Intermediate" Frequencies
(b) Vol. **85**, No. 2, p. 487–492 (1978).
See also *7*, p. *80*; 22.2.3, p. 452.

H. Jex and C. J. Maetz
*Lattice Dynamics of the Molecular Crystals KCN and NaCN in Their Disordered Cubic Phase
(b) Vol. **85**, No. 2, p. 511–517 (1978).
See also 8, p. 83; 22.9, p. 523.

S. C. Varshney and A. A. Gundjian
Internal Strain and Piezoelectricity in Zincblende Crystals
(b) Vol. **85**, No. 2, p. 733–742 (1978).
See also *12.2*, p. *138*; *14.3*, p. *217*; 22, p. 409.

V. G. Vaks, E. V. Zarochentsev, S. P. Kravchuk, V. P. Safronov, and A. V. Trefilov
*Thermal Expansion and Grüneisen Parameters in Alkali Metals
(b) Vol. **85**, No. 2, p. 749–759 (1978).
See also *8*, p. *83*; 21.2, p. 392.

V. Ramamurthy and K. K. Singh
*On the Crossing of Dispersion Curves of Alkali Metals in [ζ 00] Direction
(b) Vol. **85**, No. 2, p. 761–768 (1978).
See also 12.1, p. 134; 21.2, p. 392.

V. I. Altukhov and V. G. Kvachadze
Phonon Scattering on Small-Radius Colloids in KCl Crystals
(b) Vol. **85**, No. 2, p. 769–775 (1978).
See also 8, p. 83; *10.2*, p. *109*; 22.5.2, p. 479.

B. Prasad and R. S. Srivastava
*Thermal Properties and Grüneisen Parameters of Gold Using Toya's First Principle Approach
(b) Vol. **85**, No. 2, p. 789–793 (1978).
See also *8*, p. *83*; 21.6, p. 401.

H. H. Boghossian and K. S. Dubey
*Heat Transfer in a Lattice at Low Temperatures: Application to a Doped Sample of Al_2O_3
(b) Vol. **85**, No. 2, p. K99–K102 (1978).
See also *8*, p. *84*; 22.6, p. 491.

W. Lochmann
*Phonon-Assisted Auger Recombination in Indirect Gap Semiconductors
(a) Vol. **45**, No. 2, p. 423–432 (1978).
See also *13.1*, p. *152*; 20, p. 315; 22.1.1, p. 424; 22.1.2, p. 430.

P. Pipinys and A. Rimeika
*Current Pulses in ZnS:Cu Crystals Stimulated by Electric Flied at Low Temperatures
(a) Vol. **45**, No. 2, p. K113–K116 (1978).
See also 13.4, p. 171 ; *14.3.4*, p. *240*; 22.4.1, p. 463.

P. KONSIN
Structural Phase Transitions of Antiferroelectric and Displacive Modulated Types Caused by Electron–Phonon Interaction
(b) Vol. **86**, No.1, p. 57—66 (1978).
See also *1.2*, p. *16*; 14.4.1, p. 254; 22, p. 410.

F. SACCHETTI, P. BOSI, F. DUPRÉ, G. FROLLANI, F. MENZINGER, and M. C. SPINELLI
*Observation of Precursory Phonons of an F.C.C.–H.C.P. Transition
(b) Vol. **86**, No. 1, p. 77—81 (1978).
See also *1.2*, p. *16*; 21.1, p. 371.

J. C. CASTRO and M. DE SOUZA
*The OCN⁻Vibrational Spectrum and the Phase Transitions in KCN_xKCl_{1-x} Single Crystals
(b) Vol. **86**, No. 1, p. 137—144 (1978).
See also *1.2*, p. *16*; 12.2, p. 138; 20.1, p. 321; 22.5, p. 474; 22.5.4, p. 489.

J. BERGMANN, K. KNEIPP, and H. E. PONATH
*Raman-Induced Kerr Effect with Elliptical Polarization in Optically Active and Anisotropic Crystals
(b) Vol. **86**, No. 1, p. 215—223 (1978).
See also *20.1*, p. *322*; 22.8.1, p. 511.

K. FUKUSHI, M. NIPPUS, and R. CLAUS
*The Origin of the Anomalous TO–Doublet in Cubic CuCl Studied by Light Scattering
(b) Vol. **86**, No. 1, p. 257—262 (1978).
See also *20.1*, p. *322*; 22.5, p. 474.

W. F. W. M. VAN HEUGTEN
An Interatomic Potential for Copper from the Phonon Spectra
(b) Vol. **86**, No. 1, p. 277—281 (1978).
See also *10.1*, p. *101*; 21.1, p. 371.

R. STREHLOW
On the No–Phonon Structure in Mixed Crystals with Indirect Band Structure
(b) Vol. **86**, No. 1, p. 311—318 (1978).
See also 13.1, p. 153; 13.5.1, p. 190; *20.1*, p. *322*; 22.2.4, p. 456.

Y. L. KHAIT
*Some Observable Consequences of Kinetic Many-Body Treatment of Large Energy Fluctuations and of Rate Constants in Solids
(b) Vol. **86**, No. 1, p. 409—418 (1978).
Erratum
(b) Vol. **89**, No. 2, p. 687 (1978).
See also 8, p. 84; 9, p. 92; 10, p. 96.

I. S. GORBAN, V. A. GORINYA, R. A. DASHKOVSKAYA, V. I. LUGOVOI, A. P. MAKOVETSKAYA, and I. I. TICHINA
*One- and Two-Phonon States in Tetragonal ZnP_2 and CdP_2 Crystals
(b) Vol. **86**, No. 1, p. 419—423 (1978).
See also *20.1*, p. *323*; 22.3, p. 457.

G. M. Georgiev and J. S. Nikolov
*Spontaneous Parametric Scattering of Light in Ammonium Bioxalate Monohydrate
(b) Vol. 86, No. 1, p. 431–437 (1978).
See also *20.1*, p. *323*; 22.9, p. 523.

M. Scrocco and L. Mattioli
*The Structure of $Sr(NO_3)_2$: A Single Crystal Raman Spectra Study
(b) Vol. 86, No. 1, p. K1–K5 (1978).
See also *20.1*, p. *323*; 22.8.1, p. 511.

N. M. Gasanly, B. N. Mavrin, Kh. E. Sterin, V. I. Tagirov, and Z. D. Khalafov
*Raman Study of Layer $TlGaS_2$, β–$TlInS_2$, and $TlGaSe_2$ Crystals
(b) Vol. 86, No. 1, p. K49–K53 (1978).
See also *20.1*, p. *323*; 22.8, p. 504.

P. A. Alekseev, I. P. Sadikov, I. A. Markova, and O. D. Chistyakov
*On the Effects of the Crystalline Electric Field in $PrSn_3$
(b) Vol. 86, No. 1, p. K59–K62 (1978).
See also *18*, p. *273*; 22.7, p. 501.

J. Takahashi, K. Ozawa, S. Yamaguchi, Y. Fujino, O. Yoshinari, and M. Hirabayashi
Temperature Dependence of the Thermal Vibrations of Deuterium in a $Pd_{0.8}Au_{0.2}$ Alloy Studied by Ion Channeling
(a) Vol. 46, No. 1, p. 217–223 (1978).
See also *1.3*, p. *25*; *11*, p. *125*; 21.6, p. 401.

O. P. Sharma
*A Pseudopotential Study of the Electronic Properties of Noble Metals
(b) Vol. 86, No. 2, p. 483–492 (1978).
See also 4, p. 57; 14.1, p. 207; 21.1, p. 371; 21.6, p. 401.

S. Lefrant, J. P. Buisson, M. Ghomi, and L. Taurel
*First-Order Raman Scattering of CsF Doped with F Centres
(b) Vol. 86, No. 2, p. 493–496 (1978).
See also *20.1*, p. *324*; 22.5.2, p. 480.

B. Ulrici and E. Jahne
*Two-Phonon Infrared Absorption in $In_{1-x}Ga_xP$ Mixed Crystals
(b) Vol. 86, No. 2, p. 517–526 (1978).
See also *20.1*, p. *324*; 22.2.2, p. 450; 22.2.4, p. 456.

L. von Heimendahl
*Improvements of the Equation of Motion Technique for Disordered Systems
(b) Vol. 86, No. 2, p. 549–556 (1978).
See also *2*, p. *43*.

M. Sinyukov, R. Trommer, and M. Cardona
*Resonant Raman Scattering in InP
(b) Vol. 86, No. 2, p. 563–568 (1978).
See also *20.1*, p. *324*; 22.2.2, p. 450.

Vo Hong Anh and Nguyen Nhu Dat
*Parametric Excitation in Kane's Semiconductors
(b) Vol. **86**, No. 2, p. 585–591 (1978).
See also *13.5.2*, p. *198*; 20, p. 316; 22, p. 411.

H. Vogt and G. Neumann
Hyper-Rayleigh Scattering in Disordered Sodium Nitrate
(b) Vol. **86**, No. 2, p. 615–620 (1978).
See also 1.2, p. 17; *20.1*, p. *324*; 22.8.1, p. 512.

G. G. Nilsson, U. Karlsson, and G. Raunio
*Experimental Phonon Frequencies of RbBr at 80, 290, and 370 K
(b) Vol. **86**, No. 2, p. 653–660 (1978).
See also *8*, p. *84*; 22.5.2, p. 480.

M. B. Haxton
*The Phonon Spectra of γ- and β-CuBr
(b) Vol. **86**, No. 2, p. 661–670 (1978).
See also 22.5, p. 474.

M. Singh
*Phonon Scattering and Effect of Dislocations on the Low-Temperature Phonon Conductivity in Ge
(b) Vol. **86**, No. 2, p. 691–696 (1978).
See also 8, p. 84; 10.2, p. 111; 22.1.1, p. 425.

M. N. Iliev, E. Anastassakis, and T. Arai
*Raman Scattering of $HgCr_2Se_4$
(b) Vol. **86**, No. 2, p. 717–723 (1978).
See also 18.2, p. 285; *20.1*, p. *324*; 22.8.2, p. 518.

M. C. Al-Edani and K. S. Dubey
*Phonon Conductivity of InSb and GaAs
(b) Vol. **86**, No. 2, p. 741–750 (1978).
See also 8, p. 84; 22.2.1, p. 446; 22.2.3, p. 452.

R. Ramani, K. K. Mani†, and R. P. Singh
*Long Optical Vibrations and Elastic Constants of Hexagonal Boron Nitride
(b) Vol. **86**, No. 2, p. 759–763 (1978).
See also 12.2, p. 138; 22, p. 411.

H. A. Kassim and K. S. Dubey
*Lattice Thermal Resistivity Due to the Presence of Electrons
(b) Vol. **86**, No. 2, p. K107–K111 (1978).
See also *8*, p. *84*; 22.1.1, p. 425.

W. Gębicki and W. Nazarewicz
*Far Infrared Reflection Spectra of $Mn_xCd_{1-x}Te$ Mixed Crystals
(b) Vol. **86**, No. 2, p. K135–K138 (1978).
See also *20.1*, p. *325*; 22.4.4, p. 472.

I. V. Bodnar, A. G. Karoza, and G. F. Smirnova
*Infrared Reflectivity Spectra of $CuGa_xIn_{1-x}S_2$ Solid Solutions
(b) Vol. 86, No. 2, p. K171—K174 (1978).
See also *20.1*, p. *325*; 22.8, p. 504.

R. Kümmel, H. Rauh, and E. Bangert
Theory of Stark–Ladder Currents
(b) Vol. 87, No. 1, p. 99—110 (1978).
See also *14.3.4*, p. *242*; 22.4.1, p. 464.

M. S. Kushwaha and S. S. Kushwaha
*Unpaired Elastic Force Model in the Lattice Dynamical Study of Rare Gas Solids
(b) Vol. 87, No. 1, p. 247—253 (1978).
See also 12.2, p. 139; 23, p. 530.

Mahiradhwaj Singh
*Theory of Resonance Scattering for Electron–Phonon Interaction in Doped Semiconductors with Zincblende Structure
(b) Vol. 87, No. 1, p. 333—344 (1978).
See also 8, p. 85; 15, p. 260; 22.2.1, p. 446; 22.2.3, p. 453.

T. Soma
*Lattice Dynamics of Group IV Elements with the Local Heine-Abarenkov Model Potential
(b) Vol. 87, No. 1, p. 345—350 (1978).
See also 12.2, p. 139; 21.7, p. 406; 22.1.1, p. 425; 22.1.2, p. 432.

V.I. Zubov
*Higher Anharmonicities in the Improved Unsymmetrized Self-Consistent Field Approximation for a Crystal (I): General Consideration
(b) Vol. 87, No. 1, p. 385—394 (1978).
See also *8*, p. *85*; 23, p. 530.

V. K. Garg, D. S. Puri, and M. P. Verma
*On the Second- and Third-Order Elastic Constants of ZnS Structure Solids
(b) Vol. 87, No. 1, p. 401—407 (1978).
See also *12.2*, p. *139*; 22.2, p. 443; 22.4, p. 461; 22.5, p. 475.

Y. Thomas
*Sur la variation de la température de Debye à haute température
(On the Variation of the Debye Temperature at High Temperatures)
(b) Vol. 87, No. 1, p. K13—K16 (1978).
See also *8*, p. *85*; 21.6, p. 402.

A. Król, M. Kozielski, and W. Nazarewicz
*Localized Vibrational Modes of Al–Cu Complexes in Zinc Sulphide
(b) Vol. 87, No. 1, p. K25—K28 (1978).
See also *20.1*, p. *326*; 22.4.1, p. 464.

M. C. Al-Edani and K. S. Dubey
*Boundary Scattering and Phonon Conductivity: Application to GaAs
(b) Vol. 87, No. 1, p. K47—K52 (1978).
See also 8, p. 85; 22.2.1, p. 446.

A. YA. GORDON
Soft Mode Contribution to the Ferroelectric Heat Capacity
(b) Vol. 87, No. 1, p. K57—K60 (1978).
See also 1.2, p. 18; 8, p. 85; *14.4.1*, p. *254*.

J. CORNELIS, J. NIHOUL, F. BROSENS, and J. T. DEVREESE
Interatomic Potential for Molybdenum
(b) Vol. 87, No. 1, p. K61—K64 (1978).
See also *10.1*, p. *102*; 21, p. 358.

M. LEPPIHALME and T. TUOMI
Phonon Energies and Barrier Height from Photovoltage Spectra of Au–Intrinsic Ge Diodes
(a) Vol. 47, No. 1, p. 245—250 (1978).
See also 13.5.1, p. 191; *14.3.3*, p. *235*; *16*, p. *264*; 22.1.1, p. 425.

B. BORIE
On the Observation of Forbidden Bragg Maxima for White Tin
(a) Vol. 47, No. 1, p. K39—K42 (1978).
See also *1.1*, p. *12*; 21.7, p. 406.

D.W. FIELD
Comment on "On the Observation of Forbidden Bragg Maxima for White Tin"
(a) Vol. 47, No. 1, p. K43 (1978).
See also *1.1*, p. *12*; 21.7, p. 406.

B. I. SWANSON
Displacive Phase Transformations in K_2SnBr_6
(a) Vol. 47, No. 1, p. K95—K98 (1978).
See also *1.2*, p. *19*; 20.1, p. 327; 22.8, p. 505.

I. J. GUPTA and S. K. TRIKHA
*The Reduced Conductivity of Dielectrics when Both Boundary and Dislocation Scatterings are Important at Low Temperatures
(b) Vol. 87, No. 2, p. 513—516 (1978).
See also *8*, p. *86*; 23, p. 530.

R. KAISER, W. SPENGLER, S. SCHICKTANZ, and C. POLITIS
*Raman Spectra and Superconductivity of Various Phases of a High-T_c Superconductor:NbN
(b) Vol. 87, No. 2, p. 565—573 (1978).
See also 14.2, p. 214; *20.1*, p. *327*.

M. KALM and CH. UIHLEIN
Investigations on the Temperature-Dependent TPA Linewidth of the Z_3-Exciton in CuCl
(b) Vol. 87, No. 2, p. 575—580 (1978).
See also *13.5.1*, p. *192*; *20.1*, p. *327*; 22.5, p. 475.

R. RUPPIN
Validity Range of the Maxwell-Garnett Theory
(b) Vol. 87, No. 2, p. 619—624 (1978).
See also *20.1*, p. *328*; 22.2.1, p. 446; 22.6, p. 494.

M. Z. Zgierski and W. H. Henneker
*Vibronic Spectra of Molecular Systems
(b) Vol. 87, No. 2, p. 653—661 (1978).
See also 13.5.1, p. 192; 22.9, p. 525.

M. A. B. Whitaker
The Motional Effect in Nuclear Double-Resonance
(b) Vol. 87, No. 2, p. 689—695 (1978).
See also *19*, p. *308*; 22.8, p. 505.

V. M. Fomin and E. P. Pokatilov
On the Representation of the Non-Stationary Density Matrix of an Electron–Phonon System by a Functional Integral
(b) Vol. 87, No. 2, p. 725—732 (1978).
See also 14.3.4, p. 242; *20*, p. *316*.

R. Ramji Rao and A. Ramanand
*Lattice Dynamics, Specific Heat, and Bulk Modulus of Scandium
(b) Vol. 87, No. 2, p. 751—756 (1978).
See also 8, p. 86; 12.1, p. 135; 21.4, p. 397.

B. Prasad and R. S. Srivastava
*Thermal Properties and Grüneisen Parameters of Alkali Metals Employing the Model Potential of Krasko and Gurskii
(b) Vol. 87, No. 2, p. 771—776 (1978).
See also 8, p. 86; 21.2, p. 392.

V. L. Volkov
*The Electron–Phonon Interaction in $Pb_{1-x}Sn_xTe$
(b) Vol. 87, No. 2, p. K93—K96 (1978).
See also *13.1*, p. *155*; 22, p. 414.

K. R. Allakhverdiev, E. A. Vinogradov, N. N. Melnik, M. A. Nizametdinova, E. Yu. Salaev, and R. M. Sardarly
*Vibrational Properties of TlSe
(b) Vol. 87, No. 2, p. K115—K117 (1978).
See also *20.1*, p. *328*; 22.3, p. 458.

K. S. Dubey
*Heat Transfer in a Doped Crystal Lattice at Low Temperatures: Application to Sb-Doped Ge
(b) Vol. 87, No. 2, p. K133—K136 (1978).
See also *8*, p. *86*; 22.1.1, p. 425.

U. Kh. Kopvillem and V. M. Choodnovskii
*Phonon Avalanche in Glasses
(b) Vol. 87, No. 2, p. K145—K149 (1978).
See also *2*, p. *72*; 7, p. 81.

D. W. Field
*Anharmonicity in X-Ray Diffraction Intensities in White Tin
(a) Vol. 47, No. 2, p. 651—656 (1978).
See also 21.7, p. 406.

B. Cvikl
*On Coherent Neutron Diffraction in Nematic Liquid Crystals
(b) Vol. 88, No. 1, p. 25—36 (1978).
See also *2*, p. *46*.

V. I. Zubov
*Higher Anharmonicities in the Improved Unsymmetrized Self-Consistent Field Approximation for a Crystal (II): Comparison of Calculations with Experiment
(b) Vol. 88, No. 1, p. 43—50 (1978).
See also *8*, p. *87*; 23, p. 530.

Nguyen Van Trong
*Neutron Scattering on Anomalous Fluctuations of Phonons
(b) Vol. 88, No. 1, p. 145—148 (1978).
See also 11, p. 127; 13.5.2, p. 198.

I. P. Zvyagin
*Hopping Conductivity in the Presence of an Electromagnetic Wave
(b) Vol. 88, No. 1, p. 149—154 (1978).
See also *2*, p. *46*; *14.3*, p. *222*.

F. Bechstedt and D. Haus
*Theory of Second-Order Resonance Raman Scattering in the Case of Strong Excitonic Effects
(b) Vol. 88, No. 1, p. 163—171 (1978).
See also 13.5.1, p. 192; *20.1*, p. *329*; 22.4.1, p. 465.

V. Lemos, F. Cerdeira, and M. Cardona
Atomic Motions Corresponding to Zone Center Phonons in Paratellurite
(b) Vol. 88, No. 1, p. 199—206 (1978).
See also *20.1*, p. *329*; 22.6, p. 494.

P. Pfeffer and W. Zawadzki
Inelastic Scattering of Electrons by Optic Phonons in InSb-Type Semiconductors
(b) Vol. 88, No. 1, p. 247—260 (1978).
See also 13.1, p. 156; *14.3*, p. *222*; 22.2.3, p. 453.

Mahiradhwaj Singh
*Phonon Attenuation in n-Type Semiconductors with ZB Structure in Presence of Internal Strains and Magnetic Field
(b) Vol. 88, No. 1, p. 295—300 (1978).
See also 7, p. 81; 12.2, p. 140; 18, p. 274; 22.1.1, p. 426; 22.2.3, p. 453.

B. Prévot, C. Schwab, and B. Dorner
*Phonon Dispersion in Red HgI_2
(b) Vol. 88, No. 1, p. 327—333 (1978).
See also 7, p. 81; *20.1*, p. *329*; 22.5, p. 475.

K. Allakhverdiev, R. Sardarly, F. Wondre, and J. F. Ryan
*Raman and Infrared Spectra of $TlGaSe_2$
(b) Vol. 88, No. 1, p. K5—K9 (1978).
See also *20.1*, p. *329*; 22.8, p. 506.

E. Heinze and S. Grande
Nuclear Spin Relaxation in the Dipolar System in Solid p-Alkoxybenzylidene-p-n-Alkylanilines
(b) Vol. 88, No. 1, p. K23—K25 (1978).
See also 1.2, p. 20; *19*, p. *309*; 22.9, p. 526.

J. Cornelis, J. Nihoul, F. Brosens, and J. T. Devreese
Interatomic Potentials from Experimental Phonon Spectra. Application to Au, Cu, K, Ni, Pd, Pt, and Rb
(b) Vol. 88, No. 1, p. K93—K96 (1978).
See also *4*, p. *58*; 21, p. 360.

H. H. Boghosian and K. S. Dubey
*Role of Electron–Phonon Interaction and Peripheral Phonons in the Lattice Thermal Conductivity of a Doped Semiconductor at Low Temperatures: Application to Phosphorous-Doped Germanium
(b) Vol. 88, No. 2, p. 417—427 (1978).
See also *8*, p. *87*; 22.1.1, p. 427.

L. Graf
*Electronic Transitions in the Raman Spectra of $FeCl_2 \cdot 2\,H_2O$ and $FeCl_2 \cdot 2\,D_2O$
(b) Vol. 88, No. 2, p. 429—437 (1978).
See also *13.5.4*, p. *201*; 20.1, p. 330; 22.5, p. 475.

M. I. Dykman
*Theory of Cyclotron Resonance of Two-Dimensional Electrons Interacting with Surface and Volume Phonons
(b) Vol. 88, No. 2, p. 463—475 (1978).
See also *13.1*, p. *157*; 20.1, p. 330.

M. Wagner
*On the Non-Adiabatic Properties of Tunneling Centres
(b) Vol. 88, No. 2, p. 517—530 (1978).
See also *10.2*, p. *115*; 22.5.2, p. 482.

R. Glocker and L. Fritsche
*Pressure Dependence of the Electron–Phonon Coupling in Lanthanum
(b) Vol. 88, No. 2, p. 639—644 (1978).
See also 12.1, p. 136; *13.1*, p. *157*; 14.2, p. 215; 21.4, p. 398.

I. J. Gupta and S. K. Trikha
*Lattice Thermal Conductivity of Solid Neon in the Temperature Range 0.5 to 10 K
(b) Vol. 88, No. 2, p. 815—818 (1978).
See also *8*, p. *87*; 23, p. 530.

R. Mehrotra and M. M. Pant
*Long Wave Phonon Modes and Infrared Absorption of a NaI Slab
(b) Vol. 88, No. 2, p. K131—K133 (1978).
See also 20.1, p. 331; 22.5.2, p. 483.

V. K. Bashenov, D. I. Marvakov, and A. G. Petukhov
*Lattice Dynamics of Diamond-Type Crystals from Keating's Valence-Force Field
(b) Vol. 88, No. 2, p. K161—K164 (1978).
See also 22.1, p. 422; 22.1.1, p. 427; 22.1.2, p. 435.

E. I. Terukov, W. Reichelt, M. Wolf, H.Hemschik, and H. Oppermann
On the Influence of the Substitution of O^{16} by O^{18} in a Few Vanadium Oxides on Their Semiconductor–Metal Transition
(a) Vol. **48**, No. 2, p. 377—381 (1978).
See also *1.2*, p. *21*; 14.3, p. 223; 22.6, p. 495.

J. P. Huvenne, G. Vergoten, B. Boniface, G. Fleury et P. Legrand
*Etude vibrationelle de basse fréquence de l'oxyde de plomb Pb_3O_4. Relation entre l'évolution des spectres et les transitions ferroélectrique et ferroélastique de la maille
(Low–Frequency Vibrations of Lead Oxide Pb_3O_4. Relationhip between the Development of Spectra and Ferroelectric and Ferroelastic Phase Transitions)
(a) Vol. **48**, No. 2, p. 417—423 (1978).
See also *20.1*, p. *331*; 22.6, p. 495.

M. M. Arakelyan
*Non-Linear Ultrasound Absorption in an Electromagnetic Wave Field under Quantum Size Effect Conditions
(b) Vol. **89**, No. 1, p. 37—43 (1978).
See also 7, p. 82.

V. N. Kashcheev
*Phonon Relaxation in One- and Two-Dimensional Displacive-Type Ferroelectrics
(b) Vol. **89**, No. 1, p. 45—52 (1978).
See also 14.4.1, p. 256.

D. Wruck
*Vibrational Properties of the ⟨001⟩ Split Interstitial in Silicon
(b) Vol. **89**, No. 1, p. 177—186 (1978).
See also 20.1, p. 332; 22.1.2, p. 436.

V. C. Jain and J. Shanker
The Exchange Charge Shell Model Analysis of Strain Derivatives of Static Polarizability and Effective Charge Parameter of Alkali Halides
(b) Vol. **89**, No. 1, p. 213—219 (1978).
See also *4*, p. *59*; 14.4, p. 250; 22.5.2, p. 483.

P. Kocever and E. Fitz
*The Influence of Acoustic Phonon Disturbances on the Conductivity of Piezo-Electric Semiconductors
(b) Vol. **89**, No. 1, p. 225—233 (1978).
See also 14.3, p. 224; 22.2.1, p. 448.

Mahiradhwaj Singh
*The Influence of Internal Strain on Bound Hole–Phonon Interaction in p-Type Semiconductors with Zincblende Structure
(b) Vol. **89**, No. 1, p. 299—307 (1978).
See also 8, p. 88; 22.2, p. 444.

N. M. Gasanly, B. M. Yavadov, V. I. Tagirov, and E. A. Vinogradov
*Infrared and Raman Spectra of Layer InSe Single Crystals
(b) Vol. **89**, No. 1, p. K43—K48 (1978).
See also *20.1*, p. *332*; 22.3, p. 458.

H. H. Boghossian and K. S. Dubey
*Heat Transfer in Doped Al_2O_3 at Low Temperatures
(b) Vol. **89**, No. 1, p. K65—K67 (1978).
See also *8*, p. *88*; 22.6, p. 496.

S. Ewert, G. Heiland, H. Lüth, U. Roll, and W. Sander
Molecular Vibrations in a Polymethine Dye Layer Studied by Inelastic Electron Tunneling Spectroscopy
(b) Vol. **89**, No. 2, p. 459—465 (1978).
See also *14.3.1*, p. *228*; *20.1*, p. *334*; 22.9, p. 527.

M. S. Kushwaha and S. S. Kushwaha
*Frequency Spectra and Thermal Properties of Rare Gas Solids (Argon and Krypton)
(b) Vol. **89**, No. 2, p. 583—586 (1978).
See also 8, p. 88; 23, p. 530.

R. Enderlein
*On the Theory of Light Propagation through Disordered Solids. Scattering from Configurational Fluctuations
(b) Vol. **89**, No. 2, p. 625—632 (1978).
See also 2, p. 50; *20.1*, p. *334*; 22, p. 418.

B. Esser
*On the Theory of Resonance Raman Scattering in Disordered Semiconductors
(b) Vol. **89**, No. 2, p. 633—643 (1978).
See also 2, p. 50; 13.1, p. 159; *20.1*, p. *334*.

H. Böttger, V. V. Bryksin, and P. Kleinert
*Theory of Resonant Raman Scattering in Systems with Localized Electron States
(b) Vol. **89**, No. 2, p. 675—685 (1978).
See also *13*, p. *148*; 20.3, p. 348; 22, p. 418.

V. K. Bashenov, I. Baumann, and D. I. Marvakov
*Lattice Dynamics of Gallium Selenide in the One-Layer Approach
(b) Vol. **89**, No. 2. p. K115—K118 (1978).
See also 22.3, p. 459.

P. Franzosi
*Exciton–Phonon Interaction in CdS
(b) Vol. **89**, No. 2, p. K119—K122 (1978).
See also *13.5.1*, p. *195*; 20.1, p. 334; 22.4.1, p. 466.

L. C. Scavarda do Carmo and F. Lüty
*Librational and Phonon Sidebands of the Infrared CN^- Stretching Mode in NaCN and KCN Crystals
(b) Vol. **89**, No. 2, p. K149—K153 (1978).
See also *20.1*, p. *334*; 22.8, p. 507.

K. Bärner and H. Berg
*Elastic Anomalies at the Second-Order Phase Transition in MnAs
(a) Vol. **49**, No. 2, p. 545—554 (1978).
See also *12.1*, p. *136*; 18.1, p. 279; 18.2, p. 290; 21.1, p. 379.

W. AUCH, W. STEUDLE, J. U. VON SCHÜTZ, and H. C. WOLF
The Phase Transition in the 1:1 CT-Complex Anthracene/TCNB at 206 K. A NMR Study
(a) Vol. **49**, No. 2, p. 563—567 (1978).
See also 1.2, p. 23; *19*, p. *313*; 22.9, p. 528.

A. SIEMS
Quantitative Study of Kikuchi Structures (I): Energy Analysis of Intensity Profiles
(a) Vol. **49**, No. 2, p. 621—628 (1978).
See also *1.1*, p. *14*; 13.5.2, p. 199; 22.1.2, p. 437.

A. D. MOKRUSHIN and R. KH. SABIROV
Positron Thermalization as Quantum Statistical Process
(b) Vol. **90**, No. 1, p. 99—104 (1978).
See also *13*, p. *148*; 21, p. 364.

V. RAMAMURTHY and M. SATISHKUMAR
*Analysis of Phonon Frequencies of Alkali Metals
(b) Vol. **90**, No. 1, p. 117—122 (1978).
See also 12.1, p. 136; 21.2, p. 393.

A. MANABE, H. NOGUCHI and A. MITSUISHI
Dielectric Function in HgSe at 2, 95, and 300 K
(b) Vol. **90**, No. 1, p. 157—166 (1978).
See also 13.1, p. 159; 13.5.2, p. 199; *20.1*, p. *335*; 22.4.2, p. 469.

R. STREHLOW, W. KÜHN, H. FRIEDRICH, and H. LANGE
On the Quantitative Characterization of Chemical Disorder of $In_{1-x}Ga_xP$ from Wavelength-Modulated No-Phonon Absorption
(b) Vol. **90**, No. 1, p. 167—173 (1978).
See also 13.1, p. 159; 13.5.1, p. 195; *20.1*, p. *335*; 22.2.4, p. 456.

L. DUMOULIN, P. NÉDELLEC, C. ARZOUMANIAN, and J. P. BURGER
*Optical Phonons in PdH_x and PdD_x by Superconducting Tunneling Measurements
(b) Vol. **90**, No. 1, p. 207—210 (1978).
See also *14.2*, p. *215*; 21, p. 364.

W. F. W. M. VAN HEUGTEN and J. TH. M. DE HOSSON
Interatomic Potentials for Alkali Metals. A Comparative Study
(b) Vol. **90**, No. 1, p. 225—232 (1978).
See also *4*, p. *59*; 10.1, p. 107; 21.2, p. 393.

P. A. VAROTSOS
An Estimate of the Pressure Dependence of the Dielectric Constant in Alkali Halides
(b) Vol. **90**, No. 1, p. 339—343 (1978).
See also 4, p. 59; 12.2, p. 141; *14.4*, p. *251*; 22.5.1, p. 477; 22.5.2, p. 485.

MAHIRADHWAJ SINGH
*Electron–Phonon Interaction and Phonon Conductivity of the n-Type Semiconductors Mg_2Ge and Mg_2Sn at Low Temperatures
(b) Vol. **90**, No. 1, p. 385—389 (1978).
See also 8, p. 89; 22.7, p. 502.

V. C. SAHNI and M. L. BANSAL
*Raman Study of the Mixed System $[(NH_4)_{1-x}K_x]_2\ CuCl_4 \cdot H_2O$
(b) Vol. **90**, No. 1, p. 415—422 (1978).
See also *20.1*, p. *336*; 22.5.4, p. 490.

MAHIRADHWAJ SINGH
*Resonance Scattering of Phonons in the Thermal Conductivity of CN Doped KCl
(b) Vol. **90**, No. 1, p. K43—K47 (1978).
See also *8*, p. *89*; 22.5.2, p. 485.

A. V. GOLTSEV, I. G. LANG, and S. T. PAVLOV
*Oscillations of Secondary Radiation Intensity in Polar Semiconductors in a Magnetic Field
(b) Vol. **90**, No. 1, p. K63—K67 (1978).
See also *20.1*, p. *336*; 22, p. 420.

J. T. DEVREESE, R. EVRARD, and E. KARTHEUSER
*Note on the Solution of the Boltzmann Equation for Electron–LO Phonon Scattering
(b) Vol. **90**, No. 1, p. K73—K76 (1978).
See also *14.3.4*, p. *246*; 22.2.3, p. 454.

W. F. W. M. VAN HEUGTEN, A. VAN VEEN, and L. M. CASPERS
Molybdenium–Helium and Helium–Helium Interaction Potentials from Helium Desorp tion Experiments
(b) Vol. **90**, No. 1, p. K89—K94 (1978).
See also *10.1*, p. *107*; 21, p. 364.

H. MÖHWALD and A. THAER
Raman Study of the Phase Transition in the CT-Crystal Anthracene-TCNB
(a) Vol. **50**, No. 1, p. 131—138 (1978).
See also *1.2*, p. *24*; 20.1, p. 336; 22.9, p. 528.

A. KRÓL, M. J. KOZIELSKI, and W. NAZAREWICZ
*Infrared Studies of Al Complexes in Zinc Sulphide
(b) Vol. **90**, No. 2, p. 649—656 (1978).
See also 10.2, p. 121; *20.1*, p. *337*; 22.4.1, p. 466.

N. N. GERASIMENKO M. ROLLÉ, LI-JEN CHENG, Y. H. LEE, J. C. CORELLI, and J. W. CORBETT
*Infrared Absorption of Silicon Irradiated by Protons
(b) Vol. **90**, No. 2, p. 689—695 (1978).
See also 11, p. 131; *20.1*, p. *337*; 22.1.2, p. 439.

G. P. SRIVASTAVA
*Role of Optical Phonons in the High-Temperature Lattice Thermal Conductivity of Semiconductors
(b) Vol. **90**, No. 2, p. K125—K128 (1978).
See also *8*, p. *90*; 22.1.1, p. 428; 22.1.2, p. 439; 22.2.1, p. 449; 22.2.3, p. 455.

G. E. ALBERGA and J. T. DEVREESE
*Mobility of Electrons in n-Type InSb at 77 K in the Avalanche Regime: Theory and Experiment
(b) Vol. **90**, No. 2, p. K149—K152 (1978).
See also *14.3.4*, p. *247*; 22.2.3, p. 455.

SH. M. EFENDIEV, V. E[illegible]BAGIEV, A. CH. ZEINALLY, and V. M. SKORIKOV

Optical Transitions on $Bi_{12}SiO_{20}$

(a) Vol. **50**, No. 2, p. K141—K143 (1978).

See also 13.1, p. 161; *20.1*, p. *338*; 22.8, p. 508.

7. Acoustic Properties

Original Papers and Short Notes

I. A. Kurova and N. N. Ormont
*IR Luminescence of Acoustoelectric Domains in n-GaAs
(a) Vol. **45**, No. 1, p. 333—342 (1978).
See also *20.3*, p. *341*; 22.2.1, p. 445.

B. G. Helme and P. J. King
*The Phonon Viscosity Tensor of Si, Ge, GaAs, and InSb
(a) Vol. **45**, No. 1, p. K33—K37 (1978).
See also 6, p. 65; 22.1.1, p. 424; 22.1.2, p. 429; 22.2.1, p. 445; 22.2.3, p. 452.

Yu. V. Gulyaev, S. N. Ivanov, and A. G. Kozorezov
*On the Electronic Absorption and Amplification of Sound in Semiconductors at "Intermediate" Frequencies
(b) Vol. **85**, No. 2, p. 487—492 (1978).
See also 6, p. 66; 22.2.3, p. 452.

T. Gotô, T. Fujimura, and K. Kamiyoshi
*Ultrasonic Investigation of the Phase Transitions in $NH_4Cl_{1-x}Br_x$ Mixed Crystals
(a) Vol. **45**, No. 2, p. 513—519 (1978).
See also *1.2*, p. *16*; 14.4, p. 248; 22.5.4, p. 489.

M. K. Roy
*Coupling of Surface Acoustic Waves Propagating in Two Separated Piezoelectric Media
(a) Vol. **45**, No. 2, p. 549—552 (1978).
See also *13.3*, p. *164*; 14.3.2, p. 230; 22.4.1, p. 463; 22.6.1, p. 499.

E. I. Shtyrkov and V. V. Samartsev
*Dynamic Holograms on the Superposition States of Atoms
(a) Vol. **45**, No. 2, p. 647—655 (1978).
See also *20*, p. *316*.

K. Míšek
*Anomalous Attenuation of Ultrasound in a Niobium Single Crystal at Low Temperatures
(b) Vol. **86**, No. 1, p. K31—K33 (1978).
See also 14.2, p. 213; 21, p. 355.

V. S. Postnikov, S. A. Gridnev, B. N. Prasolov, and S. K. Turkov
*Specific Features of Low-Frequency Sound Attenuation in $LiNbO_3$ Poly- and Single-Domain Crystals
(a) Vol. **47**, No. 1, p. 65—69 (1978).
See also *12.2*, p. *139*; 14.4.1, p. 255; 22.8.1, p. 513.

A. V. Mitin and V. A. Tarasov
*Ultrasonic Spectrum in a Multidomain Structure
(b) Vol. **87**, No. 2, p. 433—440 (1978).
See also *18.2*, p. *286*.

U. KH. KOPVILLEM and V. M. CHOODNOVSKII
Phonon Avalanche in Glasses
(b) Vol. 87, No. 2, p. K145–K149 (1978).
See also *2*, p. *45*; *6*, p. *72*.

N. K. GOBRAN and N. S. GHALI
*Ultrasonic Anomalies in Iron at Elevated Temperatures
(a) Vol. 47, No. 2, p. 365–370 (1978).
See also 18.2, p. 287; 21.1.1, p. 387.

I. L. CHISTYI, L. CSILLAG, V. F. KITAEVA, N. KROÓ, and N. N. SOBOLEV
*Investigation of Molecular Light Scattering in $PbMoO_4$
(a) Vol. 47, No. 2, p. 609–615 (1978).
See also *20.1*, p. *328*; 22.8.1, p. 514.

M. KOBAYASHI, K. TAKESAWA, and I. YOKOTA
*Carrier Concentration Dependence of Ultrasonic Attenuation in Piezoelectric Semiconductors
(a) Vol. 47, No. 2, p. 657–660 (1978).
See also 22.4.1, p. 464.

D. VAJDA
*Dependence of the Threshold Electric Field of the Acoustoelectric Current Instabilities in Monocrystalline n-InSb on the Length of a Sample
(a) Vol. 47, No. 2, p. K123–K124 (1978).
See also *14.3.4*, p. *243*; 22.2.3, p. 453.

MAHIRADHWAJ SINGH
Phonon Attenuation in n-Type Semiconductors with ZB Structure in Presence of Internal Strains and Magnetic Field
(b) Vol. 88, No. 1, p. 295–300 (1978).
See also *6*, p. *73*; 12.2, p. 140; 18, p. 274; 22.1.1, p. 426; 22.2.3, p. 453.

B. PREVOT, C. SCHWAB, and B. DORNER
Phonon Dispersion in Red HgI_2
(b) Vol. 88, No. 1, p. 327–333 (1978).
See also *6*, p. *73*; *20.1*, p. *329*; 22.5, p. 475.

G. H. MELKUMYAN and A. V. SHEKOYAN
*On the Theory of Self-Action of Ultrasonic and Hypersonic Waves
(a) Vol. 48, No. 1, p. 23–29 (1978).

R. FLORIAN, J. PELZL, M. ROSENBERG, H. VARGAS, and R. WERNHARDT
*Photoacoustic Detection of Phase Transitions
(a) Vol. 48, No. 1, p. K35–K38 (1978).
See also *1.2*, p. *21*; 22.8, p. 506.

D. ROUBY and P. FLEISCHMANN
*Spectral Analysis of Acoustic Emission from Aluminium Single Crystals Undergoing Plastic Deformation
(a) Vol. 48, No. 2, p. 439–445 (1978).
See also 10.1, p. 105; 21, p. 361.

A. M. Kapitonov, E. M. Smokotin, A. P. Korenko, and G. N. Stepanov
Ultrasonic-Wave Velocities in Cr_2O_3 near the Magnetic Phase Transition
(a) Vol. **48**, No. 2, p. K203—K206 (1978).
See also *1.2*, p. *22*; *18*, p. *275*; 22.6, p. 496.

M. M. Arakelyan
Non-Linear Ultrasound Absorption in an Electromagnetic Wave Field under Quantum Size Effect Conditions
(b) Vol. **89**, No. 1, p. 37—43 (1978).
See also *6*, p. *75*.

W. Rehwald and A. Vonlanthen
Ultrasonic Studies of the Structural Phase Transition in Squaric Acid
(b) Vol. **90** No. 1 p. 61—66 (1978).
See also *1.2*, p. *23*; 12.2, p. 141; 22.9, p. 528.

R. Baubinas, B.- P. Kietis, R. Rėksnys, and A. Sakalas
*Bipolar Photoconductivity and Acoustoelectric Current in CdTe, CdS, and CdSe
(a) Vol. **50**, No. 1, p. K63—K65 (1978).
See also *16*, p. *267*; 22.4.1, p. 466; 22.4.2, p. 469; 22.4.3, p. 471.

S. S. Hsu, J. H. Chen, and P. C. Clapp
*Acoustic Absorption and Emission Associated with the Martensitic Transformation in an FeNi Alloy
(a) Vol. **50**, No. 2, p. 393—398 (1978).
See also 1.3, p. 28; 21.1, p. 382; 21.1.1, p. 390.

N. P. Kobelev and Ya. M. Soifer
Electronic Damping of Dislocations in Lead
(a) Vol. **50**, No. 2, p. K185—K188 (1978).
See also *10.1*, p. *108*; *14.2*, p. *216*; 21, p. 367.

8. Thermal Properties

Review Article

A. S. Borukhovich
Kinetic Properties of IVa and Va Subgroup Transition Metal Monocarbides
(a) Vol. **46**, No. 1, p. 11—37 (1978).
See also 13.1, p. 150; *14.3*, p. *217*; 15, p. 259; 22, p. 408; 22.3, p. 457; 22.3.1, p. 460.

Original Papers and Short Notes

V. G. Vaks, E. V. Zarochentsev, S. P. Kravchuk, V. P. Safronov, and A. V. Trefilov
*Lattice Heat Capacity and Thermal Vibrations in Alkali Metals
(b) Vol. **85**, No. 1, p. 63—74 (1978).
See also *6*, p. *64*; 21.2, p. 392.

Š. Jánoš, A. Feher, and H. U. Åström
*Thermal Conductivity of Neodymium from 0.5 to 8 K
(b) Vol. **85**, No. 1, p. K31—K32 (1978).
See also 21.4, p. 396.

H. Jex and C. J. Maetz
Lattice Dynamics of the Molecular Crystals KCN and NaCN in Their Disordered Cubic Phase
(b) Vol. **85**, No. 2, p. 511—517 (1978).
See also *6*, p. *66*; 22.9, p. 523.

K. Flachbart, A. Feher, Š. Jánoš, Z. Málek, and A. Ryska
*Thermal Conductivity of Nb–Ti Alloy in the Low-Temperature Range
(b) Vol. **85**, No. 2, p. 545—551 (1978).
See also 14.2, p. 213; 21, p. 354.

V. G. Vaks, E. V. Zarochentsev, S. P. Kravchuk, V. P. Safronov, and A. V. Trefilov
*Thermal Expansion and Grüneisen Parameters in Alkali Metals
(b) Vol. **85**, No. 2, p. 749—759 (1978).
See also *6*, p. *66*; 21.2, p. 392.

V. I. Altukhov, and V. G. Kvachadze
Phonon Scattering on Small-Radius Colloids in KCl Crystals
(b) Vol. **85**, No. 2, p. 769—775 (1978).
See also 6, p. 66; *10.2*, p. *109*; 22.5.2, p. 479.

B. Prasad and R. S. Srivastava
*Thermal Properties and Grüneisen Parameters of Gold Using Toya's First Principle Approach
(b) Vol. **85**, No. 2, p. 789—793 (1978).
See also *6*, p. *66*; 21.6, p. 401.

H. H. Boghossian and K. S. Dubey
*Heat Transfer in a Lattice at Low Temperatures: Application to a Doped Sample of Al_2O_3
(b) Vol. 85, No. 2, p. K99—K102 (1978).
See also *6*, p. *66*; 22.6, p. 491.

Y. L. Khait
Some Observable Consequences of Kinetic Many-Body Treatment of Large Energy Fluctuations and of Rate Constants in Solids
(b) Vol. 86, No. 1, p. 409—418 (1978).
Erratum
(b) Vol. 89, No. 2, p. 687 (1978).
See also *6*, p. *67*; 9, p. 92; 10, p. 96.

E. Popko and J. M. Pawlikowski
*Contribution of the Lattice Dilatation to the Temperature Shift of the Energy Gap of $Cd_xHg_{1-x}Te$
(a) Vol. 46, No. 1, p. K9—K12 (1978).
See also *13.1*, p. *153*; 22.4.4, p. 472.

A. Šlebarski
*Anomalous Change of the Curie Temperature of $GdAl_2$
(a) Vol. 46, No. 1, p. K29—K34 (1978).
See also *18.2*, p. *284*; 21, p. 356; 21.4, p. 397.

M. Drulis and Z. Biegański
*Low Temperature Phase Transition in Cerium Trihydride. Specific Heat Measurements
(a) Vol. 46, No. 1, p. K39—K41 (1978).
See also *18*, p. *274*; 21.4, p. 397.

G. G. Nilsson, U. Karlsson, and G. Raunio
*Experimental Phonon Frequencies of RbBr at 80, 290, and 370 K
(b) Vol. 86, No. 2, p. 653—660 (1978).
See also *6*, p. *69*; 22.5.2, p. 480.

M. Singh
Phonon Scattering and Effect of Dislocations on the Low-Temperature Phonon Conductivity in Ge
(b) Vol. 86, No. 2, p. 691—696 (1978).
See also *6*, p. *69*; 10.2, p. 111; 22.1.1, p. 425.

M. C. Al-Edani and K. S. Dubey
Phonon Conductivity of InSb and GaAs
(b) Vol. 86, No. 2, p. 741—750 (1978).
See also *6*, p. *69*; 22.2.1, p. 446; 22.2.3, p. 452.

H. A. Kassim and K. S. Dubey
*Lattice Thermal Resistivity Due to the Presence of Electrons
(b) Vol. 86, No. 2, p. K107—K111 (1978).
See also *6*, p. *69*; 22.1.1, p. 425.

U. ROY, K. PETROV, I. TSOLOVSKI, and P. PESHEV
Erratum to: On the Structural Phase Transition and the Thermal Expansion of Spinel Lithium Titanate
(a) Vol. **46**, No. 2, p. K171 (1978).
See also *1.2*, p. *18*; 22.8.1, p. 513.

MAHIRADHWAJ SINGH
Theory of Resonance Scattering for Electron–Phonon Interaction in Doped Semiconductors with Zincblende Structure
(b) Vol. **87**, No. 1, p. 333–344 (1978).
See also *6*, p. *70*; 15, p. 260; 22.2.1, p. 446; 22.2.3, p. 453.

V. I. ZUBOV
*Higher Anharmonicities in the Improved Unsymmetrized Self-Consistent Field Approximation for a Crystal (I): General Consideration
(b) Vol. **87**, No. 1, p. 385–394 (1978).
See also *6*, p. *70*; 23, p. 530.

Y. THOMAS
*Sur la variation de la température de Debye à haute température
(On the Variation of the Debye Temperature at High Temperatures)
(b) Vol. **87**, No. 1, p. K13–K16 (1978).
See also *6*, p. *70*; 21.6, p. 402.

M. C. AL-EDANI and K. S. DUBEY
Boundary Scattering and Phonon Conductivity: Application to GaAs
(b) Vol. **87**, No. 1, p. K47–K52 (1978).
See also *6*, p. *70*; 22.2.1, p. 446.

A. YA. GORDON
Soft Mode Contribution to the Ferroelectric Heat Capacity
(b) Vol. **87**, No. 1, p. K57–K60 (1978).
See also 1.2, p. 18; 6, p. 71; *14.1.1*, p. *254*.

A. MIGDAŁ-MIKULI, E. MIKULI, M. RACHWALSKA, and S. HODOROWICZ
Investigation of the Phase Situation in Polycrystalline $[Mg(NH_3)_6](ClO_4)_2$ by Adiabatic Calorimetry and X-Ray Diffraction Methods
(a) Vol. **47**, No. 1, p. 57–64 (1978).
See also *1.2*, p. *18*; 22.8, p. 505.

U. RINDELHARDT, S. MAGER, W. BLAU, and E. HEGENBARTH
Recovery of Low Temperature Specific Heat, Resistivity, and Coercive Force in Fe–Cr–Al Alloys
(a) Vol. **47**, No. 1, p. 157–161 (1978).
See also *10.1*, p. *103*; 14.1, p. 207; 18.2, p. 286; 21.1, p. 373; 21.1.1, p. 387.

A. JEŻOWSKI and J. RAFAŁOWICZ
*Heat Flow Asymmetry on a Junction of Quartz with Graphite
(a) Vol. **47**, No. 1, p. 229–232 (1978).
See also 22.6, p. 493.

O. Pavlov and Z. Bontcheva-Mladenova
Measurements of the Ag_2Te–Ag_2S System at 300 K
(a) Vol. 47, No. 1, p. K59—K61 (1978).
See also *1.3*, p. *26*; *14.3*, p. *221*; 15, p. 260; 22, p. 413.

M. Maszkiewicz
*Specific Heat of High Purity Nickel Single Crystal near the Critical Temperature
(a) Vol. 47, No. 1, p. K77—K80 (1978).
See also 21.1, p. 373.

W. Brückner and W. Moldenhauer
*Resistivity–Temperature Characteristic and Thermal Switching
(a) Vol. 47, No. 1, p. K99—K102 (1978).
See also *14.3.4*, p. *242*; 22.6, p. 493.

I. J. Gupta and S. K. Trikha
*The Reduced Conductivity of Dielectrics when Both Boundary and Dislocation Scatterings are Important at Low Temperatures
(b) Vol. 87, No. 2, p. 513—516 (1978).
See also *6*, p. *71*; 23, p. 530.

R. Ramji Rao and A. Ramanand
Lattice Dynamics, Specific Heat, and Bulk Modulus of Scandium
(b) Vol. 87, No. 2, p. 751—756 (1978).
See also *6*, p. *72*; 12.1, p. 135; 21.4, p. 397.

B. Prasad and R. S. Srivastava
Thermal Properties and Grüneisen Parameters of Alkali Metals Employing the Model Potential of Krasko and Gurskii
(b) Vol. 87, No. 2, p. 771—776 (1978).
See also *6*, p. *72*; 21.2, p. 392.

K. S. Dubey
*Heat Transfer in a Doped Crystal Lattice at Low Temperatures: Application to Sb-Doped Ge
(b) Vol. 87, No. 2, p. K133—K136 (1978).
See also *6*, p. *72*; 22.1.1, p. 425.

R. Capelletti, E. Okuno, G. E. Matthews, and J. H. Crawford, Jr.
*ITC Spectra of Impurity Aggregate in CaF_2 Crystals Doped with Trivalent Rare-Earth Ions
(a) Vol. 47, No. 2, p. 617—624 (1978).
See also 10.2, p. 114; 22.5.3, p. 487.

M. Godlewska and M. Rachwalska
An Adiabatic Calorimetric Study of the Phase Situation in 2,2-Dinitropropane
(a) Vol. 47, No. 2, p. 661—665 (1978).
See also *1.2*, p. *20*; 22.9, p. 525.

K. Balcerek and T. Tyc
*Heat Flux Rectification in Tin–α-Brass System
(a) Vol. 47, No. 2, p. K125—K128 (1978).
See also 21, p. 360.

H. Misiorek, T. Zakrzewski, and J. Rafałowicz
*Influence of Plastic Deformation on the Thermal Conductivity Maximum of Copper and Aluminium in the Temperature Range 4.2 to 70 K
(a) Vol. **47**, No. 2, p. K137—K140 (1978).
Erratum
(a) Vol. **50**, No. 2, p. K261 (1978).
See also *10.1*, p. *104*; 21, p. 360; 21.1, p. 375.

E. A. Pushkarev, N. S. Petrenko, and V. A. Finkel
*Thermal Expansion of the Superconduction Compound HfV_2 at Low Temperatures
(a) Vol. **47**, No. 2, p. K145—K148 (1978).
See also *14.2*, p. *214*; 21, p. 360.

V. I. Zubov
*Higher Anharmonicites in the Improved Unsymmetrized Self-Consistent Field Approximation for a Crystal (II): Comparison of Calculations with Experiment
(b) Vol. **88**, No. 1, p. 43—50 (1978).
See also *6*, p. *73*; 23, p. 530.

M. Chabin, F. Gilletta, and C. Ridou
*Thermal Properties of $TlCdF_3$ and $RbCaF_3$ near Their Phase Transitions
(a) Vol. **48**, No. 1, p. 67—70 (1978).
See also *1.2*, p. *21*; 22.8.1, p. 514.

J. Mucha and J. Rafałowicz
*Thermal Conductivity Minimum of Aluminium
(a) Vol. **48**, No. 1, p. 221—224 (1978).
See also 21, p. 360.

H. H. Boghosian and K. S. Dubey
*Role of Electron–Phonon Interaction and Peripheral Phonons in the Lattice Thermal Conductivity of a Doped Semiconductor at Low Temperatures: Application to Phosphorous-Doped Germanium
(b) Vol. **88**, No. 2, p. 417—427 (1978).
See also *6*, p. *74*; 22.1.1, p. 427.

I. J. Gupta and S. K. Trikha
*Lattice Thermal Conductivity of Solid Neon in the Temperature Range 0.5 to 10 K
(b) Vol. **88**, No. 2, p. 815—818 (1978).
See also *6*, p. *74*; 23, p. 530.

M. Rachwalska and T. Stanek
An Adiabatic Calorimetric Study of the Phase Situation in $[Ni(H_2O)_6](ClO_4)_2$
(a) Vol. **48**, No. 2, p. 297—300 (1978).
See also *1.2*, p. *21*; 22.8, p. 506.

A. T. Macrander
Density of Solid Krypton at Melting and Isochoric Equation of State of Solid Krypton and Solid Argon
(a) Vol. **48**, No. 2, p. 571—579 (1978).
See also *1.2*, p. *22*; 23, p. 530.

V. N. Andreev, F. A. Chudnovskii, A. V. Petrov, and E. I. Terukov
*Thermal Conductivity of VO_2, V_3O_5, and V_2O_3
(a) Vol. **48**, No. 2, p. K153—K156 (1978).
See also 22.6, p. 495.

S. Fakioğlu and Ş. Erkoç
*Magnetic and Thermal Properties of Nd^{3+} in Scheelite-Type Crystals
(b) Vol. **89**, No. 1, p. 289—298 (1978).
See also *18.1*, p. *279*; 22.8.1, p. 515.

Mahiradhwaj Singh
*The Influence of Internal Strain on Bound Hole–Phonon Interaction in p-Type Semiconductors with Zincblende Structure
(b) Vol. **89**, No. 1, p. 299—307 (1978).
See also *6*, p. *75*; 22.2, p. 444.

H. H. Boghossian and K. S. Dubey
*Heat Transfer in Doped Al_2O_3 at Low Temperatures
(b) Vol. **89**, No. 1, p. K65—K67 (1978).
See also *6*, p. *76*; 22.6, p. 496.

M. Meissner, A. Tausend, and D. Wobig
*Heat Capacities and Thermodynamic Properties of α-Monoclinic Selenium in the Range 3 to 300 K
(a) Vol. **49**, No. 1, p. 59—66 (1978).
See also 22.1.3, p. 442.

K. Fukamichi
Magnetic Phase Diagram of the Dilute Cr–Sb Alloy System
(a) Vol. **49**, No. 1, p. K37—K40 (1978).
See also 14.1, p. 210; *18.4*, p. *301*; 21.1, p. 378.

P. A. Varotsos
*On the Pressure Variation of the Excessive Specific Heat Associated with Vacancies
(a) Vol. **49**, No. 1, p. K57—K61 (1978).
See also *10*, p. *97*; 12, p. 133.

M. S. Kushwaha and S. S. Kushwaha
Frequency Spectra and Thermal Properties of Rare Gas Solids (Argon and Krypton)
(b) Vol. **89**, No. 2, p. 583—586 (1978).
See also *6*, p. *76*; 23, p. 530.

C. R. Brooks
*The Heat Capacity at Constant Volume of Platinum
(b) Vol. **89**, No. 2, p. K123—K125 (1978).
See also 21.6, p. 405.

E. M. Savitskii, P. V. Geld, V. E. Zinovev, N. B. Gorina, V. I. Sperelup, V. P. Polyakova, and A. L. Sokolov
*Transport Properties of Osmium Single Crystals at High Temperatures
(a) Vol. **49**, No. 2, p. K117—K120 (1978).
See also *14.1*, p. *211*; 21, p. 364.

A. Freudenhammer
*Consequences of the Linear Part of the Specific Heat of Dilute Magnetic Alloys
(b) Vol. **90**, No. 1, p. 269—275 (1978).
See also 13, p. 148; 21, p. 364.

Mahiradhwaj Singh
Electron–Phonon Interaction and Phonon Conductivity of the n-Type Semiconductors Mg_2Ge and Mg_2Sn at Low Temperatures
(b) Vol. **90**, No. 1, p. 385—389 (1978).
See also *6*, p. *77*; 22.7, p. 502.

H. Van Cong
*Influence of Band-Tail Parameters on Electronic Specific Heat and Transport Coefficients in Heavily Doped Semiconductors
(b) Vol. **90**, No. 1, p. 401—407 (1978).
See also 13.4, p. 184; *14.3*, p. *224*; 22, p. 419.

Mahiradhwaj Singh
*Resonance Scattering of Phonons in the Thermal Conductivity of CN Doped KCl
(b) Vol. **90**, No. 1, p. K43—K47 (1978).
See also *6*, p. *78*; 22.5.2, p. 485.

J. D. Patterson
*The Low Temperature Lorenz Number of Spin Glasses
(b) Vol. **90**, No. 1, p. K85—K88 (1978).
See also *2*, p. *51*; 14.1, p. 212; 21.1.1, p. 390.

B. Mrygoń
*Scaling Hypothesis for Non-Ideal System
(b) Vol. **90**, No. 1, p. K99—K101 (1978).
See also *1.2*, p. *24*; 21.1, p. 380.

G. Pompe, J. Sonntag, A. Biedermann, and S. Schiefer
*Abnormal Thermal Conductivity of Chromium–Aluminium Alloys in the Temperature Range 5 to 100 K
(a) Vol. **50**, No. 1, p. 147—152 (1978).
See also 14.1, p. 212; 21.1, p. 381.

G. Hausch and E. Török
Elastic, Magnetoelastic, and Thermal Properties of Some Ferromagnetic Metallic Glasses
(a) Vol. **50**, No. 1, p. 159—164 (1978).
See also *2*, p. *51*; 12.1, p. 136; *18.2*, p. *291*; 21.1, p. 381; 21.1.1, p. 390.

P. Macko and J. Doupovec
*Thermal and Electrical Properties of Neutron Irradiated GeS_{11} Glass
(a) Vol. **50**, No. 1, p. K51—K54 (1978).
See also 2, p. 52; 11, p. 130; *14.3*, p. *225*; 22, p. 420.

P. A. Varotsos
*A Macroscopic Calculation of the Specific Heats of a Vacancy
(b) Vol. **90**, No. 2, p. 629—633 (1978).
See also 9, p. 94; 10, p. 98.

N. N. Delyagin, V. I. Krylov, and V. I. Nesterov
*Temperature Anomalies of the Magnetic Hyperfine Field for Sn in Metallic Ferromagnets and the Thermal Expansion of the Crystal Lattice
(b) Vol. **90**, No. 2, p. 741—749 (1978).
See also *18.2*, p. *292*; 21.1, p. 382.

G. P. Srivastava
*Role of Optical Phonons in the High-Temperature Lattice Thermal Conductivity of Semiconductors
(b) Vol. **90**, No. 2, p. K125—K128 (1978).
See also *6*, p. *78*; 22.1.1, p. 428; 22.1.2, p. 439; 22.2.1, p. 449; 22.2.3, p. 455.

A. S. Bhalla, L. E. Cross, and D. A. Payne
Thermal, Optical, and Elastic Properties of Ferroelastic Isopropyl Ammonium Chloroplatinate (IPACP) and Chlorostannate (IPACS)
(a) Vol. **50**, No. 2, p. 661—667 (1978).
See also *1.2*, p. *24*; 12.2, p. 141; 20.1, p. 338; 22.9, p. 528.

F. Smutný
*Specific Heat Anomaly at the $F\bar{4}3c \rightarrow Pca2_1$ Phase Transition in Nickel Chlorine Boracite $Ni_3B_7O_{13}Cl$
(a) Vol. **50**, No. 2, p. K201—K204 (1978).
See also *14.4.1*, p. *257*; 22.8, p. 509.

C. Paorici and L. Zecchina
*Thermal Conductivity in In–Se–Te Alloys
(a) Vol. **50**, No. 2, p. K235—K238 (1978).
See also 22.3.1, p. 460.

9. Diffusion

Review Article

P. A. Thrower and R. M. Mayer
*Point Defects and Self-Diffusion in Graphite
(a) Vol. **47**, No. 1, p. 11—37 (1978).
See also *10.2*, p. *109*; 13.4, p. 169; 22.1.3, p. 441.

Original Papers and Short Notes

H. Eckseler and Ch. Herzig
*On the Mass Dependence of the Energy Factor ΔK of the Isotope Effect for Impurity Diffusion in Copper
(b) Vol. **85**, No. 1, p. 185—193 (1978).
See also 10.1, p. 99; 21.1, p. 368.

V. K. Yarmarkin, S. P. Teslenko, and A. I. Knyazev
Mössbauer Study of Sb and Fe Impurities in Semiconductining and Insulating TiO_2
(a) Vol. **45**, No. 1, p. 63—69 (1978).
See also *5*, p. *60*; 10.2, p. 109; 22.6, p. 491.

G. Rein, H. Mehrer, and K. Maier
*Diffusion of ^{197}Pt and ^{199}Au in Platinum at Low Temperatures
(a) Vol. **45**, No. 1, p. 253—261 (1978).
See also 10.1, p. 99; 21.6, p. 401.

A. N. Aleshin, V. Yu. Aristov, B. S. Bokshtein, and L. S. Shvindlerman
Kinetic Properties of ⟨111⟩ Tilt Boundaries in Aluminium
(a) Vol. **45**, No. 1, p. 359—366 (1978).
See also *1.4*, p. *29*; 21, p. 353.

P. Varotsos
*Diffusion of Monovalent Ions in an Alkali Halide Crystal
(a) Vol. **45**, No. 1, p. K9—K12 (1978).
See also 22.5.2, p. 478.

H. Runge and H. Wörl
*Autoradiographic Detection of Getter Effect of Argon-Implanted Layers in Silicon
(a) Vol. **45**, No. 2, p. 509—512 (1978).
See also *11*, p. *124*; 14.3, p. 218; 22.1.2, p. 430.

L. N. Aleksandrov, R. N. Lovyagin, P. A. Simonov, and I. S. Bzinkovskaya
p–n Junctions in the Surface Region of Silicon Obtained by Evaporation of Silicon in Ultrahigh Vacuum
(a) Vol. **45**, No. 2, p. 521—527 (1978).
See also *14.3.3*, p. *233*; 22.1.2, p. 430.

S. K. Sen, M. B. Dutt, and A. K. Barua
*The Diffusion of Iron in Copper and of Nickel in Silver
(a) Vol. **45**, No. 2, p. 657—663 (1978).
See also 21.1, p. 370; 21.1.1, p. 385; 21.6, p. 401.

H. Siethoff
*On the Creep of Germanium and Silicon
(a) Vol. **45**, No. 2, p. K135—K137 (1978).
See also *10.2*, p. *110*; 22.1.1, p. 424; 22.1.2, p. 430.

Y. L. Khait
Some Observable Consequences of Kinetic Many-Body Treatment of Large Energy Fluctuations and of Rate Constants in Solids
(b) Vol. **86**, No. 1, p. 409—418 (1978).
Erratum
(b) Vol. **89**, No. 2, p. 687 (1978).
See also *6*, p. *67*; 8, p. 84; 10, p. 96.

P. R. Jay and B. Tuck
*Mercury Probe *C–U* Measurements on GaP
(a) Vol. **46**, No. 1, p. 199—208 (1978).
See also *14.3*, p. *219*; 22.2.2, p. 450.

M. Schneider and E. Nebauer
*Diffusivity and Solubility of Copper in LEC-GaP
(a)Vol. **46**, No. 1, p. K71—K74 (1978).
See also 22.2.2, p. 450.

W. H. Tanttila
*Self-Diffusion of Solid Sodium
(b) Vol. **86**, No. 2, p. K175—K179 (1978).
Erratum
(b) Vol. **89**, No. 1, p. K109 (1978).
See also 21.2, p. 392.

H. Rauh and D. Simon
*On the Diffusion Process of Point Defects in the Stress Field of Edge Dislocations
(a) Vol. **46**, No. 2, p. 499—510 (1978).
See also 10, p. 96.

G. C. Jain, A. Prasad, and B. C. Chakravarty
*Diffusion of Ga into Si through SiO_2
(a) Vol. **46**, No. 2, p. K151—K155 (1978).
See also 22.1.2, p. 432.

B. Huber and G. Sicking
*Tritium Diffusion in Palladium Alloys with Pt and Cu
(a) Vol. **47**, No. 1, p. K85—K89 (1978).
See also 21, p. 359.

K. D. Becker
*Nuclear Spin Relaxation by Quadrupolar Interactions. The Diffusion of Substitutional Impurities in Ionic Crystals
(b) Vol. **87**, No. 2, p. 589—597 (1978).
See also *19*, p. *308*; 22.5.1, p. 477.

CH. HERZIG and H. E. PELTNER
*Electromigration and Residual Resistivity of Cobalt in Gold
(a) Vol. **47**, No. 2, p. 497–506 (1978).
See also 14.1, p. 208; 21.6, p. 403.

E. STRACKE and CH. HERZIG
*Electromigration, Thermomigration, and Solubility of Copper in Lead
(a) Vol. **47**, No. 2, p. 513–521 (1978).
See also 14.1, p. 208; 21, p. 359.

E. N. MGBENU
*Room Temperature Au–Ge Interdiffusion
(a) Vol. **47**, No. 2, p. K153–K155 (1978).
See also 14.3.1, p. 227; 22.1.1, p. 426.

O. A. TROITSKII
Fractionation of Stable Isotopes of Metals by Combined Electric Migration and Zone Melting
(a) Vol. **48**, No. 1, p. 229–234 (1978).
See also *3*, p. *55*; 21, p. 360.

R. SH. MALKOVICH and V. A. POKOEVA
*Diffusion of an Ionized Impurity in a Semi-Infinite Semiconductor
(a) Vol. **48**, No. 1, p. 241–248 (1978).
See also 22, p. 415.

W. RACZYŃSKI
*Permeability, Diffusivity, and Solubility of Hydrogen and Deuterium in Pure Iron at 10 to 60 °C
(a) Vol. **48**, No. 1, p. K27–K30 (1978).
See also 21.1.1, p. 388.

V. N. PAVLOVICH and N. V. RUDKO
*Nonadiabatic Quantum Theory of Interstitial Diffusion in Crystals
(b) Vol. **88**, No. 2, p. 407–416 (1978).
See also *10*, p. *96*.

A. BUROFF, E. NEBAUER, and P. SÜPTITZ
*Influence of an Electric DC Field on the Migration of Silver on Amorphous As_2S_3
(a) Vol. **48**, No. 2, p. K109–K112 (1978).
See also *2*, p. *48*; 14.3.4, p. 244; 22, p. 416.

J. KALUS, K. FUNKE, and A. HEIDEMANN
*Search for Quasielastic Neutron Scattering by Silver in β-AgI
(a) Vol. **48**, No. 2, p. K179–K180 (1978).
See also 22.5.1, p. 477.

R. KUTNER, I. SOSNOWSKA, and W. KLEY
*Next-Neighbour Jumps of Hydrogen in Quasi-Elastic Neutron Scattering from α-PdH_x
(b) Vol. **89**, No. 1, p. K29–K33 (1978).
See also 21.6, p. 404.

E. Nebauer
*A Comparative Study of Zinc Tracer Diffusion Profiles in LEC, SSD, and CVD GaP
(a) Vol. **49**, No. 1, p. 249–254 (1978).
See also 10.2, p. 118; 22.2.2, p. 451.

G. Müller, M. Haubold, R. Schimko, M. Trapp, and G. Schwarz
*Investigations on the Diffusion on Implanted Zinc in $GaAs_{1-x}P_x$ by Ion Microprobe
(a) Vol. **49**, No. 1, p. 279–284 (1978).
See also *11*, p. *129*; 22.2.4, p. 456.

W. Bergholz and W. Schröter
Precipitation of Cobalt in Silicon Studied by Mössbauer Spectroscopy
(a) Vol. **49**, No. 2, p. 489–498 (1978).
See also *5*, p. *62*; *10.2*, p. *119*; 22.1.2, p. 437.

G. Krautheim, A. Neidhardt, and U. Reinhold
*Impurity Diffusion of ^{110m}Ag in Cu
(a) Vol. **49**, No. 2, p. K125–K128 (1978).
See also 21.1, p. 379.

A. T. Fromhold, Jr. and A.-B. Chen
*Hopon Mechanism for Ion Transport in Ionic Compounds
(b) Vol. **90**, No. 1, p. K21–K25 (1978).
See also *10.2*, p. *120*; 22, p. 419.

J. Reichel
The Influence of White Light on the Annealing Process of Self-Interstitials in p-Type Silicon
(a) Vol. **50**, No. 1, p. 87–93 (1978).
See also *10.2*, p. *120*; 22.1.2, p. 438.

J. Bihr, H. Mehrer, and K. Maier
*A Comparison between Microsectioning Studies of Low-Temperature Self-Diffusion in Silver
(a) Vol. **50**, No. 1, p. 171–178 (1978).
See also 21.6, p. 405.

S. K. Pabi
*Mechanism of the Dissolution of Si in Al
(a) Vol. **50**, No. 1, p. K9–K12 (1978).
See also 14.3.4, p. 246; 21, p. 365; 22.1.2, p. 438.

P. A. Varotsos
A Macroscopic Calculation of the Specific Heats of a Vacancy
(b) Vol. **90**, No. 2, p. 629–633 (1978).
See also *8*, p. *89*; 10, p. 98.

V. Paidar
*Influence of Diffusion on Dislocation Motion in Solid Solutions
(a) Vol. **50**, No. 2, p. 483–490 (1978).
See also 10, p. 98.

K. Maier, H. Mehrer, and J. Bihr
*A Novel Method for the Direct Determination of the Activation Enthalpy in Tracer Diffusion Studies with Applications to "Low"-Temperature Self-Diffusion in Silver
(a) Vol. **50**, No. 2, p. 543—549 (1978).
See also 21.6, p. 405.

G. Tomasi, G. Frohberg, Th. Hehenkamp, and H. Wever
*Hall-Field Electrotransport of Carbon in Iron
(a) Vol. **50**, No. 2, p. 563—572 (1978).
See also 14.1, p. 212; 21.1.1, p. 391.

S. E. Ormrod and D. L. Kirk
*The Effect of Path Tortuosity and Anisotropy upon the Diffusion of Ions through Polycrystalline β Aluminas
(a) Vol. **50**, No. 2, p. K193—K196 (1978).
See also 22.6, p. 498.

10. Defect Properties (Irradiation Defects See 11)

Original Papers and Short Notes

B. Michel
*Dilatation Energy of Extended Defects with Arbitrary Shape and Size near Solid Surfaces
(b) Vol. **85**, No. 1, p. K49—K52 (1978).
See also *1.6*, p. *40*.

B. K. D. Gairola
*Nonlocal Theory of Elastic Interaction between Point Defects
(b) Vol. **85**, No. 2, p. 577—585 (1978).

B. A. Greenberg and M. A. Ivanov
*On the Theory of Plastic Deformation with an Account of Dislocation Transformations of Several Types
(a) Vol. **45**, No. 2, p. 403—410 (1978).

Y. L. Khait
*Some Observable Consequences of Kinetic Many-Body Treatment of Large Energy Fluctuations and of Rate Constants in Solids
(b) Vol. **86**, No. 2, p. 409—418 (1978).
Erratum
(b) Vol. **89**, No. 2, p. 687 (1978).
See also *6*, p. *67*; 8, p. 84; 9, p. 92.

H. Rauh and D. Simon
On the Diffusion Process of Point Defects in the Stress Field of Edge Dislocations
(a) Vol. **46**, No. 2, p. 499—510 (1978).
See also *9*, p. *92*.

P. Rapps, K.-H. Katerbau, H. Mughrabi, K. Urban, and M. Wilkens
*Electron Diffraction in Crystalline Specimens Containing a High Density of Lattice Defects (II): Numerical Evaluation for the Case of Narrow Dislocation Dipoles — Comparison with Experimental Data
(a) Vol. **47**, No. 2, p. 479—488 (1978).
See also *1.4*, p. *31*.

B. A. Greenberg, M. A. Ivanov, Yu. N. Gornostirev, and L. E. Karkina
*Phenomenological Theory of Plastic Deformation with Several Types of Mobile and Immobile Dislocations (I): Theory
(a) Vol. **47**, No. 2, p. 731—741 (1978).

V. N. Pavlovich and N. V. Rudko
*Nonadiabatic Quantum Theory of Interstitial Diffusion in Crystals
(b) Vol. **88**, No. 2, p. 407—416 (1978).
See also *9*, p. *93*.

T. Mura and N. Kinoshita
*The Polynomial Eigenstrain Problem for an Anisotropic Ellipsoidal Inclusion
(a) Vol. **48**, No. 2, p. 447—450 (1978).
See also 12, p. 133.

P. A. VAROTSOS
*On the Pressure Variation of the Excessive Specific Heat Associated with Vacancies
(a) Vol. **49**, No. 1, p. K57—K61 (1978).
See also *8*, p. *88*; 12, p. 133.

B. MICHEL
*A Note on the Interface-Interaction of Point Defects and Volume Defects
(b) Vol. **89**, No. 2, p. K179—K181 (1978).
See also *1.6*, p. *41*.

F. LOGES
*The Strain Energy Associated with an Extended Defect in an Isotropic Solid
(b) Vol. **89**, No. 2, p. K191—K194 (1978).
See also 12, p. 133.

B. A. GREENBERG, M. A. IVANOV, YU. N. GORNOSTIREV, and L. E. KARKINA
*Phenomenological Theory of Plastic Deformation with Several Types of Mobile and Immobile Dislocations (II): An Analysis of the Features of Plastic Deformation of Ordered Alloys with a Superstructure Ll_2
(a) Vol. **49**, No. 2, p. 517—528 (1978).

M. WILKENS and H. FÖLL
*The Black–White Vector ***l*** of Small Dislocation Loops on Transmission Electron Microscope Images
(a) Vol. **49**, No. 2, p. 555—561 (1978).
See also *1.4*, p. *33*.

K. MAEDA and S. TAKEUCHI
Computer Simulation of Deformation in Two-Dimensional Amorphous Structures
(a) Vol. **49**, No. 2, p. 685—696 (1978).
See also *2*, p. *50*; 12, p. 133.

M. J. MARCINKOWSKI
Further Generalizations of the Unified Theory of Grain Boundaries
(a) Vol. **49**, No. 2, p. 725—735 (1978).
See also *1.4*, p. *34*.

V. A. STRELTSOV
Effects of Hydrostatic Pressure of the Second Phase Induced Elastic Fields
(a) Vol. **49**, No. 2, p. K129—K133 (1978).
See also *12*, p. *133*.

V. A. SOLOVEV and V. A. STRELTSOV
*On the Crack Nucleation in Hydrostatically Compressed Crystals
(a) Vol. **49**, No. 2, p. K145—K148 (1978).
See also *12*, p. *133*.

K. JAGANNADHAM and M. J. MARCINKOWSKI
*Comparison of the Image and Surface Dislocation Models
(a) Vol. **50**, No. 1, p. 293—302 (1978).

P. A. Varotsos
A Macroscopic Calculation of the Specific Heats of a Vacancy
(b) Vol. **90**, No. 2, p. 629–633 (1978).
See also *8*, p. *89*; 9, p. 94.

V. Paidar
Influence of Diffusion on Dislocation Motion in Solid Solutions
(a) Vol. **50**, No. 2, p. 483–490 (1978).
See also *9*, p. *94*.

M. J. Marcinkowski and K. Jagannadham
*The Fine Structure and Energy of Grain Boundaries
(a) Vol. **50**, No. 2, p. 601–610 (1978).
See also *1.4*, p. *35*.

10.1 Metals

Original Papers and Short Notes

H. Eckseler and Ch. Herzig
On the Mass Dependence of the Energy Factor ΔK of the Isotope Effect for Impurity Diffusion in Copper
(b) Vol. **85**, No. 1, p. 185—193 (1978).
See also *9*, p. *91*; 21.1, p. 368.

R. Kamel, A. R. Ali, and Z. Farid
*Annealing Spectrum of Precipitating Al–Li Alloys
(a) Vol. **45**, No. 1, p. 47—52 (1978).
See also 12.1, p. 134; 21, p. 352; 21.2, p. 392.

R. Z. Valiev and O. A. Kaibyshev
*Mechanism of Superplastic Deformation in a Magnesium Alloy (III): The Deformation Model
(a) Vol. **45**, No. 1, p. 77—82 (1978).
See also 21, p. 352; 21.3, p. 394.

H. Neuhäuser and G. Schulz
*Dislocation Path Lengths in Cu Single Crystals of Low Dislocation Density after Pulse Loading
(a) Vol. **45**, No. 1, p. 89—97 (1978).
See also 21.1, p. 369.

S. Takahashi
*Compression-Induced Magnetic Anisotropy and Superlattice Dislocations in Ni_3Fe (III): The Annealing Effect
(a) Vol. **45**, No. 1, p. 133—138 (1978).
See also *18.2*, p. *281*; 21.1, p. 369; 21.1.1, p. 384.

J. P. Rivière
Electrical Resistivity of B_2-Type Fe–Al Ordered Alloys
(a) Vol. **45**, No. 1, p. 157—162 (1978).
See also *14.1*, p. *205*; 21.1.1, p. 384.

T. Hirata
*Solute Atom Clustering during Ageing an Aluminium–Magnesium–Silicon Solid Solution under Stress
(a) Vol. **45**, No. 1, p. 201—205 (1978).
See also 21, p. 353; 21.3, p. 394.

G. Rein, H. Mehrer, and K. Maier
Diffusion of ^{197}Pt and ^{199}Au in Platinum at Low Temperatures
(a) Vol. **45**, No. 1, p. 253—261 (1978).
See also *9*, p. *91*; 21.6, p. 401.

C. E. Laciana, A. J. Pedraza, and E. J. Savino
*Lattice Distortion and Migration Energy of Oxygen in Vanadium and Hydrogen in Niobium, Tantalum, and Vanadium
(a) Vol. **45**, No. 1, p. 315–326 (1978).
See also 21, p. 353; 21.1, p. 369.

A. A. Galkin, V. V. Tokii, and B. P. Filatov
*Dislocation Multiplication (II): The Effect of the Initial Defect Structure and Deformation Conditions on the Dislocation Multiplication
(a) Vol. **45**, No. 1, p. 327–332 (1978).
See also 21, p. 353.

A. G. Khachaturyan and A. F. Rumynina
*Strain Energy Controlled Shape of Crystals Forming Due to a Martensite Transformation or Decomposition
(a) Vol. **45**, No. 2, p. 393–402 (1978).
See also 21.1, p. 370; 21.1.1, p. 385.

R. Kamel, A. R. Ali, and Z. Farid
*Annealing Spectrum of Cold-Worked Dilute Al–Li Alloys
(a) Vol. **45**, No. 2, p. 419–422 (1978).
See also 12.1, p. 134; 21, p. 354.

J. Roggen, J. Nihoul, L. Stals, and J. Cornelis
*α-Particle Irradiation Damage and Stage I Recovery in Zinc
(a) Vol. **45**, No. 2, p. 439–443 (1978).
See also *11*, p. *124*; 21.1, p. 370.

J. Degauque, B. Astie, and L. P. Kubin
*Evidence for the Interaction between Magnetic Domain Walls and Dislocations in High-Purity Iron from Magnetomechanical Damping Experiments
(a) Vol. **45**, No. 2, p. 493–501 (1978).
See also *18.2*, p. *283*; 21.1.1, p. 385.

R. A. Varin, J. W. Wyrzykowski, W. Łojkowski, and M. W. Grabski
Spreading of Extrinsic Grain Boundary Dislocations in Plastically Deformed Aluminium
(a) Vol. **45**, No. 2, p. 565–569 (1978).
See also *1.4*, p. *29*; 21, p. 354.

G. Dlubek, J. Pielaszek, and O. Brümmer
*Comparison of Positron Annihilation and X-Ray Estimations of the Dislocation Density and Depth of Dislocation Profiles in Nickel Transformed into Hydride and Decomposed
(a) Vol. **45**, No. 2, p. K87–K90 (1978).
See also 21.1, p. 370.

A. Ranga Rao and C. Suryanarayana
*Solute–Vacancy Binding Energies in Magnesium Alloys
(a) Vol. **45**, No. 2, p. K131–K133 (1978).
See also 12.1, p. 134; 21.3, p. 394.

O. K. Alekseeva, U. V. Moish, V. I. Sarrak, G. A. Filippov, and V. P. Shantarovich
*Positron Annihilation for Martensite Defect Identification
(a) Vol. **45**, No. 2, p. K159—K162 (1978).
See also 21.1.1, p. 385.

K. M. Miller and P. D. Bristowe
*A Computer Simulation of Point Defects and Planar Defects in Nickel
(b) Vol. **86**, No. 1, p. 93—102 (1978).
See also 12.1, p. 134; 21.1, p. 371.

W. F. W. M. van Heugten
*An Interatomic Potential for Copper from the Phonon Spectra
(b) Vol. **86**, No. 1, p. 277—281 (1978).
See also 6, p. 67; 21.1, p. 371.

V. Hivert, P. Groh, W. Frank, I. Ritchie, and P. Moser
*Internal Friction Peaks Due to Dislocation Relaxations in Plastically Deformed and/or Irradiated High-Purity α-Iron (II): Interpretation
(a) Vol. **46**, No. 1, p. 89—98 (1978).
See also 11, p. 125; 21.1.1, p. 385.

D. Werneth
*Mechanische und magnetische Relaxationsmessungen an reinem und dotierten Nickel nach plastischer Verformung und Elektronenbestrahlung
Mechanical and Magnetic Relaxation Measurements on Pure and Doped Nickel after Plastic Deformation and Electron Irradiation
(a) Vol. **46**, No. 1, p. 99—106 (1978).
See also 12.1, p. 134; 18.2, p. 284; 21.1, p. 371.

Jong K. Lee and William C. Johnson
*Calculation of the Elastic Strain Field of a Cuboidal Precipitate in an Anisotropic Matrix
(a) Vol. **46**, No. 1, p. 267—272 (1978).
See also *12.1*, p. *134*; 21, p. 356.

J. Narayan and S. M. Ohr
*Residual Image Contrast of Dislocation Loops
(a) Vol. **46**, No. 1, p. 281—284 (1978).
See also *1.4*, p. *29*; 22.6, p. 492.

V. Raman and P. Mukhopadhyay
*Mechanical Equation of State in Zirconium Alloys
(a) Vol. **46**, No. 1, p. K5—K8 (1978).
See also 21, p. 356.

G. Ferron and M. Quintard
*Evidence of an Internal Friction Phase Transformation Peak in Niobium–Hydrogen Alloys
(a) Vol. **46**, No. 1, p. K43—K46 (1978).
See also *12.1*, p. *134*; 21, p. 356.

N. A. Tyapunina, E. K. Naimi, S. V. Gasparyan, and G. M. Zinenkova
*The Orientation Dependence of Dislocation Internal Friction in Real Crystals (II): Hexagonal Crystals
(a) Vol. **46**, No. 2, p. 411—420 (1978).
See also *12.1*, p. *135*; 21, p. 357; 21.1, p. 372; 21.3, p. 394.

P. Jung
*Subthreshold Irradiation Effects in Quenched Platinum
(a) Vol. **46**, No. 2, p. 527—531 (1978).
See also 11, p. 125; 14.1, p. 207; 21.6, p. 402.

L. M. Caspers, R. H. J. Fastenau, A. van Veen, and W. F. W. M. van Heugten
*Mutation of Vacancies to Divacancies by Helium Trapping in Molybdenum. Effect on the Onset of Percolation
(a) Vol. **46**, No. 2, p. 541—546 (1978).
See also 21, p. 357.

E. Seitz, W. Schmatz, G. S. Bauer, and W. Just
*A Trigonal Distortion Field for Substitutionally Dissolved Bi Atoms in Pb
(a) Vol. **46**, No. 2, p. 557—565 (1978).
See also 1.3, p. 25; 21, p. 357.

J. G. Erlings and F. W. Schapink
*Dislocation Networks in Twin Boundaries
(a) Vol. **46**. No. 2, p. 653—657 (1978).
See also 21.6, p. 402.

R. Kamel, A. R. Ali, and Z. Farid
*On the Recovery of Quenched and Irradiated Dilute Al–Li Alloys
(a) Vol. **46**, No. 2, p. 697—701 (1978).
See also 12.1, p. 135; 14.1, p. 207; 21, p. 357.

R. A. Varin and J. W. Wyrzykowski
*"In Situ" Observation of Dislocation Generation from Grain Boundary in Aluminium
(a) Vol. **46**, No. 2, p. K79—K81 (1978).
See also 21, p. 357.

H. J. Blythe and L. H. Dunleavy
*Point-Defect Relaxations in Dilute Iron–Nickel Alloys after Low-Temperature Neutron Irradiation
(a) Vol. **46**, No. 2, p. K95—K99 (1978).
See also 11, p. 125; *18.2*, p. *285*; 21.1, p. 372; 21.1.1, p. 386.

W. F. Huang and Y. C. Leung
*On the Analysis of the Angular Correlation Data from Positron Annihilation in Metals
(a) Vol. **46**, No. 2, p. K107—K109 (1978).
See also 21, p. 357.

J. Th. M. de Hosson
*On the Vibrational Entropy of a $\frac{1}{2}$ ⟨111⟩ {110} Edge Dislocation in B.C.C. Iron
(b) Vol. **87**, No. 1, p. 151—161 (1978).
See also 21.1.1, p. 387.

J. Cornelis, J. Nihoul, F. Brosens, and J. T. Devreese
*Interatomic Potential for Molybdenum
(b) Vol. **87**, No. 1, p. K61—K64 (1978).
See also 6, p. 71; 21, p. 358.

V. M. Chernov and M. M. Savin
*Elastic Interaction of 60° and Screw Dislocations with Point Defects in F.C.C. Crystals
(a) Vol. 47, No. 1, p. 45—55 (1978).
See also 21, p. 358.

G. Schroeder
*Investigation of the Dislocation Arrangement in Plastically Deformed Iron Single Crystals by Means of Ferromagnetic Techniques
(a) Vol. 47, No. 1, p. 145—156 (1978).
See also 18.2, p. 286; 21.1.1, p. 387.

U. Rindelhardt, S. Mager, W. Blau, and E. Hegenbarth
*Recovery of Low Temperature Specific Heat, Resistivity, and Coercive Force in Fe–Cr–Al Alloys
(a) Vol. 47, No. 1, p. 157—161 (1978).
See also 8, p. 85; 14.1, p. 207; 18.2, p. 286; 21.1, p. 373; 21.1.1, p. 387.

C. Esnouf and G. Fantozzi
*Theoretical Aspects of the Bordoni Relaxation
(a) Vol. 47, No. 1, p. 201—211 (1978).
See also *12.1*, p. *135*; 21, p. 358.

D. Varchon, C. Oytana et P. Delobelle
*Application d'une méthode d'analyse de la relaxation anélastique. Extension au fluage à haute température
(Application of a Method for Analysing Inelastic Relaxation-Extension to Creep at High Temperature)
(a) Vol. 47, No. 1, p. 259—269 (1978).
See also 21, p. 358.

A. J. Morton
Faulted Defects in γ-Brass-Type Alloys
(a) Vol. 47, No. 1, p. 293—304 (1978).
See also *1.3*, p. *26*; *1.4*, p. *30*; 21.1, p. 373.

F. Perrot
*Phenomenological Study of Vacancies in Aluminium within Density-Functional and Second-Order Perturbation Formalisms
(b) Vol. 87, No. 2, p. 707—718 (1978).
See also 14.1, p. 208; 21, p. 359.

Chang Peng-Hsiang and G. Schroeder
*Domain Patterns and Dislocations in Cyclically Deformed Ni Single Crystals and Their Behavior during Annealing
(a) Vol. 47, No. 2, p. 397—406 (1978).
See also *18.2*, p. *287*; 21.1, p. 374.

C. E. Laciana, A. J. Pedraza, and E. J. Savino
*Lattice Distortion Due to Oxygen and Nitrogen Di-Interstitial Clusters in Niobium and Vanadium
(a) Vol. 47, No. 2, p. 425—436 (1978).
See also 21, p. 359; 21.1, p. 374.

J. F. Dinhut, J. P. Rivière et J. C. Desoyer
*Défauts ponctuels produits par déformation plastique et phénomènes d'ordre dans les alliages Fe–Co et Fe–Co–2%V
(Point Defects Produced by Plastic Deformation and Ordering Phenomena in Fe–Co and Fe–Co–2%V Alloys)
(a) Vol. **47**, No. 2, p. 469—478 (1978).
See also 14.1, p. 208; 21.1, p. 374; 21.1.1, p. 387.

R. H. J. Fastenau, A. van Veen, P. Penning, and L. M. Caspers
*Monte Carlo Simulation of the Interaction of Interstitial Helium, Selfinterstitials, and Vacancies with Vacancies or Vacancy Clusters
(a) Vol. **47**, No. 2, p. 577—587 (1978).
See also 21, p. 359.

D. Kuhlmann-Wilsdorf
*Deviations from Schmid's Law in Uni-Directional Strain
(a) Vol. **47**, No. 2, p. 639—649 (1978).
See also *12.1*, p. *135*; 21, p. 359.

H. Misiorek, T. Zakrzewski, and J. Rafałowicz
*Influence of Plastic Deformation on the Thermal Conductivity Maximum of Copper and Aluminium in the Temperature Range 4.2 to 70 K
(a) Vol. **47**, No. 2, p. K137—K140 (1978).
Erratum
(a) Vol. **50**, No. 2, p. K261 (1978).
See also *8*, p. *87*; 21, p. 360; 21.1, p. 375.

S. Besse, M. Quintard, Ph. Mazot et J. de Fouquet
*Influence de l'hydrogène sur le frottement intérieur du titane
(Influence of Hydrogen on the Internal Friction of Titanium)
(a) Vol. **48**, No. 1, p. 99—103 (1978).
See also *12.1*, p. *135*; 21.1, p. 375.

D. Segers, M. Dorikens, and L. Dorikens-Vanpraet
*Temperature Dependence of Positron Annihilation Parameters in Bismuth
(a) Vol. **48**, No. 1, p. 133—136 (1978).
See also 21.7, p. 406.

P. Vostrý
*On the Influence of Silver on the Recovery Spectrum of Quenched Cadmium Foils
(a) Vol. **48**, No. 1, p. K39—K42 (1978).
See also *14.1*, p. *209*; 21, p. 361.

O. Simmich and H. Löffler
A Simple Model to Describe Isothermic Phase Transformations in Binary Alloys
(a) Vol. **48**, No. 1, p. K67—K70 (1978).
See also *1.2*, p. *21*; 21, p. 361.

K. Narayandas, M. Radhakrishnan, and C. Balasubramanian
Annealing Study of the Electrical Resistivity and Defect Density in Silver Films
(a) Vol. **48**, No. 1, p. K71—K74 (1978).
See also *14.1*, p. *209*; 21.6, p. 403.

S. Panchanadeeswaran, R. W. Ure, Jr., and J. G. Byrne
*Positron Trapping at Precipitates in Aluminum–4 wt% Copper Single Crystals
(a) Vol. **48**, No. 1, p. K83—K87 (1978).
See also *1.3*, p. *27*; 21, p. 361.

J. Thiemann
Host Nuclear Magnetic Resonance in Copper-Band Magnetic Alloys
(b) Vol. **88**, No. 2, p. 397—406 (1978).
See also *19*, p. *310*; 21.1, p. 375.

C. Koenig
On the Positron Localization in Ordered and Disordered Metallic Alloys
(b) Vol. **88**, No. 2, p. 569—579 (1978).
See also *1.3*, p. *27*; 21, p. 361.

H. Bossac, H.-G. Fabian, and H. Löffler
*A Quantitative Interpratation of the Decomposition Kinetics of Some Al–Zn–Mg Alloys
(a) Vol. **48**, No. 2, p. 369—375 (1978).
See also *1.3*, p. *27*; 14.1, p. 209; 21, p. 361; 21.1, p. 376; 21.3, p. 395.

D. Rouby and P. Fleischmann
Spectral Analysis of Acoustic Emission from Aluminium Single Crystals Undergoing Plastic Deformation
(a) Vol. **48**, No. 2, p. 439—445 (1978).
See also *7*, p. *81*; 21, p. 361.

W. H. M. Alsem, P. M. Bronsveld, and E. W. van Royen
*Superlattice Dislocations in Cu_2NiZn
(a) Vol. **48**, No. 2, p. 497—502 (1978).
See also *1.3*, p. *27*; 21.1, p. 376.

M. L. Johnson, S. Saterlie, D. Boice, and J. G. Byrne
*The Removal of Defects from Solids as Observed with Positron Annihilation
(a) Vol. **48**, No. 2, p. 551—554 (1978).
See also *1.3*, p. *27*; 21, p. 362; 21.1, p. 376.

E. Vlachá, B. Sprušil, and P. Vostrý
*Dynamic Defect Annealing during Plastic Deformation of Al
(a) Vol. **48**, No. 2, p. K113—K114 (1978).
See also 21, p. 362.

A. Bittar and D. Lesueur
*Quenching and Annealing of Defects in Bismuth and Its Dilute Alloys
(a) Vol. **48**, No. 2, p. K123—K126 (1978).
See also *14.1*, p. *209*; 21.7, p. *407*.

H. Gärtner und W. Auer
*Zum Einfluß von Nickel-Clustern auf den spezifischen elektrischen Widerstand von paramagnetischen Kupfer–Nickel-Legierungen
(The Influence of Nickel Clusters on the Specific Resistance of Paramagnetic Cu–Ni Alloys)
(a) Vol. **49**, No. 1, p. 149—152 (1978).
See also *14.1*, p. *210*; 21.1, p. 377.

H. Löffler, O. Kabisch, and R. Kroggel
*About a Relation between Microhardness and the Average Radius of the Guinier-Preston Zones in Al–Zn Alloys
(a) Vol. **49**, No. 1, p. 299—303 (1978).
See also 21, p. 363; 21.1, p. 377.

H.-J. Ullrich, U. Reinhold, S. Däbritz, P. Paufler, K. Kleinstück, and B. Pietrass
*On the Lattice Transformation in Plastically Deformed V_3Si Single Crystals Studied by the Kossel Technique
(a) Vol. **49**, No. 1, p. 323—330 (1978).
See also *1.2*, p. *22*; 14.2, p. 215; 21.1, p. 377.

J. R. Nicholls and R. D. Rawlings
*Activation Analysis of Plastic Deformation of Ni_3Al–10 at%Fe
(a) Vol. **49**, No. 1, p. K85—K89 (1978).
See also 21.1, p. 378; 21.1.1, p. 389.

R. Rodloff and H. Neuhäuser
*Mean Velocity of Edge and Screw Dislocation Groups in Neutron-Irradiated Copper Single Crystals
(a) Vol. **49**, No. 2, p. 445—454 (1978).
See also 11, p. 130; 21.1, p. 378.

M. Mayer, O. Vöhringer und E. Macherauch
*Zur quantitativen Erfassung des Portevin-Le Chatelier-Effektes bei homogenen CuNi- und CuNiZn-Legierungen
(On the Quantitative Characterization of the Portevin-Le Chatelier Effect in Homogeneous CuNi and CuNiZn Alloys)
(a) Vol. **49**, No. 2, p. 473—482 (1978).
See also 21.1, p. 378.

V. M. Aliabev and V. A. Pavlov
*Creep and Fracture of α-Iron between 77 and 673 K
(a) Vol. **49**, No. 2, p. 509—515 (1978).
See also 21.1.1, p. 389.

H. O. K. Kirchner, H. Mughrabi, F. Prinz, and H. Schmid
*Dislocation Loops in α-Iron
(a) Vol. **49**, No. 2, p. 539—543 (1978).
See also 12.1, p. 136; 21.1.1, p. 389.

H. J. Blythe and L. H. Dunleavy
Low-Temperature Magnetic Relaxation in Neutron-Irradiated α-Iron
(a) Vol. **49**, No. 2, p. 609—619 (1978).
See also *11*, p. *130*; *18.2*, p. *290*; 21.1.1, p. 389.

M. Ignat, R. Bonnet, D. Caillard, and J. L. Martin
*Creep of the Lamellar Al–$CuAl_2$ Composite (I): Microstructural Observations
(a) Vol. **49**, No. 2, p. 675—684 (1978).
See also 21, p. 364; 21.1, p. 379.

W. F. W. M. van Heugten and J. Th. M. de Hosson
Interatomic Potentials for Alkali Metals. A Comparative Study
(b) Vol. **90**, No. 1, p. 225—232 (1978).
See also *4*, p. *59*; 6, p. 77; 21.2, p. 393.

S. Tanigawa, S. Nanao, K. Kuribayashi, and M. Doyama
*Positron Annihilation in Dezincified α-Brass
(b) Vol. **90**, No. 1, p. K31—K34 (1978).
See also 21.1, p. 380.

W. F. W. M. van Heugten, A. van Veen, and L. M. Caspers
*Molybdenium–Helium and Helium–Helium Interaction Potentials from Helium Desorption Experiments
(b) Vol. **90**, No. 1, p. K89—K94 (1978).
See also 6, p. 78; 21, p. 364.

J. Fidler and P. Skalicky
TEM Investigation of Grown-in Dislocations and Stacking Faults in $SmCo_5$ Single Crystals
(a) Vol. **50**, No. 1, p. 73—81 (1978).
See also *1.4*, p. *34*; 21.1, p. 380; 21.4, p. 398.

C. Radomsky and H. Löffler
*On the Relation between X-Ray Small-Angle Scattering and Electrical Resistivity
(a) Vol. **50**, No. 1, p. 123—130 (1978).
See also 14.1, p. 212; 21.1, p. 381.

M. I. Darby, M. N. Read, and K. N. R. Taylor
*The Energy of Interstitial Hydrogen in α-Zirconium
(a) Vol. **50**, No. 1, p. 203—211 (1978).
See also *13.4*, p. *184*; 21.1, p. 381.

Y. Motohashi and S. Ohtake
*Elastic Energy, Stability, and Interaction of Slip Dislocations and the Choice of Slip Systems in Rhombohedral Crystals, Bi, Sb, As, and Hg
(a) Vol. **50**, No. 1, p. 335—347 (1978).
See also 21.7, p. 407.

G. Ya. Akimov, V. I. Zaitsev, and V. A. Streltsov
*Stress Relaxation in Hydrostatically Compressed Zinc Single Crystals
(a) Vol. **50**, No. 1, p. K33—K36 (1978).
See also *12.1*, p. *136*; 21.1, p. 381.

W. F. W. M. van Heugten, F. van den Berg, L. M. Caspers, A. van Veen, and J. Th. M. de Hosson
*The Influence of Interatomic Potentials on the Interaction of He with a ½ ⟨111⟩ {110} Edge Dislocation in Molybdenum
(b) Vol. **90**, No. 2, p. 643—648 (1978).
See also 21, p. 365.

R. I. Kuznetsova and N. N. Zhukov
*Structural Peculiarities of the Superplastic Behaviour of the Alloy Al–4 wt%Ge Due to Porosity
(a) Vol. **50**, No. 2, p. 385–391 (1978).
See also 1.4, p. 34; 21, p. 366.

Y. Motohashi and S. Ohtake
*The Effect of Self-Energy of Dislocations on the Choice of Slip Systems in Crystals
(a) Vol. **50**, No. 2, p. 449–458 (1978).
See also *12.1*, p. *136*; 21, p. 366.

G. J. den Otter and R. Vetter
*A Model Comprising Work Hardening and Solid Solution Hardening
(a) Vol. **50**, No. 2, p. 529–536 (1978).
See also *12.1*, p. *137*; 21.1, p. 382.

M. van Sande, R. de Ridder, J. van Landuyt, and S. Amelinckx
A Study by Means of Electron Microscopy and Electron Diffraction of Vacancy Ordering in Ternary Alloys of the Systems AlCuNi
(a) Vol. **50**, No. 2, p. 587–599 (1978).
See also *1.3*, p. *28*; *1.4*, p. *34*; 21, p. 366; 21.1, p. 382.

I. G. Ritchie, J. F. Dufresne, and P. Moser
*Internal Friction in Deformed Pure Iron
(a) Vol. **50**, No. 2, p. 617–626 (1978).
See also *12.1*, p. *137*; 21.1.1, p. 391.

V. L. Sedov and L. V. Solomatina
*Structural Inhomogeneity in Invar Alloys Studied by Positron Annihilation
(a) Vol. **50**, No. 2, p. 757–765 (1978).
See also 21.1, p. 382; 21.1.1, p. 391.

N. P. Kobelev and Ya. M. Soifer
*Electronic Damping of Dislocations in Lead
(a) Vol. **50**, No. 2, p. K185–K188 (1978).
See also 7, p. 82; *14.2*, p. *216*; 21, p. 367.

10.2 Non-Metals

Review Article

P. A. Thrower and R. M. Mayer
*Point Defects and Self-Diffusion in Graphite
(a) Vol. **47**, No. 1, p. 11—37 (1978).
See also *9*, p. *91*; 13.4, p. 169; 22.1.3, p. 441.

Original Papers and Short Notes

R. Alcalá, V. M. Orera, and J. Beamonte
*A New Type of F_3-Centre in Additively Coloured CaF_2
(b) Vol. **85**, No. 1, p. 283—290 (1978).
See also *20.1*, p. *320*; 22.5.3, p. 487.

L. Bosi and M. Nimis
*New Lifetime Data for the Excited State of the M-Centre in Alkali Halides
(b) Vol. **85**, No. 1, p. K27—K29 (1978).
Erratum
(b) Vol. **89**, No. 1, p. K107 (1978).
See also *13.4*, p. *170*; 22.5.2, p. 478.

V. K. Yarmarkin, S. P. Teslenko, and A. I. Knyazev
Mössbauer Study of Sb and Fe Impurities in Semiconducting and Insulating TiO_2
(a) Vol. **45**, No. 1, p. 63—69 (1978).
See also *5*, p. *60*; 9, p. 91; 22.6, p. 491.

W. Bollmann
*Association Parameters of $[Me^{3+}-F_i^-]$ Dipoles in CaF_2 Crystals
(a) Vol. **45**, No. 1, p. 187—193 (1978).
See also 14.4, p. 248; 22.5.3, p. 487.

H. Gottschalk, G. Patzer, and H. Alexander
*Stacking Fault Energy and Ionicity of Cubic III–V Compounds
(a) Vol. **45**, No. 1, p. 207—217 (1978).
See also 4, p. 57; 22.2.1, p. 445; 22.2.2, p. 450; 22.2.3, p. 452.

P. Schrey, R. Balzer, and H. Peisl
*Release of Stored Energy of X-Irradiated KCl and CsBr
(b) Vol. **85**, No. 2, p. 553—559 (1978).
See also *11*, p. *123*; 22.5.2, p. 478.

F. Lüty and A. Diaz-Gongora
Electro-Birefringence from Paraelectric Defects in Cubic Crystals
(b) Vol. **85**, No. 2, p. 693—706 (1978).
See also *14*, p. *203*; *20*, p. *315*; 22.5.2, p. 479.

V. I. Altukhov and V. G. Kvachadze
*Phonon Scattering on Small-Radius Colloids in KCl Crystals
(b) Vol. **85**, No. 2, p. 769—775 (1978).
See also 6, p. 66; 8, p. 83; 22.5.2, p. 479.

S. WINTER
A Comment on the Optical Absorption in Germanium with 60°-Dislocations
(b) Vol. **85**, No. 2, p. K95—K98 (1978).
See also *13.4*, p. *171*; *20.1*, p. *321*; 22.1.1, p. 424.

V. V. KALININ, N. N. GERASIMENKO, and S. I. STENIN
*The Effect of Implantation Temperature on the Mechanism of Misfit Dislocation Formation
(a) Vol. **45**, No. 2, p. 377—385 (1978).
See also 1.4, p. 29; *11*, p. *124*; 22.1.2, p. 429.

P. BRAILLON, L. KUBIN, and J. SERUGHETTI
*Plastic Deformation of Calcite Single Crystals Deformed in Compression Parallel to [111]
(a) Vol. **45**, No. 2, p. 453—462 (1978).
See also 22.8.1, p. 511.

V. T. BUBLIK
*The Mean Square Atomic Displacements and Enthalpies of Vacancy Formation in Some Semiconductors
(a) Vol. **45**, No. 2, p. 543—548 (1978).
See also 22, p. 410; 22.2, p. 443; 22.4, p. 461.

J. NARAYAN
**a* ⟨100⟩ Dislocation Loops in Magnesium Oxide
(a) Vol. **45**, No. 2, p. 625—630 (1978).
See also 22.6, p. 491.

B. STROCKA
*Imaging of Photorefractive Effects in $LiNbO_3$ by Double Crystal X-Ray Reflection Topography
(a) Vol. **45**, No. 2, p. K83—K86 (1978).
See also 14.4.1, p. 253; *16*, p. *262*; 22.8.1, p. 511.

H. SIETHOFF
*On the Creep of Germanium and Silicon
(a) Vol. **45**, No. 2, p. K135—K137 (1978).
See also *9*, p. *92*; 22.1.1, p. 424; 22.1.2, p. 430.

A. DIAZ-GONGORA and F. LÜTY
*Kerr Effect and Birefringence from Paraelectric CN^- Defects in Nine Alkali Halides
(b) Vol. **86**, No. 1, p. 127—136 (1978).
See also *14*, p. *203*; *20*, p. *316*; 22.5.2, p. 479.

R. STEINBRECH, W. SKROTZKI, and P. HAASEN
*Latent Hardening of Doped Alkali Halide Single Crystals
(a) Vol. **46**, No. 1, p. 107—115 (1978).
See also 22.5.2, p. 479.

S. UNTERRICKER and J. HAUSBRAND
PAC Examination of Radiation Damaged $CdSiP_2$
(a) Vol. **46**, No. 1, p. 125—130 (1978).
See also *11*, p. *125*; 22.8, p. 504.

P. Hennl
*Optical Properties of Chalcogen–Vacancy Centres in Alkali Halides
(a) Vol. **46**, No. 1, p. 147–152 (1978).
See also *20.1*, p. *323*; 20.3, p. **343**; 22.5.2, p. 479.

E. Hadjoudis, F. Milia, J. Seliger, and R. Blinc
*Temperature Dependence of the ^{14}N Quadrupole Resonance in Some Thermochromic and Photochromic N-Salicylideneanilines
(a) Vol. **46**, No. 1, p. 239–244 (1978).
See also *19*, p. *306*; 22.9, p. 524.

G. A. Andreev, M. Hartmanová, and V. A. Klimov
*Precipitation of Pb Impurity from the Lattice of NaCl Crystals
(a) Vol. **46**, No. 1, p. 259–265 (1978).
See also 20.1, p. 324; 22.5.2, p. 479.

N. A. Tyapunina, G. M. Zinenkova, and E. V. Shtrom
*Dislocation Multiplication in Alkali Halide Crystals Exposed to Ultrasonic Waves. The Original Stage
(a) Vol. **46**, No. 1, p. 327–336 (1978).
See also 22.5.2, p. 479.

N. A. Tyapunina, E. K. Naimi, and S. V. Gasparyan
*The Orientation Dependence of Dislocation Internal Friction in Real Crystals (I): Alkali Halide Crystals
(a) Vol. **46**, No. 1, p. 351–356 (1978).
See also *12.2*, p. *138*; 22.5.2, p. 480.

T. M. Bulatova and E. M. Nadgornyi
*Internal Stresses and Dislocation Mobility in NaCl Single Crystals
(a) Vol. **46**, No. 1, p. 357–364 (1978).
See also 22.5.2, p. 480.

V. L. Indenbom, V. I. Nikitenko, E. V. Suvorov, and V. M. Kaganer
*Section Topography as Single-Crystal Interferometry
(a) Vol. **46**, No. 1, p. 379–386 (1978).
See also 22.1.2, p. 431.

R. J. Gaboriaud, M. F. Denanot, M. Boisson et J. Grilhé
*Etude des dislocations par microscopie électronique dans l'oxyde d'yttrium (Y_2O_3) monocristallin déformé plastiquement
(Study of Dislocations by Electron Microscopy in Plastically Deformed Single-Crystalline Yttrium Oxide (Y_2O_3))
(a) Vol. **46**, No. 1, p. 387–396 (1978).
See also 22.6, p. 492.

M. Singh
Phonon Scattering and Effect of Dislocations on the Low-Temperature Phonon Conductivity in Ge
(b) Vol. **86**, No. 2, p. 691–696 (1978).
See also *6*, p. *69*; 8, p. 84; 22.1.1, p. 425.

K. D. Nierzewski, T. Todorov, and M. Georgiev
*Room Temperature F–F′ Conversion in KCl Crystals of Various Purity and Method of Coloration
(b) Vol. **86**, No. 2, p. 697–704 (1978).
See also 20.1, p. 324; 22.5.2, p. 480.

G. A. Andreev and V. A. Klimov
*Flotation Study of Ca^{2+} and Sr^{2+} Precipitation in KCl Crystals
(a) Vol. **46**, No. 2, p. 421–426 (1978).
See also *1.4*, p. *30*; 22.5.2, p. 480.

A. Bohg
Measurement of Stresses in Thin Films on Single Crystalline Substrates
(a) Vol. **46**, No. 2, p. 445–450 (1978).
See also *1.5*, p. *36*; 22.1.2, p. 432.

B. Hughes and G. H. Narayanan
*On Annealing-Induced Prismatic Dislocation Loops and Electrical Changes in Heavily Te-Doped GaAs
(a) Vol. **46**, No. 2, p. 627–637 (1978).
See also *1.4*, p. *30*; 14.3, p. 220; 22.2.1, p. 446.

J. H. Basson and C. A. B. Ball
*On the Density of Misfit Dislocations in Double- and Triple-Layer Heterojunction Devices
(a) Vol. **46**, No. 2, p. 707–715 (1978).
See also *1.5*, p. *36*.

A. Martinaitis and A. Sakalas
High Temperature Investigations of the Hall Effect in p-CdSe in Selenium Vapour Atmosphere
(a) Vol. **46**, No. 2, p. K141–K144 (1978).
See also *14.3*, p. *220*; 22.4.2, p. 468.

V. K. Agrawal
*Rotation Twinning in Cadmium Iodide Crystals
(a) Vol. **46**, No. 2, p. K145–K146 (1978).
See also 22.5, p. 475.

L. P. Khiznichenko, E. Oteniyazov, and N. S. Tulyaganova
*Internal Friction in Neutron-Irradiated Silicon
(a) Vol. **46**, No. 2, p. K147–K149 (1978).
See also 11, p. 126; *12.2*, p. *139*; 22.1.2, p. 432.

L. E. Oliveira, P. Murilo Oliveira, and B. Maffeo
*F Centers in Alkaline-Earth Fluorides. Inadequacy of the Muffin-Tin Approximation
(b) Vol. **87**, No. 1, p. 25–30 (1978).
See also *20.1*, p. *325*; 22.5.3, p. 487.

Hj. Bernhardt
*Bound Polarons in YAG Crystals
(b) Vol. **87**, No. 1, p. 213–219 (1978).
See also *13.5.3*, p. *200*; 22.8.2, p. 519.

Ch. Diaine, J. L. Gisclon, and J. Dupuy
*Colloid Centres in AgCl Induced by Heavy Ion Implantation
(b) Vol. **87**, No. 1, p. 293–300 (1978).
See also *11*, p. *126*; 20.1, p. 326; 22.5.1, p. 477.

Ch. S. Kuo and D. Schroeer
Mössbauer Study of Dy^{3+} in CaF_2
(b) Vol. **87**, No. 1, p. 325–331 (1978).
See also *5*, p. *61*; 22.5.3, p. 487.

E. Beyer, R. Biederbick, A. Hofstaetter, R. Oeder, A. Scharmann, D. Schwabe, and B. Vitt
*Some Features of the Creation of Intrinsic Hole Centres in Scheelite Type Crystals
(b) Vol. **87**, No. 1, p. K35–K38 (1978).
See also 11, p. 126; *19*, p. *307*; 22.8.1, p. 513.

R. Wernicke
*The Influence of Kinetic Processes on the Electrical Conductivity of Donor-Doped $BaTiO_3$ Ceramics
(a) Vol. **47**, No. 1, p. 139–144 (1978).
See also *14.3*, p. *220*; 22.8.1, p. 513.

A. V. Martinaitis, A. P. Sakalas, Z. V. Januškevičius, and J. K. Viščakas
*Defect Structure of CdSe in Selenium Vapour
(a) Vol. **47**, No. 1, p. 187–194 (1978).
See also *14.3*, p. *220*; 22.4.2, p. 468.

L. Nikolova and T. Todorov
*Photoinduced Reorientation of F_A Centres in KCl: Na near Room Temperature
(a) Vol. **47**, No. 1, p. 213–217 (1978).
See also *20.1*, p. *326*; 22.5.2, p. 481.

F.-G. Kirscht, J. Doerschel, and C. Hansch
*Some Observations Concerning the Initial Stage of Plastic Flow in Silicon
(a) Vol. **47**, No. 1, p. K17–K20 (1978).
See also 22.1.2, p. 433.

G. C. Kuczynski, P. E. Dietz, W. H. Hamill, and J. H. Jischke
*H_2O Related Centers in Potassium Chloride
(b) Vol. **87**, No. 2, p. 581–588 (1978).
See also *20.1*, p. *327*; 22.5.2, p. 481.

M. Vanhaelst and P. Clauws
EPR Spectrum of the Oxygen Vacancy in Single Crystals V_2O_5
(b) Vol. **87**, No. 2, p. 719–723 (1978).
See also *19*, p. *308*; 22.6, p. 494.

K. O. Kechechyan and A. A. Kirakossyan
*Impurity Centres in Semiconductors Containing Screw Dislocations
(b) Vol. **87**, No. 2, p. K123–K126 (1978).
See also *13.4*, p. *176*; 22, p. 414.

R. Hultzsch
Passive Q-Switching of the Ruby Laser by Means of Colour Centres in SrF_2
(a) Vol. **47**, No. 2, p. 415–423 (1978).
See also *20.2*, p. *339*; 22.5.3, p. 487.

K. C. Bleijenberg and C. W. M. Timmermans
Luminescence and Electrical Conductivity of Uranium-Activated Sodium Fluoride Crystals
(a) Vol. **47**, No. 2, p. 589–595 (1978).
See also *14.4*, p. *249*; *20.3*, p. *345*; 22.5.2, p. 481.

R. Capelletti, E. Okuno, G. E. Matthews, and J. H. Crawford, Jr.
ITC Spectra of Impurity Aggregate in CaF_2 Crystals Doped with Trivalent Rare-Earth Ions
(a) Vol. **47**, No. 2, p. 617–624 (1978).
See also *8*, p. *86*; 22.5.3, p. 487.

D. E. Aboltin, V. J. Grabovskis, A. R. Kangro, Ch. Lushchik, A. A. O'Konnel-Bronin, I. K. Vitol, and V. E. Zirap
*Thermally Stimulated and Tunneling Luminescence and Frenkel Defect Recombination in KCl and KBr at 4.2 to 77 K
(a) Vol. **47**, No. 2, p. 667–675 (1978).
See also *20.3*, p. *345*; 22.5.2, p. 481.

V. Platikanova and J. Malinowski
*Neutralization of Trapped Photoholes in AgBr
(a) Vol. **47**, No. 2, p. 683–690 (1978).
See also 13.4, p. 176; 22.5.1, p. 477.

K. S. Bhatki, S. D. Pradhan, and B. V. Thosar
*Further Studies on Positron Lifetimes in γ-Irradiated Teflon in Low Dose Range
(a) Vol. **47**, No. 2, p. 691–698 (1978).
See also 11, p. 127; 14.4, p. 249; 19, p. 309; 22.9, p. 526.

S. B. S. Sastry and K. Balasubramanyam
*Studies on Tin Centres in RbCl Crystals
(a) Vol. **47**, No. 2, p. 711–716 (1978).
See also 11, p. 127; *20.1*, p. *328*; 22.5.2, p. 481.

P.A. Varotsos and K. Alexopoulos
*Calculation of the Migration Volume of Vacancies in Ionic Solids from Macroscopic Parameters
(a) Vol. **47**, No. 2, p. K133–K136 (1978).
See also 22.5.2, p. 482.

K. P. Arefiev, A. A. Tsoi, and S. A. Vorobiev
Influence of High-Temperature Annealing on Positron Annihilation in Electron-Irradiated Silicon
(a) Vol. **47**, No. 2, p. K149–K152 (1978).
See also *11*, p. *127*; 22.1.2, p. 433.

R. K. Gartia and B. S. Acharya
*On the Production of Z_1 Centres
(a) Vol. **47**, No. 2, p. K165–K168 (1978).
See also 20.3, p. 345; 22.5.2, p. 482.

A. A. Kirakossyan and K. O. Kechechyan
*Interdislocation Interaction on Account of Carriers Bound with Dislocations
(b) Vol. 88, No. 1, p. K15—K18 (1978).
See also *13.4*, p. *177*.

A. Watterich
*Z_4 Like Centre in NaCl Doped by Divalent Metals: Mg, Mn, Ni, Cd, or Pb
(b) Vol. 88, No. 1, p. K51—K54 (1978).
See also 20.1, p. 330; 22.5.2, p. 482.

J. R. Hanscomb and P. Chantikul
The Effect of Strontium Chloride as Impurity on the Thermoelectric Power of Potassium Chloride
(a) Vol. 48, No. 1, p. 79—87 (1978).
See also *15*, p. *260*; 22.5.2, p. 482.

W. H. Robinson, A. J. Glover, and A. Wolfenden
*Electrical–Mechanical Coupling of Dislocations in KCl, NaCl, LiF, and CaF_2
(a) Vol. 48, No. 1, p. 155—163 (1978).
See also 22.5.2, p. 482; 22.5.3, p. 488.

R. K. Gartia, B. S. Acharya, and V. V. Ratnam
*Effect of Re-Irradiation on Z_1 Centres
(a) Vol. 48, No. 1, p. 235—239 (1978).
See also *20.3*, p. *345*; 22.5.2, p. 482.

D. E. Ioannou and S. M. Davidson
*SEM Observation of Dislocations in Boron Implanted Silicon Using Schottky Barrier EBIC Technique
(a) Vol. 48, No. 1, p. K1—K4 (1978).
See also 22.1.2, p. 434.

T. Cornelissens, G. van Tendeloo, J. van Landuyt, and S. Amelinckx
Peierls Distortion, Chain Polytypism, and Dislocation Coupling in NbS_3
(a) Vol. 48, No. 1, p. K5—K9 (1978).
Erratum
(a) Vol. 49, No. 2, p. K217 (1978).
See also *1.4*, p. *31*; 22, p. 415.

F.-G. Kirscht, C. Hansch, and J. Doerschel
*Dislocation Reactions in Plastically Deformed Silicon Observed by X-Ray Topography
(a) Vol. 48, No. 1, p. K11—K14 (1978).
See also 22.1.2, p. 435.

M. Wagner
*On the Non-Adiabatic Properties of Tunneling Centres
(b) Vol. 88, No. 2, p. 517—530 (1978).
See also *6*, p. *74*; 22.5.2, p. 482.

E. Goovaerts and D. Schoemaker
*Inelastic Light Scattering of the V_K Centers in the Alkali Halides
(b) Vol. 88, No. 2, p. 615–621 (1978).
See also *20.1*, p. *330*; 22.5.2, p. 482.

R. S. Eachus, R. E. Graves, and M. T. Olm
*Electron Spin Resonance of Mobile Electrons in AgCl
(b) Vol. 88, No. 2, p. 705–711 (1978).
See also 16, p. 265; *19*, p. *310*; 22.5.1, p. 477.

P. Nath, J.-E. Sundgren, and H. T. G. Nilsson
Structural and Mechanical Properties of Composite Films of Iron Particles in an Oxide Matrix
(a) Vol. 48, No. 2, p. 345–355 (1978).
See also *1.5*, p. *37*; 22.6, p. 495.

M. B. O'Reilly
*A Phenomenological Theory for Ionic Conductivity in Solid Electrolytes
(a) Vol. 48, No. 2, p. 489–496 (1978).
See also *14.4*, p. *250*; 22.5.1, p. 477.

L. Baldi, G. F. Cerofolini, G. Ferla, and G. Frigerio
*Gold Solubility in Silicon and Gettering by Phosphorus
(a) Vol. 48, No. 2, p. 523–532 (1978).
See also 14.3.3, p. 235; 22.1.2, p. 435.

Ya. L. Shaiovich and R. A. Zvinchuk
Conditions of Magnetite Precipitation during Decomposition of Wüstite $Fe_{1-x}O$
(a) Vol. 48, No. 2, p. 543–550 (1978).
See also *1.2*, p. *21*; 22.8.2, p. 519.

M. Lefeld-Sosnowska
*X-Ray Diffraction Contrast of a 60° Dislocation Investigated by Means of Section Topography
(a) Vol. 48, No. 2, p. 565–569 (1978).
See also *1.4*, p. *32*; 22.1.2, p. 435.

K. Maeda, K. Nakagawa, and S. Takeuchi
*Activation Analysis of Plastic Deformation of CdTe below Room Temperature
(a) Vol. 48, No. 2, p. 587–591 (1978).
See also *12.2*, p. *140*; 22.4.3, p. 471.

E. V. Shvedov and V. M. Ievlev
*Nucleus Saturation Density and Epitaxy in the Condensation of Metals on Alkali-Halide Crystals (II): Oriented Growth of Islands
(a) Vol. 48, No. 2, p. 603–608 (1978).
See also *3*, p. *55*; 22.5.2, p. 483.

J. H. von Bardeleben, A. Goltzene, B. Meyer, and C. Schwab
*Effects of Iron Content and Stoichiometry on the Coloration of $CuGaS_2$
(a) Vol. 48, No. 2, p. K145–K148 (1978).
See also 19, p. 311; 20.1, p. 331; 22.8, p. 506.

O. K. Alekseeva, V. I. Mikhailov, and V. P. Shantarovich
*Positron Annihilation in Point Defects of the Glassy As–Se System
(a) Vol. **48**, No. 2, p. K169—K173 (1978).
See also 2, p. *48*; 22, p. 416.

S. B. S. Sastry and S. Sapru
*Thermoluminescence of γ-Irradiated Rubidium Chloride Doped withEuropium
(a) Vol. **48**, No. 2, p. K189—K191 (1978).
See also *20.3*, p. *347*; 22.5.2, p. 483.

W. Fuhs, U. Holzhauer, S. Mantl, F. W. Richter, and R. Sturm
*Annihilation of Positrons in Electron-Irradiated Silicon Crystals
(b) Vol. **89**, No. 1, p. 69—75 (1978).
See also 11, p. 128; 22.1.2, p. 436.

C. S. Lam and Y. P. Varshni
*Binding Energies of Positron to F and F′ Colour Centres in Alkali Halides
(b) Vol. **89**, No. 1, p. 103—108 (1978).
See also, *13.4*, p. *179*; 22.5.2, p. 483.

F. Lohse, G. Reuter, and J. M. Spaeth
*Optical Investigation of Atomic Hydrogen Centres in Alkali Chlorides Doped with I^- or Br^-
(b) Vol. **89**, No. 1, p. 109—116 (1978).
See also 13.4, p. 180; *20.1*, p. *332*; 22.5.2, p. 483.

L. Bosi, A. Longoni, and M. Nimis
*Lifetime of Color Centers in NaBr
(b) Vol. **89**, No. 1, p. 221—224 (1978).
See also 13.4, p. 180; *20.3*, p. *347*; 22.5.2, p. 484.

T. Goworek, C. Rybka, and J. Wawryszczuk
*Positronium in Mixed Molecular Crystals
(b) Vol. **89**, No. 1, p. 253—259 (1978).
See also 22.9, p. 526.

A. R. Forouhi and I. Bloomer
*A Quantum Mechanical Approach to the Velocity of Dislocations in Ice
(b) Vol. **89**, No. 1, p. 309—312 (1978).
See also 22, p. 417.

G. Guillot and A. Nouailhat
*Influence of the Purity of Crystals on the Stability of F-Centers and Interstitials in Alkali Halides
(b) Vol. **89**, No. 1, p. K85—K89 (1978).
See also *20.1*, p. *333*; 22.5.2, p. 484.

I. A. Parfianovich, E. E. Penzina, V. M. Khulugurov, V. M. Metzik, and L. M. Sobolev
*Luminescence and Absorption of Colour Centres in RbCl: Eu and KCl: Eu Crystals
(b) Vol. **89**, No. 1, p. K91—K94 (1978).
See also 20.1, p. 333; *20.3*, p. *347*; 22.5.2, p. 484.

F. Carlier and G. Jacobs
*Transient Infrared Absorption of Bound Polarons in KBr
(b) Vol. **89**, No. 1, p. K95–K98 (1978).
See also 13.5.3, p. 200; *20.1*, p. *333*; 22.5.2, p. 484.

H. Steinhardt and P. Haasen
*Creep and Dislocation Velocities in Gallium Arsenide
(a) Vol. **49**, No. 1, p. 93–101 (1978).
See also 22.2.1, p. 448.

K. Graff and H. Pieper
Degradation of Carrier Lifetime in Silicon Crystals at Room Temperature
(a) Vol. **49**, No. 1, p. 137–144 (1978).
See also *14.3*, p. *224*; 22.1.2, p. 436.

N. Terao
*Experimental Study of Crack Propagation in Brittle Amorphous Solids. Refraction of Propagating Crack by Stress Waves
(a) Vol. **49**, No. 1, p. 237–248 (1978).
See also 2, p. 49; *12.2*, p. *140*.

E. Nebauer
A Comparative Study of Zinc Tracer Diffusion Profiles in LEC, SSD, and CVD GaP
(a) Vol. **49**, No. 1, p. 249–254 (1978).
See also *9*, p. *94*; 22.2.2, p. 451.

S. Myhra
*Defect Annealing and Transport Properties in High-Purity InSb Irradiated at 80 K
(a) Vol. **49**, No. 1, p. 285–292 (1978).
See also *14.3*, p. *224*; 15, p. 261; 22.2.3, p. 454 .

A. L. Guerrero, S. C. Jain, and P. L. Pratt
*The Role of Dimers and Trimers in the Aggregation of Impurity–Vacancy Dipoles in Alkali Halides
(a) Vol. **49**, No. 1, p. 353–362 (1978).
See also *14.4*, p. *250*; 22.5.2, p. 484.

Z. Liliental and G. Kästner
*Voltage Dependence on Precipitation Contrast
(a) Vol. **49**, No. 1, p. K1–K2 (1978).
See also 22.4.3, p. 471.

J. Llopis and J. Piqueras
*Cathodoluminescence from Grain Boundaries in MgO and GaP
(a) Vol. **49**, No. 1, p. K9–K12 (1978).
See also *20.3*, p. *348*; 22.2.2, p. 451; 22.6, p. 496.

A. Ioan, V. Topa, and M. Giurgea
* New Colour Centres in KCl: ($Tl^+ + Ca^{2+}$) and KCl: ($Tl^+ + Sr^{2+}$) Crystals
(b) Vol. **89**, No. 2, p. 513–520 (1978).
See also 20.1, p. 334; 22.5.2, p. 485.

I. K. AMANIS, J. G. KLIAVA et J. J. PURĀNS
*Nouveaux centres paramagnétiques dans les halogénures alcalins mixtes: $Ag_c^0(Br_1)$ dans KCl: KBr
New Paramagnetic Centres in Mixed Alkali Halides: $Ag_c^0(Br_1)$ in KCl: KBr
(b) Vol. **89**, No. 2, p. 531–535 (1978).
See also 19, p. 312; 22.5.2, p. 485.

J. POŹNIAK
*Dipolar Relaxation Induced by an AC Field in Various Crystallographic Directions of Me^{2+}-Doped Alkali Halide Crystals (I): Nearest-Neighbour Approach
(b) Vol. **89**, No. 2, p. 577–582 (1978).
See also 14.4, p. 251; 22.5.2, p. 485.

M. DUBIEL, G. BERG, and F. FRÖHLICH
*Aggregation of Point Defects in Highly Doped $NaCl:Ca^{++}$ Crystals
(b) Vol. **89**, No. 2, p. 595–602 (1978).
See also 19, p. 312; 22.5.2, p. 485.

W. BERGHOLZ and W. SCHRÖTER
*Precipitation of Cobalt in Silicon Studied by Mössbauer Spectroscopy
(a) Vol. **49**, No. 2, p. 489–498 (1978).
See also *5*, p. *62*; 9, p. 94; 22.1.2, p. 437.

E. JOHNSON, J. FERRER, and L. T. CHADDERTON
Radiolytic Radiation Damage of Sodalite
(a) Vol. **49**, No. 2, p. 585–591 (1978).
See also *11*, p. *130*; 22.8, p. 507.

D. SHAW and E. WATSON
The Shift of Optical Absorption Edge in CdTe with Temperature and Doping Level
(a) Vol. **49**, No. 2, p. 593–602 (1978).
See also *20.1*, p. *335*; 22.4.3, p. 471.

L. TRÉPIED and J.-C. DOUKHAN
*Dissociated ***a***-Dislocations in Quartz. Influence on Plastic Deformation
(a) Vol. **49**, No. 2, p. 713–724 (1978).
See also *1.4*, p. *33*; 22.6, p. 496.

D. FERRÉ and J. L. FARVACQUE
*Asymmetrical Behaviour of "60°" Dislocation Glide in Indium Antimonide
(a) Vol. **49**, No. 2, p. 737–745 (1978).
See also 22.2.3, p. 454.

J. P. ESTIENNE and J. L. FARVACQUE
*Occurrence of an "***a***" Screw Dislocation Substructure in Tellurium by a Torsion Experiment
(a) Vol. **49**, No. 2, p. K109–K111 (1978).
See also 22.1.3, p. 442.

J. POŹNIAK
*Dipolar Relaxation Induced by an AC Field in Various Crystallographic Directions of Me^{2+} Doped Alkali Halide Crystals (II): Next-Nearest Neighbour Approach
(b) Vol. **90**, No. 1, p. 261–267(1978).
See also 14.4, p. 251; 22.5.2, p. 485.

K. Thörmer and F. Lüty
A New Off-Center Defect System, RbCl:Li$^+$, Studied with Dielectric Techniques under Hydrostatic Pressure
(b) Vol. **90**, No. 1, p. 277—287 (1978).
See also *14.4*, p. *251*; 22.5.3, p. 488.

S. B. S. Sastry and K. Balasubramanyam
*Thermoluminescence of Irradiated RbCl and RbCl:Sn Crystals
(b) Vol. **90**, No. 1, p. 375—383 (1978).
See also *20.3*, p. *349*; 22.5.2, p. 485.

A. T. Fromhold, Jr. and A.-B. Chen
*Hopon Mechanism for Ion Transport in Ionic Compounds
(b) Vol. **90**, No. 1, p. K21—K25 (1978).
See also *9*, p. *94*; 22, p. 419.

W. Szkiełko
*Spin-Dependent Recombination at Dislocations in Silicon Revealed at Reverse Current in a p–n Junction
(b) Vol. **90**, No. 1, p. K81—K83 (1978).
See also *14.3.3*, p. *237*; 16, p. 267; 19, p. 314; 22.1.2, p. 438.

J. Reichel
*Der Einfluß von weißem Licht auf den Ausheilprozeß von Eigen-Zwischengitteratomen in p-Silizium
(The Influence of White Light on the Annealing Process of Self-Interstitials in p-Type Silicon)
(a) Vol. **50**, No. 1, p. 87—93 (1978).
See also 9, p. 94; 22.1.2, p. 438.

I. A. Maltseva, A. Mamedov, Yu. V. Rud, and Yu. K. Undalov
*Luminescence of In-Doped $CdGeP_2$ Crystals
(a) Vol. **50**, No. 1, p. 139—146 (1978).
See also *20.3*, p. *350*; 22.8, p. 508.

E. V. Suvorov, V. I. Nikitenko, and K. Yu. Mukhin
*On the Possibilities of X-Ray Dynamic Focussing to Study Dislocation Microstress Fields
(a) Vol. **50**, No. 1, p. 213—219 (1978).
See also 22.1.2, p. 438.

K. H. Yang, H. F. Kappert, and G. H. Schwuttke
*Minority Carrier Lifetime in Annealed Silicon Crystals Containing Oxygen
(a) Vol. **50**, No. 1, p. 221—235 (1978).
See also *14.3.3*, p. *238*; 22.1.2, p. 438.

Y. H. Lee, P. R. Brosious, and J. W. Corbett
High-Temperature Ion Implantation in Diamond
(a) Vol. **50**, No. 1, p. 237—242 (1978).
See also *11*, p. *130*; 19, p. 314; 22.1, p. 423.

P. Iacconi, D. Lapraz, and R. Caruba
Traps and Emission Centres in Thermoluminescent ZrO_2
(a) Vol. **50**, No. 1, p. 275–283 (1978).
See also 13.4, p. 184; *20.3*, p. *350*; 22.6, p. 497.

Z. Janáček, J. Kuběna, and V. Holý
X-Ray Laue Diffraction from a Bent Crystal. Integrated Intensity
(a) Vol. **50**, No. 1, p. 285–291 (1978).
See also *1.1*, p. *14*; 22.1.2, p. 438.

V. K. Jain and S. P. Kathuria
*Z_3 Centre Thermoluminescence in LiF TLD Phosphor
(a) Vol. **50**, No. 1, p. 329–333 (1978).
See also 20.1, p. 336; *20.3*, p. *350*; 22.5.2, p. 485.

K. Atobe
*The Role of F-Centers in Thermoluminescence Processes in Reactor Irradiated SrF_2 Crystals
(a) Vol. **50**, No. 1, p. K77–K80 (1978).
See also 11, p. 130; *20.3*, p. *350*; 22.5.3, p. 488.

W. Raupach
*Atomistic Calculation of the Core Configuration of Screw Dislocations in Rock Salt Crystals
(a) Vol. **50**, No. 1, p. K81–K84 (1978).
See also 22.5.2, p. 485.

S. Mantovani, E. Mazzega, S. Valeri, and U. del Pennino
*Edge Dislocation Energy Level in Silicon
(a) Vol. **50**, No. 1, p. K123–K126 (1978).
See also *13.4*, p. *184*; 22.1.2, p. 438.

B. Rambabu, C. Ramasastry, and B. V. R. Chowdari
*Optical Absorption of Off-Centre Cu^+ in CsBr
(b) Vol. **90**, No. 2, p. 465–473 (1978).
See also *20.1*, p. *337*; 22.5.2, p. 486.

A. Król, M. J. Kozielski, and W. Nazarewicz
*Infrared Studies of Al Complexes in Zinc Sulphide
(b) Vol. **90**, No. 2, p. 649–656 (1978).
See also *6*, p. *78*; *20.1*, p. *337*; 22.4.1, p. 466.

J. de Oliveira, O. de O. Damasceno, J. R. F. Ferreira, A. L. de Oliveira, J. D. Fabris, P. Boyer, and A. Baudry
*Electric Field Gradient Measurements at ^{181}Ta Nuclei in HfO_2–CaO and HfO_2–MgO Systems
(b) Vol. **90**, No. 2, p. K169–K174 (1978).
See also *13*, p. *149*; 14.4, p. 252 ; 22.6.1, p. 500.

L. C. Jen and J. F. Merklin
*Optically Induced Dichroism of Z_2-Centers in LiF:Mg, Ti (TLD-100)
(a) Vol. **50**, No. 2, p. 469–474 (1978).
See also *20.1*, p. *338*; 22.5.2, p. 486.

G. L. Bleris, Th. Karakostas, N. A. Economou, and R. de Ridder
The Ordered State of In_2Te_3 and Its Relation to the Transition State
(a) Vol. **50**, No. 2, p. 579—586 (1978).
See also *1.2*, p. *24*; *1.3*, p. *28*; 22.3, p. 459.

I. Yonenaga and K. Sumino
*Dislocation Dynamics in the Plastic Deformation of Silicon Crystals (I) : Experiments
(a) Vol. **50**, No. 2, p. 685—693 (1978).
See also 22.1.2, p. 439.

S. Tsunekawa and H. Takei
Twinning Structure of Ferroelastic $LaNbO_4$ and $NdNbO_4$ Crystals
(a) Vol. **50**, No. 2, p. 695—702 (1978).
See also *1*, p. *10*; 14.4.1, p. 257; 22.8.1, p. 516.

A. P. Dolgolenko and I. I. Fishchuk
Defect Clusters and Simple Defect Build-Up Kinetics in Fast-Neutron Irradiated n-Si
(a) Vol. **50**, No. 2, p. 751—755 (1978).
See also *11*, p. *131*; 22.1.2, p. 439.

J. A. A. Khan and J. Leteurtre
*An Experimental Investigation of the Stress Oriented Dislocation Loop Nucleation in Whiskers
(a) Vol. **50**, No. 2, p. K161—K164 (1978).
See also 11, p. 131; 22.5.2, p. 486.

M. E. van Hulle and W. Maenhout-van der Vorst
*Ionic Conductivity of β-AgI Microcrystals
(a) Vol. **50**, No. 2, p. K175—K180 (1978).
See also *14.4*, p. *252*; 22.5.1, p. 477.

L. G. Aravin, R. A. Salukvadze, V. P. Shantarovich, and A. A. Khomenko
Positron Annihilation in Carbon Structures
(a) Vol. **50**, No. 2, p. K213—K217 (1978).
See also *1.4*, p. *35*; 22.1, p. 423.

H. Hesse and S. Kapphan
Doping of $SrTiO_3$ with Hydrogen and Deuterium
(a) Vol. **50**, No. 2, p. K243—K245 (1978).
See also *14.4.1*, p. *257*; *20.1*, p. *338*; 22.8.1, p. 517.

11. Irradiation Effects (X-Ray Diffraction Investigations See 1 and 10)

Review Articles

P. P. Seregin, F. S. Nasredinov, and L. N. Vasilev
*A Study of Radiation Defects in Solids by Means of Mössbauer Spectroscopy
(a) Vol. **45**, No. 1, p. 11—45 (1978).
See also *5*, p. *60*; 22.5, p. 474; 22.8, p. 503; 22.9, p. 522.

V. V. Titov
*Ion Implantation — Problems and Perspectives
(a) Vol. **48**, No. 1, p. 13—22 (1978).
See also *3*, p. *54*; 22.1.2, p. 429.

Original Papers and Short Notes

P. J. Alonso and R. Alcalá
*Optical Absorption of X-Irradiated CaF_2:Co
(b) Vol. **85**, No. 1, p. K23—K26 (1978).
See also *20.1*, p. *320*; 22.5.3, p. 487.

K. Ellmer and R. Wedell
*Erratum to: Total Energy Loss Distributions of Low Energy Heavy Particles after Penetration through Matter
(b) Vol. **85**, No. 1, p. K67 (1978).

J. Gazsó and J. Hajtó
Self-Controlled Laser Beam Chopping Effect in $GeSe_2$ Thick Films
(a) Vol. **45**, No. 1, p. 181—186 (1978).
See also 2, p. 42; *20.1*, p. *320*; 22.7, p. 501.

I. M. Belii, F. F. Komarov, V. S. Tishkov, and V. M. Yankovskii
Formation of Chemical Compounds by Ion Bombardment of Thin Transition Metal Films
(a) Vol. **45**, No. 1, p. 343—352 (1978).
See also *1.2*, p. *15*; *1.5*, p. *36*; 21.1, p. 369.

A. Bhalla, L. E. Cross, and L. Tongson
*Passivation of the *c* Surfaces of Single Crystal Gadolinium Molybdate ($Gd_2(MoO_4)_3$) against Attack by Hydrofluoric Acid by Inert Ion Beam Irradiation
(a) Vol. **45**, No. 1, p. K1—K4 (1978).
See also 14.4.1, p. 253; 22.8.1, p. 510.

A. Bessis, Y. Bismuth et J. Bigot
*Etude de l'endommagement du chrome au cours de l'irradiation aux neutrons à 24,5K
(Study of Damage in Chromium by Neutron Irradiation at 24.5 K)
(a) Vol. **45**, No. 1, p. K71—K76 (1978).
See also *14.1*, p. *205*; 21.1, p. 369.

P. Schrey, R. Balzer, and H. Peisl
*Release of Stored Energy of X-Irradiated KCl and CsBr
(b) Vol. **85**, No. 2, p. 553—559 (1978).
See also *10.2*, p. *109*; 22.5.2, p. 478.

V. V. Kalinin, N. N. Gerasimenko, and S. I. Stenin
*The Effect of Implantation Temperature on the Mechanism of Misfit Dislocation Formation
(a) Vol. **45**, No. 2, p. 377–385 (1978).
See also 1.4, p. 29; *10.2*, p. *110*; 22.1.2, p. 429.

F. Fujimoto, N. Sumida, H. Fujita, and Y. Uchida
*Electron Channeling in ⟨111⟩ Direction of Tungsten Crystals (II): Energy Range of 1000 to 2000 keV
(a) Vol. **45**, No. 2, p. 411–418 (1978).
See also 13, p. 143; 21, p. 354.

J. Roggen, J. Nihoul, L. Stals, and J. Cornelis
*α-Particle Irradiation Damage and Stage I Recovery in Zinc
(a) Vol. **45**, No. 2, p. 439–443 (1978).
See also *10.1*, p. *100*; 21.1, p. 370.

H. Runge and H. Wörl
*Autoradiographic Detection of Getter Effect of Argon-Implanted Layers in Silicon
(a) Vol. **45**, No. 2, p. 509–512 (1978).
See also *9*, p. *91*; 14.3, p. 218; 22.1.2, p. 430.

H. D. Carstanjen, J. Dünstl, G. Löbl, and R. Sizmann
Lattice Location and Determination of Thermal Amplitudes of Deuterium in α-$PdD_{0.007}$ by Channeling
(a) Vol. **45**, No. 2, p. 529–536 (1978).
See also *1.1*, p. *11*; 21, p. 354.

G. Mende
*A Comparison of Calculated Arsenic Implantation Profiles in Silicon with Experimental Results
(a) Vol. **45**, No. 2, p. K121–K123 (1978).
See also 22.1.2, p. 430.

D. Landgraf-Dietz
*A Simple Method for the Determination of Parameters of Ion Implanted Doping Profiles by Means of Threshold Voltage Measurements of MOSFET's
(a) Vol. **45**, No. 2, p. K125–K129 (1978).
See also *14.3.3*, p. *233*; 22.1.2, p. 430.

M. Lindén, E. Ejder, and R. Hellborg
*Channeling Studies of As-Grown GaN
(a) Vol. **45**, No. 2, p. K167–K170 (1978).
See also 22.2, p. 443.

V. V. Kaplin, D. E. Popov, and S. A. Vorobev
*Quantum-Mechanical Treatment of Anomalous Rutherford Scattering of Swift Electrons in Crystals
(b) Vol. **86**, No. 1, p. K35–K38 (1978).
See also 13, p. 143; 22.1.2, p. 431.

V. HIVERT, P. GROH, W. FRANK, I. RITCHIE, and P. MOSER
*Internal Friction Peaks Due to Dislocation Relaxations in Plastically Deformed and/or Irradiated High-Purity α-Iron (II): Interpretation
(a) Vol. **46**, No. 1, p. 89—98 (1978).
See also *10.1*, p. *101*; 21.1.1, p. 385.

S. UNTERRICKER and J. HAUSBRAND
*PAC Examination of Radiation Damaged $CdSiP_2$
(a) Vol. **46**, No. 1, p. 125—130 (1978).
See also 10.2, p. 110; 22.8, p. 504.

J. TAKAHASHI, K. OZAWA, S. YAMAGUCHI, Y. FUJINO, O. YOSHINARI, and M. HIRABAYASHI
*Temperature Dependence of the Thermal Vibrations of Deuterium in a $Pd_{0.8}Au_{0.2}$ Alloy Studied by Ion Channelling
(a) Vol. **46**, No. 1, p. 217—223 (1978).
See also *1.3*, p. *25*; 6, p. 68; 21.6, p. 401.

P. L. F. HEMMENT and R. W. BICKNELL
*1.5 MeV H^+ and He^+ Irradiations of Arsenic Diffused Layers in ⟨100⟩ Silicon
(a) Vol. **46**, No. 1, p. 225—231 (1978).
See also 13.4, p. 173; 14.3, p. 219; 22.1.2, p. 431.

S. KOPTA, E. MAYDELL-ONDRUSZ, and B. RAJCHEL
*Backscattering Measurements of Postimplantation Damage Profiles in Silicon Single Crystals
(a) Vol. **46**, No. 1, p. 285—290 (1978).
See also 22.1.2, p. 431.

ANAND P. PATHAK
*Stopping Power of Solids in Planar Channelling
(b) Vol. **86**, No. 2, p. 751—758 (1978).
See also 13, p. 144; 21.1, p. 372; 21.6, p. 402.

P. JUNG
Subthreshold Irradiation Effects in Quenched Platinum
(a) Vol. **46**, No. 2, p. 527—531 (1978).
See also *10.1*, p. *102*; 14.1, p. 207; 21.6, p. 402.

P. K. MADDEN and J. E. HARBOTTLE
The Imaging of Slip Bands in Nickel
(a) Vol. **46**, No. 2, p. 639—652 (1978).
See also *1.5*, p. *36*; 21.1, p. 372.

A. E. KARKIN, B. N. GOSHCHITSKII, V. E. ARKHIPOV, E. E. VALIEV, and S. K. SIDOROV
Electrical Resistivity and Transition Temperature of Highly Disordered Nb_3Sn
(a) Vol. **46**, No. 2, p. K87—K90 (1978).
See also *14.2*, p. *214*; 21, p. 357.

H. J. BLYTHE and L. H. DUNLEAVY
Point-Defect Relaxations in Dilute Iron–Nickel Alloys after Low-Temperature Neutron Irradiation
(a) Vol. **46**, No. 2, p. K95—K99 (1978).
See also *10.1*, p. *102*; *18.2*, p. *285*; 21.1, p. 372; 21.1.1, p. 386.

L. P. Khiznichenko, E. Oteniyazov, and N. S. Tulyaganova
Internal Friction in Neutron-Irradiated Silicon
(a) Vol. **46**, No. 2, p. K147—K149 (1978).
See also *10.2*, p. *112*; *12.2*, p. *139*; 22.1.2, p. 432.

M. Hou, P. Delavignette, and A. Art
*Detection of Implanted Ions by X-Ray Emission Analysis in TEM
(a) Vol. **46**, No. 2, p. K167—K169 (1978).
See also 22, p. 412.

Ch. Diaine, J. L. Gisclon, and J. Dupuy
*Colloid Centres in AgCl Induced by Heavy Ion Implantation
(b) Vol. **87**, No. 1, p. 293—300 (1978).
See also *10.2*, p. *113*; 20.1, p. 326; 22.5.1, p. 477.

E. Beyer, R. Biederbick, A. Hofstaetter, R. Oeder, A. Scharmann, D. Schwabe, and B. Vitt
Some Features of the Creation of Intrinsic Hole Centres in Scheelite Type Crystals
(b) Vol. **87**, No. 1, p. K35—K38 (1978).
See also *10.2*, p. *113*; *19*, p. *307*; 22.8.1, p. 513.

B. Rauschenbach and W. Hinz
*Sputtering of Flat Glass Surfaces with Noble Gas Ions
(a) Vol. **47**, No. 1, p. 79—84 (1978).
See also *2*, p. *45*.

C. Wagner and F. Burkhardt
*The Annealing Characteristics of Phosphorus-Implanted Silicon Investigated at Low Temperatures
(a) Vol. **47**, No. 1, p. 131—138 (1978).
See also *14.3*, p. *220*; 22.1.2, p. 432.

N. Y. Chernysheva, G. A. Kachurin, and V. A. Bogatyriov
*Amorphisation and Recrystallisation of InSb Ion Implanted Layers
(a) Vol. **47**, No. 1, p. K5—K8 (1978).
See also 1.5, p. 37; 22.2.3, p. 453.

P. Erlenwein
On a Relationship between the Symmetries of Faceted Cones on Ion Bombarded Single-Crystal Faces and Ejection Patterns
(a) Vol. **47**, No. 1, p. K9—K10 (1978).
See also *1.4*, p. *30*; 21, p. 358; 21.6, p. 402.

J. Koshy
*Effect of Neutron Bombardment on the Structure of Barium Sulphate Single Crystals
(a) Vol. **47**, No. 1, p. K21—K22 (1978).
See also 22.8.1, p. 513.

W. Wanarski and Z. Zagórski
*Influence of γ-Irradiation on the Electric Permittivity of Thiourea Single Crystal
(a) Vol. **47**, No. 1, p. K45—K46 (1978).
See also *14.4.1*, p. *255*; 22.9, p. 525.

N. N. Gerasimenko, P. P. Dobrovolskii, N. P. Kibalina, and V. F. Stas
Investigation of a High-Resistivity Surface Layer in Electron Irradiated Silicon by Schottky Barrier Capacitance Measurements
(a) Vol. 47, No. 1, p. K91—K94 (1978).
See also *14.3.2*, p. *230*; 22.1.2, p. 433.

R. Wedell
*Radiative Damping Effect for Ultrarelativistic Channelled Particles
(b) Vol. 87, No. 2, p. 491—496 (1978).
See also 13, p. 145; 22.1.2, p. 433.

G. Foti, E. Rimini, and S. U. Campisano
*Laser Annealing of Pb–Implanted Silicon
(a) Vol. 47, No. 2, p. 533—538 (1978).
See also 22.1.2, p. 433.

N. N. Sirota, V. I. Lukashova, and L. A. Kurochkin
*Irradiation Effect on Cyclotron Resonance in Germanium
(a) Vol. 47, No. 2, p. 677—682 (1978).
See also 19, p. 309; 22.1.1, p. 426.

K. S. Bhatki, S. D. Pradhan, and B. V. Thosar
Further Studies on Positron Lifetimes in γ-Irradiated Teflon in Low Dose Range
(a) Vol. 47, No. 2, p. 691—698 (1978).
See also *10.2*, p. *114*; 14.4, p. 249; 19, p. 309; 22.9, p. 526.

S. B. S. Sastry and K. Balasubramanyam
*Studies on Tin Centres in RbCl Crystals
(a) Vol. 47, No. 2, p. 711—716 (1978).
See also *10.2*, p. *114*; *20.1*, p. *328*; 22.5.2, p. 481.

H. F. Kappert, K. F. Heidemann, B. Grabe, and E. te Kaat
*Range and Damage Profiling after Heavy Ion Implantation in the MeV Region
(a) Vol. 47, No. 2, p. 751—762 (1978).
See also 22.1.2, p. 433.

K. P. Arefiev, A. A. Tsoi, and S. A. Vorobiev
*Influence of High-Temperature Annealing on Positron Annihilation in Electron-Irradiated Silicon
(a) Vol. 47, No. 2, p. K149—K152 (1978).
See also 10.2, p. 114; 22.1.2, p. 433.

Nguyen Van Trong
Neutron Scattering on Anomalous Fluctuations of Phonons
(b) Vol. 88, No. 1, p. 145—148 (1978).
See also *6*, p. *73*; 13.5.2, p. 198.

V. G. Baryshevskii, A. O. Grubich, and I. Ya. Dubrovskaya
*Diffraction of Radiation from Channelled Charged Particles
(b) Vol. 88, No. 1, p. 351—358 (1978).
See also *1.1*, p. *13*.

T. I. Kolchenko and V. M. Lomako
*Removal and Scattering of Charge Carriers by Defect Clusters in Semiconductors
(a) Vol. 48, No. 1, p. 263—269 (1978).
See also 13.4, p. 178; 14.3, p. 222; 22.2.1, p. 447.

C. S. Sunandana
ESR Saturation of SO_3^- in X-Irradiated $KHSO_4$
(a) Vol. 48, No. 1, p. K19—K21 (1978).
See also *19*, p. *310*; 22.8.1, p. 514.

P. M. Mooney, R. T. Young, J. Karins, Y. H. Lee, and J. W. Corbett
*Defects in Laser Damaged Silicon Observed by DLTS
(a) Vol. 48, No. 1, p. K31—K34 (1978).
See also *13.4*, p. *178*; 22.1.2, p. 435.

A. F. Akkerman
Reflection of Slow Hydrogen and Helium Ions from Solid Surfaces
(a) Vol. 48, No. 1, p. K47—K50 (1978).
See also 21, p. 361; 21.1.1, p. 388.

O. Yu. Borkovskaya, N. L. Dmitruk, R. V. Konakova, and Yu. A. Tkhorik
*Investigation of Radiation Defects in GaAs by Means of Schottky Diode Characteristics
(a) Vol. 48, No. 1, p. K55—K58 (1978).
See also 14.3.3, p. 235; 22.2.1, p. 447.

O. K. Alekseeva, V. N. Bykov, V. A. Levdik, N. F. Miron, and V. P. Shantarovich
*Positron Study of Electron-Irradiated Vanadium
(a) Vol. 48, No. 1, p. K97—K99 (1978).
See also 21.1, p. 375.

M. Treilleux, P. Thévenard, G. Chassagne, and L. W. Hobbs
Observation of Implanted Potassium Aggregates in MgO Single Crystals
(a) Vol. 48, No. 2, p. 425—430 (1978).
See also *1.4*, p. *32*; 22.6, p. 495.

Yu. V. Bogatyrev and F. P. Korshunov
*Investigation of Properties of the Si–SiO_2 Interface in MOS Structures Irradiated by Fast Electrons
(a) Vol. 48, No. 2, p. K127—K130 (1978).
See also *14.3.2*, p. *231*; 22.1.2, p. 436; 22.6, p. 495.

S. V. Plotnikov, E. I. Rozum, V. A. Kuzminikh, and S. A. Vorobiev
Channeling Effect of Fast Electrons on the Luminescence Yield of NaCl:TI Crystals
(a) Vol. 48, No. 2, p. K175—K178 (1978).
See also *20.3*, p. *346*; 22.5.2, p. 483.

W. Fuhs, U. Holzhauer, S. Mantl, F. W. Richter, and R. Sturm
Annihilation of Positrons in Electron-Irradiated Silicon Crystals
(b) Vol. 89, No. 1, p. 69—75 (1978).
See also *10.2*, p. *117*; 22.1.2, p. 436.

M. WAUTELET and L. D. LAUDE
Electronic Structure of Solids under Laser Irradiation
(b) Vol. **89**, No. 1, p. 275–279 (1978).
See also *13.1*, p. *158*; 20.2, p. 339; 22, p. 417.

A. F. AKKERMAN and G. YA. CHERNOV
*Mean Free Paths by Inelastic Interactions, Stopping Powers, and Energy Straggling for Electrons of Energies up to 20 keV in Various Solids
(b) Vol. **89**, No. 1, p. 329–333 (1978).
See also 13, p. 147.

T. D. DZHAFAROV, M. N. KOLOTOV, and A. A. LITVIN
The Influence of γ-Irradiation on the Degradation of Red-Emitting GaP LED's
(a) Vol. **49**, No. 1, p. 211–215 (1978).
See also *14.3.3*, p. *236*; *20.3*, p. *347*; 22.2.2, p. 451.

M. T. PHAM
*A Very Simple Method for Profiling the Ion-Implanted Si-Surface
(a) Vol. **49**, No. 1, p. 261–265 (1978).
See also *14.3.2*, p. *231*; 22.1.2, p. 436.

G. MÜLLER, M. HAUBOLD, R. SCHIMKO, M. TRAPP, and G. SCHWARZ
*Investigations on the Diffusion on Implanted Zinc in $GaAs_{1-x}P_x$ by Ion Microprobe
(a) Vol. **49**, No. 1, p. 279–284 (1978).
See also *9*, p. *94*; 22.2.4, p. 456.

G. BATTAGLIN, G. DELLA MEA, A. V. DRIGO, G. FOTI, G. G. BENTINI, and M. SERVIDORI
*Two-Stage Laser Annealing of Lattice Disorder in Phosphorus Implanted Silicon
(a) Vol. **49**, No. 1, p. 347–352 (1978).
See also 22.1.2, p. 437.

V. S. LYSENKO and A. N. NAZAROV
Effect of Laser Beam Irradiation on the Electrophysical Properties of the Interface in Boron Ion Implanted MIS Structures
(a) Vol. **49**, No. 1, p. 405–411 (1978).
See also 13.3, p. 167; *14.3.3*, p. *236*; 22.1.2, p. 437.

G. STETTER, H. COUFAL, and E. LÜSCHER
*EPR Study of the Annealing Behaviour of Neutron-Transmutation-Doped Silicon
(a) Vol. **49**, No. 1, p. K17–K20 (1978).
See also 14.3, p. 224; *19*, p. *312*; 22.1.2, p. 437.

R. KOSICIK, P. DUDA, W. KEILIG, and R. MATTHEIS
The Influence of Ion Bombardment on the Magnetic Anisotropy in Evaporated Amorphous Gd–Co Films
(a) Vol. **49**, No. 1, p. K67–K69 (1978).
See also *2*, p. *50*; *18.3*, p. *296*; 21.1, p. 378; 21.4, p. 398.

K. ELLMER and J. STURM
*Energy–Angle Distributions of ^{14}N and ^{20}Ne Ions after Passage through Thin Carbon Foils
(b) Vol. **89**, No. 2, p. 611–618 (1978).
See also 22.1, p. 422.

H. KUZMANY
Instability of the Raman Spectra of Tetrathiofulvalene-Tetracyanoquinodimethan (TTF-TCNQ) and Tetrathiofulvalene (TTF)
(b) Vol. **89**, No. 2, p. K139—K144 (1978).
See also *20.1*, p. *334*; 22.9, p. 527.

A. WATTERICH and K. POLÁK
ESR Investigation of Isolated V^{2+} Ions in NaCl
(b) Vol. **89**, No. 2, p. K187—K190 (1978).
See also *19*, p. *313*; 22.5.2, p. 485.

R. RODLOFF and H. NEUHÄUSER
Mean Velocity of Edge and Screw Dislocation Groups in Neutron-Irradiated Copper Single Crystals
(a) Vol. **49**, No. 2, p. 445—454 (1978).
See also *10.1*, p. *106*; 21.1, p. 378.

E. JOHNSON, J. FERRER, and L. T. CHADDERTON
*Radiolytic Radiation Damage of Sodalite
(a) Vol. **49**, No. 2, p. 585—591 (1978).
See also 10.2, p. 119; 22.8, p. 507.

H. J. BLYTHE and L. H. DUNLEAVY
*Low-Temperature Magnetic Relaxation in Neutron-Irradiated α-Iron
(a) Vol. **49**, No. 2, p. 609—619 (1978).
See also 10.1, p. 106; *18.2*, p. *290*; 21.1.1, p. 389.

A. DUNLOP, N. LORENZELLI et J. C. JOUSSET
*Décanalisation des particules α par le carbone en solution solide dans le palladium (Dechanneling of α Particles by Carbon Dissolved in Palladium)
(a) Vol. **49**, No. 2, p. 643—650 (1978).
See also 14.1, p. 211; 21, p. 363.

V. N. BRUDNYI, M. A. KIROV, A. I. POTAPOV, I. K. POLUSHINA, V. D. PROCHUKHAN, and YU. V. RUD
*Change of Electrical Properties in Electron Irradiated $CdGeAs_2$ Crystals
(a) Vol. **49**, No. 2, p. 761—765 (1978).
See also 13.4, p. 182; *14.3*, p. *224*; 22.8, p. 507.

Y. H. LEE, P. R. BROSIOUS, and J. W. CORBETT
*High-Temperature Ion Implantation in Diamond
(a) Vol. **50**, No. 1, p. 237—242 (1978).
See also 10.2, p. 120; 19, p. 314; 22.1, p. 423.

P. MACKO and J. DOUPOVEC
Thermal and Electrical Properties of Neutron Irradiated GeS_{11} Glass
(a) Vol. **50**, No. 1, p. K51—K54 (1978).
See also 2, p. 52; *8*, p. *89*; *14.3*, p. *225*; 22, p. 420.

K. ATOBE
The Role of F-Centers in Thermoluminescence Processes in Reactor Irradiated SrF_2 Crystals
(a) Vol. **50**, No. 1, p. K77—K80 (1978).
See also *10.2*, p. *121*; *20.3*, p. *350*; 22.5.3, p. 488.

N. N. GERASIMENKO, M. ROLLÉ, LI-JEN CHENG, Y. H. LEE, J. C. CORELLI, and J. W. CORBETT
Infrared Absorption of Silicon Irradiated by Protons
(b) Vol. **90**, No. 2, p. 689—695 (1978).
See also *6*, p. *78*; *20.1*, p. *337*; 22.1.2, p. 439.

A. P. PATHAK and M. P. SRIVASTAVA
*Channeling and a Semistatistical Model for the Interatomic Potential
(b) Vol. **90**, No. 2, p. 703—709 (1978).

V. N. BRUDNYI, D. L. BUDNITSKII, M. A. KRIVOV, R. V. MASAGUTOVA, V. D. PROCHUKHAN, and YU. V. RUD
The Electrical and Optical Properties of 2.0 MeV Electron-Irradiated $ZnGeP_2$
(a) Vol. **50**, No. 2, p. 379—384 (1978).
See also 13.4, p. 185; *14.3*, p. *225*; *20.1*, p. *337*; 22.8, p. 508.

W. R. FAHRNER and E. KLAUSMANN
Characterization of MOS Structures with Buried Layers
(a) Vol. **50**, No. 2, p. 433—444 (1978).
See also *14.3.2*, p. *232*; *14.3.3*, p. *238*; 22.1.2, p. 439.

R. E. BLAHA and W. R. FAHRNER
Technology and Reliability of Thermal SiO_2 Layers as Used for Passivating Silicon Power Devices
(a) Vol. **50**, No. 2, p. 551—562 (1978).
See also *14.3.1*, p. *229*; *14.3.3*, p. *238*; 22.6, p. 497.

V. LANTTO
*Contribution of High-Angle Deflections to the Fast-Electron Transmission
(a) Vol. **50**, No. 2, p. 611—616 (1978).
See also 21.1, p. 382.

A. P. DOLGOLENKO and I. I. FISHCHUK
*Defect Clusters and Simple Defect Build-Up Kinetics in Fast-Neutron Irradiated n-Si
(a) Vol. **50**, No. 2, p. 751—755 (1978).
See also 10.2, p. 122; 22.1.2, p. 439.

J. A. A. KHAN and J. LETEURTRE
An Experimental Investigation of the Stress Oriented Dislocation Loop Nucleation in Whiskers
(a) Vol. **50**, No. 2, p. K161—K164 (1978).
See also *10.2*, p. *122*; 22.5.2, p. 486.

H.-J. KOS and S. MIEKE
Effect of Deformation on the Thermoluminescence of Titanium Doped Lithium Fluoride
(a) Vol. **50**, No. 2, p. K165—K168 (1978).
See also *12.2*, p. *141*; *20.3*, p. *351*; 22.5.2, p. 486.

A. K. SEMENYUK and M. D. VARENTSOV
Anisotropy of Galvanomagnetic Effects in Irradiated Silicon with Layer Inhomogeneity
(a) Vol. **50**, No. 2, p. K169—K173 (1978).
See also *14.3*, p. *225*; 22.1.2, p. 440.

G. Götz, W. Karthe, B. Schnabel, and N. Sobolev
ESR Line Shape of Amorphous Centres in Ion Implanted Silicon
(a) Vol. **50**, No. 2, p. K209—K212 (1978).
See also *19*, p. *314*; 22.1.2, p. 440.

A. Kalz, F. Storbeck, and G. Blasek
Adsorption of H_2O on α-Fe(111) under the Influence of 2.5 keV Electrons
(a) Vol. **50**, No. 2, p. K247—K250 (1978).
See also *1.6*, p. *41*; 21.1.1, p. 391.

12. Mechanical Properties (Plastic Deformations See 10)

Original Papers and Short Notes

Y. Li and D. Y. Chung
*The Anisotropic Behavior of Poisson's Ratio, Young's Modulus, and Shear Modulus in Crystals with Different Symmetries
(a) Vol. **46**, No. 2, p. 603—610 (1978).

T. Mura and N. Kinoshita
The Polynomial Eigenstrain Problem for an Anisotropic Ellipsoidal Inclusion
(a) Vol. **48**, No. 2, p. 447—450 (1978).
See also *10*, p. *96*.

P. A. Varotsos
On the Pressure Variation of the Excessive Specific Heat Associated with Vacancies
(a) Vol. **49**, No. 1, p. K57—K61 (1978).
See also *8*, p. *88*; *10*, p. *97*.

F. Loges
*The Strain Energy Associated with an Extended Defect in an Isotropic Solid
(b) Vol. **89**, No. 2, p. K191—K194 (1978).
See also *10*, p. *97*.

K. Maeda and S. Takeuchi
Computer Simulation of Deformation in Two-Dimensional Amorphous Structures
(a) Vol. **49**, No. 2, p. 685—696 (1978).
See also *2*, p. *50*; 10, p. 97.

V. A. Streltsov
*Effects of Hydrostatic Pressure of the Second Phase Induced Elastic Fields
(a) Vol. **49**, No. 2, p. K129—K133 (1978).
See also 10, p. 97.

V. A. Solovev and V. A. Streltsov
*On the Crack Nucleation in Hydrostatically Compressed Crystals
(a) Vol. **49**, No. 2, p. K145—K148 (1978).
See also *10*, p. *97*.

J. Kubát, M. Rigdahl, and R. Seldén
*On the Pre-Exponential Factor in Stress Relaxation
(a) Vol. **50**, No. 1, p. 117—122 (1978).

12.1 Metals

Original Papers and Short Notes

R. Kamel, A. R. Ali, and Z. Farid
Annealing Spectrum of Precipitating Al–Li Alloys
(a) Vol. **45**, No. 1, p. 47–52 (1978).
See also *10.1*, p. *99*; 21, p. 352; 21.2, p. 392.

P. Turchi, Y. Calvayrac, and F. Plicque
Effects of Configurational Order in a Ni_3Fe Single Crystal on Debye Temperature Deduced from Elastic Constants and X-Ray Measurements
(a) Vol. **45**, No. 1, p. 229–238 (1978).
See also *1.3*, p. *25*; 6, p. 65; 21.1, p. 369; 21.1.1, p. 384.

V. Ramamurthy and K. K. Singh
On the Crossing of Dispersion Curves of Alkali Metals in [ζ00] Direction
(b) Vol. **85**, No. 2, p. 761–768 (1978).
See also *6*, p. *66*; 21.2, p. 392.

R. Kamel, A. R. Ali, and Z. Farid
Annealing Spectrum of Cold-Worked Dilute Al–Li Alloys
(a) Vol. **45**, No. 2, p. 419–422 (1978).
See also *10.1*, p. *100*; 21, p. 354.

A. Ranga Rao and C. Suryanarayana
Solute–Vacancy Binding Energies in Magnesium Alloys
(a) Vol. **45**, No. 2, p. K131–K133 (1978).
See also *10.1*, p. *100*; 21.3, p. 394.

K. M. Miller and P. D. Bristowe
A Computer Simulation of Point Defects and Planar Defects in Nickel
(b) Vol. **86**, No. 1, p. 93–102 (1978).
See also *10.1*, p. *101*; 21.1, p. 371.

D. Werneth
Mechanical and Magnetic Relaxation Menasurements on Pure and Doped Nickel after Plastic Deformation and Electron Irradiation
(a) Vol. **46**, No. 1, p. 99–106 (1978).
See also *10.1*, p. *101*; 18.2, p. 284; 21.1, p. 371.

Jong K. Lee and William C. Johnson
*Calculation of the Elastic Strain Field of a Cuboidal Precipitate in an Anisotropic Matrix
(a) Vol. **46**, No. 1, p. 267–272 (1978).
See also *10.1*, p. *101*; 21, p. 356.

G. Ferron and M. Quintard
*Evidence of an Internal Friction Phase Transformation Peak in Niobium–Hydrogen Alloys
(a) Vol. **46**, No. 1, p. K43–K46 (1978).
See also *10.1*, p. *101*; 21, p. 356.

N. A. Tyapunina, E. K. Naimi, S. V. Gasparyan, and G. M. Zinenkova
*The Orientation Dependence of Dislocation Internal Friction in Real Crystals (II): Hexagonal Crystals
(a) Vol. **46**, No. 2, p. 411—420 (1978).
See also *10.1*, p. *101*; 21, p. 357; 21.1, p. 372; 21.3, p. 394.

R. Yamamoto, T. Hoshiai, H. Matsuoka, K. Taira, and M. Doyama
*The Calculations of the Heat of Crystallization and Bulk Modulus of Glassy Iron
(a) Vol. **46**, No. 2, p. 671—674 (1978).
See also *2*, p. *44*; 21.1.1, p. 386.

R. Kamel, A. R. Ali, and Z. Farid
On the Recovery of Quenched and Irradiated Dilute Al–Li Alloys
(a) Vol. **46**, No. 2, p. 697—701 (1978).
See also *10.1*, p. *102*; 14.1, p. 207; 21, p. 357.

I. O. Bashkin, E. G. Ponyatovskii, and M. E. Kost
Phase Transformations in Hydrides of Rare-Earth Metals under Pressure (II): The Sc–H and Yb–H Systems under High Hydrogen Pressure
(b) Vol. **87**, No. 1, p. 369—372 (1978).
See also *1.2*, p. *18*; 21.4, p. 397.

C. Esnouf and G. Fantozzi
*Theoretical Aspects of the Bordoni Relaxation
(a) Vol. **47**, No. 1, p. 201—211 (1978).
See also *10.1*, p. *103*; 21, p. 358.

R. Ramji Rao and A. Ramanand
Lattice Dynamics, Specific Heat, and Bulk Modulus of Scandium
(b) Vol. **87**, No. 2, p. 751—756 (1978).
See also *6*, p. *72*; 8, p. 86; 21.4, p. 397.

D. Kuhlmann-Wilsdorf
*Deviations from Schmid's Law in Uni-Directional Strain
(a) Vol. **47**, No. 2, p. 639—649 (1978).
See also *10.1*, p. *104*; 21, p. 359.

W. Gustaw, T. Goworek, and J. Wawryszczuk
*Influence of Fatigue on Positron Annihilation in Steel
(a) Vol. **47**, No. 2, p. K141—K144 (1978).
See also 21.1.1, p. 388.

S. Besse, M. Quintard, Ph. Mazot et J. de Fouquet
*Influence de l'hydrogène sur le frottement intérieur du titane
Influence of Hydrogen on the Internal Friction of Titanium
(a) Vol. **48**, No. 1, p. 99—103 (1978).
See also *10.1*, p. *104*; 21.1, p. 375.

V. A. Streltsov
*Shear Stresses Induced by Hydrostatic Pressure near the Surface of an Inclusion
(a) Vol. **48**, No. 1, p. K23—K25 (1978).
See also *1.4*, p. *32*; 21, p. 360.

R. Glocker and L. Fritsche
Pressure Dependence of the Electron–Phonon Coupling in Lanthanum
(b) Vol. **88**, No. 2, p. 639—644 (1978).
See also *6*, p. *74*; *13.1*, p. *157*; 14.2, p. 2145; 21.4, p. 398.

K. N. Khanna and P. K. Sharma
*A Model Potential and Its Calculations
(b) Vol. **88**, No. 2, p. 809—814 (1978).
See also 13.2, p. 162; 18.1, p. 278; 21.2, p. 393.

H. O. K. Kirchner, H. Mughrabi, F. Prinz, and H. Schmid
Dislocation Loops in α-Iron
(a) Vol. **49**, No. 2, p. 539—543 (1978).
See also *10.1*, p. *106*; 21.1.1, p. 389.

K. Bärner and H. Berg
*Elastic Anomalies at the Second-Order Phase Transition in MnAs
(a) Vol. **49**, No. 2, p. 545—554 (1978).
See also 6, p. 76; 18.1, p. 279; 18.2, p. 290; 21.1, p. 379.

B. B. Sahay and P. L. Srivastava
*Isothermal Compressibility for Various Liquid Metals by Harmonic Model Potential
(b) Vol. **90**, No. 1, p. 93—97 (1978).
See also 2, p. 51; 21, p. 364; 21.1, p. 379; 21.2, p. 393.

V. Ramamurthy and M. Satishkumar
Analysis of Phonon Frequencies of Alkali Metals
(b) Vol. **90**, No. 1, p. 117—122 (1978).
See also *6*, p. *77*; 21.2, p. 393.

G. Hausch and E. Török
Elastic, Magnetoelastic, and Thermal Properties of Some Ferromagnetic Metallic Glasses
(a) Vol. **50**, No. 1, p. 159—164 (1978).
See also *2*, p. *51*; 8, p. 89; *18.2*, p. *291*; 21.1, p. 381; 21.1.1, p. 390.

G. Ya. Akimov, V. I. Zaitsev, and V. A. Streltsov
*Stress Relaxation in Hydrostatically Compressed Zinc Single Crystals
(a) Vol. **50**, No. 1, p. K33—K36 (1978).
See also *10.1*, p. *107*; 21.1, p. 381.

T. Soma
*Equation of State under Pressure for Metallic β-Sn
(b) Vol. **90**, No. 2, p. K129—K131 (1978).
See also 1.2, p. 24; *4*, p. *59*; 21, p. 366.

Y. Motohashi and S. Ohtake
*The Effect of Self-Energy of Dislocations on the Choice of Slip Systems in Crystals
(a) Vol. **50**, No. 2, p. 449—458 (1978).
See also *10.1*, p. *108*; 21, p. 366.

G. J. den Otter and R. Vetter
*A Model Comprising Work Hardening and Solid Solution Hardening
(a) Vol. **50**, No. 2, p. 529—536 (1978).
See also *10.1*, p. *108*; 21.1, p. 382.

I. G. Ritchie, J. F. Dufresne, and P. Moser
*Internal Friction in Deformed Pure Iron
(a) Vol. **50**, No. 2, p. 617—626 (1978).
See also *10.1*, p. *108*; 21.1.1, p. 391.

12.2 Non-Metals

Original Papers and Short Notes

V. K. Garg and M. P. Verma
Erratum to: Evaluation of Some Important Latice Sums on NaCl and CsCl Structures
(b) Vol. **85**, No. 1, p. 417 (1978).
See also *1.1*, p. *11*; 22.5.2, p. 478.

R. Ethiraj, V. G. Krishna Murty, and K. G. Bansigir
Piezo-Optic Birefringence in Mixed Crystals of KCl–KBr
(a) Vol. **45**, No. 1, p. 83—87 (1978).
See also *20*, p. *315*; 22.5.4, p. 489.

S. C. Goyal, R. Prakash, and S. P. Tripathi
Many-Body Lattice Theory of Elastic Dielectric and Its Application to Solids with NaCl Structure
(b) Vol. **85**, No. 2, p. 477—486 (1978).
Erratum
(b) Vol. **89**, No. 1, p. 335 (1978).
See also *6*, p. *65*; 22.5.2, p. 478.

S. C. Varshney and A. A. Gundjian
*Internal Strain and Piezoelectricity in Zincblende Crystals
(b) Vol. **85**, No. 2, p. 733—742 (1978).
See also 6, p. 66; *14.3*, p. *217*; 22, p. 409.

J. C. Castro and M. de Souza
The OCN^- Vibrational Spectrum and the Phase Transitions in KCN_xKCl_{1-x}Single Crystals
(b) Vol. **86**, No. 1, p. 137—144 (1978).
See also *1.2*, p. *16*; *6*, p. *67*; 20.1, p. 321; 22.5, p. 474; 22.5.4, p. 489.

S. H. El-Mahalawy and B. L. Evans
Pressure Dependence of the Electrical Conductivity in 2H-MoS_2 and 2H–WSe_2
(b) Vol. **86**, No. 1, p. 151—157 (1978).
See also *14.3*, p. *218*; 22, p. 410.

V. K. Bashenov, D. I. Marvakov, and A. M. Mutal
Internal Strain Parameter in Diamond-Type Crystals
(b) Vol. **86**, No. 1, p. K7—K10 (1978).
See also *4*, p. *57*; 22.1, p. 422; 22.1.1, p. 424; 22.1.2, p. 431.

N. A. Tyapunina, E. K. Naimi, and S. V. Gasparyan
*The Orientation Dependence of Dislocation Internal Friction in Real Crystals (I): Alkali Halide Crystals
(a) Vol. **46**, No. 1, p. 351—356 (1978).
See also *10.2*, p. *111*; 22.5.2, p. 480.

R. Ramani, K. K. Mani†, and R. P. Singh
Long Optical Vibrations and Elastic Constants of Hexagonal Boron Nitride
(b) Vol. **86**, No. 2, p. 759—763 (1978).
See also *6*, p. *69*; 22, p. 411.

I. Rattan and N. C. Srivastava
*Magnetoelastic Interaction Energy of a Cubic Ferromagnet in an Arbitrary Coordinate System
(a) Vol. **46**, No. 2, p. 475—478 (1978).
See also *18.2*, p. *285*; 22.8.2, p. 519.

L. P. Khiznichenko, E. Oteniyazov, and N. S. Tulyaganova
*Internal Friction in Neutron-Irradiated Silicon
(a) Vol. **46**, No. 2, p. K147—K149 (1978).
See also *10.2*, p. *112*; 11, p. 126; 22.1.2, p. 432.

M. S. Kushwaha and S. S. Kushwaha
Unpaired Elastic Force Model in the Lattice Dynamical Study of Rare Gas Solids
(b) Vol. **87**, No. 1, p. 247—253 (1978).
See also *6*, p. *70*; 23, p. 530.

T. Soma
Lattice Dynamics of Group IV Elements with the Local Heine-Abarenkov Model Potential
(b) Vol. **87**, No. 1, p. 345—350 (1978).
See also *6*, p. *70*; 21.7, p. 406; 22.1.1, p. 425; 22.1.2, p. 432.

V. K. Garg, D. S. Puri, and M. P. Verma
*On the Second- and Third-Order Elastic Constants of ZnS Structure Solids
(b) Vol. **87**, No. 1, p. 401—407 (1978).
See also *6*, p. *70*; 22.2, p. 443; 22.4, p. 461; 22.5, p. 475.

V. S. Postnikov, S. A. Gridnev, B. N. Prasolov, and S. K. Turkov
*Specific Features of Low-Frequency Sound Attenuation in $LiNbO_3$ Poly- and Single-Domain Crystals
(a) Vol. **47**, No. 1, p. 65—69 (1978).
See also *7*, p. *80*; 14.4.1, p. 255; 22.8.1, p. 513.

S. K. Sarkar and S. Sengupta
Lundqvist Three-Body Interaction and Lattice Dynamics of Ionic Crystals
(b) Vol. **87**, No. 2, p. 517—525 (1978).
See also *4*, p. *58*; 22.5.2, p. 481.

E. Wiszt
A Contribution to the Phenomenological Theory of the Anomalies of the Elastic Constants near a Structural Phase Transition
(b) Vol. **87**, No. 2, p. 675—688 (1978).
See also *1.2*, p. *19*; 22.5, p. 475.

R. Prakash, L. K. Bansal, and S. C. Goyal
*Photoelastic Constants of the Alkaline Earth Oxides
(b) Vol. **87**, No. 2, p. K127—K131 (1978).
See also *4*, p. *58*; 22.6, p. 494.

S. A. Gridnev and B. M. Darinskii
*Attenuation of Low-Frequency Elastic Oscillations in KH_2PO_4-Type Ferroelectric Crystals
(a) Vol. **47**, No. 2, p. 379—384 (1978).
See also 14.4.1, p. 255; 22.8.1, p. 514.

N. Terao
*Experimental Study of Crack Propagation in Brittle Amorphous Solids. Bifurcation of a Crack in Glass
(a) Vol. **47**, No. 2, p. 565—575 (1978).
See also *2*, p. *45*; 22.8, p. 505.

Mahiradhwaj Singh
Phonon Attenuation in n-Type Semiconductors with ZB Structure in Presence of Internal Strains and Magnetic Field
(b) Vol. **88**, No. 1, p. 295—300 (1978).
See also *6*, p. *73*; 7, p. 81; 18, p. 274; 22.1.1, p. 426; 22.2.3, p. 453.

Z. M. Alekseeva, A. P. Vyatkin, G. F. Karavaev, and N. P. Krivorotov
*Effect of Pressure on Hole Effective Masses in GaAs
(b) Vol. **88**, No. 1, p. 321—326 (1978).
See also *13.1*, p. *156*; 14.3, p. 222; 22.2.1, p. 447.

H. Konwent and J. Lorenc
*A Thermodynamic Theory of the Improper Antiferroelectric Phase Transition in ADP
(b) Vol. **88**, No. 2, p. 747—755 (1978).
See also 1.2, p. 21; *14.4.1*, p. *256*; 22.8.1, p. 515.

K. Maeda, K. Nakagawa, and S. Takeuchi
*Activation Analysis of Plastic Deformation of CdTe below Room Temperature
(a) Vol. **48**, No. 2, p. 587—591 (1978).
See also *10.2*, p. *116*; 22.4.3, p. 471.

V. A. Savastenko and A. U. Sheleg
*Study of the Elastic Properties of Gallium Nitride
(a) Vol. **48**, No. 2, p. K135—K139 (1978).
See also 22.2, p. 443.

Debjani Bose and A. N. Basu
*Second-Neighbour Interaction Potential in Alkali Halide Crystals
(b) Vol. **89**, No. 1, p. K35—K38 (1978).
See also *4*, p. *59*; 22.5.2, p. 484.

U. Holland and F. Lüty
Hydrostatic Pressure Effects on the Absorption and Emission of F_A(Li) Centers in KCl
(b) Vol. **89**, No. 1, p. K73—K77 (1978).
See also *13.4*, p. *180*; *20.3*, p. *347*; 22.5.2, p. 484.

N. Terao
*Experimental Study of Crack Propagation in Brittle Amorphous Solids. Refraction of Propagating Crack by Stress Waves
(a) Vol. **49**, No. 1, p. 237—248 (1978).
See also 2, p. 49; *10.2*, p. *118*.

B. Pertzsch and U. Rössler
Stress Dependence of Shallow Acceptor States in Semiconductors with Zincblende Symmetry
(b) Vol. **89**, No. 2, p. 475—487 (1978).
See also *13.4*, p. *181*; 22.2, p. 444.

N. Terao
*Experimental Study of Crack Propagation in Brittle Amorphous Solids: Cracks Formed in Glassy Carbon
(a) Vol. **49**, No. 2, p. K101—K102 (1978).
See also *2*, p. *51*; 22.1, p. 422.

W. Rehwald and A. Vonlanthen
Ultrasonic Studies of the Structural Phase Transition in Squaric Acid
(b) Vol. **90**, No. 1, p. 61—66 (1978).
See also *1.2*, p. *23*; 7, p. 82; 22.9, p. 528.

P. A. Varotsos
An Estimate of the Pressure Dependence of the Dielectric Constant in Alkali Halides
(b) Vol. **90**, No. 1, p. 339—343 (1978).
See also 4, p. 59; 6, p. 77; *14.4*, p. *251*; 22.5.1, p. 477; 22.5.2, p. 485.

Z. Dobrovolskis, A. Krotkus, and J. Požela
Negative Differential Conductivity in Hydrostatically Pressed Indium Antimonide at Room Temperature
(a) Vol. **50**, No. 1, p. K73—K75 (1978).
See also *14.3.4*, p. *246*; 22.2.3, p. 454.

G. Schmidt and I. Boczek
Pseudoelasticity and Shape Memory of PLZT Ceramic
(a) Vol. **50**, No. 1, p. K109—K111 (1978).
See also *14.4.1*, p. *257*; 22.8.1, p. 516.

R. A. Noack
Inversion of the γ and L Conduction Bands of GaSb under Uniaxial Stress
(b) Vol. **90**, No. 2, p. 615—620 (1978).
See also *13.1*, p. *161*; *20.3*, p. *350*; 22.2.3, p. 454.

A. S. Bhalla, L. E. Cross, and D. A. Payne
Thermal, Optical, and Elastic Properties of Ferroelastic Isopropyl Ammonium Chloroplatinate (IPACP) and Chlorostannate (IPACS)
(a) Vol. **50**, No. 2, p. 661—667 (1978).
See also *1.2*, p. *24*; 8, p. 90; 20.1, p. 338; 22.9, p. 528.

H.-J. Kos and S. Mieke
*Effect of Deformation on the Thermoluminescence of Titanium Doped Lithium Fluoride
(a) Vol. **50**, No. 2, p. K165—K168 (1978).
See also 11, p. 131; *20.3*, p. *351*; 22.5.2, p. 486.

13. Electron States

Review Article

V. I. Ivanov-Omskii, L. I. Korovin, and E. M. Shereghii
Phonon-Assisted Cyclotron Resonance in Semiconductors
(b) Vol. **90**, No. 1, p. 11—32 (1978).
See also *6*, p. *64*; *19*, p. *303*; 22, p. 408; 22.2.3, p. 452; 22.4.1, p. 463.

Original Papers and Short Notes

A. M. Berezhkovskii and A. A. Ovchinnikov
*Electron Energy Spectrum of a Finite Crystal in a Uniform Electric Field
(b) Vol. **85**, No. 1, p. 121—126 (1978).

V. I. Grebennikov, O. B. Sokolov, and E. A. Turov
*Effects of Correlation Interaction of Electrons with Deep Holes in X-Ray Spectra of Metals
(b) Vol. **85**, No. 1, p. 127—134 (1978).
See also 20, p. 315; 21, p. 352.

C.-O. Almbladh and P. Minnhagen
*On the Thermal Broadening of Core Levels in Metals Due to Many-Electron Effects
(b) Vol. **85**, No. 1, p. 135—143 (1978).
See also 17, p. 269; 20, p. 315; 21, p. 352.

N. Kumar and J. Heinrichs
*Electron Localization and Statistical Mechanics of a Zero-Component Classical Field
(a) Vol. **85**, No. 1, p. 163—170 (1978).

O. Bílek and P. Kadura
*On the Tight Binding Eigenvalues of Finite S.C. and F.C.C. Crystallites with (100) Surfaces
(b) Vol. **85**, No. 1, p. 225—231 (1978).

J. Richter and S. Kobe
Density of States and Itinerant Ferromagnetism in Amorphous Systems with Dense Random Packed Hard Sphere Structure
(b) Vol. **85**, No. 1, p. 233—238 (1978).
See also *2*, p. *42*; *18.2*, p. *281*.

G. V. Ionova, E. F. Makarov, O. M. Pachev, and S. P. Ionov
*Electronic Ordering in Crystals
(b) Vol. **85**, No. 1, p. 359—370 (1978).
See also *18*, p. *273*.

V. S. Sardaryan and M. M. Arakelyan
*Bound States of Electrons and Optical Phonons under Conditions of One-Dimensional Quantizing
(b) Vol. **85**, No. 2, p. 459—463 (1978).
See also *6*, p. *65*.

R. J. Jelitto and M. Heise
*On the Dynamics of Electrons in the Strong-Coupling Hubbard Chain
(b) Vol. **85**, No. 2, p. 535—543 (1978).
See also *18.2*, p. *282*.

G. V. Ionova, E. F. Makarov, O. M. Pachev, and S. P. Ionov
*Analysis of Charge–Orbital Ordering in Compounds with Mixed Valence
(b) Vol. **85**, No. 2, p. 683—691 (1978).
See also *18*, p. *273*.

W. I. Friesen, D. B. Litvin, and B. Bergersen
*Theory of the X-Ray and Optical Properties of Polysulphur Nitride
(b) Vol. **85**, No. 2, p. 715—722 (1978).
See also *20*, p. *315*; 22, p. 409.

M. V. Eremin and Yu. V. Rakitin
*Channel Model in Isotropic Exchange Theory (III)
(b) Vol. **85**, No. 1, p. 783—788 (1978).
See also 1, p. 9; 22.5.4, p. 489.

F. Fujimoto, N. Sumida, H. Fujita, and Y. Uchida
Electron Channeling in ⟨111⟩ Direction of Tungsten Crystals (II) Energy Range of 1000 to 2000 keV
(a) Vol. **45**, No. 2, p. 411—418 (1978).
See also *11*, p. *124*; 21, p. 354.

H. Gollisch and L. Fritsche
*Relativistic One-Particle Equation for Electron States of Heavy Metals
(b) Vol. **86**, No. 1, p. 145—150 (1978).
See also 21, p. 355.

S. Zawisz
*An Exact Calculation of the Average Density of States in Disordered Systems
(b) Vol. **86**, No. 1, p. 367—372 (1978).

E. Kolley and W. Kolley
*Singularities in the Paramagnetic Susceptibility of Random Hubbard Alloys
(b) Vol. **86**, No. 1, p. 397—408 (1978).
See also *18.1*, p. *277*; 18.2, p. 284; 21, p. 355.

V. V. Kaplin, D. E. Popov, and S. A. Vorobev
Quantum-Mechanical Treatment of Anomalous Rutherford Scattering of Swift Electrons in Crystals
(b) Vol. **86**, No. 1, p. K35—K38 (1978).
See also *11*, p. *124*; 22.1.2, p. 431.

Th. Eifrig, E. Kolley, and W. Kolley
*Electron Correlations and Ferromagnetism in Disordered Alloys
(b) Vol. **86**, No. 1, p. K79—K83 (1978).
See also 18.2, p. 284; 21, p. 356.

A. P. Dmitriev, A. E. Stefanovich, and L. D. Tsendin
*Discontinuities in Drift Solutions Due to Reversal in Ambipolar Drift. Injection and Accumulation of Hot Carriers
(a) Vol. **46**, No. 1, p. 45–53 (1978).
See also *14.3.4*, p. *241*; 22, p. 411.

A. Freudenhammer
*On the RKKY Interaction in Spin Glass Systems
(b) Vol. **86**, No. 2, p. 579–584 (1978).
See also *18*, p. *274*; 21.1, p. 371.

R. Lenk
Multiple Scattering in Condensed Materials (I): The Embedding Field and the Effective Medium
(b) Vol. **86**, No. 2, p. 603–613 (1978).
See also *1*, p. *9*.

F. M. Galperin
*Atomic Magnetic Moments of Fe, Co, Ni and Hyperfine Fields at Impurities in Ferromagnetic Iron
(b) Vol. **86**, No. 2, p. 679–684 (1978).
See also *18.2*, p. *285*; 21.1, p. 372; 21.1.1, p. 386.

Anand P. Pathak
Stopping Power of Solids in Planar Channelling
(b) Vol. **86**, No. 2, p. 751–758 (1978).
See also *11*, p. *125*; 21.1, p. 372; 21.6, p. 402.

P. I. Baranskii, L. S. Demidenko, S. L. Korolyuk, and A. G. Samoilovich
Impurity Scattering of Electrons in Quantizing Magnetic Fields
(b) Vol. **86**, No. 2, p. 765–772 (1978).
See also *14.3*, p. *219*; 22.2.1, p. 446.

I. M. Dykman and P. M. Tomchuk
*Electromagnetic Waves in Inhomogeneous Semiconductors with Non-Equilibrium Carriers
(b) Vol. **86**, No. 2, p. 773–780 (1978).
See also *14.3*, p. *220*; 20, p. 316; 22, p. 412.

B. V. Egorov and M. A. Krivoglaz
*The Theory of Autolocalized Electron States in Ferromagnetic Semiconductors at Arbitrary Relation between the Band Width and the Spin–Electron Interaction
(b) Vol. **86**, No. 2, p. 781–794 (1978).
See also 13.5.3, p. 200; 18.2, p. 285; 22, p. 412.

K. Berndt, O. Brümmer, and U. Marx
*Study of the Fe 2p States by Means of Photoelectron Spectroscopy in Iron and Iron Alloys
(b) Vol. **86**, No. 2, p. K93–K95 (1978).
See also 20, p. 316; 21.1, p. 372; 21.1.1, p. 386.

D. D. Richardson
*Use of Finite Atomic Radius in Dispersion Force Calculations
(b) Vol. **86**, No. 2, p. K97–K100 (1978).
See also *4*, p. *57*.

C. Tsallis
*Influence of Long-Range Forces in Isotropic Two-Dimensional Magnetism
(b) Vol. **86**, No. 2, p. K101—K105 (1978).
See also *18*, p. *274*.

D. Grensing and W.-H. Steeb
*Free Lattice Electron Gas with Nearest and Second Nearest Neighbour Hopping
(b) Vol. **86**, No. 2, p. K131—K134 (1978).

L. P. Mokhracheva, V. A. Tskhai, and P. V. Geld
*Charge Distribution and Ionicity of the Chemical Bonds in TiC, VC, and TiN
(b) Vol. **87**, No. 1, p. 49—52 (1978).
See also *4*, p. *58*; 22.3, p. **457**.

M. A. Ball and M. M. Islam
*Effects of Core and Correlation Corrections on the Electrical Resistivity of Lithium
(b) Vol. **87**, No. 1, p. 145—150 (1978).
See also *14.1*, p. *207*; 21.2, p. **392**.

A. Milchev and K. Unger
Third-Order Contribution to the Effective Ion–Ion Interaction and Formation of Angular Forces in Crystalline Solids (I): General Formulation of the Expression for the Total Energy
(b) Vol. **87**, No. 1, p. 227—237 (1978).
See also *4*, p. *58*; 22.2, p. **443**.

G. Kamieniarz
*On the Theory of the Ising Model with a Transverse Field
(b) Vol. **87**, No. 2, p. 455—459 (1978).
See also *18.1*, p. *278*.

A. Milchev and K. Unger
Third-Order Contribution to the Effective Ion–Ion Interaction and Formation of Angular Forces in Crystalline Solids (II): Application to Tetrahedral Solids
(b) Vol. **87**, No. 2, p. 473—478 (1978).
See also *4*, p. *58*; 22.2, p. **443**.

R. Wedell
Radiative Damping Effect for Ultrarelativistic Channelled Particles
(b) Vol. **87**, No. 2, p. 491—496 (1978).
See also *11*, p. *127*; 22.1.2, p. **433**.

J. Richter
*Electronic Structure of Amorphous Alloys
(b) Vol. **87**, No. 2, p. K89—K92 (1978).
See also *2*, p. *45*; 18.2, p. 287.

D. Dădârlat and M. Crişan
*Dispersion Theory Approach to the Scaling Function for a Random System
(b) Vol. **87**, No. 2, p. K151—K154 (1978).

A. Richter, G. Röpke, and F. Goedsche
*Functional Integral Approach for the Hubbard Model with Arbitrary Electron Density
(b) Vol. 88, No. 1, p. 189—198 (1978).
See also *18*, p. *274*; 18.1, p. 278; 18.2, p. 288; 18.4, p. 300.

Yu. E. Lozovik, V. N. Nishanov, and E. P. Kaminskaya
*Boundary Effects of Transition into Excitonic Phase
(b) Vol. 88, No. 1, p. 359—365 (1978).
See also *14*, p. *203*.

K. I. Papunashvili
**H*-Linear "Planar" Hall Effects of Hot Electrons in n-Ge
(b) Vol. 88, No. 1, p. K27—K30 (1978).
See also *14.3.4*, p. *243*; 22.1.1, p. 426.

V. L. Aksionov, H. Konvent, and J. Schreiber
*Commutator and Anticommutator Green Functions in the Ising Model with Transverse Field
(b) Vol. 88, No. 1, p. K43—K45 (1978).
See also *18.2*, p. *288*.

G. Lautz and K. Sachse
*Noise Spectra of Hot Carriers in n–Germanium
(a) Vol. 48, No. 1, p. 121—131 (1978).
See also *14.3.4*, p. *243*; 22.1.1, p. 426.

I. M. Dykman, V. M. Rosenbaum, and F. T. Vasko
*Hot Electrons in Semiconductors with Quasi-Relativistic Band Structure
(b) Vol. 88, No. 2, p. 385—395 (1978).
See also *14.3.4*, p. *243*; 16, p. 265; 22, p. 415.

Th. Frauenheim and G. Röpke
*Magnetic Behaviour of the s–f Two-Band Hubbard Model
(b) Vol. 88, No. 2, p. 457—462 (1978).
See also *18*, p. *274*.

Farid A. Khwaja, A. A. Katsnelson, and V. M. Silonov
*Many-Electron Effects on the Characteristics of Short-Range Order in the Pseudopotential Approximation
(b) Vol. 88, No. 2, p. 477—483 (1978).
See also *1.3*, p. *27*; 21.1, p. 375.

P. Kadura and L. Künne
*From Small to Infinite F.C.C. Clusters
(b) Vol. 88, No. 2, p. 537—544 (1978).
See also 4, p. 59.

C. Papatriantafillou
*NMP-TCNQ: A Realization of the Disordered Hubbard Model
(b) Vol. 88, No. 2, p. 663—674 (1978).
See also 14.4, p. 249; 22.9, p. 526.

U. BEHN

*On the Ward Identities for the Hubbard Model. Particle-Number Conservation and the Analytical Structure of the Self-Energy
(b) Vol. 88, No. 2, p. 699—704 (1978).
See also 14, p. 204.

G. H. WANNIER

*A Result Not Dependent on Rationality for Bloch Electrons in a Magnetic Field
(b) Vol. 88, No. 2, p. 757—765 (1978).

J.-P. GALLINAR

*On the Optical Conductivity in a Strongly Attractive Hubbard Model Chain
(b) Vol. 88, No. 2, p. K143—K146 (1978).
See also *14*, p. *204*.

G. HILSCHER and E. GRATZ

*An Estimation of the Density of States from Magnetization and Resistivity Measurements in $Ti(Fe_{1-x}Co_x)$ and $Zr(Fe_{1-x}Co_x)_2$
(a) Vol. 48, No. 2, p. 473—480 (1978).
See also 14.1, p. 209; 18.1, p. 279; 18.2, p. 289; 21.1, p. 376; 21.1.1, p. 388.

R. RICHTER, F. GOEDSCHE, and G. VOJTA

Statistical Theory of Electrical Resistivity of Magnetic Ternary Alloys with Partial Long-Range Order
(b) Vol. 89, No. 1, p. 77—83 (1978).
See also *14.1*, p. *209*; 18.4, p. 301; 21, p. 362.

P. BLAUDECK and R. LENK

Multiple Scattering in Condensed Materials (II): Calculation of the Density of States for Non-Crystalline Copper and Iron
(b) Vol. 89, No. 1, p. 199—202 (1978).
See also *2*, p. *49*; 21.1, p. 377; 21.1.1, p. 389.

A. F. AKKERMAN and G. YA. CHERNOV

Mean Free Paths by Inelastic Interactions, Stopping Powers, and Energy Straggling for Electrons of Energies up to 20 keV in Various Solids
(b) Vol. 89, No. 1, p. 329—333 (1978).
See also *11*, p. *129*.

R. N. SINGH and P. L. SRIVASTAVA

*Model Potential Calculation of Binding Energy and Compressibility of Alkali Metals
(b) Vol. 89, No. 1, p. K13—K16 (1978).
See also *4*, p. *59*; 21.2, p. 393.

M. V. EREMIN, V. N. KALINENKOV, and YU. V. RAKITIN

*Two-Centre Exchange Interactions between Orbitally Degenerate Ions (I): One Ion in Orbitally Degenerate State
(b) Vol. 89, No. 2, p. 503—512 (1978).
See also 1, p. 10.

G. SEIFERT, E. MROSAN, and H. MÜLLER
*Electronic Structure of 4d and 5d Transition Metal Clusters
(b) Vol. **89**, No. 2, p. 553—560 (1978).
See also 21, p. 363.

E. WIMMER, P. WEINBERGER, A. KOSAKOV, and G. LEONHARDT
*On the Electronic Structure of As_2O_3, Sb_2O_3, and As_2S_3
(b) Vol. **89**, No. 2, p. 619—624 (1978).
See also 2, p. 50; *20*, p. *317*; 22.6, p. 496.

H. BÖTTGER, V. V. BRYKSIN, and P. KLEINERT
*Theory of Resonant Raman Scattering in Systems with Localized Electron States
(b) Vol. **89**, No. 2, p. 675—685 (1978).
See also *6*, p. *76*; 20.3, p. 348; 22, p. 418.

A. ČÍŽEK, M. ŠOB, E. DĘBOWSKA, and B. ROZENFELD
*Positron Annihilation in Some Uranium and Thorium Compounds
(b) Vol. **89**, No. 2, p. K159—K161 (1978).
See also 21.5, p. 400.

A. D. MOKRUSHIN and R. KH. SABIROV
*Positron Thermalization as Quantum Statistical Process
(b) Vol. **90**, No. 1, p. 99—104 (1978).
See also 6, p. 77; 21, p. 364.

TH. NATTERMANN
*Low-Temperature Renormalization Group and Symmetry Breaking Perturbations
(b) Vol. **90**, No. 1, p. 105—115 (1978).
See also *18.2*, p. *291*.

M. V. EREMIN, V. N. KALINENKOV, and YU. V. RAKITIN
*Two-Centre Exchange Interactions between Orbitally Degenerate Ions (II): Two Ions in Orbitally Degenerate States
(b) Vol. **90**, No. 1, p. 123—134 (1978).
See also 1, p. 10.

P. M. VAN DEN BROEK and J. F. CORNWELL
*Clebsch-Gordan Coefficients of Symmetry Groups
(b) Vol. **90**, No. 1, p. 211—224 (1978).
See also *1*, p. *10*.

A. FREUDENHAMMER
Consequences of the Linear Part of the Specific Heat of Dilute Magnetic Alloys
(b) Vol. **90**, No. 1, p. 269—275 (1978).
See also *8*, p. *89*; 21, p. 364.

V. I. LITVINOV, V. M. BAGINSKII, and A. A. MITROFANOV
*The Phase Diagram of a System with Two Order Parameters. Mott Transition and Ferromagnetism
(b) Vol. **90**, No. 1, p. 423—427 (1978).
See also *1.2*, p. *23*; 14, p. 204; 18, p. 275.

F. A. Zhivopistsev and F. C. Hernández
*Fast Electron Scattering in Metals with the Excitation of a Core Electron
(b) Vol. **90** No. 2 p. 441—447 (1978).
See also 13.5.2, p. 199; 21.2, p. 393.

J. P. Leburton and R. Evrard
*Oscillatory Electron Drift Velocity in Polar Semiconductor in Non-Ohmic Regime at Low Temperature
(b) Vol. **90**, No. 2, p. K137—K139 (1978).
See also *14.3.4*, p. *246*; 22.2.1, p. 449; 22.2.3, p. 455.

D. Kapor and A. Ivić
*The Coefficients in a Boson Series Expansion of Spin 1/2 Operators
(b) Vol. **90** No. 2 p. K157—K159 (1978).
See also *18.2*, p. *292*.

J. de Oliveira, O. de O. Damasceno, J. R. F. Ferreira, A. L. de Oliveira, J. D. Fabris, P. Boyer, and A. Baudry
*Electric Field Gradient Measurements at ^{181}Ta Nuclei in HfO_2–CaO and HfO_2–MgO Systems
(b) Vol. **90**, No. 2, p. K169—K174 (1978).
See also *10.2*, p. *121*; 14.4, p. 252; 22.6.1, p. 500.

13.1 Band Structure

Review Articles

R. Suryanarayanan
Valence Related Optical and Other Studies of Sm and Tm Chalcogenides
(b) Vol. **85**, No. 1, p. 9—43 (1978).
See also 14.1, p. 205; 14.3, p. 217; *20.1*, p. *319*; 22.7, p. 501.

J. A. Wilson
Modelling the Contrasting Semimetallic Characters of TiS_2 and $TiSe_2$
(b) Vol. **86**, No. 1, p. 11—36 (1978).
See also 6, p. 64; *14.1*, p. *205*; 22, p. 408.

A. S. Borukhovich
Kinetic Properties of IVa and Va Subgroup Transition Metal Monocarbides
(a) Vol. **46**, No. 1, p. 11—37 (1978).
See also 8, p. 83; *14.3*, p. *217*; 15, p. 259; 22, p. 408; 22.3, p. 457; 22.3.1, p. 460.

M. Tomášek and Š. Pick
Projected Surface Energy Bands and Shockley Surface States of Transition Metals. General Considerations and the Example of the B.C.C. Lattice
(b) Vol. **89**, No. 1, p. 11—26 (1978).
See also *13.3*, p. *164*; 21.1, p. 368; 21.1.1, p. 384.

P. C. Mathur, R. Shyam, and Sushil Jain
*Transport Phenomena in III–V Compound Semiconductors
(a) Vol. **50**, No. 1, p. 11—40 (1978).
See also 6, p. 64; 13.4, p. 169; *14.3*, p. *217*; 20.1, p. 319; 22.2, p. 443.

Original Papers and Short Notes

V. N. Kuznetsov and A. N. Men
*Symmetry and Analytical Expressions of Matrix Components in the Tight-Binding Method
(b) Vol. **85**, No. 1, p. 95—104 (1978).

E. Krüger
Superconductivity Originating from Quasi-Orbital Electrons (I): The Model of Quasi-Orbital Conduction Electrons
(b) Vol. **85**, No. 1, p. 261—270 (1978).
See also *14.2*, p. *213*; 21, p. 352.

H. Neumann and A. Kosakov
*Tight-Binding Calculation of the Density of Valence States of SnS, SnSe, and SnTe
(b) Vol. **85**, No. 1, p. K11—K14 (1978).
See also 17, p. 269; 22.7, p. 501.

W. Hörig, H. Neumann, B. Schumann, and G. Kühn
*Optical Properties of $CuGaSe_2$ near and above the Fundamental Absorption Edge
(b) Vol. **85**, No. 1, p. K57—K61 (1978).
See also *20.1*, p. *320*; 22.8, p. 503.

R. ROSEI, E. COLAVITA, S. MODESTI, and N. NISTICÒ
*Electronic Structure of Dilute α-Phase Ag–Cd Alloys from Thermoreflectance and XPS Measurements
(b) Vol. **85**, No. 2, p. 445—452 (1978).
See also 20.1, p. 321; 21, p. 354; 21.6, p. 401.

A. I. BELKIND, A. M. BRODSKII, and V. V. GRECHOV
Theory of Photoemission from Molecular Crystals
(b) Vol. **85**, No. 2, p. 465—472 (1978).
See also *17*, p. *269*; 22.9, p. 522.

E. KRÜGER
Superconductivity Originating from Quasi-Orbital Electrons (II): The Superconducting Ground State of Quasi Orbital Conduction Electrons
(b) Vol. **85**, No. 2, p. 493—504 (1978).
See also *14.2*, p. *213*; 21, p. 354.

M. BRAUWERS and A. K. BHATTACHARYA
Temperature Dependence of the Electrical Resistivity and Hall Coefficient in a Model of Concentrated Alloy
(b) Vol. **85**, No. 2, p. 519—524 (1978).
See also *14.1*, p. *206*; 21, p. 354.

R. ENDERLEIN, B. SEIFERT, and F. BECHSTEDT
*Electronic Elementary Excitations in the Far UV-Spectral Region. Core Level Exciton and Core Hole Polarization
(b) Vol. **85**, No. 2, p. 569—576 (1978).
See also *13.5.1*, p. *190*; 20, p. 315; 22.1, p. 422.

L. LEIBLER
Theory of Electron Spin Resonance in Graded Mixed Semiconductors
(b) Vol. **85**, No. 2, p. 611—619 (1978).
See also *19*, p. *304*; 22.2.4, p. 456; 22.4.4, p. 472.

R. M. M. FONVILLE, W. GEERTSMA, and C. HAAS
*On the Nature of the Top of the Valence Band in Layered Mo and W Dichalcogenides
(b) Vol. **85**, No. 2, p. 621—627 (1978).
See also 14.3, p. 217; 18.1, p. 277; 19, p. 304; 22, p. 409.

J. J. DUBOWSKI
*Disorder Scattering in $Cd_xHg_{1-x}Te$ Mixed Crystals
(b) Vol. **85**, No. 2, p. 663—672 (1978).
See also *14.3*, p. *217*; 15, p. 259; 22.4.4, p. 472.

Y. YAMADA
*Band Structure Calculation of ZnSb
(b) Vol. **85**, No. 2, p. 723—732 (1978).
See also 22.3, p. 457.

P. J. LIN-CHUNG
*Many Special Point Scheme for Noncubic Lattices
(b) Vol. **85**, No. 2, p. 743—748 (1978).
See also 22, p. 409.

G. P. Srivastava
*Electronic and Optical Properties of Crystalline and Amorphous Si
(b) Vol. 85, No. 2, p. K121—K126 (1978).
See also 2, p. 43; *20.1*, p. *321*; 22.1.2, p. 429.

A. Aladgadgyan and A. Toneva
*Study of the Valence Band Structure of $Pb_{1-x}Sn_xTe$
(b) Vol. 85, No. 2, p. K131—K134 (1978).
See also 14.3, p. 218; 22.7, p. 501.

W. Lochmann
*Phonon-Assisted Auger Recombination in Indirect Gap Semiconductors
(a) Vol. 45, No. 2, p. 423—432 (1978).
See also *6*, p. *66*; 20, p. 315; 22.1.1, p. 424; 22.1.2, p. 430.

R. A. Evarestov and A. N. Ermoshkin
*Large Unit Cell Calculations of the Band Structure of Ionic Crystals Using the Mulliken-Ruedenberg Approximation
(b) Vol. 86, No. 1, p. 47—55 (1978).
See also 22.5.2, p. 479.

A. N. Trukhin
*Investigation of the Photoelectric and Photoluminescent Properties of Crystalline Quartz and Vitreous Silica in the Fundamental Absorption Region. A Model for Electronic Structure and Migration of Energy in SiO_2
(b) Vol. 86, No. 1, p. 67—75 (1978).
See also 2, p. 43; 16, p. 263; 20.3, p. 342; 22.6, p. 491.

N. I. Kulikov, V. N. Borzunov, and A. D. Zvonkov
*The Electronic Band Structure and Interatomic Bond in Nickel and Titanium Hydrides
(b) Vol. 86, No. 1, p. 83—91 (1978).
See also 4, p. 57; 21.1, p. 371.

S. Matsuo, T. Nishimoto, S. Hayashi, and S. Noguchi
*Magnetothermal Oscillation and Spin Splitting of Electrons in Dilute Bismuth–Antimony Alloys
(b) Vol. 86, No. 1, p. 169—176 (1978).
See also 14.1, p. 206; 21.7, p. 406.

E. Majerníková and Š. Barta
*Magnetic Freeze-Out Effect in Noncrystalline Semiconductors
(b) Vol. 86, No. 1, p. 183—194 (1978).
See also 2, p. 43; *13.5.1*, p. *190*; 18, p. 273; 22, p. 410.

W. Müller, M. Overhamm, G. Bauer, E. Gornik, P. Grosse, H. W. Pötzl, and M. Rautenberg
On the Influence of Background Radiation on Landau Emission Spectra
(b) Vol. 86, No. 1, p. 205—213 (1978).
See also *14.3.4*, p. *241*; 20.1, p. 322; *20.3*, p. *343*; 22.2.3, p. 452.

T. Ikari and K. Hashimoto
Electrical Properties of TlTe Single Crystals
(b) Vol. **86**, No. 1, p. 239—248 (1978).
Erratum
(b) Vol. **90**, No. 2, p. 751 (1978).
See also *14.3*, p. *218*; *15*, p. *259*; 22.3, p. 457.

T. Soma
Local Heine-Abarenkov Model Potential for Covalent Crystals
(b) Vol. **86**, No. 1, p. 263—268 (1978).
See also *4*, p. *57*; 22.1, p. 422; 22.1.1, p. 424; 22.1.2, p. 430.

D. Ihle
Electrical Conductivity in the Strong-Coupling Non-Half-Filled Band Hubbard Model
(b) Vol. **86**, No. 1, p. 283—289 (1978).
See also *14*, p. *203*.

J. Humlíček
On the Nature of E_1-Transitions in Germanium
(b) Vol. **86**, No. 1, p. 303—310 (1978).
See also *14.3.4*, p. *241*; *20.1*, p. *322*; 22.1.1, p. 424.

R. Strehlow
On the No-Phonon Structure in Mixed Crystals with Indirect Band Structure
(b) Vol. **86** No. 1 p. 311—318 (1978).
See also 6, p. 67; 13.5.1, p. 190; *20.1*, p. *322*; 22.2.4, p. 456.

V. P. Oleinik and V. A. Sinyak
*To the Theory of the Tunnel Effect in a Semiconductor in Constant Electric and Magnetic Fields
(b) Vol. **86**, No. 1, p. 319—327 (1978).
See also 14.3, p. 218; 22, p. 410.

R. Gründler, K. Breuer, and W. Tews
Optical Properties of Amorphous and Polycrystalline BeO Thin Films from Electron Energy Loss Measurements
(b) Vol. **86**, No. 1, p. 329—338 (1978).
See also 2, p. 43; *14.4*, p. *248*; *20.1*, p. *322*; 22.6, p. 492.

N. N. Tyutyulkov, J. I. Petkov und P. Schuster
*Bandstruktur von Bis (Catena-Poly(Methylarsen))
(Band Structure of Bis (Catena-Poly (Methylarsen)))
(b) Vol. **86**, No. 1, p. 385—395 (1978).
See also 22.9, p. 523.

E. Popko and J. M. Pawlikowski
*Contribution of the Lattice Dilatation to the Temperature Shift of the Energy Gap of $Cd_xHg_{1-x}Te$
(a) Vol. **46**, No. 1, p. K9—K12 (1978).
See also *8*, p. *84*; 22.4.4, p. 472.

R. Car, G. Ciucci, and L. Quartapelle
*Electronic Band Structure of SnSe
(b) Vol. **86**, No. 2, p. 471—478 (1978).
See also 22, p. 411.

W. JOHN, D. HAMANN, and P. URWANK
*Partial Densities of States and Electron–Phonon Interaction in Transition Metals
(b) Vol. 86, No. 2, p. 569–577 (1978).
See also 14.2, p. 214; 21.1, p. 371; 21.1.1, p. 386.

J. KOSSUT
*The Disorder Scattering in Zincblende Narrow-Gap Semiconducting Mixed Crystals
(b) Vol. 86, No. 2, p. 593–601 (1978).
See also *13.4*, p. *174*; 14.3, p. 219; 22.4.4, p. 472.

B. I. RESER and V. V. DYAKIN
*Instability of the OPW Method in Relation to the Error in Core Eigenvalues
(b) Vol. 87, No. 1, p. 41–47 (1978).
See also 22.1.2, p. 432.

A. KUROWSKA
*Theory of Free Carrier Absorption in Mixed Semiconductors with Composition Fluctuations
(b) Vol. 87, No. 1, p. 69–78 (1978).
See also *20.1*, p. *325*; 22.4.4, p. 472.

V. M. SHATALOV, V. A. TELEZHKIN, and K. B. TOLPYGO
*Inadequacy of the Direct Interband Transition Scheme Applied to Valence Semiconductor Fundamental Absorption Spectra
(b) Vol. 87, No. 1, p. 79–86 (1978).
See also 13.5.1, p. 191; 20.1, p. 325; 22, p. 412.

H. L. ALTHAUS, G. WEISER, and S. NAGEL
*Optical Spectra and Band Structure of As_2Se_3
(b) Vol. 87, No. 1, p. 117–128 (1978).
See also *20.1*, p. *326*; 22, p. 413.

P. ZIESCHE, H. WONN, CH. MÜLLER, V. V. NEMOSHKALENKO, and V. P. KRIVITSKII
*Theoretical and Experimental Investigation of the Electronic Structure of the B2-Compound FeAl
(b) Vol. 87, No. 1, p. 129–134 (1978).
See also 20, p. 316; 21, p. 357; 21.1.1, p. 386.

P. RENNERT, TH. DÖRRE, and U. GLÄSER
*Theoretical Investigation of the Anisotropy of Beryllium by Polarized K-Emission, APS, and Compton Profiles
(b) Vol. 87, No. 1, p. 221–226 (1978).
See also 20, p. 316; 21.3, p. 394.

D. M. EAGLES
*Widths of 5d Energy Bands in the Heavy Rare-Earth Metals
(b) Vol. 87, No. 1, p. 301–307 (1978).
See also 21.4, p. 397.

B. DOWGIAŁŁO-PLENKIEWICZ and P. PLENKIEWICZ
*Spin–Orbit and Crystal Field Splittings of the Valence Band in II_3V_2 Semiconducting Compounds
(b) Vol. 87, No. 1, p. 309–315 (1978).
See also 22.3, p. 458.

A. S. Filipchenko, L. P. Bolshakov, and A. Naurizbaev
The Moss-Burstein Effect in n-Type Indium Arsenide Crystals Doped with Different Donor Impurities
(b) Vol. 87, No. 1, p. K1—K3 (1978).
See also *20.1*, p. *326*; 22.2.1, p. 446.

L. Pasemann
*On the Screening of the Ions in Solids. Band Structure Calculations of Si with Several Screening Variants
(b) Vol. 87, No. 2, p. 553—564 (1978).
See also 22.1.2, p. 433.

J. Mertsching
*Peierls Distortion and Antiferromagnetism in the Quasi-One-Dimensional Peierls-Hubbard Model
(b) Vol. 87, No. 2, p. 599—606 (1978).
See also *18.4*, p. *300*.

J. Spałek, A. M. Oleś, and K. A. Chao
The Effective Magnetic Interactions between Impurity and Conduction Electrons for the Wolff Model
(b) Vol. 87, No. 2, p. 625—635 (1978).
See also *13.4*, p. *176*; *18.4*, p. *300*.

W. Szuszkiewicz, A. M. Witowski, and M. Grynberg
*The Dynamic Dielectric Function in HgSe and HgTe
(b) Vol. 87, No. 2, p. 637—645 (1978).
See also 14.4, p. 249; 20.1, p. 328; 22.4.2, p. 468; 22.4.3, p. 470.

C. K. Kim and A. B. Kunz
*Electronic Structure of Silicon with Non-Spherical Potentials for the Two Basis Atoms
(b) Vol. 87, No. 2, p. 765—769 (1978).
See also 22.1.2, p. 433.

V. K. Bashenov, Y. K. Markolenko, and V. V. Timofeenko
*Semiconductor–Metal Transition of Diamond-Type Crystals
(b) Vol. 87, No. 2, p. K73—K75 (1978).
See also *1.2*, p. *19*; 22.1, p. 422.

V. K. Bashenov, A. M. Mutal, and V. V. Timofeenko
*Valence-Band Density of States for Mg_2Si from Pseudopotential Calculation
(b) Vol. 87, No. 2, p. K77—K79 (1978).
See also 22.7, p. 502.

V. L. Volkov
*The Electron–Phonon Interaction in $Pb_{1-x}Sn_xTe$
(b) Vol. 87, No. 2, p. K93—K96 (1978).
See also *6*, p. *72*; 22, p. 414.

V. M. Shatalov
*Quasi-Molecular Model Interpretation of the Resonances in the Fundamental Absorption Region of Valence Semiconductors
(b) Vol. 88, No. 1, p. 129—133 (1978).
See also 13.5.1, p. 192; 20.1, p. 328; 22, p. 414.

W. Eberhardt, G. Kalkoffen, C. Kunz, D. Aspnes, and M. Cardona
*Photoemission Studies of 2p Core Levels of Pure and Heavily Doped Silicon
(b) Vol. 88, No. 1, p. 135—143 (1978).
See also 17, p. 270; 20.1, p. 328; 22.1.2, p. 434.

R. Enderlein, R. E. Hummel, J. B. Andrews, R. J. Nastasi-Andrews, and C. W. Shanley
*Interpretation of Compositional Modulation Spectra to Determine Some Optical Properties of Alloys
(b) Vol. 88, No. 1, p. 173—178 (1978).
See also *20.1*, p. *329*; 21.1, p. 375.

A. A. Cottey
*Band Theory of Size-Quantised Electron States in Thin Crystalline Films
(b) Vol. 88, No. 1, p. 207—219 (1978).
See also 13.3, p. 166; *14.3.1*, p. *227*.

S. D. Beneslavskii and O. Ziep
*Hot Electrons in Zero-Gap Semiconductors with Non-Parabolic Electronic Band Structure
(b) Vol. 88, No. 1, p. 221—230 (1978).
See also *14.3.4*, p. *243*; 14.4, p. 249; 22.4.4, p. 473.

P. Pfeffer and W. Zawadzki
Inelastic Scattering of Electrons by Optic Phonons in InSb-Type Semiconductors
(b) Vol. 88, No. 1, p. 247—260 (1978).
See also 6, p. 73; *14.3*, p. *222*; 22.2.3, p. 453.

S. B. Shrivastva and H. P. Bonde
*Positron Annihilation Rates in the Metals of Ag and Au Series
(b) Vol. 88, No. 1, p. 269—272 (1978).
See also 21.6, p. 403.

B. K. Agarwal and R. K. Johri
*X-Ray K-Absorption Structure of Zirconium Metal and Its Compounds
(b) Vol. 88, No. 1, p. 309—313 (1978).
See also *20*, p. *317*; 21, p. 360.

Z. M. Alekseeva, A. P. Vyatkin, G. F. Karavaev, and N. P. Krivorotov
*Effect of Pressure on Hole Effective Masses in GaAs
(b) Vol. 88, No. 1, p. 321—326 (1978).
See also *12.2*, p. *140*; 14.3, p. 222; 22.2.1, p. 447.

M. Baleva
*Thermoreflectance of PbTe
(b) Vol. 88, No. 1, p. 335—339 (1978).
See also *20.1*, p. *329*; 22, p. 414.

A. D. Jannussis, V. Papatheou, and P. Tsilimigras
*The Motion of an Electron in One-Dimensional Periodic Potential under the Influence of Uniform Electric and Magnetic Fields
(b) Vol. 88, No. 1, p. 341—350 (1978).
See also 18, p. 274.

I. Ya. Dekhtyar, V. T. Adonkin, S. P. Likhtorovich, and Sh. A. Rustamov
*Positronium-Like States in GaAs at Low Temperature
(b) Vol. 88, No. 1, p. K55—K58 (1978).
See also 22.2.1, p. 447.

V. V. Nemoshkalenko, Vl. N. Antonov, V. N. Antonov, P. Ziesche, and H. Wonn
*On the Influence of Relativistic Effects on the Electronic Structure of Luthecium
(b) Vol. 88, No. 1, p. K83—K86 (1978).
See also 21.4, p. 398.

K. Hübner
*Chemical Bond and Related Properties of SiO_2 (V): Pseudopotential Study of Bonding and Structure
(a) Vol. 48, No. 1, p. 147—154 (1978).
See also 2, p. 47; *4*, p. *59*; 14.4, p. 249; 22.6, p. 494.

M. I. Dykman
*Theory of Cyclotron Resonance of Two-Dimensional Electrons Interacting with Surface and Volume Phonons
(b) Vol. 88, No. 2, p. 463—475 (1978).
See also *6*, p. *74*; 20.1, p. 330.

R. Richter and F. Goedsche
*Complex Band Structure and Density of States for a One-Dimensional Model of Transition Metal Alloys
(b) Vol. 88, No. 2, p. 485—491 (1978).
See also 21.1, p. 376.

K. Saiki, M. Yoshimi, and S. Tanaka
Modulation Spectroscopy on the Group IV and VI Transition-Metal Dichalcogenides
(b) Vol. 88, No. 2, p. 607—614 (1978).
See also *13.5.1*, p. *193*; *20.1*, p. *330*; 22, p. 415.

R. Glocker and L. Fritsche
*Pressure Dependence of the Electron–Phonon Coupling in Lanthanum
(b) Vol. 88, No. 2, p. 639—644 (1978).
See also *6*, p. *74*; 12.1, p. 136; 14.2, p. 215; 21.4, p. 398.

K. Lochner, H. Bässler, B. Tieke, and G. Wegner
*Photoconduction in Polydiacetylene Multilayer Structures and Single Crystals. Evidence for Band-to-Band Excitation
(b) Vol. 88, No. 2, p. 653—661 (1978).
See also *16*, p. *265*; 22.9, p. 526.

E. N. Mgbenu and A. Maduemezia
*X-Ray Emission Spectrum of Metallic Lithium
(b) Vol. 88, No. 2, p. 805—808 (1978).
See also 13.2, p. 162; *20*, p. *317*; 21.2, p. 393.

C. K. Sarkar and D. Chattopadhyay
Effect of the Electron Spin on the Magneto-Seebeck Coefficient of Polar Semiconductors at Large Magnetic Fields
(b) Vol. 88, No. 2, p. K109—K112 (1978).
See also *15*, p. *260*; 22.2.3, p. 453.

F. Claro and G. H. Wannier
*Closure of Bands for Bloch Electrons in a Magnetic Field
(b) Vol. 88, No. 2, p. K147—K151 (1978).

J. M. Pawlikowski, J. Misiewicz, N. Mirowska, and P. Becla
*Absorption Edge of Zn_3P_2 at 5, 80, and 300 K
(b) Vol. 88, No. 2, p. K153—K156 (1978).
See also *20.1*, p. *331*; 22.3, p. 458.

R. H. Williams
*The Electronic Structure of Tin Diselenide Containing Cyclopropylamine
(b) Vol. 88, No. 2, p. K157—K159 (1978).
See also 17, p. 270; 22, p. 416.

V. P. Dolgopolskii, R. P. Seisyan, and Á. L. Efros, Jr.
*The Ground State of a Wannier-Mott Exciton in Diamond-Like Semiconductors in "Intermediate" Magnetic Fields
(b) Vol. 88, No. 2, p. K177—K180 (1978).
See also *13.5.1*, p. *194*; 22.2.1, p. 447.

W. Braune, G. Kuka, H.-J. Gollnest, and R. Herrmann
*Microwave Spectroscopy in Semimetallic Bismuth–Antimony Alloys
(b) Vol. 89, No. 1, p. 95—101 (1978).
See also *19*, p. *311*; 21.7, p. 407.

F. R. Kessler, J. Metzdorf, and W. Zurek
Interband Faraday Ellipticity in Germanium
(b) Vol. 89, No. 1, p. 203—211 (1978).
See also 13.5.1, p. 194; *18*, p. *275*; *20*, p. *317*; 22.1.1, p. 427.

M. Wautelet and L. D. Laude
*Electronic Structure of Solids under Laser Irradiation
(b) Vol. 89, No. 1, p. 275—279 (1978).
See also 11, p. 129; 20.2, p. 339; 22, p. 417.

P. Świątek, A. M. Witowski, and M. Grynberg
*Experimental and Theoretical Study of the Dynamic Dielectric Function in p-HgTe
(b) Vol. 89, No. 1, p. K1—K4 (1978).
See also 14.4, p. 250; 20.1, p. 332; 22.4.3, p. 471.

V. K. Bashenov, D. I. Marvakov, and A. G. Petukhov
*Average Structure Factors and Stability of Binary Crystals
(b) Vol. 89, No. 1, p. K5—K8 (1978).
See also *4*, p. *59*; 22, p. 417.

D. R. Mašović and S. Zeković
*Model Pseudopotential for Aluminium
(b) Vol. 89, No. 1, p. K57—K60 (1978).
See also 21, p. 362.

U. Saxena and O. N. Srivastava
Effect of Nonstoichiometry on the Electrical Conductivity of Erbium Oxide Films
(a) Vol. 49, No. 1, p. 371—378 (1978).
See also *1.5*, p. *38*; 2, p. 49; *14.3.1*, p. *228*; 22.6, p. 496.

A. HAUG, D. KERKHOFF, and W. LOCHMANN
*Calculation of Auger Coefficients for III–V Semiconductors with Emphasis on GaSb
(b) Vol. **89**, No. 2, p. 357—365 (1978).
See also 20, p. 317; 22.2, p. 444; 22.2.3, p. 454.

P. KACMAN, J. BLINOWSKI, and M. GRYNBERG
*Dynamic Dielectric Function in the Three-Band Model for Zero-Gap Semiconductors in a Magnetic Field
(b) Vol. **89**, No. 2, p. 521—530 (1978).
See also 14.4, p. 251; 18, p. 275; 22, p. 418.

K. ELK, J. RICHTER, and V. CHRISTOPH
Electrical Conductivity of Disordered Alloys with Strong Hybridization
(b) Vol. **89**, No. 2, p. 537—540 (1978).
See also *14.1*, p. *211*; 21, p. 363.

B. ESSER
On the Theory of Resonance Raman Scattering in Disordered Semiconductors
(b) Vol. **89**, No. 2, p. 633—643 (1978).
See also 2, p. 50; *6*, p. *76*; *20.1*, p. *334*; 22, p. 418.

J. A. GAJ, J. GINTER, and R. R. GAŁĄZKA
*Exchange Interaction of Manganese $3d^5$ States with Band Electrons in $Cd_{1-x}Mn_xTe$
(b) Vol. **89**, No. 2, p. 655—662 (1978).
See also 18.2, p. 290; 22.8, p. 507.

G. SEIFERT, E. MROSAN, H. MÜLLER, and P. ZIESCHE
*SW-Xα Calculations of Small Niobium Clusters
(b) Vol. **89**, No. 2, p. K175—K178 (1978).
See also 21, p. 363.

P. STEINER, F. J. REITER, H. HÖCHST, and S. HÜFNER
*The KLL Auger Spectra of Na and Mg Metal and Their Plasmon Structure
(b) Vol. **90**, No. 1, p. 45—51 (1978).
See also *13.5.2*, p. *199*; 21.2, p. 393; 21.3, p. 395.

A. MANABE, H. NOGUCHI, and A. MITSUISHI
Dielectric Function in HgSe at 2, 95, and 300 K
(b) Vol. **90**, No. 1, p. 157—166 (1978).
See also 6, p. 77; 13.5.2, p. 199; *20.1*, p. *335*; 22.4.2, p. 469.

R. STREHLOW, W. KÜHN, H. FRIEDRICH, and H. LANGE
On the Quantitative Characterization of Chemical Disorder of $In_{1-x}Ga_xP$ from Wavelength-Modulated No-Phonon Absorption
(b) Vol. **90**, No. 1, p. 167—173 (1978).
See also 6, p. 77; 13.5.1, p. 195; *20.1*, p. *335*; 22.2.4, p. 456.

O. ZIEP, M. MOCKER, D. GENZOW, and K. H. HERRMANN
*Auger Recombination in PbSnTe-Like Semiconductors
(b) Vol. **90**, No. 1, p. 197—205 (1978).
See also 22, p. 419.

E. Marschall and H. Bross
*Ferromagnetic Nickel: Self-Consistent Calculation of Energy Bands
(b) Vol. **90**, No. 1, p. 241–250 (1978).
See also 18.1, p. 280; 18.2, p. 291; 21.1, p. 380.

G. B. Dubrovskii, S. T. Pavlov, and M. S. Vershinin
*Interband Optical Absorption in Semiconductors with One-Dimensional Electronic Superlattice
(b) Vol. **90**, No. 1, p. 429–432 (1978).
See also *20.1*, p. *336*; 22.3, p. 459.

V. K. Bashenov, D. I. Marvakov, and A. G. Petukhov
*Bond-Orbital Model of Layer Gallium Selenide
(b) Vol. **90**, No. 1, p. K5–K8 (1978).
See also *4*, p. *59*; 22.3, p. 459.

O. Hunderi and P. O. Nilsson
*The Electronic Band Structure of Zinc: A Theoretical Note
(b) Vol. **90**, No. 1, p. K69–K72 (1978).
See also 21.1, p. 380.

K. Berndt, U. Marx, and O. Brümmer
*Electronic Structure of FeNi Alloys Studied by Means of X-Ray Photoelectron Spectroscopy (I): Valence Bands
(b) Vol. **90**, No. 2, p. 487–496 (1978).
See also *20*, p. *318*; 21.1, p. 381; 21.1.1, p. 390.

H. Stöhr and H. Bross
*Solid Silicon Selfconsistently with X_α Potentials
(b) Vol. **90**, No. 2, p. 497–507 (1978).
See also 22.1.2, p. 439.

H. Köhler and W. Ossau
*Magneto-Seebeck and Shubnikov-de Haas Effect in n-Type $(Bi_{1-x}Sb_x)_2Te_3$
(b) Vol. **90**, No. 2, p. 509–515 (1978).
See also 14.3, p. 225; *15*, p. *261*; 22.7, p. 502.

W. Braun, M. Neumann, M. Iwan, and E. E. Koch
*UPS Measurements of CO Chemisorbed and Condensed on Rh (111) Using Synchrotron Radiation
(b) Vol. **90**, No. 2, p. 525–533 (1978).
See also *17*, p. *270*; 21, p. 365.

D. Genzow, A. G. Mironov, and O. Ziep
*On the Interband Absorption in Lead Chalcogenides
(b) Vol. **90**, No. 2, p. 535–542 (1978).
See also *20.1*, p. *337*; 22, p. 420.

Nguyen Vinh Quang
*The Influence of an Electromagnetic Wave on the Electronic Spectrum of Crystals
(b) Vol. **90**, No. 2, p. 597–608 (1978).
See also *20.2*, p. *340*.

M. Raychaudhury, P. Chatterjee, and S. Chatterjee
*Electron–Phonon Coupling Constant and Critical Temperature of Some A15 Superconducting Compounds of Nb and V
(b) Vol. **90**, No. 2, p. 609—613 (1978).
See also *14.2*, p. *216*.

R. A. Noack
*Inversion of the Γ and L Conduction Bands of GaSb under Uniaxial Stress
(b) Vol. **90**, No. 2, p. 615—620 (1978).
See also 12.2, p. 141; *20.3*, p. *350*; 22.2.3, p. 454.

H. Eschrig and I. Bergert
*An Optimized LCAO Version for Band Structure Calculations. Application to Copper
(b) Vol. **90**, No. 2, p. 621—628 (1978).
See also 21.1, p. 381.

A. Milchev
*Frequency- and Wave-Vector-Dependent Dielectric Function of a Model Semiconductor
(b) Vol. **90**, No. 2, p. 679—688 (1978).
See also *14.4*, p. *252*; 22.1.1, p. 428; 22.1.2, p. 439; 22.2.1, p. 449; 22.4.2, p. 469.

E. Krüger
Superconductivity Originating from Quasi-Orbital Electrons (III): Quasi-Orbital Conduction Electrons in Non-Adiabatic Systems
(b) Vol. **90**, No. 2, p. 719—731 (1978).
See also *14.2*, p. *216*; 21, p. 366.

A. M. Altshuler, Yu. Ku. Vekilov, and G. R. Umarov
*Characteristic Features of the Electronic Spectrum and the Stability of Ternary Diamond-Like Semiconductors
(b) Vol. **90**, No. 2, p. 733—740 (1978).
See also 1, p. 10; 22, p. 421.

E. S. Alekseev, R. G. Arkhipov, and S. V. Popova
*Band Structure of the Metastable F.C.C. Tantalum Nitride
(b) Vol. **90**, No. 2, p. K133—K136 (1978).
See also 22, p. 421.

A. F. Plotnikov, V. A. Tolokonnikov, and V. E. Shubin
*Investigation of the Energy Band Diagram of the Anodic Oxide–Indium Antimonide System by Internal Photoemission
(a) Vol. **50**, No. 2, p. 733—738 (1978).
See also *14.3.3*, p. *239*; 16, p. 268; 22.2.3, p. 455.

Sh. M. Efendiev, V. E. Bagiev, A. Ch. Zeinally, and V. M. Skorikov
Optical Transitions in $Bi_{12}SiO_{20}$
(a) Vol. **50**, No. 2, p. K141—K143 (1978).
See also 6, p. 79; *20.1*, p. *338*; 22.8, p. 508.

13.2 Fermi Surfaces

Original Papers and Short Notes

B. G. Mulimani and M. A. Angadi
*A Note on the Analysis of the dHvA Effect in Zinc Based Kondo Alloys with Magnetic Breakdown
(b) Vol. 85, No. 1, p. K33—K38 (1978).
See also 21.1, p. 368.

J. Shy-Yih Wang and M. Rasolt
*Effect of Nonlocal Potential on the Fermi Surface of Metals and the Phase Shift Parametrization Scheme
(b) Vol. 86, No. 1, p. 37—40 (1978).
See also 21, p. 355.

R. Stegmann and G. Oelgart
*Hybrid Resonance in a Quantizing Magnetic Field in BiSb
(b) Vol. 87, No. 2, p. 485—489 (1978).
See also *19*, p. *308*; 22.7, p. 502.

W. Braune, J. Lebech, and K. Saermark
*Propagation of High-Frequency Electromagnetic Waves in the Voigt Configuration through Samples of Bi at 45, 70, and 300 GHz
(b) Vol. 87, No. 2, p. 527—533 (1978).
See also *14.1*, p. *208*; 21.6, p. 402.

M. Jaczyński, J. Kossut, and R. R. Gałązka
*Influence of Exchange Interaction on the Quantum Transport Phenomena in $Hg_{1-x}Mn_xTe$
(b) Vol. 88, No. 1, p. 73—85 (1978).
See also *14.3*, p. *221*; 15, p. 260; 22.4.4, p. 473.

K. Yonemitsu, K. Takano, and T. Matsuda
Hall Coefficient of Dilute Aluminium Alloys (I): Calculation by Four-OPW Approximation
(b) Vol. 88, No. 1, p. 273—282 (1978).
See also *14.1*, p. *209*; 21, p. 360.

G. Oelgart, R. Stegmann, and R. Herrmann
*Propagation of Electromagnetic Waves in a Direction Parallel to the Magnetic Field in Semiconductive Bi–Sb Alloys
(b) Vol. 88, No. 2, p. 507—516 (1978).
See also 14.3, p. 223; 19, p. 310; 21.7, p. 407.

E. N. Mgbenu and A. Maduemezia
X-Ray Emission Spectrum of Metallic Lithium
(b) Vol. 88, No. 2, p. 805—808 (1978).
See also *13.1*, p. *157*; *20*, p. *317*; 21.2, p. 393.

K. N. Khanna and P. K. Sharma
A Model Potential and Its Calculations
(b) Vol. 88, No. 2, p. 809—814 (1978).
See also *12.1*, p. *136*; 18.1, p. 278; 21.2, p. 393.

K. SAERMARK
*Azbel-Kaner Cyclotron Resonance as a Long-Wavelength Limit of an Exact Expression for the Surface Impedance of a Metal in a Magnetic Field
(b) Vol. **89**, No. 1, p. 117–121 (1978).
See also 14.1, p. 210; 21, p. 362.

J. REICHELT and K. WINZER
*Magnetoresistance and de Haas-van Alphen Effect in $LaAl_2$
(b) Vol. **89**, No. 2, p. 489–494 (1978).
See also 21, p. 363.

13.3 Surface and Interface States

Review Article

M. Tomášek and Š. Pick
*Projected Surface Energy Bands and Shockley Surface States of Transition Metals. General Considerations and the Example of the B.C.C. Lattice
(b) Vol. **89**, No. 1, p. 11—26 (1978).
See also 13.1, p. 150; 21.1, p. 368; 21.1.1, p. 384.

Original Papers and Short Notes

G. S. Agarwal and H. D. Vollmer
*Surface Polariton Effects in Spontaneous Emission (II): Effects of Spatial Dispersion
(b) Vol. **85**, No. 1, p. 301—310 (1978).
See also 13.5.2, p. 198; 20.1, p. 320; 21, p. 352; 22.4.1, p. 463.

M. Nishida
*Cluster Model Calculation for Threefold Coordinated Site Chemisorption of Cl on the Ge(111) Surface by the Extended Hückel Theory
(b) Vol. **85**, No. 2, p. K69—K72 (1978).
See also 22.1.1, p. 424.

I. Bartoš
*Surface Green Functions for *N* Interfaces
(b) Vol. **85**, No. 2, p. K127—K130 (1978).

M. K. Roy
*Coupling of Surface Acoustic Waves Propagating in Two Separated Piezoelectric Media
(a) Vol. **45**, No. 2, p. 549—552 (1978).
See also *7*, p. *80*; 14.3.2, p. 230; 22.4.1, p. 463; 22.6 .1, p. 499.

W. C. Meixner and P. R. Antoniewicz
*Effect of Atomic Size on the Effective Polarizability of Physisorbed Atoms
(b) Vol. **86**, No. 1, p. 339—343 (1978).

B. D. Kandilarov, M. G. Tashkova, P. C. Petrova, and V. Detcheva[1]
*Influence of the Position of the Interface Boundary on the Existence of Interface States
(b) Vol. **86**, No. 1, p. 425—430 (1978).
See also 14.3.3, p. 233.

J. Wortmann
*Optisch stimulierte Elektronenemission von tribomechanisch aktivierten Aluminium (Optically Stimulated Exoelectron Emmission of Tribomechanically Activated Aluminium)
(a) Vol. **46**, No. 1, p. 69—76 (1978).
See also *17*, p. *269*; 21, p. 356.

G. Sarrabayrouse, J. Buxo, and D. Esteve
*The Influence of Surface States on the Electrical Characteristics of Tunneling MIS Schottky Barriers
(a) Vol. **46**, No. 1, p. 185—190 (1978).
See also *14.3.3*, p. *234*.

M. Samoć and Z. Zboiński
*Trapping of Charge Carriers in the Deformed Surface Region of Anthracene Single Crystals
(a) Vol. **46**, No. 1, p. 251–257 (1978).
See also **14.3**, p. **219**; **16**, p. **263**; **22.9**, p. **524**.

S. Cristoloveanu, A. Chovet, and G. Kamarinos
*Some Comments on "The Effect of Effective Lifetime Modulation in Semiconductors"
(a) Vol. **46**, No. 1, p. K13–K16 (1978).
See also **14.3.3**, p. **234**; **22**, p. **411**.

I. Munteanu
*On the Effective Lifetime Modulation in Semiconductors
(a) Vol. **46**, No. 1, p. K17–K20 (1978).
See also 14.3.3, p. **234**; 22, p. **411**.

P. R. Antoniewicz
*The Depolarization Interaction of Absorbed Non-Polar Molecules
(b) Vol. **86**, No. 2, p. 645–652 (1978).
See also 21, p. 357.

I. Bartoš, M. Hietschold, and G. Paasch
*On Integral Characteristics of the Electron Structure of Crystalline Surfaces
(b) Vol. **87**, No. 1, p. 111–115 (1978).

B. D. Kandilarov, M. G. Tashkova, and P. C. Petrova
*Interface States in One-Dimensional Models of Deformed Heterojunctions
(b) Vol. **87**, No. 1, p. 361–368 (1978).
See also *14.3.3*, p. *235*.

T. Takebe, J. Saraie, and T. Tanaka
*Study of the Surface Barrier of the Metal–n-CdTe Contact
(a) Vol. **47**, No. 1, p. 123–130 (1978).
See also *14.3.2*, p. *230*; 14.3.4, p. 242; 22.4.3, p. 470.

K. A. Mirza, P. D. Townsend, and G. L. Destefanis
Light Emission from the Surface of $LiNbO_3$
(a) Vol. **47**, No. 1, p. K63–K66 (1978).
See also 14.4.1, p. 255; *20.3*, p. *344*; 22.8.1, p. 513.

V. A. Tyagai, O. V. Snitko, A. M. Evstigneev, and A. N. Krasiko
Surface Phenomena in the Polarization-Dependent Electroreflectance Spectra of Si
(b) Vol. **87**, No. 2, p. 613–618 (1978).
See also *14.3.4*, p. *242*; *20.1*, p. *327*; 22.1.2, p. 433.

K. Masuda
*Chemisorption on a Model Transition Metal. Coverage Dependence of the Absorption Behaviour
(b) Vol. **87**, No. 2, p. 739–749 (1978).
See also 21.1, p. 373.

S. Horiguchi, K. Koyama, and Y. H. Ohtsuki
*Auger Neutralization of Slow Protons at Solid Surfaces
(b) Vol. 87, No. 2, p. 757—763 (1978).
See also 21.6, p. 402.

L. Risch
*The Influence of Surface Layers on Photoreflectance in the Exciton Region of CdS
(b) Vol. 88, No. 1, p. 111—116 (1978).
See also 13.5.1, p. 192; 20.1, p. 328; 22.4.1, p. 464.

A. A. Cottey
Band Theory of Size-Quantised Electron States in Thin Crystalline Films
(b) Vol. 88, No. 1, p. 207—219 (1978).
See also *13.1*, p. *156*; *14.3.1*, p. *227*.

P. Viktorovitch, G. Kamarinos, P. Even, and E. Fabre
Correlation between Interface States and MIS Silicon Solar Cell Performances
(a) Vol. 48, No. 1, p. 137—145 (1978).
See also *14.3.3*, p. *235*; *16*, p. *264*; 22.1.2, p. 434.

P. Schlotter and H. Bässler
*Sensitized Injection into Anthracene Crystals from Surface Oxidation Products
(a) Vol. 48, No. 1, p. K15—K18 (1978).
See also *16*, p. *264*; 22.9, p. 526.

D. Heitmann
*Theory and Experiment of the Light Emission of Surface Plasmons Induced by Fast Electrons on Modulated Surfaces
(b) Vol. 88, No. 2, p. 493—506 (1978).
See also *13.5.2*, p. *198*; 20.3, p. 346; 21.6, p. 403.

F. Flores, C. Tejedor, and A. Martín-Rodero
*Surface States in the (111) and $(\bar{1}\bar{1}\bar{1})$ Faces of Zincblende Compounds
(b) Vol. 88, No. 2, p. 591—597 (1978).
See also 22.2, p. 443.

A. V. Sachenko, V. A. Tyagai, and A. G. Kundzich
Exciton Luminescence in Semiconductors. Surface Recombination and Space-Charge-Layer Effects
(b) Vol. 88, No. 2, p. 797—804 (1978).
See also *13.5.1*, p. *193*; *20.3*, p. *346*; 22.4.1, p. 465.

J. von Wienskowski
*Carbon Monoxide Adsorption on Cleaved Si (111) and Ge (111) Surfaces
(b) Vol. 88, No. 2, p. K169—K172 (1978).
See also 22.1.1, p. 427; 22.1.2, p. 435.

H. D. Dimitrov
*On the Surface Electrical Resistivity of Thin Semiconducting and Semimetallic Films
(b) Vol. 89, No. 1, p. 319—328 (1978).
See also *14.3.1*, p. *228*; 14.3.2, p. 231; 22, p. 417.

C. Mande and A. Pimpale
*Evidence for Surface Plasmons in X-Ray Emission Bands
(b) Vol. **89**, No. 1, p. K21—K24 (1978).
See also *13.5.2*, p. *198*; 20, p. 317; 21.6, p. 404.

G. Heiland and D. Kohl
Interpretation of Surface Phenomena on ZnO by the Compensation Model
(a) Vol. **49**, No. 1, p. 27—37 (1978).
See also 13.4, p. 180; *14.3.2*, p. *231*; 14.3.4, p. 244; *16*, p. *265*; 22.4, p. 462.

S. Lehr and H. Pagnia
*Work Function of Irradiated Real and Cleaned Indium Antimonide Surfaces
(a) Vol. **49**, No. 1, p. 83—91 (1978).
See also 16, p. 265; 17, p. 270; 22.2.3, p. 453.

L. Risch
*Investigation of Surface Properties of CdS by Means of Photoreflectance in the Exciton Region
(a) Vol. **49**, No. 1, p. 177—182 (1978).
See also 16, p. 266; 20.1, p. 333; 22.4.1, p. 465.

V. S. Lysenko and A. N. Nazarov
Effect of Laser Beam Irradiation on the Electrophysical Properties of the Interface in Boron Ion Implanted MIS Structures
(a) Vol. **49**, No. 1, p. 405—411 (1978).
See also 11, p. 129; *14.3.3*, p. *236*; 22.1.2, p. 437.

L. Fritsche, H. Gollisch, and J. Noffke
*Dipole Barriers at Structured Metal Surfaces
(b) Vol. **89**, No. 2, p. 367—373 (1978).
See also 21, p. 363.

G. Kühnel, W. Siegel, and E. Ziegler
Metal–n-$ZnSiP_2$ Schottky Contacts
(a) Vol. **49**, No. 2, p. 705—711 (1978).
See also *14.3.4*, p. *245*; 22.8, p. 507.

K. I. Kirov and S. P. Alexandrova
*Some Sources of Error in Interface State Density Evaluation from Q-U Data in MOS Systems
(a) Vol. **49**, No. 2, p. 781—788 (1978).
See also *14.3.2*, p. *231*.

F. Schauer, V. Šmíd, and V. Žaludek
On the Contact Capacitance and the Field Effect in Chalcogenide Glasses
(a) Vol. **49**, No. 2, p. K157—K161 (1978).
See also *2*, p. *51*; *14.3.4*, p. *245*.

H. J. de Bruin and S. P. S. Badwal
*Faradaic Reaction Kinetics in Solid Electrolytes by Impedance Dispersion Analysis
(a) Vol. **49**, No. 2, p. K181—K184 (1978).
See also *14.3.4*, p. *246*; 22.6, p. 497.

A. Čížek, M. Šob, and I. Ya. Dekhtyar
*Investigation of the Surface of Vanadium Single Crystals by Positron Annihilation
(b) Vol. **90**, No. 1, p. K1–K3 (1978).
See also 21.1, p. 380.

J. von Wienskowski
*Oxidation of Cleaved Si(111) Surfaces
(b) Vol. **90**, No. 1, p. K9–K12 (1978).
See also 22.1.2, p. 438.

A. N. Cherkasov, A. A. Galkin, and L. T. Tsymbal
*W and Mo RF Surface Impedance. New Type of Oscillations
(b) Vol. **90**. No. 1, p. K17–K20 (1978).
See also *14.1*, p. *211*; 19, p. 313; 21, p. 364.

H. Flietner
*Schottky-Barrier Theory and the Complex Band Structure
(b) Vol. **90**, No. 1, p. K27–K30 (1978).
See also *14.3.2*, p. *231*; 22, p. 419.

A. Wolkenberg
Energetic Barrier Height Hypothesis on the Silicon–Electrolyte and Gallium Arsenide–Electrolyte Interfaces
(a) Vol. **50**, No. 1, p. 249–256 (1978).
See also *14.3.2*, p. *232*; *14.3.3*, p. *238*; 22.1.2, p. 438; 22.2.1, p. 449.

L. Rothkirch, R. Herrmann, H. Dwelk, and U. Preppernau
Surface Impedance Investigations of Superconducting Niobium
(b) Vol. **90**, No. 2, p. 517–524 (1978).
See also *14.2*, p. *216*; 21, p. 365.

L. Rothkirch
*Magnetic Surface States in Superconductors
(b) Vol. **90**. No. 2, p. K115–K119 (1978).
See also *14.2*, p. *216*; 21, p. 366.

K. Masuda
*On the Electronic Structure of Ordered Adlayers
(b) Vol. **90**, No. 2, p. K179–K183 (1978).
See also 21, p. 366.

J. Wortmann
*Upon the Photostimulated Electron Emission from Tribomechanically Activated Aluminium
(a) Vol. **50**, No. 2, p. 509–516 (1978).
See also *17*, p. *271*; 21, p. 366.

V. F. Kiselev, V. A. Matveev, and R. V. Prudnikov
*The Proton Mechanism of Neutralization of Surface Recombination and Fast Capture Centres on Germanium
(a) Vol. **50**, No. 2, p. 739–744 (1978).
See also 22.1.1, p. 428.

13.4 Impurity and Defect States

Review Articles

P. A. Thrower and R. M. Mayer
Point Defects and Self-Diffusion in Graphite
(a) Vol. **47**, No. 1, p. 11—37 (1978).
See also *9*, p. *91*; *10.2*, p. *109*; 22.1.3, p. 441.

P. C. Mathur, R. Shyam, and Sushil Jain
Transport Phenomena in III–V Compound Semiconductors
(a) Vol. **50**, No. 1, p. 11—40 (1978).
See also 6, p. 64; *13.1*, p. *150*; *14.3*, p. *217*; 20.1, p. 319; 22.2, p. 443.

Original Papers and Short Notes

I. P. Zvyagin
Comments on the Theory of Hopping Conduction
(b) Vol. **85**, No. 1, p. 105—109 (1978).
See also *14.3*, p. *217*; 22, p. 408.

H. Michel and M. Wagner
*On the Use of Generalized Langevin Equations in Phonon Transport Problems
(b) Vol. **85**, No. 1, p. 195—201 (1978).
See also *6*, p. *64*.

R. Pässler
*Temperature Dependences of the Nonradiative Multiphonon Carrier Capture and Ejection Properties of Deep Traps in Semiconductors (I): Theoretical Results
(b) Vol. **85**, No. 1, p. 203—215 (1978).
See also 22, p. 408.

D. Suisky, R. Boyn, R. Enderlein, and B. Biering
First-Harmonic Electromodulation in Uniaxial Crystals
(b) Vol. **85**, No. 1, p. 217—224 (1978).
See also *14.3.4*, p. *240*; *20.1*, p. *319*; 22.4.1, p. 463.

C. Norris and G. P. Williams
*Photoelectron Measurements of α-Phase Alloys CuZn, CuGa, and CuGe
(b) Vol. **85**, No. 1, p. 325—330 (1978).
See also *17*, p. *269*; 21.1, p. 368.

A. V. Shendrik and D. M. Yudin
*A New Concept of the Model for the Paramagnetic E′-Centre in the SiO_{2-x} System
(b) Vol. **85**, No. 1, p. 343—349 (1978).
See also 2, p. 42; *19*, p. *303*; 22.8.1, p. 510.

G. Lautenschläger and E. Mrosan
*Change in Electronic Charge Density around 3d-Transition Metal Impurities in Al and Cu
(b) Vol. **85**, No. 1, p. K19—K21 (1978).
See also 21, p. 352; 21.1, p. 368.

L. Bosi and M. Nimis
*New Lifetime Data for the Excited State of the M-Centre in Alkali Halides
(b) Vol. **85**, No. 1, p. K27—K29 (1978).
Erratum
(b) Vol. **89**, No. 1, p. K107 (1978).
See also *10.2*, p. *109*; 22.5.2, p. 478.

K. Hoffmann and D. Pöss
Electron Spin Resonance Investigations on Lithium Fluoride Doped with Titanium
(a) Vol. **45**, No. 1, p. 263—268 (1978).
See also *19*, p. *303*; 22.5.2, p. 478.

Hj. Bernhardt
Studies of the Colour of Lead Molybdate Crystals
(a) Vol. **45**, No. 1, p. 353—357 (1978).
See also *20.1*, p. *321*; *20.3*, p. *341*; 22.8.1, p. 510.

I. B. Barkan, S. I. Marennikov, and M. V. Entin
*Dynamic Optical Storage in $LiNbO_3$ Crystals
(a) Vol. **45**, No. 1, p. K17—K22 (1978).
See also 14.4.1, p. 253; 16, p. 262; *20.1*, p. *321*; 22.8.1, p. 510.

R. M. Mehra, A. K. Kathuria, Radhey Shyam, A. L. Dawar, and P. C. Mathur
*Determination of the Density of Localized States in a-Se Films
(a) Vol. **45**, No. 1, p. K59—K62 (1978).
See also *2*, p. *43*; 14.3.1, p. 226; 22.1.3, p. 441.

L. Mărculescu, N. Vlahovici, and T. Velula
*On the Decay of the Luminescence Emission of Pb^{2+} Centres in Alkali Halide Crystals
(b) Vol. **85**, No. 2, p. 439—443 (1978).
Erratum
(b) Vol. **89**, No. 1, p. 335 (1978).
See also *20.3*, p. *341*; 22.5.2, p. 478.

M. Höhne
EPR of Lithium in Mechanically Affected Silicon
(b) Vol. **85**, No. 2, p. 525—534 (1978).
See also 1.6, p. 40; *19*, p. *304*; 22.1.2, p. 429.

D. Boulanger and M.-S. Martin
Crystal Growth and EPR Study of Cr^+-Doped MgTe Single Crystals
(b) Vol. **85** No. 2, p. 597—601 (1978).
See also 3, p. 54; *19*, p. *304*; 22.4.3, p. 470.

M. Hamera
*Calculation of the g-Shift and Superhyperfine Structure Tensor of Mn^{2+} Ion in $A_{II}B_{IV}$ Tetrahedral Semiconductors
(b) Vol. **85**, No. 2, p. 645—655 (1978).
See also 19, p. 304; 22.4, p. 461.

S. Marklund
*Electron States Associated with the Core Region of the 60° Dislocation in Silicon and Germanium
(b) Vol. **85**, No. 2, p. 673—681 (1978).
See also 22.1.1, p. 424; 22.1.2, p. 429.

E. Mrosan and G. Lehmann
*Charge Density Perturbation around Impurities in Metals
(b) Vol. **85**, No. 2, p. K85—K88 (1978).
See also 21, p. 354.

S. Winter
*A Comment on the Optical Absorption in Germanium with 60°-Dislocations
(b) Vol. **85**, No. 2, p. K95—K98 (1978).
See also 10.2, p. 110; *20.1*, p. *321*; 22.1.1, p. 424.

A. Frey
*Composition Dependence of Nitrogen and Tellurium Bound State Energies in $In_{1-x}Ga_xP$
(b) Vol. **85**, No. 2, p. K107—K110 (1978).
See also *20.3*, p. *342*; 22.2.4, p. 456.

H. J. Eichler, Ch. Hartig, and J. Knof
Laser Induced Gratings in CdS
(a) Vol. **45**, No. 2, p. 433—438 (1978).
See also *20.2*, p. *339*; 22.4.1, p. 463.

G. C. Stevens and D. Bloor
ESR Studies of a Diacetylene Polymer (I): Partially Crystalline Polymer Extracts
(a) Vol. **45**, No. 2, p. 483—491 (1978).
See also *19*, p. *304*; 22.9, p. 523.

V. C. Kieu
*Current Controlled Negative Resistance in GaAs at Low Temperatures
(a) Vol. **45**, No. 2, p. 571—578 (1978).
See also *14.3.4*, p. *240*; 22.2.1, p. 445.

A. N. Zyuganov, Yu. G. Pismennyi, S. V. Svechnikov, and P. S. Smertenko
*Regime of Contact Emission Limited Current with Weak Harmonic Distortion of Steady Bias
(a) Vol. **45**, No. 2, p. 631—638 (1978).
See also *14.3.4*, p. *240*; 22, p. 410.

K. D. Glinchuk and A. V. Prokhorovich
*The Role of Different Local Centres in the Determination of the Concentration Dependence of the Intrinsic Emission Intensity in n-Type GaAs
(a) Vol. **45**, No. 2, p. K91—K94 (1978).
See also *20.3*, p. *342*; 22.2.1, p. 445.

P. Pipinys and A. Rimeika
*Current Pulses in ZnS:Cu Crystals Stimulated by Electric Field at Low Temperatures
(a) Vol. **45**, No. 2, p. K113—K116 (1978).
See also *6*, p. *66*; *14.3.4*, p. *240*; 22.4.1, p. 463.

Ch. Schnittler
Some New Aspects of the SCLC Characteristic of Homogeneous Insulators with Traps
(a) Vol. **45**, No. 2, p. K179—K182 (1978).
See also *14.3.4*, p. *241*; 22, p. 410.

A. P. Radlinski

*On the Properties of Cobalt Impurities in Zinc Selenide Crystals (II): The Optical Properties in Near-Infrared. $^4A_2(^4F) \leftrightarrow {}^4T_2(^4F)$ Transitions
(b) Vol. **86**, No. 1, p. 41–46 (1978).
See also 20.1, p. 321; 20.3, p. 342; 22.4.2, p. 468.

M. Höhne

Norandom Strain in "Crushed" Silicon. EPR of Thermally Excited Lithium Donors
(b) Vol. **86**, No. 1, p. 119–126 (1978).
See also 1.6, p. 40; *19*, p. *305*; 22.1.2, p. 430.

M. Schwab

Phase Transition in the Mixed Crystals $Tb_pY_{1-p}AsO_4$
(b) Vol. **86**, No. 1, p. 195–203 (1978).
See also *1.2*, p. *16*; 18.4, p. 298; 20.1, p. 322; 22.8.1, p. 511.

E. Mrosan, Ch. Schober, and G. Lehmann

Multiple Scattering Theory for Magnetoresistance of Localized Impurities in Metals
(b) Vol. **86**, No. 1, p. 249–256 (1978).
See also *14.1*, p. *206*; 18, p. 273; 21, p. 355.

H. Neubrand

ESR from Boron in Silicon at Zero and Small External Stress (I): Line Positions and Line Structure
(b) Vol. **86**, No. 1, p. 269–275 (1978).
See also *19*, p. *305*; 22.1.2, p. 430.

V. Gomathy, C. Basu, and U. S. Ghosh

*Configuration Interaction and Intensity of Crystal Field Spectra of the Tetrahedral Ni^{2+} Ion
(b) Vol. **86**, No. 1, p. 379–384 (1978).
See also 20.1, p. 322; 22.4, p. 461.

R. Pässler

*Intermediate-Temperature Dependences of the Nonradiative Multiphonon Carrier-Capture Cross-Section of Deep Traps in Semiconductors
(b) Vol. **86**, No. 1, p. K39–K44 (1978).
See also 22, p. 410; 22.2.1, p. 445.

R. Pässler

*A Simple Approximation for the Temperature Dependences of the Nonradiative Multiphonon Carrier–Capture Cross-Section of Deep Traps in Semiconductors
(b) Vol. **86**, No. 1, p. K45–K47 (1978).
See also 22, p. 411.

V. Drchal and J. Kudrnovský

*The T-Matrix Approximation for the Anderson Model: A Self-Consistent Solution
(b) Vol. **86**, No. 1, p. K89–K92 (1978).
See also 21, p. 356.

V. I. Ivanov-Omskii, V. A. Maltseva, A. D. Britov, and S. D. Sivachenko
*Photo- and Cathodoluminescence of $Cd_{0.3}Hg_{0.7}Te$ Alloys
(a) Vol. **46**, No. 1, p. 77—80 (1978).
See also *20.3*, p. *343*; 22.4.4, p. 472.

P. Gaworzewski and H. Riemann
On the Effect of Heat Treatment upon the Charge Carrier Concentration of Dislocation-Free Floating-Zoned Silicon Crystals
(a) Vol. **46**, No. 1, p. 117—124 (1978).
See also *14.3*, p. *219*; 22.1.2, p. 431.

G. C. Stevens and D. Bloor
ESR Studies of a Diacetylene Polymer (II): Spin Concentrations and *g*-Values for Single Crystal Samples
(a) Vol. **46**, No. 1, p. 141—146 (1978).
See also *19*, p. *305*; 22.9, p. 523.

R. W. A. Hillhouse and J. Woods
Plasma Growth of Rutile Crystals and Their Photoelectronic Properties
(a) Vol. **46**, No. 1, p. 163—172 (1978).
See also *3*, p. *54*; 16, p. 263; 22.6, p. 492.

P. L. F. Hemment and R. W. Bicknell
1.5 MeV H^+ and He^+ Irradiations of Arsenic Diffused Layers in ⟨100⟩ Silicon
(a) Vol. **46**, No. 1, p. 225—231 (1978).
See also *11*, p. *125*; 14.3, p. 219; 22.1.2, p. 431.

C. Lawther and J. Woods
*The Calculation of Donor Densities from *C–U* Measurements on MIS ZnSe Diodes
(a) Vol. **46**, No. 1, p. 245—250 (1978).
See also *14.3.1*, p. *226*; 22.4.2, p. 468.

S. Nešpůrek, J. Obrda, and J. Sworakowski
A Study of Traps for Current Carriers in Organic Solids. N,N′-diphenyl-p-phenylenediamine
(a) Vol. **46**, No. 1, p. 273—280 (1978).
See also *14.3.4*, p. *241*; 22.9, p. 524.

E. A. Tsvetkov
*Calculation of the Optical Fluorescence Intensities and Crystal Field Parameters of Nd^{3+} in Calcium Fluorophosphate
(a) Vol. **46**, No. 1, p. 311—314 (1978).
See also 20.2, p. 339; 20.3, p. 343; 22.8, p. 504.

V. N. Kadchenko and M. Elango
*Diffusion Parameters of Hot Holes Created by Ionizing Radiation in NaCl(Ag)
(a) Vol. **46**, No. 1, p. 315—319 (1978).
See also *20.3*, p. *343*; 22.5.2, p. 479.

I. M. Kotina, S. R. Novikov, and T. I. Pirozhkova
*On the Determination of the Parameters of Impurity Centres from the Thermally Stimulated Current of p–n Junctions
(a) Vol. **46**, No. 1, p. K21—K23 (1978).
See also *14.3.3*, p. *234*; 22, p. 411.

M. E. Foglio
*Statistical Mechanics of Impurity States (I): General Treatment
(b) Vol. **86**, No. 2, p. 459–469 (1978).

E. Gerlach and M. Rautenberg
*Ionized Impurity Scattering in Semiconductors
(b) Vol. **86**, No. 2, p. 479–482 (1978).
See also *14.3*, p. *219*; 22, p. 411.

G. Appold, H. Pascher, R. Ebert, U. Steigenberger, and M. von Ortenberg
*Determination of the Effective g-Value of Electrons in n-Type InSb by Spin-Flip Raman Scattering and Electric-Dipole-Excited Electron Spin Resonance
(b) Vol. **86**, No. 2, p. 557–561 (1978).
See also 19, p. 306; 20.1, p. 324; 22.2.3, p. 452.

J. Kossut
*The Disorder Scattering in Zincblende Narrow-Gap Semiconducting Mixed Crystals
(b) Vol. **86**, No. 2, p. 593–601 (1978).
See also *13.1*, p. *154*; 14.3, p. 219; 22.4.4, p. 472.

V. Hizhnyakov and S. Zazubovich
*Vacancy-Induced Splitting of Excited States and the Structure of $Sn^{2+}v_c^-$ Centres in Alkali Halides
(b) Vol. **86**, No. 2, p. 733–739 (1978).
See also *20.1*, p. *324*; 22.5.2, p. 480.

G. C. Stevens and D. Bloor
ESR Studies of a Diacetylene Polymer (III): Thermal History Effects and Temperature Dependence of Polycrystalline Samples
(a) Vol. **46**, No. 2, p. 619–626 (1978).
See also *19*, p. *306*; 22.9, p. 524.

B. R. Sethi and P. C. Mathur
*Effect of Heavy Doping of ZnSe Crystals with Indium in Creating Compensating Acceptors
(a) Vol. **46**, No. 2, p. 717–721 (1978).
See also *14.3*, p. *220*; 22.4.2, p. 468.

M. E. Foglio
*Statistical Mechanics of Impurity States (II): Applications
(b) Vol. **87**, No. 1, p. 87–98 (1978).

R. Mansfield and G. Azhdarov
*Negative Magnetoresistance in Indium Arsenide in the Extreme Quantum Limit
(b) Vol. **87**, No. 1, p. 163–167 (1978).
See also *14.3.4*, p. *242*; 22.2.1, p. 446.

R. Bindemann, R. Schwabe, and T. Hänsel
*The "Quasidirect" Radiative Recombination of Free Holes at Neutral Shallow Donors in GaP
(b) Vol. **87**, No. 1, p. 169–177 (1978).
See also *20.3*, p. *344*; 22.2.2, p. 450.

C. PEDRINI
*Cu^+ Luminescent Centre in LiCl Single Crystals
(b) Vol. **87**, No. 1, p. 273–286 (1978).
See also *20.3*, p. *344*; 22.5.2, p. 481.

V. NEKVASIL
*The Crystal Field for Nd^{3+} in Garnets
(b) Vol. **87**, No. 1, p. 317–323 (1978).
See also *20.1*, p. *326*; 22.8.2, p. 519.

S. L. A. DE QUEIROZ, B. KOILLER, B. MAFFEO, and H. S. BRANDI
*A Cluster-Bethe Lattice Treatment for the F-Center in Alkali Halides
(b) Vol. **87**, No. 1, p. 351–359 (1978).
See also *20.1*, p. *326*; 22.5.2, p. 481.

E. MROSAN and G. LEHMANN
*Electronic States of 3d-Transition Metal Impurities in Aluminium
(b) Vol. **87**, No. 1, p. K21–K24 (1978).
See also 21, p. 358.

P. KRISPIN and J. MAEGE
Capacitance–Frequency Dispersion and Electroluminescence Efficiency of GaP p–n Junctions
(a) Vol. **47**, No. 1, p. 91–97 (1978).
See also *14.3.3*, p. *235*; *20.3*, p. *344*; 22.2.2, p. 450.

B. BAILLAND and P. PINARD
*Interfacial Tunnel Interaction in MOS Elements in Depletion
(a) Vol. **47**, No. 1, p. 251–258 (1978).
See also *14.3.2*, p. *230*; 22.6, p. 493.

C. E. WICKERSHAM and J. E. GREENE
The Effect of Substrate Bias on the Electrical and Optical Properties of In_2O_3 Films Grown by RF Sputtering
(a) Vol. **47**, No. 1, p. 329–337 (1978).
See also *14.3.1*, p. *226*; *20.1*, p. *327*; 22.6, p. 493.

M. C. WINTERSGILL and P. D. TOWNSEND
The Effectiveness of the Implantation of Oxygen in Increasing the Luminescence from LiF: Mg:Ti
(a) Vol. **47**, No. 1, p. K67–K69 (1978).
See also *20.3*, p. *344*; 22.5.2, p. 481.

M. A. MEKHTIEV and S. M. SEID-RZAEVA
*Quasi-Local Acceptor Levels and Electron Mobility in $Hg_{1-x}Cd_xTe$
(b) Vol. **87**, No. 2, p. 447–454 (1978).
See also *14.3*, p. *221*; 22.4.4, p. 473.

C. JACCARD and M. ECABERT
Optical Detection of Nuclear Magnetic Resonance with Distant F-Centre Pairs in KCl
(b) Vol. **87**, No. 2, p. 497–505 (1978).
See also *19*, p. *308*; *20.3*, p. *344*; 22.5.2, p. 481.

A. Andreeff, L. P. Kaun, B. Lippold, W. Matz, N. I. Moreva, and K. Walther
*Crystal Field Levels of Pr^{3+} in $PrAl_3$ Studied by Inelastic Neutron Scattering
(b) Vol. 87, No. 2, p. 535—541 (1978).
See also 21.4, p. 397.

R. Droste, R. Gmeiner, and R. Geick
*Electronic Absorption of $SrCl_2$:Fe^{2+} and $SrCl_2$:Co^{2+}
(b) Vol. 87, No. 2, p. 543—552 (1978).
See also *20.1*, p. *327*; 22.5.3, p. 487.

S. W. Biernacki
*Jahn-Teller Coupling of Cr^{2+} Ion in ZnS, ZnSe, and ZnTe
(b) Vol. 87, No. 2, p. 607—612 (1978).
See also *20.1*, p. *327*; 22.4.1, p. 464; 22.4.2, p. 468; 22.4.3, p. 470.

J. Spałek, A. M. Oleś, and K. A. Chao
*The Effective Magnetic Interactions between Impurity and Conduction Electrons for the Wolff Model
(b) Vol. 87, No. 2, p. 625—635 (1978).
See also 13.1, p. 155; *18.4*, p. *300*.

K. O. Kechechyan and A. A. Kirakossyan
*Impurity Centres in Semiconductors Containing Screw Dislocations
(b) Vol. 87, No. 2, p. K123—K126 (1978).
See also *10.2*, p. *113*; 22, p. 414.

A. L. Dawar, K. V. Krishna, O. P. Taneja, and P. C. Mathur
*Electrical Transport Properties of Polycrystalline p-Type InSb Films
(a) Vol. 47, No. 2, p. 375—378 (1978).
See also *14.3.1*, p. *227*; 22.2.3, p. 453.

W. Ecke, A. A. Lebedev, and A. T. Mamadalimov
*Deep Defect States in Evaporated Heteroepitaxial Silicon Films
(a) Vol. 47, No. 2, p. 507—512 (1978).
See also *14.3.1*, p. *227*; 22.1.2, p. 433.

V. Platikanova and J. Malinowski
Neutralization of Trapped Photoholes in AgBr
(a) Vol. 47, No. 2, p. 683—690 (1978).
See also *10.2*, p. *114*; 22.5.1, p. 477.

B. R. Sethi, P. L. Talwar, O. P. Sharma, and P. C. Mathur
*Effect of Indium Doping and Zinc Annealing on the Electrical Transport Properties of Zinc Selenide Single Crystals
(a) Vol. 47, No. 2, p. 699—702 (1978).
See also *14.3*, p. *221*; 22.4.2, p. 468.

S. Radhakrishna and T. Bhaskar Rao
Optical Absorption and ESR Studies of Nickel Ions in Sodium Zinc Sulphate Tetrahydrate
(a) Vol. 47, No. 2, p. 723—729 (1978).
See also *19*, p. *309*; *20.1*, p. *328*; 22.8.1, p. 514.

B. Pohoryles and R. Piotrzkowski
Effect of Hydrostatic Pressure on the Photoconductivity of Plastically Deformed n-Type Germanium
(a) Vol. **47**, No. 2, p. K115–K117 (1978).
See also *16*, p. *264*; 22.1.1, p. 426.

E. Mlitzke
*On the Original Method of Balarin and Zetzsche
(a) Vol. **47**, No. 2, p. K181–K182 (1978).
See also *20.3*, p. *345*; 22, p. 414.

N. B. Yunusov and V. P. Zentsov
*Electronic Structure and ESR of Ti^{2+} in Fluorite Type Crystals
(b) Vol. **88**, No. 1, p. 87–97 (1978).
See also 19, p. 309; 22.5.3, p. 488.

W. Kraak, G. Oelgart, G. Schneider, and R. Herrmann
*The Semiconductor–Semimetal Transition in $Bi_{1-x}Sb_x$ Alloys with $x \geqq 0.22$
(b) Vol. **88**, No. 1, p. 105–110 (1978).
See also 14.1, p. 208; *14.3*, p. *221*, 22.7, p. 502.

N. P. Gavaleshko, P. N. Gorley, P. V. Kavyuk, L. S. Solonchuk, I. I. Tarasyuk, and V. A. Shenderovskii
*Transverse and Longitudinal Nernst-Ettingshausen Effect in p-Te
(b) Vol. **88**, No. 1, p. 179–188 (1978).
See also 14.3, p. 222; *15*, p. *260*; 22.1.3, p. 441.

D. P. Bhattacharya and Z. S. Kachlishvili
*Faraday Rotation in Presence of Attractive Traps
(b) Vol. **88**, No. 1, p. 283–287 (1978).
See also *14.3.4*, p. *243*; 18, p. 274; 22.1.1, p. 426.

S. Fakioğlu
Susceptibility Calculations on Spin-Doped Crystals
(b) Vol. **88**, No. 1, p. 301–308 (1978).
See also *18.1*, p. *278*; 22.8.1, p. 514.

A. A. Kirakossyan and K. O. Kechechyan
*Interdislocation Interaction on Account of Carriers Bound with Dislocations
(b) Vol. **88**, No. 1, p. K15–K18 (1978).
See also *10.2*, p. *115*.

M. Egée
Effect of a Direct Electric Field on the Photoluminescence Processes and Model of the Luminogen Centre in ZnS:Cu
(a) Vol. **48**, No. 1, p. 89–98 (1978).
See also 14.3, p. 222; *20.3*, p. *345*; 22.4.1, p. 465.

F. Holuj and A. Jesmanowicz
EPR of Fe^{3+} in Andalusite and Kyanite at V-Band and the Pair Spectra
(a) Vol. **48**, No. 1, p. 191–198 (1978).
See also *19*, p. *310*; 22.8, p. 506.

A. V. Bobyl, G. S. Pekar, and M. K. Sheinkman
*Dark Conductivity and Photoconductivity of Nickel-Doped CdS Single Crystals
(a) Vol. 48, No. 1, p. 249—256 (1978).
See also *16*, p. *264*; 22.4.1, p. 465.

T. I. Kolchenko and V. M. Lomako
Removal and Scattering of Charge Carriers by Defect Clusters in Semiconductors
(a) Vol. 48, No. 1, p. 263—269 (1978).
See also *11*, p. *128*; 14.3, p. 222; 22.2.1, p. 447.

P. M. Mooney, R. T. Young, J. Karins, Y. H. Lee, and J. W. Corbett
*Defects in Laser Damaged Silicon Observed by DLTS
(a) Vol. 48, No. 1, p. K31—K34 (1978).
See also *11*, p. *128*; 22.1.2, p. 435.

P. Höschl, P. Moravec, A. Martinaitis, and A. Sakalas
*Study of Photo-Hall Effect on Single Crystals of Cadmium Telluride Doped with Chlorine
(a) Vol. 48, No. 1, p. K43—K46 (1978).
See also *16*, p. *265*; 22.4.3, p. 470.

E. Ziegler, W. Siegel, G. Kühnel, and E. Buhrig
Incorporation of Gallium in $ZnSiP_2$
(a) Vol. 48, No. 1, p. K63—K65 (1978).
See also *14.3*, p. *222*; 22.8, p. 506.

V. E. Petrashen, Yu. V. Yablokov, and R. L. Davidovich
*Electron Paramagnetic Resonance Study of Cu^{2+} in $K_2Zn(ZrF_6)_2 \cdot 6\,H_2O$
(b) Vol. 88, No. 2, p. 439—443 (1978).
See also *19*, p. *310*; 22.8, p. 506.

A. E. Kalnin and M. F. Trinkler
*Spectroscopy of Mixed Alkali Halide Crystals Activated by Thallium
(b) Vol. 88, No. 2, p. 445—455 (1978).
See also 20.1, p. 330; *20.3*, p. *346*; 22.5.4, p. 489.

Ch. Hoentzsch and J. M. Spaeth
*Localized Vibrations of Substitutional Atomic Hydrogen in KCl and RbCl Studied by ENDOR
(b) Vol. 88, No. 2, p. 581—589 (1978).
See also *19*, p. *310*; 22.5.2, p. 482.

F. Galeskii, A. E. Yunovich, K. H. Herrmann, H. Kostial, I. Rechenberg, and P. Schäfer
Stimulated Emission, Absorption Spectra, and Recombination in Epitaxial PbTe
(b) Vol. 88, No. 2, p. 675—681 (1978).
See also 20.1, p. 330; *20.2*, p. *339*; 22, p. 415.

B. Stébé and G. Munschy
*Binding of an Exciton to a Neutral Acceptor
(b) Vol. 88, No. 2, p. 713—725 (1978).
See also *13.5.1*, p. *193*; 14.3, p. 223; 22.2.1, p. 447.

S. D. Zvonkov, V. D. Sedykh, Yu. S. Stark, and A. Sh. Chavchanidze
*Electronic Structure and Hyperfine Interactions in α-Fe-Based Dilute Solid Solutions
(b) Vol. **88**, No. 2, p. K103—K107 (1978).
See also 21.1.1, p. **388**.

N. A. Poklonskii, V. F. Stelmakh, V. D. Tkachev, and S. V. Voitikov
*Screening of Electrostatic Fields in Semiconductors by Multicharged Defects
(b) Vol. **88**, No. 2, p. K165—K168 (1978).
See also *14.3*, p. *223*; 22, p. **416**.

A. S. Filipchenko, T. B. Kurenkeev, L. P. Bolshakov, and E. I. Chaikina
Photoluminescence of Heavily Doped n-Type Gallium Antimonide
(a) Vol. **48**, No. 2, p. 281—285 (1978).
See also *20.3*, p. *346*; 22.2.3, p. **453**.

C. Manfredotti, A. M. Mancini, A. Rizzo, R. Murri, and L. Vasanelli
*Electrical Properties of GaS_xSe_{1-x} Solid Solutions
(a) Vol. **48**, No. 2, p. 293—296 (1978).
See also *14.3*, p. *223*; 22.3.1, p. **460**.

I. S. Gorban, G. A. Grishchenko, A. P. Sakalas, A. S. Sodeika, I. I. Tychina, and A. K. Tkachenko
*Electrical Properties of ZnP_2 Single Crystals of Tetragonal and Monoclinic Modifications
(a) Vol. **48**, No. 2, p. 329—334 (1978).
See also *14.3*, p. *223*; 22.3, p. **458**.

Ch. Schnittler
A Simplified Approach to the SCLC Characteristic of Homogeneous Solids with Traps Distributed in Energy
(a) Vol. **48**, No. 2, p. 357—368 (1978).
See also *14.3.4*, p. *244*; 22, p. **416**.

K. D. Glinchuk, A. V. Prokhorovich, V. E. Rodionov, and V. I. Vovnenko
*Study of Non-Linear Extrinsic Luminescence in GaAs
(a) Vol. **48**, No. 2, p. 593—602 (1978).
See also *20.3*, p. *346*; 22.2.1, p. **448**.

A. S. Filipchenko, L. P. Bolshakov, A. Naurizbaev, A. G. Braginskaya, and A. N. Popkov
*Moss-Burstein Effect in n-Type Gallium Antimonide Crystals
(a) Vol. **48**, No. 2, p. K115—K117 (1978).
See also 20.1, p. **331**; 22.2.3, p. **453**.

C. A. J. Ammerlaan and J. C. Wolfrat
*Extended-Hückel-Theory Calculations for the Positive Divacancy in Silicon
(b) Vol. **89**, No. 1, p. 85—94 (1978).
See also 22.1.2, p. **436**.

C. S. Lam and Y. P. Varshni
*Binding Energies of Positron to F and F′ Colour Centres in Alkali Halides
(b) Vol. **89**, No. 1, p. 103—108 (1978).
See also *10.2*, p. *117*; 22.5.2, p. **483**.

F. Lohse, G. Reuter, and J. M. Spaeth
Optical Investigation of Atomic Hydrogen Centres in Alkali Chlorides Doped with I^- or Br^-
(b) Vol. **89**, No. 1, p. 109–116 (1978).
See also *10.2*, p. *117*; *20.1*, p. *332*; 22.5.2, p. 483.

L. Bosi, A. Longoni, and M. Nimis
Lifetime of Color Centers in NaBr
(b) Vol. **89**, No. 1, p. 221–224 (1978).
See also *10.2*, p. *117*; *20.3*, p. *347*; 22.5.2, p. 484.

B. Milsch and W. Windsch
Electron Paramagnetic Resonance of Mn^{2+} in Guanidinium Aluminium Sulfate Hexahydrate
(b) Vol. **89**, No. 1, p. 241–251 (1978).
See also *14.4.1*, p. *256*; *19*, p. *311*; 22.8.1, p. 515.

R. A. Serway, F. H. Yang, and S. A. Marshall
Line Broadening in the ESR Absoption Spectra of Fe^{3+}, Cr^{3+}, and Gd^{3+} in the Oxide YAlG
(b) Vol. **89**, No. 1, p. 267–273 (1978).
See also *19*, p. *311*; 22.8.2, p. 520.

H. Bill
*Parametrization of the Measured Temperature Dependence of the Crystal Field Splitting in Axial ^{8}S State Ion Centers
(b) Vol. **89**, No. 1, p. K49–K52 (1978).
See also 19, p. 311; 22.5.4, p. 489.

U. Holland and F. Lüty
*Hydrostatic Pressure Effects on the Absorption and Emission of F_A(Li) Centers in KCl
(b) Vol. **89**, No. 1, p. K73–K77 (1978).
See also 12.2, p. 140; *20.3*, p. *347*; 22.5.2, p. 484.

G. Heiland and D. Kohl
Interpretation of Surface Phenomena on ZnO by the Compensation Model
(a) Vol. **49**, No. 1, p. 27–37 (1978).
See also 13.3, p. 167; *14.3.2*, p. *231*; 14.3.4, p. 244; *16*, p. *265*; 22.4, p. 462.

W. Hoyer, G. Lautz, and M. Heyen
Low-Frequency Current Noise of GaAs Epitaxial Layers between 4.4 and 300 K
(a) Vol. **49**, No. 1, p. 111–117 (1978).
Erratum
(a) Vol. **50**, No. 2, p. 771 (1978).
See also *14.3.1*, p. *228*; *14.3.3*, p. *236*; 22.2.1, p. 448.

I. Dudeck and R. Kassing
*Determination of the Temperature Dependence of the Capture Cross-Sections of the Gold Acceptor Level and of the Temperature of Current Filaments in Silicon pin Diodes
(a) Vol. **49**, No. 1, p. 153–161 (1978).
See also 14.3.3, p. 236; *14.3.4*, p. *244*; 22.1.2, p. 436.

A. S. Filipchenko, L. P. Bolshakov, and A. Naurizbaev
*Absorption in Indium Antimonide and Arsenide Crystals Doped with Various Admixtures
(a) Vol. **49**, No. 1, p. 183–188 (1978).
See also *20.1*, p. *333*; 22.2.1, p. 448; 22.2.3, p. 454.

A. V. MOSKALANAV and J. R. ZAKIS
*X-Ray Induced Paramagnetic States and Luminescence in As_2S_3 and As_2Se_3
(a) Vol. **49**, No. 1, p. 231–235 (1978).
See also 2, p. 49; 19, p. 312; *20.3*, p. *347*; 22, p. 417.

R. BINDEMANN, H. FISCHER, and K. KREHER
*The Spectral Dependence of the Photoneutralization Cross-Section of Oxygen in GaP Near the Threshold
(a) Vol. **49**, No. 1, p. 331–336 (1978).
See also 13.5.1, p. 194; *20.3*, p. *347*; 22.2.2, p. 451.

W. MYCIELSKI and A. LIPIŃSKI
*SCL Currents and Trap Distribution in Polycrystalline Tetracene
(a) Vol. **49**, No. 1, p. K41–K42 (1978).
See also *14.3.4*, p. *245*; 22.9, p. 527.

H. ATZMÜLLER and U. SCHRÖDER
*Theoretical Investigations of Haynes' Rule
(b) Vol. **89**, No. 2, p. 349–356 (1978).
See also *13.5.1*, p. *194*; 22, p. 418.

F. G. GÄRTNER and E. MOLLWO
IR Absorption of OH and OD Centres and OH/OH, OD/OD and OH/OD Complexes in Cu-Doped ZnO Single Crystals (I): Experimental Results
(b) Vol. **89**, No. 2, p. 381–388 (1978).
See also *20.1*, p. *333*; 22.4, p. 462.

R. BIEDERBICK, A. HOFSTAETTER, and A. SCHARMANN
*Temperature-Dependent Mn^{2+}-^6S State Splitting in Scheelites
(b) Vol. **89**, No. 2, p. 449–458 (1978).
See also *19*, p. *312*; 22.8.1, p. 516.

W. SENSKE, M. SONDERGELD, and H. J. QUEISSER
*Impurity Interaction on Excited Acceptor States in GaP
(b) Vol. **89**, No. 2, p. 467–474 (1978).
See also *20.3*, p. *348*; 22.2.2, p. 451.

B. PERTZSCH and U. RÖSSLER
*Stress Dependence of Shallow Acceptor States in Semiconductors with Zincblende Symmetry
(b) Vol. **89**, No. 2, p. 475–487 (1978).
See also 12.2, p. 140; 22.2, p. 444.

S. ASANO, N. YAMASHITA, and Y. NAKAO
*Luminescence of the Pb^{2+}-Ion Dimer Center in CaS and CaSe Phosphors
(b) Vol. **89**, No. 2, p. 663–673 (1978).
See also *20.3*, p. *348*; 22.4.1, p. 466; 22.4.2, p. 469.

S. ÖBERG
*Electronic States Associated with 60° Dislocations in Gallium Arsenide
(b) Vol. **89**, No. 2, p. K127–K130 (1978).
See also 22.2.1, p. 448.

I. K. KRISTENSEN
Decay of Photoluminescence in GaSe
(b) Vol. **89**, No. 2, p. K155—K157 (1978).
See also *20.3*, p. *348*; 22.3, p. 459.

V. KUMAR and R. KUMAR
*On the *g* and Hyperfine Interaction Parameters of Er^{3+} in Three Hydroxides
(b) Vol. **89**, No. 2, p. K195—K197 (1978).
See also 18.1, p. 279; 22.8, p. 507.

D. JÄGER, E. KÄHLER, and R. HEIDEMANN
Negative Conductance in SCLC Diodes with Traps
(a) Vol. **49**, No. 2, p. 431—435 (1978).
See also *14.3.3*, p. *236*; *14.3.4*, p. *245*; 22.1.2, p. 437.

B. RABIN, H. TABATABAI, and P. SIFFERT
*Capacitance of Cadmium Telluride Schottky Diodes
(a) Vol. **49**, No. 2, p. 577—584 (1978).
See also *14.3.3*, p. *237*; 22.4.3, p. 471.

B. BARANOV, L. DÄWERITZ, V. B. GUTAN, G. JUNGK, H. NEUMANN, and H. RAIDT
Growth and Properties of $Al_xGa_{1-x}N$ Epitaxial Layers
(a) Vol. **49**, No. 2, p. 629—636 (1978).
See also *1.5*, p. *39*; 14.3.1, p. 228; 20.1, p. 335; 20.3, p. 348; 22.2.4, p. 456.

R. MACH, W. GERICKE, H. TREPTOW, and W. LUDWIG
Direct Impact Excitation of Luminescent Centres in ZnSe Single Crystals
(a) Vol. **49**, No. 2, p. 667—674 (1978).
See also *14.3.3*, p. *237*; *20.3*, p. *349*; 22.4.2, p. 469.

V. N. BRUDNYI, M. A. KIROV, A. I. POTAPOV, I. K. POLUSHINA, V. D. PROCHUKHAN, and YU. V. RUD
Change of Electrical Properties in Electron Irradiated $CdGeAs_2$ Crystals
(a) Vol. **49**, No. 2, p. 761—765 (1978).
See also *11*, p. *130*; *14.3*, p. *224*; 22.8, p. 507.

V. M. HARUTYUNYAN and F. V. GASPARYAN
*Impedance of p^+–n–n^+ Structures Based on Semiconductors Compensated by Deep Doubly Ionized Centres
(a) Vol. **49**, No. 2, p. 797—801 (1978).
See also *14.3.3*, p. *237*; 22.1.2, p. 437.

G. RUDLOF, J. BECHERER, and H. GLAEFEKE
*Behaviour of the Fractional Glow Technique with First-Order Detrapping Processes, Traps Distributed in Energy or Frequency Factor
(a) Vol. **49**, No. 2, p. K121—K124 (1978).
See also *17*, p. *270*; 22, p. 419.

Z. S. KACHLISHVILI
*On a Certain Feature of Nonlinear Galvanomagnetic Characteristics
(a) Vol. **49**, No. 2, p. K135—K137 (1978).
See also *14.3.4*, p. *245*; 22, p. 419.

S. NEŠPŮREK and J. SWORAKOWSKI
A Differential Method of Analysis of Steady-State Space-Charge-Limited Currents: An Extension
(a) Vol. **49**, No. 2, p. K149—K152 (1978).
See also *14.3.4*, p. *245*; 22, p. 419.

E. ZIEGLER, W. SIEGEL, and G. KÜHNEL
Potential Barrier Scattering in n-Type Single Crystals of $ZnSiP_2$
(a) Vol. **49**, No. 2, p. K205—K208 (1978).
See also *14.3*, p. *224*; 22.8, p. 507.

F. G. GÄRTNER and E. MOLLWO
IR-Absorption of OH and OD Centres and OH/OH, OD/OD, and OH/OD Complexes in Cu-Doped ZnO Single Crystals (II): Discussion and Quantitative Interpretation of Models
(b) Vol. **90**, No. 1, p. 33—44 (1978).
See also *20.1*, p. *335*; 22.4, p. 462.

H. SCHREY and C. KLINGSHIRN
Biexcitons, Impurity Transitions, and Electron–Hole Plasma in Highly Excited II–VI Compounds?
(b) Vol. **90**, No. 1, p. 67—76 (1978).
See also *13.5.1*, p. *195*; *20.3*, p. *349*; 22.4.1, p. 466.

S. WINTER
*Electron States below the Conduction Band in Germanium Originating from Dissociated 60°-Dislocations
(b) Vol. **90**, No. 1, p. 289—293 (1978).
See also 22.1.1, p. 428.

H. NEUBRAND
ESR from Boron in Silicon at Zero and Small External Stress (II): Linewidth and Crystal Defects
(b) Vol. **90**, No. 1, p. 301—308 (1978).
See also *19*, p. *313*; 22.1.2, p. 437.

G. GOBSCH, H. HABERLANDT, H.-J. WECKNER, and J. REINHOLD
*Calculation of the g-Tensor and ^{29}Si Hyperfine Tensors of the E_1' Centre in Silicon Dioxide
(b) Vol. **90**, No. 1, p. 309—317 (1978).
See also *19*, p. *313*; 22.6, p. 497.

S. LIGENZA and H. PTASIEWICZ-BĄK
*Magneto-Electric Splitting of the Crystal Field Γ_2-Level of Cr^{3+} Ions in $MgCr_2O_4$ Studied by Neutron Spectroscopy
(b) Vol. **90**, No. 1, p. 319—326 (1978).
See also *18*, p. *275*; 22.8.2, p. 520.

YU. D. AREPEV, V. P. OLEINIK, E. A. SALKOV, and K. B. TOLPYGO
*The Anomalous Stark Shift of the Ground State Energy of a Donor–Acceptor Pair in an Electric Field
(b) Vol. **90**, No. 1, p. 355—360 (1978).
See also *20.3*, p. *349*; 22.4.1, p. 466.

J. K. Srivastava and B. W. Dale
Cross Relaxation Studies in the Cr^{3+}-Doped Magnetically Dilute Paramagnetic Al_2O_3–Fe_2O_3 System (I): Effect of Cr^{3+} Concentration
(b) Vol. **90**, No. 1, p. 391—399 (1978).
See also *5*, p. *62*; 22.6.1, p. 499.

H. Van Cong
Influence of Band-Tail Parameters on Electronic Specific Heat and Transport Coefficients in Heavily Doped Semiconductors
(b) Vol. **90**, No. 1, p. 401—407 (1978).
See also *8*, p. *89*; *14.3*, p. *224*; 22, p. 419.

V. I. Sokolov, N. A. Moskvina, and B. V. Melnikov
*Discovery of Exciton–Impurity Complexes in Zinc Selenide by Electroabsorption
(b) Vol. **90**, No. 1, p. K39—K42 (1978).
See also *13.5.1*, p. *196*; 14.3.4, p. 246; 20.1, p. 336; 22.4.2, p. 469.

R. Schöpke and E. Mrosan
*Influence of Multiple Scattering Effects on the Residual Resistivity of 3d-Transition Metal Impurities in Aluminium
(b) Vol. **90**, No. 1, p. K95—K98 (1978).
See also *14.1*, p. *212*; 21, p. 365.

M. I. Darby, M. N. Read, and K. N. R. Taylor
*The Energy of Interstitial Hydrogen in α-Zirconium
(a) Vol. **50**, No. 1, p. 203—211 (1978).
See also *10.1*, p. *107*; 21.1, p. 381.

P. Iacconi, D. Lapraz, and R. Caruba
Traps and Emission Centres in Thermoluminescent ZrO_2
(a) Vol. **50**, No. 1, p. 275—283 (1978).
See also 10.2, p. 121; *20.3*, p. *350*; 22.6, p. 497.

A. le Bloa, P. N. Favennec, and Y. Colin
*Determination of Complementary Transition Energy Threshold by Photocapacitance Measurements. Application to Oxygen Implanted GaAs
(a) Vol. **50**, No. 1, p. 311—322 (1978).
See also 14.3.3, p. 238; 22.2.1, p. 449.

V. B. Sharan, D. N. S. Srivastwa, and S. C. Sen
*Diffuse Absorption Spectrum of NH_4Cl + TlCl Mixed Phosphors
(a) Vol. **50**, No. 1, p. 323—327 (1978).
See also *20.1*, p. *336*; 22.5.4, p. 490.

Ľ. Hrivnák
*On the Switching Effect in Crystalline Semiconductors
(a) Vol. **50**, No. 1, p. K97—K99 (1978).
See also *14.3.4*, p. *246*; 22, p. 420.

S. Mantovani, E. Mazzega, S. Valeri, and U. del Pennino
*Edge Dislocation Energy Level in Silicon
(a) Vol. **50**, No. 1, p. K123—K126 (1978).
See also *10.2*, p. *121*; 22.1.2, p. 438.

J. K. Srivastava and B. W. Dale

Cross Relaxation Studies in the Cr^{3+}-Doped Magnetically Dilute Paramagnetic Al_2O_3–Fe_2O_3 System (II): Effect of Temperature and External Magnetic Field

(b) Vol. **90**, No. 2, p. 571—574 (1978).

See also *5*, p. *62*; 22.6.1, p. 499.

P. Petelenz

*Effect of the Mixing between Frenkel Excitons and Charge Transfer States on Charge Carrier Transfer in a Linear Molecular Crystal

(b) Vol. **90**, No. 2, p. 635—642 (1978).

See also *13.5.1*, p. *197*.

S. Benci, R. Capelletti, F. Fermi, M. Manfredi, J. Z. Damm, and E. Mugeński

Spectral and Time Decay Analysis of Photoluminescence Excited in the A Band Region in KCl:Pb

(b) Vol. **90**, No. 2, p. 657—666 (1978).

See also *20.1*, p. *337*; *20.3*, p. *350*; 22.5.2, p. 486.

V. B. Sharan, D. N. S. Srivastwa, and S. C. Sen

*Absorption, Excitation, and Fluorescence Spectra of Thallium-Activated Ammonium Bromide

(b) Vol. **90**, No. 2, p. 671—678 (1978).

See also 20.1, p. 337; 20.3, p. 350; 22.5, p. 476.

N. A. Kulagin and V. I. Tutlis

*Change of the Praseodymium Spectrum under the Free State-to-Crystal Transition

(b) Vol. **90**, No. 2, p. K109—K113 (1978).

See also 20.1, p. 337; 22, p. 421.

K.-A. Chao and A. Ferreira da Silva

*Matrix Elements in the Matsubara-Toyozawa Theory of Impurity Band

(b) Vol. **90**, No. 2, p. K153—K155 (1978).

See also 22, p. 421.

V. N. Brudnyi, D. I. Budnitskii, M. A. Krivov, R. V. Masagutova, V. D. Prochukhan, and Yu. V. Rud

*The Electrical and Optical Properties of 2.0 MeV Electron-Irradiated $ZnGeP_2$

(a) Vol. **50**, No. 2, p. 379—384 (1978).

See also 11, p. 131; *14.3*, p. *225*; *20.1*, p. *337*; 22.8, p. 508.

A. Roizès and R. Schuttler

Galvanomagnetic Effects in p-Type Silicon. Application to Electrically Active Impurity Contents

(a) Vol. **50**, No. 2, p. 399—409 (1978).

See also *14.3*, p. *225*; 22.1.2, p. 439.

P. T. Landsberg, D. J. Robbins, and E. Schöll

Threshold Switching as a Generation–Recombination Induced Non-Equilibrium Phase Transition

(a) Vol. **50**, No. 2, p. 423—426 (1978).

See also 2, p. 52; *14.3.4*, p. *247*; 22.1, p. 423.

A. Anedda, I. Garbato, F. Raga, and A. Serpi
Photoconductivity and Trap Distribution in $CdIn_2S_4$
(a) Vol. **50**, No. 2, p. 643–650 (1978).
See also *16*, p. *267*; 22.8, p. 508.

C. Lhermitte, D. Carles, C. Viger, and C. Vautier
Transient Photoconductivity Effects on n-Type Cadmium Telluride Films in Presence of a Uniform Distribution of Traps. Theoretical Model and Experimental Results
(a) Vol. **50**, No. 2, p. 651–659 (1978).
See also *16*, p. *267*; 22.4.3, p. 471.

S. Kawamura and B. S. H. Royce
*Thermally Stimulated Current Studies of Electron and Hole Traps in Single Crystal Al_2O_3
(a) Vol. **50**, No. 2, p. 669–677 (1978).
See also *14.3*, p. *225*; 22.6, p. 497.

A. A. Bukharaev and N. R. Yafaev
*Laser and Thermal Bleaching of Colour Centres in Sodium Borate Glasses
(a) Vol. **50**, No. 2, p. 711–716 (1978).
See also 2, p. 53; *20.1*, p. *338*; 22.8.1, p. 516.

V. V. Serdyuk, A. P. Chebanenko and G. G. Chemeresyuk
*Stromumschaltmechanismus in inhomogenen Zinkselenidkristallen
(Current Switching Mechanism in Inhomogeneous Zinc Selenide Crystals)
(a) Vol. **50**, No. 2, p. 717–722 (1978).
See also *14.3.4*, p. *247*; 16, p. 267; 22.4.2, p. 469.

A. A. Kaminskii, S. E. Sarkisov, Tran Ngoc, B. I. Denker, V. V. Osiko, and A. M. Prokhorov
Stimulated Emission Spectroscopy of Concentrated Lithium Neodymium Phosphate Glasses in $^4F_{3/2} \rightarrow {}^4I_{11/2}$ and $^4F_{3/2} \rightarrow {}^4I_{13/2}$ Transitions
(a) Vol. **50**, No. 2, p. 745–750 (1978).
See also *20.1*, p. *338*; *20.2*, p. *340*; 22.8.1, p. 516.

N. E. Korsunskaya, I. V. Markevich, T. V. Torchinskaya, and M. K. Sheinkman
*Some Peculiarities of Thermostimulated Conductivity and Optical Quenching of the Photocurrent in Crystals with the Auger Excitation of Recombination Centres
(a) Vol. **50**, No. 2, p. 767–770 (1978).
See also *14.3*, p. *225*; 16, p. 268; 22.4.1, p. 467.

A. S. Filipchenko, T. B. Kurenkeev, and T. Seisenbaev
Photoluminescence of Doped n-Type Indium Arsenide and Indium Antimonide Crystals
(a) Vol. **50**, No. 2, p. K251–K254 (1978).
See also *20.3*, p. *351*; 22.2.1, p. 449; 22.2.3, p. 455.

13.5 Elementary Excitations (Phonons See 6)

*Original Papers and Short Notes**

G. C. Aers, A. D. Boardman, and P. Clark
A Computational Study of the Correlation of Attenuated Total Reflection Minima with Surface Polariton Modes
(b) Vol. **85**, No. 1, p. 171—178 (1978).
See also *18*, p. *273*; *20.1*, p. *319*; 21.1, p. 368.

R. Zimmermann
*Final State Interactions in the Gain and Absorption Spectra of Electron–Hole Liquids
(b) Vol. **86**, No. 1, p. K63—K66 (1978).
See also *20.1*, p. *323*; 22.2.1, p. 445.

V. G. Plekhanov and A. A. O'Connel-Bronin
*Polariton Emission in LiH, LiD, and NaI Single Crystals
(b) Vol. **86**, No. 2, p. K123—K125 (1978).
See also 20.3, p. 343; 22, p. 412; 22.5.2, p. 480.

B. L. Gelmont, V. I. Ivanov-Omskii, and E. N. Ukraintsev
*The Frequency Dependent Dielectric Function of Gapless Semiconductors in a Magnetic Field
(b) Vol. **87**, No. 1, p. 381—383 (1978).
See also 14.4, p. 248; 22.4.3, p. 470.

J. Popielawski
*On the Optical Criterion for the Mott Transition
(b) Vol. **88**, No. 1, p. 241—245 (1978).
See also 20.1, p. 329.

G. Röpke, K. Kilimann, D. Kremp, W. D. Kraeft, and R. Zimmermann
*The Influence of Dynamical Effects on the Two-Particle States (Excitons) in the Electron–Hole Plasma
(b) Vol. **88**, No. 1, p. K59—K63 (1978).
See also *13.5.1*, p. *193*; 22, p. 415.

E. O. Goebel and O. Hildebrand
*Thermalization of the Electron–Hole Plasma in GaAs. The Case of Resonant Excitation
(b) Vol. **88**, No. 2, p. 645—652 (1978).
See also *20.3*, p. *346*; 22.2.1, p. 447.

R. Claus
*Polariton Dispersion and Crystal Optics in Monoclinic Materials
(b) Vol. **88**, No. 2, p. 683—688 (1978).
See also *14.4*, p. *250*.

R. Claus and G. Borstel
*Polariton Dispersion in Triclinic Crystals
(b) Vol. **88**, No. 2, p. K123—K125 (1978).

G. A. Puchkovskaya, V. L. Strizhevskii, Yu. A. Frolkov, N. M. Chepilko, and Yu. N. Yashkir
*Crystal Optics of Surface Polaritons in Anisotropic Crystals
(b) Vol. **89**, No. 1, p. 27–36 (1978).
See also 20.1, p. 331; 22.6, p. 496.

U. Lindner and V. K. Fedyanin
*Solitons in a One-Dimensional Modified Hubbard Model
(b) Vol. **89**, No. 1, p. 123–129 (1978).
See also 18.4, p. 301.

L. Merten and R. Claus
*Dispersion of Dielectric Axes in Monoclinic and Triclinic Crystals Related to Lattice Dynamics
(b) Vol. **89**, No. 1, p. 159–163 (1978).
See also 14.4, p. 250.

R. Zimmermann, K. Kilimann, W. D. Kraeft, D. Kremp, and G. Röpke
*Dynamical Screening and Self-Energy of Excitons in the Electron–Hole Plasma
(b) Vol. **90**, No. 1, p. 175–187 (1978).
See also *13.5.1*, p. *196*; 22, p. 419.

13.5.1 Excitons

Original Papers and Short Notes

M. E. Michel-Beyerle, R. Haberkorn, J. Kinder, and H. Seidlitz
*Direct Evidence for the Singlet–Triplet Exciton Annihilation in Anthracene Crystals
(b) Vol. **85**, No. 1, p. 45–49 (1978).
See also 20.3, p. 341; 22.9, p. 522.

V. V. Hizhnyakov and A. V. Sherman
*On the Theory of Resonant Secondary Radiation of Excitons Weakly Interacting with Phonons
(b) Vol. **85**, No. 1, p. 51–61 (1978).
See also *6*, p. *64*; 20.1, p. 319; 20.3, p. 341; 22, p. 408.

G. A. Boiko, V. S. Dneprovskii, M. V. Kraevskii, K. Marinova, S. M. Oak, E. K. Silina, and V. S. Fokin
*A Study of the Kinetics of Recombination Radiation of CdS and CdSe Crystals
(b) Vol. **85**, No. 1, p. 111–119 (1978).
See also *20.3*, p. *341*; 22.4.1, p. 463; 22.4.2, p. 468.

O. Goede, H.-J. Krug, and A. Horst
*Deformation Effects on Exciton Excitation Spectra in CdS
(b) Vol. **85**, No. 1, p. 239–245 (1978).
See also *20.1*, p. *319*; 22.4.1, p. 463.

K. Hümmer and P. Gebhardt
*Angular Dependence of the Reflection Spectra and Directional Dispersion of the Anisotropic Exciton Polaritons in ZnO
(b) Vol. **85**, No. 1, p. 271–282 (1978).
See also *20.1* p. *320*; 22.4, p. 461.

K. A. Aristova, C. Trallero Giner, I. G. Lang, and S. T. Pavlov
*Scattering and Decay of Hot Excitons Interacting with Longitudinal Optical Phonons
(b) Vol. **85**, No. 1, p. 351–357 (1978).
See also *6*, p. *65*; 20.1, p. 320.

H.-J. Wünsche, V. E. Khartsiev, and K. Henneberger
*Binding-Energy Calculations for Multiple Bound Excitons in Silicon
(b) Vol. **85**, No. 1, p. K53–K56 (1978).
See also 22.1.2, p. 429.

G. L. Belenkii and M. O. Godzhaev
*Excitons in Gallium Sulphide
(b) Vol. **85**, No. 2, p. 453–458 (1978).
See also *20.3*, p. *341*; 22.3, p. 457.

M. E. Michel-Beyerle and R. Haberkorn
*Temperature-Independent Intersystem Crossing in Anthracene Crystals
(b) Vol. **85**, No. 2, p. 473–475 (1978).
See also *20.3*, p. *342*; 22.9, p. 522.

H. Haug and D. B. Tran Thoai
*Dynamical Screening of Excitons by Free Carriers
(b) Vol. **85**, No. 2, p. 561—568 (1978).
See also 13.5.2, p. 198; 22, p. 409.

R. Enderlein, B. Seifert, and F. Bechstedt
*Electronic Elementary Excitations in the Far UV-Spectral Region. Core Level Exciton and Core Hole Polarization
(b) Vol. **85**, No. 2, p. 569—576 (1978).
See also *13.1*, p. *151*; 20, p. 315; 22.1, p. 422.

J. Nahum and F. J. Bryant
*Exciton Emission of Ytterbium-Implanted Cadmium Telluride
(b) Vol. **85**, No. 2, p. 629—637 (1978).
See also *20.3*, p. *342*; 22.4.3, p. 470.

Vu Duy Phach, A. Bivas, B. Hönerlage, and J. B. Grun
*Two-Photon Resonant Raman Scattering via Biexcitons
(b) Vol. **86**, No. 1, p. 159—168 (1978).
See also 20.1, p. 321; 20.3, p. 342; 22.5, p. 474.

E. Majerníková and Š. Barta
*Magnetic Freeze-Out Effect in Noncrystalline Semiconductors
(b) Vol. **86**, No. 1, p. 183—194 (1978).
See also 2, p. 43; *13.1*, p. *152*; 18, p. 273; 22, p. 410.

M. Pöhler and G. Staupendahl
*Investigation of the Simultaneous Absorption of Three CO_2-Laser Photons by Tellurium
(b) Vol. **86**, No. 1, p. 225—230 (1978).
See also *20.1*, p. *322*; 22.1.3, p. 441.

R. Strehlow
On the No-Phonon Structure in Mixed Crystals with Indirect Band Structure
(b) Vol. **86**, No. 1, p. 311—318 (1978).
See also 6, p. 67; 13.1, p. 153; *20.1*, p. *322*; 22.2.4, p. 456.

M. Combescot
*Thermodynamics of an Electron–Hole System in Semiconductors
(b) Vol. **86**, No. 1, p. 349—358 (1978).
See also 22, p. 410.

V. A. Benderskii, V. Kh. Brikenshtein, A. G. Lavrushko, and P. G. Filippov
*Non-Linear Fluorescence Quenching in Molecular Crystals (I): Recombination of Localized Excitons
(b) Vol. **86**, No. 2, p. 449—458 (1978).
See also *20.3*, p. *343*; 22.9, p. 524.

H.-J. Wünsche, K. Henneberger, and V. E. Khartsiev
*Binding-Energy Calculations for Multiple Bound Excitons
(b) Vol. **86**, No. 2, p. 505—515 (1978).
See also 22.1.2, p. 431.

K. Hümmer, R. Helbig, and M. Baumgärtner
*Manifestation of *k*-Linear Term Effects on the Zeeman Splitting of the A1-Exciton Emission Lines in ZnO
(b) Vol. **86**, No. 2, p. 527—533 (1978).
See also *20.3*, p. *343*; 22.4, p. 461.

W. Ekardt and M. I. Sheboul
*Many-Exciton Effects in the Theory of Biexcitons
(b) Vol. **86**, No. 2, p. 535—540 (1978).
See also 22.5, p. 474.

R. Resta
*Quantum Defect Theory of Excitons. Solid Neon
(b) Vol. **86**, No. 2, p. 627—633 (1978).
See also 23, p. 530.

V. A. Korneichuk, T. I. Mitrofanova, P. E. Mozol, and G. S. Pekar
Photoconductivity of CdS:Cu under Ruby Laser Excitation
(b) Vol. **86**, No. 2, p. K155—K158 (1978).
See also *16*, p. *263*; 22.4.1, p. 464.

A. R. Hassan
*Exciton Molecule Creation by Pure and Phonon-Assisted Two-Photon Transitions in Solids
(b) Vol. **87**, No. 1, p. 31—40 (1978).
See also 20.1, p. 325; 22.5, p. 475.

V. M. Shatalov, V. A. Telezhkin, and K. B. Tolpygo
Inadequacy of the Direct Interband Transition Scheme Applied to Valence Semiconductor Fundamental Absorption Spectra
(b) Vol. **87**, No. 1, p. 79—86 (1978).
See also *13.1*, p. *154*; 20.1, p. 325; 22, p. 412.

A. K. Kostin, V. V. Savelev, and A. V. Vannikov
*Free Charges and Excitons in Stilbene Single Crystals Excited by Electron Pulses
(b) Vol. **87**, No. 1, p. 255—259 (1978).
See also 20.1, p. 326; 20.3, p. 344; 22.9, p. 524.

S. Nikitine, S. G. Elkomoss, and C. Schwab
*Study of Linewidth and Exciton–Impurity Scattering in Cu_2O
(b) Vol. **87**, No. 1, p. 261—272 (1978).
See also *20.1*, p. *326*; 22.6, p. 493.

E. L. Frankevich, M. M. Tribel, I. A. Sokolik, and A. I. Pristupa
Magnetic-Resonant Modulation of Photoconductivity of Crystalline Charge Transfer Complexes. Anthracene–Tetracyanbenzene
(b) Vol. **87**, No. 1, p. 373—379 (1978).
See also *16*, p. *264*; 20.3, p. 344; 22.9, p. 524.

M. Leppihalme and T. Tuomi
Phonon Energies and Barrier Height from Photovoltage Spectra of Au–Intrinsic Ge Diodes
(a) Vol. **47**, No. 1, p. 245—250 (1978).
See also 6, p. 71; *14.3.3*, p. *235*; *16*, p. *264*; 22.1.1, p. 425.

M. Kalm and Ch. Uihlein
*Investigations on the Temperature-Dependent TPA Linewidth of the Z_3-Exciton in CuCl
(b) Vol. 87, No. 2, p. 575—580 (1978).
See also 6, p. 71; *20.1*, p. *327*; 22.5, p. 475.

V. Šubrta
*Exciton Propagation and Memory Effect in Molecular Crystals at Low Temperatures
(b) Vol. 87, No. 2, p. 647—652 (1978).
See also 22.9, p. 525.

M. Z. Zgierski and W. H. Henneker
Vibronic Spectra of Molecular Systems
(b) Vol. 87, No. 2, p. 653—661 (1978).
See also *6*, p. *72*; 22.9, p. 525.

Z. Zboiński
*A Kinetic Model of Triplet Exciton Dynamics in Organic Crystals with Discrete Trapping Levels
(b) Vol. 87, No. 2, p. 663—674 (1978).
See also 20.3, p. 344; 22.9, p. 525.

I. Rückmann, J. Puls, and J. Voigt
*Induced Absorption of CdS Single Crystals in the Energy Ranges of Exciton and Exciton–Biexciton Transitions
(b) Vol. 87, No. 2, p. K111—K114 (1978).
See also *20.1*, p. *328*; 22.4.1, p. 464.

L. Risch
The Influence of Surface Layers on Photoreflectance in the Exciton Region of CdS
(b) Vol. 88, No. 1, p. 111—116 (1978).
See also *13.3*, p. *166*; 20.1, p. 328; 22.4.1, p. 464.

V. M. Shatalov
Quasi-Molecular Model Interpretation of the Resonances in the Fundamental Absorption Region of Valence Semiconductors
(b) Vol. 88, No. 1, p. 129—133 (1978).
See also *13.1*, p. *155*; 20.1, p. 328; 22, p. 414.

I. Balslev
*The Excitonic Surface Potential in Semiconductors
(b) Vol. 88, No. 1, p. 155—161 (1978).
See also 20.1, p. 329; 22.4, p. 461.

F. Bechstedt and D. Haus
Theory of Second-Order Resonance Raman Scattering in the Case of Strong Excitonic Effects
(b) Vol. 88, No. 1, p. 163—171 (1978).
See also *6*, p. *73*; *20.1*, p. *329*; 22.4.1, p. 465.

H. Maaref, J. Barrau, M. Brousseau, J. Collet, J. Mazzaschi, and M. Pugnet
*First-Order Mott Transition of the Exciton Gas in GaP
(b) Vol. 88, No. 1, p. 261—268 (1978).
See also 20.3, p. 345; 22.2.2, p. 450.

T. Iwasaki, T. Goto, and Y. Nishina
*Oscillator Strength of Excitons in Highly Excited PbI_2
(b) Vol. 88, No. 1, p. 289—293 (1978).
See also 20.1, p. 329; 20.3, p. 345; 22.5, p. 475.

A. I. Bobrysheva and V. I. Vybornov
*The Biexciton in Polar Crystals
(b) Vol. 88, No. 1, p. 315—319 (1978).
See also 22.2.1, p. 447.

K. Germanova, K. Marinova, and S. Stoyanov
*Effect of Surface Layers on the Free Exciton Reflectance of ZnTe
(b) Vol. 88, No. 1, p. K19—K22 (1978).
See also 20.1, p. 329; 22.4.3, p. 470.

G. Röpke, K. Kilimann, D. Kremp, W. D. Kraeft, and R. Zimmermann
*The Influence of Dynamical Effects on the Two-Particle States (Excitons) in the Electron-Hole Plasma
(b) Vol. 88, No. 1, p. K59—K63 (1978).
See also *13.5*, p. *187*; 22, p. 415.

B. Hönerlage, Vu Duy Phach, and J. B. Grun
*Polarization Properties of Biexciton Luminescence and Two-Photon Raman Emission in CuCl
(b) Vol. 88, No. 2, p. 545—553 (1978).
See also *20.3*, p. *346*; 22.5, p. 475.

K. Saiki, M. Yoshimi, and S. Tanaka
*Modulation Spectroscopy on the Group IV and VI Transition-Metal Dichalcogenides
(b) Vol. 88, No. 2, p. 607—614 (1978).
See also 13.1, p. 157; *20.1*, p. *330*; 22, p. 415.

H. Asami, T. Nakayama, T. Chong, and N. Itoh
*Reaction Rate Constants of Triplet Excitons in Naphthalene Single Crystals
(b) Vol. 88, No. 2, p. 623—631 (1978).
See also 20.3, p. 346; 22.9, p. 526.

P. Petelenz
*Effect of the Mixing between Frenkel Excitons and Charge Transfer States on Charge Carrier Transfer in Molecular Crystals (II): Physical Consequences
(b) Vol. 88, No. 2, p. 689—697 (1978).
See also 22.9, p. 526.

B. Stébé and G. Munschy
Binding of an Exciton to a Neutral Acceptor
(b) Vol. 88, No. 2, p. 713—725 (1978).
See also *13.4*, p. *178*; 14.3, p. 223; 22.2.1, p. 447.

A. V. Sachenko, V. A. Tyagai, and A. G. Kundzich
*Exciton Luminescence in Semiconductors. Surface Recombination and Space-Charge-Layer Effects
(b) Vol. 88, No. 2, p. 797—804 (1978).
See also 13.3, p. 166; *20.3*, p. *346*; 22.4.1, p. 465.

V. P. Dolgopolskii, R. P. Seisyan, and A. L. Efros, Jr.
*The Ground State of a Wannier-Mott Exciton in Diamond-Like Semiconductors in "Intermediate" Magnetic Fields
(b) Vol. **88**, No. 2, p. K177—K180 (1978).
See also *13.1*, p. *158*; 22.2.1, p. 447.

A. Haydar and A. Coret
*Excitons and Structural Defects in Cu_2O at Low Temperature
(a) Vol. **48**, No. 2, p. 465—471 (1978).
See also *16*, p. *265*; 22.6, p. 495.

R. Kühne and P. Reineker
*Exact Derivation of a Generalized Master Equation for the Motion of Excitons
(b) Vol. **89**, No. 1, p. 131—141 (1978).

F. R. Kessler, J. Metzdorf, and W. Zurek
*Interband Faraday Ellipticity in Germanium
(b) Vol. **89**, No. 1, p. 203—211 (1978).
See also 13.1, p. 158; *18*, p. *275*; *20*, p. *317*; 22.1.1, p. 427.

W. G. Albrecht, H. Coufal, R. Haberkorn, and M. E. Michel-Beyerle
*Excitation Spectra of Exciton Fission in Organic Crystals
(b) Vol. **89**, No. 1, p. 261—265 (1978).
See also 20.1, p. 332; 22.9, p. 527.

A. Kozanecki, Z. Werner, H. Rzewuski, and J. J. Loferski
*A Study of Excitonic Spectra in CdS_xSe_{1-x} Mixed Crystals
(b) Vol. **89**, No. 1, p. 313—317 (1978).
See also *20.3*, p. *347*; 22.4.4, p. 473.

R. Bindemann, H. Fischer, and K. Kreher
The Spectral Dependence of the Photoneutralization Cross-Section of Oxygen in GaP Near the Threshold
(a) Vol. **49**, No. 1, p. 331—336 (1978).
See also *13.4*, p. *181*; *20.3*, p. *347*; 22.2.2, p. 451.

H. Atzmüller and U. Schröder
*Theoretical Investigations of Haynes' Rule
(b) Vol. **89**, No. 2, p. 349—356 (1978).
See also *13.4*, p. *181*; 22, p. 418.

F. R. Kessler and W. Olsowski
*Exciton Contribution to Oscillatory Magnetophotoconductivity in Ge
(b) Vol. **89**, No. 2, p. 389—396 (1978).
See also *16*, p. *266*; 22.1.1, p. 428.

W. Stössel and H. J. Wagner
*Comparison of Additional Boundary Conditions Based on Thermoreflectance Spectra of the $A_{n=1}$-Exciton in CdS Crystals at Oblique Incidence
(b) Vol. **89**, No. 2, p. 403—417 (1978).
See also *20.1*, p. *333*; 22.4.1, p. 465.

S. W. Koch, H. Haug, G. Schmieder, W. Bohnert, and C. Klingshirn
Stimulated Intrinsic Recombination Processes in II–VI Compounds
(b) Vol. **89**, No. 2, p. 431—440 (1978).
See also *20.2*, p. *339*; *20.3*, p. *348*; 22.4, p. 462; 22.4.1, p. 465; 22.4.3, p. 471.

R. Schwabe, F. Thuselt, H. Weinert, R. Bindemann, and K. Unger
*Luminescence from Electron–Hole Drops in Undoped GaP
(b) Vol. **89**, No. 2, p. 561—570 (1978).
See also *20.3*, p. *348*; 22.2.2, p. 451.

H. Friedrich, Yu. V. Ozerov, R. Strehlow, and A. N. Georgobiani
*On the Electron–Phonon and Exciton–Phonon Interaction in ZnTe
(b) Vol. **89**, No. 2, p. 603—609 (1978).
See also *20.1*, p. *334*; 22.4.3, p. 471.

P. Franzosi
*Exciton–Phonon Interaction in CdS
(b) Vol. **89**, No. 2, p. K119—K122 (1978).
See also *6*, p. *76*; 20.1, p. 334; 22.4.1, p. 466.

N. F. Berk, L. Yarmus, W. Bizzaro, and J. Rosenthal
*The Source of "Residual" Linewidth in EPR of Triplet Excitons in Molecular Crystals
(b) Vol. **89**, No. 2, p. K167—K170 (1978).
See also *19*, p. *313*; 22, p. 418.

O. Goede, L. John, and D. Hennig
*Compositional Disorder-Induced Broadening for Free Excitons in II–VI Semiconducting Mixed Crystals
(b) Vol. **89**, No. 2, p. K183—K186 (1978).
See also *20.1*, p. *334*; 22.4.4, p. 473.

G. Behnke and H. Büttner
*The Exciton Dispersion in Polar Semiconductors
(b) Vol. **90**, No. 1, p. 53—59 (1978).
See also 13.5.3, p. 200; 22.5, p. 476.

H. Schrey and C. Klingshirn
*Biexcitons, Impurity Transitions, and Electron–Hole Plasma in Highly Excited II–VI Compounds?
(b) Vol. **90**, No. 1, p. 67—76 (1978).
See also 13.4, p. 183; *20.3*, p. *349*; 22.4.1, p. 466.

I. Broser, M. Rosenzweig, R. Broser, M. Richard, and E. Birkicht
*A Quantitative Study of Excitonic Polariton Reflectance in CdS]
(b) Vol. **90**, No. 1, p. 77—91 (1978).
See also *20.1*, p. *335*; 22.4.1, p. 466.

R. Strehlow, W. Kühn, H. Friedrich, and H. Lange
On the Quantitative Characterization of Chemical Disorder of $In_{1-x}Ga_xP$ from Wavelength Modulated No-Phonon Absorption
(b) Vol. **90**, No. 1, p. 167—173 (1978).
See also 6, p. 77; 13.1, p. 159; *20.1*, p. *335*; 22.2.4, p. 456.

R. Zimmermann, K. Kilimann, W. D. Kraeft, D. Kremp, and G. Röpke
*Dynamical Screening and Self-Energy of Excitons in the Electron–Hole Plasma
(b) Vol. **90**, No. 1, p. 175—187 (1978).
See also *13.5*, p. *188*; 22, p. 419.

E. Schuberth and M. Creuzburg
*Luminescence from Rare-Gas-Doped Solid Neon
(b) Vol. **90**, No. 1, p. 189—196 (1978).
See also *20.3*, p. *349*; 23, p. 531.

R. Schwabe, F. Thuselt, T. Hänsel, and F. Lévy
*Electron Beam Excited Free Exciton Luminescence of GaSe
(b) Vol. **90**, No. 1, p. 233—239 (1978).
See also *20.3*, p. *349*; 22.3, p. 459.

V. Šubrta and V. Čápek
*Fading Memory as Insufficient Condition for the Exciton Diffusion
(b) Vol. **90**, No. 1, p. 295—299 (1978).
See also 22.9, p. 528.

V. I. Lesin
*Analytical Form of the Magnetic Field Dependences of Rate Constants of Triplet Exciton Annihilation and Triplet Exciton Quenching by Free Radicals in Molecular Crystals
(b) Vol. **90**, No. 1, p. K13—K16 (1978).
See also 22.9, p. 528.

V. I. Sokolov, N. A. Moskvina, and B. V. Melnikov
*Discovery of Exciton–Impurity Complexes in Zinc Selenide by Electroabsorption
(b) Vol. **90**, No. 1, p. K39—K42 (1978).
See also *13.4*, p. *184*; 14.3.4, p. 246; 20.1, p. 336; 22.4.2, p. 469.

S. A. Boiko, D. I. Bletskan, and S. F. Terekhova
*Exciton Absorption of Germanium Diselenide
(b) Vol. **90**, No. 1, p. K49—K52 (1978).
See also *20.1*, p. *336*; 22, p. 420.

V. L. Broude, N. A. Vidmont, and V. V. Korshunov
*Singlet Exciton Annihilation Kinetics in Anthracene Crystals
(b) Vol. **90**, No. 1, p. K53—K58 (1978).
See also *20.3*, p. *349*; 22.9, p. 528.

V. A. Gadziev, S. G. Abdullaeva, F. K. Aliev, F. K. Allakhverdiev, and A. Z. Badalov
Electroabsorption in GaTe
(b) Vol. **90**, No. 1, p. K77—K79 (1978).
See also *14.3.4*, p. *246*; *20.1*, p. *336*; 22.3, p. 459.

R. M. Westervelt, J. L. Staehli, and E. E. Haller
*Nucleation Phenomena in Electron–Hole Drop Formation in Ge and Si (III): Experimental Results for Ge
(b) Vol. **90**, No. 2, p. 557—569 (1978).
See also 22.1.1, p. 428.

R. S. Markiewicz

*Acoustic Phonon Scattering by Electron–Hole Droplets in Ge (II): Total Scattering Rate and Anisotropy of Scattering, as Reflected in the Shape of a Droplet Cloud
(b) Vol. **90**, No. 2, p. 585—596 (1978).
See also 14.3, p. 225; 22.1.1, p. 428.

P. Petelenz

*Effect of the Mixing between Frenkel Excitons and Charge Transfer States on Charge Carrier Transfer in a Linear Molecular Crystal
(b) Vol. **90**, No. 2, p. 635—642 (1978).
See also *13.4*, p. *185*.

13.5.2 Plasmons

Original Papers and Short Notes

G. S. Agarwal and H. D. Vollmer
Surface Polariton Effects in Spontaneous Emission (II): Effects of Spatial Dispersion
(b) Vol. **85**, No. 1, p. 301–310 (1978).
See also *13.3*, p. *164*; 20.1, p. 320; 21, p. 352; 22.4.1, p. 463.

C. J. Rossouw and M. J. Whelan
Thickness Determination of Thin Graphite and Aluminium Nitride Foils by Transmission Electron Microscopy Using Diffraction and Energy Loss Analysis
(a) Vol. **45**, No. 1, p. 277–288 (1978).
See also *1.5*, p. *36*; 22, p. 409; 22.1, p. 422.

H. Haug and D. B. Tran Thoai
Dynamical Screening of Excitons by Free Carriers
(b) Vol. **85**, No. 2, p. 561–568 (1978).
See also *13.5.1*, p. *190*; 22, p. 409.

Vo Hong Anh and Nguyen Nhu Dat
Parametric Excitation in Kane's Semiconductors
(b) Vol. **86**, No. 2, p. 585–591 (1978).
See also *6*, p. *69*; 20, p. 316; 22, p. 411.

W. Gasser
*Dynamical Electrical Conductivity, Dielectric Function, and Plasma Oscillations of the Hubbard Model
(b) Vol. **87**, No. 1, p. 185–190 (1978).
See also *14*, p. *203*; 18, p. 274.

Nguyen Van Trong
Neutron Scattering on Anomalous Fluctuations of Phonons
(b) Vol. **88**, No. 1, p. 145–148 (1978).
See also *6*, p. *73*; 11, p. 127.

D. Heitmann
*Theory and Experiment of the Light Emission of Surface Plasmons Induced by Fast Electrons on Modulated Surfaces
(b) Vol. **88**, No. 2, p. 493–506 (1978).
See also *13.3*, p. *166*; 20.3, p. 346; 21.6, p. 403.

W. Gasser
*On the Dielectric Function of the Hubbard Model
(b) Vol. **89**, No. 1, p. 165–169 (1978).
See also 14, p. 204; 18.1, p. 279.

C. Mande and A. Pimpale
*Evidence for Surface Plasmons in X-Ray Emission Bands
(b) Vol. **89**, No. 1, p. K21–K24 (1978).
See also *13.3*, p. *167*; 20, p. 317; 21.6, p. 404.

T. Puritis
Plasmon Mechanism for Secondary Breakdown
(a) Vol. **49**, No. 1, p. 225–230 (1978).
See also *14.3.4*, p. *244*; 22.1.2, p. 436.

A. Siems
Quantitative Study of Kikuchi Structures (I): Energy Analysis of Intensity Profiles
(a) Vol. **49**, No. 2, p. 621–628 (1978).
See also *1.1*, p. *14*; 6, p. 77; 22.1.2, p. 437.

P. Steiner, F. J. Reiter, H. Höchst, and S. Hüfner
*The KLL Auger Spectra of Na and Mg Metal and Their Plasmon Structure
(b) Vol. **90**, No. 1, p. 45–51 (1978).
See also *13.1*, p. *159*; 21.2, p. 393; 21.3, p. 395.

A. Manabe, H. Noguchi, and A. Mitsuishi
Dielectric Function in HgSe at 2, 95, and 300 K
(b) Vol. **90**, No. 1, p. 157–166 (1978).
See also 6, p. 77; 13.1, p. 159; *20.1*, p. *335*; 22.4.2, p. 469.

F. A. Zhivopistsev and F. C. Hernández
Fast Electron Scattering in Metals with the Excitation of a Core Electron
(b) Vol. **90**, No. 2, p. 441–447 (1978).
See also *13*, p. *149*; 21.2, p. 393.

13.5.3 Polarons

Original Papers and Short Notes

V. Čápek and J. Příívratská
*Spin Polarons in the s–d Model with Off-Diagonal Exchange
(b) Vol. **85**, No. 1, p. K1–K4 (1978).
See also *18.2*, p. *281*; 22, p. 409.

B. V. Egorov and M. A. Krivoglaz
The Theory of Autolocalized Electron States in Ferromagnetic Semiconductors at Arbitrary Relation between the Band Width and the Spin–Electron Interaction
(b) Vol. **86**, No. 2, p. 781–794 (1978).
See also *13*, p. *144*; 18.2, p. 285; 22, p. 412.

R. Mańka
*The Singularities of the Small Polaron Effective Mass
(b) Vol. **87**, No. 1, p. 53–60 (1978).

Hj. Bernhardt
*Bound Polarons in YAG Crystals
(b) Vol. **87**, No. 1, p. 213–219 (1978).
See also *10.2*, p. *112*; 22.8.2, p. 519.

V. L. Vinetskii and T. A. Kudykina
*Localized Bipolarons in Ionic Crystals
(b) Vol. **87**, No. 2, p. 507–511 (1978).
See also 22, p. 414.

Y. Lépine, G. Bader, and D. Matz
*Ground State Energy of Bound Polarons
(b) Vol. **89**, No. 1, p. 53–59 (1978).
See also 22, p. 416.

F. Carlier and G. Jacobs
*Transient Infrared Absorption of Bound Polarons in NaCl
(b) Vol. **89**, No. 1, p. 193–198 (1978).
See also *20.1*, p. *332*; 22.5.2, p. 483.

F. Carlier and G. Jacobs
Transient Infrared Absorption of Bound Polarons in KBr
(b) Vol. **89**, No. 1, p. K95–K98 (1978).
See also *10.2*, p. *118*; *20.1*, p. *333*; 22.5.2, p. 484.

G. Behnke and H. Büttner
The Exciton Dispersion in Polar Semiconductors
(b) Vol. **90**, No. 1, p. 53–59 (1978).
See also *13.5.1*, p. *195*; 22.5, p. 476.

E. G. Batyev and A. V. Efanov
*Polaron State of an Electron Near a Structural Instability
(b) Vol. **90**, No. 2, p. 449–455 (1978).
See also *1.2*, p. *24*.

13.5.4 Magnons

Original Papers and Short Notes

G. Georgiev and I. Gochev
*Localized Magnons around an Impurity in a Spin Chain with Next-Nearest Neighbour Interaction
(b) Vol. **85**, No. 2, p. 657—662 (1978).
See also *18.2*, p. *282*.

D. I. Pushkarov and Kh. I. Pushkarov
*Quasi-Classical Description of the Heisenberg Ferromagnet: Soliton Solutions
(b) Vol. **85**, No. 2, p. K89—K93 (1978).
See also *18.2*, p. *283*.

S. Ligenza
Exchange Splitting of Ionic S-State in $MnFe_2O_4$ Studied by Neutron Spectroscopy
(b) Vol. **86**, No. 2, p. 635—643 (1978).
See also *18.3*, p. *295*; 22.8.2, p. 518.

C. G. Patil and B. S. Krishnamurthy
Carrier Mobility in Ferromagnetic Semiconductors
(b) Vol. **86**, No. 2, p. 725—732 (1978).
See also *14.3*, p. *219*; 18.2, p. 285; 22.4.4, p. 472.

Diep-The-Hung
*Erratum to: Effect of Induced Surface Magnetic Moment on Spin-Wave Spectrum
(b) Vol. **86**, No. 2, p. K181 (1978).
See also 1.6, p. 40; 18.4, p. 299.

Kh. I. Pushkarov, D. I. Pushkarov, and J. P. Vlahov
*Solitary Excitations in One-Dimensional Ferromagnets with Biquadratic Exchange
(b) Vol. **88**, No. 1, p. 65—71 (1978).
See also *18.2*, p. *287*.

M. Matlak and W. Nolting
*Influence of Finite Carrier Concentrations on the Electronic Spectrum of a Ferromagnetic Semiconductor
(b) Vol. **88**, No. 1, p. 231—240 (1978).
See also 18.2, p. 288; 22, p. 414.

W. Borgieł
*On the Longitudinal Correlation Functions of the Heisenberg Spin System with Single-Ion Anisotropy
(b) Vol. **88**, No. 1, p. K47—K50 (1978).
See also 18.1, p. 278; 18.2, p. 288.

L. Graf
*Electronic Transitions in the Raman Spectra of $FeCl_2 \cdot 2\,H_2O$ and $FeCl_2 \cdot 2\,D_2O$
(b) Vol. **88**, No. 2, p. 429—437 (1978).
See also *6*, p. *74*; 20.1, p. 330; 22.5, p. 475.

C. G. Patil and B. S. Krishnamurthy
*Magnon Drag Thermoelectric Power in Ferromagnetic Semiconductors
(b) Vol. 88, No. 2, p. K117—K122 (1978).
See also *15*, p. *260*; 18.2, p. 289; 22, p. 415.

Kh. I. Pushkarov, D. I. Pushkarov, and J. P. Vlahov
*Solitary Bound States of Magnons and Lattice Deformation in Ferromagnets with Biquadratic Exchange
(b) Vol. 90, No. 1, p. 361—368 (1978).
See also *18.2*, p. *291*.

E. Kolley and W. Kolley
*Electron Correlation Effects on the Spin Wave Stiffness Constant in Itinerant Ferromagnets
(b) Vol. 90, No. 1, p. K103—K107 (1978).
See also 18.2, p. 291; 21, p. 365.

14. Electrical Properties. Transport Phenomena

Original Papers and Short Notes

Č. Koňák, L. Budík, and L. Novák
Electro-Optical Properties of NH_4Cl Single Crystals near Their Order–Disorder Phase Transition
(a) Vol. **45**, No. 1, p. K63—K66 (1978).
See also *1.2*, p. *16*; 20, p. 315; 22.5, p. 474.

F. Lüty and A. Diaz-Gongora
*Electro-Birefringence from Paraelectric Defects in Cubic Crystals
(b) Vol. **85**, No. 2, p. 693—706 (1978).
See also 10.2, p. 109; *20*, p. *315*; 22.5.2, p. 479.

J. L. Jacquemin and G. Bordure
*Numerical Simulation of the AC Electrical Properties of Random Inhomogeneous Systems. Application to Heat-Treated Polymers
(a) Vol. **45**, No. 2, p. 579—584 (1978).
See also 14.4, p. 248; 22.9, p. 523.

A. Diaz-Gongora and F. Lüty
*Kerr Effect and Birefringence from Paraelectric CN^- Defects in Nine Alkali Halides
(b) Vol. **86**, No. 1, p. 127—136 (1978).
See also 10.2, p. 110; *20*, p. *316*; 22.5.2, p. 479.

D. Ihle
*Electrical Conductivity in the Strong-Coupling Non-Half-Filled Band Hubbard Model
(b) Vol. **86**, No. 1, p. 283—289 (1978).
See also 13.1, p. 153.

G. J. Ashwell
Organic Metals: Chain Stabilisation in Bipyridinium–TCNQ Salts
(b) Vol. **86**, No. 2, p. 705—715 (1978).
See also *1.1*, p. *11*; 22.9, p. 524.

W. C. Kok
*Resistivity of Spin Glasses below the Ordering Temperature
(b) Vol. **86**, No. 2, p. K113—K117 (1978).
See also *2*, p. *44*.

W. Gasser
*Dynamical Electrical Conductivity, Dielectric Function, and Plasma Oscillations of the Hubbard Model
(b) Vol. **87**, No. 1, p. 185—190 (1978).
See also *13.5.2*, p. *198*; 18, p. 274.

Yu. E. Lozovik, V. N. Nishanov, and E. P. Kaminskaya
*Boundary Effects of Transition into Excitonic Phase
(b) Vol. **88**, No. 1, p. 359—365 (1978).
See also *13*, p. *146*.

A. A. GOGOLIN and V. I. MELNIKOV
*Frequency Dependences of Conductivity and Dielectric Permeability of a One-Dimensional Disordered Conductor
(b) Vol. **88**, No. 2, p. 377–383 (1978).
See also 22.9, p. 526.

U. BEHN
On the Ward Identities for the Hubbard Model. Particle-Number Conservation and the Analytical Structure of the Self-Energy
(b) Vol. **88**, No. 2, p. 699–704 (1978).
See also *13*, p. *147*.

J.-P. GALLINAR
*On the Optical Conductivity in a Strongly Attractive Hubbard Model Chain
(b) Vol. **88**, No. 2, p. K143–K146 (1978).
See also *13*, p. *147*.

A. V. GAISANYUK, K. I. KUGEL, V. A. PETROV, and V. V. SOKOLOV
*Resistivity of the ZrC–C System: An Example of Percolation Behaviour
(a) Vol. **48**, No. 2, p. K131–K133 (1978).
See also *1.2*, p. *22*; 22.3, p. 458.

I. I. FISHCHUK and V. N. RUDKO
*Theory of the Phononless High Frequency Hall Effect in Disordered Systems
(b) Vol. **89**, No. 1, p. 61–68 (1978).
See also *2*, p. *48*.

W. GASSER
On the Dielectric Function of the Hubbard Model
(b) Vol. **89**, No. 1, p. 165–169 (1978).
See also *13.5.2*, p. *198*; 18.1, p. 279.

V. I. LITVINOV, V. M. BAGINSKII, and A. A. MITROFANOV
The Phase Diagram of a System with Two Order Parameters. Mott Transition and Ferromagnetism
(b) Vol. **90**, No. 1, p. 423–427 (1978).
See also *1.2*, p. *23*; *13*, p. *148*; 18, p. 275.

M. PRZYBYLSKI, A. GRAJA, A. RAJCHEL, M. GAWRON, and T. BOROWIAK
*DC Conductivity of Complex Salts of TCNQ with N-Methyl-Pyridines
(a) Vol. **50**, No. 1, p. K105–K107 (1978).
See also 22.9, p. 528.

A. S. DAVYDOV
*The Role of Proteins in Electron Transport at Large Distances
(b) Vol. **90**, No. 2, p. 457–464 (1978).
See also 22.9, p. 528.

I. V. KRIVOSHEI, V. P. BABIYCZUK, I. M. GUELLA, I. F. GOLOVKINA, N. V. MANSIA, V. A. STARODUB, and S. A. CHUEVA
*A New Conducting Compound: Tetrafluoroborate of Dibenztetrathiofulvalene; Electrical and Magnetic Properties
(a) Vol. **50**, No. 2, p. K197–K200 (1978).
See also *18*, p. *276*; 22.9, p. 529.

14.1 Metals. Semi-Metals

Review Articles

R. Suryanarayanan
Valence Related Optical and Other Studies of Sm and Tm Chalcogenides
(b) Vol. **85**, No. 1, p. 9–43 (1978).
See also 13.1, p. 150; 14.3, p. 217; *20.1*, p. *319*; 22.7, p. 501.

J. A. Wilson
*Modelling the Contrasting Semimetallic Characters of TiS_2 and $TiSe_2$
(b) Vol. **86**, No. 1, p. 11–36 (1978).
See also 6, p. 64; 13.1, p. 150; 22, p. 408.

Original Papers and Short Notes

G. J. Ashwell
*A Comment upon the Metallic and Semiconducting Phases of $(DEPE)^{2+}$ $(TCNQ)_4^{2-}$
(b) Vol. **85**, No. 1, p. K7–K9 (1978).
See also *14.3*, p. *217*; 22.9, p. 522.

J. P. Rivière
*Electrical Resistivity of B_2-Type Fe–Al Ordered Alloys
(a) Vol. **45**, No. 1, p. 157–162 (1978).
See also 10.1, p. 99; 21.1.1, p. 384.

R. Mattheis
*Electrical Resistivity of PdH_x Films
(a) Vol. **45**, No. 1, p. 219–227 (1978).
See also *14.2*, p. *213*; 21, p. 353.

P. Svoboda
*Size Effect Correction to the Longitudinal Magnetoresistance of Thin Wires
(a) Vol. **45**, No. 1, p. K43–K46 (1978).
See also 21, p. 353; 21.6, p. 401.

A. Bessis, Y. Bismuth et J. Bigot
*Etude de l'endommagement du chrome au cours de l'irradiation aux neutrons à 24,5 K
(Study of Damage in Chromium by Neutron Irradiation at 24.5 K)
(a) Vol. **45**, No. 1, p. K71–K76 (1978).
See also *11*, p. *123*; 21.1, p. 369.

E. I. Terukov, W. Reichelt, and H. Oppermann
*Zum Einfluß von Dotierungen auf den Phasenübergang des V_3O_5
(On the Influence of Doping on the Phase Transition of V_3O_5)
(a) Vol. **45**, No. 1, p. K77–K79 (1978).
See also *14.3.1*, p. *226*; 22.6, p. 491.

M. Brauwers and A. K. Bhattacharya
*Temperature Dependence of the Electrical Resistivity and Hall Coefficient in a Model of Concentrated Alloy
(b) Vol. **85**, No. 2, p. 519—524 (1978).
See also 13.1, p. 151; 21, p. 354.

F. Goedsche and G. Vojta
*Spin-Disorder Resistivity of Ternary Alloys in the CPA Weak Scattering Limit
(b) Vol. **85**, No. 2, p. K81—K84 (1978).
See also 18.2, p. 282; 21, p. 354; 21.1, p. 370.

B. I. Shklovskii
*Anisotropy of Percolation Conduction
(b) Vol. **85**, No. 2, p. K111—K114 (1978).
See also *14.4*, p. *248*.

S. Matsuo, T. Nishimoto, S. Hayashi, and S. Noguchi
Magnetothermal Oscillation and Spin Splitting of Electrons in Dilute Bismuth–Antimony Alloys
(b) Vol. **86**, No. 1, p. 169—176 (1978).
See also *13.1*, p. *152*; 21.7, p. 406.

F. Goedsche, R. Richter, and G. Vojta
*Coherent Potential Approximation for the Electrical Resistivity of Partially Ordered Binary Magnetic Alloys
(b) Vol. **86**, No. 1, p. 177—182 (1978).
See also 18.2, p. 283; 21, p. 355.

V. Christoph and W. Schiller
*Force-Correlation Function Method Applied to the Electrical Resistivity of Many-Band Metals
(b) Vol. **86**, No. 1, p. 231—237 (1978).
See also 21, p. 355.

E. Mrosan, Ch. Schober, and G. Lehmann
*Multiple Scattering Theory for Magnetoresistance of Localized Impurities in Metals
(b) Vol. **86**, No. 1, p. 249—256 (1978).
See also 13.4, p. 172; 18, p. 273; 21, p. 355.

H. Krizek, K. N. R. Taylor, and W. D. Corner
*Resistivity of Some Light–Heavy Rare-Earth Alloys (II): Magnetic Contributions to ϱ_{res}, ϱ_{s-f} and T_N
(a) Vol. **46**, No. 1, p. 39—43 (1978).
See also 21.4, p. 396.

G. C. Vezzoli
Optical and Electrical Effects during the Solidification of Mercury
(a) Vol. **46**, No. 1, p. K49—K53 (1978).
See also *1.2*, p. *17*; 20.1, p. 324; 21, p. 356.

O. P. Sharma
A Pseudopotential Study of the Electronic Properties of Noble Metals
(b) Vol. **86**, No. 2, p. 483—492 (1978).
See also 4, p. 57; *6*, p. *68*; 21.1, p. 371; 21.6, p. 401.

R. S. Sorbello
*Electromigration in Liquid Metal Alloys
(b) Vol. **86**, No. 2, p. 671—678 (1978).
See also *2*, p. *44*; 21.2, p. 392.

P. Jung
Subthreshold Irradiation Effects in Quenched Platinum
(a) Vol. **46**, No. 2, p. 527—531 (1978).
See also *10.1*, p. *102*; 11, p. 125; 21.6, p. 402.

R. Kamel, A. R. Ali, and Z. Farid
On the Recovery of Quenched and Irradiated Dilute Al–Li Alloys
(a) Vol. **46**, No. 2, p. 697—701 (1978).
See also *10.1*, p. *102*; 12.1, p. 135; 21, p. 357.

M. A. Ball and M. M. Islam
*Effects of Core and Correlation Corrections on the Electrical Resistivity of Lithium
(b) Vol. **87**, No. 1, p. 145—150 (1978).
See also *13*, p. *145*; 21.2, p. 392.

A. Eroglu, S. Arajs, C. A. Moyer, and K. V. Rao
*Electrical Resistivity of Chromium–Gold Alloys between 77 and 700 K
(b) Vol. **87**, No. 1, p. 287—291 (1978).
See also 21.1, p. 372; 21.6, p. 402.

U. Rindelhardt, S. Mager, W. Blau, and E. Hegenbarth
Recovery of Low Temperature Specific Heat, Resistivity, and Coercive Force in Fe–Cr–Al Alloys
(a) Vol. **47**, No. 1, p. 157—161 (1978).
See also 8, p. 85; *10.1*, p. *103*; 18.2, p. 286; 21.1, p. 373; 21.1.1, p. 387.

C. L. Bauer and A. G. Jordan
Study of Interfacial Reactions in Bimetallic Thin-Film Couples by Contact Resistance Measurements
(a) Vol. **47**, No. 1, p. 321—328 (1978).
See also *1.5*, p. *37*; 21.1, p. 373.

K. Melidis
*On the Temperature Dependence of the Hall Effect and Electrical Resistivity of Cd, Zn, and Mg Single Crystals
(a) Vol. **47**, No. 1, p. K27—K30 (1978).
See also 21, p. 358; 21.1, p. 373; 21.3, p. 394.

I. M. Chapnik
Atomic Volume and Semiconducting Properties in Sulfides of Non-Magnetic Transition Metals
(a) Vol. **47**, No. 1, p. K71—K75 (1978).
See also *1.3*, p. *26*; *14.3*, p. *221*; 22, p. 413.

YA. A. KRAFTMAKHER and T. YU. PINEGINA
*Anomalies of Resistivity and Thermopower of Samarium at High Temperatures
(a) Vol. 47, No. 1, p. K81—K83 (1978).
See also 1.2, p. 18; *15*, p. *260*; 21.4, p. 397.

W. BRAUNE, J. LEBECH, and K. SAERMARK
*Propagation of High-Frequency Electromagnetic Waves in the Voigt Configuration through Samples of Bi at 45, 70, and 300 GHz
(b) Vol. 87, No. 2, p. 527—533 (1978).
See also *13.2*, p. *162*; 21.6, p. 402.

F. PERROT
Phenomenological Study of Vacancies in Aluminium within Density-Functional and Second-Order Perturbation Formalisms
(b) Vol. 87, No. 2, p. 707—718 (1978).
See also *10.1*, p. *103*; 21, p. 359.

Y. WASEDA and H. S. CHEN
*Calculation of Electrical Resistivity and Its Temperature Coefficient of Amorphous Cu_{60}–Zr_{40} Alloy Using *t*-Matrix
(b) Vol. 87, No. 2, p. 777—782 (1978).
See also *2*, p. *45*; 21.1, p. 373.

J. A. ASHBY, H. M. FLOWER, and R. D. RAWLINGS
*Correlation of Electrical Resistivity with Microstructure in an Fe–Co–2% V Alloy
(a) Vol. 47, No. 2, p. 407—414 (1978).
See also *1.4*, p. *31*; 21.1, p. 374; 21.1.1, p. 387.

J. F. DINHUT, J. P. RIVIÈRE, and J. C. DESOYER
Point Defects Produced by Plastic Deformation and Ordering Phenomena in Fe–Co and Fe–Co–2% V Alloys
(a) Vol. 47, No. 2, p. 469—478 (1978).
See also *10.1*, p. *104*; 21.1, p. 374; 21.1.1, p. 387.

CH. HERZIG and H. E. PELTNER
Electromigration and Residual Resistivity of Cobalt in Gold
(a) Vol. 47, No. 2, p. 497—506 (1978).
See also *9*, p. *93*; 21.6, p. 403.

E. STRACKE and CH. HERZIG
*Electromigration, Thermomigration, and Solubility of Copper in Lead
(a) Vol. 47, No. 2, p. 513—521 (1978).
See also *9*, p. *93*; 21, p. 359.

Y. KIMURA and M. UMEDA
*Temperature Dependence of the Electrical Resistivity of Laves Phase Compounds $V_2(Hf, Zr)$
(a) Vol. 47, No. 2, p. K111—K113 (1978).
See also *1.2*, p. *20*; 21, p. 359.

W. KRAAK, G. OELGART, G. SCHNEIDER, and R. HERRMANN
The Semiconductor–Semimetal Transition in $Bi_{1-x}Sb_x$ Alloys with $x \geqq 0.22$
(b) Vol. 88, No. 1, p. 105—110 (1978).
See also *13.4*, p. *177*; *14.3*, p. *221*; 22.7, p. 502.

K. Yonemitsu, K. Takano, and T. Matsuda
*Hall Coefficient of Dilute Aluminum Alloys (I): Calculation by Four–OPW Approximation
(b) Vol. **88**, No. 1, p. 273—282 (1978).
See also 13.2, p. 162; 21, p. 360.

K. Bartkowski, D. Włosewicz, and J. Rafałowicz
*Electrical Resistivity Anitsotropy of Tin Monocrystals of Different Concentration of Chemical Impurities in the Temperature Range 5 to 70 K
(a) Vol. **48**, No. 1, p. 215—219 (1978).
See also 21.7, p. 407.

P. Vostrý
*On the Influence of Silver on the Recovery Spectrum of Quenched Cadmium Foils
(a) Vol. **48**, No. 1, p. K39—K42 (1978).
See also *10.1*, p. *104*; 21, p. 361.

K. Narayandas, M. Radhakrishnan, and C. Balasubramanian
*Annealing Study of the Electrical Resistivity and Defect Density in Silver Films
(a) Vol. **48**, No. 1, p. K71—K74 (1978).
See also 10.1, p. 104; 21.6, p. 403.

G. C. Vezzoli
*Peaked Electrical Conductivity in Liquid Bismuth and Associated Optical Scattering
(a) Vol. **48**, No. 1, p. K75—K77 (1978).
See also *2*, p. *47*; 20.1, p. 330; 21.7, p. 407.

H. Bossac, H.-G. Fabian, and H. Löffler
A Quantitative Interpretation of the Decompostition Kinetics of Some Al–Zn–Mg Alloys
(a) Vol. **48**, No. 2, p. 369—375 (1978).
See also *1.3*, p. *27*; *10.1*, p. *105*; 21, p. 361; 21.1, p. 376; 21.3, p. 395.

G. Hilscher and E. Gratz
An Estimation of the Density of Stastes from Magnetization and Resistivity Measurements in $Ti(Fe_{1-x}Co_x)$ and $Zr(Fe_{1-x}Co_x)_2$
(a) Vol. **48**, No. 2, p. 473—480 (1978).
See also *13*, p. *147*; 18.1, p. 279; 18.2, p. 289; 21.1, p. 376; 21.1.1, p. 388.

A. Bittar and D. Lesueur
*Quenching and Annealing of Defects in Bismuth and Its Dilute Alloys
(a) Vol. **48**, No. 2, p. K123—K126 (1978).
See also *10.1*, p. *105*; 21.7, p. 407.

M. Ohta, T. Kanadani, A. Sakakibara, and M. Yamada
*Aging and Reversion in Low-Concentrated Al–Zn Alloys
(a) Vol. **48**, No. 2, p. K141—K144 (1978).
See also 21, p. 362.

R. Richter, F. Goedsche, and G. Vojta
*Statistical Theory of Electrical Resistivity of Magnetic Ternary Alloys with Partial Long-Range Order
(b) Vol. **89**, No. 1, p. 77—83 (1978).
See also 13, p. 147; 18.4, p. 301; 21, p. 362.

K. Saermark
*Azbel-Kaner Cyclotron Resonance as a Long-Wavelength Limit of an Exact Expression for the Surface Impedance of a Metal in a Magnetic Field
(b) Vol. **89**, No. 1, p. 117–121 (1978).
See also *13.2*, p. *163*; 21, p. 362.

J. Richter and W. Schiller
*Residual Resistivity of Disordered Systems
(b) Vol. **89**, No. 1, p. 143–150 (1978).
See also *2*, p. *48*; 21, p. 362.

Th. Eifrig
*On the Electronic Properites of Disordered Quasi-One-Dimensional Systems with Electron Correlation
(b) Vol. **89**, No. 1, p. 171–175 (1978).
See also *2*, p. *48*; 18.1, p. 279; 22.9, p. 526.

A. Philipp and K. Seeger
*Influence of Halogens on the DC Conductivity and Magnetoresistance of $(SN)_x$
(b) Vol. **89**, No. 1, p. 187–192 (1978).
See also 22, p. 417.

N. V. Volkenshtein, V. E. Startsev, and V. I. Cherepanov
*Low Temperature Anomalies of the Hall Effect in Tungsten
(b) Vol. **89**, No. 1, p. K53–K56 (1978).
See also 21, p. 362.

D. Möckel, W. Rössler, and F. Baumann
*Electrical Resistivity of Amorphous Sb Films and Sb–Au Films
(a) Vol. **49**, No. 1, p. 103–110 (1978).
See also 2, p. 49; *14.3.1*, p. *228*; 21.6, p. 404; 21.7, p. 407.

H. Gärtner und W. Auer
*Zum Einfluß von Nickel-Clustern auf den spezifischen elektrischen Widerstand von paramagnetischen Kupfer–Nickel-Legierungen
(The Influence of Nickel Clusters on the Specific Resistance of Paramagnetic Cu–Ni Alloys)
(a) Vol. **49**, No. 1, p. 149–152 (1978).
See also *10.1*, p. *105*; 21.1, p. 377.

K. Fukamichi
Magnetic Phase Diagramm of the Dilute Cr–Sb Alloy System
(a) Vol. **49**, No. 1, p. K37–K40 (1978).
See also 8, p. 88; *18.4*, p. *301*; 21.1, p. 378.

W. Andrä, H. Danan, R. Mattheis, and U. Zeimer
*On the Electrical Resistivity of PdH_x
(a) Vol. **49**, No. 1, p. K97–K100 (1978).
See also 21, p. 363.

W. A. Theiner and H. P. Geserich
Anderson Transition in $FeSi_2$ Films
(b) Vol. **89**, No. 2, p. 441–447 (1978).
See also *1.2*, p. *22*; *1.5*, p. *39*; 2, p. 50; 20.1, p. 334; 21.1.1, p. 389.

K. Elk, J. Richter, and V. Christoph
*Electrical Conductivity of Disordered Alloys with Strong Hybridization
(b) Vol. **89**, No. 2, p. 537—540 (1978).
See also 13.1, p. 159; 21, p. 363.

G. Kuka, W. Kraak, H.-J. Gollnest, and R. Herrmann
*Temperature Dependence of the Resistivity in Semimetals of the Bismuth Type
(b) Vol. **89**, No. 2, p. 547—551 (1978).
See also 21.7, p. 407.

H. Sato, T. Babauchi, and K. Yonemitsu
*Hall Coefficient of Dilute Aluminium Alloys (II): Application
(b) Vol. **89**, No. 2, p. 571—576 (1978).
See also 21, p. 363.

Y. D. Yao and S. Arajs
*Determination of the Curie Temperature of Ferromagnetic Alloys Using the Deviation from Matthiessen's Rule
(b) Vol. **89**, No. 2, p. K201—K205 (1978).
See also 18.2, p. 290; 21.1, p. 378.

A. Dunlop, N. Lorenzelli and J. C. Jousset
Dechanneling of α Particles by Carbon Dissolved in Palladium
(a) Vol. **49**, No. 2, p. 643—650 (1978).
See also *11*, p. *130*; 21, p. 363.

E. M. Savitskii, P. V. Geld, V. E. Zinovev, N. B. Gorina, V. I. Sperelup, V. P. Polyakova, and A. L. Sokolov
*Transport Properties of Osmium Single Crystals at High Temperatures
(a) Vol. **49**, No. 2, p. K117—K120 (1978).
See also *8*, p. *88*; 21, p. 364.

K. Narayandas, M. Radhakrishnan, and C. Balasubramanian
*Annealing Effect of Silver–Tellurium Films
(a) Vol. **49**, No. 2, p. K213—K215 (1978).
See also 1.2, p. 23; 2, p. 51; 21.6, p. 405; 22.1.3, p. 442.

H. P. Aubauer
*Residual Electrical Resistivity of Alloys
(b) Vol. **90**, No. 1, p. 345—354 (1978).
See also 21, p. 364.

N. N. Sinha and P. L. Srivastava
*Resistivity and Thermo-Electric Power of Monovalent, Divalent, and Trivalent Liquid Metals
(b) Vol. **90**, No. 1, p. 369—374 (1978).
See also 2, p. 51; *15*, p. *261*; 21, p. 364; 21.1, p. 380; 21.2, p. 393.

A. N. Cherkasov, A. A. Galkin, and L. T. Tsymbal
*W and Mo RF Surface Impedance. New Type of Oscillations
(b) Vol. **90**, No. 1, p. K17—K20 (1978).
See also *13.3*, p. *168*; 19, p. 313; 21, p. 364.

J. D. Patterson
The Low Temperature Lorenz Number of Spin Glasses
(b) Vol. **90**, No. 1, p. K85—K88 (1978).
See also *2*, p. *51*; *8*, p. *89*; 21.1.1, p. 390.

R. Schöpke and E. Mrosan
*Influence of Multiple Scattering Effects on the Residual Resistivity of 3d-Transition Metal Impurities in Aluminium
(b) Vol. **90**, No. 1, p. K95—K98 (1978).
See also *13.4*, p. *184*; 21, p. 365.

J. M. Roland, X. Quillard, A. Moreau and R. Gay
Analysis of Electron Diffraction Micrographs of Cu–Al Alloys by the Patterson Function
(a) Vol. **50**, No. 1, p. 41—52 (1978).
See also *1.3*, p. *28*; 21.1, p. 380.

E. M. Schulson and R. B. Turner
*The Electrical Resistivity of Ordered Zr_3Al
(a) Vol. **50**, No. 1, p. 83—86 (1978).
See also 21.1, p. 380.

C. Radomsky and H. Löffler
On the Relation between X-Ray Small-Angle Scattering and Electrical Resistivity
(a) Vol. **50**, No. 1, p. 123—130 (1978).
See also *10.1*, p. *107*; 21.1, p. 381.

G. Pompe, J. Sonntag, A. Biedermann, and S. Schiefer
Abnormal Thermal Conductivity of Chromium–Aluminium Alloys in the Temperature Range 5 to 100 K
(a) Vol. **50**, No. 1, p. 147—152 (1978).
See also *8*, p. *89*; 21.1, p. 381.

H. Puslat
*On the Temperature Dependence of the Hall Coefficient in Cr–Al Alloys
(b) Vol. **90**, No. 2, p. K121—K123 (1978).
See also 21.1, p. 382.

G. Frommeyer
*High-Frequency Conductivity, Classical and Anomalous Skin Effect of Eutectic Silver–Copper Composite Wires
(a) Vol. **50**, No. 2, p. 517—528 (1978).
See also 21.1, p. 382; 21.6, p. 405.

G. Tomasi, G. Frohberg, Th. Hehenkamp, and H. Wever
*Hall-Field Electrotransport of Carbon in Iron
(a) Vol. **50**, No. 2, p. 563—572 (1978).
See also *9*, p. *95*; 21.1.1, p. 391.

14.2 Superconductivity. Superconducting Materials and Devices

Review Article

V. S. Postnikov, V. V. Postnikov, and V. M. Fedorov
*Metastability and Superconductivity in Thin Films
(b) Vol. **87**, No. 1, p. 11—23 (1978).
See also 1.2, p. 15; *1.6*, p. *40*; 21, p. 352; 22, p. 408.

Original Papers and Short Notes

E. Krüger
*Superconductivity Originating from Quasi-Orbital Electrons (I): The Model of Quasi-Orbital Conduction Electrons
(b) Vol. **85**, No. 1, p. 261—270 (1978).
See also 13.1, p. 150; 21, p. 352.

E. K. Naimi
*The Quantum Polarization Effect of Uniform Superconductors in the Electromagnetic Field Having a Critical Frequency
(b) Vol. **85**, No. 1, p. K43—K47 (1978).

R. Mattheis
*Electricel Resistivity of PdH_x Films
(a) Vol. **45**, No. 1, p. 219—227 (1978).
See also *14.1*, p. *205*; 21, p. 353.

G. Fuchs
*Scaling Law in Superconducting NbTi Alloys
(a) Vol. **45**, No. 1, p. 239—247 (1978).
See also 21, p. 353.

E. Krüger
*Superconductivity Originating from Quasi-Orbital Electrons (II): The Superconducting Ground State of Quasi-Orbital Conduction Electron
(b) Vol. **85**, No. 2, p. 493—504 (1978).
See also 13.1, p. 151; 21, p. 354.

K. Flachbart, A. Feher, Š. Jánoš, Z. Málek, and A. Ryska
Thermal Conductivity of Nb–Ti Alloy in the Low-Temperature Range
(b) Vol. **85**, No. 2, p. 545—551 (1978).
See also *8*, p. *83*; 21, p. 354.

V. S. Postnikov, V. V. Postnikov, and V. M. Fedorov
*Instability and Superconductivity in Pd–Ag–D and Pd–H Systems
(b) Vol. **85**, No. 2, p. K115—K117 (1978).
See also 21.6, p. 401.

K. Míšek
Anomalous Attenuation of Ultrasound in a Niobium Single Crystal at Low Temperatures
(b) Vol. **86**, No. 1, p. K31—K33 (1978).
See also *7*, p. *80*; 21, p. 355.

B. K. Chakrabarti, B. P. Barua, and S. K. Sinha
*Coupling of the Martensitic Transformation with the Superconducting Transition
(b) Vol. **86**, No. 1, p. K55—K58 (1978).
See also *1.2*, p. *17*; 21, p. 356.

D. R. Krahn, R. L. Henry, D. B. Tanner, and P. E. Wigen
*The Critical Fields of Superconducting Palladium Hydride
(a) Vol. **46**, No. 1, p. 209—212 (1978).
See also 21, p. 356.

P. Weber, W. Richter, and A. Feuerpfeil
*Improved Sensitivity of Thin Film SQUIDs with Permalloy Films
(a) Vol. **46**, No. 1, p. K67—K69 (1978).
See also 18.2.1, p. 293; 21.1.1, p. 386.

W. John, D. Hamann, and P. Urwank
Partial Densities of States and Electron–Phonon Interaction in Transition Metals
(b) Vol. **86**, No. 2, p. 569—577 (1978).
See also *13.1*, p. *154*; 21.1, p. 371; 21.1.1, p. 386.

A. E. Karkin, B. N. Goshchitskii, V. E. Arkhipov, E. E. Valiev, and S. K. Sidorov
*Electrical Resistivity and Transition Temperature of Highly Disordered Nb_3Sn
(a) Vol. **46**, No. 2, p. K87—K90 (1978).
See also 11, p. 125; 21, p. 357.

J. Sitek, J. Kružliak, M. Tomašich, J. Cirák, and M. Prejsa
*Phase Analysis of Superconducting Nb–Sn Materials by Mössbauer Spectroscopy
(a) Vol. **47**, No. 1, p. 287—291 (1978).
See also 5, p. 61.

R. Kaiser, W. Sprengler, S. Schicktanz, and C. Politis
Raman Spectra and Superconductivity of Various Phases of a High-T_c Superconductor: NbN
(b) Vol. **87**, No. 2, p. 565—573 (1978).
See also *6*, p. *71*; *20.1*, p. *327*.

G. Ilonca and T. W. Barbee, Jr.
*Superconductivity in V–Si and V–Si–C Alloy Films
(a) Vol. **47**, No. 2, p. 357—364 (1978).

E. A. Pushkarev, N. S. Petrenko, and V. A. Finkel
*Thermal Expansion of the Superconducting Compound HfV_2 at Low Temperatures
(a) Vol. **47**, No. 2, p. K145—K148 (1978).
See also *8*, p. *87*; 21, p. 360.

W. Haberkorn, H. Knauer, and J. Richter
*A Theoretical Study of the Current–Phase Relation in Josephson Contacts
(a) Vol. **47**, No. 2, p. K161—K164 (1978).

G. Dharmadurai and B. A. Ratnam
*A Non-Ohmic Potential Transverse to a Long Superconducting Weak-Link
(a) Vol. **47**, No. 2, p. K187—K190 (1978).
See also 21, p. 360.

P. Otschik
*The Effect of Oxygen Impurities on the Electron Tunneling
(b) Vol. 88, No. 2, p. 563—568 (1978).

R. Glocker and L. Fritsche
Pressure Dependence of the Electron–Phonon Coupling in Lanthanum
(b) Vol. 88, No. 2, p. 639—644 (1978).
See also *6*, p. *74*; 12.1, p. 136; *13.1*, p. *157*; 21.4, p. 398.

W. Richter and R. Diller
*Flat Thin Film SQUIDs with Current Transformers
(a) Vol. 48, No. 2, p. 513—521 (1978).
See also 21, p. 362.

T. Skośkiewicz
*Preparation of Superconducting Hydrides of Palladium–Nickel Alloys under High Hydrogen Pressures
(a) Vol. 48, No. 2, p. K165—K168 (1978).
See also 21.1, p. 377; 21.6, p. 404.

F. Dettmann
*Properties of Quantum Interference Devices with V–VO_x–Pb Tunnel Junctions
(a) Vol. 49, No. 1, p. 293—298 (1978).
See also 22.6, p. 496.

H.-J. Ullrich, U. Reinhold, S. Däbritz, P. Paufler, K. Kleinstück, and B. Pietrass
On the Lattice Transformation in Plastically Deformed V_3Si Single Crystals Studied by the Kossel Technique
(a) Vol. 49, No. 1, p. 323—330 (1978).
See also *1.2*, p. *22*; *10.1*, p. *106*; 21.1, p. 377.

P. Müller
*Preparation of Superconducting Nb_3Au Diffusion Layers
(a) Vol. 49, No. 1, p. K75—K79 (1978).
See also 1.2, p. 22.

H.-J. Köhler, K. Blüthner, S. Linke, P. Weber, and G. Albrecht
*Some Properties of Miniature Nb–NbO_x–Pb Tunnel Junctions
(a) Vol. 49, No. 2, p. 569—572 (1978).

W. Zahn and F. Dettmann
*Resonance Effects in Quantum Interference Devices with Tunnel Junctions
(a) Vol. 49, No. 2, p. 573—576 (1978).

W. Richter and F. Dettmann
*Thin-Film DC SQID Gradiometer
(a) Vol. 49, No. 2, p. K209—K211 (1978).

L. Dumoulin, P. Nédellec, C. Arzoumanian, and J. P. Burger
*Optical Phonons in PdH_x and PdD_x by Superconducting Tunneling Measurements
(b) Vol. 90, No. 1, p. 207—210 (1978).
See also *6*, p. *77*; 21, p. 364.

W. N. Mathews Jr.
*Quasiparticle, Charge, and Energy Conservation in Weak-Coupling Superconductors
(b) Vol. **90**, No. 1, p. 327—338 (1978).

H. U. Habermeier, R. Aoki, and H. Kronmüller
*The Quantitative Determination of Flux-Density Gradients in Type-II Superconductors by Means of the Magneto-Optical Faraday Effect
(a) Vol. **50**, No. 1, p. 187—194 (1978).
See also 18, p. 275; 20, p. 318; 21, p. 365.

L. Rothkirch, R. Herrmann, H. Dwelk, and U. Preppernau
*Surface Impedance Investigations of Superconducting Niobium
(b) Vol. **90**, No. 2, p. 517—524 (1978).
See also 13.3, p. 168; 21, p. 365.

M. Raychaudhury, P. Chatterjee, and S. Chatterjee
*Electron–Phonon Coupling Constant and Critical Temperature of Some A15 Superconducting Compounds of Nb and V
(b) Vol. **90**, No. 2, p. 609—613 (1978).
See also *13.1*, p. *161*.

E. Krüger
*Superconductivity Originating from Quasi-Orbital Electrons (III): Quasi-Orbital Conduction Electrons in Non-Adiabatic Systems
(b) Vol. **90**, No. 2, p. 719—731 (1978).
See also 13.1, p. 161; 21, p. 366.

L. Rothkirch
*Magnetic Surface States in Superconductors
(b) Vol. **90**, No. 2, p. K115—K119 (1978).
See also *13.3*, p. *168*; 21, p. 366.

F. Dettmann, H.-J. Köhler, and W. Hild
*Static Characteristics of Tunneling Cryotrons with Nb–NbO_x–Pb Junctions
(a) Vol. **50**, No. 2, p. 417—422 (1978).

N. P. Kobelev and Ya. M. Soifer
*Electronic Damping of Dislocations in Lead
(a) Vol. **50**, No. 2, p. K185—K188 (1978).
See also 7, p. 82; *10.1*, p. *108*; 21, p. 367.

14.3 Semiconductors

Review Articles

R. Suryanarayanan
Valence Related Optical and Other Studies of Sm and Tm Chalcogenides
(b) Vol. **85**, No. 1, p. 9—43 (1978).
See also 13.1, p. 150; 14.1, p. 205; *20.1*, p. *319*; 22.7, p. 501.

A. S. Borukhovich
*Kinetic Properties of IVa and Va Subgroup Transition Metal Monocarbides
(a) Vol. **46**, No. 1, p. 11—37 (1978).
See also 8, p. 83; 13.1, p. 150; 15, p. 259; 22, p. 408; 22.3, p. 457; 22.3.1, p. 460.

P. C. Mathur, R. Shyam, and Sushil Jain
*Transport Phenomena in III–V Compound Semiconductors
(a) Vol. **50**, No. 1, p. 11—40 (1978).
See also 6, p. 64; *13.1*, p. *150*; 13.4, p. 169; 20.1, p. 319; 22.2, p. 443.

Original Papers and Short Notes

I. P. Zvyagin
*Comments on the Theory of Hopping Conduction
(b) Vol. **85**, No. 1, p. 105—109 (1978).
See also 13.4, p. 169; 22, p. 408.

G. J. Ashwell
*A Comment upon the Metallic and Semiconducting Phases of $(DEPE)^{2+}$ $(TCNQ)_4^{2-}$
(b) Vol. **85**, No. 1, p. K7—K9 (1978).
See also *14.1*, p. *205*; 22.9, p. 522.

W. Eichler
*On the Analysis of Galvano- and Thermomagnetic Effects in Pure Perfect Tellurium Crystals
(a) Vol. **45**, No. 1, p. 195—199 (1978).
See also *15*, p. *259*; 22.1.3, p. 441.

R. M. M. Fonville, W. Geertsma, and C. Haas
On the Nature of the Top of the Valence Band in Layered Mo and W Dichalcogenides
(b) Vol. **85**, No. 2, p. 621—627 (1978).
See also *13.1*, p. *151*; 18.1, p. 277; 19, p. 304; 22, p. 409.

J. J. Dubowski
*Disorder Scattering in $Cd_xHg_{1-x}Te$ Mixed Crystals
(b) Vol. **85**, No. 2, p. 663—672 (1978).
See also *13.1*, p. *151*; 15, p. 259; 22.4.4, p. 472.

S. C. Varshney and A. A. Gundjian
*Internal Strain and Piezoelectricity in Zincblende Crystals
(b) Vol. **85**, No. 2, p. 733—742 (1978).
See also 6, p. 66; *12.2*, p. *138*; 22, p. 409.

A. ALADGADGYAN and A. TONEVA
Study of the Valence Band Structure of $Pb_{1-x}Sn_xTe$
(b) Vol. **85**, No. 2, p. K131–K134 (1978).
See also *13.1*, p. *152*; 22.7, p. 501.

H. RUNGE and H. WÖRL
Autoradiographic Detection of Getter Effect of Argon-Implanted Layers in Silicon
(a) Vol. **45**, No. 2, p. 509–512 (1978).
See also *9*, p. *91*; *11*, p. *124*; 22.1.2, p. 430.

J. KOČKA and J. KRIŠTOFIK
*AC Conductivity of $CdAs_2$-Based Glasses
(a) Vol. **45**, No. 2, p. 559–564 (1978).
See also *2*, p. *43*; 22.3, p. 457.

V. CRISTEA and V. BABES
The Seebeck Effect in Nb-Doped TiO_2 Rutile
(a) Vol. **45**, No. 2, p. 617–624 (1978).
See also *15*, p. *259*; 22.6, p. 491.

A. ZARĘBA
*Piezoconductivity of β-Rhombohedral Boron
(a) Vol. **45**, No. 2, p. K183–K187 (1978).
See also 22.1, p. 422.

S. H. EL-MAHALAWY and B. L. EVANS
*Pressure Dependence of the Electrical Conductivity in 2H-MoS_2 and 2H–WSe_2
(b) Vol. **86**, No. 1, p. 151–157 (1978).
See also 12.2, p. 138; 22, p. 410.

T. IKARI and K. HASHIMOTO
*Electrical Properties of TlTe Single Crystals
(b) Vol. **86**, No. 1, p. 239–248 (1978).
Erratum
(b) Vol. **90**, No. 2, p. 751 (1978).
See also 13.1, p. 153; *15*, p. *259*; 22.3, p. 457.

B. JENSEN
*The Infrared Absorption Spectrum of n-Type GaAs
(b) Vol. **86**, No. 1, p. 291–301 (1978).
See also *20.1*, p. *322*; 22.2.1, p. 445.

V. P. OLEINIK and V. A. SINYAK
To the Theory of the Tunnel Effect in a Semiconductor in Constant Electric and Magnetic Fields
(b) Vol. **86**, No. 1, p. 319–327 (1978).
See also *13.1*, p. *153*; 22, p. 410.

W. EICHLER and G. SIMON
*Galvanomagnetic Transport Properties of Sb_2Te_3
(b) Vol. **86**, No. 1, p. K85–K88 (1978).
See also 22.7, p. 501.

P. Gaworzewski and H. Riemann
*On the Effect of Heat Treatment upon the Charge Carrier Concentration of Dislocation-Free Floating-Zoned Silicon Crystals
(a) Vol. **46**, No. 1, p. 117—124 (1978).
See also 13.4, p. 173; 22.1.2, p. 431.

P. R. Jay and B. Tuck
*Mercury Probe C–U Measurements on GaP
(a) Vol. **46**, No. 1, p. 199—208 (1978).
See also *9*, p. *92*; 22.2.2, p. 450.

P. L. F. Hemment and R. W. Bicknell
1.5 MeV H^+ and He^+ Irradiations of Arsenic Diffused Layers in ⟨100⟩ Silicon
(a) Vol. **46**, No. 1, p. 225—231 (1978).
See also *11*, p. *125*; 13.4, p. 173; 22.1.2, p. 431.

M. Samoć and Z. Zboiński
Trapping of Charge Carriers in the Deformed Surface Region of Anthracene Single Crystals
(a) Vol. **46**, No. 1, p. 251—257 (1978).
See also *13.3*, p. *165*; 16, p. 263; 22.9, p. 524.

S. Prasad and N. C. Goel
*DC Resistivity of Doped Mn–Zn Polycrystalline Ferrites
(a) Vol. **46**, No. 1, p. 373—377 (1978).
See also 22.8.2, p. 518.

E. Gerlach and M. Rautenberg
*Ionized Impurity Scattering in Semiconductors
(b) Vol. **86**, No. 2, p. 479—482 (1978).
See also *13.4*, p. *174*; 22, p. 411.

P. Thomas and D. Wuertz
*Low Temperature Polaronic Hopping Conduction
(b) Vol. **86**, No. 2, p. 541—548 (1978).
See also *2*, p. *43*; 15, p. 260; 22.1.1, p. 425; 22.1.2, p. 432.

J. Kossut
The Disorder Scattering in Zincblende Narrow-Gap Semiconducting Mixed Crystals
(b) Vol. **86**, No. 2, p. 593—601 (1978).
See also *13.1*, p. *154*; *13.4*, p. *174*; 22.4.4, p. 472.

C. G. Patil and B. S. Krishnamurthy
*Carrier Mobility in Ferromagnetic Semiconductors
(b) Vol. **86**, No. 2, p. 725—732 (1978).
See also 13.5.4, p. 201; 18.2, p. 285; 22.4.4, p. 472.

P. I. Baranskii, L. S. Demidenko, S. L. Korolyuk, and A. G. Samoilovich
*Impurity Scattering of Electrons in Quantizing Magnetic Fields
(b) Vol. **86**, No. 2, p. 765—772 (1978).
See also 13, p. 144; 22.2.1, p. 446.

I. M. Dykman and P. M. Tomchuk
*Electromagnetic Waves in Inhomogeneous Semiconductors with Non-Equilibrium Carriers
(b) Vol. **86**, No. 2, p. 773—780 (1978).
See also *13*, p. *144*; 20, p. 316; 22, p. 412.

B. Hughes and G. H. Narayanan
On Annealing-Induced Prismatic Dislocation Loops and Electrical Changes in Heavily Te-Doped GaAs
(a) Vol. **46**, No. 2, p. 627—637 (1978).
See also *1.4*, p. *30*; *10.2*, p. *112*; 22.2.1, p. 446.

B. R. Sethi and P. C. Mathur
*Effect of Heavy Doping of ZnSe Crystals with Indium in Creating Compensating Acceptors
(a) Vol. **46**, No. 2, p. 717—721 (1978).
See also *13.4*, p. *174*; 22.4.2, p. 468.

A. Martinaitis and A. Sakalas
*High Temperature Investigations of the Hall Effect in p-CdSe in Selenium Vapour Atmosphere
(a) Vol. **46**, No. 2, p. K141—K144 (1978).
See also 10.2, p. 112; 22.4.2, p. 468.

S. Guha and S. Ghosh
*Hybrid Mode Interaction in Inhomogeneous Semiconductor Plasmas
(b) Vol. **87**, No. 1, p. K9—K12 (1978).
See also *20*, p. *316*; 22, p. 413.

S. Roth and J. Sonntag
*Temperature Dependence of the Electrical Resistivity of Semiconducting CrAl Solid Solutions at High Temperatures
(b) Vol. **87**, No. 1, p. K69—K72 (1978).
See also *18.4*, p. *300*; 21, p. 358; 21.1, p. 372.

C. Wagner and F. Burkhardt
*The Annealing Characteristics of Phosphorus-Implanted Silicon Investigated at Low Temperatures
(a) Vol. **47**, No. 1, p. 131—138 (1978).
See also *11*, p. *126*; 22.1.2, p. 432.

R. Wernicke
*The Influence of Kinetic Processes on the Electrical Conductivity of Donor-Doped $BaTiO_3$ Ceramics
(a) Vol. **47**, No. 1, p. 139—144 (1978).
See also 10.2, p. 113; 22.8.1, p. 513.

A. V. Martinaitis, A. P. Sakalas, Z. V. Januškevičius, and J. K. Viščakas
*Defect Structure of CdSe in Selenium Vapour
(a) Vol. **47**, No. 1, p. 187—194 (1978).
See also *10.2*, p. *113*; 22.4.2, p. 468.

V. CHÁB, L. ŠTOURAČ, and B. VELICKÝ
*The Composition Dependence of Activation Energy in the As_x–Te_{100-x} and $As_2Te_3 + Tl_x$ Glassy Systems
(a) Vol. 47, No. 1, p. 271—276 (1978).
See also *2*, p. *45*; 15, p. 260; 22, p. 413.

O. PAVLOV and Z. BONTCHEVA-MLADENOVA
*Measurements of the Ag_2Te–Ag_2S System at 300 K
(a) Vol. 47, No. 1, p. K59—K61 (1978).
See also *1.3*, p. *26*; 8, p. 86; 15, p. 260; 22, p. 413.

I. M. CHAPNIK
*Atomic Volume and Semiconducting Properties in Sulfides of Non-Magnetic Transition Metals
(a) Vol. 47, No. 1, p. K71—K75 (1978).
See also *1.3*, p. *26*; 14.1, p. 207; 22, p. 413.

M. A. MEKHTIEV and S. M. SEID-RZAEVA
*Quasi-Local Acceptor Levels and Electron Mobility in $Hg_{1-x}Cd_xTe$
(b) Vol. 87, No. 2, p. 447—454 (1978).
See also *13.4*, p. *175*; 22.4.4, p. 473.

A. I. MERKULOV, S. I. RADAUTSAN, and V. E. TEZLEVAN
*The Effect of the Degree of Doping on Magneto-Electrical Properties of n-$CdCr_2Se_4$ Monocrystals
(b) Vol. 87, No. 2, p. K141—K143 (1978).
See also 22.8, p. 505.

B. R. SETHI, P. L. TALWAR, O. P. SHARMA, and P. C. MATHUR
*Effect of Indium Doping and Zinc Annealing on the Electrical Transport Properties of Zinc Selenide Single Crystals
(a) Vol. 47, No. 2, p. 699—702 (1978).
See also *13.4*, p. *176*; 22.4.2, p. 468.

W. BRÜCKNER, W. MOLDENHAUER, and H. OPPERMANN
*Conductivity Transitions in (Ga + Mo)-Doped VO_2
(a) Vol. 47, No. 2, p. K129—K132 (1978).
See also *1.2*, p. *20*; 22.6, p. 494.

GH. ILONCA
*Electrical Resistivity and Magnetoresistivity of Amorphous $Cr_xPd_{82-x}Ge_{18}$ Alloys
(b) Vol. 88, No. 1, p. 37—41 (1978).
See also *2*, p. *46*; 22, p. 414.

M. JACZYŃSKI, J. KOSSUT, and R. R. GAŁĄZKA
*Influence of Exchange Interaction on the Quantum Transport Phenomena in $Hg_{1-x}Mn_xTe$
(b) Vol. 88, No. 1, p. 73—85 (1978).
See also *13.2*, p. *162*; 15, p. 260; 22.4.4, p. 473.

W. KRAAK, G. OELGART, G. SCHNEIDER, and R. HERRMANN
*The Semiconductor–Semimetal Transition in $Bi_{1-x}Sb_x$ Alloys with $x \geqq 0.22$
(b) Vol. 88, No. 1, p. 105—110 (1978).
See also *13.4*, p. *177*; 14.1, p. 208; 22.7, p. 502.

I. P. Zvyagin
*Hopping Conductivity in the Presence of an Electromagnetic Wave
(b) Vol. 88, No. 1, p. 149–154 (1978).
See also *2*, p. *46*; 6, p. 73.

N. P. Gavaleshko, P. N. Gorley, P. V. Kavyuk, L. S. Solonchuk, I. I. Tarasyuk, and V. A. Shenderovskii
Transverse and Longitudinal Nernst-Ettingshausen Effect in p-Te
(b) Vol. 88, No. 1, p. 179–188 (1978).
See also *13.4*, p. *177*; *15*, p. *260*; 22.1.3, p. 441.

P. Peepper and W. Zawadzki
*Inelastic Scattering of Electrons by Optic Phonons in InSb-Type Semiconductors
(b) Vol. 88, No. 1, p. 247–260 (1978).
See also 6, p. 73; 13.1, p. 156; 22.2.3, p. 453.

Z. M. Alekseeva, A. P. Vyatkin, G. F. Karavaea, and N. P. Krivorotov
Effect of Pressure on Hole Effective Masses in GaAs
(b) Vol. 88, No. 1, p. 321–326 (1978).
See also *12.2*, p. *140*; *13.1*, p. *156*; 22.2.1, p. 447.

D. Wuertz and P. Thomas
*The Pre-Exponential in DC-Hopping Conduction
(b) Vol. 88, No. 1, p. K73–K77 (1978).
See also *2*, p. *46*; 22.1.2, p. 434.

M. Egée
Effect of a Direct Electric Field on the Photoluminesćene Processes and Model of the Luminogen Centre in ZnS:Cu
(a) Vol. 48, No. 1, p. 89–98 (1978).
See also 13.4, p. 177; *20.3*, p. *345*; 22.4.1, p. 465.

E. Kuroda, M. Matsuda, and M. Maki
Growth and Characterization of Silicon Ribbon Crystals Grown with Wetting and Non-Wetting Dies
(a) Vol. 48, No. 1, p. 105–111 (1978).
See also *3*, p. *55*; 22.1.2, p. 434.

T. I. Kolchenko and V. M. Lomako
Removal and Scattering of Charge Carriers by Defect Clusters in Semiconductors
(a) Vol. 48, No. 1, p. 263–269 (1978).
See also *11*, p. *128*; 13.4, p. 178; 22.2.1, p. 447.

J. Kočka
*Influence of Contacts on the AC Conductivity of Glasses
(a) Vol. 48, No. 1, p. K59–K62 (1978).
See also *2*, p. *47*; 14.3.4, p. 243; 22, p. 415.

E. Ziegler, W. Siegel, G. Kühnel, and E. Buhrig
*Incorporation of Gallium in $ZnSiP_2$
(a) Vol. 48, No. 1, p. K63–K65 (1978).
See also 13.4, p. 178; 22.8, p. 506.

G. Oelgart, R. Stegmann, and R. Herrmann
*Propagation of Electromagnetic Waves in a Direction Parallel to the Magnetic Field in Semiconductive Bi–Sb Alloys
(b) Vol. 88, No. 2, p. 507—516 (1978).
See also *13.2*, p. *162*; 19, p. 310; 21.7, p. 407.

B. Stébé and G. Munschy
Binding of an Excition to a Neutral Acceptor
(b) Vol. 88, No. 2, p. 713—725 (1978).
See also *13.4*, p. *178*; *13.5.1*, p. *193*; 22.2.1, p. 447.

S. Jandl, G. Perluzzo, and P. E. Girard
*Mobility of Electrons in SnSSe Single Crystals
(b) Vol. 88, No. 2, p. K135—K137 (1978).
See also 22, p. 416.

N. A. Poklonskii, V. F. Stelmakh, V. D. Tkachev, and S. V. Voitikov
*Screening of Electrostatic Fields in Semiconductors by Multicharged Defects
(b) Vol. 88, No. 2, p. K165—K168 (1978).
See also *13.4*, p. *179*; 22, p. 416.

C. Manfredotti, A. M. Mancini, A. Rizzo, R. Murri, and L. Vasanelli
*Electrical Properties of GaS_xSe_{1-x} Solid Solutions
(a) Vol. 48, No. 2, p. 293—296 (1978).
See also *13.4*, p. *179*; 22.3.1, p. 460.

I. S. Gorban, G. A. Grishchenko, A. P. Sakalas, A. S. Sodeika, I. I. Tychina, and A. K. Tkachenko
*Electrical Properties of ZnP_2 Single Crystals of Tetragonal and Monoclinic Modifications
(a) Vol. 48, No. 2, p. 329—334 (1978).
See also *13.4*, p. *179*; 22.3, p. 458.

E. I. Terukov, W. Reichelt, M. Wolf, H. Hemschik, and H. Oppermann
On the Influence of the Substitution of O^{16} by O^{18} in a Few Vanadium Oxides on Their Semiconductor–Metal Transition
(a) Vol. 48, No. 2, p. 377—381 (1978).
See also *1.2*, p. *21*; 6, p. 75; 22.6, p. 495.

G. Juška, S. Vengris, and K. Arlauskas
*Electron Lifetime in Amorphous Selenium
(a) Vol. 48, No. 2, p. K149—K151 (1978).
See also *2*, p. *48*; 22.1.3, p. 441.

M. Moldovanova and R. Assenov
*A New Interpretation of the Temperature Dependence of the Hall Coefficient in "Pure" and Doped SnTe
(a) Vol. 48, No. 2, p. K193—K197 (1978).
See also 15, p. 261; 22, p. 416.

P. Kocever and E. Fitz
The Influence of Acoustic Phonon Disturbances on the Conductivity of Piezo-Electric Semiconductors
(b) Vol. **89**, No. 1, p. 225—233 (1978).
See also *6*, p. *75*; 22.2.1, p. 448.

K. Graff and H. Pieper
*Degradation of Carrier Lifetime in Silicon Crystals at Room Temperature
(a) Vol. **49**, No. 1, p. 137—144 (1978).
See also 10.2, p. 118; 22.1.2, p. 436.

S. Myhra
*Defect Annealing and Transport Properties in High-Purity InSb Irradiated at 80 K
(a) Vol. **49**, No. 1, p. 285—292 (1978).
See also *10.2*, p. *118*; 15, p. 261; 22.2.3, p. 454.

G. Stetter, H. Coufal, and E. Lüscher
EPR Study of the Annealing Behaviour of Neutron-Transmutation-Doped Silicon
(a) Vol. **49**, No. 1, p. K17—K20 (1978).
See also *11*, p. *129*; *19*, p. *312*; 22.1.2, p. 437.

S. Guha and S. Ghosh
*Drift Wave Instability in Solid-State Plasmas
(b) Vol. **89**, No. 2, p. K171—K174 (1978).
See also *20*, p. *317*; 22, p. 418.

V. N. Brudnyi, M. A. Kirov, A. I. Potapov, I. K. Polushina, V. D. Prochukhan, and Yu. V. Rud
*Change of Electrical Properties in Electron Irradiated $CdGeAs_2$ Crystals
(a) Vol. **49**, No. 2, p. 761—765 (1978).
See also *11*, p. *130*; 13.4, p. 182; 22.8, p. 507.

E. Ziegler, W. Siegel, and G. Kühnel
*Potential Barrier Scattering in n-Type Single Crystals of $ZnSiP_2$
(a) Vol. **49**, No. 2, p. K205—K208 (1978).
See also 13.4, p. 183; 22.8, p. 507.

H. Van Cong
*Influence of Band-Tail Parameters on Electronic Specific Heat and Transport Coefficients in Heavily Doped Semiconductors
(b) Vol. **90**, No. 1, p. 401—407 (1978).
See also *8*, p. *89*; 13.4, p. 184; 22, p. 419.

B. Gillot, F. Bouton, F. Chassagneux, and A. Rousset
*Electrical Conductivity of $(Fe^{2+}Al^{3+}_{2-x}Cr^{3+}_x)O_4^{2-}$ Spinels and Defect Phases $\gamma(Fe^{3+}_{1/3}Al^{3+}_{2/3-y}Cr^{3+}_y)_2O_3^{2-}$
(a) Vol. **50**, No. 1, p. 109—116 (1978).
See also 18.3, p. 296; 22.8.2, p. 520.

N. G. Nilsson
*Empirical Approximations Applied to the Generalized Einstein Relation for Degenerate Semiconductors
(a) Vol. **50**, No. 1, p. K43—K45 (1978).
See also 22, p. 420.

P. MACKO and J. DOUPOVEC
*Thermal and Electrical Properties of Neutron Irradiated GeS_{11} Glass
(a) Vol. **50**, No. 1, p. K51—K54 (1978).
See also 2, p. 52; *8*, p. *89*; 11, p. 130; 22, p. 420.

H. KÖHLER and W. OSSAU
Magneto-Seebeck and Shubnikov-de Haas Effect in n-Type $(Bi_{1-x}Sb_x)_2Te_3$
(b) Vol. **90**, No. 2, p. 509—515 (1978).
See also *13.1*, p. *160*; *15*, p. *261*; 22.7, p. 502.

R. S. MARKIEWICZ
Acoustic Phonon Scattering by Electron–Hole Droplets in Ge (II): Total Scattering Rate and Anisotropy of Scattering, as Reflected in the Shape of a Droplet Cloud
(b) Vol. **90**, No. 2, p. 585—596 (1978).
See also *13.5.1*, p. *197*; 22.1.1, p. 428.

V. N. BRUDNYI, D. L. BUDNITSKII, M. A. KRIVOV, R. V. MASAGUTOVA, V. D. PROCHUKHAN, and YU. V. RUD
*The Electrical and Optical Properties of 2.0 MeV Electron-Irradiated $ZnGeP_2$
(a) Vol. **50**, No. 2, p. 379—384 (1978).
See also 11, p. 131; 13.4, p. 185; *20.1*, p. *337*; 22.8, p. 508.

A. ROIZÈS and R. SCHUTTLER
*Galvanomagnetic Effects in p-Type Silicon. Application to Electrically Active Impurity Contents
(a) Vol. **50**, No. 2, p. 399—409 (1978).
See also 13.4, p. 185; 22.1.2, p. 439.

I. A. MALTSEVA, A. MAMEDOV, YU. V. RUD, and YU. K. UNDALOV
The Photosensitivity of Diodes Based on $CdGeP_2$:In Crystals
(a) Vol. **50**, No. 2, p. 445—448 (1978).
See also *16*, p. *267*; 22.8, p. 508.

S. KAWAMURA and B. S. H. ROYCE
*Thermally Stimulated Current Studies of Electron and Hole Traps in Single Crystal Al_2O_3
(a) Vol. **50**, No. 2, p. 669—677 (1978).
See also *13.4*, p. *186*; 22.6, p. 497.

V. K. MALYUTENKO, S. S. BOLGOV, and YU. M. MALOZOVSKY
*Magnetoconcentration Effect at Nonlinear Recombination of Current Carriers
(a) Vol. **50**, No. 2, p. 723—732 (1978).
See also 22.2.3, p. 455.

N. E. KORSUNSKAYA, I. V. MARKEVICH, T. V. TORCHINSKAYA, and M. K. SHEINKMAN
*Some Peculiarities of Thermostimulated Conductivity and Optical Quenching of the Photocurrent in Crystals with the Auger Excitation of Recombination Centres
(a) Vol. **50**, No. 2, p. 767—770 (1978).
See also *13.4*, p. *186*; 16, p. 268; 22.4.1, p. 467.

A. K. SEMENYUK and M. D. VARENTSOV
*Anisotropy of Galvanomagnetic Effects in Irradiated Silicon with Layer Inhomogeneity
(a) Vol. **50**, No. 2, p. K169—K173 (1978).
See also 11, p. 131; 22.1.2, p. 440.

14.3.1 Films

Original Papers and Short Notes

R. M. Mehra, A. K. Kathuria, Radhey Shyam, A. L. Dawar, and P. C. Mathur
Determination of the Density of Localized States in a-Se Films
(a) Vol. **45**, No. 1, p. K59—K62 (1978).
See also *2*, p. *43*; *13.4*, p. *170*; 22.1.3, p. 441.

E. I. Terukov, W. Reichelt, and H. Oppermann
*Zum Einfluß von Dotierungen auf den Phasenübergang des V_3O_5
(On the Influence of Doping on the Phase Transition of V_3O_5)
(a) Vol. **45**, No. 1, p. K77—K79 (1978).
See also *14.1*, p. *205*; 22.6, p. 491.

V. Dolocan
*Some Electrical Properties of Bi_2O_3 Thin Films
(a) Vol. **45**, No. 2, p. K155—K157 (1978).
See also 14.3.4, p. 241; 22.6, p. 491.

S. D. Phadke and R. N. Karekar
Photocurrents through Polythiourea and Polymalononitrile Thin Films
(a) Vol. **45**, No. 2, p. K163—K165 (1978).
See also *16*, p. *263*; 22.9, p. 523.

M. Hirose, A. Fischer, and K. Ploog
*Growth of Al_2O_3 Layer on MBE GaAs
(a) Vol. **45**, No. 1, p. K175—K177 (1978).
See also *3*, p. *54*; 22.6, p. 491.

B. M. Askerov, B. I. Kuliev, S. R. Figarova, and V. Ya. Shteinshraiber
Theory of the Thermomagnetic Effects in Thin Films in a Magnetic Field of Arbitrary Magnitude
(b) Vol. **86**, No. 1, p. 103—108 (1978).
See also *15*, p. *259*.

C. Lawther and J. Woods
*The Calculation of Donor Densities from *C–U* Measurements on MIS ZnSe Diodes
(a) Vol. **46**, No. 1, p. 245—250 (1978).
See also *13.4*, p. *173*; 22.4.2, p. 468.

A. E. Gershinskii, L. V. Mironova, and E. I. Cherepov
*On the Conductivity of the Si–Thermal SiO_2–Electrolyte System in Superhigh Electric Fields
(a) Vol. **46**, No. 2, p. 467—473 (1978).
See also *14.3.4*, p. *241*; 22.1.2, p. 432; 22.6, p. 492.

A. Banerjee, Prem Nath, V. D. Vankar, and K. L. Chopra
*Properties of $Zn_xCd_{1-x}S$ Films Prepared by Solution Spray Technique
(a) Vol. **46**, No. 2, p. 723—728 (1978).
See also *1.5*, p. *36*; 16, p. 263; 22.4.4, p. 472.

C. E. Wickersham and J. E. Greene
*The Effect of Substrate Bias on the Electrical and Optical Properties of In_2O_3 Films Grown by RF Sputtering
(a) Vol. **47**, No. 1, p. 329—337 (1978).
See also 13.4, p. 175; *20.1*, p. *327*; 22.6, p. 493.

C. Albers, Ch. Walkowiak, P. Schäfer, and I. Rechenberg
*Time Dependent Negative Weak-Field Magnetoresistance in n-Type Lead Telluride Epitaxial Films
(a) Vol. **47**, No. 1, p. K31—K34 (1978).
See also 22, p. 413.

K. O. Lee
*The Meyer-Neldel Rule and the DC Conductivity in Amorphous Ge Films
(a) Vol. **47**, No. 1, p. K47—K49 (1978).
See also *2*, p. *45*; 22.1.1, p. 425.

N. Romeo, G. Sberveglieri, L. Tarricone, J. Vidal, and A. Wojtowicz
*Electrical Properties of Sb-Doped ZnTe Thin Films
(a) Vol. **47**, No. 2, p. 371—374 (1978).
See also 22.4.3, p. 470.

A. L. Dawar, K. V. Krishna, O. P. Taneja, and P. C. Mathur
*Electrical Transport Properties of Polycrystalline p-Type InSb Films
(a) Vol. **47**, No. 2, p. 375—378 (1978).
See also 13.4, p. 176; 22.2.3, p. 453.

W. Ecke, A. A. Lebedev, and A. T. Mamadalimov
*Deep Defect States in Evaporated Heteroepitaxial Silicon Films
(a) Vol. **47**, No. 2, p. 507—512 (1978).
See also *13.4*, p. *176*; 22.1.2, p. 433.

E. N. Mgbenu
Room Temperature Au–Ge Interdiffusion
(a) Vol. **47**, No. 2, p. K153—K155 (1978).
See also *9*, p. *93*; 22.1.1, p. 426.

A. A. Cottey
*Band Theory of Size-Quantised Electron States in Thin Crystalline Films
(b) Vol. **88**, No. 1, p. 207—219 (1978).
See also *13.1*, p. *156*; 13.3, p. 166.

V. Dutta, P. Nath, and K. L. Chopra
*Structural and Electrical Properties of Polycrystalline Ge Films
(a) Vol. **48**, No. 1, p. 257—262 (1978).
See also *1.5*, p. *37*; 22.1.1, p. 427.

Nguyen van Dong and Tran Quoc Hai
*Electronic Transport in Doped a-Ge and a-Si Prepared by DC Cathodic Sputtering
(b) Vol. **88**, No. 2, p. 555—561 (1978).
See also *2*, p. *47*; 22.1.1, p. 427; 22.1.2, p. 435.

W. R. Fahrner, D. Bräunig, R. Ferretti, and B. Müller
*A Rapid Destruction-Free Determination of Breakdown Voltage of Thin Insulating Layers
(a) Vol. **48**, No. 2, p. 533—541 (1978).
See also *14.3.4*, p. *244*; 22, p. 416.

K. I. KIROV, S. S. GEORGIEV, E. V. GEROVA, and S. P. ALEKSANDROVA
*Investigation of SiO_2 Layers Deposited by Plasma Decomposition of Tetra-Ethoxy Silane in a Planar Reactor
(a) Vol. **48**, No. 2, p. 609—614 (1978).
See also *1.5*, p. *38*; 22.6, p. 495.

H. D. DIMITROV
*On the Surface Electrical Resistivity of Thin Semiconducting and Semimetallic Films
(b) Vol. **89**, No. 1, p. 319—328 (1978).
See also *13.3*, p. *166*; 14.3.2, p. 231; 22, p. 417.

M. ALTWEIN, H. FINKENRATH, and D. GERHARDT
*Temperature Dependence of the Magnetoresistance in CdO
(a) Vol. **49**, No. 1, p. 67—71 (1978).
See also 22.4, p. 462.

D. MÖCKEL, W. RÖSSLER, and F. BAUMANN
*Electrical Resistivity of Amorphous Sb Films and Sb–Au Films
(a) Vol. **49**, No. 1, p. 103—110 (1978).
See also 2, p. 49; *14.1*, p. *210*; 21.6, p. 404; 21.7, p. 407.

W. HOYER, G. LAUTZ, and M. HEYEN
*Low-Frequency Current Noise of GaAs Epitaxial Layers between 4.4 and 300 K
(a) Vol. **49**, No. 1, p. 111—117 (1978).
Erratum
(a) Vol. **50**, No. 2, p. 771 (1978).
See also 13.4, p. 180; *14.3.3*, p. *236*; 22.2.1, p. 448.

U. SAXENA and O. N. SRIVASTAVA
*Effect of Nonstoichiometry on the Electrical Conductivity of Erbium Oxide Films
(a) Vol. **49**, No. 1, p. 371—378 (1978).
See also *1.5*, p. *38*; 2, p. 49; 13.1, p. 158; 22.6, p. 496.

S. EWERT, G. HEILAND, H. LÜTH, U. ROLL, and W. SANDER
*Molecular Vibrations in a Polymethine Dye Layer Studied by Inelastic Electron Tunneling Spectroscopy
(b) Vol. **89**, No. 2, p. 459—465 (1978).
See also 6, p. 76; *20.1*, p. *334*; 22.9, p. 527.

W. FUHS, M. MILLEVILLE, and J. STUKE
*Drift Mobility and Photoconductivity in Amorphous Silicon
(b) Vol. **89**, No. 2, p. 495—502 (1978).
See also 2, p. 50; *16*, p. *266*; 22.1.2, p. 437.

B. BARANOV, L. DÄWERITZ, V. B. GUTAN, G. JUNGK, H. NEUMANN, and H. RAIDT
Growth and Properties of $Al_xGa_{1-x}N$ Epitaxial Layers
(a) Vol. **49**, No. 2, p. 629—636 (1978).
See also *1.5*, p. *39*; 13.4, p. 182; 20.1, p. 335; 20.3, p. 348; 22.2.4, p. 456.

S. D. Shutov and M. S. Iovu
*Dispersive Charge Transport in Vitreous 0.55 $As_2S_3 \cdot 0.45\ Sb_2S_3$
(a) Vol. **50**, No. 1, p. K59—K61 (1978).
See also *2*, p. *52*; 22, p. 420.

R. E. Blaha and W. R. Fahrner
*Technology and Reliability of Thermal SiO_2 Layers as Used for Passivating Silicon Power Devices
(a) Vol. **50**, No. 2, p. 551—562 (1978).
See also 11, p. 131; *14.3.3*, p. *238*; 22.6, p. 497.

14.3.2 Surfaces and Interfaces

Original Papers and Short Notes

B. Joseph
*Influence of Insulator Surface Conduction on the Transient Response of MIS Capacitors
(a) Vol. **45**, No. 1, p. K25—K28 (1978).
See also 14.3.3, p. 233; 22.1.2, p. 429.

M. K. Roy
Coupling of Surface Acoustic Waves Propagating in Two Separated Piezoelectric Media
(a) Vol. **45**, No. 2, p. 549—552 (1978).
See also *7*, p. *80*; *13.3*, p. *164*; 22.4.1, p. 463; 22.6.1, p. 499.

R. S. Nakhmanson and L. K. Popov
*Measurements of Small-Signal Photo-EMF of Si and Ge MIS Structures Using Scanning Light Probe
(a) Vol. **46**, No. 1, p. 59—68 (1978).
See also *14.3.3*, p. *233*; 16, p. 263; 22.1.1, p. 425; 22.1.2, p. 431.

T. Takebe, J. Saraie, and T. Tanaka
*Study of the Surface Barrier of the Metal–n-CdTe Contact
(a) Vol. **47**, No. 1, p. 123—130 (1978).
See also 13.3, p. 165; 14.3.4, p. 242; 22.4.3, p. 470.

B. Balland and P. Pinard
*Interfacial Tunnel Interaction in MOS Elements in Depletion
(a) Vol. **47**, No. 1, p. 251—258 (1978).
See also *13.4*, p. *175*; 22.6, p. 493.

N. N. Gerasimenko, P. P. Dobrovolskii, N. P. Kibalina, and V. F. Stas
*Investigation of a High-Resistivity Surface Layer in Electron Irradiated Silicon by Schottky Capacitance Measurements
(a) Vol. **47**, No. 1, p. K91—K94 (1978).
See also 11, p. 127; 22.1.2, p. 433.

V. A. Gritsenko, E. E. Meerson, and S. P. Sinitsa
*Unsteady Silicon Nitride Conductivity in High Electric Fields
(a) Vol. **48**, No. 1, p. 31—37 (1978).
See also 2, p. 47; *14.3.4*, p. *243*; 22, p. 415.

A. Wolkenberg
*Capacitance Measurements of the Silicon–Silicon Dioxide–Potassium Chloride Water Solution Interface as a Method to Determine the Fixed Charge Value and Its Centroid
(a) Vol. **48**, No. 1, p. 203—208 (1978).
See also *14.3.3*, p. *235*; 22.1.2, p. 434.

H. Menninger, G. Voigt, H. Raidt, H. Fabig, and J. Maege
*Scanning Electron Microscopic Investigations of GaP
(a) Vol. **48**, No. 2, p. 407—415 (1978).
See also *1.4*, p. *32*; 20.3, p. 346; 22.2.2, p. 451.

Yu. V. Bogatyrev and F. P. Korshunov
*Investigation of Properties of the Si–SiO_2 Interface in MOS Structures Irradiated by Fast Electrons
(a) Vol. **48**, No. 2, p. K127—K130 (1978).
See also *11*, p. *128*; 22.1.2, p. 436; 22.6, p. 495.

W. Brodkorb
*The Influence of Surface Recombination on the Double-Injection Negative Resistance
(a) Vol. **48**, No. 2, p. K157—K159 (1978).
See also *14.3.4*, p. *244*.

M. Manzel and W. Ecke
*The Influence of Surface Recombination on the Double-Injection Switching
(a) Vol. **48**, No. 2, p. K161—K163 (1978).
See also *14.3.4*, p. *244*; 22.1.2, p. 436.

H. D. Dimitrov
On the Surface Electrical Resistivity of Thin Semiconducting and Semimetallic Films
(b) Vol. **89**, No. 1, p. 319—328 (1978).
See also *13.3*, p. *166*; *14.3.1*, p. *228*; 22, p. 417.

G. Heiland and D. Kohl
*Interpretation of Surface Phenomena on ZnO by the Compensation Model
(a) Vol. **49**, No. 1, p. 27—37 (1978).
See also 13.3, p. 167; 13.4, p. 180; 14.3.4, p. 244; *16*, p. *265*; 22.4, p. 462.

H. P. Bauer and U. Birkholz
*Electrical Conductivity of Passivated Si_2Te_3 Single Crystals
(a) Vol. **49**, No. 1, p. 127—131 (1978).
See also *14.3.4*, p. *244*; 22, p. 417.

M. T. Pham
*A Very Simple Method for Profiling the Ion-Implanted Si-Surface
(a) Vol. **49**, No. 1, p. 261—265 (1978).
See also *11*, p. *129*; 22.1.2, p. 436.

T. J. Magee and J. Peng
*Investigation of Solid–Solid Reactions of Au Films on Silicon
(a) Vol. **49**, No. 1, p. 313—322 (1978).
See also *1.3*, p. *28*; 1.5, p. 38; 21.6, p. 404; 22.1.2, p. 437.

K. I. Kirov and S. P. Alexandrova
*Some Sources of Error in Interface State Density Evaluation from Q–U Data in MOS Systems
(a) Vol. **49**, No. 2, p. 781—788 (1978).
See also *13.3*, p. *167*.

H. Flietner
*Schottky-Barrier Theory and the Complex Band Structure
(b) Vol. **90**, No. 1, p. K27—K30 (1978).
See also *13.3*, p. *168*; 22, p. 419.

A. Wolkenberg
*Energetic Barrier Height Hypothesis on the Silicon–Electrolyte and Gallium Arsenide–Electrolyte Interfaces
(a) Vol. **50**, No. 1, p. 249–256 (1978).
See also 13.3, p. 168; *14.3.3*, p. *238*; 22.1.2, p. 438; 22.2.1, p. 449.

W. R. Fahrner and E. Klausmann
*Characterization of MOS Structures with Buried Layers
(a) Vol. **50**, No. 2, p. 433–444 (1978).
See also 11, p. 131; *14.3.3*, p. *238*; 22.1.2, p. 439.

14.3.3 Devices. Junctions (Contact Problems See 14.3.4)

Original Papers and Short Notes

G. H. Döhler
*Analytical Solution for the Junction Behaviour of Relaxation Semiconductors
(a) Vol. **45**, No. 1, p. 113–116 (1978).
See also *14.3.4*, p. *240*; 22, p. 409.

A. Orlik
*Numerical Calculation of the Capacitance of the Planar p–n Junction
(a) Vol. **45**, No. 1, p. 139–147 (1978).

K. N. Ratnakumar and D. Kakati
*A Design Technique for Specific Reverse Bias Characteristics of a p–i–n Diode
(a) Vol. **45**, No. 1, p. K13–K16 (1978).

B. Joseph
Influence of Insulator Surface Conduction on the Transient Response of MIS Capacitors
(a) Vol. **45**, No. 1, p. K25–K28 (1978).
See also *14.3.2*, p. *230*; 22.1.2, p. 429.

L. N. Aleksandrov, R. N. Lovyagin, P. A. Simonov, and I. S. Bzinkovskaya
*p–n Junctions in the Surface Region of Silicon Obtained by Evaporation of Silicon in Ultrahigh Vacuum
(a) Vol. **45**, No. 2, p. 521–527 (1978).
See also 9, p. 91; 22.1.2, p. 430.

P. M. Batev, M. D. Ivanovitch, E. I. Ķafediiska, and S. S. Simeonov
*Schottky Barrier on W–GaAs Contact
(a) Vol. **45**, No. 2, p. 671–675 (1978).
See also 14.3.4, p. 240; 22.2.1, p. 445.

D. Landgraf-Dietz
*A Simple Method for the Determination of Parameters of Ion Implanted Doping Profiles by Means of Threshold Voltage Measurements of MOSFET's
(a) Vol. **45**, No. 2, p. K125–K129 (1978).
See also *11*, p. *124*; 22.1.2, p. 430.

B. D. Kandilarov, M. G. Tashkova, P. C. Petrova, and V. Detcheva
Influence of the Position of the Interface Boundary on the Existence of Interface States
(b) Vol. **86**, No. 1, p. 425–430 (1978).
See also *13.3*, p. *164*.

R. S. Nakhmanson and L. K. Popov
*Measurements of Small-Signal Photo-EMF of Si and Ge MIS Structures Using Scanning Light Probe
(a) Vol. **46**, No. 1, p. 59–68 (1978).
See also *14.3.2*, p. *230*; 16, p. 263; 22.1.1, p. 425; 22.1.2, p. 431.

G. Sarrabayrouse, J. Buxo, and D. Esteve
*The Influence of Surface States on the Electrical Characteristics of Tunneling MIS Schottky Barriers
(a) Vol. **46**, No. 1, p. 185—190 (1978).
See also *13.3*, p. *164*.

P. C. Mathur, Vijender Sharma, and Praveen Saxena
*A Small Signal Noise Analysis for Realistic GaAs IMPATT Diode
(a) Vol. **46**, No. 1, p. 321—326 (1978).
See also 22.2.1, p. 445.

S. Cristoloveanu, A. Chovet, and G. Kamarinos
Some Comments on "The Effect of Effective Lifetime Modulation in Semiconductors"
(a) Vol. **46**, No. 1, p. K13—K16 (1978).
See also *13.3*, p. *165*; 22, p. 411.

I. Munteanu
On the Effective Lifetime Modulation in Semiconductors
(a) Vol. **46**, No. 1, p. K17—K20 (1978).
See also *13.3*, p. *165*; 22, p. 411.

I. M. Kotina, S. R. Novikov, and T. I. Pirozhkova
*On the Determination of the Parameters of Impurity Centres from the Thermally Stimulated Current of p–n Junctions
(a) Vol. **46**, No. 1, p. K21—K23 (1978).
See also *13.4*, p. *173*; 22, p. 411.

J. Richter and P. Seidel
*An Analytic Expression for the Zero-Bias Offset in the Conductance–Voltage Characteristic of Metal–Insulator–Metal Tunnel Junctions
(a) Vol. **46**, No. 1, p. K25—K28 (1978).
See also 22, p. 411.

J. M. Hess, P. H. Nguyen, B. Lepley, and S. Ravelet
*Barrier Height Study on Au–InP Schottky Diodes
(a) Vol. **46**, No. 1, p. K55—K59 (1978).
See also 14.3.4, p. 241; 22.2.2, p. 450.

G. Weinhausen
*Theoretical Study of the Depletion-Layer Capacitance in Abrupt p–n Semiconductor Junctions
(a) Vol. **46**, No. 2, p. 517—525 (1978).
See also 14.3.4, p. 241; 22, p. 412.

F. M. Livingstone, R. M. De La Rue, and W. Duncan
*pSi–nCdS Heterojunction Photodetectors
(a) Vol. **46**, No. 2, p. 595—601 (1978).
See also 16, p. 263; 22.1.2, p. 432; 22.4.1, p. 464.

A. D. Buonaquisti and R. A. Collins
Energy Distribution of Electrons Emited from MIM Thin-Film Sandwich Cathodes
(a) Vol. **46**, No. 2, p. K91—K93 (1978).
See also *17*, p. *269*; 22.6, p. 492.

B. D. Kandilarov, M. G. Tashkova, and P. C. Petrova
*Interface States in One-Dimensional Models of Deformed Heterojunctions
(b) Vol. **87**, No. 1, p. 361—368 (1978).
See also *13.3*, p. *165*.

P. Krispin and J. Maege
*Capacitance–Frequency Dispersion and Electroluminescence Efficiency of GaP p–n Junctions
(a) Vol. **47**, No. 1, p. 91—97 (1978).
See also 13.4, p. 175; *20.3*, p. *344*; 22.2.2, p. 450.

L. Gouskov, A. Gouskov, and B. Girault
*Electrooptical Properties of SnO_2 Spray–GaSe and SnO_2 Spray–GaTe Heterojunctions
(a) Vol. **47**, No. 1, p. 195—199 (1978).
See also 14.3.4, p. 242; 16, p. 264; 20.3, p. 344; 22.3, p. 458.

M. Leppihalme and T. Tuomi
*Phonon Energies and Barrier Height from Photovoltage Spectra of Au–Intrinsic Ge Diodes
(a) Vol. **47**, No. 1, p. 245—250 (1978).
See also 6, p. 71; 13.5.1, p. 191; *16*, p. *264*; 22.1.1, p. 425.

H. Reichl, D. Huber, and J. Müller
*Photoluminescence Investigations of Diffused GaAs Light Emitting Diodes
(a) Vol. **47**, No. 2, p. 489—495 (1978).
See also *20.3*, p. *345*; 22.2.1, p. 447.

P. Viktorovitch, G. Kamarinos, P. Even, and E. Fabre
*Correlation between Interface States and MIS Silicon Solar Cell Performances
(a) Vol. **48**, No. 1, p. 137—145 (1978).
See also 13.3, p. 166; *16*, p. *264*; 22.1.2, p. 434.

A. Wolkenberg
*Capacitance Measurements of the Silicon–Silicon Dioxide–Potassium Chloride Water Solution Interface as a Method to Determine the Fixed Charge Value and Its Centroid
(a) Vol. **48**, No. 1, p. 203—208 (1978).
See also *14.3.2*, p. *230*; 22.1.2, p. 434.

O. Yu. Borkovskaya, N. L. Dmitruk, R. V. Konakova, and Yu. A. Tkhorik
Investigation of Radiation Defects in GaAs by Means of Schottky Diode Characteristics
(a) Vol. **48**, No. 1, p. K55—K58 (1978).
See also *11*, p. *128*; 22.2.1, p. 447.

L. Baldi, G. F. Cerofolini, G. Ferla, and G. Frigerio
Gold Solubility in Silicon and Gettering by Phosphorus
(a) Vol. **48**, No. 2, p. 523—532 (1978).
See also *10.2*, p. *116*; 22.1.2, p. 435.

K. W. Böer
*The Physics of the Photovoltaic Effect
(a) Vol. **49**, No. 1, p. 13–26 (1978).
See also *16*, p. *265*; 22, p. 417; 22.4.1, p. 465.

W. Hoyer, G. Lautz, and M. Heyen
*Low-Frequency Current Noise of GaAs Epitaxial Layers between 4.4 and 300 K
(a) Vol. **49**, No. 1, p. 111–117 (1978).
Erratum
(a) Vol. **50**, No. 2, p. 771 (1978).
See also 13.4, p. 180; *14.3.1*, p. *228*; 22.2.1, p. 448.

I. Dudeck and R. Kassing
Determination of the Temperature Dependence of the Capture Cross-Sections of the Gold Acceptor Level and of the Temperature of Current Filaments in Silicon pin Diodes
(a) Vol. **49**, No. 1, p. 153–161 (1978).
See also *13.4*, p. *180*; *14.3.4*, p. *244*; 22.1.2, p. 436.

T. D. Dzhafarov, M. N. Kolotov, and A. A. Litvin
*The Influence of γ-Irradiation on the Degradation of Red-Emitting GaP LED's
(a) Vol. **49**, No. 1, p. 211–215 (1978).
See also 11, p. 129; *20.3*, p. *347*; 22.2.2, p. 451.

A. Soliman and N. Saleh
*Photovoltaic Operation of Nonuniformly Doped Intrinsic Photoconductors
(a) Vol. **49**, No. 1, p. 393–403 (1978).
See also *16*, p. *266*; 22.2.3, p. 454.

V. S. Lysenko and A. N. Nazarov
*Effect of Laser Beam Irradiation on the Electrophysical Properties of the Interface in Boron Ion Implanted MIS Structures
(a) Vol. **49**, No. 1, p. 405–411 (1978).
See also 11, p. 129; 13.3, p. 167; 22.1.2, p. 437.

H. Nelkowski and K. Petermann
*Preparation and Investigation of ZnS–$CuGaS_2$ Heterojunctions
(a) Vol. **49**, No. 1, p. K13–K16 (1978).
See also 1.5, p. 39; 3, p. 56; *20.3*, p. *348*; 22.4.1, p. 465; 22.8, p. 507.

M. Vodenicharova
*Thermionic-Field Emission from a Metal into Organic Polymer Semiconductor Films
(a) Vol. **49**, No. 1, p. K49–K53 (1978).
See also *14.3.4*, p. *245*; 22.9, p. 527.

D. Jäger, E. Kähler, and R. Heidemann
*Negative Conductance in SCLC Diodes with Traps
(a) Vol. **49**, No. 2, p. 431–435 (1978).
See also 13.4, p. 182; *14.3.4*, p. *245*; 22.1.2, p. 437.

K. W. Böer
*Heterojunction Interface Double Layer and Consequences for Photovoltaic Cells, Specifically $Cd_zZn_{1-z}S/Cu_2S$
(a) Vol. **49**, No. 2, p. 455—462 (1978).
See also *16*, p. *266*; 22, p. 418; 22.4.1, p. 466.

B. Rabin, H. Tabatabai, and P. Siffert
*Capacitance of Cadmium Telluride Schottky Diodes
(a) Vol. **49**, No. 2, p. 577—584 (1978).
See also *13.4*, p. *182*; 22.4.3, p. 471.

G. Sarrabayrouse and J. Buxo
*The Reversely Biased MIS Tunneling Schottky Barrier. Application to Photodetection
(a) Vol. **49**, No. 2, p. 603—608 (1978).
See also *16*, p. *267*.

P. P. Tirodkar and R. Pratap
*Equivalent Rectangular Shape of a Potential Barrier in a Tunnel Junction
(a) Vol. **49**, No. 2, p. 637—641 (1978).
See also 14.4, p. 251; 22.6, p. 496.

R. Mach, W. Gericke, H. Treptow, and W. Ludwig
*Direct Impact Excitation of Luminescen t Centres in ZnSe Single Crystals
(a) Vol. **49**, No. 2, p. 667—674 (1978).
See also 13.4, p. 182; *20.3*, p. *349*; 22.4.2, p. 469.

V. M. Harutyunyan and F. V. Gasparyan
*Impedance of p^+–n–n^+ Structures Based o n Semiconductors Compensated by Deep Doubly Ionized Centres
(a) Vol. **49**, No. 2, p. 797—801 (1978).
See also *13.4*, p. *182*; 22.1.2, p. 437.

H. Neumann and B. Jacobs
*Electrical Properties of Liquid Phase Epitaxial InP Grown from Tin Solution
(a) Vol. **49**, No. 2, p. K139—K144 (1978).
See also 22.2.2, p. 451.

W. Szkiełko
*Spin-Dependent Recombination at Dislocations in Silicon Revealed at Reverse Current in a p–n Junction
(b) Vol. **90**, No. 1, p. K81—K83 (1978).
See also *10.2*, p. *120*; 16, p. 267; 19, p. 314; 22.1.2, p. 438.

S. Ristić and V. Cvekić
*Some Effects of Localized Stress on Silicon Planar Transistors
(a) Vol. **50**, No. 1, p. 153—157 (1978).
See also 22.1.2, p. 438.

D. I. Tsiulyanu, A. M. Andriesh, and E. P. Kolomeyko
*Arsenic Trisulphide in Isotype Amorphous-Crystalline Heterojunctions
(a) Vol. **50**, No. 1, p. 195—202 (1978).
See also *2*, p. *52*; 14.3.4, p. 246; 22, p. 420.

K. H. Yang, H. F. Kappert, and G. H. Schwuttke
*Minority Carrier Lifetime in Annealed Silicon Crystals Containing Oxygen
(a) Vol. **50**, No. 1, p. 221—235 (1978).
See also *10.2*, p. *120*; 22.1.2, p. 438.

E. Kaminska et A. Piotrowska
*Stabilité thermique sous vide du CdF_2
(Thermal Stability of CdF_2 in Vacuum)
(a) Vol. **50**, No. 1, p. 243—247 (1978).
See also 22.5.3, p. 488.

A. Wolkenberg
*Energetic Barrier Height Hypothesis on the Silicon–Electrolyte and Gallium Arsenide–Electrolyte Interfaces
(a) Vol. **50**, No. 1, p. 249—256 (1978).
See also 13.3, p. 168; *14.3.2*, p. *232*; 22.1.2, p. 438; 22.2.1, p. 449.

A. le Bloa, P. N. Favennec, and Y. Colin
Determination of Complementary Transition Energy Threshold by Photocapacitance Measurements. Application to Oxygen Implanted GaAs
(a) Vol. **50**, No. 1, p. 311—322 (1978).
See also *13.4*, p. *184*; 22.2.1, p. 449.

K. I. Kirov, E. D. Atanasova, and N. A. Ivanov
*Investigation of Al_2O_3 Layers Obtained by DC Planar-Magnetron Sputtering
(a) Vol. **50**, No. 1, p. 363—368 (1978).
See also *1.5*, p. *39*; 14.4, p. 252; 22.6, p. 497.

J. K. Wysocki and C. M. Vodenicharov
*Approximate Formula for the Emission Current Density in the Second Intermediate Region of Christov
(a) Vol. **50**, No. 2, p. 411—415 (1978).
See also *17*, p. *271*; 22.6, p. 497.

W. R. Fahrner and E. Klausmann
*Characterization of MOS Structures with Buried Layers
(a) Vol. **50**, No. 2, p. 433—444 (1978).
See also 11, p. 131; *14.3.2*, p. *232*; 22.1.2, p. 430.

C. Lawther and J. Woods
*Blue Light Emission in Forward-Biased ZnS MIS Diodes
(a) Vol. **50**, No. 2, p. 491—502 (1978).
See also *20.3*, p. *350*; 22.4.1, p. 467.

R. E. Blaha and W. R. Fahrner
*Technology and Reliability of Thermal SiO_2 Layers as Used for Passivating Silicon Power Devices
(a) Vol. **50**, No. 2, p. 551—562 (1978).
See also 11, p. 131; *14.3.1*, p. *229*; 22.6, p. 497.

A. F. Plotnikov, V. A. Tolokonnikov, and V. E. Shubin
*Investigation of the Energy Band Diagram of the Anodic Oxide–Indium Antimonide System by Internal Photoemission
(a) Vol. **50**, No. 2, p. 733—738 (1978).
See also *13.1*, p. *161*; 16, p. 268; 22.2.3, p. 455.

C. Hamann, H.-J. Höhne, F. Kersten, M. Müller, F. Przyborowski, and M. Starke
*Switching Effects on Polycrystalline Films of Lead Phthalocyanine (PbPc)
(a) Vol. **50**, No. 2, p. K189—K192 (1978).
See also 22.9, p. 529.

14.3.4 High-Field Phenomena, Space-Charge Effects, Inhomogeneities, Injected Carriers (Electroluminescence See 20.3; Junctions See 14.3.3)

Original Papers and Short Notes

D. Suisky, R. Boyn, R. Enderlein, and B. Biering
*First-Harmonic Electromodulation in Uniaxial Crystals
(b) Vol. **85**, No. 1, p. 217–224 (1978).
See also 13.4, p. 169; *20.1*, p. *319*; 22.4.1, p. 463.

V. M. Fomin and E. P. Pokatilov
*Non-Linear Optical Properties of the Band Charge Carrier Gas
(b) Vol. **85**, No. 1, p. 403–415 (1978).
See also *20.2*, p. *339*; 22, p. 409.

G. H. Döhler
*Analytical Solution for the Junction Behaviour of Relaxation Semiconductors
(a) Vol. **45**, No. 1, p. 113–116 (1978).
See also *14.3.3*, p. *233*; 22, p. 409.

E. M. Baskin, M. D. Blokh, M. V. Entin, and L. I. Magarill
Erratum to: Current Quadratic in Field and Photogalvanic Effect in Crystals without Inversion Centre
(b) Vol. **85**, No. 2, p. K135 (1978).
See also 16, p. 262; 22.2, p. 443.

A. Dargys, R. Sedrakyan, and J. Požela
*Measurement of the Electron Drift Velocity in InSb up to Fields of 800 V/cm in the Presence of Impact Ionization
(a) Vol. **45**, No. 2, p. 387–392 (1978).
See also 22.2.3, p. 452.

V. C. Kieu
*Current Controlled Negative Resistance in GaAs at Low Temperatures
(a) Vol. **45**, No. 2, p. 571–578 (1978).
See also *13.4*, p. *171*; 22.2.1, p. 445.

A. N. Zyuganov, Yu. G. Pismennyi, S. V. Svechnikov, and P. S. Smertenko
*Regime of Contact Emission Limited Current with Weak Harmonic Distortion of Steady Bias
(a) Vol. **45**, No. 2, p. 631–638 (1978).
See also 13.4, p. 171; 22, p. 410.

P. M. Batev, M. D. Ivanovitch, E. I. Kafedjiska, and S. S. Simeonov
Schottky Barrier on W–GaAs Contact
(a) Vol. **45**, No. 2, p. 671–675 (1978).
See also *14.3.3*, p. *233*; 22.2.1, p. 445.

P. Pipinys and A. Rimeika
*Current Pulses in ZnS:Cu Crystals Stimulated by Electric Field at Low Temperatures
(a) Vol. **45**, No. 2, p. K113–K116 (1978).
See also *6*, p. *66*; 13.4, p. 171; 22.4.1, p. 463.

V. Dolocan
Some Electrical Properties of Bi_2O_3 Thin Films
(a) Vol. **45**, No. 2, p. K155—K157 (1978).
See also *14.3.1*, p. *226*; 22.6, p. 491.

Ch. Schnittler
*Some New Aspects of the SCLC Characteristic of Homogeneous Insulators with Traps
(a) Vol. **45**, No. 2, p. K179—K182 (1978).
See also 13.4, p. 171; 22, p. 410.

W. Müller, M. Overhamm, G. Bauer, E. Gornik, P. Grosse, H. W. Pötzl, and M. Rautenberg
*On the Influence of Background Radiation on Landau Emission Spectra
(b) Vol. **86**, No. 1, p. 205—213 (1978).
See also 13.1, p. 152; 20.1, p. 322; *20.3*, p. *343*; 22.2.3, p. 452.

J. Humlíček
*On the Nature of E_1-Transitions in Germanium
(b) Vol. **86**, No. 1, p. 303—310 (1978).
See also 13.1, p. 153; *20.1*, p. *322*; 22.1.1, p. 424.

A. P. Dmitriev, A. E. Stefanovich, and L. D. Tsendin
*Discontinuities in Drift Solutions Due to Reversal in Ambipolar Drift. Injection and Accumulation of Hot Carriers
(a) Vol. **46**, No. 1, p. 45—53 (1978).
See also *13*, p. *144*; 22, p. 411.

S. Nešpůrek, J. Obrda, and J. Sworakowski
*A Study of Traps for Current Carriers in Organic Solids. N,N′-diphenyl-p-phenylenediamine
(a) Vol. **46**, No. 1, p. 273—280 (1978).
See also 13.4, p. 173; 22.9, p. 524.

J. M. Hess, P. H. Nguyen, B. Lepley, and S. Ravelet
Barrier Height Study on Au–InP Schottky Diodes
(a) Vol. **46**, No. 1, p. K55—K59 (1978).
See also *14.3.3*, p. *234*; 22.2.2, p. 450.

A. E. Gershinskii, L. V. Mironova, and E. I. Cherepov
*On the Conductivity of the Si–Thermal SiO_2–Electrolyte System in Superhigh Electric Fields
(a) Vol. **46**, No. 2, p. 467—473 (1978).
See also *14.3.1*, p. *226*; 22.1.2, p. 432; 22.6, p. 492.

G. Weinhausen
Theoretical Study of the Depletion-Layer Capacitance in Abrupt p–n Semiconductor Junctions
(a) Vol. **46**, No. 2, p. 517—525 (1978).
See also *14.3.3*, p. *234*; 22, p. 412.

L. Baldassarre, V. Capozzi, G. Maggipinto, and A. Minafra
*Space Charge Limited Current Measurements in $ZnIn_2S_4$
(a) Vol. **46**, No. 2, p. 589—593 (1978).
See also 22.8, p. 504.

R. Kümmel, H. Rauh, and E. Bangert
*Theory of Stark-Ladder Currents
(b) Vol. 87, No. 1, p. 99—110 (1978).
See also 6, p. 70; 22.4.1, p. 464.

R. Mansfield and G. Azhdarov
*Negative Magnetoresistance in Indium Arsenide in the Extreme Quantum Limit
(b) Vol. 87, No. 1, p. 163—167 (1978).
See also *13.4*, p. *174*; 22.2.1, p. 446.

T. Takebe, J. Saraie, and T. Tanaka
*Study of the Surface Barrier of the Metal–n-CdTe Contact
(a) Vol. 47, No. 1, p. 123—130 (1978).
See also 13.3, p. 165; *14.3.2*, p. *230*; 22.4.3, p. 470.

L. Gouskov, A. Gouskov, and B. Girault
Electrooptical Properties of SnO_2 Spray–GaSe and SnO_2 Spray–GaTe Heterojunctions
(a) Vol. 47, No. 1, p. 195—199 (1978).
See also *14.3.3*, p. *235*; 16, p. 264; 20.3, p. 344; 22.3, p. 458.

J. M. Dusseau, L. Ensuque, Im-Saroeun, and T. P. Lepetre
*High Electric Field Effect in Silicon–Boron Compounds such as $B_{14}Si$
(a) Vol. 47, No. 1, p. K11—K15 (1978).
See also 22, p. 413.

A. Kobayashi and K. Suzuki
*High Field Transport Properties of a Magnetic Semiconductor
(a) Vol. 47, No. 1, p. K51—K53 (1978).
See also *18.2*, p. *286*; 22.8, p. 505.

W. Brückner and W. Moldenhauer
*Resistivity–Temperature Characteristic and Thermal Switching
(a) Vol. 47, No. 1, p. K99—K102 (1978).
See also *8*, p. *86*; 22.6, p. 493.

V. A. Tyagai, O. V. Snitko, A. M. Evstigneev, and A. N. Krasiko
*Surface Phenomena in the Polarization-Dependent Electroreflectance Spectra of Si
(b) Vol. 87, No. 2, p. 613—618 (1978).
See also 13.3, p. 165; *20.1*, p. *327*; 22.1.2, p. 433.

V. M. Fomin and E. P. Pokatilov
On the Representation of the Non-Stationary Density Matrix of an Electron–Phonon System by a Functional Integral
(b) Vol. 87, No. 2, p. 725—732 (1978).
See also 6, p. 72; *20*, p. *316*.

V. M. Bondar, V. S. Radchenko, and L. S. Solonchuk
*Conductivity of Warm Holes in Tellurium
(b) Vol. 87, No. 2, p. K97—K100 (1978).
See also 22.1.3, p. 441.

D. Vajda
*Dependence of the Threshold Electric Field of the Acoustoelectric Current Instabilities in Monocrystalline n-InSb on the Length of a Sample
(a) Vol. **47**, No. 2, p. K123–K124 (1978).
See also *7*, p. *81*; 22.2.3, p. 453.

I. Mojzes
*Formation of AuGe Contacts to n-GaAs
(a) Vol. **47**, No. 2, p. K183–K185 (1978).
See also 22.2.1, p. 447.

S. D. Beneslavskii and O. Ziep
*Hot Electrons in Zero-Gap Semiconductors with Non-Parabolic Electronic Band Structure
(b) Vol. **88**, No. 1, p. 221–230 (1978).
See also *13.1*, p. *156*; 14.4, p. 249; 22.4.4, p. 473.

D. P. Bhattacharya and Z. S. Kachlishvili
*Faraday Rotation in Presence of Attractive Traps
(b) Vol. **88**, No. 1, p. 283–287 (1978).
See also *13.4*, p. *177*; 18, p. 274; 22.1.1, p. 426.

K. I. Papunashvili
**H*-Linear "Planar" Hall Effects of Hot Electrons in n-Ge
(b) Vol. **88**, No. 1, p. K27–K30 (1978).
See also *13*, p. *146*; 22.1.1, p. 426.

V. A. Gritsenko, E. E. Meerson, and S. P. Sinitsa
*Unsteady Silicon Nitride Conductivity in High Electric Fields
(a) Vol. **48**, No. 1, p. 31–37 (1978).
See also 2, p. 47; *14.3.2*, p. *230*; 22, p. 415.

G. Lautz and K. Sachse
*Noise Spectra of Hot Carriers in n-Germanium
(a) Vol. **48**, No. 1, p. 121–131 (1978).
See also *13*, p. *146*; 22.1.1, p. 426.

J. Kočka
Influence of Contacts on the AC Conductivity of Glasses
(a) Vol. **48**, No. 1, p. K59–K62 (1978).
See also *2*, p. *47*; *14.3*, p. *222*; 22, p. 415.

I. M. Dykman, V. M. Rosenbaum, and F. T. Vasko
*Hot Electrons in Semiconductors with Quasi-Relativistic Band Structure
(b) Vol. **88**, No. 2, p. 385–395 (1978).
See also *13*, p. *146*; 16, p. 265; 22, p. 415.

L. F. Kurtenok, E. A. Movchan, O. G. Sarbey, V. V. Mitin, and M. Asche
*Static High Field Domains in n-Si
(a) Vol. **48**, No. 2, p. 323–328 (1978).
See also 22.1.2, p. 435.

CH. SCHNITTLER
*A Simplified Approach to the SCLC Characteristic of Homogeneous Solids with Traps Distributed in Energy
(a) Vol. **48**, No. 2, p. 357–368 (1978).
See also 13.4, p. 179; 22, p. 416.

M. H. EL-FOULY and J. T. EDMOND
*Electrical Conductivity of Some Chalcogenide Glasses as a Function of Electric Field and Temperature
(a) Vol. **48**, No. 2, p. 395–405 (1978).
See also *2*, p. *48*; 22, p. 416.

W. R. FAHRNER, D. BRÄUNIG, R. FERRETTI, and B. MÜLLER
*A Rapid Destruction-Free Determination of Breakdown Voltage of Thin Insulating Layers
(a) Vol. **48**, No. 2, p. 533–541 (1978).
See also *14.3.1*, p. *227*; 22, p. 416.

A. BUROFF, E. NEBAUER, and P. SÜPTITZ
Influence of an Electric DC Field on the Migration of Silver in Amorphous As_2S_3
(a) Vol. **48**, No. 2, p. K109–K112 (1978).
See also *2*, p. *48*; *9*, p. *93*; 22, p. 416.

W. BRODKORB
*The Influence of Surface Recombination on the Double-Injection Negative Resistance
(a) Vol. **48**, No. 2, p. K157–K159 (1978).
See also *14.3.2*, p. *231*.

M. MANZEL and W. ECKE
*The Influence of Surface Recombination on the Double-Injection Switching
(a) Vol. **48**, No. 2, p. K161–K163 (1978).
See also *14.3.2*, p. *231*; 22.1.2, p. 436.

G. HEILAND and D. KOHL
Interpretation of Surface Phenomena on ZnO by the Compensation Model
(a) Vol. **49**, No. 1, p. 27–37 (1978).
See also 13.3, p. 167; 13.4, p. 180; *14.3.2*, p. *231*; *16*, p. *265*; 22.4, p. 462.

H. P. BAUER and U. BIRKHOLZ
*Electrical Conductivity of Passivated Si_2Te_3 Single Crystals
(a) Vol. **49**, No. 1, p. 127–131 (1978).
See also 14.3.2, p. 231; 22, p. 417.

I. DUDECK and R. KASSING
*Determination of the Temperature Dependence of the Capture Cross-Sections of the Gold Acceptor Level and of the Temperature of Current Filaments in Silicon pin Diodes
(a) Vol. **49**, No. 1, p. 153–161 (1978).
See also *13.4*, p. *180*; 14.3.3, p. 236; 22.1.2, p. 436.

T. PURITIS
*Plasmon Mechanism for Secondary Breakdown
(a) Vol. **49**, No. 1, p. 225–230 (1978).
See also *13.5.2*, p. *199*; 22.1.2, p. 436.

S. GUHA and S. GHOSH
*Effect of a Density Gradient on the Electromagnetic Wave Instability in Semiconductor Plasmas
(a) Vol. **49**, No. 1, p. 363—370 (1978).
See also 22.2.3, p. 454.

W. MYCIELSKI and A. LIPIŃSKI
*SCL Currents and Trap Distribution in Polycrystalline Tetracene
(a) Vol. **49**, No. 1, p. K41—K42 (1978).
See also *13.4*, p. *181*; 22.9, p. 527.

M. VODENICHAROVA
*Thermionic-Field Emission from a Metal into Organic Polymer Semiconductor Films
(a) Vol. **49**, No. 1, p. K49—K53 (1978).
See also *14.3.3*, p. *236*; 22.9, p. 527.

D. JÄGER, E. KÄHLER, and R. HEIDEMANN
*Negative Conductance in SCLC Diodes with Traps
(a) Vol. **49**, No. 2, p. 431—435 (1978).
See also 13.4, p. 182; *14.3.3*, p. *236*; 22.1.2, p. 437.

J. BAARS, D. BASSETT, and M. SCHULZ
*Metal–Semiconductor Barrier Studies of PbTe
(a) Vol. **49**, No. 2, p. 483—488 (1978).
See also 22, p. 419.

K. G. GERMANOVA
*On Some Stationary and Non-Stationary Temperature-Electric Instabilities in Photoconducting CdS Crystals
(a) Vol. **49**, No. 2, p. 499—507 (1978).
See also 16, p. 266; 22.4.1, p. 466.

G. KÜHNEL, W. SIEGEL, and E. ZIEGLER
*Metal–n-$ZnSiP_2$ Schottky Contacts
(a) Vol. **49**, No. 2, p. 705—711 (1978).
See also 13.3, p. 167; 22.8, p. 507.

Z. S. KACHLISHVILI
*On a Certain Feature of Nonlinear Galvanomagnetic Characteristics
(a) Vol. **49**, No. 2, p. K135—K137 (1978).
See also *13.4*, p. *182*; 22, p. 419.

S. NEŠPŮREK and J. SWORAKOWSKI
*A Differential Method of Analysis of Steady-State Space-Charge-Limited Currents: An Extension
(a) Vol. **49**, No. 2, p. K149—K152 (1978).
See also 13.4, p. 183; 22, p. 419.

F. SCHAUER, V. ŠMÍD, and V. ŽALUDEK
*On the Contact Capacitance and the Field Effect in Chalcogenide Glasses
(a) Vol. **49**, No. 2, p. K157—K161 (1978).
See also *2*, p. *51*; 13.3, p. 167.

H. J. DE BRUIN and S. P. S. BADWAL
*Faradaic Reaction Kinetics in Solid Electrolytes by Impedance Dispersion Analysis
(a) Vol. **49**, No. 2, p. K181—K184 (1978).
See also *13.3*, p. *167*; 22.6, p. 497.

K. G. NIKIFOROV, S. I. RADAUTSAN, and V. E. TEZLEVAN
*Negative Magnetoresistance in the Ga-Doped Magnetic Semiconductor $CdCr_2S_4$
(a) Vol. **49**, No. 2, p. K197—K200 (1978).
See also 22.8.2, p. 520.

V. I. SOKOLOV, N. A. MOSKVINA, and B. V. MELNIKOV
Discovery of Exciton–Impurity Complexes in Zinc Selenide by Electroabsorption
(b) Vol. **90**, No. 1, p. K39—K42 (1978).
See also *13.4*, p. *184*; *13.5.1*, p. *196*; 20.1, p. 336; 22.4.2, p. 469.

J. T. DEVREESE, R. EVRARD, and E. KARTHEUSER
*Note on the Solution of the Boltzmann Equation for Electron–LO Phonon Scattering
(b) Vol. **90**, No. 1, p. K73—K76 (1978).
See also *6*, p. *78*; 22.2.3, p. 454.

V. A. GADZHIEV, S. G. ABDULLAEVA, F. K. ALIEV, F. K. ALLAKHVERDIEV, and A. Z. BADALOV
*Electroabsorption in GaTe
(b) Vol. **90**, No. 1, p. K77—K79 (1978).
See also 13.5.1, p. 196; *20.1*, p. *336*; 22.3, p. 459.

D. I. TSIULYANU, A. M. ANDRIESH, and E. P. KOLOMEYKO
Arsenic Trisulphide in Isotype Amorphous–Crystalline Heterojunctions
(a) Vol. **50**, No. 1 p. 195—202 (1978).
See also *2*, p. *52*; *14.3.3*, p. *237*; 22, p. 420.

S. K. PABI
Mechanism of the Dissolution of Si in Al
(a) Vol. **50**, No. 1, p. K9—K12 (1978).
See also *9*, p. *94*; 21, p. 365; 22.1.2, p. 438.

Z. DOBROVOLSKIS, A. KROTKUS, and J. POŽELA
*Negative Differential Conductivity in Hydrostatically Pressed Indium Antimonide at Room Temperature
(a) Vol. **50**, No. 1, p. K73—K75 (1978).
See also 12.2, p. 141; 22.2.3, p. 454.

Ľ. HRIVNÁK
*On the Switching Effect in Crystalline Semiconductors
(a) Vol. **50**, No. 1, p. K97—K99 (1978).
See also *13.4*, p. *184*; 22, p. 420.

J. P. LEBURTON and R. EVRARD
*Oscillatory Electron Drift Velocity in Polar Semiconductor in Non-Ohmic Regime at Low Temperature
(b) Vol. **90**, No. 2, p. K137—K139 (1978).
See also *13*, p. *149*; 22.2.1, p. 449; 22.2.3, p. 455.

G. E. Alberga and J. T. Devreese
*Mobility of Electrons in n-Type InSb at 77 K in the Avalanche Regime: Theory and Experiment
(b) Vol. **90**, No. 2, p. K149—K152 (1978).
See also *6*, p. *78*; 22.2.3, p. 455.

P. T. Landsberg, D. J. Robbins, and E. Schöll
*Threshold Switching as a Generation–Recombination Induced Non-Equilibrium Phase Transition
(a) Vol. **50**, No. 2, p. 423—426 (1978).
See also 2, p. 52; 13.4, p. 185; 22.1, p. 423.

V. V. Serdyuk, A. P. Chebanenko und G. G. Chemeresyuk
*Stromumschaltmechanismus in inhomogenen Zinkselenidkristallen
(Current Switching Mechanism in Inhomogeneous Zinc Selenide Crystals)
(a) Vol. **50**, No. 2, p. 717—722 (1978).
See also *13.4*, p. *186*; 16, p. 267; 22.4.2, p. 469.

G. Weimann and W. Schlapp
*Ohmic Contacts on Indium Phosphide
(a) Vol. **50**, No. 2, p. K219—K223 (1978).
See also 22.2.2, p. 451.

L. Brehmer and M. Pinnow
*Estimation of the Charge Carrier Mobility in Low Density Polyethylene by Means of Current Oscillations
(a) Vol. **50**, No. 2, p. K239—K242 (1978).
See also 22.9, p. 529.

14.4 Dielectrics

Original Papers and Short Notes

W. Bollmann
Association Parameters of [Me^{3+}–F_1^-] Dipoles in CaF_2 Crystals
(a) Vol. **45**, No. 1, p. 187–193 (1978).
See also *10.2*, p. *109*; 22.5.3, p. 487.

B. I. Shklovskii
*Anisotropy of Percolation Conduction
(b) Vol. **85**, No. 2, p. K111–K114 (1978).
See also *14.1*, p. *206*.

T. Gotô, T. Fujimura, and K. Kamiyoshi
Ultrasonic Investigation of the Phase Transitions in $NH_4Cl_{1-x}Br_x$ Mixed Crystals
(a) Vol. **45**, No. 2, p. 513–519 (1978).
See also *1.2*, p. *16*; *7*, p. *80*; 22.5.4, p. 489.

J. L. Jacquemin and G. Bordure
Numerical Simulation of the AC Electrical Properties of Random Inhomogeneous Systems. Application to Heat-Treated Polymers
(a) Vol. **45**, No. 2, p. 579–584 (1978).
See also *14*, p. *203*; 22.9, p. 523.

R. Gründler, K. Breuer, and W. Tews
*Optical Properties of Amorphous and Polycrystalline BeO Thin Films from Electron Energy Loss Measurements
(b) Vol. **86**, No. 1, p. 329–338 (1978).
See also 2, p. 43; 13.1, p. 153; *20.1*, p. *322*; 22.6, p. 492.

C. R. A. Catlow
*Ion Distribution Functions for Complex Solids and Their Application to the Conductivity of Glasses
(a) Vol. **46**, No. 1, p. 191–198 (1978).
See also 22.8.1, p. 511.

R. Kirsch
An Automated FSA Ellipsometer System for in Situ Investigations
(a) Vol. **46**, No. 2, p. 459–466 (1978).
See also *20.1*, p. *325*; 21.6, p. 402.

P. K. C. Pillai, R. G. Mendiratta, C. K. Pillai, and R. Nath
*Photo-Induced AC Impedance Measurements in Single Crystal Photoconductors
(a) Vol. **46**, No. 2, p. K127–K130 (1978).
See also *16*, p. *264*; 22.4, p. 461; 22.5, p. 475.

B. L. Gelmont, V. I. Ivanov-Omskii, and E. N. Ukraintsev
The Frequency Dependent Dielectric Function of Gapless Semiconductors in a Magnetic Field
(b) Vol. **87**, No. 1, p. 381–383 (1978).
See also *13.5*, p. *187*; 22.4.3, p. 470.

L. Benguigui and D. Cabib
*Determination of the Tilt Angle in the Smectic C Phase of Bis-(Heptyloxy)-Azoxybenzene by Means of Dielectric Measurements
(a) Vol. **47**, No. 1, p. 71—78 (1978).
See also 1.4, p. 30; 22.9, p. 525.

W. Szuszkiewicz, A. M. Witowski, and M. Grynberg
The Dynamic Dielectric Function in HgSe and HgTe
(b) Vol. **87**, No. 2, p. 637—645 (1978).
See also *13.1*, p. *155*; 20.1, p. 328; 22.4.2, p. 468; 22.4.3, p. 470.

R. Blinc and F. C. de Sa Barreto
Electric Field Induced Lifshits Point in Smectic Liquid Crystals
(b) Vol. **87**, No. 2, p. K105—K109 (1978).
See also *1.2*, p. *19*; *2*, p. *45*.

R. P. Goyal and S. C. Goyal
*Dielectric Behaviour of Transition Metal Oxides
(b) Vol. **87**, No. 2, p K119—K122 (1978).
See also 22.6, p. 494.

K. C. Bleijenberg and C. W. M. Timmermans
*Luminescence and Electrical Conductivity of Uranium-Activated Sodium Fluoride Crystals
(a) Vol. **47**, No. 2, p. 589—595 (1978).
See also 10.2, p. 114; *20.3*, p. *345*; 22.5.2, p. 481.

K. S. Bhatki, S. D. Pradhan, and B. V. Thosar
Further Studies on Positron Lifetimes in γ-Irradiated Teflon in Low Dose Range
(a) Vol. **47**, No. 2, p. 691—698 (1978).
See also *10.2*, p. *114*; 11, p. 127; 19, p. 309; 22.9, p. 526.

F. Krok, W. Bogusz, B. Jarosz, and W. Jakubowski
*Electrode Polarization in CdF_2 Crystals
(a) Vol. **47**, No. 2, p. K103—K105 (1978).
See also 22.5.3, p. 487.

S. D. Beneslavskii and O. Ziep
Hot Electrons in Zero-Gap Semiconductors with Non-Parabolic Electronic Band Structure
(b) Vol. **88**, No. 1, p. 221—230 (1978).
See also *13.1*, p. *156*; *14.3.4*, p. *243*; 22.4.4, p. 473.

K. Hübner
Chemical Bond and Related Properties of SiO_2 (V): Pseudopotential Study of Bonding and Structure
(a) Vol. **48**, No. 1, p. 147—154 (1978).
See also 2, p. 47; *4*, p. *59*; *13.1*, p. *157*; 22.6, p. 494.

C. Papatriantafillou
NMP-TCNQ: A Realization of the Disordered Hubbard Model
(b) Vol. **88**, No. 2, p. 663—674 (1978).
See also *13*, p. *146*; 22.9, p. 526.

R. Claus
*Polariton Dispersion and Crystal Optics in Monoclinic Materials
(b) Vol. **88**, No. 2, p. 683—688 (1978).
See also *13.5*, p. *187*.

M. B. O'Reilly
*A Phenomenological Theory for Ionic Conductivity in Solid Electrolytes
(a) Vol. **48**, No. 2, p. 489—496 (1978).
See also *10.2*, p. *116*; 22.5.1, p. 477.

L. Merten and R. Claus
*Dispersion of Dielectric Axes in Monoclinic and Triclinic Crystals Related to Lattice Dynamics
(b) Vol. **89**, No. 1, p. 159—163 (1978).
See also *13.5*, p. *188*.

V. C. Jain and J. Shanker
The Exchange Charge Shell Model Analysis of Strain Derivatives of Static Polarizability and Effective Charge Parameter of Alkali Halides
(b) Vol. **89**, No. 1, p. 213—219 (1978).
See also *4*, p. *59*; 6, p. 75; 22.5.2, p. 483.

I. Földvári, R. Voszka, and Z. Morlin
*The Properties of Ni Ions in NaCl Single Crystals (I): Vacuum Ultraviolet, Ionic Conductivity, and X-Ray Diffraction Studies
(b) Vol. **89**, No. 1, p. 235—240 (1978).
See also 1.2, p. 22; *20.1*, p. *332*; 22.5.2, p. 484; 22.5.4, p. 489.

P. Świątek, A. M. Witowski, and M. Grynberg
Experimental and Theoretical Study of the Dynamic Dielectric Function in p-HgTe
(b) Vol. **89**, No. 1, p. K1—K4 (1978).
See also *13.1*, p. *158*; 20.1, p. 332; 22.4.3, p. 471.

H. Kiess and R. Clarke
*Pyroelectric Effect on PTS
(a) Vol. **49**, No. 1, p. 133—136 (1978).
See also *1*, p. *9*; 15, p. 261; 22.9, p. 527.

V. M. Efimov, V. A. Kolosanov, and S. P. Sinitsa
*Electron and Hole Conduction in Silicon Nitride at Moderate Electric Fields
(a) Vol. **49**, No. 1, p. 217—223 (1978).
See also 22, p. 417.

A. L. Guerrero, S. C. Jain, and P. L. Pratt
*The Role of Dimers and Trimers in the Aggregation of Impurity-Vacancy Dipoles in Alkali Halides
(a) Vol. **49**, No. 1, p. 353—362 (1978).
See also *10.2*, p. *118*; 22.5.2, p. 484.

P. Kacman, J. Blinowski, and M. Grynberg
Dynamic Dielectric Function in the Three-Band Model for Zero-Gap Semiconductors in a Magnetic Field
(b) Vol. **89**, No. 2, p. 521–530 (1978).
See also *13.1*, p. *159*; 18, p. 275; 22, p. 418.

J. Poźniak
Dipolar Relaxation Induced by an AC Field in Various Crystallographic Directions of Me^{2+}-Doped Alkali Halide Crystals (I): Nearest-Neighbour Approach
(b) Vol. **89**, No. 2, p. 577–582 (1978).
See also *10.2*, p. *119*; 22.5.2, p. 485.

H.-D. Maier, D. Müller, and J. Petersson
*Temperature and Frequency Dependence of the Dielectric Constant and Proton NMR near the Structural Phase Transition of Squaric Acid
(b) Vol. **89**, No. 2, p. 587–593 (1978).
See also *19*, p. *312*; 22.9, p. 527.

P. P. Tirodkar and R. Pratap
Equivalent Rectangular Shape of a Potential Barrier in a Tunnel Junction
(a) Vol. **49**, No. 2, p. 637–641 (1978).
See also *14.3.3*, p. *237*; 22.6, p. 496.

V. P. Arora and V. K. Agarwal
*Dielectric Behaviour of Nematic Phase OHMBBA at Radio Frequencies
(a) Vol. **49**, No. 2, p. 757–760 (1978).
See also *2*, p. *51*; 22.9, p. 528.

A. Mansingh, R. Singh, and M. Sayer
*Dielectric Behaviour of Vanadium Dioxide
(a) Vol. **49**, No. 2, p. 773–779 (1978).
See also 22.6, p. 496.

J. Poźniak
Dipolar Relaxation Induced by an AC Field in Various Crystallographic Directions of Me^{2+}-Doped Alkali Halide Crystals (II): Next-Nearest Neighbour Approach
(b) Vol. **90**, No. 1, p. 261–267 (1978).
See also *10.2*, p. *119*; 22.5.2, p. 485.

K. Thörmer and F. Lüty
*A New Off-Center Defect System, RbCl:Li^{+}, Studied with Dielectric Techniques under Hydrostatic Pressure
(b) Vol. **90**, No. 1, p. 277–287 (1978).
See also 10.2, p. 120; 22.5.3, p. 488.

P. A. Varotsos
*An Estimate of the Pressure Dependence of the Dielectric Constant in Alkali Halides
(b) Vol. **90**, No. 1, p. 339–343 (1978).
See also 4, p. 59; 6, p. 77; 12.2, p. 141; 22.5.1, p. 477; 22.5.2, p. 485.

K. I. Kirov, E. D. Atanasova, and N. A. Ivanov
Investigation of Al_2O_3 Layers Obtained by DC Planar-Magnetron Sputtering
(a) Vol. **50**, No. 1, p. 363–368 (1978).
See also *1.5*, p. *39*; *14.3.3*, p. *238*; 22.6, p. 497.

A. Milchev
*Frequency- and Wave-Vector-Dependent Dielectric Function of a Model Semiconductor
(b) Vol. **90**, No. 2, p. 679–688 (1978).
See also *13.1*, p. *161*; 22.1.1, p. 428; 22.1.2, p. 439; 22.2.1, p. 449; 22.4.2, p. 469.

J. de Oliveira, O. de O. Damasceno, J. R. F. Ferreira, A. L. de Oliveira, J. D. Fabris, P. Boyer, and A. Baudry
Electric Field Gradient Measurements at ^{181}Ta Nuclei in HfO_2–CaO and HfO_2–MgO Systems
(b) Vol. **90**, No. 2, p. K169–K174 (1978).
See also *10.2*, p. *121*; *13*, p. *149*; 22.6.1, p. 500.

M. E. van Hulle and W. Maenhout-van der Vorst
*Ionic Conductivity of β-AgI Microcrystals
(a) Vol. **50**, No. 2, p. K175–K178 (1978).
See also *10.2*, p. *122*; 22.5.1, p. 477.

H. Kresse, D. Demus, and A. Wiegeleben
*Dielectric Relaxation of Liquid Crystals with Nematic, Smectic A, and Smectic C Phases
(a) Vol. **50**, No. 2, p. K181–183 (1978).
See also *2*, p. *53*; 22.9, p. 529.

14.4.1 Ferroelectrics

Original Papers and Short Notes

TH. NATTERMANN
*Static and Dynamic Critical Behaviour of Uniaxial Ferroelectrics and the Phase Transition in TGS
(b) Vol. **85**, No. 1, p. 291–300 (1978).
See also 1.2, p. 15; 22.9, p. 522.

S. FRANCK, A. KÜHNEL, and S. WENDT
*On the Microscopic Mechanism of the Ferroelectric Phase Transition of TSCC: Mn^{2+}
(b) Vol. **85**, No. 1, p. 319–323 (1978).
See also 19, p. 303; 22.9, p. 522.

K. S. ALEKSANDROV, I. P. ALEKSANDROVA, V. F. SHABANOV, V. I. YUZVAK, Y. Z. NOZIK, and L. I. FIKIN
*Mechanism of Ferroelectric Phase Transition on $NaNH_4SeO_4 \cdot 2H_2O$
(a) Vol. **45**, No. 1, p. 53–62 (1978).
See also 1.2, p. 15; 19, p. 303; 20, p. 315; 22.8.1, p. 510.

V. I. KOVALEVICH, L. A. SHUVALOV, and T. R. VOLK
*Spontaneous Polarization Reversal and Photorefractive Effect in Single-Domain Iron-Doped Lithium Niobate Crystals
(a) Vol. **45**, No. 1, p. 249–252 (1978).
See also 22.8.1, p. 510.

A. BHALLA, L. E. CROSS, and L. TONGSON
Passivation of the *c* Surfaces of Single Crystal Gadolinium Molybdate ($Gd_2(MoO_4)_3$) against Attack by Hydrofluoric Acid by Inert Ion Beam Irradiation
(a) Vol. **45**, No. 1, p. K1–K4 (1978).
See also *11*, p. *123*; 22.8.1, p. 510.

I. B. BARKAN, S. I. MARENNIKOV, and M. V. ENTIN
Dynamic Optical Storage in $LiNbO_3$ Crystals
(a) Vol. **45**, No. 1, p. K17–K22 (1978).
See also *13.4*, p. *170*; 16, p. 262; *20.1*, p. *321*; 22.8.1, p. 510.

Z. WRÓBEL
*Effect of Proportional Compositions of $Pb(Zr_xTi_{1-x})O_3$ Solid Solutions on Their Electric Properties with Particular Regard to the Morphotropic Phase Boundary
(a) Vol. **45**, No. 1, p. K67–K69 (1978).
See also 22.8.1, p. 511.

B. STROCKA
Imaging of Photorefractive Effects in $LiNbO_3$ by Double Crystal X-Ray Reflection Topography
(a) Vol. **45**, No. 2, p. K83–K86 (1978).
See also *10.2*, p. *110*; *16*, p. *262*; 22.8.1, p. 511.

P. Konsin
Structural Phase Transitions of Antiferroelectric and Displacive Modulated Types Caused by Electron–Phonon Interaction
(b) Vol. **86**, No. 1, p. 57–66 (1978).
See also *1.2*, p. *16*; 6, p. 67; 22, p. 410.

C. A. S. Lima, A. Scalabrin, L. C. M. Miranda, H. Vargas, and S. P. S. Porto
*Temperature Behaviour of the Dielectric Constant of Tetragonal $BaTiO_3$
(b) Vol. **86**, No. 1, p. 373–378 (1978).
See also 22.8.1, p. 511.

L. Benguigui
*Disordered Ferroelectrics: $Ba_xSr_{1-x}TiO_3$ Single Crystals
(a) Vol. **46**, No. 1, p. 337–342 (1978).
See also 22.8.1, p. 512.

F. Milia, M. Voudouris, A. Kandušer, A. Levstik, and R. Blinc
^{75}As Nuclear Quadrupole Resonance Study of the Phase Transition in Na_2HAsO_4
(b) Vol. **86**, No. 2, p. K119–K121 (1978).
See also *1.2*, p. *17*; *19*, p. *306*; 22.8.1, p. 512.

R. M. Henson, R. R. Zeyfang, and E. Linhart
*Pyroelectric Properties of Pb(Ti, Zr, $[Fe_{1/2}Ta_{1/2}])O_3$ Polycrystalline Solid Solutions
(a) Vol. **46**, No. 2, p. 511–515 (1978).
See also 22.8, p. 504.

H. Arndt, G. Schmidt, N. R. Ivanov, and L. A. Shuvalov
*Electrooptic Switching Anomalies in Ferroelectric TGS
(a) Vol. **46**, No. 2, p. 533–540 (1978).
See also *20.1*, p. *325*; 22.9, p. 524.

V. M. Fridkin and B. N. Popov
*The Anomalous Photovoltaic Effect and Photoconductivity in Ferroelectrics
(a) Vol. **46**, No. 2, p. 729–733 (1978).
See also *16*, p. *263*; 22.8.1, p. 512.

F. Prokert and R. Schalge
*Neutron and Light Scattering from Ferroelectric Fluctuations and Domain Walls of Strontium Barium Niobate
(b) Vol. **87**, No. 1, p. 179–184 (1978).
See also 22.8.1, p. 513.

C. V. Manjunath and R. Srinivasan
*ESR Study of a Cr^{5+} Centre in Ferroelectric Ammonium Sulphate
(b) Vol. **87**, No. 1, p. 395–400 (1978).
See also *19*, p. *307*; 22.8.1, p. 513.

A. Ya. Gordon
*Soft Mode Contribution to the Ferroelectric Heat Capacity
(b) Vol. **87**, No. 1, p. K57–K60 (1978).
See also 1.2, p. 18; 6, p. 71; 8, p. 85.

V. S. Postnikov, S. A. Gridnev, B. N. Prasolov, and S. K. Turkov
Specific Features of Low-Frequency Sound Attenuation in $LiNbO_3$ Poly- and Single-Domain Crystals
(a) Vol. 47, No. 1, p. 65–69 (1978).
See also *7*, p. *80*; *12.2*, p. *139*; 22.8.1, p. 513.

W. Wanarski and Z. Zagórski
*Influence of γ-Irradiation on the Electric Permittivity of Thiourea Single Crystal
(a) Vol. 47, No. 1, p. K45–K46 (1978).
See also *11*, p. *126*; 22.9, p. 525.

K. A. Mirza, P. D. Townsend, and G. L. Destefanis
Light Emission from the Surface of $LiNbO_3$
(a) Vol. 47, No. 1, p. K63–K66 (1978).
See also 13.3, p. 165; *20.3*, p. *344*; 22.8.1, p. 513.

S. A. Gridnev and B. M. Darinskii
Attenuation of Low-Frequency Elastic Oscillations in KH_2PO_4-Type Ferroelectric Crystals
(a) Vol. 47, No. 2, p. 379–384 (1978).
See also *12.2*, p. *139*; 22.8.1, p. 514.

H. Ihrig
*The Positive Temperature Coefficient Resistivity of $BaTiO_3$ Ceramics as a Function of the Amount of Titanium-Rich Second Phase
(a) Vol. 47, No. 2, p. 437–444 (1978).
See also 22.8.1, p. 514.

H. Pykacz
*Pyroelectric Properties of TGS with Internal Bias Field
(a) Vol. 47, No. 2, p. 625–630 (1978).
See also 22.9, p. 525.

R. R. Levitskii, N. A. Korynevskii, and I. V. Stasyuk
*Distribution Functions and Thermodynamical Properties of KD_2PO_4 and $ND_4D_2PO_4$ Type Crystals
(b) Vol. 88, No. 1, p. 51–63 (1978).
See also *1.2*, p. *20*; 22.8.1, p. 514.

J. Grossmann, D. Müller, and J. Petersson
*Study of the Ultraslow Critical Dynamics of the Order Parameter of the Ferroelectric $AgNa(NO_2)_2$ by the Spin Relaxation Times T_1, $T_{1\varrho}$, T_2 of the ^{23}Na Nucleus
(b) Vol. 88, No. 1, p. 99–103 (1978).
See also *19*, p. *309*; 22.8.1, p. 514.

V. V. Gagulin, N. V. Fadeeva, A. G. Belous, A. V. Titov, K. P. Mitrofanov, M. V. Plotnikova, S. P. Soloviev, and Yu. N. Venevtsev
*New Seignette-Magnets with Hexagonal Barium Titanate Structure
(a) Vol. 48, No. 1, p. 183–189 (1978).
See also *1*, p. *9*; 5, p. 61; 18.2, p. 289; 18.4, p. 301; 19, p. 310; 22.8.1, p. 514.

H. Konwent and J. Lorenc
*A Thermodynamic Theory of the Improper Antiferroelectric Phase Transition in ADP
(b) Vol. **88**, No. 2, p. 747–755 (1978).
See also 1.2, p. 21; *12.2*, p. *140*; 22.8.1, p. 515.

L. E. Cross and A. Bhalla
*Domain "Decoration" in $Gd_2(MoO_4)_3$
(a) Vol. **48**, No. 2, p. 431–437 (1978).
See also *1.6*, p. *41*; 22.8.1, p. 515.

C. J. de Pater
*Dielectric Constant Ferroelectric Hysteresis of Rb_2ZnBr_4
(a) Vol. **48**, No. 2, p. 503–512 (1978).
See also 22.5.4, p. 489.

V. N. Kashcheev
Phonon Relaxation in One- and Two-Dimensional Displacive-Type Ferroelectrics
(b) Vol. **89**, No. 1, p. 45–52 (1978).
See also *6*, p. *75*.

B. Milsch and W. Windisch
*Electron Paramagnetic Resonance of Mn^{2+} in Guanidinium Aluminium Sulfate Hexahydrate
(b) Vol. **89**, No. 1, p. 241–251 (1978).
See also 13.4, p. 180; *19*, p. *311*; 22.8.1, p. 515.

A. Ya. Gordon
*On Phase Transitions in Two-Dimensional Ferroelectrics
(b) Vol. **89**, No. 1, p. K69–K71 (1978).
See also *1.2*, p. *22*.

H. E. Müser, W. Kuhn, and J. Albers
*Optical Investigation of the Cubic–Tetragonal Phase Transition in $BaTiO_3$ Single Crystals
(a) Vol. **49**, No. 1, p. 51–58 (1978).
See also *1.2*, p. *22*; 20, p. 317; 22.8.1, p. 515.

R. von Baltz
*Theory of the Anomalous Bulk Photovoltaic Effect in Ferroelectrics
(b) Vol. **89**, No. 2, p. 419–429 (1978).
See also *16*, p. *266*; 22.8.1, p. 516.

V. A. Izvozchikov, G. A. Bordovskii, V. T. Avanesyan, and V. A. Bordovskii
*On the Problem of Ferroelectricity in Photoconducting Lead Oxides
(a) Vol. **49**, No. 2, p. K173–175 (1978).
See also 22.6, p. 497.

T. I. Chekmasova and I. P. Aleksandrova
*NMR Investigation of the High-Pressure Phase in $LiNH_4SO_4$
(a) Vol. **49**, No. 2, p. K185–K188 (1978).
See also *1.2*, p. *23*; 19, p. 313; 22.8.1, p. 516.

N. A. Zakharov, V. S. Krikorov, E. F. Kustov, and S. Yu. Stefanovich
*New Non-Linear Crystals in the $A_2B_2O_7$ Series
(a) Vol. **50**, No. 1, p. K13—K16 (1978).
See also 3, p. 56; *20*, p. *318*; 22.8, p. 508.

G. Schmidt and I. Boczek
*Pseudoelasticity and Shape Memory of PLZT Ceramic
(a) Vol. **50**, No. 1, p. K109—K111 (1978).
See also 12.2, p. 141; 22.8.1, p. 516.

M. A. F. Scarparo, R. S. Katiyar, R. Srivastava, and S. P. S. Porto
*Evidence of Non-Linear Behaviour of Soft Modes in Crystals with KDP Structure
(b) Vol. **90**, No. 2, p. 543—556 (1978).
See also 20.1, p. 337; 22.8.1, p. 516.

G. Luther, H. E. Müser, and A. Levstik
*Dielectric Behaviour of Ferroelectric $Na(H_{1-x}D_x)_3(SeO_3)_2$ at RF and Microwave Frequencies
(a) Vol. **50**, No. 2, p. 459—468 (1978).
See also 22.8, p. 508.

S. Tsunekawa and H. Takei
Twinning Structure of Ferroelastic $LaNbO_4$ and $NbNbO_4$ Crystals
(a) Vol. **50**, No. 2, p. 695—702 (1978).
See also *1*, p. *10*; 10.2, p. 122; 22.8.1, p. 516.

R. Jakubas, Z. Czapla, L. Sobczyk, and S. Olejnik
*Ferroelectric Properties of Selenate Alums
(a) Vol. **50**, No. 2, p. K137—K140 (1978).
See also 22.8, p. 508.

Z. Surowiak
*The Effect of a Stable Electric Field on the Properties of $BaTiO_3$ Films
(a) Vol. **50**, No. 2, p. K145—K148 (1978).
See also 22.8.1, p. 517.

A. K. Moskalev, I. A. Belobrova, I. P. Aleksandrova, Sh. Sawada, and Y. Shiroishi
*Investigation of the Incommensurate Phase of Rb_2ZnCl_4 by the NQR Method
(a) Vol. **50**, No. 2, p. K157—K159 (1978).
See also *19*, p. *314*; 22.8, p. 509.

F. Smutný
*Specific Heat Anomaly at the $F\bar{4}3c \rightarrow Pca2_1$ Phase Transition in Nickel Chlorine Boracite, $Ni_3B_7O_{13}Cl$
(a) Vol. **50**, No. 2, p. K201—K204 (1978).
See also *8*, p. *90*; 22.8, p. 509.

H. Hesse and S. Kapphan
*Doping of $SrTiO_3$ with Hydrogen and Deuterium
(a) Vol. **50**, No. 2, p. K243—K245 (1978).
See also 10.2, p. 122; *20.1*, p. *338*; 22.8.1, p. 517.

V. M. Fridkin, E. I. Boikova, B. N. Popov, and G. I. Rosenman
The Influence of the Anomalous Photovoltaic Effect on the Exoelectron Emission in $LiNbO_3$
(a) Vol. **50**, No. 2, p. K255—K259 (1978).
See also *16*, p. *268*; *17*, p. *271*; 22.8.1, p. 517.

15. Thermoelectric and Thermomagnetic Properties

Review Article

A. S. BORUKHOVICH
Kinetic Properties of IVa and Va Subgroup Transition Metal Monocarbides
(a) Vol. **46**, No. 1, p. 11–37 (1978).
See also 8, p. 83; 13.1, p. 150; *14.3*, p. *217*; 22, p. 408; 22.3, p. 457; 22.3.1, p. 460.

Original Papers and Short Notes

W. EICHLER
*On the Analysis of Galvano- and Thermomagnetic Effects in Pure Perfect Tellurium Crystals
(a) Vol. **45**, No. 1, p. 195–199 (1978).
See also *14.3*, p. *217*; 22.1.3, p. 441.

J. J. DUBOWSKI
Disorder Scattering in $Cd_xHg_{1-x}Te$ Mixed Crystals
(b) Vol. **85**, No. 2, p. 663–672 (1978).
See also *13.1*, p. *151*; *14.3*, p. *217*; 22.4.4, p. 472.

V. CRISTEA and V. BABES
*The Seebeck Effect in Nb-Doped TiO_2 Rutile
(a) Vol. **45**, No. 2, p. 617–624 (1978).
See also 14.3, p. 218; 22.6, p. 491.

B. M. ASKEROV, B. I. KULIEV, S. R. FIGAROVA, and V. YA. SHTEINSHRAIBER
*Theory of the Thermomagnetic Effects in Thin Films in a Magnetic Field of Arbitrary Magnitude
(b) Vol. **86**, No. 1, p. 103–108 (1978).
See also 14.3.1, p. 226.

T. IKARI and K. HASHIMOTO
*Electrical Properties of TlTe Single Crystals
(b) Vol. **86**, No. 1, p. 239–248 (1978).
Erratum
(b) Vol. **90**, No. 2, p. 751 (1978).
See also 13.1, p. 153; *14.3*, p. *218*; 22.3, p. 457.

D. V. GITSU, F. M. MUNTYANU, M. I. ONU, and A. I. RUSSU
*Peculiarities of the Thermomagnetic Power in $Bi_{0.91}Sb_{0.09}$ Alloy in Strong Magnetic Field
(b) Vol. **86**, No. 1, p. K71–K73 (1978).
See also 21.7, p. 406.

L. G. KHVOSTANTSEV and V. A. SIDOROV
*Influence of Pressure up to 90 kbar on the Thermoelectric Properties of Antimony
(a) Vol. **46**, No. 1, p. 305–309 (1978).
See also 21.7, p. 406.

P. THOMAS and D. WUERTZ
Low Temperature Polaronic Hopping Conduction
(b) Vol. 86, No. 2, p. 541—548 (1978).
See also *2*, p. *43*; *14.3*, p. *219*; 22.1.1, p. 425; 22.1.2, p. 432.

MAHIRADHWAJ SINGH
Theory of Resonance Scattering for Electron–Phonon Interaction in Doped Semiconductors with Zincblende Structure
(b) Vol. 87, No. 1, p. 333—344 (1978).
See also *6*, p. *70*; 8, p. 85; 22.2.1, p. 446; 22.2.3, p. 453.

V. CHÁB, L. ŠTOURAČ, and B. VELICKÝ
The Composition Dependence of Activation Energy in the As_x–Te_{100-x} and As_2Te_3+Tl_x Glassy Systems
(a) Vol. 47, No. 1, p. 271—276 (1978).
See also *2*, p. *45*; *14.3*, p. *221*; 22, p. 413.

O. PAVLOV and Z. BONTCHEVA-MLADENOVA
Measurements of the Ag_2Te–Ag_2S System at 300 K
(a) Vol. 47, No. 1, p. K59—K61 (1978).
See also *1.3*, p. *26*; 8, p. 86; *14.3*, p. *221*; 22, p. 413.

YA. A. KRAFTMAKHER and T. YU. PINEGINA
*Anomalies of Resistivity and Thermopower of Samarium at High Temperatures
(a) Vol. 47, No. 1, p. K81—K83 (1978).
See also 1.2, p. 18; *14.1*, p. *208*; 21.4, p. 397.

M. JACZYŃSKI, J. KOSSUT, and R. R. GAŁĄZKA
Influence of Exchange Interaction on the Quantum Transport Phenomena in $Hg_{1-x}Mn_xTe$
(b) Vol. 88, No. 1, p. 73—85 (1978).
See also *13.2*, p. *162*; *14.3*, p. *221*; 22.4.4, p. 473.

N. P. GAVALESHKO, P. N. GORLEY, P. V. KAVYUK, L. S. SOLONCHUK, I. I. TARASYUK, and V. A. SHENDEROVSKII
*Transverse and Longitudinal Nernst-Ettingshausen Effect in p-Te
(b) Vol. 88, No. 1, p. 179—188 (1978).
See also *13.4*, p. *177*; 14.3, p. 222; 22.1.3, p. 441.

J. R. HANSCOMB and P. CHANTIKUL
*The Effect of Strontium Chloride as Impurity on the Thermoelectric Power of Potassium Chloride
(a) Vol. 48, No. 1, p. 79—87 (1978).
See also 10.2, p. 115; 22.5.2, p. 482.

C. K. SARKAR and D. CHATTOPADHYAY
*Effect of the Electron Spin on the Magneto-Seebeck Coefficient of Polar Semiconductors at Large Magnetic Fields
(b) Vol. 88, No. 2, p. K109—K112 (1978).
See also 13.1, p. 157; 22.2.3, p. 453.

C. G. PATIL and B. S. KRISHNAMURTHY
*Magnon Drag Thermoelectric Power in Ferromagnetic Semiconductors
(b) Vol. 88, No. 2, p. K117—K122 (1978).
See also *13.5.4*, p. *202*; 18.2, p. 289; 22, p. 415.

M. Moldovanova and R. Assenov
*A New Interpretation of the Temperature Dependence of the Hall Coefficient in "Pure" and Doped SnTe
(a) Vol. **48**, No. 2, p. K193–K197 (1978).
See also *14.3*, p. *223*; 22, p. 416.

H. Kiess and R. Clarke
Pyroelectric Effect on PTS
(a) Vol. **49**, No. 1, p. 133–136 (1978).
See also *1*, p. *9*; *14.4*, p. *250*; 22.9, p. 527.

S. Myhra
Defect Annealing and Transport Properties in High-Purity InSb Irradiated at 80 K
(a) Vol. **49**, No. 1, p. 285–292 (1978).
See also *10.2*, p. *118*; *14.3*, p. *224*; 22.2.3, p. 454.

N. N. Sinha and P. L. Srivastava
*Resistivity and Thermo-Electric Power of Monovalent, Divalent, and Trivalent Liquid Metals
(b) Vol. **90**, No. 1, p. 369–374 (1978).
See also 2, p. 51; *14.1*, p. *211*; 21, p. 364; 21.1, p. 380; 21.2, p. 393.

H. Köhler and W. Ossau
*Magneto-Seebeck and Shubnikov-de Haas Effect in n-Type $(Bi_{1-x}Sb_x)_2Te_3$
(b) Vol. **90**, No. 2, p. 509–515 (1978).
See also *13.1*, p. *160*; 14.3, p. 225; 22.7, p. 502.

16. Photoconductivity. Photovoltaic Effects

Original Papers and Short Notes

B. Reimer and H. Bässler
*Transient Photoconduction in a Polydiacetylene Single Crystal. Determination of Transit Time, Free Lifetime, and Recombination Time of Charge Carriers
(b) Vol. **85**, No. 1, p. 145—153 (1978).
See also 22.9, p. 522.

R. Keiper and R. Schuchardt
*On the Theory of Photo-Induced DC Hopping Conductivity in Disordered Semiconductors
(b) Vol. **85**, No. 1, p. 155—162 (1978).
See also 22, p. 408.

V. G. Golubev, E. I. Georgitse, and V. I. Ivanov-Omskii
*Photoconductivity of HgTe in Magnetic Field at 337 μm
(b) Vol. **85**, No. 1, p. K39—K41 (1978).
See also 18, p. 273; *20.1*, p. *320*; 22.4.3, p. 470.

M. Kitao, H. Ikeda, H. Hasegawa, and S. Yamada
*Photoconductivity of Amorphous As_2Se_3
(a) Vol. **45**, No. 1, p. 107—111 (1978).
See also 2, p. 42; 22, p. 409.

I. B. Barkan, S. I. Marennikov, and M. V. Entin
Dynamic Optical Storage in $LiNbO_3$ Crystals
(a) Vol. **45**, No. 1, p. K17—K22 (1978).
See also *13.4*, p. *170*; 14.4.1, p. 253; *20.1*, p. *321*; 22.8.1, p. 510.

E. M. Baskin, M. D. Blokh, M. V. Entin, and L. I. Magarill
*Erratum to: Current Quadratic in Field and Photogalvanic Effect in Crystals without Inversion Centre
(b) Vol. **85**, No. 2, p. K135 (1978).
See also *14.3.4*, p. *240*; 22.2, p. 443.

G. Jacobsen and W. Maenhout-van der Vorst
*Photoconductivity and the Negative Differential Photo-Effect in Sulfur-Annealed ZnS Single Crystals
(a) Vol. **45**, No. 2, p. 463—468 (1978).
See also 22.4.1, p. 463.

B. Strocka
*Imaging of Photorefractive Effects in $LiNbO_3$ by Double Crystal X-Ray Reflection Topography
(a) Vol. **45**, No. 2, p. K83—K86 (1978).
See also *10.2*, p. *110*; 14.4.1, p. 253; 22.8.1, p. 511.

G. A. Medvedkin, Yu. V. Rud, Yu. A. Valov, and V. I. Sokolova
*Photoconductivity of p-Type $CdSnP_2$ Single Crystals
(a) Vol. **45**, No. 2, p. K95—K99 (1978).
See also 22.8, p. 503.

S. D. PHADKE and R. N. KAREKAR
*Photocurrents through Polythiourea and Polymalononitrile Thin Films
(a) Vol. **45**, No. 2, p. K163–K165 (1978).
See also 14.3.1, p. 226; 22.9, p. 523.

A. N. TRUKHIN
Investigation of the Photoelectric and Photoluminescent Properties of Crystalline Quartz and Vitreous Silica in the Fundamental Absorption Region. A Model for Electronic Structure and Migration of Energy in SiO_2
(b) Vol. **86**, No. 1, p. 67–75 (1978).
See also 2, p. 43; *13.1*, p. *152*; 20.3, p. 342; 22.6, p. 491.

R. S. NAKHMANSON and L. K. POPOV
Measurements of Small-Signal Photo-EMF of Si and Ge MIS Structures Using Scanning Light Probe
(a) Vol. **46**, No. 1, p. 59–68 (1978).
See also *14.3.2*, p. *230*; *14.3.3*, p. *233*; 22.1.1, p. 425; 22.1.2, p. 431.

R. W. A. HILLHOUSE and J. WOODS
Plasma Growth of Rutile Crystals and Their Photoelectronic Properties
(a) Vol. **46**, No. 1, p. 163–172 (1978).
See also *3*, p. *54*; 13.4, p. 173; 22.6, p. 492.

M. SAMOĆ and Z. ZBOIŃSKI
Trapping of Charge Carriers in the Deformed Surface Region of Anthracene Single Crystals
(a) Vol. **46**, No. 1, p. 251–257 (1978).
See also *13.3*, p. *165*; 14.3, p. 219; 22.9, p. 524.

D. GENZOW
*Anisotropy of Photoconductivity in Lead Chalcogenides
(b) Vol. **86**, No. 2, p. K143–K146 (1978).
See also 22, p. 412.

V. A. KORNEICHUK, T. I. MITROFANOVA, P. E. MOZOL, and G. S. PEKAR
*Photoconductivity of CdS:Cu under Ruby Laser Excitation
(b) Vol. **86**, No. 2, p. K155–K158 (1978).
See also 13.5.1, p. 191; 22.4.1, p. 464.

F. M. LIVINGSTONE, R. M. DE LA RUE, and W. DUNCAN
pSi–nCdS Heterojunction Photodetectors
(a) Vol. **46**, No. 2, p. 595–601 (1978).
See also *14.3.3*, p. *234*; 22.1.2, p. 432; 22.4.1, p. 464.

A. BANERJEE, PREM NATH, V. D. VANKAR, and K. L. CHOPRA
Properties of $Zn_xCd_{1-x}S$ Films Prepared by Solution Spray Technique
(a) Vol. **46**, No. 2, p. 723–728 (1978).
See also *1.5*, p. *36*; *14.3.1*, p. *226*; 22.4.4, p. 472.

V. M. FRIDKIN and B. N. POPOV
*The Anomalous Photovoltaic Effect and Photoconductivity in Ferroelectrics
(a) Vol. **46**, No. 2, p. 729–733 (1978).
See also *14.4.1*, p. *254*; 22.8.1, p. 512.

P. K. C. Pillai, R. G. Mendiratta, C. K. Pillai, and R. Nath
*Photo-Induced AC Impedance Measurements in Single Crystal Photoconductors
(a) Vol. 46, No. 2, p. K127—K130 (1978).
See also *14.4*, p. *248*; 22.4, p. 461; 22.5, p. 475.

L. Vescan, H. Stötzel, T. Stoica, and A. Kottwitz
*Photoconductivity in a-GeSe
(b) Vol. 87, No. 1, p. 239—246 (1978).
See also 2, p. 44; 22.7, p. 501.

E. L. Frankevich, M. M. Tribel, I. A. Sokolik, and A. I. Pristupa
*Magnetic-Resonant Modulation of Photoconductivity of Crystalline Charge Transfer Complexes. Anthracene–Tetracyanbenzene
(b) Vol. 87, No. 1, p. 373—379 (1978).
See also 13.5.1, p. 191; 20.3, p. 344; 22.9, p. 524.

L. Gouskov, A. Gouskov, and B. Girault
Electrooptical Properties of SnO_2 Spray–GaSe and SnO_2 Spray–GaTe Heterojunctions
(a) Vol. 47, No. 1, p. 195—199 (1978).
See also *14.3.3*, p. *235*; 14.3.4, p. 242; 20.3, p. 344; 22.3, p. 458.

M. Geddo, A. Stella, and G. Soncini
*Photoelectron–Grain Boundary Scattering in Al Films of MOS Structures
(a) Vol. 47, No. 1, p. 223—227 (1978).
See also 21, p. 358.

M. Leppihalme and T. Tuomi
*Phonon Energies and Barrier Height from Photovoltage Spectra of Au–Intrinsic Ge Diodes
(a) Vol. 47, No. 1, p. 245—250 (1978).
See also 6, p. 71; 13.5.1, p. 191; *14.3.3*, p. *235*; 22.1.1, p. 425.

B. Pohoryles and R. Piotrzkowski
*Effect of Hydrostatic Pressure on the Photoconductivity of Plastically Deformed n-Type Germanium
(a) Vol. 47, No. 2, p. K115—K117 (1978).
See also 13.4, p. 177; 22.1.1, p. 426.

P. Viktorovitch, G. Kamarinos, P. Even, and E. Fabre
*Correlation between Interface States and MIS Silicon Solar Cell Performances
(a) Vol. 48, No. 1, p. 137—145 (1978).
See also 13.3, p. 166; *14.3.3*, p. *235*; 22.1.2, p. 434.

A. V. Bobyl, G. S. Pekar, and M. K. Sheinkman
*Dark Conductivity and Photoconductivity of Nickel-Doped CdS Single Crystals
(a) Vol. 48, No. 1, p. 249—256 (1978).
See also *13.4*, p. *178*; 22.4.1, p. 465.

P. Schlotter and H. Bässler
*Sensitized Injection into Anthracene Crystals from Surface Oxidation Products
(a) Vol. 48, No. 1, p. K15—K18 (1978).
See also *13.3*, p. *166*; 22.9, p. 526.

P. Höschl, P. Moravec, A. Martinaitis, and A. Sakalas
*Study of Photo-Hall Effect on Single Crystals of Cadmium Telluride Doped with Chlorine
(a) Vol. **48**, No. 1, p. K43—K46 (1978).
See also *13.4*, p. *178*; 22.4.3, p. 470.

P. Doubrava, V. L. Averyanov, and J. Škácha
Spectral Characteristics of Photodarkening of a-$As_{60}Se_{40}$ Films
(a) Vol. **48**, No. 1, p. K93—K96 (1978).
See also *2*, p. *47*; *20.1*, p. *330*; 22, p. 415.

I. M. Dykman, V. M. Rosenbaum, and F. T. Vasko
*Hot Electrons in Semiconductors with Quasi-Relativistic Band Structure
(b) Vol. **88**, No. 2, p. 385—395 (1978).
See also *13*, p. *146*; *14.3.4*, p. *243*; 22, p. 415.

H. Mell, B. Movaghar, and L. Schweitzer
*Magnetic Field Dependence of the Photoconductivity in Amorphous Silicon
(b) Vol. **88**, No. 2, p. 531—535 (1978).
See also *2*, p. *47*; 22.1.2, p. 435.

K. Lochner, H. Bässler, B. Tieke, and G. Wegner
*Photoconduction in Polydiacetylene Multilayer Structures and Single Crystals. Evidence for Band-to-Band Excitation
(b) Vol. **88**, No. 2, p. 653—661 (1978).
See also *13.1*, p. *157*; 22.9, p. 526.

R. S. Eachus, R. E. Graves, and M. T. Olm
Electron Spin Resonance of Mobile Electrons in AgCl
(b) Vol. **88**, No. 2, p. 705—711 (1978).
See also *10.2*, p. *116*; *19*, p. *310*; 22.5.1, p. 477.

A. Haydar and A. Coret
*Excitons and Structural Defects in Cu_2O at Low Temperature
(a) Vol. **48**, No. 2, p. 465—471 (1978).
See also *13.5.1*, p. *194*; 22.6, p. 495.

K. W. Böer
*The Physics of the Photovoltaic Effect
(a) Vol. **49**, No. 1, p. 13—26 (1978).
See also *14.3.3*, p. *236*; 22, p. 417; 22.4.1, p. 465.

G. Heiland and D. Kohl
*Interpretation of Surface Phenomena on ZnO by the Compensation Model
(a) Vol. **49**, No. 1, p. 27—37 (1978).
See also 13.3, p. 167; 13.4, p. 180; *14.3.2*, p. *231*; 14.3.4, p. 244; 22.4, p. 462.

S. Lehr and H. Pagnia
Work Function of Irradiated Real and Cleaned Indium Antimonide Surfaces
(a) Vol. **49**, No. 1, p. 83—91 (1978).
See also *13.3*, p. *167*; 17, p. 270; 22.2.3, p. 453.

K. Mettler
*Effect of Reabsorbed Recombination Radiation on Photoluminescene and Photoconductivity in a Semi-Infinite Direct-Gap Semiconductor
(a) Vol. **49**, No. 1, p. 163—168 (1978).
See also *20.3*, p. *347*; 22.2.1, p. 448.

R. Butendeich and W. Ruppel
*Photovoltage at the Metal–CdS Schottky Contact
(a) Vol. **49**, No. 1, p. 169—175 (1978).
See also 22.4.1, p. 465.

L. Risch
Investigation of Surface Properties of CdS by Means of Photoreflectance in the Exciton Region
(a) Vol. **49**, No. 1, p. 177—182 (1978).
See also *13.3*, p. *167*; 20.1, p. 333; 22.4.1, p. 465.

A. Soliman and N. Saleh
*Photovoltaic Operation of Nonuniformly Doped Intrinsic Photoconductors
(a) Vol. **49**, No. 1, p. 393—403 (1978).
See also *14.3.3*, p. *236*; 22.2.3, p. 454.

F. R. Kessler and W. Olsowski
*Exciton Contribution to Oscillatory Magnetophotoconductivity in Ge
(b) Vol. **89**, No. 2, p. 389—396 (1978).
See also *13.5.1*, p. *194*; 22.1.1, p. 428.

R. von Baltz
*Theory of the Anomalous Bulk Photovoltaic Effect in Ferroelectrics
(b) Vol. **89**, No. 2, p. 419—429 (1978).
See also *14.4.1*, p. *256*; 22.8.1, p. 516.

W. Fuhs, M. Milleville, and J. Stuke
*Drift Mobility and Photoconductivity in Amorphous Silicon
(b) Vol. **89**, No. 2, p. 495—502 (1978).
See also 2, p. 50; *14.3.1*, p. *228*; 22.1.2, p. 437.

C. Clemen, X. I. Saldaña, P. Munz, and E. Bucher
*Photovoltaic Properties of Some Semiconducting Layer Structures
(a) Vol. **49**, No. 2, p. 437—443 (1978).
See also 22, p. 418.

K. W. Böer
*Heterojunction Interface Double Layer and Consequences for Photovoltaic Cells, Specifically $Cd_zZn_{1-z}S/Cu_2S$
(a) Vol. **49**, No. 2, p. 455—462 (1978).
See also *14.3.3*, p. *237*; 22, p. 418; 22.4.1, p. 466.

K. G. Germanova
On Some Stationary and Non-Stationary Temperature-Electric Instabilities in Photoconducting CdS Crystals
(a) Vol. **49**, No. 2, p. 499—507 (1978).
See also *14.3.4*, p. *245*; 22.4.1, p. 466.

G. Sarrabayrouse and J. Buxo
*The Reversely Biased MIS Tunneling Schottky Barrier. Application to Photodetection
(a) Vol. 49, No. 2, p. 603—608 (1978).
See also *14.3.3*, p. *237*.

J. A. Groenink and D. A. van Wezep
*Photoconductivity of $PbMoO_4$ and $PbWO_4$ Single Crystals
(a) Vol. 49, No. 2, p. 651—654 (1978).
See also 22.8.1, p. 516.

W. Szkiełko
Spin-Dependent Recombination at Dislocations in Silicon Revealed at Reverse Current in a p–n Junction
(b) Vol. 90, No. 1, p. K81—K83 (1978).
See also *10.2*, p. *120*; *14.3.3*, p. *237*; 19, p. 314; 22.1.2, p. 438.

F. Adduci, L. Baldassarre, G. Maggipinto, A. Minafra, and F. Lévy
*Photoelectromagnetic Investigation of PbI_2
(a) Vol. 50, No. 1, p. 257—262 (1978).
See also 22.5, p. 476.

R. Baubinas, B.-P. Kietis, R. Reksnys, and A. Sakalas
*Bipolar Photoconductivity and Acoustoelectric Current in CdTe, CdS, and CdSe
(a) Vol. 50, No. 1, p. K63—K65 (1978).
See also *7*, p. *82*; 22.4.1, p. 466; 22.4.2, p. 469; 22.4.3, p. 471.

I. A. Maltseva, A. Mamedov, Yu. V. Rud, and Yu. K. Undalov
*The Photosensitivity of Diodes Based on $CdGeP_2$:In Crystals
(a) Vol. 50, No. 2, p. 445—448 (1978).
See also 14.3, p. 225; 22.8, p. 508.

A. Anedda, L. Garbato, F. Raga, and A. Serpi
*Photoconductivity and Trap Distribution in $CdIn_2S_4$
(a) Vol. 50, No. 2, p. 643—650 (1978).
See also 13.4, p. 186; 22.8, p. 508.

C. Lhermitte, D. Carles, C. Viger, and C. Vautier
*Transient Photoconductivity Effects on n-Type Cadmium Telluride Films in Presence of a Uniform Distribution of Traps. Theoretical Model and Experimental Results
(a) Vol. 50, No. 2, p. 651—659 (1978).
See also 13.4, p. 186; 22.4.3, p. 471.

V. V. Serdyuk, A. P. Chebanenko and G. G. Chemeresyuk
Current Switching Mechanism in Inhomogeneous Zinc Selenide Crystals
(a) Vol. 50, No. 2, p. 717—722 (1978).
See also *13.4*, p. *186*; *14.3.4*, p. *247*; 22.4.2, p. 469.

A. F. Plotnikov, V. A. Tolokonnikov, and V. E. Shubin
Investigation of the Energy Band Diagram of the Anodic Oxide–Indium Antimonide System by Internal Photoemission
(a) Vol. **50**, No. 2, p. 733—738 (1978).
See also *13.1*, p. *161*; *14.3.3*, p. *239*; 22.2.3, p. 455.

N. E. Korsunskaya, I. V. Markevich, T. V. Torchinskaya, and M. K. Sheinkman
Some Peculiarities of Thermostimulated Conductivity and Optical Quenching of the Photocurrent in Crystals with the Auger Excitation of Recombination Centres
(a) Vol. **50**, No. 2, p. 767—770 (1978).
See also *13.4*, p. *186*; *14.3*, p. *225*; 22.4.1, p. 467.

V. M. Fridkin, E. I. Boikova, B. N. Popov, and G. I. Rosenman
*The Influence of the Anomalous Photovoltaic Effect on the Exoelectron Emission in $LiNbO_3$
(a) Vol. **50**, No. 2, p. K255—K259 (1978).
See also 14.4.1, p. 258; *17*, p. *271*; 22.8.1, p. 517.

17. Emission of Electrons and Ions

Original Papers and Short Notes

C.-O. Almbladh and P. Minnhagen
On the Thermal Broadening of Core Levels in Metals Due to Many-Electron Effects
(b) Vol. **85**, No. 1, p. 135–143 (1978).
See also *13*, p. *142*; 20, p. 315; 21, p. 352.

C. Norris and G. P. Williams
*Photoelectron Measurements of α-Phase Alloys CuZn, CuGa, and CuGe
(b) Vol. **85**, No. 1, p. 325–330 (1978).
See also *13.4*, p. *169*; 21.1, p. 368.

H. Neumann and A. Kosakov
Tight-Binding Calculation of the Density of Valence States of SnS, SnSe, and SnTe
(b) Vol. **85**, No. 1, p. K11–K14 (1978).
See also *13.1*, p. *150*; 22.7, p. 501.

A. I. Belkind, A. M. Brodskii, and V. V. Grechov
*Theory of Photoemission from Molecular Crystals
(b) Vol. **85**, No. 2, p. 465–472 (1978).
See also 13.1, p. 151; 22.9, p. 522.

J. Wortmann
*Optisch stimulierte Elektronenemission von tribomechanisch aktiviertem Aluminium (Optically Stimulated Exoelectron Emission of Tribomechanically Activated Aluminium)
(a) Vol. **46**, No. 1, p. 69–76 (1978).
See also *13.3*, p. *164*; 21, p. 356.

M. V. Kruglov, V. N. Shchemelev, and G. G. Kareva
Detection of Atomic Plane Displacements in Near-Surface Layers of Crystals Using the Angular Dependence of Photoemission during Bragg Diffraction of X-Ray
(a) Vol. **46**, No. 1, p. 343–350 (1978).
See also *1.1*, p. *11*; 1.6, p. 40; 22.1.2, p. 431.

V. N. Kadchenko and M. Elango
*Diffusion Approach to the Photostimulated Electron Emission of Ionic Crystals after Photoexcitation
(a) Vol. **46**, No. 1, p. K61–K65 (1978).
See also 22.5.2, p. 480.

H. Hora
*Strong Coupling of Electrons to Black Body Radiation at High Temperatures
(b) Vol. **86**, No. 2, p. 685–690 (1978).

A. D. Buonaquisti and R. A. Collins
*Energy Distribution of Electrons Emited from MIM Thin-Film Sandwich Cathodes
(a) Vol. **46**, No. 2, p. K91–K93 (1978).
See also 14.3.3, p. 235; 22.6, p. 492.

G. PAASCH
*Theory of Angle-Resolved Photoemission. The Role of Electron Diffraction in the XPS Spectra
(b) Vol. **87**, No. 1, p. 191–201 (1978).

W. EBERHARDT, G. KALKOFEN, C. KUNZ, D. ASPNES, and M. CARDONA
Photoemission Studies of 2p Core Levels of Pure and Heavily Doped Silicon
(b) Vol. **88**, No. 1, p. 135–143 (1978).
See also *13.1*, p. *156*; 20.1, p. 328; 22.1.2, p. 434.

R. H. WILLIAMS
The Electronic Structure of Tin Diselenide Containing Cyclopropylamine
(b) Vol. **88**, No. 2, p. K157–K159 (1978).
See also *13.1*, p. *158*; 22, p. 416.

R. GAUTHIER, M. BABOUT, and C. GUITTARD
*Laser Induced Desorption Studied with Mirror Electron Microscopy
(a) Vol. **48**, No. 2, p. 459–464 (1978).
See also 21, p. 361.

S. LEHR and H. PAGNIA
Work Function of Irradiated Real and Cleaned Indium Antimonide Surfaces
(a) Vol. **49**, No. 1, p. 83–91 (1978).
See also *13.3*, p. *167*; 16, p. 265; 22.2.3, p. 453.

H. HUMBERG and H. MERZ
*Polarization Dependence of the Photoelectron Energy Distribution Curves of Cesium
(b) Vol. **89**, No. 2, p. 397–402 (1978).
See also 21.2, p. 393.

ATSUO KASUYA, S. TOGASHI, T. GOTO, and Y. NISHINA
*Two-Photon Photoemission in PbI_2 and GaAs
(b) Vol. **89**, No. 2, p. K145–K147 (1978).
See also 22.2.1, p. 448; 22.5, p. 476.

G. RUDLOF, J. BECHERER, and H. GLAEFEKE
*Behaviour of the Fractional Glow Technique with First-Order Detrapping Processes, Traps Distributed in Energy or Frequency Factor
(a) Vol. **49**, No. 2, p. K121–K124 (1978).
See also *13.4*, p. *182*; 22, p. 419.

V. T. CHEREPIN, A. A. KOSYACHKOV, and M. A. VASILEV
*Secondary Ion Emission of Transition Metal Carbides
(a) Vol. **50**, No. 1, p. K113–K116 (1978).
See also 22.3, p. 459.

W. BRAUN, M. NEUMANN, M. IWAN, and E. E. KOCH
*UPS Measurements of CO Chemisorbed and Condensed on Rh(111) Using Synchrotron Radiation
(b) Vol. **90**, No. 2, p. 525–533 (1978).
See also *13.1*, p. *160*; 21, p. 365.

J. K. Wysocki and C. M. Vodenicharov
*Approximate Formula for the Emission Current Density in the Second Intermediate Region of Christov
(a) Vol. **50**, No. 2, p. 411—415 (1978).
See also *14.3.3*, p. *238*; 22.6, p. 497.

H. Düsterhöft
*Dependence of Secondary Ion Emission on the Mass of the Bombarding Ions (Z_1-Dependence)
(a) Vol. **50**, No. 2, p. 503—507 (1978).
See also 21, p. 366.

J. Wortmann
*Upon the Photostimulated Electron Emission from Tribomechanically Activated Aluminium
(a) Vol. **50**, No. 2, p. 509—516 (1978).
See also *13.3*, p. *168*; 21, p. 366.

V. M. Fridkin, E. I. Boikova, B. N. Popov, and G. I. Rosenman
*The Influence of the Anomalous Photovoltaic Effect on the Exoelectron Emission in $LiNbO_3$
(a) Vol. **50**, No. 2, p. K255—K259 (1978).
See also 14.4.1, p. 258; *16*, p. *268*; 22.8.1, p. 517.

17.1 Field Emission Microscope Investigations

Original Papers and Short Notes

W. Zingg and H. Warlimont
*On Imaging of Aluminium by Field Ion Microscopy
(a) Vol. **45**, No. 1, p. 117–122 (1978).
See also 21, p. 353.

J.-M. Derochette
*Adsorption of Nitrogen on Iridium. Influence of Steps on the Surface Potential Studied by Field Emission
(a) Vol. **45**, No. 1, p. 163–169 (1978).
See also *1.6*, p. *40*; 21, p. 353.

N. Rihon
*Energy Distribution Study of Field-Emitted Electrons from (100) and (111) Nickel Planes
(a) Vol. **49**, No. 2, p. 697–703 (1978).
See also 21.1, p. 379.

18. Magnetic Properties

Original Papers and Short Notes

G. C. Aers, A. D. Boardman, and P. Clark
*A Computational Study of the Correlation of Attenuated Total Reflection Minima with Surface Polariton Modes
(b) Vol. 85, No. 1, p. 171–178 (1978).
See also 13.5, p. 187; *20.1*, p. *319*; 21.1, p. 368.

G. V. Ionova, E. F. Makarov, O. M. Pachev, and S. P. Ionov
*Electronic Ordering in Crystals
(b) Vol. 85, No. 1, p. 359–370 (1978).
See also *13*, p. *142*.

V. G. Golubev, E. I. Georgitse, and V. I. Ivanov-Omskii
Photoconductivity of HgTe in Magnetic Field at 337 μm
(b) Vol. 85, No. 1, p. K39–K41 (1978).
See also *16*, p. *262*; *20.1*, p. *320*; 22.4.3, p. 470.

R. Huster and E. Huster
*Hydrogen Absorption and Magnetic Behaviour of Palladium–Iron Alloys
(a) Vol. 45, No. 1, p. K23 (1978).
See also 21.1.1, p. 384.

J. Alizon, G. Berthet, J. P. Blanc, J. Gallice, H. Robert, J. M. Fabre, and L. Giral
*Dynamic Nuclear Polarization in TCNQ Salts
(b) Vol. 85, No. 2, p. 603–610 (1978).
See also 19, p. 304; 22.9, p. 523.

G. V. Ionova, E. F. Makarov, O. M. Pachev, and S. P. Ionov
*Analysis of Charge–Orbital Ordering in Compounds with Mixed Valence
(b) Vol. 85, No. 2, p. 683–691 (1978).
See also *13*, p. *143*.

E. Majerníková and Š. Barta
Magnetic Freeze-Out Effect in Noncrystalline Semiconductors
(b) Vol. 86, No. 1, p. 183–194 (1978).
See also 2, p. 43; *13.1*, p. *152*; *13.5.1*, p. *190*; 22, p. 410.

E. Mrosan, Ch. Schober, and G. Lehmann
Multiple Scattering Theory for Magnetoresistance of Localized Impurities in Metals
(b) Vol. 86, No. 1, p. 249–256 (1978).
See also 13.4, p. 172; *14.1*, p. *206*; 21, p. 355.

P. A. Alekseev, I. P. Sadikov, I. A. Markova, and O. D. Chistyakov
*On the Effects of the Crystalline Electric Field in $PrSn_3$
(b) Vol. 86, No. 1, p. K59–K62 (1978).
See also *6*, p. *68*; 22.7, p. 501.

M. Drulis and Z. Biegański
*Low Temperature Phase Transition in Cerium Trihydride. Specific Heat Measurements
(a) Vol. **46**, No. 1, p. K39—K41 (1978).
See also *8*, p. *84*; 21.4, p. 397.

A. Freudenhammer
*On the RKKY Interaction in Spin Glass Systems
(b) Vol. **86**, No. 2, p. 579—584 (1978).
See also *13*, p. *144*; 21.1, p. 371.

C. Tsallis
*Influence of Long-Range Forces in Isotropic Two-Dimensional Magnetism
(b) Vol. **86**, No. 2, p. K101—K105 (1978).
See also *13*, p. *145*.

W. Gasser
Dynamical Electrical Conductivity, Dielectric Function, and Plasma Oscillations of the Hubbard Model
(b) Vol. **87**, No. 1, p. 185—190 (1978).
See also *13.5.2*, p. *198*; *14*, p. *203*.

A. M. Moisin
*Neutron Scattering Study of Local Ordering in Fe–Ni and Fe–Al Alloys
(a) Vol. **47**, No. 1, p. 305—309 (1978).
See also 21.1, p. 373; 21.1.1, p. 387.

A. Richter, G. Röpke, and F. Goedsche
*Functional Integral Approach for the Hubbard Model with Arbitrary Electron Density
(b) Vol. **88**, No. 1, p. 189—198 (1978).
See also *13*, p. *146*; 18.1, p. 278; 18.2, p. 288; 18.4, p. 300.

D. P. Bhattacharya and Z. S. Kachlishvili
Faraday Rotation in Presence of Attractive Traps
(b) Vol. **88**, No. 1, p. 283—287 (1978).
See also *13.4*, p. *177*; *14.3.4*, p. *243*; 22.1.1, p. 426.

Mahiradhwaj Singh
Phonon Attenuation in n-Type Semiconductors with ZB Structure in Presence of Internal Strains and Magnetic Field
(b) Vol. **88**, No. 1, p. 295—300 (1978).
See also *6*, p. *73*; 7, p. 81; 12.2, p. 140; 22.1.1, p. 426; 22.2.3, p. 453.

A. D. Jannussis, V. Papatheou, and P. Tsilimigras
The Motion of an Electron in One-Dimensional Periodic Potential under the Influence of Uniform Electric and Magnetic Fields
(b) Vol. **88**, No. 1, p. 341—350 (1978).
See also *13.1*, p. *156*.

Th. Frauenheim and G. Röpke
*Magnetic Behaviour of the s–f Two-Band Hubbard Model
(b) Vol. **88**, No. 2, p. 457—462 (1978).
See also *13*, p. *146*.

T. Wydro
*High-Temperature Series in the Quantum-Classical Heisenberg Model
(b) Vol. **88**, No. 2, p. 787–796 (1978).

H. Szymczak
*On Magnetostriction in Amorphous Thin Films
(b) Vol. **88**, No. 2, p. K97–K101 (1978).
See also *2*, p. *47*; 21, p. 361.

A. M. Kapitonov, E. M. Smokotin, A. P. Korenko, and G. N. Stepanov
*Ultrasonic-Wave Velocities in Cr_2O_3 near the Magnetic Phase Transition
(a) Vol. **48**, No. 2, p. K203–K206 (1978).
See also *1.2*, p. *22*; 7, p. 82; 22.6, p. 496.

F. R. Kessler, J. Metzdorf, and W. Zurek
*Interband Faraday Ellipticity in Germanium
(b) Vol. **89**, No. 1, p. 203–211 (1978).
See also 13.1, p. 158; 13.5.1, p. 194; *20*, p. *317*; 22.1.1, p. 427.

P. Kacman, J. Blinowski, and M. Grynberg
Dynamic Dielectric Function in the Three-Band Model for Zero-Gap Semiconductors in a Magnetic Field
(b) Vol. **89**, No. 2, p. 521–530 (1978).
See also *13.1*, p. *159*; 14.4, p. 251; 22, p. 418.

Z. Onyszkiewicz
*Hysteresis of the Metamagnetic Phase Transitions
(b) Vol. **89**, No. 2, p. K131–K134 (1978).

S. Ligenza and H. Ptasiewicz-Bąk
*Magneto-Electric Splitting of the Crystal Field Γ_2-Level of Cr^{3+} Ions in $MgCr_2O_4$ Studied by Neutron Spectroscopy
(b) Vol. **90**, No. 1, p. 319–326 (1978).
See also *13.4*, p. *183*; 22.8.2, p. 520.

V. I. Litvinov, V. M. Baginskii, and A. A. Mitrofanov
The Phase Diagram of a System with Two Order Parameters. Mott Transition and Ferromagnetism
(b) Vol. **90**, No. 1, p. 423–427 (1978).
See also *1.2*, p. *23*; *13*, p. *148*; 14, p. 204.

H. U. Habermeier, R. Aoki, and H. Kronmüller
The Quantitative Determination of Flux-Density Gradients in Type-II Superconductors by Means of the Magneto-Optical Faraday Effect
(a) Vol. **50**, No. 1, p. 187–194 (1978).
See also *14.2*, p. *215*; 20, p. 318; 21, p. 365.

D. Deutges, A. Freudenhammer, and G. Heber
*Spin Glass Transition. A Dynamical Instability
(b) Vol. **90**, No. 2, p. 475–485 (1978).
See also 21.1, p. 381.

I. V. Krivoshei, V. P. Babiyczuk, I. M. Guella, I. F. Golovkina, N. V. Mansia, V. A. Starodub, and S. A. Chueva
*A New Conducting Compound: Tetrafluoroborate of Dibenztetrathiofulvalene; Electrical and Magnetic Properties
(a) Vol. **50**, No. 2, p. K197–K200 (1978).
See also *14*, p. *204*; 22.9, p. 529.

18.1 Paramagnetic Properties

Review Article

K. Bärner
Exchange Interactions and the Coupled Spin–Lattice Phase Transitions in Manganese Pnictides
(b) Vol. 88, No. 1, p. 13—23 (1978).
See also *18.2*, p. *281*; 18.3, p. 295; *18.4*, p. *298*; 21.1, p. 368.

Original Papers and Short Notes

E. Wieser, W. Brückner, B. Thuss, and U. Gerlach
Mössbauer Effect Studies on Modified VO_2. $V_{1-x}Fe_xO_{2-y}$ and $V_{1-x-z}Fe_xMo_zO_2$
(a) Vol. 45, No. 1, p. 123—131 (1978).
See also *1.2*, p. *15*; *5*, p. *60*; 22.6.1, p. 499.

R. M. M. Fonville, W. Geertsma, and C. Haas
On the Nature of the Top of the Valance Band in Layered Mo and W Dichalcogenides
(b) Vol. 85, No. 2, p. 621—627 (1978).
See also *13.1*, p. *151*; 14.3, p. 217; 19, p. 304; 22, p. 409.

K. Kawasaki and R. A. Tahir-Kheli
*Effects of Bond Dilution on Dynamics of Anisotropic Paramagnets at High Temperatures
(b) Vol. 85, No. 2, p. 707—713 (1978).
See also 18.2, p. 282; 18.4, p. 298.

E. Burzo, D. P. Lazăr, and L. Palii
*Paramagnetic Behaviour of $Gd(Fe_xCo_{1-x})_2$ Compounds
(a) Vol. 45, No. 2, p. K145—K148 (1978).
See also 18.3, p. 295; 21.1, p. 370; 21.1.1, p. 385; 21.4, p. 396.

M. V. Medvedev
Phase Diagrams of an Ising Magnet with Random Exchange Bonds
(b) Vol. 86, No. 1, p. 109—118 (1978).
See also *18.2*, p. *283*; *18.4*, p. *298*.

E. Kolley and W. Kolley
*Singularities in the Paramagnetic Susceptibility of Random Hubbard Alloys
(b) Vol. 86, No. 1, p. 397—408 (1978).
See also *13*, p. *143*; 18.2, p. 284; 21, p. 355.

J. Sznajd
Cross-Over from Discontinuous to Continuous Phase Transition under an External Magnetic Field
(b) Vol. 86, No. 1, p. K11—K14 (1978).
See also *1.2*, p. *16*; 18.4, p. 298; 22.6, p. 492.

S. Lahiry, J. Sharma, G. D. Sootha, and H. O. Gupta
*Cu^{2+} in Pseudotetrahedral Symmetry: Magnetic Anisotropy, Susceptibility, and EPR of Bis(Trimethyl- and Triethyl Benzyl Ammonium) Tetrachlorocuprate(II) from 300 to 77 K
(a) Vol. 46, No. 1, p. 153—162 (1978).
See also *19*, p. *305*; 22.9, p. 523.

K. N. Khanna and P. K. Sharma
*Susceptibility of Alkali Metals
(b) Vol. **86**, No. 2, p. K167–K170 (1978).
See also 21.1, p. 392.

H. H. Heikens, R. S. Kuindersma, C. F. van Bruggen, and C. Haas
*Magnetic Properties of $MnEr_2S_4$, $MgEr_2S_4$, and $Mn_xMg_{1-x}Y_2S_4$
(a) Vol. **46**, No. 2, p. 687–695 (1978).
See also *18.4*, p. *299*; 22.8, p. 504.

G. Kamieniarz
*On the Theory of the Ising Model with a Transverse Field
(b) Vol. **87**, No. 2, p. 455–459 (1978).
See also *13*, p. *145*.

H. K. Diemer, E. J. Bijvank, and H. W. den Hartog
*EPR Investigations of ADP:Cr^{3+} above and below T_C
(b) Vol. **87**, No. 2, p. 697–706 (1978).
See also *19*, p. *308*; 22.8.1, p. 513.

A. Richter, G. Röpke, and F. Goedsche
Functional Integral Approach for the Hubbard Model with Arbitrary Electron Density
(b) Vol. **88**, No. 1, p. 189–198 (1978).
See also *13*, p. *146*; *18*, p. *274*; 18.2, p. 288; 18.4, p. 300.

S. Fakioğlu
*Susceptility Calculations on Spin-Doped Crystals
(b) Vol. **88**, No. 1, p. 301–308 (1978).
See also 13.4, p. 177; 22.8.1, p. 514.

W. Borgieł
On the Longitudinal Correlation Functions of the Heisenberg Spin System with Single-Ion Anisotropy
(b) Vol. **88**, No. 1, p. K47–K50 (1978).
See also *13.5.4*, p. *201*; 18.2, p. 288.

G. Ya. Selyutin
Anomaly of Width and Lineshape of Light Absorption in Magnetic Crystals
(b) Vol. **88**, No. 1, p. K79–K81 (1978).
See also 18.2, p. 288; *20.1*, p. *330*.

K. N. Khanna and P. K. Sharma
A Model Potential and Its Calculations
(b) Vol. **88**, No. 2, p. 809–814 (1978).
See also *12.1*, p. *136*; 13.2, p. 162; 21.2, p. 393.

I. Pop, A. Giurgiu, E. Pop, and V. Iuşan
*Coexistence of Itinerant Antiferromagnetism with Paramagnetism of Localized Magnetic Moments in Low Concentrated Cr–Fe Alloys
(b) Vol. **88**, No. 2, p. K181–K185 (1978).
See also *18.4*, p. *301*; 21.1, p. 376; 21.1.1, p. 388.

G. Hilscher and E. Gratz
An Estimation of the Density of States from Magnetization and Resistivity Measurements $Ti(Fe_{1-x}Co_x)$ and $Zr(Fe_{1-x}Co_x)_2$
(a) Vol. **48**, No. 2, p. 473—480 (1978).
See also *13*, p. *147*; 14.1, p. 209; 18.2, p. 289; 21.1, p. 376; 21.1.1, p. 388.

W. Gasser
On the Dielectric Function of the Hubbard Model
(b) Vol. **89**, No. 1, p. 165—169 (1978).
See also *13.5.2*, p. *198*; 14, p. 204.

Th. Eifrig
On the Electronic Properties of Disordered Quasi-One-Dimensional Systems with Electron Correlation
(b) Vol. **89**, No. 1, p. 171—175 (1978).
See also *2*, p. *48*; *14.1*, p. *210*; 22.9, p. 526.

S. Fakioğlu and Ş. Erkoç
*Magnetic and Thermal Properties of Nd^{3+} in Scheelite-Type Crystals
(b) Vol. **89**, No. 1, p. 289—298 (1978).
See also *8*, p. *88*; 22.8.1, p. 515.

R. Nagarajan and J. K. Srivastava
*Anomalous Mössbauer Lineshapes in a Magnetically Ordered Ga_2O_3–Fe_2O_3 System in Presence of an External Electric Field Applied during the Solid Solution Formation
(b) Vol. **89**, No. 1, p. K79—K83 (1978).
See also *5*, p. *62*; 22.6.1, p. 499.

B. Lorenz
*On the Magnetic Susceptibility of $(Ti_{1-x}V_x)_4O_7$
(b) Vol. **89**, No. 1, p. K99—K101 (1978).
See also 22.6.1, p. 499.

I. Pop, M. Coldea, and V. U. S. Rao
*NMR and Magnetic Susceptibility of Intermetallic Compounds ZrV_2, $Zr_{0.99}Pt_{0.01}V_2$, and $Zr_{0.9}Fe_{0.1}V_2$
(a) Vol. **49**, No. 1, p. 207—210 (1978).
See also *19*, p. *311*; 21.1, p. 377.

C. A. J. Ammerlaan and J. C. Wolfrat
*Magnetic Dipole–Dipole Hyperfine Integrals for Slater-Type Orbitals
(b) Vol. **89**, No. 2, p. 541—546 (1978).

V. Kumar and R. Kumar
On the g and Hyperfine Interaction Parameters of Er^{3+} in Three Hydroxides
(b) Vol. **89**, No. 2, p. K195—K197 (1978).
See also *13.4*, p. *182*; 22.8, p. 507.

K. Bärner and H. Berg
Elastic Anomalies at the Second-Order Phase Transition in MnAs
(a) Vol. **49**, No. 2, p. 545—554 (1978).
See also 6, p. 76; *12.1*, p. *136*; 18.2, p. 290; 21.1, p. 379.

E. Marschall and H. Bross
Ferromagnetic Nickel: Self-Consistent Calculation of Energy Bands
(b) Vol. **90**, No. 1, p. 241–250 (1978).
See also *13.1*, p. *160*; 18.2, p. 291; 21.1, p. 380.

L. M. Noskova, L. A. Ugodnikova, and S. G. Chuchman
*The Néel Points of Yttrium–Rare Earth Metal Alloys
(a) Vol. **50**, No. 1, p. K47–K49 (1978).
See also *18.4*, p. *301*; 21.4, p. 399.

B. E. Vugmeister
*Spin Diffusion and Spin–Lattice Relaxation in Paramagnetic Crystals
(b) Vol. **90**, No. 2, p. 711–718 (1978).
See also *19*, p. *314*.

18.2 Ferromagnetic Properties

Review Article

K. Bärner
*Exchange Interactions and the Coupled Spin–Lattice Phase Transitions in Manganese Pnictides
(b) Vol. 88, No. 1, p. 13–23 (1978).
See also 18.1, p. 277; 18.3, p. 295; *18.4*, p. *298*; 21.1, p. 368.

Original Papers and Short Notes

J. Richter and S. Kobe
*Density of States and Itinerant Ferromagnetism in Amorphous Systems with Dense Random Packed Hard Sphere Structure
(b) Vol. 85, No. 1, p. 233–238 (1978).
See also *2*, p. *42*; 13, p. 142.

W. Treplin and R. J. Jelitto
*Long-Range Order in Nearly One- and Nearly Two-Dimensional Heisenberg Ferromagnets. An Upper Bound for the Magnetization
(b) Vol. 85, No. 1, p. 247–252 (1978).

B. Mrygoń and A. Urbaniak
*Temperature Shift of the Maximum of Neutron Critical Scattering in the Second Born Approximation
(b) Vol. 85, No. 1, p. 387–391 (1978).

V. Čápek and J. Přívratská
*Spin Polarons in the s–d Model with Off-Diagonal Exchange
(b) Vol. 85, No. 1, p. K1–K4 (1978).
See also *13.5.3*, p. *200*; 22, p. 409.

A. V. Deryagin, N. V. Kudrevatykh, and V. N. Moskalev
*Magnetic Anisotropy and Intersublattice Exchange Interaction in $Er_2(Co_{1-x}M_x)_{17}$ Intermetallic Compounds
(a) Vol. 45, No. 1, p. 71–76 (1978).
See also 21.1, p. 368; 21.4, p. 396.

S. Takahashi
*Compression-Induced Magnetic Anisotropy and Superlattice Dislocations in Ni_3Fe(III): The Annealing Effect
(a) Vol. 45, No. 1, p. 133–138 (1978).
See also *10.1*, p. *99*; 21.1, p. 369; 21.1.1, p. 384.

S. A. Mahmoud, M. A. Semary, and F. Mohsen
*Reduction of the Magnetization Curves for Ni5%Mn Alloy
(a) Vol. 45, No. 1, p. K5–K7 (1978).
See also 21.1, p. 369.

C. J. Schinkel, Z. Żołnierek, and R. Troć
*High Field Magnetization of U_2N_2Te
(a) Vol. **45**, No. 1, p. K39—K41 (1978).
See also 22.8, p. 503.

J. Schneider, A. Handstein, and R. Hesske
*Mictomagnetic to Ferromagnetic Transition in Amorphous Ni–Fe Based Alloys
(a) Vol. **45**, No. 1, p. K47—K50 (1978).
See also *2*, p. *42*; 21.1, p. 369; 21.1.1, p. 384.

M. V. Medvedev and E. L. Rumyantsev
*On the Existence of a Spin-Glass State in an Ising Alloy Model with Mixed Nearest-Neighbour Exchange Interactions
(b) Vol. **85**, No. 2, p. 427—438 (1978).
See also *18.4*, p. *298*; 21.1, p. 369; 21.1.1, p. 384.

R. J. Jelitto and M. Heise
*On the Dynamics of Electrons in the Strong-Coupling Hubbard Chain
(b) Vol. **85**, No. 2, p. 535—543 (1978).
See also *13*, p. *143*.

G. Georgiev and I. Gochev
*Localized Magnons around an Impurity in a Spin Chain with Next-Nearest Neighbour Interaction
(b) Vol. **85**, No. 2, p. 657—662 (1978).
See also *13.5.4*, p. *201*.

K. Kawasaki and R. A. Tahir-Kheli
Effects of Bond Dilution on Dynamics of Anisotropic Paramagnets at High Temperatures
(b) Vol. **85**, No. 2, p. 707—713 (1978).
See also *18.1*, p. *277*; 18.4, p. 298.

G. Chaddha
*Isotropic Ferromagnets with Biquadratic Interactions
(b) Vol. **85**, No. 2, p. 777—781 (1978).

R. Ciszewski, A. Jabłonka, A. Sobaszek, and D. Sikorska
Temperature Shift of the Maximum of the Neutron Critical Scattering in an Iron Monocrystal
(b) Vol. **85**, No. 2, p. K73—K75 (1978).
See also *1*, p. *9*; 21.1.1, p. 385.

R. Ciszewski, A. Jabłonka, and A. Sobaszek
Experimental Anisotropy of the Neutron Critical Scattering from a Ferromagnet Monocrystal
(b) Vol. **85**, No. 2, p. K77—K79 (1978).
See also *1*, p. *9*; 21.1.1, p. 385.

F. Goedsche and G. Vojta
Spin-Disorder Resistivity of Ternary Alloys in the CPA Weak Scattering Limit
(b) Vol. **85**, No. 2, p. K81—K84 (1978).
See also *14.1*, p. *206*; 21, p. 354; 21.1, p. 370.

D. I. PUSHKAROV and KH. I. PUSHKAROV
*Quasi-Classical Description of the Heisenberg Ferromagnet: Soliton Solutions
(b) Vol. **85**, No. 2, p. K89—K93 (1978).
See also *13.5.4*, p. *201*.

K. GŁOWACKI
*Differential Cross Section for Magnetic Scattering in the Critical Region
(b) Vol. **85**, No. 2, p. K119—K120 (1978).

H. MAEDA
*Magnetocrystalline Anisotropy of Pr $(Co_{1-x}Cu_x)_5$ Compounds. Experiment and Theoretical Analysis
(a) Vol. **45**, No. 2, p. 445—452 (1978).
See also 21.1, p. 370; 21.4, p. 396.

J. DEGAUQUE, B. ASTIE, and L. P. KUBIN
*Evidence for the Interaction between Magnetic Domain Walls and Dislocations in High-Purity Iron from Magnetomechanical Damping Experiments
(a) Vol. **45**, No. 2, p. 493—501 (1978).
See also *10.1*, p. *100*; 21.1.1, p. 385.

S. NIZIOŁ, H. BIŃCZYCKA, A. SZYTUŁA, J. TODOROVIĆ, R. FRUCHART, J. P. SENATEUR et D. FRUCHART
*Structure magnétique des MnCoSi
(Magnetic Structure of MnCoSi)
(a) Vol. **45**, No. 2, p. 591—597 (1978).
See also *18.4*, p. *298*; 22.8, p. 503.

R. C. O'HANDLEY and D. S. BOUDREAUX
*Magnetic Properties of Transition Metal–Metalloid Glasses. A Charge Transfer Model
(a) Vol. **45**, No. 2, p. 607—615 (1978).
See also *2*, p. *43*; 21.1, p. 370.

J. BARTON and E. WIESER
*Investigation of Magnetic Preferred Directions in an Fe–Mn Base Alloy
(a) Vol. **45**, No. 2, p. K105—K108 (1978).
See also *5*, p. *60*; 21.1, p. 370; 21.1.1, p. 385.

R. G. PILLAY and P. N. TANDON
*Hyperfine Magnetic Field at Sn in Heusler Alloys Rh_2MnSn and Rh_2NiSn
(a) Vol. **45**, No. 2, p. K109—K112 (1978).
See also *5*, p. *60*; 21, p. 355.

M. V. MEDVEDEV
*Phase Diagrams of an Ising Magnet with Random Exchange Bonds
(b) Vol. **86**, No. 1, p. 109—118 (1978).
See also 18.1, p. 277; *18.4*, p. *298*.

F. GOEDSCHE, R. RICHTER, and G. VOJTA
Coherent Potential Approximation for the Electrical Resistivity of Partially Ordered Binary Magnetic Alloys
(b) Vol. **86**, No. 1, p. 177—182 (1978).
See also *14.1*, p. *206*; 21, p. 355.

V. N. Kashcheev
*Some Aspects of Spin Relaxation in Ferromagnets in a Magnetic Field
(b) Vol. **86**, No. 1, p. 359—365 (1978).
See also 19, p. 305.

E. Kolley and W. Kolley
Singularities in the Paramagnetic Susceptibility of Random Hubbard Alloys
(b) Vol. **86**, No. 1, p. 397—408 (1978).
See also *13*, p. *143*; *18.1*, p. *277*; 21, p. 355.

Th. Eifrig, E. Kolley, and W. Kolley
Electron Correlations and Ferromagnetism in Disordered Alloys
(b) Vol. **86**, No. 1, p. K79—K83 (1978).
See also *13*, p. *143*; 21, p. 356.

D. Werneth
Mechanical and Magnetic Relaxation Measurements on Pure and Doped Nickel after Plastic Deformation and Electron Irradiation
(a) Vol. **46**, No. 1, p. 99—106 (1978).
See also *10.1*, p. *101*; 12.1, p. 134; 21.1, p. 371.

G. N. Belozerskii
*Superparamagnetism–Determination of the Diffusion Coefficient of the Vector of Magnetization, of Anisotropy Energy, and of Volume Distribution of Particles by Mössbauer Effect
(a) Vol. **46**, No. 1, p. 131—140 (1978).
See also *5*, p. *61*.

A. Tari
*The Quenching Effect of Iron Substitution on the Holmium Moment in the $Ho_{0.01}Gd_{0.99}(Rh_{1-x}Fe_x)_2$ Compounds
(a) Vol. **46**, No. 1, p. 173—177 (1978).
See also 19, p. 305; 21.1.1, p. 385; 21.4, p. 396.

H. Maeda
*Anisotropy of Saturation Magnetization in $Pr(Co_{1-x}Cu_x)_5$ Compounds
(a) Vol. **46**, No. 1, p. 233—238 (1978).
See also 21.1, p. 371; 21.4, p. 396.

J. P. Jakubovics and D. J. Fathers
*Image Profiles and Resolution of Magnetic Domains in the Scanning Electron Microscope
(a) Vol. **46**, No. 1, p. 291—303 (1978).
See also 21.1.1, p. 386; 21.4, p. 397.

A. Ślebarski
*Anomalous Change of the Curie Temperature of $GdAl_2$
(a) Vol. **46**, No. 1, p. K29—K34 (1978).
See also *8*, p. *84*; 21, p. 356; 21.4, p. 397.

A. Murasik, H. Ptasiewicz-Bąk, and A. Zygmunt
Magnetic Structure and Crystal-Field Effects in CeZn
(a) Vol. **46**, No. 1, p. K75—K78 (1978).
See also *18.4*, p. *299*; 21, p. 356; 21.4, p. 397.

W. Schumacher, U. Lindner, and R. Jauch
*Magnetic Solutions of the Hubbard Model
(b) Vol. **86**, No. 2, p. 621—625 (1978).
See also *18.4*, p. *299*.

F. M. Galperin
*Atomic Magnetic Moments of Fe, Co, Ni and Hyperfine Fields at Impurities in Ferromagnetic Iron
(b) Vol. **86**, No. 2, p. 679—684 (1978).
See also *13*, p. *144*; 21.1, p. 372; 21.1.1, p. 386.

M. N. Iliev E. Anastassakis and T. Aräi
Raman Scattering of $HgCr_2Se_4$
(b) Vol. **86**, No. 2, p. 717—723 (1978).
See also *6*, p. *69*; *20.1*, p. *324*; 22.8.2, p. 518.

C. G. Patil and B. S. Krishnamurthy
Carrier Mobility in Ferromagnetic Semiconductors
(b) Vol. **86**, No. 2, p. 725—732 (1978).
See also 13.5.4, p. 201; *14.3*, p. *219*; 22.4.4, p. 472.

B. V. Egorov and M. A. Krivoglaz
The Theory of Autolocalized Electron States in Ferromagnetic Semiconductors at Arbitrary Relation between the Band Width and the Spin–Electron Interaction
(b) Vol. **86**, No. 2, p. 781—794 (1978).
See also *13*, p. *144*; 13.5.3, p. 200; 22, p. 412.

V. K. Saxena and R. Kishore
*Disorder Induced Behaviour of Magnetisation at Low Temperatures
(b) Vol. **86**, No. 2, p. K163—K166 (1978).
See also *2*, p. *44*.

I. Rattan and N. C. Srivastava
*Magnetoelastic Interaction Energy of a Cubic Ferromagnet in an Arbitrary Coordinate System
(a) Vol. **46**, No. 2, p. 475—478 (1978).
See also *12.2*, p. *139*; 22.8.2, p. 519.

M. Kopcewicz
*Mössbauer Study of the Fast Magnetization Reversal Forced in Permalloy and Invar by an External RF Magnetic Field
(a) Vol. **46**, No. 2, p. 675—685 (1978).
See also *5*, p. *61*; 21.1, p. 372; 21.1.1, p. 386.

H. J. Blythe and L. H. Dunleavy
*Point-Defect Relaxations in Dilute Iron–Nickel Alloys after Low-Temperature Neutron Irradiation
(a) Vol. **46**, No. 2, p. K95—K99 (1978).
See also *10.1*, p. *102*; 11, p. 125; 21.1, p. 372; 21.1.1, p. 386.

C. P. van E. Oosthuizen and P. de V. du Plessis
*Uniaxial Magnetic Anisotropy Induced in a Compression Deformed Iron Single Crystal
(a) Vol. **46**, No. 2, p. K111—K113 (1978).
See also 21.1.1, p. 386.

P. PAWLICKI
*Heat Magnetization in the Singlet–Triplet Model
(b) Vol. 87, No. 1, p. K17—K20 (1978).
See also 21, p. 357.

V. A. FEIGIN
*Longitudinal Dynamic Susceptibility of the Heisenberg Ferromagnet
(b) Vol. 87, No. 1, p. K33—K34 (1978).

G. SCHROEDER
Investigation of the Dislocation Arrangement in Plastically Deformed Iron Single Crystals by Means of Ferromagnetic Techniques
(a) Vol. 47, No. 1, p. 145—156 (1978).
See also *10.1*, p. *103*; 21.1.1, p. 387.

U. RINDELHARDT, S. MAGER, W. BLAU, and E. HEGENBARTH
Recovery of Low Temperature Specific Heat, Resistivity, and Coercive Force in Fe–Cr–Al Alloys
(a) Vol. 47, No. 1, p. 157—161 (1978).
See also 8, p. 85; *10.1*, p. *103*; 14.1, p. 207; 21.1, p. 373; 21.1.1, p. 387.

M. MARYŠKO
*The Influence of Electromagnetic Propagation on Ferromagnetic Resonance in an Infinite Cylinder
(a) Vol. 47, No. 1, p. 277—286 (1978).
See also 18.3, p. 295; *19*, p. *307*; 22.8.2, p. 519.

J. STASZ
*ESR Line Intensity near the Curie Point in $CdCr_2Se_4$, $HgCr_2Se_4$, and $CdCr_2S_4$ Single Crystals
(a) Vol. 47, No. 1, p. K23—K26 (1978).
See also *19*, p. *308*; 22.8, p. 505.

A. KOBAYASHI and K. SUZUKI
*High Field Transport Properties of a Magnetic Semiconductor
(a) Vol. 47, No. 1, p. K51—K53 (1978).
See also *14.3.4*, p. *242*; 22.8, p. 505.

E. A. VASILEV and A. E. GELYASIN
*Magnetic Properties of the $Mn_{1+x}Sb_{1-x}Sn_x$ System
(a) Vol. 47, No. 1, p. K55—K57 (1978).
See also *18.3*, p. *295*; 21, p. 359.

A. V. MITIN and V. A. TARASOV
*Ultrasonic Spectrum in a Multidomain Structure
(b) Vol. 87, No. 2, p. 433—440 (1978).
See also 7, p. *80*.

YU. A. IZYUMOV, YU. N. SKRYABIN, and V. M. LAPTEV
*Critical Behaviour near the Intersection of Second-Order Phase Transition Lines in a Random System
(b) Vol. 87, No. 2, p. 441—445 (1978).

N. K. JAGGI
*Dependence of the Spin Glass Transition on the Details of the Disorder
(b) Vol. 87, No. 2, p. K81—K84 (1978).
See also *18.4*, p. *300*.

K. J. MICHALSKI
*Magnetic Symmetry of Ordered Phases in Mn_3GaC
(b) Vol. 87, No. 2, p. K85—K87 (1978).
See also 18.4, p. 300; 22.8, p. 505.

J. RICHTER
Electronic Structure of Amorphous Alloys
(b) Vol. 87, No. 2, p. K89—K92 (1978).
See also *2*, p. *45*; *13*, p. *145*.

H. PTASIEWICZ-BĄK, J. LECIEJEWICZ, and A. ZYGMUNT
*Magnetic Ordering in UGeY (Y = S, Se, Te) by Neutron Diffraction
(a) Vol. 47, No. 2, p. 349—356 (1978).
See also *18.4*, p. *300*; 22.8, p. 505.

N. K. GOBRAN and N. S. GHALI
Ultrasonic Anomalies in Iron at Elevated Temperatures
(a) Vol. 47, No. 2, p. 365—370 (1978).
See also *7*, p. *81*; 21.1.1, p. 387.

CHANG PENG-HSIANG and G. SCHROEDER
*Domain Patterns and Dislocations in Cyclically Deformed Ni Single Crystals and Their Behaviour during Annealing
(a) Vol. 47, No. 2, p. 397—406 (1978).
See also *10.1*, p. *103*; 21.1, p. 374.

H. J. BAUER, M. BECKER, H. PRETSCH, and M. ZWICK
Magnetic Study of the Activation Energy of the Phase Transition Kinetics between Nickel and Nickel Hydride
(a) Vol. 47, No. 2, p. 445—450 (1978).
See also *1.2*, p. *19*; 21.1, p. 374.

R. HERZ and H. KRONMÜLLER
*Field-Induced Phase Transitions in the Helical State of Dysprosium
(a) Vol. 47, No. 2, p. 451—458 (1978).
See also 21.4, p. 398.

P. L. ROSSITER and M. E. HOUGHTON
*Magnetic Properties and Microstructure of an Fe–27.5 Cr–17.5 Co–0.5Al Alloy
(a) Vol. 47, No. 2, p. 597—608 (1978).
See also *1.4*, p. *31*; 21.1, p. 374; 21.1.1, p. 387.

KH. I. PUSHKAROV, D. I. PUSHKAROV, and J. P. VLAHOV
*Solitary Excitations in One-Dimensional Ferromagnets with Biquadratic Exchange
(b) Vol. 88, No. 1, p. 65—71 (1978).
See also *13.5.4*, p. *201*.

M. V. Medvedev
*Spin Waves in a Uniaxial Ferromagnet with Random Exchange Bonds of Different Signs
(b) Vol. 88, No. 1, p. 117—127 (1978).

A. Richter, G. Röpke, and F. Goedsche
Functional Integral Approach for the Hubbard Model with Arbitrary Electron Density
(b) Vol. 88, No. 1, p. 189—198 (1978).
See also *13*, p. *146*; *18*, p. *274*; 18.1, p. 278; 18.4, p. 300.

M. Matlak and W. Nolting
Influence of Finite Carrier Concentrations on the Electronic Spectrum of a Ferromagnetic Semiconductor
(b) Vol. 88, No. 1, p. 231—240 (1978).
See also *13.5.4*, p. *201*; 22, p. 414.

H. A. Brown
*The Bethe-Peierls-Weiss Approximation for Amorphous Magnets
(b) Vol. 88, No. 1, p. K31—K34 (1978).
See also *2*, p. *46*.

M. Jursich and H. A. Brown
*Note on the Constant Coupling Approximation
(b) Vol. 88, No. 1, p. K39—K42 (1978).

V. L. Aksionov, H. Konvent, and J. Schreiber
*Commutator and Anticommutator Green Functions in the Ising Model with Transverse Field
(b) Vol. 88, No. 1, p. K43—K45 (1978).
See also *13*, p. *146*.

W. Borgieł
On the Longitudinal Correlation Functions of the Heisenberg Spin System with Single-Ion Anisotropy
(b) Vol. 88, No. 1, p. K47—K50 (1978).
See also *13.5.4*, p. *201*; 18.1, p. 278.

G. Ya. Selyutin
Anomaly of Width and Lineshape of Light Absorption in Magnetic Crystals
(b) Vol. 88, No. 1, p. K79—K81 (1978).
See also 18.1, p. 278; *20.1*, p. *330*.

V. K. Saxena
*Constant Coupling Approximation for the Curie Temperature of Ferromagnets with Random Anisotropy
(b) Vol. 88, No. 1, p. K87—K91 (1978).
See also *2*, p. *46*.

T. H. O'Dell
*Static Wall Structure in Bubble Films under Large In-Plane Fields
(a) Vol. 48, No. 1, p. 59—66 (1978).

M. E. Houghton and P. L. Rossiter
*Induced Anisotropy in an Fe–27.5 Cr–17.5 Co–0.5 Al Alloy
(a) Vol. 48, No. 1, p. 71—77 (1978).
See also *1.4*, p. *31*; 21.1, p. 375; 21.1.1, p. 388.

M. Kikuchi, K. Fukamichi, T. Masumoto, T. Jagielinski, K. I. Arai, and N. Tsuya
*Giant ΔE Effect and Elinvar Characteristics in Amorphous Fe–B Binary Alloys
(a) Vol. 48, No. 1, p. 175—181 (1978).
See also 2, p. 47; 21.1.1, p. 388.

V. V. Gagulin, N. V. Fadeeva, A. G. Belons, A. V. Titov, K. P. Mitrofanov, M. V. Plotnikova S. P. Soloviev, and Yu. N. Venevtsev
New Seignette-Magnets with Hexagonal Barium Titanate Structure
(a) Vol. 48, No. 1, p. 183—189 (1978).
See also *1*, p. *9*; 5, p. 61; *14.4.1*, p. *255*; 18.4, p. 301; 19, p. 310; 22.8.1, p. 514.

W. I. Khan and D. Melville
*Magnetic Anisotropy Associated with Iron Substitution in $Y_2(Co_xFe_{1-x})_{17}$
(a) Vol. 48, No. 1, p. 209—214 (1978).
See also 21.1, p. 375.

W. Nawrocka and R. Zossel
*Higher-Order Corrections to the Spectrum of the Easy-Axis Heisenberg Ferromagnet
(b) Vol. 88, No. 2, p. 727—733 (1978).

C. G. Patil and B. S. Krishnamurthy
Magnon Drag Thermoelectric Power in Ferromagnetic Semiconductors
(b) Vol. 88, No. 2, p. K117—K122 (1978).
See also *13.5.4*, p. *202*; *15*, p. *260*; 22, p. 415.

J. D. Patterson, G. R. Gruzalski, and D. J. Sellmyer
*Finite Temperature Calculations for the Random Anisotropy Model for Amorphous Magnets
(b) Vol. 88, No. 2, p. K127—K130 (1978).
See also *2*, p. *47*.

B. K. Ponomarev and V. G. Thiessen
*Band Ferromagnetism in Nickel–Palladium Alloys
(b) Vol. 88, No. 2, p. K139—K142 (1978).
See also 18.4, p. 301; 21.1, p. 376.

G. Hilscher and E. Gratz
An Estimation of the Density of States from Magnetization and Resistivity Measurements in $Ti(Fe_{1-x}Co_x)$ and $Zr(Fe_{1-x}Co_x)_2$
(a) Vol. 48, No. 2, p. 473—480 (1978).
See also *13*, p. *147*; 14.1, p. 209; 18.1, p. 279; 21.1, p. 376; 21.1.1, p. 388.

C. Bansal
*Concentration Dependence of the Average Magnetic Moment for F.C.C. Co–Fe Alloys
(a) Vol. 48, No. 2, p. K119—K121 (1978).
See also 21.1, p. 376; 21.1.1, p. 389.

V. G. Thiessen, V. E. Antonov, I. T. Belash, B. K. Ponomarev, and E. G. Ponyatovskii
*Magnetization of Co–H Solid Solutions
(a) Vol. **48**, No. 2, p. K185—K187 (1978).
See also 21.1, p. 377.

M. Tanaka and K. Takahashi
*Correlation Functions of a Spin-One Ising System with Bilinear and Biquadratic Interactions
(b) Vol. **89**, No. 1, p. K39—K42 (1978).

V. Ilkovič
*Variational Method for the Cluster Model of the Structurally Disordered Heisenberg Ferromagnet
(b) Vol. **89**, No. 1, p. K103—K106 (1978).

H. Bittel, W. Grosse-Nobis, and Th. Jannemann
*On the Magnetostriction of Grain-Oriented Ferromagnetic Material
(a) Vol. **49**, No. 1, p. K21—K25 (1978).
See also 21.1.1, p. 389.

P. Raj, S. K. Kulshreshtha, and J. V. Yakhmi
Mössbauer and X-Ray Studies of $Fe_{1.67}Ge$
(a) Vol. **49**, No. 1, p. K91—K95 (1978).
See also *1.3*, p. *28*; *5*, p. *62*; 21.1.1, p. 389.

G. F. Kventsel and J. Katriel
*An Exactly Soluble Model of Singlet-Ground-State Ferromagnetism
(b) Vol. **89**, No. 2, p. 645—650 (1978).

J. A. Gaj, J. Ginter, and R. R. Gałązka
Exchange Interaction of Manganese $3d^5$ States with Band Electrons in $Cd_{1-x}Mn_xTe$
(b) Vol. **89**, No. 2, p. 655—662 (1978).
See also *13.1*, p. *159*; 22.8, p. 507.

Y. D. Yao and S. Arajs
Determination of the Curie Temperature of Ferromagnetic Alloys Using the Deviation from Matthiessen's Rule
(b) Vol. **89**, No. 2, p. K201—K205 (1978).
See also *14.1*, p. *211*; 21.1, p. 378.

K. Bärner and H. Berg
Elastic Anomalies at the Second-Order Phase Transition in MnAs
(a) Vol. **49**, No. 2, p. 545—554 (1978).
See also 6, p. 76; *12.1*, p. *136*; 18.1, p. 279; 21.1, p. 379.

H. J. Blythe and L. H. Dunleavy
*Low-Temperature Magnetic Relaxation in Neutron-Irradiated α-Iron
(a) Vol. **49**, No. 2, p. 609—619 (1978).
See also 10.1, p. 106; *11*, p. *130*; 21.1.1, p. 389.

I. TOMÁŠ
*Interaction of Bubble Domains with a Concentric Hard Overlay
(a) Vol. **49**, No. 2, p. 747—755 (1978).

Y. ANDERSSON, S. RUNDQVIST, O. BECKMAN, L. LUNDGREN, and P. NORDBLAD
Properties of Fe_2P Crystals Prepared from a Liquid Copper Medium
(a) Vol. **49**, No. 2, p. K153—K156 (1978).
See also *3*, p. *56*; 21.1, p. 379; 21.1.1, p. 389.

M. MITERA, M. NAKA, T. MASUMOTO, N. KAZAMA, and K. WATANABE
*Effects of Metalloids on the Magnetic Properties of Iron Based Amorphous Alloys
(a) Vol. **49**, No. 2, p. K163—K166 (1978).
See also *2*, p. *51*; 21.1.1, p. 390.

TH. NATTERMANN
*Low-Temperature Renormalization Group and Symmetry Breaking Perturbations
(b) Vol. **90**, No. 1, p. 105—115 (1978).
See also *13*, p. *148*.

W. TREPLIN and R. J. JELITTO
*A Green's Function Investigation of Nearly Two-Dimensional and Nearly One-Dimensional Heisenberg Ferromagnets
(b) Vol. **90**, No. 1, p. 135—145 (1978).

E. MARSCHALL and H. BROSS
Ferromagnetic Nickel: Self-Consistent Calculation of Energy Bands
(b) Vol. **90**, No. 1, p. 241—250 (1978).
See also *13.1*, p. *160*; 18.1, p. 280; 21.1, p. 380.

KH. I. PUSHKAROV, D. I. PUSHKAROV, and J. P. VLAHOV
*Solitary Bound States of Magnons and Lattice Deformation in Ferromagnets with Biquadratic Exchange
(b) Vol. **90**, No. 1, p. 361—368 (1978).
See also *13.5.4*, p. *202*.

E. KOLLEY and W. KOLLEY
Electron Correlation Effects on the Spin Wave Stiffness Constant in Itinerant Ferromagnets
(b) Vol. **90**, No. 1, p. K103—K107 (1978).
See also *13.5.4*, p. *202*; 21, p. 365.

G. HAUSCH and E. TÖRÖK
*Elastic, Magnetoelastic, and Thermal Properties of Some Ferromagnetic Metallic Glasses
(a) Vol. **50**, No. 1, p. 159—164 (1978).
See also *2*, p. *51*; 8, p. 89; 12.1, p. 136; 21.1, p. 381; 21.1.1, p. 390.

Y. AOKI and K. FUKAMICHI
*Ferromagnetic Behaviour of Co-Rich Co–V Alloys Annealead at 600 °C
(a) Vol. **50**, No. 1, p. 263—267 (1978).
See also 21.1, p. 381.

T. RAJASEKHARAN and K. V. S. RAMA RAO
*Nuclear Magnetic Resonance Study of Ferromagnetic $Mn_{1-x}Co_xSb$
(a) Vol. **50**, No. 1, p. 303—310 (1978).
See also *19*, p. *314*; 22.8, p. 508.

E. A. Dorofeeva, V. P. Makarov, B. V. Molotilov, and L. P. Babichev
*On the Nature of Additional Losses in Fe–3% Si Single Crystals
(a) Vol. **50**, No. 1, p. K5—K7 (1978).
See also 21.1.1, p. 390.

T. Jagieliński, K. I. Arai, N. Tsuya, and K. Fukamichi
*Magnetoelastic Effects in Invar-Type Amorphous Fe–B Alloys
(a) Vol. **50**, No. 1, p. K25—K28 (1978).
See also *2*, p. *52*; 21.1.1, p. 390.

N. N. Delyagin, V. I. Krylov, and V. I. Nesterov
*Temperature Anomalies of the Magnetic Hyperfine Field for Sn in Metallic Ferromagnets and the Thermal Expansion of the Crystal Lattice
(b) Vol. **90**, No. 2, p. 741—749 (1978).
See also *8*, p. *90*; 21.1, p. 382.

K. J. Michalski and W. Reich
*Metamagnet with Temperature-Dependent Anisotropy in the Classical Approximation
(b) Vol. **90**, No. 2, p. K141—K143 (1978).
See also 18.4, p. 302.

G. Chaddha
*Heisenberg Ferromagnet with Biquadratic Exchange
(b) Vol. **90**, No. 2, p. K145—K148 (1978).

D. Kapor and A. Ivić
*The Coefficients in a Boson Series Expansion of Spin 1/2 Operators
(b) Vol. **90**, No. 2, p. K157—K159 (1978).
See also *13*, p. *149*.

H. A. Brown
*On the Classical-Spin Model of Amorphous Ferromagnets
(b) Vol. **90**, No. 2, p. K165—K168 (1978).
See also *2*, p. *52*.

G. Schroeder, R. Schäfer, and H. Kronmüller
*Magneto-Optical Investigation of the Domain Structure of Amorphous $Fe_{80}B_{20}$ Alloys
(a) Vol. **50**, No. 2, p. 475—481 (1978).
See also 2, p. 52; *20*, p. *318*; 21.1.1, p. 390.

J. Jelonek and M. Szczygielski
*Dispersion of Initial Magnetic Permeability for $HgCr_2Se_4$
(a) Vol. **50**, No. 2, p. 679—683 (1978).
See also 22.8.2, p. 521.

Yu. M. Vysotskii, E. A. Vasilev, A. E. Gelyasin, and A. V. Shablovskii
*Nuclear Zeeman Effect in the $Mn_{1+x}Sb_{1-x}Sn_x$ System
(a) Vol. **50**, No. 2, p. K205—K208 (1978).
See also 5, p. 63; *18.3*, p. *297*; 21, p. 367.

18.2.1 Ferromagnetic Films

Original Papers and Short Notes

W. KÜMMERLE and U. GRADMANN
*Magnetic Properties of Ferromagnetic γ-Fe Films on Cu (111)
(a) Vol. **45**, No. 1, p. 171—180 (1978).
See also 21.1.1, p. 384.

P. WEBER, W. RICHTER, and A. FEUERPFEIL
Improved Sensitivity of Thin Film SQUIDs with Permalloy Films
(a) Vol. **46**, No. 1, p. K67—K69 (1978).
See also *14.2*, p. *214*; 21.1.1, p. 386.

I. S. EDELMAN and V. D. BAURIN
*Influence of Cr^{3+} Ions on the Magnetooptical Properties of Cobalt Ferrite
(a) Vol. **46**, No. 2, p. K83—K86 (1978).
See also *20.1*, p. *325*; 22.8.2, p. 519.

A. CORCIOVEI, D. MAZILU, and D. MIHALACHE
*Quadrupolar–Paramagnetic Critical Temperature of a Spin-One Isotropic Heisenberg Film with Biquadratic Interactions
(b) Vol. **87**, No. 1, p. 61—67 (1978).

H.-R. MÜLLER and R. PERTHEL
*Model Calculations for the Anisotropy Formation by Resputtering Processes during the Growth of Amorphous Gd–Co Films
(b) Vol. **87**, No. 1, p. 203—211 (1978).
See also *2*, p. *44*; 3, p. 54; 21.1, p. 372; 21.4, p. 397.

Y. SHIMADA and H. KOJIMA
*Thermal Stability of Amorphous Ferromagnetic Thin Films Produced by Sputtering
(a) Vol. **47**, No. 2, p. K119—K121 (1978).
See also *2*, p. *46*; 21.1.1, p. 387.

A. DUDA and W. MACIEJEWSKI
*Spin-Wave Resonance Spectrum in Thin Films with Internal Isotropically Coupled Impurity Layer
(b) Vol. **88**, No. 2, p. 735—746 (1978).

H. PUSZKARSKI
*Elliptical Spin Precession in Microscopic Spin-Wave Resonance Theory
(b) Vol. **88**, No. 2, p. K113—K116 (1978).
See also *19*, p. *311*.

A. BOSTANJOGLO and H. P. GEMÜND
*Extraordinary Viscous Magnetic Wall Motion in a Wide Range of Velocities
(a) Vol. **48**, No. 2, p. 481—488 (1978).
See also 21.1, p. 376; 21.1.1, p. 388.

YA. M. POGOSYAN, S. D. GEVORKYAN, and R. STRAUBEL
*Submicroscopical Magnetization Structures in Permalloy Films Induced by Fields Normal to the Film Plane
(a) Vol. **49**, No. 1, p. K3—K4 (1978).
See also 21.1, p. 378; 21.1.1, p. 389.

J. ČERMÁK, V. TVAROŽEK, R. HARMAN, and J. KACZÉR
*Contribution to the Technology of Gd–Co Films
(a) Vol. **49**, No. 1, p. K81—K83 (1978).
See also 2, p. 50; 21.1, p. 378; 21.4, p. 398.

S. MARTELLI, G. MATTEUCCI, and M. V. ANTISARI
*Shadow and Interference Electron Microscopy of Magnetic Domain Walls
(a) Vol. **49**, No. 2, p. K103—K107 (1978).
See also 21.1, p. 379.

T. H. O'DELL
*The Peak Velocity of a Domain Wall in a Magnetic Bubble Film under a Large In-Plane Field
(a) Vol. **50**, No. 1, p. 179—186 (1978).

A. G. SHISHKOV, E. N. ILICHEVA, N. G. KANAVINA, O. S. KOLOTOV, and R. V. TELESNIN
*Bloch-Line Mobility in Permalloy Films in a Wide Thickness Range
(a) Vol. **50**, No. 1, p. K101—K104 (1978).
See also 21.1, p. 381; 21.1.1, p. 390.

N. N. EVTIKHIEV, N. A. EKONOMOV, A. R. KREBS, N. A. ZAMYATINA, A. S. KOMALOV, V. G. PINKO, and L. V. IVAEVA
*FePt and CoPt Alloy Films for Magneto-Optic Storage
(a) Vol. **50**, No. 2, p. K153—K155 (1978).
See also *20*, p. *318*; 21.1, p. 383; 21.1.1, p. 391.

18.3 Ferrimagnetic Properties

Review Article

K. Bärner
Exchange Interactions and the Coupled Spin–Lattice Phase Transitions in Manganese Pnictides
(b) Vol. **88**, No. 1, p. 13–23 (1978).
See also 18.1, p. 277; *18.2*, p. *281*; *18.4*, p. *298*; 21.1, p. 368.

Original Papers and Short Notes

T. Merceron, A. Marais, M. Porte et L. Brossard
*Etude au moyen de l'ordre directionnel d'une série de ferrites de manganèse à excès de manganèse. Valence des cations et répartition de ces ions sur les sites A et B
(Study of a Series of Manganese Ferrites with Manganese Excess by Directional Ordering. Cation Valency and Repartition of Ions on Sites A and B)
(a) Vol. **45**, No. 2, p. 503–507 (1978).
See also 22.8.2, p. 518.

E. Burzo, D. P. Lazăr, and L. Palii
Paramagnetic Behaviour of $Gd(Fe_xCo_{1-x})_2$ Compounds
(a) Vol. **45**, No. 2, p. K145–K148 (1978).
See also *18.1*, p. *277*; 21.1, p. 370; 21.1.1, p. 385; 21.4, p. 396.

M. W. Muller
*Site Ordering and Magnetostriction in Garnet (II): Iron
(b) Vol. **86**, No. 1, p. 345–348 (1978).
See also *1*, p. *9*; 22.8.2, p. 518.

S. Ligenza
Exchange Splitting of Ionic S-State in $MnFe_2O_4$ Studied by Neutron Spectroscopy
(b) Vol. **86**, No. 2, p. 635–643 (1978).
See also 13.5.4, p. 201; 22.8.2, p. 518.

M. Maryško
The Influence of Electromagnetic Propagation on Ferromagnetic Resonance in an Infinite Cylinder
(a) Vol. **47**, No. 1, p. 277–286 (1978).
See also *18.2*, p. *286*; *19*, p. *307*; 22.8.2, p. 519.

E. A. Vasilev and A. E. Gelyasin
*Magnetic Properties of the $Mn_{1+x}Sb_{1-x}Sn_x$ System
(a) Vol. **47**, No. 1, p. K55–K57 (1978).
See also *18.2*, p. *286*; 21, p. 359.

V. A. Povitskii, A. N. Salugin, Yu. V. Baldokhin, E. F. Makarov, and N. V. Elistratov
Spin Reorientation in Hematite (α-Fe_2O_3) with Al Impurity
(b) Vol. **87**, No. 2, p. K101–K104 (1978).
See also *5*, p. *61*; 22.8.2, p. 519.

B. Knappe, H.-R. Müller, and B. Springmann
Honogeneity of Amorphous Gd–Co Films Prepared by Bias Sputtering
(a) Vol. **47**, No. 2, p. 523—532 (1978).
See also *1.5*, p. *37*; *2*, p. *45*; 21.1, p. 374; 21.4, p. 398.

V. M. Ryzhkovskii
*On Structural and Magnetic Properties of the Mn_2As–Mn_2Sb Quasibinary System
(a) Vol. **47**, No. 2, p. K177—K180 (1978).
See also 1.2, p. 20; *18.4*, p. *300*; 22.8, p. 505.

H.-R. Müller
*Influence of Anisotropic Short-Range Order on Magnetic Properties of the Amorphous Ferrimagnet Gd–Co
(a) Vol. **49**, No. 1, p. K63—K66 (1978).
See also *2*, p. *50*; 21.1, p. 378; 21.4, p. 398.

R. Kosicik, P. Duda, W. Keilig, and R. Mattheis
*The Influence of Ion Bombardment on the Magnetic Anisotropy in Evaporated Amorphous Gd–Co Films
(a) Vol. **49**, No. 1, p. K67—K69 (1978).
See also *2*, p. *50*; 11, p. 129; 21.1, p. 378; 21.4, p. 398.

M. Maryško
Note on FMR Absorption in a Low Resistivity Ferrite Cylinder
(a) Vol. **49**, No. 1, p. K71—K73 (1978).
See also *19*, p. *312*; 22.8.2, p. 520.

J. D. Browne, N. J. Hance, G. B. Johnston, and C. F. Sampson
*The Magnetic Behaviour of Mn_3In
(a) Vol. **49**, No. 2, p. K177—K179 (1978).
See also 18.4, p. 301; 21.1, p. 379.

P. Hendy and E. W. Lee
*A Powder Neutron Diffraction Study of Some Rare-Earth–Co_2 Compounds
(a) Vol. **50**, No. 1, p. 101—107 (1978).
See also 21.1, p. 381; 21.4, p. 398.

B. Gillot, F. Bouton, F. Chassagneux, and A. Rousset
Electrical Conductivity of $(Fe^{2+}Al^{3+}_{2-x}Cr^{3+}_{x})O^{2-}_{4}$ Spinels and Defect Phases $\gamma(Fe^{3+}_{1/3}Al^{3+}_{2/3-y}Cr^{3+}_{y})_2O^{2-}_{3}$
(a) Vol. **50**, No. 1, p. 109—116 (1978).
See also *14.3*, p. *224*; 22.8.2, p. 520.

E. M. Smokotin, G. F. Gusyatskii, L. M. Protopopova, and A. M. Kapitonov
*The Effect of Inhomogeneous Stress on the FMR in Ferrites
(a) Vol. **50**, No. 1, p. 269—274 (1978).
See also *19*, p. *314*; 22.8.2, p. 520.

A. Bombik, B. Lesniewska, and A. Oleś
*Atomic and Magnetic Structures of $HoVO_3$
(a) Vol. **50**, No. 1, p. K17—K20 (1978).
See also *1.1*, p. *14*; 22.8.2, p. 520.

M. URNER-WILLE, T. S. TE VELDE, and P. G. VAN ENGEN
*Kerr-Effect Enhancement in Amorphous GdFe Films
(a) Vol. **50**, No. 1, p. K29—K31 (1978).
See also 2, p. 52; *20*, p. *318*; 21.1.1, p. 390; 21.4, p. 398.

Y. WATANABE, K. URADE, and S. SAITO
*Contribution of the Fe^{2+} Ion to the Magnetic Anisotropy Constant in Ferrites
(b) Vol. **90**, No. 2, p. 697—702 (1978).
See also 22.8.2, p. 521.

H.-R. MÜLLER, W. KEILIG, R. KOSICIK, P. ROSEMANN, and Z. FRAIT
*Significance and Interpretation of Magnetic Anisotropy Measurements in Amorphous Gd–Co Films
(a) Vol. **50**, No. 2, p. 537—542 (1978).
See also 21.1, p. 382; 21.4, p. 399.

Z. JIRÁK, S. VRATISLAV, and J. ZAJÍČEK
*The Magnetic Structure of the Tetragonal $Mn_{2.5}Cr_{0.5}O_4$
(a) Vol. **50**, No. 2, p. K131—K135 (1978).
See also 22.8.1, p. 517.

YU. M. VYSOTSKII, E. A. VASILEV, A. E. GELYASIN, and A. V. SHABLOVSKII
*Nuclear Zeeman Effect in the $Mn_{1+x}Sb_{1-x}Sn_x$ System
(a) Vol. **50**, No. 2, p. K205—K208 (1978).
See also 5, p. 63; *18.2*, p. *292*; 21, p. 367.

M.-L. CRAUS and N. REZLESCU
*The Influence of the Composition and Thermal Treatment on the Cation Redistribution in the $Mg_xFe_{3-x}O_4$ System
(a) Vol. **50**, No. 2, p. K225—K229 (1978).
See also 22.8.2, p. 521.

18.4 Antiferromagnetic Properties

Review Article

K. Bärner
*Exchange Interactions and the Coupled Spin–Lattice Phase Transitions in Manganese Pnictides
(b) Vol. **88**, No. 1, p. 13–23 (1978).
See also 18.1, p. 277; *18.2*, p. *281*; 18.3, p. 295; 21.1, p. 368.

Original Papers and Short Notes

M. V. Medvedev and E. L. Rumyantsev
*On the Existence of a Spin-Glass State in an Ising Alloy Model with Mixed Nearest-Neighbour Exchange Interactions
(b) Vol. **85**, No. 2, p. 427–438 (1978).
See also *18.2*, p. *282*; 21.1, p. 369; 21.1.1, p. 384.

K. Kawasaki and R. A. Tahir-Kheli
Effects of Bond Dilution on Dynamics of Anisotropic Paramagnets at High Temperatures
(b) Vol. **85**, No. 2, p. 707–713 (1978).
See also *18.1*, p. *277*; 18.2, p. 282.

S. Niziol, H. Bińczycka, A. Szytuła, J. Todorović, R. Fruchart, J. P. Senateur et D. Fruchart
*Structure magnétique des MnCoSi
(Magnetic Structure of MnCoSi)
(a) Vol. **45**, No. 2, p. 591–597 (1978).
See also *18.2*, p. *283*; 22.8, p. 503.

D. Gignoux and H. Asmat
*Magnetic Properties and Structures of TbAlGa and HoAlGa
(a) Vol. **45**, No. 2, p. K149–K153 (1978).
See also *1.1*, p. *11*; 21, p. 355; 21.4, p. 396.

M. V. Medvedev
*Phase Diagrams of an Ising Magnet with Random Exchange Bonds
(b) Vol. **86**, No. 1, p. 109–118 (1978).
See also 18.1, p. 277; *18.2*, p. *283*.

M. Schwab
Phase Transition in the Mixed Crystals $Tb_pY_{1-p}AsO_4$
(b) Vol. **86**, No. 1, p. 195–203 (1978).
See also *1.2*, p. *16*; 13.4, p. 172; 20.1, p. 322; 22.8.1, p. 511.

J. Sznajd
Cross-Over from Discontinuous to Continuous Phase Transition under an External Magnetic Field
(b) Vol. **86**, No. 1, p. K11–K14 (1978).
See also *1.2*, p. *16*; 18.1, p. 277; 22.6, p. 492.

P. Meriel, A. Herr, and A. Meyer
*Neutron Diffraction Study of the Magnetic Structure of $TbBe_{13}$ at 4.2 K
(a) Vol. **46**, No. 1, p. K35—K38 (1978).
See also 21.3, p. 394; 21.4, p. 397.

S. Nadolski, A. M. Balbashov, and A. Ya. Chervonenkis
*Influence of Photomagnetic Effects on the ^{57}Fe NMR in $YFeO_3$
(a) Vol. **46**, No. 1, p. K47—K48 (1978).
See also *19*, p. *306*; 22.8.1, p. 512.

A. Murasik, H. Ptasiewicz-Bąk, and A. Zygmunt
Magnetic Structure and Crystal-Field Effects in CeZn
(a) Vol. **46**, No. 1, p. K75—K78 (1978).
See also *18.2*, p. *284*; 21, p. 356; 21.4, p. 397.

M.-F. Charlot, J.-J. Girerd, and O. Kahn
*Molecular Orbital Approach of Antiferromagnetic Coupling in Linear Chains $AMCl_3$ with M = V, Mn, and Ni and A = Cs or $N(CH_3)_4$
(b) Vol. **86**, No. 2, p. 497—504 (1978).
See also 22.8, p. 504.

W. Schumacher, U. Lindner, and R. Jauch
*Magnetic Solutions of the Hubbard Model
(b) Vol. **86**, No. 2, p. 621—625 (1978).
See also *18.2*, p. *285*.

Diep-The-Hung
Erratum to: Effect of Induced Surface Magnetic Moment on Spin-Wave Spectrum
(b) Vol. **86**, No. 2, p. K181 (1978).
See also 1.6, p. 40; *13.5.4*, p. *201*.

D. Samaras, A. Collomb, et J. C. Joubert
*Etude des propriétés magnétiques du composé $SrNd_2Fe_2O_7$ — Détermination des structures magnétiques par diffraction neutronique
(Study of Magnetic Properties of the Compound $SrNd_2Fe_2O_7$ — Determination of Magnetic Structures by Neutron Diffraction)
(a) Vol. **46**, No. 2, p. 493—498 (1978).
See also 22.8.1, p. 512.

H. H. Heikens, R. S. Kuindersma, C. F. van Bruggen, and C. Haas
*Magnetic Properties of $MnEr_2S_4$, $MgEr_2S_4$, and $Mn_xMg_{1-x}Y_2S_4$
(a) Vol. **46**, No. 2, p. 687—695 (1978).
See also *18.1*, p. *278*; 22.8, p. 504.

S. Siek, A. Szytuła, and J. Leciejewicz
*The Crystal and Magnetic Structure of $CeMn_2Si_2$
(a) Vol. **46**, No. 2, p. K101—K105 (1978).
See also *1.1*, p. *12*; 22.8, p. 504.

H. Kalbfleisch and K. Melchior
*Susceptibility of $DyPO_4$ near the Magnetic Phase Boundaries
(a) Vol. **46**, No. 2, p. K119—K121 (1978).
See also 22.8.1, p. 512.

S. Roth and J. Sonntag
*Temperature Dependence of the Electrical Resistivity of Semiconducting CrAl Solid Solutions at High Temperatures
(b) Vol. 87, No. 1, p. K69—K72 (1978).
See also *14.3*, p. *220*; 21, p. 358; 21.1, p. 372.

S. V. Koptsik, V. A. Povitskii, M. Yu. Messinev, Yu. V. Baldokhin, and N. F. Shchukin
Mössbauer Study on Shocked Hematite
(a) Vol. 47, No. 1, p. 39—43 (1978).
See also *5*, p. *61*; 22.6, p. 493.

G. I. Makovetskii and G. M. Shakhlevich
*Magnetic Properties of the $CrS_{1-x}Se_x$ System
(a) Vol. 47, No. 1, p. 219—222 (1978).
See also 22, p. 413.

J. Mertsching
*Peierls Distortion and Antiferromagnetism in the Quasi-One-Dimensional Peierls-Hubbard Model
(b) Vol. 87, No. 2, p. 599—606 (1978).
See also *13.1*, p. *155*.

J. Spałek, A. M. Oleś, and K. A. Chao
*The Effective Magnetic Interactions between Impurity and Conduction Electrons for the Wolff Model
(b) Vol. 87, No. 2, p. 625—635 (1978).
See also 13.1, p. 155; *13.4*, p. *176*.

N. K. Jaggi
*Dependence of the Spin Glass Transition on the Details of the Disorder
(b) Vol. 87, No. 2, p. K81—K84 (1978).
See also *18.2*, p. *287*.

K. J. Michalski
Magnetic Symmetry of Ordered Phases in Mn_3GaC
(b) Vol. 87, No. 2, p. K85—K87 (1978).
See also *18.2*, p. *287*; 22.8, p. 505.

H. Ptasiewicz-Bąk, J. Leciejewicz, and A. Zygmunt
*Magnetic Ordering in UGeY (Y = S, Se, Te) by Neutron Diffraction
(a) Vol. 47, No. 2, p. 349—356 (1978).
See also *18.2*, p. *287*; 22.8, p. 505.

V. M. Ryzhkovskii
*On Structural and Magnetic Properties of the Mn_2As–Mn_2Sb Quasibinary System
(a) Vol. 47, No. 2, p. K177—K180 (1978).
See also 1.2, p. 20; *18.3*, p. *296*; 22.8, p. 505.

A. Richter, G. Röpke, and F. Goedsche
Functional Integral Approach for the Hubbard Model with Arbitrary Electron Density
(b) Vol. 88, No. 1, p. 189—198 (1978).
See also *13*, p. *146*; *18*, p. *274*; 18.1, p. 278; 18.2, p. 288.

V. V. Gagulin, N. V. Fadeeva, A. G. Belous, A. V. Titov, K. P. Mitrofanov, M. V. Plotnikova, S. P. Soloviev, and Yu. N. Venevtsev
New Seignette-Magnets with Hexagonal Barium Titanate Structure
(a) Vol. **48**, No. 1, p. 183—189 (1978).
See also *1*, p. *9*; 5, p. 61; *14.4.1*, p. *255*; 18.2, p. 289; 19, p. 310; 22.8.1, p. 514.

W. K. Unger
*Search for Light-Induced Magnetization in Magnetic Semiconductors
(a) Vol. **48**, No. 1, p. K89—K91 (1978).
See also 22.8.2, p. 519.

B. K. Ponomarev and V. G. Thiessen
Band Ferromagnetism in Nickel–Palladium Alloys
(b) Vol. **88**, No. 2, p. K139—K142 (1978).
See also *18.2*, p. *289*; 21.1, p. 376.

I. Pop, A. Giurgiu, E. Pop, and V. Iuşan
Coexistence of Itinerant Antiferromagnetism with Paramagnetism of Localized Magnetic Moments in Low Concentrated Cr–Fe Alloys
(b) Vol. **88**, No. 2, p. K181—K185 (1978).
See also *18.1*, p. *278*; 21.1, p. 376; 21.1.1, p. 388.

R. Richter, F. Goedsche, and G. Vojta
Statistical Theory of Electrical Resistivity of Magnetic Ternary Alloys with Partial Long-Range Order
(b) Vol. **89**, No. 1, p. 77—83 (1978).
See also 13, p. 147; *14.1*, p. *209*; 21, p. 362.

U. Lindner and V. K. Fedyanin
Solitons in a One-Dimensional Modified Hubbard Model
(b) Vol. **89**, No. 1, p. 123—129 (1978).
See also *13.5*, p. *188*.

K. Fukamichi
*Magnetic Phase Diagram of the Dilute Cr–Sb Alloy System
(a) Vol. **49**, No. 1, p. K37—K40 (1978).
See also 8, p. 88; 14.1, p. 210; 21.1, p. 378.

J. D. Browne, N. J. Hance, G. B. Johnston, and C. F. Sampson
The Magnetic Behaviour of Mn_3In
(a) Vol. **49**, No. 2, p. K177—K179 (1978).
See also *18.3*, p. *296*; 21.1, p. 379.

S. Nadolski, E. Jędryka, K. Piotrowski, and B. Wanklyn
*Cr^{53} Nuclear Magnetic Resonance in Domain Walls of Yttrium Orthochromite
(a) Vol. **49**, No. 2, p. K193—K196 (1978).
See also *19*, p. *313*; 22.8.2, p. 520.

L. M. Noskova, L. A. Ugodnikova, and S. G. Chuchman
*The Néel Points of Yttrium–Rare Earth Metal Alloys
(a) Vol. **50**, No. 1, p. K47—K49 (1978).
See also *18.1*, p. *280*; 21.4, p. 399.

D. E. G. Williams, H. W. G. Spraget, and R. Perrin
*The Magnetic Moment of $Cr_{0.6}Mn_{0.4}$
(b) Vol. **90**, No. 2, p. 667—670 (1978).
See also 21.1, p. 382.

K. J. Michalski and W. Reich
Metamagnet with Temperature-Dependent Anisotropy in the Classical Approximation
(b) Vol. **90**, No. 2, p. K141—K143 (1978).
See also *18.2*, p. *292*.

A. Collomb, D. Samaras et J. C. Joubert
*Détermination par diffraction neutronique des structures magnétiques des composés $SrTCrO_4$ (T = La, Nd)
(Neutron Diffraction Determination of Magnetic Structures of $SrTCrO_4$ Compounds (T = La, Nd))
(a) Vol. **50**, No. 2, p. 635—642 (1978).
See also 22.8.1, p. 516.

19. Magnetic Resonance

Review Article

V. I. Ivanov-Omskii, L. I. Korovin, and E. M. Shereghii
*Phonon-Assisted Cyclotron Resonance in Semiconductors
(b) Vol. **90**, No. 1, p. 11—32 (1978).
See also *6*, p. *64*; 13, p. 142; 22, p. 408; 22.2.3, p. 452; 22.4.1, p. 463.

Original Papers and Short Notes

S. I. Fărcaş, A. Darabont, and A. Nicula
*Theoretical Study of the Forbidden ESR Transitions of Cr^{3+} and Fe^{3+} in Alums
(b) Vol. **85**, No. 1, p. 75—79 (1978).
See also 22.8.1, p. 510.

S. Franck, A. Kühnel, and S. Wendt
On the Microscopic Mechanism of the Ferroelectric Phase Transition of TSCC:Mn^{2+}
(b) Vol. **85**, No. 1, p. 319—323 (1978).
See also *14.4.1*, p. *253*; 22.9, p. 522.

A. V. Shendrik and D. M. Yudin
*A New Concept of the Model for the Paramagnetic E′-Centre in the SiO_{2-x} System
(b) Vol. **85**, No. 1, p. 343—349 (1978).
See also 2, p. 42; *13.4*, p. *169*; 22.8.1, p. 510.

I. N. Geifman
*Electron Paramagnetic Resonance of Ti^{3+} in $KTaO_3$
(b) Vol. **85**, No. 1, p. K5—K6 (1978).
See also 22.8.1, p. 510.

K. S. Aleksandrov, I. P. Aleksandrova, V. F. Shabanov, V. I. Yuzvak, Y. Z. Nozik, and L. I. Fikin
Mechanism of Ferroelectric Phase Transition in $NaNH_4SeO_4 \cdot 2\,H_2O$
(a) Vol. **45**, No. 1, p. 53—62 (1978).
See also 1.2, p. 15; *14.4.1*, p. *253*; 20, p. 315; 22.8.1, p. 510.

K. Hoffmann and D. Pöss
*Electron Spin Resonance Investigations on Lithium Fluoride Doped with Titanium
(a) Vol. **45**, No. 1, p. 263—268 (1978).
See also 13.4, p. 170; 22.5.2, p. 478.

S. Grande, H.-D. Mecke, and St. Idziak
*High-Resolution Proton Magnetic Resonance in Single Crystals of Potassium Trihydrogen Selenite
(a) Vol. **45**, No. 1, p. K29—K32 (1978).
See also *4*, p. *57*; 22.8, p. 503.

K. V. S. Rama Rao, T. Rajasekharan, V. Seshubai, and C. Ramasastry
*Ferromagnetic Resonance Study in Thin Films of $Mn_{1-x}Cu_xSb$
(a) Vol. **45**, No. 1, p. K51—K53 (1978).
See also 22.7, p. 501.

M. Höhne
*EPR of Lithium in Mechanically Affected Silicon
(b) Vol. **85**, No. 2, p. 525–534 (1978).
See also 1.6, p. 40; 13.4, p. 170; 22.1.2, p. 429.

D. Boulanger and M.-S. Martin
*Crystal Growth and EPR Study of Cr^{+}-Doped MgTe Single Crystals
(b) Vol. **85**, No. 2, p. 597–601 (1978).
See also 3, p. 54; 13.4, p. 170; 22.4.3, p. 470.

J. Alizon, G. Berthet, J. P. Blanc, J. Gallice, H. Robert, J. M. Fabre, and L. Giral
Dynamic Nuclear Polarization in TCNQ Salts
(b) Vol. **85**, No. 2, p. 603–610 (1978).
See also *18*, p. *273*; 22.9, p. 523.

L. Leibler
*Theory of Electron Spin Resonance in Graded Mixed Semiconductors
(b) Vol. **85**, No. 2, p. 611–619 (1978).
See also 13.1, p. 151; 22.2.4, p. 456; 22.4.4, p. 472.

R. M. M. Fonville, W. Geertsma, and C. Haas
On the Nature of the Top of the Valence Band in Layered Mo and W Dichalcogenides
(b) Vol. **85**, No. 2, p. 621–627 (1978).
See also *13.1*, p. *151*; 14.3, p. 217; 18.1, p. 277; 22, p. 409.

M. Vanhaelst, P. Matthys, and E. Boesman
*The Phonon Induced Zero Field Splitting in Vanadium Doped Alkali Chlorides
(b) Vol. **85**, No. 2, p. 639–644 (1978).
See also 22.5.2, p. 479.

M. Hamera
Calculation of the *g*-Shift and Superhyperfine Structure Tensor of Mn^{2+} Ion in $A_{II}B_{IV}$ Tetrahedral Semiconductors
(b) Vol. **85**, No. 2, p. 645–655 (1978).
See also *13.4*, p. *170*; 22.4, p. 461.

S. R. Shukla and S. D. Pandey
*EPR of Mn^{2+} in Cobalt Potassium Selenate Hexahydrate Tutton Salt
(b) Vol. **85**, No. 2, p. K103–K105 (1978).
See also 22.8.1, p. 511. *3*

G. C. Stevens and D. Bloor
*ESR Studies of a Diacetylene Polymer (I): Partially Crystalline Polymer Extracts
(a) Vol. **45**, No. 2, p. 483–491 (1978).
See also 13.4, p. 171; 22.9, p. 523.

M. Krupski and J. Stankowski
New High-Pressure Phase in $[Ni(NH_3)_6](ClO_4)_2$ Detected by EPR
(a) Vol. **45**, No. 2, p. 585–589 (1978).
See also *1.2*, p. *16*; 22.8, p. 503.

D. GOURIER, D. VIVIEN, J. THERY, J. LIVAGE, and R. COLLONGUES
*ESR of Copper-Doped β-Alumina. Localization of Cu^{2+} and Linewidth Effect
(a) Vol. **45**, No. 2, p. 599–606 (1978).
See also 22.6.1, p. 499.

Y. OCHIAI and E. MATSUURA
*Spin–Lattice Relaxation at High Temperatures in Heavily Doped n-Type Silicon
(a) Vol. **45**, No. 2, p. K101–K104 (1978).
See also 22.1.2, p. 430.

M. HÖHNE
*Nonrandom Strain in "Crushed" Silicon. EPR of Thermally Excited Lithium Donors
(b) Vol. **86**, No. 1, p. 119–126 (1978).
See also 1.6, p. 40; 13.4, p. 172; 22.1.2, p. 430.

H. NEUBRAND
*ESR from Boron in Silicon at Zero and Small External Stress (I): Line Positions and Line Structure
(b) Vol. **86**, No. 1, p. 269–275 (1978).
See also 13.4, p. 172; 22.1.2, p. 430.

V. N. KASHCHEEV
Some Aspects of Spin Relaxation in Ferromagnets in a Magnetic Field
(b) Vol. **86**, No. 1, p. 359–365 (1978).
See also *18.2*, p. *284*.

P. G. BARANOV, R. A. ZHITNIKOV, and V. A. KHRAMTSOV
*EPR and Optical Absorption of Zn^+ Ions in NaCl Crystals
(b) Vol. **86**, No. 1, p. K67–K70 (1978).
See also *20.1*, p. *323*; 22.5.2, p. 479.

J. DUMAS
*Analysis of the Intensity of the ESR Line in Amorphous Si–Ge Films
(b) Vol. **86**, No. 1, p. K75–K77 (1978).
See also *2*, p. *43*; 22.1.1, p. 424; 22.1.2, p. 431.

G. C. STEVENS and D. BLOOR
*ESR Studies of a Diacetylene Polymer (II): Spin Concentrations and g-Values for Single Crystal Samples
(a) Vol. **46**, No. 1, p. 141–146 (1978).
See also 13.4, p. 173; 22.9, p. 523.

S. LAHIRY, J. SHARMA, G. D. SOOTHA, and H. O. GUPTA
*Cu^{2+} in Pseudotetrahedral Symmetry: Magnetic Anisotropy, Susceptibility, and EPR of Bis (Trimethyl- and Triethyl Benzyl Ammonium) Tetrachlorocuprate (II) from 300 to 77 K
(a) Vol. **46**, No. 1, p. 153–162 (1978).
See also *18.1*, p. *277*; 22.9, p. 523.

A. TARI
The Quenching Effect of Iron Substitution on the Holmium Moment in the $Ho_{0.01}Gd_{0.99}(Rh_{1-x}Fe_x)_2$ Compounds
(a) Vol. **46**, No. 1, p. 173–177 (1978).
See also *18.2*, p. *284*; 21.1.1, p. 385; 21.4, p. 396.

M. Punkkinen and L. P. Ingman
*Molecular Motion and Spin–Lattice Relaxation in Polyethylene
(a) Vol. **46**, No. 1, p. 213—216 (1978).
See also 22.9, p. 524.

E. Hadjoudis, F. Milia, J. Seliger, and R. Blinc
*Temperature Dependence of the ^{14}N Quadrupole Resonance in Some Thermochromic and Photochromic N-Salicylideneanilines
(a) Vol. **46**, No. 1, p. 239—244 (1978).
See also *10.2*, p. *111*; 22.9, p. 524.

S. Nadolski, A. M. Balbashov, and A. Ya. Chervonenkis
*Influence of Photomagnetic Effects on the ^{57}Fe MNR in $YFeO_3$
(a) Vol. **46**, No. 1, p. K47—K48 (1978).
See also *18.4*, p. *299*; 22.8.1, p. 512.

G. Appold, H. Pascher, R. Ebert, U. Steigenberger, and M. von Ortenberg
Determination of the Effective g-Value of Electrons in n-Type InSb by Spin-Flip Raman Scattering and Electric-Dipole-Excited Electron Spin Resonance
(b) Vol. **86**, No. 2, p. 557—561 (1978).
See also *13.4*, p. *174*; 20.1, p. 324; 22.2.3, p. 452.

F. Milia, M. Voudouris, A. Kandušer, A. Levstik, and R. Blinc
*^{75}As Nuclear Quadrupole Resonance Study of the Phase Transition in Na_2HAsO_4
(b) Vol. **86**, No. 2, p. K119—K121 (1978).
See also *1.2*, p. *17*; 14.4.1, p. 254; 22.8.1, p. 512.

V. Chandra Mouli and G. Sivarama Sastry
*ESR Studies of Cu^{2+} Doped $Na_2HAsO_4 \cdot 7\ H_2O$ Single Crystals
(b) Vol. **86**, No. 2, p. K127—K130 (1978).
See also 22.8.1, p. 512.

T. Taki
*^{31}P Multiple-Pulse Nuclear Magnetic Resonance Experiments in Inorganic Solids
(b) Vol. **86**, No. 2, p. K147—K150 (1978).
See also 22, p. 412.

M. Adachi and N. Takeuchi
*ESR Study of Copper(II) Ions in NaCl X-Irradiated at Room Temperature
(b) Vol. **86**, No. 2, p. K151—K153 (1978).
See also 22.5.2, p. 480.

S. Grande, H.-D. Mecke, and L. A. Shuvalov
Deuterium Resonance of $KD_3(SeO_3)_2$ Single Crystals above and below Phase Transition
(a) Vol. **46**, No. 2, p. 547—555 (1978).
See also *1.2*, p. *17*; *4*, p. *57*; 22.8, p. 504.

G. C. Stevens and D. Bloor
*ESR Studies of a Diacetylene Polymer (III): Thermal History Effects and Temperature Dependence of Polycrystalline Samples
(a) Vol. **46**, No. 2, p. 619—626 (1978).
See also 13.4, p. 174; 22.9, p. 524.

A. GOLTZENE and C. SCHWAB
*EPR Investigation of the Photomemory Effect of Cu_2O
(a) Vol. **46**, No. 2, p. K157—K161 (1978).
See also 22.6, p. 493.

H. J. L. VAN DER VALK, P. MEERTENS, and C. HAAS
*Electron Paramagnetic Resonance of Divalent V, Mn, and Co in Single Crystals of CdI_2 and PbI_2
(b) Vol. **87**, No. 1, p. 135—144 (1978).
See also 22.5, p. 475.

C. V. MANJUNATH and R. SRINIVASAN
*ESR Study of a Cr^{5+} Centre in Ferroelectric Ammonium Sulphate
(b) Vol. **87**, No. 1, p. 395—400 (1978).
See also *14.4.1*, p. *254*; 22.8.1, p. 513.

R. S. DE BIASI
*Forbidden $\Delta m_s = 2$ Transitions in the EPR Spectrum of Fe^{3+} in MgO
(b) Vol. **87**, No. 1, p. K29—K31 (1978).
See also 22.6, p. 493.

E. BEYER, R. BIEDERBICK, A. HOFSTAETTER, R. OEDER, A. SCHARMANN, D. SCHWABE, and B. VITT
*Some Features of the Creation of Intrinsic Hole Centres in Scheelite Type Crystals
(b) Vol. **87**, No. 1, p. K35—K38 (1978).
See also *10.2*, p. *113*; 11, p. 126; 22.8.1, p. 513.

R. S. SARASWAT and G. C. UPRETI
*Estimation of T_1 of Fast Relaxing Co^{2+} Ions from the Observed EPR Linewidth of Mn^{2+} and VO^{2+} Impurities in Some Cobalt Salt Single Crystals
(b) Vol. **87**, No. 1, p. K39—K42 (1978).
See also 22.8, p. 505.

E. HEINZE and S. GRANDE
*Nuclear Spin–Lattice Relaxation in the Solid State of Mesogenic p-Alkoxybenzylidene-Alkylanilines
(b) Vol. **87**, No. 1, p. K43—K45 (1978).
See also 22.9, p. 525.

R. MÜLLER and G. SCHELER
*Reorientation of Water Molecules in Monocrystals of Oxalic Acid Dihydrate ($(COOH)_2 \cdot 2\,H_2O$) Investigated by High Resolution NMR in Solids
(b) Vol. **87**, No. 1, p. K53—K55 (1978).
See also 22.9, p. 525.

R. KIND, S. PLESKO, and J. ROOS
*Structural Phase Transitions in the Perovskite-Type Layer Compound $NH_3(CH_2)_3NH_3MnCl_4$
(a) Vol. **47**, No. 1, p. 233—240 (1978).
See also *1.2*, p. *18*; 22.9, p. 525.

M. MARYŠKO
*The Influence of Electromagnetic Propagation on Ferromagnetic Resonance in an Infinite Cylinder
(a) Vol. **47**, No. 1, p. 277—286 (1978).
See also *18.2*, p. *286*; 18.3, p. 295; 22.8.2, p. 519.

J. Stasz
*ESR Line Intensity near the Curie Point in $CdCr_2Se_4$, $HgCr_2Se_4$, and $CdCr_2S_4$ Single Crystals
(a) Vol. 47, No. 1, p. K23–K26 (1978).
See also *18.2*, p. *286*; 22.8, p. 505.

R. Stegmann and G. Oelgart
*Hybrid Resonance in a Quantizing Magnetic Field in BiSb
(b) Vol. 87, No. 2, p. 485–489 (1978).
See also *13.2*, p. *162*; 22.7, p. 502.

C. Jaccard and M. Ecabert
*Optical Detection of Nuclear Magnetic Resonance with Distant F-Centre Pairs in KCl
(b) Vol. 87, No. 2, p. 497–505 (1978).
See also 13.4, p. 175; *20.3*, p. *344*; 22.5.2, p. 481.

K. D. Becker
*Nuclear Spin Relaxation by Quadrupolar Interactions. The Diffusion of Substitutional Impurities in Ionic Crystals
(b) Vol. 87, No. 2, p. 589–597 (1978).
See also *9*, p. *92*; 22.5.1, p. 477.

M. A. B. Whitaker
*The Motional Effect in Nuclear Double-Resonance
(b) Vol. 87, No. 2, p. 689–695 (1978).
See also 6, p. 72; 22.8, p. 505.

H. K. Diemer, E. J. Bijvank, and H. W. den Hartog
*EPR Investigations of ADP:Cr^{3+} above and below T_c
(b) Vol. 87, No. 2, p. 697–706 (1978).
See also *18.1*, p. *278*; 22.8.1, p. 513.

M. Vanhaelst and P. Clauws
*EPR Spectrum of the Oxygen Vacancy in Single Crystals V_2O_5
(b) Vol. 87, No. 2, p. 719–723 (1978).
See also 10.2, p. 113; 22.6, p. 494.

E. Burzo and I. Ardelean
*EPR Study of Fe^{3+} Ions in Lead Borate Glasses
(b) Vol. 87, No. 2, p. K137–K140 (1978).
See also 22.8.1, p. 513.

N. Guskos, J. Kuriata, and T. Rewaj
*Spin–Lattice Coefficients of Gd^{3+} in SrF_2 Single Crystals
(b) Vol. 87, No. 2, p. K155–K158 (1978).
See also 22.5.3, p. 487.

A. A. Klyukanov, E. P. Pokatilov, V. N. Gladilin, and Tr. H. Wu
*Pinning Structure of Cyclotron Resonance Lines
(b) Vol. 87, No. 2, p. K159–K162 (1978).
See also 22.4.1, p. 464.

N. N. Sirota, V. I. Lukashova, and L. A. Kurochkin
Irradiation Effect on Cyclotron Resonance in Germanium
(a) Vol. 47, No. 2, p. 677—682 (1978).
See also *11*, p. *127*; 22.1.1, p. 426.

K. S. Bhatki, S. D. Pradhan, and B. V. Thosar
Further Studies on Positron Lifetimes in γ-Irradiated Teflon in Low Dose Range
(a) Vol. 47, No. 2, p. 691—698 (1978).
See also *10.2*, p. *114*; 11, p. 127; 14.4, p. 249; 22.9, p. 526.

S. Radhakrishna and T. Bhaskar Rao
*Optical Absorption and ESR Studies of Nickel Ions in Sodium Zinc Sulphate Tetrahydrate
(a) Vol. 47, No. 2, p. 723—729 (1978).
See also 13.4, p. 176; *20.1*, p. *328*; 22.8.1, p. 514.

T. Grochulski, K. Leibler, and A. Sienkiewicz
*Magnetic Resonance in MnTe at High Hydrostatic Pressures
(a) Vol. 47, No. 2, p. K169—K171 (1978).
See also 22.7, p. 502.

N. B. Yunusov and V. P. Zentsov
Electronic Structure and ESR of Ti^{2+} in Fluorite Type Crystals
(b) Vol. 88, No. 1, p. 87—97 (1978).
See also *13.4*, p. *177*; 22.5.3, p. 488.

J. Grossmann, D. Müller, and J. Petersson
*Study of the Ultraslow Critical Dynamics of the Order Parameter of the Ferroelectric $AgNa(NO_2)_2$ by the Spin Relaxation Times T_1, $T_{1\varrho}$, T_2 of the ^{23}Na Nucleus
(b) Vol. 88, No. 1, p. 99—103 (1978).
See also *14.4.1*, p. *255*; 22.8.1, p. 514.

S. Ueda and T. Shimizu
*Nuclar Magnetic Resonance of ^{77}Se in Amorphous $Cu(As_{0.4}Se_{0.6})_{1-x}$
(b) Vol. 88, No. 1, p. K1—K3 (1978).
See also *2*, p. *46*; 22, p. 414.

M. Godlewski and A. M. Hennel
*EPR Measurements on GaAs:Co
(b) Vol. 88, No. 1, p. K11—K13 (1978).
See also 22.2.1, p. 447.

E. Heinze and S. Grande
*Nuclear Spin Relaxation in the Dipolar System in Solid p-Alkoxybenzylidene-p-n-Alkylanilines
(b) Vol. 88, No. 1, p. K23—K25 (1978).
See also 1.2, p. 20; 6, p. 74; 22.9, p. 526.

T. Rewaj and M. Krupski
*The Effect of Hydrostatic Stress and Temperature on the EPR Spectrum of Gd^{3+} Ion in CdF_2:$-b_4^0$ as a Function of Interionic Distances
(b) Vol. 88, No. 1, p. K65—K68 (1978).
See also 22.5.3, p. 488.

S. A. Marshall, T. Marshall, and R. A. Serway
*Electron Spin Resonance Absorption Spectrum of Trivalent Gadolinium in the Oxide YAlG
(a) Vol. 48, No. 1, p. 165—173 (1978).
See also 22.8.2, p. 519.

V. V. Gagulin, N. V. Fadeeva, A. G. Belous, A. V. Titov, K. P. Mitrofanov, M. V. Plotnikova, S. P. Soloviev, and Yu. N. Venevtsev
New Seignette-Magnets with Hexagonal Barium Titanate Structure
(a) Vol. 48, No. 1, p. 183—189 (1978).
See also *1*, p. *9*; 5, p. 61; *14.4.1*, p. *255*; 18.2, p. 289; 18.4, p. 301; 22.8.1, p. 514.

F. Holuj and A. Jesmanowicz
*EPR of Fe^{3+} in Andalusite and Kyanite at V-Band and the Pair Spectra
(a) Vol. 48, No. 1, p. 191—198 (1978).
See also 13.4, p. 177; 22.8, p. 506.

C. S. Sunandana
*ESR Saturation of SO_3^- in X-Irradiated $KHSO_4$
(a) Vol. 48, No. 1, p. K19—K21 (1978).
See also 11, p. 128; 22.8.1, p. 514.

J. Thiemann
*Host Nuclear Magnetic Resonance in Copper-Band Magnetic Alloys
(b) Vol. 88, No. 2, p. 397—406 (1978).
See also 10.1, p. 105; 21.1, p. 375.

V. E. Petrashen, Yu. V. Yablokov, and R. L. Davidovich
*Electron Paramagnetic Resonance Study of Cu^{2+} in $K_2Zn(ZrF_6)_2 \cdot 6\,H_2O$
(b) Vol. 88, No. 2, p. 439—443 (1978).
See also *13.4*, p. *178*; 22.8, p. 506.

G. Oelgart, R. Stegmann, and R. Herrmann
Propagation of Electromagnetic Waves in a Direction Parallel to the Magnetic Field in Semiconductive Bi–Sb Alloys
(b) Vol. 88, No. 2, p. 507—516 (1978).
See also *13.2*, p. *162*; 14.3, p. 223; 21.7, p. 407.

Ch. Hoentzsch and J. M. Spaeth
*Localized Vibrations of Substitutional Atomic Hydrogen in KCl and RbCl Studied by ENDOR
(b) Vol. 88, No. 2, p. 581—589 (1978).
See also *13.4*, p. *178*; 22.5.2, p. 482.

W. Gehlhoff
*On the Calculation of Superhyperfine Structures
(b) Vol. 88, No. 2, p. 599—605 (1978).

R. S. Eachus, R. E. Graves, and M. T. Olm
*Electron Spin Resonance of Mobile Electrons in AgCl
(b) Vol. 88, No. 2, p. 705—711 (1978).
See also *10.2*, p. *116*; 16, p. 265; 22.5.1, p. 477.

H. Puszkarski
*Elliptical Spin Precession in Microscopic Spin-Wave Resonance Theory
(b) Vol. 88, No. 2, p. K113—K116 (1978).
See also *18.2.1*, p. *293*.

V. K. Jain and P. Venkateswarlu
*EPR of Mn^{2+} in $MgSeO_4 \cdot 6\,H_2O$ Single Crystals
(b) Vol. 88, No. 2, p. K173—K175 (1978).
See also 22.8.1, p. 515.

J. H. von Bardeleben, A. Goltzene, B. Meyer, and C. Schwab
Effects of Iron Content and Stoichiometry on the Coloration of $CuGaS_2$
(a) Vol. 48, No. 2, p. K145—K148 (1978).
See also *10.2*, p. *116*; 20.1, p. 331; 22.8, p. 506.

W. Braune, G. Kuka, H.-J. Gollnest, and R. Herrmann
*Microwave Spectroscopy in Semimetallic Bismuth–Antimony Alloys
(b) Vol. 89, No. 1, p. 95—101 (1978).
See also *13.1*, p. *158*; 21.7, p. 407.

M. Vanhaelst, P. Matthys, and E. Boesman
*Electron Spin Transfer in Alkali Chlorides and Alkali Bromides Doped with Divalent Vanadium
(b) Vol. 89, No. 1, p. 151—157 (1978).
See also *20.1*, p. *332*; 22.5.2, p. 483.

B. Milsch and W. Windsch
*Electron Paramagnetic Resonance of Mn^{2+} in Guanidinium Aluminium Sulfate Hexahydrate
(b) Vol. 89, No. 1, p. 241—251 (1978).
See also 13.4, p. 180; *14.4.1*, p. *256*; 22.8.1, p. 515.

R. A. Serway, F. H. Yang, and S. A. Marshall
*Line Broadening in the ESP Absorption Spectra of Fe^{3+}, Cr^{3+}, and Gd^{3+} in the Oxide YAlG
(b) Vol. 89, No. 1, p. 267—273 (1978).
See also 13.4, p. 180; 22.8.2, p. 520.

M. Vanhaelst, J. Broeckx, P. Matthys, and E. Boesman
*Spin–Lattice Relaxation of the V^{2+} Centre in NaCl, KCl, and RbCl
(b) Vol. 89, No. 1, p. K25—K28 (1978).
See also 22.5.2, p. 484.

H. Bill
Parametrization of the Measured Temperature Dependence of the Crystal Field Splitting in Axial ^8S State Ion Centers
(b) Vol. 89, No. 1, p. K49—K52 (1978).
See also *13.4*, p. *180*; 22.5.4, p. 489.

I. Pop, M. Coldea, and V. U. S. Rao
*NMR and Magnetic Susceptibility of Intermetallic Compounds ZrV_2, $Zr_{0.99}Pt_{0.01}V_2$, and $Zr_{0.9}Fe_{0.1}V_2$
(a) Vol. 49, No. 1, p. 207—210 (1978).
See also *18.1*, p. *279*; 21.1, p. 377.

A. V. Moskalonov and J. R. Zakis
X-Ray Induced Paramagnetic States and Luminescence in As_2S_3 and As_2Se_3
(a) Vol. **49**, No. 1, p. 231—235 (1978).
See also 2, p. 49; *13.4*, p. *181*; *20.3*, p. *347*; 22, p. 417.

G. Stetter, H. Coufal, and E. Lüscher
*EPR Study of the Annealing Behaviour of Neutron-Transmutation-Doped Silicon
(a) Vol. **49**, No. 1, p. K17—K20 (1978).
See also *11*, p. *129*; 14.3, p. 224; 22.1.2, p. 437.

J. G. Kliava and J. J. Purāns
*Analysis of Mn^{2+} EPR Spectra in Polycrystals Using Bir's Method to Account for the Transition Probability
(a) Vol. **49**, No. 1, p. K43—K47 (1978).
See also 22.8.1, p. 515.

J.-M. Moret and R. Lacroix
*Paramagnetic Resonance of Mn^{2+} in Na_2S Single Crystals
(a) Vol. **49**, No. 1, p. K55—K56 (1978).
See also 22, p. 417.

M. Maryško
*Note on FMR Absorption in a Low Resistivity Ferrite Cylinder
(a) Vol. **49**, No. 1, p K71—K73 (1978).
See also 18.3, p. 296; 22.8.2, p. 520.

A. Hofstaetter, R. Oeder, A. Scharmann, D. Schwabe, and B. Vitt
*Paramagnetic Resonance and Thermoluminescence of the $PbWO_4/PbMoO_4$ Mixed Crystal System
(b) Vol. **89**, No. 2, p. 375—380 (1978).
See also 20.3, p. 348; 22.8.1, p. 515.

R. Biederbick, A. Hofstaetter, and A. Scharmann
Temperature-Dependent Mn^{2+}-^6S State Splitting in Scheelites
(b) Vol. **89**, No. 2, p. 449—458 (1978).
See also *13.4*, p. *181*; 22.8.1, p. 516.

I. K. Amanis, J. G. Kliava, and J. J. Purans
New Paramagnetic Centres in Mixed Alkali Halides: $Ag^0_C(Br_1)$ in KCl:KBr
(b) Vol. **89**, No. 2, p. 531—535 (1978).
See also *10.2*, p. *119*; 22.5.2, p. 485.

H.-D. Maier, D. Müller, and J. Petersson
*Temperature and Frequency Dependence of the Dielectric Constant and Proton NMR near the Structural Phase Transition of Squaric Acid
(b) Vol. **89**, No. 2, p. 587—593 (1978).
See also *14.4*, p. *251*; 22.9, p. 527.

M. Dubiel, G. Berg, and F. Fröhlich
Aggregation of Point Defects in Highly Doped NaCl:Ca^{++} Crystals
(b) Vol. **89**, No. 2, p. 595—602 (1978).
See also *10.2*, p. *119*; 22.5.2, p. 485.

M. Razeghi and B. Houlier
*EPR Study of Gd^{3+} around the Ferroelastic Transition Point of $Pb_3(PO_4)_2$
(b) Vol. **89**, No. 2, p. K135—K137 (1978).
See also *1.2*, p. *23*; 22.8.1, p. 516.

N. F. Berk, L. Yarmus, W. Bizzaro, and J. Rosenthal
*The Source of "Residual" Linewidth in EPR of Triplet Excitons in Molecular Crystals
(b) Vol. **89**, No. 2, p. K167—K170 (1978).
See also *13.5.1*, p. *195*; 22, p. 418.

A. Watterich and K. Polák
*ESR Investigation of Isolated V^{2+} Ions in NaCl
(b) Vol. **89**, No. 2, p. K187—K190 (1978).
See also 11, p. 130; 22.5.2, p. 485.

W. Auch, W. Steudle, J. U von Schütz, and H. C. Wolf
*The Phase Transition in the 1:1 CT-Complex Anthracene/TCNB at 206 K. A NMR Study
(a) Vol. **49**, No. 2, p. 563—567 (1978).
See also 1.2, p. 23; 6, p. 77; 22.9, p. 528.

R. Durný
*EPR of Mn^{2+} in Ge–S Glasses
(a) Vol. **49**, No. 2, p. K113—K116 (1978).
See also *2*, p. *51*; 22, p. 419.

T. I. Chekmasova and I. P. Aleksandrova
NMR Investigation of the High-Pressure Phase in $LiNH_4SO_4$
(a) Vol. **49**, No. 2, p. K185—K188 (1978).
See also *1.2*, p. *23*; *14.4.1*, p. *256*; 22.8.1, p. 516.

S. Nadolski, E. Jedryka, K. Piotrowski, and B. Wanklyn
*Cr^{53} Nuclear Magnetic Resonance in Domain Walls of Yttrium Orthochromite
(a) Vol. **49**, No. 2, p. K193—K196 (1978).
See also *18.4*, p. *301*; 22.8.2, p. 520.

H. Neubrand
*ESR from Boron in Silicon at Zero and Small External Stress (II): Linewidth and Crystal Defects
(b) Vol. **90**, No. 1, p. 301—308 (1978).
See also 13.4, p. 183; 22.1.2, p. 437.

G. Gobsch, H. Haberlandt, H.-J. Weckner, and J. Reinhold
*Calculation of the g-Tensor and ^{29}Si Hyperfine Tensors of the E'_1 Centre in Silicon Dioxide
(b) Vol. **90**, No. 1, p. 309—317 (1978).
See also *13.4*, p. *183*; 22.6, p. 497.

A. N. Cherkasov, A. A. Galkin, and L. T. Tsymbal
W and Mo RF Surface Impedance. New Type of Oscillations
(b) Vol. **90**, No. 1, p. K17—K20 (1978).
See also *13.3*, p. *168*; *14.1*, p. *211*; 21, p. 364.

S. A. Seryshev, A. G. Lundin, and M. L. Afanasev
*NMR in Tungsten Hexafluoride at High Hydrostatic Pressure
(b) Vol. **90**, No. 1, p. K35—K37 (1978).
See also 22.5, p. 476.

M. Baran, H. Szymczak, W. Wardzyński, and B. M. Wanklyn
*EPR and Optical Absorption Spectra of $K_2Cr_5F_{17}$ Monocrystals
(b) Vol. **90**, No. 1, p. K59—K61 (1978).
See also *20.1*, p. *336*; 22.8, p. 507.

W. Szkiełko
Spin-Dependent Recombination at Dislocations in Silicon Revealed at Reverse Current in p–n Junction
(b) Vol. **90**, No. 1, p. K81—K83 (1978).
See also *10.2*, p. *120*; *14.3.3*, p. *237*; 16, p. 267; 22.1.2, p. 438.

Y. H. Lee, P. R. Brosious, and J. W. Corbett
*High Temperature Ion Implantation in Diamond
(a) Vol. **50**, No. 1, p. 237—242 (1978).
See also 10.2, p. 120; *11*, p. *130*; 22.1, p. 423.

E. M. Smokotin, G. F. Gusyatskii, L. M. Protopopova, and A. M. Kapitonov
*The Effect of Inhomogeneous Stress on the FMR in Ferrites
(a) Vol. **50**, No. 1, p. 269—274 (1978).
See also *18.3*, p. *296*; 22.8.2, p. 520.

T. Rajasekharan and K. V. S. Rama Rao
*Nuclear Magnetic Resonance Study of Ferromagnetic $Mn_{1-x}Co_xSb$
(a) Vol. **50**, No. 1, p. 303—310 (1978).
See also *18.2*, p. *291*; 22.8, p. 508.

B. E. Vugmeister
*Spin Diffusion and Spin Lattice Relaxation in Paramagnetic Crystals
(b) Vol. **90**, No. 2, p. 711—718 (1978).
See also *18.1*, p. *280*.

J.-J. Pilloud and C. Jaccard
*ESR of (X_2^-) Centers in CsBr:Tl^+ and CsCl:Tl^+
(b) Vol. **90**, No. 2, p. K161—K164 (1978).
See also 22.5.2, p. 486.

A. K. Moskalev, I. A. Belobrova, I. P. Aleksandrova, Sh. Sawada, and Y. Shiroishi
*Investigation of the Incommensurate Phase of Rb_2ZnCl_4 by the NQR Method
(a) Vol. **50**, No. 2, p. K157—K159 (1978).
See also *14.4.1*, p. *257*; 22.8, p. 509.

G. Götz, W. Karthe, B. Schnabel, and N. Sobolev
*ESR Line Shape of Amorphous Centres in Ion Implanted Silicon
(a) Vol. **50**, No. 2, p. K209—K212 (1978).
See also 11, p. 132; 22.1.2, p. 440.

20. Optical Properties

Original Papers and Short Notes

V. I. GREBENNIKOV, O. B. SOKOLOV, and E. A. TUROV
Effects of Correlation Interaction of Electrons with Deep Holes in X-Ray Spectra of Metals
(b) Vol. **85**, No. 1, p. 127–134 (1978).
See also *13*, p. *142*; 21, p. 352.

C.-O. ALMBLADH and P. MINNHAGEN
On the Thermal Broadening of Core Levels in Metals Due to Many-Electron Effects
(b) Vol. **85**, No. 1, p. 135–143 (1978).
See also *13*, p. *142*; 17, p. 269; 21, p. 352.

K. S. ALEKSANDROV, I. P. ALEKSANDROVA, V. F. SHABANOV, V. I. YUZVAK, Y. Z. NOZIK, and L. I. FIKIN
Mechanism of Ferroelectric Phase Transition in $NaNH_4SeO_4 \cdot 2\ H_2O$
(a) Vol. **45**, No. 1, p. 53–62 (1978).
See also 1.2, p. 15; *14.4.1*, p. *253*; 19, p. 303; 22.8.1, p. 510.

R. ETHIRAJ, V. G. KRISHNA MURTY, and K. G. BANSIGIR
*Piezo-Optic Birefringence in Mixed Crystals of KCl–KBr
(a) Vol. **45**, No. 1, p. 83–87 (1978).
See also 12.2, p. 138; 22.5.4, p. 489.

Č. KOŇÁK, L. BUDÍK, and L. NOVÁK
Electro-Optical Properties of NH_4Cl Single Crystals near Their Order–Disorder Phase Transition
(a) Vol. **45**, No. 1, p. K63–K66 (1978).
See also *1.2*, p. *16*; 14, p. 203; 22.5, p. 474.

R. ENDERLEIN, B. SEIFERT, and F. BECHSTEDT
Electronic Elementary Excitations in the Far UV-Spectral Region. Core Level Exciton and Core Hole Polarization
(b) Vol. **85**, No. 2, p. 569–576 (1978).
See also *13.1*, p. *151*; *13.5.1*, p. *190*; 22.1, p. 422.

F. LÜTY and A. DIAZ-GONGORA
*Electro-Birefringence from Paraelectric Defects in Cubic Crystals
(b) Vol. **85**, No. 2, p. 693–706 (1978).
See also 10.2, p. 109; *14*, p. *203*; 22.5.2, p. 479.

W. I. FRIESEN, D. B. LITVIN, and B. BERGERSEN
*Theory of the X-Ray and Optical Properties of Polysulphur Nitride
(b) Vol. **85**, No. 2, p. 715–722 (1978).
See also *13*, p. *143*; 22, p. 409.

W. LOCHMANN
Phonon-Assisted Auger Recombination in Indirect Gap Semiconductors
(a) Vol. **45**, No. 2, p. 423–432 (1978).
See also *6*, p. *66*; *13.1*, p. *152*; 22.1.1, p. 424; 22.1.2, p. 430.

E. I. Shtyrkov and V. V. Samartsev
*Dynamic Holograms on the Superposition States of Atoms
(a) Vol. **45**, No. 2, p. 647—655 (1978).
See also *7*, p. *80*.

A. Diaz-Gongora and F. Lüty
*Kerr Effect and Birefringence from Paraelectric CN^- Defects in Nine Alkali Halides
(b) Vol. **86**, No. 1, p. 127—136 (1978).
See also 10.2, p. 110; *14*, p. *203*; 22.5.2, p. 479.

V. A. Antonov, P. A. Arsenev, and D. S. Petrova
Analysis of Spectral and Structural Regularities of Neodymium Ions in Garnet Monocrystals
(a) Vol. **46**, No. 1, p. 55—58 (1978).
See also *1.4*, p. *29*; 22.8.2, p. 518.

Vo Hong Anh and Nguyen Nhu Dat
Parametric Excitation in Kane's Semiconductors
(b) Vol. **86**, No. 2, p. 585—591 (1978).
See also *6*, p. *69*; *13.5.2*, p. *198*; 22, p. 411.

I. M. Dykman and P. M. Tomchuk
Electromagnetic Waves in Inhomogeneous Semiconductors with Non-Equilibrium Carriers
(b) Vol. **86**, No. 2, p. 773—780 (1978).
See also *13*, p. *144*; *14.3*, p. *220*; 22, p. 412.

K. Berndt, O. Brümmer, and U. Marx
Study of the Fe 2p States by Means of Photoelectron Spectroscopy in Iron and Iron Alloys
(b) Vol. **86**, No. 2, p. K93—K95 (1978).
See also *13*, p. *144*; 21.1, p. 372; 21.1.1, p. 386.

P. Ziesche, H. Wonn, Ch. Müller, V. V. Nemoshkalenko, and V. P. Krivitskii
Theoretical and Experimental Investigation of the Electronic Structure of the B2-Compound FeAl
(b) Vol. **87**, No. 1, p. 129—134 (1978).
See also *13.1*, p. *154*; 21, p. 357; 21.1.1, p. 386.

P. Rennert, Th. Dörre, and U. Gläser
Theoretical Investigation of the Anisotropy of Beryllium by Polarized K-Emission, APS, and Compton Profiles
(b) Vol. **87**, No. 1, p. 221—226 (1978).
See also *13.1*, p. *154*; 21.3, p. 394.

S. Guha and S. Ghosh
*Hybrid Mode Interaction in Inhomogeneous Semiconductor Plasmas
(b) Vol. **87**, No. 1, p. K9—K12 (1978).
See also *14.3*, p. *220*; 22, p. 413.

V. M. Fomin and E. P. Pokatilov
*On the Representation of the Non-Stationary Density Matrix of an Electron–Phonon System by a Functional Integral
(b) Vol. **87**, No. 2, p. 725—732 (1978).
See also 6, p. 72; 14.3.4, p. 242.

B. K. Agarwal and R. K. Johri
*X-Ray K-Absorption Structure of Zirconium Metal and Its Compounds
(b) Vol. **88**, No. 1, p. 309—313 (1978).
See also *13.1*, p. *156*; 21, p. 360.

P. R. Sarode
*Correlation between the Chemical Shifts in X-Ray Absorption Spectra of Ionic Solids and Contribution to the Chemical Bond
(b) Vol. **88**, No. 1, p. K35—K38 (1978).
See also *4*, p. *58*; 22.5.2, p. 482.

E. N. Mgbenu and A. Maduemezia
*X-Ray Emission Spectrum of Metallic Lithium
(b) Vol. **88**, No. 2, p. 805—808 (1978).
See also *13.1*, p. *157*; 13.2, p. 162; 21.2, p. 393.

F. R. Kessler, J. Metzdorf, and W. Zurek
*Interband Faraday Ellipticity in Germanium
(b) Vol. **89**, No. 1, p. 203—211 (1978).
See also 13.1, p. 158; 13.5.1, p. 194; *18*, p. *275*; 22.1.1, p. 427.

C. Mande and A. Pimpale
Evidence for Surface Plasmons in X-Ray Emission Bands
(b) Vol. **89**, No. 1, p. K21—K24 (1978).
See also *13.3*, p. *167*; *13.5.2*, p. *198*; 21.6, p. 404.

H. E. Müser, W. Kuhn, and J. Albers
Optical Investigation of the Cubic–Tetragonal Phase Transition in $BaTiO_3$ Single Crystals
(a) Vol. **49**, No. 1, p. 51—58 (1978).
See also *1.2*, p. *22*; *14.4.1*, p. *256*; 22.8.1, p. 515.

A. Haug, D. Kerkhoff, and W. Lochmann
Calculation of Auger Coefficients for III–V Semiconductors with Emphasis on GaSb
(b) Vol. **89**, No. 2, p. 357—365 (1978).
See also *13.1*, p. *159*; 22.2, p. 444; 22.2.3, p. 454.

E. Wimmer, P. Weinberger, A. Kosakov, and G. Leonhardt
*On the Electronic Structure of As_2O_3, Sb_2O_3, and As_2S_3
(b) Vol. **89**, No. 2, p. 619—624 (1978).
See also 2, p. 50; *13*, p. *148*; 22.6, p. 496.

S. Guha and S. Ghosh
*Drift Wave Instability in Solid-State Plasmas
(b) Vol. **89**, No. 2, p. K171—K174 (1978).
See also *14.3*, p. *224*; 22, p. 418.

V. V. Samartsev and R. G. Usmanov
*Dependence of the Primary and Stimulated Photon Echo Intensity on the Pulse Parameters and the Sample Temperature
(a) Vol. **49**, No. 2, p. 789—795 (1978).

H. U. Habermeier, R. Aoki, and H. Kronmüller
The Quantitative Determination of Flux-Density Gradients in Type II Superconductors by Means of the Magneto-Optical Faraday Effect
(a) Vol. **50**, No. 1, p. 187—194 (1978).
See also *14.2*, p. *215*; 18, p. 275; 21, p. 365.

N. A. Zakharov, V. S. Krikorov, E. F. Kustov, and S. Yu. Stefanovich
*New Non-Linear Crystals in the $A_2B_2O_7$ Series
(a) Vol. **50**, No. 1, p. K13—K16 (1978).
See also 3, p. 56; *14.4.1*, p. *257*; 22.8, p. 508.

M. Urner-Wille, T. S. te Velde, and P. G. van Engen
*Kerr-Effect Enhancement in Amorphous GdFe Films
(a) Vol. **50**, No. 1, p. K29—K31 (1978).
See also 2, p. 52; *18.3*, p. *297*; 21.1.1, p. 390; 21.4, p. 398.

K. Berndt, U. Marx, and O. Brümmer
*Electronic Structure of FeNi Alloys Studied by Means of X-Ray Photoelectron Spectroscopy (I): Valence Bands
(b) Vol. **90**, No. 2, p. 487—496 (1978).
See also *13.1*, p. *160*; 21.1, p. 381; 21.1.1, p. 390.

I. R. Entin
*Theoretical and Experimental Study of X-Ray Acoustic Resonance in Perfect Silicon Crystals
(b) Vol. **90**, No. 2, p. 575—584 (1978).
See also 22.1.2, p. 439.

G. Schroeder, R. Schäfer, and H. Kronmüller
*Magneto-Optical Investigation of the Domain Structure of Amorphous $Fe_{80}B_{20}$ Alloys
(a) Vol. **50**, No. 2, p. 475—481 (1978).
See also 2, p. 52; *18.2*, p. *292*; 21.1.1, p. 390.

N. N. Evtikhiev, N. A. Ekonomov, A. R. Krebs, N. A. Zamyatina, A. S. Komalov, V. G. Pinko, and L. V. Ivaeva
*FePt and CoPt Alloy Films for Magneto-Optic Storage
(a) Vol. **50**, No. 2, p. K153—K155 (1978).
See also *18.2.1*, p. *294*; 21.1, p. 383; 21.1.1, p. 391.

20.1 Spectra. Optical Constants (X-Ray Spectra See 20)

Review Articles

R. Suryanarayanan
*Valence Related Optical and Other Studies of Sm and Tm Chalcogenides
(b) Vol. **85**, No. 1, p. 9—43 (1978).
See also 13.1, p. 150; 14.1, p. 205; 14.3, p. 217; 22.7, p. 501.

P. C. Mathur, R. Shyam, and Sushil Jain
Transport Phenomena in III–V Compound Semiconductors
(a) Vol. **50**, No. 1, p. 11—40 (1978).
See also 6, p. 64; *13.1*, p. *150*; 13.4, p. 169; *14.3*, p. *217*; 22.2, p. 443.

Original Papers and Short Notes

V. V. Hizhnyakov and A. V. Sherman
On the Theory of Resonant Secondary Radiation of Excitons Weakly Interacting with Phonons
(b) Vol. **85**, No. 1, p. 51—61 (1978).
See also *6*, p. *64*; *13.5.1*, p. *189*; 20.3, p. 341; 22, p. 408.

G. C. Aers, A. D. Boardman, and P. Clark
*A Computational Study of the Correlation of Attenuated Total Reflection Minima with Surface Polariton Modes
(b) Vol. **85**, No. 1, p. 171—178 (1978).
See also 13.5, p. 187; *18*, p. *273*; 21.1, p. 368.

G. Neumann and H. Vogt
Rayleigh Wing Scattering in Disordered Sodium Nitrate
(b) Vol. **85**, No. 1, p. 179—184 (1978).
See also *1.2*, p. *15*; 2, p. 42; *6*, p. *64*; 22.8.1, p. 510.

D. Suisky, R. Boyn, R. Enderlein, and B. Biering
*First-Harmonic Electromodulation in Uniaxial Crystals
(b) Vol. **85**, No. 1, p. 217—224 (1978).
See also 13.4, p. 169; *14.3.4*, p. *240*; 22.4.1, p. 463.

O. Goede, H.-J. Krug, and A. Horst
*Deformation Effect on Exciton Excitation Spectra in CdS
(b) Vol. **85**, No. 1, p. 239—245 (1978).
See also *13.5.1*, p. *189*; 22.4.1, p. 463.

P. Kleinert and F. Bechstedt
*Influence of Mass Defect Disorder on Resonance Raman Scattering
(b) Vol. **85**, No. 1, p. 253—260 (1978).
See also *6*, p. *64*; 22, p. 409.

K. Hümmer and P. Gebhardt
*Angular Dependence of the Reflection Spectra and Directional Dispersion of the Anisotropic Exciton Polaritons in ZnO
(b) Vol. **85**, No. 1, p. 271–282 (1978).
See also *13.5.1*, p. *189*; 22.4, p. 461.

R. Alcalá, V. M. Orera, and J. Beamonte
*A New Type of F_3-Centre in Additively Coloured CaF_2
(b) Vol. **85**, No. 1, p. 283–290 (1978).
See also *10.2*, p. *109*; 22.5.3, p. 487.

G. S. Agarwal and H. D. Vollmer
Surface Polariton Effects in Spontaneous Emission (II): Effects of Spatial Dispersion
(b) Vol. **85**, No. 1, p. 301–310 (1978).
See also *13.3*, p. *164*; 13.5.2, p. 198; 21, p. 352; 22.4.1, p. 463.

V. Wagner, G. Dolling, B. M. Powell, and G. Landwehr
*Lattice Vibrations of Bi_2Te_3
(b) Vol. **85**, No. 1, p. 311–317 (1978).
See also *6*, p. *65*; 22.7, p. 501.

K. A. Aristova, C. Trallero Giner, I. G. Lang, and S. T. Pavlov
Scattering and Decay of Hot Excitons Interacting with Longitudinal Optical Phonons
(b) Vol. **85**, No. 1, p. 351–357 (1978).
See also *6*, p. *65*; *13.5.1*, p. *189*.

F. E. Faradev, N. M. Gasanly, B. N. Mavrin, and N. N. Melnik
*Raman Scattering in Some III–VI Layer Single Crystals
(b) Vol. **85**, No. 1, p. 381–386 (1978).
See also *6*, p. *65*; 22.3, p. 457.

P. J. Alonso and R. Alcalá
*Optical Absorption of X-Irradiated CaF_2: Co
(b) Vol. **85**, No. 1, p. K23–K26 (1978).
See also *11*, p. *123*; 22.5.3, p. 487.

V. G. Golubev, E. I. Georgitse, and V. I. Ivanov-Omskii
*Photoconductivity of HgTe in Magnetic Field at 337 μm
(b) Vol. **85**, No. 1, p. K39–K41 (1978).
See also *16*, p. *262*; 18, p. 273; 22.4.3, p. 470.

W. Hörig, H. Neumann, B. Schumann, and G. Kühn
*Optical Properties of $CuGaSe_2$ near and above the Fundamental Absorption Edge
(b) Vol. **85**, No. 1, p. K57–K61 (1978).
See also *13.1*, p. *150*; 22.8, p. 503.

J. Gazsó and J. Hajtó
*Self-Controlled Laser Beam Chopping Effect in $GeSe_2$ Thick Films
(a) Vol. **45**, No. 1, p. 181–186 (1978).
See also 2, p. 42; 11, p. 123; 22.7, p. 501.

HJ. BERNHARDT
*Studies of the Colour of Lead Molybdate Crystals
(a) Vol. 45, No. 1, p. 353—357 (1978).
See also 13.4, p. 170; *20.3*, p. *341*; 22.8.1, p. 510.

I. B. BARKAN, S. I. MARENNIKOV, and M. V. ENTIN
*Dynamic Optical Storage in $LiNbO_3$ Crystals
(a) Vol. 45, No. 1, p. K17—K22 (1978).
See also *13.4*, p. *170*; 14.4.1, p. 253; 16, p. 262; 22.8.1, p. 510.

R. ROSEI, E. COLAVITA, S. MODESTI, and N. NISTICÒ
Electronic Structure of Dilute α-Phase Ag–Cd Alloys from Thermoreflectance and XPS Measurements
(b) Vol. 85, No. 2, p. 445—452 (1978).
See also *13.1*, p. *151*; 21, p. 354; 21.6, p. 401.

M. C. BONNET, E. ANTIC-FIDANCEV, P. PORCHER, M. LEMAITRE-BLAISE, and P. CARO
*The Optical Spectra of Rare-Earth Cobalticyanides
(b) Vol. 85, No. 2, p. 587—596 (1978).
See also *20.3*, p. *342*; 22.8, p. 503.

S. WINTER
*A Comment on the Optical Absorption in Germanium with 60°-Dislocations
(b) Vol. 85, No. 2, p. K95—K98 (1978).
See also 10.2, p. 110; *13.4*, p. *171*; 22.1.1, p. 424.

G. P. SRIVASTAVA
*Electronic and Optical Properties of Crystalline and Amorphous Si
(b) Vol. 85, No. 2, p. K121—K126 (1978).
See also 2, p. 43; *13.1*, p. *152*; 22.1.2, p. 429.

A. CZAPLA, M. JACHIMOWSKI, E. KUSIOR, and J. SZCZYRBOWSKI
*Optical Properties of $Pb_xSn_{1-x}O_2$ Films
(a) Vol. 45, No. 2, p. 537—542 (1978).
See also 22.6.1, p. 499.

A. P. RADLINSKI
On the Properties of Cobalt Impurities in Zinc Selenide Crystals (II): The Optical Properties in Near-Infrared. $^4A_2(^4F) \leftrightarrow {}^4T_2(^4F)$ Transitions
(b) Vol. 86, No. 1, p. 41—46 (1978).
See also *13.4*, p. *172*; 20.3, p. 342; 22.4.2, p. 468.

J. C. CASTRO and M. DE SOUZA
The OCN^- Vibrational Spectrum and the Phase Transitions in KCN_xKCl_{1-x} Single Crystals
(b) Vol. 86, No. 1, p. 137—144 (1978).
See also *1.2*, p. *16*; *6*, p. *67*; 12.2, p. 138; 22.5, p. 474; 22.5.4, p. 489.

VU DUY PHACH, A. BIVAS, B. HÖNERLAGE, and J. B. GRUN
Two-Photon Resonant Raman Scattering via Biexcitons
(b) Vol. 86, No. 1, p. 159—168 (1978).
See also *13.5.1*, p. *190*; 20.3, p. 342; 22.5, p. 474.

M. SCHWAB
Phase Transition in the Mixed Crystals $Tb_pY_{1-p}AsO_4$
(b) Vol. **86**, No. 1, p. 195—203 (1978).
See also *1.2*, p. *16*; 13.4, p. 172; 18.4, p. 298; 22.8.1, p. 511.

W. MÜLLER, M. OVERHAMM, G. BAUER, E. GORNIK, P. GROSSE, H. W. PÖTZL, and M. RAUTENBERG
*On the Influence of Background Radiation on Landau Emission Spectra
(b) Vol. **86**, No. 1, p. 205—213 (1978).
See also 13.1, p. 152; *14.3.4*, p. *241*; *20.3*, p. *343*; 22.2.3, p. 452.

J. BERGMANN, K. KNEIPP, and H. E. PONATH
*Raman-Induced Kerr Effect with Elliptical Polarization in Optically Active and Anisotropic Crystals
(b) Vol. **86**, No. 1, p. 215—223 (1978).
See also *6*, p. *67*; 22.8.1, p. 511.

M. PÖHLER and G. STAUPENDAHL
*Investigation of the Simultaneous Absorption of Three CO_2-Laser Photons by Tellurium
(b) Vol. **86**, No. 1, p. 225—230 (1978).
See also *13.5.1*, p. *190*; 22.1.3, p. 441.

K. FUKUSHI, M. NIPPUS, and R. CLAUS
*The Origin of the Anomalous TO-Doublet in Cubic CuCl Studied by Light Scattering
(b) Vol. **86**, No. 1, p. 257—262 (1978).
See also *6*, p. *67*; 22.5, p. 474.

B. JENSEN
*The Infrared Absorption Spectrum of n-Type GaAs
(b) Vol. **86**, No. 1, p. 291—301 (1978).
See also *14.3*, p. *218*; 22.2.1, p. 445.

J. HUMLÍČEK
*On the Nature of E_1-Transitions in Germanium
(b) Vol. **86**, No. 1, p. 303—310 (1978).
See also 13.1, p. 153; *14.3.4*, p. *241*; 22.1.1, p. 424.

R. STREHLOW
*On the No-Phonon Structure in Mixed Crystals with Indirect Band Structure
(b) Vol. **86**, No. 1, p. 311—318 (1978).
See also 6, p. 67; 13.1, p. 153; 13.5.1, p. 190; 22.2.4, p. 456.

R. GRÜNDLER, K. BREUTER, and W. TEWS
*Optical Properties of Amorphous and Polycrystalline BeO Thin Films from Electron Energy Loss Measurements
(b) Vol. **86**, No. 1, p. 329—338 (1978).
See also 2, p. 43; 13.1, p. 153; *14.4*, p. *248*; 22.6, p. 492.

V. GOMATHY, C. BASU, and U. S. GHOSH
Configuration Interaction and Intensity of Crystal Field Spectra of the Tetrahedral Ni^{2+} Ion
(b) Vol. **86**, No. 1, p. 379—384 (1978).
See also *13.4*, p. *172*; 22.4, p. 461.

I. S. Gorban, V. A. Gorinya, R. A. Dashkovskaya, V. I. Lugovoi, A. P. Makovetskaya, and I. I. Tichina
*One- and Two-Phonon States in Tetragonal ZnP_2 and CdP_2 Crystals
(b) Vol. **86**, No. 1, p. 419—423 (1978).
See also *6*, p. *67*; 22.3, p. 457.

G. M. Georgiev and J. S. Nikolov
*Spontaneous Parametric Scattering of Light in Ammonium Bioxalate Monohydrate
(b) Vol. **86**, No. 1, p. 431—437 (1978).
See also *6*, p. *68*; 22.9, p. 523.

M. Scrocco and L. Mattioli
*The Structure of $Sr(NO_3)_2$: A Single Crystal Raman Spectra Study
(b) Vol. **86**, No. 1, p. K1—K5 (1978).
See also *6*, p. *68*; 22.8.1, p. 511.

D. Genzow, K. H. Herrmann, H. Kostial, I. Rechenberg, and A. E. Yunovich
*Interband Absorption Edge in $Pb_{1-x}Sn_xTe$ Epitaxial Layers
(b) Vol. **86**, No. 1, p. K21—K25 (1978).
See also 22, p. 410.

I. J. Dayawansa and C. F. Bohren
*The Effect of Substrate and Aggregation on Infrared Extinction Spectra of MgO Particles
(b) Vol. **86**, No. 1, p. K27—K30 (1978).
See also 22.6, p. 492.

N. M. Gasanly, B. N. Mavrin, Kh. E. Sterin, V. I. Tagirov, and Z. D. Khalafov
*Raman Study of Layer $TlGaS_2$, β-$TlInS_2$, and $TlGaSe_2$ Crystals
(b) Vol. **86**, No. 1, p. K49—K53 (1978).
See also *6*, p. *68*; 22.8, p. 504.

R. Zimmermann
*Final State Interactions in the Gain and Absorption Spectra of Electron–Hole Liquids
(b) Vol. **86**, No. 1, p. K63—K66 (1978).
See also *13.5*, p. *187*; 22.2.1, p. 445.

P. G. Basanov, R. A. Zhitnikov, and V. A. Khramtsov
*EPR and Optical Absorption of Zn^+ Ions in NaCl Crystals
(b) Vol. **86**, No. 1, p. K67—K70 (1978).
See also *19*, p. *305*; 22.5.2, p. 479.

W. K. Unger, H. Meuth, J. C. Irwin, and H. Pink
*Photoluminescence and Resonant Raman Scattering in $CdIn_2S_4$ and $ZnIn_2S_4$
(a) Vol. **46**, No. 1, p. 81—87 (1978).
See also *20.3*, p. *343*; 22.8, p. 504.

P. Hennl
*Optical Properties of Chalcogen–Vacancy Centres in Alkali Halides
(a) Vol. **46**, No. 1, p. 147—152 (1978).
See also *10.2*, p. *111*; 20.3, p. 343; 22.5.2, p. 479.

G. A. ANDREEV, M. HARTMANOVÁ, and V. A. KLIMOV
Precipitation of Pb Impurity from the Lattice of NaCl Crystals
(a) Vol. **46**, No. 1, p. 259—265 (1978).
See also *10.2*, p. *111*; 22.5.2, p. 479.

G. C. VEZZOLI
Optical and Electrical Effects during the Solidification of Mercury
(a) Vol. **46**, No. 1, p. K49—K53 (1978).
See also *1.2*, p. *17*; 14.1, p. 206; 21, p. 356.

S. LEFRANT, J. P. BUISSON, M. GHOMI, and L. TAUREL
*First-Order Raman Scattering of CsF Doped with F Centres
(b) Vol. **86**, No. 2, p. 493—496 (1978).
See also *6*, p. *68*; 22.5.2, p. 480.

B. ULRICI and E. JAHNE
*Two-Phonon Infrared Absorption in $In_{1-x}Ga_xP$ Mixed Crystals
(b) Vol. **86**, No. 2, p. 517—526 (1978).
See also *6*, p. *68*; 22.2.2, p. 450; 22.2.4, p. 456.

G. APPOLD, H. PASCHER, R. EBERT, U. STEIGENBERGER, and M. VON ORTENBERG
Determination of the Effective *g*-Value of Electrons in n-Type InSb by Spin-Flip Raman Scattering and Electric-Dipole-Excited Electron Spin Resonance
(b) Vol. **86**, No. 2, p. 557—561 (1978).
See also *13.4*, p. *174*; 19, p. 306; 22.2.3, p. 452.

M. SINYUKOV, R. TROMMER, and M. CARDONA
*Resonant Raman Scattering in InP
(b) Vol. **86**, No. 2, p. 563—568 (1978).
See also *6*, p. *68*; 22.2.2, p. 450.

H. VOGT and G. NEUMANN
*Hyper-Rayleigh Scattering in Disordered Sodium Nitrate
(b) Vol. **86**, No. 2, p. 615—620 (1978).
See also 1.2, p. 17; 6, p. 69; 22.8.1, p. 512.

K. D. NIERZEWSKI, T. TODOROV, and M. GEORGIEV
Room Temperature F–F′ Conversion in KCl Crystals of Various Purity and Method of Coloration
(b) Vol. **86**, No. 2, p. 697—704 (1978).
See also *10.2*, p. *112*; 22.5.2, p. 480.

M. N. ILIEV, E. ANASTASSAKIS, and T. ARAI
*Raman Scattering of $HgCr_2Se_4$
(b) Vol. **86**, No. 2, p. 717—723 (1978).
See also *6*, p. *69*; 18.2, p. 285; 22.8.2, p. 518.

V. HIZHNYAKOV and S. ZAZUBOVICH
*Vacancy-Induced Splitting of Excited States and the Structure of $Sn^{2+}v_c^-$ Centres in Alkali Halides
(b) Vol. **86**, No. 2, p. 733—739 (1978).
See also *13.4*, p. *174*; 22.5.2, p. 480.

W. GĘBICKI and W. NAZAREWICZ
*Far Infrared Reflection Spectra of $Mn_xCd_{1-x}Te$ Mixed Crystals
(b) Vol. **86**, No. 2, p. K135—K138 (1978).
See also *6*, p. *69*; 22.4.4, p. 472.

I. V. BODNAR, A. G. KAROZA, and G. F. SMIRNOVA
*Infrared Reflectivity Spectra of $CuGa_xIn_{1-x}S_2$ Solid Solutions
(b) Vol. **86**, No. 2, p. K171—K174 (1978).
See also *6*, p. *70*; 22.8, p. 504.

R. KIRSCH
*An Automated FSA Ellipsometer System for in Situ Investigations
(a) Vol. **46**, No. 2, p. 459—466 (1978).
See also 14.4, p. 248; 21.6, p. 402.

H. ARNDT, G. SCHMIDT, N. R. IVANOV, and L. A. SHUVALOV
*Electrooptic Switching Anomalies in Ferroelectric TGS
(a) Vol. **46**, No. 2, p. 533—540 (1978).
See also *14.4.1*, p. *254*; 22.9, p. 524.

I. S. EDELMAN and V. D. BAURIN
*Influence of Cr^{3+} Ions on the Magnetooptical Properties of Cobalt Ferrite
(a) Vol. **46**, No. 2, p. K83—K86 (1978).
See also *18.2.1*, p. *293*; 22.8.2, p. 519.

G. SMITH, E. R. VANCE, Z. HASAN, A. EDGAR, and W. A. RUNCIMAN
*A Charge Transfer Mechanism for the Colour of Rose Quartz
(a) Vol. **46**, No. 2, p. K135—K140 (1978).
See also 22.6, p. 492.

L. E. OLIVEIRA, P. MURILO OLIVEIRA, and B. MAFFEO
*F Centers in Alkaline-Earth Fluorides. Inadequacy of the Muffin-Tin Approximation
(b) Vol. **87**, No. 1, p. 25—30 (1978).
See also *10.2*, p. *112*; 22.5.3, p. 487.

A. R. HASSAN
Exciton Molecule Creation by Pure and Phonon-Assisted Two-Photon Transitions in Solids
(b) Vol. **87**, No. 1, p. 31—40 (1978).
See also *13.5.1*, p. *191*; 22.5, p. 475.

A. KUROWSKA
*Theory of Free Carrier Absorption in Mixed Semiconductors with Composition Fluctuations
(b) Vol. **87**, No. 1, p. 69—78 (1978).
See also *13.1*, p. *154*; 22.4.4, p. 472.

V. M. SHATALOV, V. A. TELEZHKIN, and K. B. TOLPYGO
Inadequacy of the Direct Interband Transition Scheme Applied to Valence Semiconductor Fundamental Absorption Spectra
(b) Vol. **87**, No. 1, p. 79—86 (1978).
See also *13.1*, p. *154*; 13.5.1, p. 191; 22, p. 412.

H. L. Althaus, G. Weiser, and S. Nagel
*Optical Spectra and Band Structure of As_2Se_3
(b) Vol. 87, No. 1, p. 117—128 (1978).
See also *13.1*, p. *154*; 22, p. 413.

A. K. Kostin, V. V. Savelev, and A. V. Vannikov
Free Charges and Excitons in Stilbene Single Crystals Excited by Electron Pulses
(b) Vol. 87, No. 1, p. 255—259 (1978).
See also *13.5.1*, p. *191*; 20.3, p. 344; 22.9, p. 524.

S. Nikitine, S. G. Elkomoss, and C. Schwab
*Study of Linewidth and Exciton–Impurity Scattering in Cu_2O
(b) Vol. 87, No. 1, p. 261—272 (1978).
See also *13.5.1*, p. *191*; 22.6, p. 493.

Ch. Diaine, J. L. Gisclon, and J. Dupuy
Colloid Centres in AgCl Induced by Heavy Ion Implantation
(b) Vol. 87, No. 1, p. 293—300 (1978).
See also *10.2*, p. *113*; *11*, p. *126*; 22.5.1, p. 477.

V. Nekvasil
*The Crystal Field for Nd^{3+} in Garnets
(b) Vol. 87, No. 1, p. 317—323 (1978).
See also *13.4*, p. *175*; 22.8.2, p. 519.

S. L. A. de Queiroz, B. Koiller, B. Maffeo, and H. S. Brandi
*A Cluster-Bethe Lattice Treatment for the F-Center in Alkali Halides
(b) Vol. 87, No. 1, p. 351—359 (1978).
See also *13.4*, p. *175*; 22.5.2, p. 481.

A. S. Filipchenko, L. P. Bolshakov, and A. Naurizbaev
*The Moss-Burstein Effect in n-Type Indium Arsenide Crystals Doped with Different Donor Impurities
(b) Vol. 87, No. 1, p. K1—K3 (1978).
See also 13.1, p. 155; 22.2.1, p. 446.

A. Król, M. Kozielski, and W. Nazarewicz
*Localized Vibrational Modes of Al–Cu Complexes in Zinc Sulphide
(b) Vol. 87, No. 1, p. K25—K28 (1978).
See also *6*, p. *70*; 22.4.1, p. 464.

V. Riede, Hoang Xuan Nguyen, H. Neumann, H. Sobotta, and F. Levy
*Refractive Index of GaS in the Far Infrared
(b) Vol. 87, No. 1, p. K65—K67 (1978).
See also 22.3, p. 458.

L. Nikolova and T. Todorov
*Photoinduced Reorientation of F_A Centres in KCl:Na near Room Temperature
(a) Vol. 47, No. 1, p. 213—217 (1978).
See also *10.2*, p. *113*; 22.5.2, p. 481.

C. E. Wickersham and J. E. Greene
*The Effect of Substrate Bias on the Electrical and Optical Properties of In_2O_3 Films Grown by RF Sputtering
(a) Vol. **47**, No. 1, p. 329–337 (1978).
See also 13.4, p. 175; *14.3.1*, p. *226*; 22.6, p. 93.

W. Thielemann, P. Walther, and H. Neumann
*Evidence for Different Absorption Tails in the Surface Depletion Layer and in the Bulk of n-Type GaAs
(a) Vol. **47**, No. 1, p. K35–K38 (1978).
See also 22.2.1, p. 446.

B. I. Swanson
*Displacive Phase Transformations in K_2SnBr_6
(a) Vol. **47**, No. 1, p. K95–K98 (1978).
See also *1.2*, p. *19*; 6, p. 71; 22.8, p. 505.

C. Razzetti and P. P. Lottici
*Resonant Raman Scattering in As_2S_3 Amorphous Thin Films
(b) Vol. **87**, No. 2, p. 479–483 (1978).
See also *2*, p. *45*; 22, p. 413.

R. Droste, R. Gmeiner, and R. Geick
*Electronic Absorption of $SrCl_2:Fe^{2+}$ and $SrCl_2:Co^{2+}$
(b) Vol. **87**, No. 2, p. 543–552 (1978).
See also *13.4*, p. *176*; 22.5.3, p. 487.

R. Kaiser, W. Spengler, S. Schicktanz, and C. Politis
*Raman Spectra and Superconductivity of Various Phases of a High-T_c Superconductor: NbN
(b) Vol. **87**, No. 2, p. 565–573 (1978).
See also *6*, p. *71*; 14.2, p. 214.

M. Kalm and Ch. Uihlein
*Investigations on the Temperature-Dependent TPA Linewidth of the Z_3-Exciton in CuCl
(b) Vol. **87**, No. 2, p. 575–580 (1978).
See also 6, p. 71; *13.5.1*, p. *192*; 22.5, p. 475.

G. C. Kuczynski, P. E. Dietz, W. H. Hamill, and J. H. Jischke
*H_2O Related Centers in Potassium Chloride
(b) Vol. **87**, No. 2, p. 581–588 (1978).
See also *10.2*, p. *113*; 22.5.2, p. 481.

S. W. Biernacki
*Jahn-Teller Coupling of Cr^{2+} Ion in ZnS, ZnSe, and ZnTe
(b) Vol. **87**, No. 2, p. 607–612 (1978).
See also *13.4*, p. *176*; 22.4.1, p. 464; 22.4.2, p. 468; 22.4.3, p. 470.

V. A. Tyagai, O. V. Snitko, A. M. Evstigneev, and A. N. Krasiko
*Surface Phenomena in the Polarization-Dependent Electroreflectance Spectra of Si
(b) Vol. **87**, No. 2, p. 613–618 (1978).
See also 13.3, p. 165; *14.3.4*, p. *242*; 22.1.2, p. 433.

R. Ruppin
*Validity Range of the Maxwell-Garnett Theory
(b) Vol. 87, No. 2, p. 619—624 (1978).
See also 6, p. 71; 22.2.1, p. 446; 22.6, p. 494.

W. Szuszkiewicz, A. M. Witowski, and M. Grynberg
The Dynamic Dielectric Function in HgSe and HgTe
(b) Vol. 87, No. 2, p. 637—645 (1978).
See also *13.1*, p. *155*; 14.4, p. 249; 22.4.2, p. 468; 22.4.3, p. 470.

I. Rückmann, J. Puls, and J. Voigt
*Induced Absorption of CdS Single Crystals in the Energy Ranges of Exciton and Exciton–Biexciton Transitions
(b) Vol. 87, No. 2, p. K111—K114 (1978).
See also *13.5.1*, p. *192*; 22.4.1, p. 464.

K. R. Allakhverdiev, E. A. Vinogradov, N. N. Melnik, M. A. Nizametdinova, E. Yu. Salaev, and R. M. Sardarly
*Vibrational Properties of TlSe
(b) Vol. 87, No. 2, p. K115—K117 (1978).
See also *6*, p. *72*; 22.3, p. 458.

I. L. Chistyi, L. Csillag, V. F. Kitaeva, N. Kroó, and N. N. Sobolev
*Investigation of Molecular Light Scattering in $PbMoO_4$
(a) Vol. 47, No. 2, p. 609—615 (1978).
See also *7*, p. *81*; 22.8.1, p. 514.

S. B. S. Sastry and K. Balasubramanyam
*Studies on Tin Centres in RbCl Crystals
(a) Vol. 47, No. 2, p. 711—716 (1978).
See also *10.2*, p. *114*; 11, p. 127; 22.5.2, p. 481.

S. Radhakrishna and T. Bhaskar Rao
*Optical Absorption and ESR Studies of Nickel Ions in Sodium Zinc Sulphate Tetrahydrate
(a) Vol. 47, No. 2, p. 723—729 (1978).
See also 13.4, p. 176; *19*, p. *309*; 22.8.1, p. 514.

L. Risch
The Influence of Surface Layers on Photoreflectance in the Exciton Region of CdS
(b) Vol. 88, No. 1, p. 111—116 (1978).
See also *13.3*, p. *166*; 13.5.1, p. 192; 22.4.1, p. 464.

V. M. Shatalov
Quasi-Molecular Model Interpretation of the Resonances in the Fundamental Absorption Region of Valence Semiconductors
(b) Vol. 88, No. 1, p. 129—133 (1978).
See also *13.1*, p. *155*; 13.5.1, p. 192; 22, p. 414.

W. Eberhardt G. Kalkoffen, C. Kunz, D. Aspnes, and M. Cardona
Photoemission Studies of 2p Core Levels of Pure and Heavily Doped Silicon
(b) Vol. 88, No. 1, p. 135—143 (1978).
See also *13.1*, p. *156*; 17, p. 270; 22.1.2, p. 434.

I. Balslev
The Excitonic Surface Potential in Semiconductors
(b) Vol. 88, No. 1, p. 155—161 (1978).
See also *13.5.1*, p. *192*; 22.4, p. 461.

F. Bechstedt and D. Haus
*Theory of Second-Order Resonance Raman Scattering in the Case of Strong Excitonic Effects
(b) Vol. 88, No. 1, p. 163—171 (1978).
See also *6*, p. *73*; 13.5.1, p. 192; 22.4.1, p. 465.

R. Enderlein, R. E. Hummel, J. B. Andrews, R. J. Nastasi-Andrews, and C. W. Shanley
*Interpretation of Compositional Modulation Spectra to Determine Some Optical Properties of Alloys
(b) Vol. 88, No. 1, p. 173—178 (1978).
See also *13.1*, p. *156*; 21.1, p. 375.

V. Lemos, F. Cerdeira, and M. Cardona
*Atomic Motions Corresponding to Zone Center Phonons in Paratellurite
(b) Vol. 88, No. 1, p. 199—206 (1978).
See also *6*, p. *73*; 22.6, p. 494.

J. Popielawski
On the Optical Criterion for the Mott Transition
(b) Vol. 88, No. 1, p. 241—245 (1978).
See also *13.5* ,p. *187*.

T. Iwasaki, T. Goto, and Y. Nishina
Oscillator Strength of Excitons in Highly Excited PbI_2
(b) Vol. 88, No. 1, p. 289—293 (1978).
See also *13.5.1*, p. *193*; 20.3, p. 345; 22.5, p. 475.

B. Prevot, C. Schwab, and B. Dorner
*Phonon Dispersion in Red HgI_2
(b) Vol. 88, No. 1, p. 327—333 (1978).
See also *6*, p. *73*; 7, p. 81; 22.5, p. 475.

M. Baleva
*Thermoreflectance of PbTe
(b) Vol. 88, No. 1, p. 335—339 (1978).
See also *13.1*, p. *156*; 22, p. 414.

K. Allakhverdiev, R. Sardarly, F. Wondre, and J. F. Ryan
*Raman and Infrared Spectra of $TlGaSe_2$
(b) Vol. 88, No. 1, p. K5—K9 (1978).
See also *6*, p. *73*; 22.8, p. 506.

K. Germanova, K. Marinova, and S. Stoyanov
Effect of Surface Layers on the Free Exciton Reflectance of ZnTe
(b) Vol. 88, No. 1, p. K19—K22 (1978).
See also *13.5.1*, p. *193*; 22.4.3, p. 470.

A. Watterich
Z_4 Like Centre in NaCl Doped by Divalent Metals: Mg, Mn, Ni, Cd, or Pb
(b) Vol. 88, No. 1, p. K51—K54 (1978).
See also *10.2*, p. *115*; 22.5.2, p. 482.

G. Ya. Selyutin
*Anomaly of Width and Lineshape of Light Absorption in Magnetic Crystals
(b) Vol. 88, No. 1, p. K79—K81 (1978).
See also 18.1, p. 278; 18.2, p. 288.

G. C. Vezzoli
Peaked Electrical Conductivity in Liquid Bismuth and Associated Optical Scattering
(a) Vol. 48, No. 1, p. K75—K77 (1978).
See also *2*, p. *47*; *14.1*, p. *209*; 21.7, p. 407.

P. Doubrava, V. L. Averyanov, and J. Škácha
*Spectral Characteristics of Photodarkening of a-$As_{60}Se_{40}$ Films
(a) Vol. 48, No. 1, p. K93—K96 (1978).
See also *2*, p. *47*; 16, p. 265; 22, p. 415.

L. Graf
Electronic Transitions in the Raman Spectra of $FeCl_2 \cdot 2\,H_2O$ and $FeCl_2 \cdot 2\,D_2O$
(b) Vol. 88, No. 2, p. 429—437 (1978).
See also *6*, p. *74*; *13.5.4*, p. *201*; 22.5, p. 475.

A. E. Kalnin and M. F. Trinkler
Spectroscopy of Mixed Alkali Halide Crystals Activated by Thallium
(b) Vol. 88, No. 2, p. 445—455 (1978).
See also *13.4*, p. *178*; *20.3*, p. *346*; 22.5.4, p. 489.

M. I. Dykman
Theory of Cyclotron Resonance of Two-Dimensional Electrons Interacting with Surface and Volume Phonons
(b) Vol. 88, No. 2, p. 463—475 (1978).
See also *6*, p. *74*; *13.1*, p. *157*.

K. Saiki, M. Yoshimi, and S. Tanaka
*Modulation Spectroscopy on the Group IV and VI Transition-Metal Dichalcogenides
(b) Vol. 88, No. 2, p. 607—614 (1978).
See also 13.1, p. 157; *13.5.1*, p. *193*; 22, p. 415.

E. Goovaerts and D. Schoemaker
*Inelastic Light Scattering of the V_k Centers in the Alkali Halides
(b) Vol. 88, No. 2, p. 615—621 (1978).
See also *10.2*, p. *116*; 22.5.2, p. 482.

F. Galeskii, A. E. Yunovich, K. H. Herrmann, H. Kostial, I. Rechenberg, and P. Schäfer
Stimulated Emission, Absorption Spectra, and Recombination in Epitaxial PbTe
(b) Vol. 88, No. 2, p. 675—681 (1978).
See also 13.4, p. 178; *20.2*, p. *339*; 22, p. 415.

A. G. Coutts and P. Day
*Charge Transfer Spectra of Transition Metal Ions in Cadmium Halide Lattices
(b) Vol. 88, No. 2, p. 767—772 (1978).
See also 22.5.4, p. 489.

E. Doni, R. Girlanda, and G. Pastori Parravicini
*A Note on the Polarization Dependence of the Two-Photon Absorption Coefficient in Crystals. Extension to Circularly Polarized Light
(b) Vol. 88, No. 2, p. 773—786 (1978).
See also 22.5.2, p. 483.

R. Mehrotra and M. M. Pant
*Long Wave Phonon Modes and Infrared Absorption of a NaI Slab
(b) Vol. 88, No. 2, p. K131—K133 (1978).
See also *6*, p. *74*; 22.5.2, p. 483.

J. M. Pawlikowski, J. Misiewicz, N. Mirowska, and P. Becla
*Absorption Edge of Zn_3P_2 at 5, 80, and 300 K
(b) Vol. 88, No. 2, p. K153—K156 (1978).
See also *13.1*, p. *158*; 22.3, p. 458.

V. N. Abramov, B. G. Ivanov, A. I. Kuznetsov, I. A. Meriloo, and M. I. Musatov
*On the Vacuum Ultraviolet Transparency of Sapphire
(a) Vol. 48, No. 2, p. 287—292 (1978).
See also 22.6, p. 495.

J. P. Huvenne, G. Vergoten, B. Boniface, G. Fleury et P. Legrand
*Etude vibrationelle de basse fréquence de l'oxyde de plomb Pb_3O_4. Relation entre l'évolution des spectres et les transitions ferroélectrique et ferroélastique de la maille
(Low-Frequency Vibrations of Lead Oxide Pb_3O_4. Relationship between the Development of Spectra and Ferroelectric and Ferroelastic Phase Transitions)
(a) Vol. 48, No. 2, p. 417—423 (1978).
See also *6*, p. *75*; 22.6, p. 495.

A. S. Filipchenko, L. P. Bolshakov, A. Naurizbaev, A. G. Braginskaya, and A. N. Popkov
Moss-Burstein Effect in n-Type Gallium Antimonide Crystals
(a) Vol. 48, No. 2, p. K115—K117 (1978).
See also *13.4*, p. *179*; 22.2.3, p. 453.

J. H. von Bardeleben, A. Goltzene, B. Meyer, and C. Schwab
Effects of Iron Content and Stoichiometry on the Coloration of $CuGaS_2$
(a) Vol. 48, No. 2, p. K145—K148 (1978).
See also *10.2*, p. *116*; 19, p. 311; 22.8, p. 506.

G. A. Puchkovskaya, V. L. Strizhevskii, Yu. A. Frolkov, N. M. Chepilko, and Yu. N. Yashkir
Crystal Optics of Surface Polaritons in Anisotropic Crystals
(b) Vol. 89, No. 1, p. 27—36 (1978).
See also *13.5*, p. *188*; 22.6, p. 496.

F. Lohse, G. Reuter, and J. M. Spaeth
*Optical Investigation of Atomic Hydrogen Centres in Alkali Chlorides Doped with I^- or Br^-
(b) Vol. **89**, No. 1, p. 109—116 (1978).
See also *10.2*, p. *117*; 13.4, p. 180; 22.5.2, p. 483.

M. Vanhaelst, P. Matthys, and E. Boesman
*Electron Spin Transfer in Alkali Chlorides and Alkali Bromides Doped with Divalent Vanadium
(b) Vol. **89**, No. 1, p. 151—157 (1978).
See also *19*, p. *311*; 22.5.2, p. 483.

D. Wruck
Vibrational Properties of the ⟨001⟩ Split Interstitial in Silicon
(b) Vol. **89**, No. 1, p. 177—186 (1978).
See also *6*, p. *75*; 22.1.2, p. 436.

F. Carlier and G. Jacobs
*Transient Infrared Absorption of Bound Polarons in NaCl
(b) Vol. **89**, No. 1, p. 193—198 (1978).
See also *13.5.3*, p. *200*; 22.5.2, p. 483.

I. Földvári, R. Voszka, and Z. Morlin
*The Properties of Ni Ions in NaCl Single Crystals (I): Vacuum Ultraviolet, Ionic Conductivity, and X-Ray Diffraction Studies
(b) Vol. **89**, No. 1, p. 235—240 (1978).
See also 1.2, p. 22; *14.4*, p. *250*; 22.5.2, p. 484; 22.5.4, p. 489.

W. G. Albrecht, H. Coufal, R. Haberkorn, and M. E. Michel-Beyerle
Excitation Spectra of Exciton Fission in Organic Crystals
(b) Vol. **89**, No. 1, p. 261—265 (1978).
See also *13.5.1*, p. *194*; 22.9, p. 527.

F. Meseguer and C. Sánchez
*Piezo-Optical Properties of Alkali Halides and Their Temperature Dependence
(b) Vol. **89**, No. 1, p. 281—288 (1978).
See also 22.5.2, p. 484.

P. Świątek, A. M. Witowski, and M. Grynberg
Experimental and Theoretical Study of the Dynamic Dielectric Function in p-HgTe
(b) Vol. **89**. No. 1, p. K1—K4 (1978).
See also *13.1*, p. *158*; 14.4, p. 250; 22.4.3, p. 471.

D. Genzow
*On the Interband Absorption in Lead-Chalcogenides under Linearly Polarized Light Excitation
(b) Vol. **89**, No. 1, p. K9—K11 (1978).
See also 22, p. 417.

N. M. Gasanly, B. M. Yavadov, V. I. Tagirov, and E. A. Vinogradov
*Infrared and Raman Spectra of Layer InSe Single Crystals
(b) Vol. **89**, No. 1, p. K43—K48 (1978).
See also *6*, p. *75*; 22.3, p. 458.

P. P. Reddy and Y. P. Reddy
*Electronic Absorption Spectrum of Co^{2+} in Strontium Tartrate Tetrahydrate Single Crystals
(b) Vol. **89**, No. 1, p. K61–K63 (1978).
See also 22.9, p. 527.

G. Guillot and A. Nouailhat
*Influence of the Purity of Crystals on the Stability of F-Centers and Interstitials in Alkali Halides
(b) Vol. **89**, No. 1, p. K85–K89 (1978).
See also *10.2*, p. *117*; 22.5.2, p. 484.

I. A. Parfianovich, E. E. Penzina, V. M. Khulugurov, V. M. Metzik, and L. M. Sobolev
Luminescence and Absorption of Colour Centres in RbCl:Eu and KCl:Eu Crystals
(b) Vol. **89**, No. 1, p. K91–K94 (1978).
See also *10.2*, p. *117*; *20.3*, p. *347*; 22.5.2, p. 484.

F. Carlier and G. Jacobs
*Transient Infrared Absorption of Bound Polarons in KBr
(b) Vol. **89**, No. 1, p. K95–K98 (1978).
See also *10.2*, p. *118*; 13.5.3, p. 200; 22.5.2, p. 484.

L. Risch
Investigation of Surface Properties of CdS by Means of Photoreflectance in the Exciton Region
(a) Vol. **49**, No. 1, p. 177–182 (1978).
See also *13.3*, p. *167*; 16, p. 266; 22.4.1, p. 465.

A. S. Filipchenko, L. P. Bolshakov, and A. Naurizbaev
*Absorption in Indium Antimonide and Arsenide Crystals Doped with Various Admixtures
(a) Vol. **49**, No. 1, p. 183–188 (1978).
See also *13.4*, p. *180*; 22.2.1, p. 448; 22.2.3, p. 454.

A. A. Kaminskii, V. V. Osiko, S. E. Sarkisov, M. T. Timoshechkin, E. V. Zharikov, J. Bohm, P. Reiche, and D. Schultze
Growth, Spectroscopic Investigations, and Some New Stimulated Emission Data of $Gd_3Ga_5O_{12}:Nd^{3+}$ Single Crystals
(a) Vol. **49**, No. 1, p. 305–311 (1978).
See also 3, p. 56; *20.2*, p. *339*; *20.3*, p. *347*; 22.8.2, p. 520.

F. G. Gärtner and E. Mollwo
*IR Absorption of OH and OD Centres and OH/OH, OD/OD and OH/OD Complexes in Cu-Doped ZnO Single Crystals (I): Experimental Results
(b) Vol. **89**, No. 2, p. 381–388 (1978).
See also 13.4, p. 181; 22.4, p. 462.

W. Stössel and H. J. Wagner
*Comparison of Additional Boundary Conditions Based on Thermoreflectance Spectra of the $A_{n=1}$-Exciton in CdS Crystals at Oblique Incidence
(b) Vol. **89**, No. 2, p. 403–417 (1978).
See also *13.5.1*, p. *194*; 22.4.1, p. 465.

W. A. Theiner and H. P. Geserich
Anderson Transition in $FeSi_2$ Films
(b) Vol. **89**, No. 2, p. 441—447 (1978).
See also *1.2*, p. *22*; *1.5*, p. *39*; 2, p. 50; 14.1, p. 210; 21.1.1, p. 389.

S. Ewert, G. Heiland, H. Lüth, U. Roll, and W. Sander
*Molecular Vibrations in a Polymethine Dye Layer Studied by Inelastic Electron Tunneling Spectroscopy
(b) Vol. **89**, No. 2, p. 459—465 (1978).
See also 6, p. 76; *14.3.1*, p. *228*; 22.9, p. 527.

A. Ioan, V. Topa, and M. Giurgea
New Colour Centres in KCl:(Tl^+ + Ca^{2+}) and KCl:(Tl^+ + Sr^{2+}) Crystals
(b) Vol. **89**, No. 2, p. 513—520 (1978).
See also *10.2*, p. *118*; 22.5.2, p. 485.

H. Friedrich, Yu. V. Ozerov, R. Strehlow, and A. N. Georgobiani
*On the Electron–Phonon and Exciton–Phonon Interaction in ZnTe
(b) Vol. **89**, No. 2, p. 603—609 (1978).
See also *13.5.1*, p. *195*; 22.4.3, p. 471.

R. Enderlein
*On the Theory of Light Propagation through Disordered Solids. Scattering from Configurational Fluctuations
(b) Vol. **89**, No. 2, p. 625—632 (1978).
See also 2, p. 50; *6*, p. *76*; 22, p. 418.

B. Esser
*On the Theory of Resonance Raman Scatterihg in Disordered Semiconductors
(b) Vol. **89**, No. 2, p. 633—643 (1978).
See also 2, p. 50; *6*, p. *76*; 13.1, p. 159; 22, p. 418.

P. Franzosi
Exciton–Phonon Interaction in CdS
(b) Vol. **89**, No. 2, p. K119—K122 (1978).
See also *6*, p. *76*; *13.5.1*, p. *195*; 22.4.1, p. 466.

H. Kuzmany
*Instability of the Raman Spectra of Tetrathiofulvalene-Tetracyanoquinodimethan (TTF-TCNQ) and Tetrathiofulvalene (TTF)
(b) Vol. **89**, No. 2, p. K139—K144 (1978).
See also 11, p. 130; 22.9, p. 527.

L. C. Scavarda do Carmo and F. Lüty
*Librational and Phonon Sidebands of the Infrared CN^- Stretching Mode in NaCN and KCN Crystals
(b) Vol. **89**, No. 2, p. K149—K153 (1978).
See also *6*, p. *76*; 22.8, p. 507.

O. Goede, L. John, and D. Hennig
*Compositional Disorder-Induced Broadening for Free Excitons in II–VI Semiconducting Mixed Crystals
(b) Vol. **89**, No. 2, p. K183—K186 (1978).
See also *13.5.1*, p. *195*; 22.4.4, p. 473.

D. Shaw and E. Watson
*The Shift of Optical Absorption Edge in CdTe with Temperature and Doping Level
(a) Vol. **49**, No. 2, p. 593—602 (1978).
See also 10.2, p. 119; 22.4.3, p. 471.

B. Baranov, L. Däweritz, V. B. Gutan, G. Jungk, H. Neumann, and H. Raidt
Growth and Properties of $Al_xGa_{1-x}N$ Epitaxial Layers
(a) Vol. **49**, No. 2, p. 629—636 (1978).
See also *1.5*, p. *39*; 13.4, p. 182; 14.3.1, p. 228; 20.3, p. 348; 22.2.4, p. 456.

F. G. Gärtner and E. Mollwo
*IR-Absorption of OH and OD Centres and OH/OH, OD/OD, and OH/OD Complexes in Cu Doped ZnO Single Crystals (II): Discussion and Quantitative Interpretation of Models
(b) Vol. **90**, No. 1, p. 33—44 (1978).
See also 13.4, p. 183; 22.4, p. 462.

I. Broser, M. Rosenzweig, R. Broser, M. Richard, and E. Birkicht
*A Quantitative Study of Excitonic Polariton Reflectance in CdS
(b) Vol. **90**, No. 1, p. 77—91 (1978).
See also *13.5.1*, p. *195*; 22.4.1, p. 466.

A. Lachter, J. Salardenne, and A. S. Barrière
*Optical Absorption Study of Iron Trifluoride Thin Films
(b) Vol. **90**, No. 1, p. 147—150 (1978).
See also 22.5, p. 476.

M. Zanini, D. Grubisic, and J. E. Fischer
*Optical Anisotropy of Highly Oriented Pyrolytic Graphite. Effect of Surface Preparation
(b) Vol. **90**, No. 1, p. 151—156 (1978).
See also 22.1, p. 423.

A. Manabe, H. Noguchi, and A. Mitsuishi
*Dielectric Function in HgSe at 2, 95, and 300 K
(b) Vol. **90**, No. 1, p. 157—166 (1978).
See also 6, p. 77; 13.1, p. 159; 13.5.2, p. 199; 22.4.2, p. 469.

R. Strehlow, W. Kühn, H. Friedrich, and H. Lange
*On the Quantitative Characterization of Chemical Disorder of $In_{1-x}Ga_xP$ from Wavelength-Modulated No-Phonon Absorption
(b) Vol. **90**, No. 1, p. 167—173 (1978).
See also 6, p. 77; 13.1, p. 159; 13.5.1, p. 195; 22.2.4, p. 456.

P. J. Flatau
*Kinetics of the Intraband Absorption and Magnetoabsorption Coefficients in Mixed Semiconductors with Composition Fluctuations
(b) Vol. **90**, No. 1, p. 251—260 (1978).
See also *1.3*, p. *28*; 22.4.4, p. 473.

E. Vigil, J. A. Rodríguez, and R. Pérez Alvarez
*Optical Constants of n- and p-Type GaAs between 2.5 and 3.5 eV
(b) Vol. **90**, No. 1, p. 409—414 (1978).
See also 22.2.1, p. 448.

V. C. Sahni and M. L. Bansal
*Raman Study of the Mixed System $[(NH_4)_{1-x}K_x]_2 CuCl_4 \cdot H_2O$
(b) Vol. **90**, No. 1, p. 415—422 (1978).
See also *6*, p. *78*; 22.5.4, p. 490.

G. B. Dubrovskii, S. T. Pavlov, and M. S. Vershinin
*Interband Optical Absorption in Semiconductors with One-Dimensional Electronic Superlattice
(b) Vol. **90**, No. 1, p. 429—432 (1978).
See also *13.1*, p. *160*; 22.3, p. 459.

V. I. Sokolov, N. A. Moskvina, and B. V. Melnikov
Discovery of Exciton–Impurity Complexes in Zinc Selenide by Electroabsorption
(b) Vol. **90**, No. 1, p. K39—K42 (1978).
See also *13.4*, p. *184*; *13.5.1*, p. *196*; 14.3.4, p. 246; 22.4.2, p. 469.

S. A. Boiko, D. I. Bletskan, and S. F. Terekhova
*Exciton Absorption of Germanium Diselenide
(b) Vol. **90**, No. 1, p. K49—K52 (1978).
See also *13.5.1*, p. *196*; 22, p. 420.

M. Baran, H. Szymczak, W. Wardzyński, and B. M. Wanklyn
*EPR and Optical Absorption Spectra of $K_2Cr_5F_{17}$ Monocrystals
(b) Vol. **90**, No. 1, p. K59—K61 (1978).
See also *19*, p. *314*; 22.8, p. 507.

A. V. Goltsev, I. G. Lang, and S. T. Pavlov
*Oscillations of Secondary Radiation Intensity in Polar Semiconductors in a Magnetic Field
(b) Vol. **90**, No. 1, p. K63—K67 (1978).
See also *6*, p. *78*; 22, p. 420.

V. A. Gadzhiev, S. G. Abdullaeva, F. K. Aliev, F. K. Allakhverdiev, and A. Z. Badalov
*Electroabsorption in GaTe
(b) Vol. **90**, No. 1, p. K77—K79 (1978).
See also 13.5.1, p. 196; *14.3.4*, p. *246*; 22.3, p. 459.

H. Möhwald and A. Thaer
Raman Study of the Phase Transition in the CT-Crystal Anthracene-TCNB
(a) Vol. **50**, No. 1, p. 131—138 (1978).
See also *1.2*, p. *24*; 6, p. 78; 22.9, p. 528.

V. B. Sharan, D. N. S. Srivastwa, and S. C. Sen
*Diffuse Absorption Spectrum of NH_4Cl + TlCl Mixed Phosphors
(a) Vol. **50**, No. 1, p. 323—327 (1978).
See also *13.4*, p. *184*; 22.5.4, p. 490.

V. K. Jain and S. P. Kathuria
Z_3 Centre Thermoluminescence in LiF TLD Phosphor
(a) Vol. **50**, No. 1, p. 329—333 (1978).
See also *10.2*, p. *121*; *20.3*, p. *350*; 22.5.2, p. 485.

I. A. Oxton and O. Knop
*Infrared Spectra of $SiHD_3$ in Crystalline Germane
(a) Vol. **50**, No. 1, p. K55—K58 (1978).
See also *1.2*, p. *24*; 23, p. 531.

B. Rambabu, C. Ramasastry, and B. V. R. Chowdari
*Optical Absorption of Off-Centre Cu^+ in CsBr
(b) Vol. **90**, No. 2, p. 465—473 (1978).
See also *10.2*, p. *121*; 22.5.2, p. 486.

D. Genzow, A. G. Mironov, and O. Ziep
*On the Interband Absorption in Lead Chalcogenides
(b) Vol. **90**, No. 2, p. 535—542 (1978).
See also *13.1*, p. *160*; 22, p. 420.

M. A. F. Scarparo, R. S. Katiyar, R. Srivastava, and S. P. S. Porto
Evidence of Non-Linear Behaviour of Soft Modes in Crystals with KDP Structure
(b) Vol. **90**, No. 2, p. 543—556 (1978).
See also *14.4.1*, p. *257*; 22.8.1, p. 516.

A. Król, M. J. Kozielski, and W. Nazarewicz
*Infrared Studies of Al Complexes in Zinc Sulphide
(b) Vol. **90**, No. 2, p. 649—656 (1978).
See also *6*, p. *78*; 10.2, p. 121; 22.4.1, p. 466.

S. Benci, R. Capelletti, F. Fermi, M. Manfredi, J. Z. Damm, and E. Mugeński
*Spectral and Time Decay Analysis of Photoluminescence Excited in the A Band Region in KCl:Pb
(b) Vol. **90**, No. 2, p. 657—666 (1978).
See also 13.4, p. 185; *20.3*, p. *350*; 22.5.2, p. 486.

V. B. Sharan, D. N. S. Srivastwa, and S. C. Sen
Absorption, Excitation, and Fluorescence Spectra of Thallium-Activated Ammonium Bromide
(b) Vol. **90**, No. 2, p. 671—678 (1978).
See also *13.4*, p. *185*; 20.3, p. 350; 22.5, p. 476.

N. N. Gerasimenko, M. Rollé, Li-Jen Cheng, Y. H. Lee, J. C. Corelli, and J. W. Corrett
*Infrared Absorption of Silicon Irradiated by Protons
(b) Vol. **90**, No. 2, p. 689—695 (1978).
See also *6*, p. *78*; 11, p. 131; 22.1.2, p. 439.

N. A. Kulagin and V. I. Tutlis
Change of the Praseodymium Spectrum under the Free State-to-Crystal Transition
(b) Vol. **90**, No. 2, p. K109—K113 (1978).
See also *13.4*, p. *185*; 22, p. 421.

V. N. Brudnyi, D. L. Budnitskii, M. A. Krivov, R. V. Masagutova, V. D. Prochukhan, and Yu. V. Rud
*The Electrical and Optical Properties of 2.0 MeV Electron-Irradiated $ZnGeP_2$
(a) Vol. **50**, No. 2, p. 379—384 (1978).
See also 11, p. 131; 13.4, p. 185; *14.3*, p. *225*; 22.8, p. 508.

L. C. Jen and J. F. Merklin
*Optically Induced Dichroism of Z_2-Centers in LiF:Mg, Ti (TLD-100)
(a) Vol. **50**, No. 2, p. 469—474 (1978).
See also *10.2*, p. *121*; 22.5.2, p. 486.

A. S. Bhalla, L. E. Cross, and D. A. Payne
*Thermal, Optical, and Elastic Properties of Ferroelastic Isopropyl Ammonium Chloroplatinate (IPACP) and Chlorostannate (IPACS)
(a) Vol. **50**, No. 2, p. 661—667 (1978).
See also *1.2*, p. *24*; 8, p. 90; 12.2, p. 141; 22.9, p. 528.

A. A. Bukharaev and N. R. Yafaev
*Laser and Thermal Bleaching of Colour Centres in Sodium Borate Glasses
(a) Vol. **50**, No. 2, p. 711—716 (1978).
See also 2, p. 53; *13.4* p. *186*; 22.8.1, p. 516.

A. A. Kaminskii, S. E. Sarkisov, Tran Ngoc, B. I. Denker, V. V. Osiko, and A. M. Prokhorov
*Stimulated Emission Spectroscopy of Concentrated Lithium Neodymium Phosphate Glasses in $^4F_{3/2} \rightarrow {}^4I_{11/2}$ and $^4F_{3/2} \rightarrow {}^4I_{13/2}$ Transitions
(a) Vol. **50**, No. 2, p. 745—750 (1978).
See also 13.4, p. 186; *20.2*, p. *340*; 22.8.1, p. 516.

Sh. M. Efendiev, V. E. Bagiev, A. Ch. Zeinally, and V. M. Skorikov
*Optical Transitions in $Bi_{12}SiO_{20}$
(a) Vol. **50**, No. 2, p. K141—K143 (1978).
See also 6, p. 79; 13.1, p. 161; 22.8, p. 508.

H. Hesse and S. Kapphan
*Doping of $SrTiO_3$ with Hydrogen and Deuterium
(a) Vol. **50**, No. 2, p. K243—K245 (1978).
See also 10.2, p. 122; *14.4.1*, p. *257*; 22.8.1, p. 517.

20.2 Lasers

Original Papers and Short Notes

V. M. Fomin and E. P. Pokatilov
Non-Linear Optical Properties of the Band Charge Carrier Gas
(b) Vol. **85**, No. 1, p. 403–415 (1978).
See also *14.3.4*, p. *240*; 22, p. 409.

H. J. Eichler, Ch. Hartig, and J. Knof
*Laser Induced Gratings in CdS
(a) Vol. **45**, No. 2, p. 433–438 (1978).
See also 13.4, p. 171; 22.4.1, p. 463.

E. A. Tsvetkov
Calculation of the Optical Fluorescence Intensities and Crystal Field Parameters of Nd^{3+} in Calcium Fluorophosphate
(a) Vol. **46**, No. 1, p. 311–314 (1978).
See also *13.4*, p. *173*; 20.3, p. 343; 22.8, p. 504.

R. Hultzsch
*Passive Q-Switching of the Ruby Laser by Means of Colour Centres in SrF_2
(a) Vol. **47**, No. 2, p. 415–423 (1978).
See also 10.2, p. 114; 22.5.3, p. 487.

F. Galeskii, A. E. Yunovich, K. H. Herrmann, H. Kostial, I. Rechenberg, and P. Schäfer
*Stimulated Emission, Absorption Spectra, and Recombination in Epitaxial PbTe
(b) Vol. **88**, No. 2, p. 675–681 (1978).
See also 13.4, p. 178; 20.1, p. 330; 22, p. 415.

G. Staupendahl and K. Schindler
*Self-Defocusing of CO_2-Laser Radiation in Tellurium
(a) Vol. **48**, No. 2, p. K199–K201 (1978).
See also 22.1.3, p. 441.

M. Wautelet and L. D. Laude
Electronic Structure of Solids under Laser Irradiation
(b) Vol. **89**, No. 1, p. 275–279 (1978).
See also 11, p. 129; *13.1*, p. *158*; 22, p. 417.

A. A. Kaminskii, V. V. Osiko, S. E. Sarkisov, M. T. Timoshechkin, E. V. Zharikov, J. Bohm, P. Reiche, and D. Schultze
*Growth, Spectroscopic Investigations, and Some New Stimulated Emission Data of $Gd_3Ga_5O_{12}$:Nd^{3+} Single Crystals
(a) Vol. **49**, No. 1, p. 305–311 (1978).
See also 3, p. 56; 20.1, p. 333; *20.3*, p. *347*; 22.8.2, p. 520.

S. W. Koch, H. Haug, G. Schmieder, W. Bohnert, and C. Klingshirn
*Stimulated Intrinsic Recombination Processes in II–VI Compounds
(b) Vol. **89**, No. 2, p. 431–440 (1978).
See also 13.5.1, p. 195; *20.3*, p. *348*; 22.4, p. 462; 22.4.1, p. 465; 22.4.3, p. 471.

NGUYEN VINH QUANG
*The Influence of an Electromagnetic Wave on the Electronic Spectrum of Crystals
(b) Vol. **90**, No. 2, p. 597—608 (1978).
See also *13.1*, p. *160*.

A. A. KAMINSKII, S. E. SARKISOV, TRAN NGOC, B. I. DENKER, V. V. OSIKO, and A. M. PROKHOROV
*Stimulated Emission Spectroscopy of Concentrated Lithium Neodymium Phosphate Glasses in $^4F_{3/2} \rightarrow {}^4I_{11/2}$ and $^4F_{3/2} \rightarrow {}^4I_{13/2}$ Transitions
(a) Vol. **50**, No. 2, p. 745—750 (1978).
See also 13.4, p. 186; *20.1*, p. *338*; 22.8.1, p. 516.

20.3 Luminescence (X-Ray Spectra See 20)

Original Papers and Short Notes

M. E. Michel-Beyerle, R. Haberkorn, J. Kinder, and H. Seidlitz
Direct Evidence for the Singlet–Triplet Exciton Annihilation in Anthracene Crystals
(b) Vol. **85**, No. 1, p. 45—49 (1978).
See also *13.5.1*, p. *189*; 22.9, p. 522.

V. V. Hizhnyakov and A. V. Sherman
On the Theory of Resonant Secondary Radiation of Excitons Weakly Interacting with Phonons
(b) Vol. **85**, No. 1, p. 51—61 (1978).
See also *6*, p. *64*; *13.5.1*, p. *189*; 20.1, p. 319; 22, p. 408.

E. L. Ivchenko, I. G. Lang, and Ş. T. Pavlov
*Resonant Secondary Radiation in Polar Semiconductors
(b) Vol. **85**, No. 1, p. 81—85 (1978).
See also *6*, p. *64*; 22, p. 408.

G. A. Boiko, V. S. Dneprovskii, M. V. Kraevskii, K. Marinova, S. M. Oak, E. K. Silina, and V. S. Fokin
*A Study of the Kinetics of Recombination Radiation of CdS and CdSe Crystals
(b) Vol. **85**, No. 1, p. 111—119 (1978).
See also *13.5.1*, p. *189*; 22.4.1, p. 463; 22.4.2, p. 468.

S. B. S. Sastry, A. Scharmann, and D. Schwabe
*Luminescence of $RbI:Pb^{2+}$ under B-Band Excitation
(b) Vol. **85**, No. 1, p. K15—K17 (1978).
Erratum
(b) Vol. **87**, No. 2, p. K163 (1978).
See also 22.5.2, p. 478.

I. A. Kurova and N. N. Ormont
*IR Luminescence of Acoustoelectric Domains in n-GaAs
(a) Vol. **45**, No. 1, p. 333—342 (1978).
See also *7*, p. *80*; 22.2.1, p. 445.

Hj. Bernhardt
*Studies of the Colour of Lead Molybdate Crystals
(a) Vol. **45**, No. 1, p. 353—357 (1978).
See also 13.4, p. 170; *20.1*, p. *321*; 22.8.1, p. 510.

L. Mărculescu, N. Vlahovici, and T. Velula
*On the Decay of the Luminescence Emission of Pb^{2+} Centres in Alkali Halide Crystals
(b) Vol. **85**, No. 2, p. 439—443 (1978).
Erratum
(b) Vol. **89**, No. 1, p. 335 (1978).
See also *13.4*, p. *170*; 22.5.2, p. 478.

G. L. Belenkii and M. O. Godzhaev
*Excitons in Gallium Sulphide
(b) Vol. **85**, No. 2, p. 453—458 (1978).
See also *13.5.1*, p. *189*; 22.3, p. 457.

M. E. Michel-Beyerle and R. Haberkorn
*Temperature-Independent Intersystem Crossing in Anthracene Crystals
(b) Vol. **85**, No. 2, p. 473—475 (1978).
See also *13.5.1*, p. *189*; 22.9, p. 522.

N. Graber and I. Zschokke-Gränacher
*Delayed Fluorescence Excitation Spectra in Doped Anthracene Crystals
(b) Vol. **85**, No. 2, p. 505—510 (1978).
See also 22.9, p. 523.

M. C. Bonnet, E. Antic-Fidancev, P. Porcher, M. Lemaitre-Blaise, and P. Caro
*The Optical Spectra of Rare-Earth Cobalticyanides
(b) Vol. **85**, No. 2, p. 587—596 (1978).
See also *20.1*, p. *321*; 22.8, p. 503.

J. Nahum and F. J. Bryant
*Exciton Emission of Ytterbium-Implanted Cadmium Telluride
(b) Vol. **85**, No. 2, p. 629—637 (1978).
See also *13.5.1*, p. *190*; 22.4.3, p. 470.

A. Frey
*Composition Dependence of Nitrogen and Tellurium Bound State Energies in $In_{1-x}Ga_xP$
(b) Vol. **85**, No. 2, p. K107—K110 (1978).
See also *13.4*, p. *171*; 22.2.4, p. 456.

K. D. Glinchuk and A. V. Prokhorovich
*The Role of Different Local Centres in the Determination of the Concentration Dependence of the Intrinsic Emission Intensity in n-Type GaAs
(a) Vol. **45**, No. 2, p. K91—K94 (1978).
See also *13.4*, p. *171*; 22.2.1, p. 445.

L. Sodomka
*Mechanoluminescent Spectra of Silicon Carbide Powders
(a) Vol. **45**, No. 2, p. K171—K174 (1978).
See also 22.3, p. 457.

A. P. Radlinski
On the Properties of Cobalt Impurities in Zinc Selenide Crystals (II): The Optical Properties in Near-Infrared. $^4A_2(^4F) \leftrightarrow {}^4T_2(^4F)$ Transitions
(b) Vol. **86**, No. 1, p. 41—46, (1978).
See also *13.4*, p. *172*; 20.1, p. 321; 22.4.2, p. 468.

A. N. Trukhin
Investigation of the Photoelectric and Photoluminescent Properties of Crystalline Quartz and Vitreous Silica in the Fundamental Absorption Region. A Model for Electronic Structure and Migration of Energy in SiO_2
(b) Vol. **86**, No. 1, p. 67—75 (1978).
See also 2, p. 43; *13.1*, p. *152*; 16, p. 263; 22.6, p. 491.

Vu Duy Phach, A. Bivas, B. Hönerlage, and J. B. Grun
Two-Photon Resonant Raman Scattering via Biexcitons
(b) Vol. **86**, No. 1, p. 159—168 (1978).
See also *13.5.1*, p. *190*; 20.1, p. 321; 22.5, p. 474.

W. MÜLLER, M. OVERHAMM, G. BAUER, E. GORNIK, P. GROSSE, H. W. PÖTZL, and M. RAUTENBERG
*On the Influence of Background Radiation on Landau Emission Spectra
(b) Vol. **86**, No. 1, p. 205–213 (1978).
See also 13.1, p. 152; *14.3.4*, p. *241*; 20.1, p. 322; 22.2.3, p. 452.

V. I. IVANOV-OMSKII, V. A. MALTSEVA, A. D. BRITOV, and S. D. SIVACHENKO
*Photo- and Cathodoluminescence of $Cd_{0.3}Hg_{0.7}Te$ Alloys
(a) Vol. **46**, No. 1, p. 77–80 (1978).
See also *13.4*, p. *173*; 22.4.4, p. 472.

W. K. UNGER, H. MEUTH, J. C. IRWIN, and H. PINK
*Photoluminescence and Resonant Raman Scattering in $CdIn_2S_4$ and $ZnIn_2S_4$
(a) Vol. **46**, No. 1, p. 81–87 (1978).
See also *20.1*, p. *323*; 22.8, p. 504.

P. HENNL
Optical Properties of Chalcogen–Vacancy Centres in Alkali Halides
(a) Vol. **46**, No. 1, p. 147–152 (1978).
See also *10.2*, p. *111*; *20.1*, p. *323*; 22.5.2, p. 479.

E. A. TSVETKOV
Calculation of the Optical Fluorescence Intensities and Crystal Field Parameters of Nd^{3+} in Calcium Fluorophosphate
(a) Vol. **46**, No. 1, p. 311–314 (1978).
See also *13.4*, p. *173*; 20.2, p. 339; 22.8, p. 504.

V. N. KADCHENKO and M. ELANGO
*Diffusion Parameters of Hot Holes Created by Ionizing Radiation in NaCl(Ag)
(a) Vol. **46**, No. 1, p. 315–319 (1978).
See also *13.4*, p. *173*; 22.5.2, p. 479.

V. A. BENDERSKII, V. KH. BRIKENSHTEIN, A. G. LAVRUSHKO, and P. G. FILIPPOV
*Non-Linear Fluorescence Quenching in Molecular Crystals (I): Recombination of Localiced Excitons
(b) Vol. **86**, No. 2, p. 449–458 (1978).
See also *13.5.1*, p. *190*; 22.9, p. 524.

K. HÜMMER, R. HELBIG, and M. BAUMGÄRTNER
*Manifestation of k-Linear Term Effects on the Zeeman Splitting of the A1-Exciton Emission Lines in ZnO
(b) Vol. **86**, No. 2, p. 527–533 (1978).
See also *13.5.1*, p. *191*; 22.4, p. 461.

V. G. PLEKHANOV and A. A. O'CONNEL-BRONIN
Polariton Emission in LiH, LiD, and NaI Single Crystals
(b) Vol. **86**, No. 2. p. K123–K125 (1978).
See also *13.5*, p. *187*; 22, p. 412; 22.5.2, p. 480.

S. BHUSHAN, A. N. PANDEY, and BALAKRISHNA RAO KAZA
*Brightness Waves of Electroluminescence in ZnO:La Phosphors
(a) Vol. **46**, No. 2, p. K123–K126 (1978).
See also 22.4, p. 461.

R. Bindemann, R. Schwabe, and T. Hänsel
*The "Quasidirect" Radiative Recombination of Free Holes at Neutral Shallow Donors in GaP
(b) Vol. 87, No. 1, p. 169—177 (1978).
See also *13.4*, p. *174*; 22.2.2, p. 450.

A. K. Kostin, V. V. Savelev, and A. V. Vannikov
Free Charges and Excitons in Stilbene Single Crystals Excited by Electron Pulses
(b) Vol. 87, No. 1, p. 255—259 (1978).
See also *13.5.1*, p. *191*; 20.1, p. 326; 22.9, p. 524.

C. Pedrini
*Cu^+ Luminescent Centre in LiCl Single Crystals
(b) Vol. 87, No. 1, p. 273—286 (1978).
See also *13.4*, p. *175*; 22.5.2, p. 481.

E. L. Frankevich, M. M. Tribel, I. A. Sokolik, and A. I. Pristupa
Magnetic-Resonant Modulation of Photoconductivity of Crystalline Charge Transfer Complexes. Anthracene–Tetracyanbenzene
(b) Vol. 87, No. 1, p. 373—379 (1978).
See also 13.5.1, p. 191; *16*, p. *264*; 22.9, p. 524.

P. Krispin and J. Maege
*Capacitance–Frequency Dispersion and Electroluminescence Efficiency of GaP p–n Junctions
(a) Vol. 47, No. 1, p. 91—97 (1978).
See also 13.4, p. 175; *14.3.3*, p. *235*; 22.2.2, p. 450.

L. Gouskov, A. Gouskov, and B. Girault
Electrooptical Properties of SnO_2 Spray–GaSe and SnO_2 Spray–GaTe Heterojunctions
(a) Vol. 47, No. 1, p. 195—199 (1978).
See also *14.3.3*, p. *235*; 14.3.4, p. 242; 16, p. 264; 22.3, p. 458.

K. A. Mirza. P. D. Townsend, and G. L. Destefanis
*Light Emission from the Surface of $LiNbO_3$
(a) Vol. 47, No. 1, p. K63—K66 (1978).
See also 13.3, p. 165; 14.4.1, p. 255; 22.8.1, p. 513.

M. C. Wintersgill and P. D. Townsend
*The Effectiveness of the Implantation of Oxygen in Increasing the Luminescence from LiF:Mg:Ti
(a) Vol. 47, No. 1, p. K67—K69 (1978).
See also 13.4, p. 175; 22.5.2, p. 481.

C. Jaccard and M. Ecabert
*Optical Detection of Nuclear Magnetic Resonance with Distant F-Centre Pairs in KCl
(b) Vol. 87, No. 2, p. 497—505 (1978).
See also 13.4, p. 175; *19*, p. *308*; 22.5.2, p. 481.

Z. Zboiński
A Kinetic Model of Triplet Exciton Dynamics in Organic Crystals with Discrete Trapping Levels
(b) Vol. 87, No. 2, p. 663—674 (1978).
See also *13.5.1*, p. *192*; 22.9, p. 525.

H. Reichl, D. Huber, and J. Müller
*Photoluminescence Investigations of Diffused GaAs Light Emitting Diodes
(a) Vol. 47, No. 2, p. 489—495 (1978).
See also *14.3.3*, p. *235*; 22.2.1, p. 447.

K. C. Bleijenberg and C. W. M. Timmermans
*Luminescence and Electrical Conductivity of Uranium-Activated Sodium Fluoride Crystals
(a) Vol. 47, No. 2, p. 589—595 (1978).
See also 10.2, p. 114; *14.4*, p. *249*; 22.5.2, p. 481.

D. E. Aboltin, V. J. Grabovskis, A. R. Kangro, Ch. Lushchik, A. A. O'Konnel-Bronin, I. K. Vitol, and V. E. Zirap
*Thermally Stimulated and Tunneling Luminescence and Frenkel Defect Recombination in KCl and KBr at 4.2 to 77 K
(a) Vol. 47, No. 2, p. 667—675 (1978).
See also *10.2*, p. *114*; 22.5.2, p. 481.

V. I. Tagirov, N. M. Gasanly, B. M. Yavadov, M. A. Sobeikh, and V. M. Salmanov
*Photoluminescence of Layered $Ga_{1-x}In_xSe$ Crystals at Two-Photon Optical Excitation
(a) Vol. 47, No. 2, p. K157—K160 (1978).
See also 22.3.1, p. 460.

R. K. Gartia and B. S. Acharya
On the Production of Z_1 Centres
(a) Vol. 47, No. 2, p. K165—K168 (1978).
See also *10.2*, p. *114*; 22.5.2, p. 482.

E. Mlitzke
*On the Original Method of Balarin and Zetzsche
(a) Vol. 47, No. 2, p. K181—K182 (1978).
See also *13.4*, p. *177*; 22, p. 414.

H. Maaref, J. Barrau, M. Brousseau, J. Collet, J. Mazzaschi, and M. Pugnet
First-Order Mott Transition of the Exciton Gas in GaP
(b) Vol. 88, No. 1, p. 261—268 (1978).
See also *13.5.1*, p. *192*; 22.2.2, p. 450.

T. Iwasaki, T. Goto, and Y. Nishina
Oscillator Strength of Excitons in Highly Excited PbI_2
(b) Vol. 88, No. 1, p. 289—293 (1978).
See also *13.5.1*, p. *193*; 20.1, p. 329; 22.5, p. 475.

M. Egée
*Effect of a Direct Electric Field on the Photoluminescence Processes and Model of the Luminogen Centre in ZnS:Cu
(a) Vol. 48, No. 1, p. 89—98 (1978).
See also 13.4, p. 177; 14.3, p. 222; 22.4.1, p. 465.

R. K. Gartia, B. S. Acharya, and V. V. Ratnam
*Effect of Re-Irradiation on Z_1 Centres
(a) Vol. 48, No. 1, p. 235—239 (1978).
See also *10.2*, p. *115*; 22.5.2, p. 482.

A. E. Kalnin and M. F. Trinkler
*Spectroscopy of Mixed Alkali Halide Crystals Activated by Thallium
(b) Vol. 88, No. 2, p. 445–455 (1978).
See also *13.4*, p. *178*; 20.1, p. 330; 22.5.4, p. 489.

D. Heitmann
Theory and Experiment of the Light Emission of Surface Plasmons Induced by Fast Electrons on Modulated Surfaces
(b) Vol. 88, No. 2, p. 493–506 (1978).
See also *13.3*, p. *166*; *13.5.2*, p. *198*; 21.6, p. 403.

B. Hönerlage, Vu Duy Phach, and J. B. Grun
*Polarization Properties of Biexciton Luminescence and Two-Photon Raman Emission in CuCl
(b) Vol. 88, No. 2, p. 545–553 (1978).
See also *13.5.1*, p. *193*; 22.5, p. 475.

H. Asami, T. Nakayama, T. Chong, and N. Itoh
Reaction Rate Constants of Triplet Excitons in Naphthalene Single Crystals
(b) Vol. 88, No. 2, p. 623–631 (1978).
See also *13.5.1*, p. *193*; 22.9, p. 526.

E. O. Goebel and O. Hildebrand
*Thermalization of the Electron–Hole Plasma in GaAs. The Case of Resonant Excitation
(b) Vol. 88, No. 2, p. 645–652 (1978).
See also *13.5*, p. *187*; 22.2.1, p. 447.

A. V. Sachenko, V. A. Tyagai, and A. G. Kundzich
*Exciton Luminescence in Semiconductors. Surface Recombination and Space-Charge-Layer Effects
(b) Vol. 88, No. 2, p. 797–804 (1978).
See also 13.3, p. 166; *13.5.1*, p. *193*; 22.4.1, p. 465.

A. S. Filipchenko, T. B. Kurenkeev, L. P. Bolshakov, and E. I. Chaikina
*Photoluminescence of Heavily Doped n-Type Gallium Antimonide
(a) Vol. 48, No. 2, p. 281–285 (1978).
See also 13.4, p. 179; 22.2.3, p. 453.

H. Menniger, G. Voigt, H. Raidt, H. Fabig, and J. Maege
Scanning Electron Microscopic Investigations of GaP
(a) Vol. 48, No. 2, p. 407–415 (1978).
See also *1.4*, p. *32*; 14.3.2, p. 230; 22.2.2, p. 451.

K. D. Glinchuk, A. V. Prokhorovich, V. E. Rodionov, and V. I. Vovnenko
*Study of Non-Linear Extrinsic Luminescence in GaAs
(a) Vol. 48, No. 2, p. 593–602 (1978).
See also *13.4*, p. *179*; 22.2.1, p. 448.

S. V. Plotnikov, E. I. Rozum, V. A. Kuzminikh, and S. A. Vorobiev
*Channeling Effect of Fast Electrons on the Luminescence Yield of NaCl:Tl Crystals
(a) Vol. 48, No. 2, p. K175–K178 (1978).
See also 11, p. 128; 22.5.2, p. 483.

S. B. S. SASTRY and S. SAPRU
*Thermoluminescence of γ-Irradiated Rubidium Chloride Doped with Europium
(a) Vol. **48**, No. 2, p. K189—K191 (1978).
See also *10.2*, p. *117*; 22.5.2, p. 483.

L. BOSI, A. LONGONI, and M. NIMIS
*Lifetime of Color Centers in NaBr
(b) Vol. **89**, No. 1, p. 221—224 (1978).
See also *10.2*, p. *117*; 13.4, p. 180; 22.5.2, p. 484.

A. KOZANECKI, Z. WERNER, H. RZEWUSKI, and J. J. LOFERSKI
*A Study of Excitonic Spectra in CdS_xSe_{1-x} Mixed Crystals
(b) Vol. **89**, No. 1, p. 313—317 (1978).
See also *13.5.1*, p. *194*; 22.4.4, p. 473.

U. HOLLAND and F. LÜTY
*Hydrostatic Pressure Effects on the Absorption and Emission of F_A(Li) Centers in KCl
(b) Vol. **89**, No. 1, p. K73—K77 (1978).
See also 12.2, p. 140; *13.4*, p. *180*; 22.5.2, p. 484.

I. A. PARFIANOVICH, E. E. PENZINA, V. M. KHULUGUROV, V. M. METZIK, and L. M. SOBOLEV
*Luminescence and Absorption of Colour Centres in RbCl:Eu and KCl:Eu Crystals
(b) Vol. **89**, No. 1, p. K91—K94 (1978).
See also *10.2*, p. *117*; 20.1, p. 333; 22.5.2, p. 484.

K. METTLER
*Effect of Reabsorbed Recombination Radiation on Photoluminescence and Photoconductivity in a Semi-Infinite Direct-Gap Semiconductor
(a) Vol. **49**, No. 1, p. 163—168 (1978).
See also *16*, p. *266*; 22.2.1, p. 448.

T. D. DZHAFAROV, M. N. KOLOTOV, and A. A. LITVIN
*The Influence of γ-Irradiation on the Degradation of Red-Emitting GaP LED's
(a) Vol. **49**, No. 1, p. 211—215 (1978).
See also 11, p. 129; *14.3.3*, p. *236*; 22.2.2, p. 451.

A. V. MOSKALONOV and J. R. ZAKIS
*X-Ray Induced Paramagnetic States and Luminescence in As_2S_3 and As_2Se_3
(a) Vol. **49**, No. 1, p. 231—235 (1978).
See also 2, p. 49; *13.4*, p. *181*; 19, p. 312; 22, p. 417.

A. A. KAMINSKII, V. V. OSIKO, S. E. SARKISOV, M. T. TIMOSHECHKIN, E. V. ZHARIKOV, J. BOHM, P. REICHE, and D. SCHULTZE
*Growth, Spectroscopic Investigations, and Some New Stimulated Emission Data of $Gd_3Ga_5O_{12}$:Nd^{3+} Single Crystals
(a) Vol. **49**, No. 1, p. 305—311 (1978).
See also 3, p. 56; 20.1, p. 333; *20.2*, p. *339*; 22.8.2, p. 520.

R. BINDEMANN, H. FISCHER, and K. KREHER
*The Spectral Dependence of the Photoneutralization Cross-Section of Oxygen in GaP Near the Threshold
(a) Vol. **49**, No. 1, p. 331—336 (1978).
See also *13.4*, p. *181*; 13.5.1, p. 194; 22.2.2, p. 451.

J. Llopis and J. Piqueras
*Cathodoluminescence from Grain Boundaries in MgO and GaP
(a) Vol. **49**, No. 1, p. K9—K12 (1978).
See also *10.2*, p. *118*; 22.2.2, p. 451; 22.6, p. 496.

H. Nelkowski and K. Petermann
*Preparation and Investigation of ZnS–$CuGaS_2$ Heterojunctions
(a) Vol. **49**, No. 1, p. K13—K16 (1978).
See also 1.5, p. 39; 3, p. 56; *14.3.3*, p. *236*; 22.4.1, p. 465; 22.8, p. 507.

A. Hofstaetter, R. Oeder, A. Scharmann, D. Schwabe, and B. Vitt
Paramagnetic Resonance and Thermoluminescence of the $PbWO_4$/$PbMoO_4$ Mixed Crystal System
(b) Vol. **89**, No. 2, p. 375—380 (1978).
See also *19*, p. *312*; 22.8.1, p. 515.

S. W. Koch, H. Haug, G. Schmieder, W. Bohnert, and C. Klingshirn
*Stimulated Intrinsic Recombination Processes in II–VI Compounds
(b) Vol. **89**, No. 2, p. 431—440 (1978).
See also 13.5.1, p. 195; *20.2*, p. *339*; 22.4, p. 462; 22.4.1, p. 465; 22.4.3, p. 471.

W. Senske, M. Sondergeld, and H. J. Queisser
*Impurity Interaction on Excited Acceptor States in GaP
(b) Vol. **89**, No. 2, p. 467—474 (1978).
See also *13.4*, p. *181*; 22.2.2, p. 451.

R. Schwabe, F. Thuselt, H. Weinert, R. Bindemann, and K. Unger
*Luminescence from Electron–Hole Drops in Undoped GaP
(b) Vol. **89**, No. 2, p. 561—570 (1978).
See also *13.5.1*, p. *195*; 22.2.2, p. 451.

V. M. Kenkre
*Model for Trapping Rates for Sensitized Fluorescence in Molecular Crystals
(b) Vol. **89**, No. 2, p. 651—654 (1978).
See also 22, p. 418.

S. Asano, N. Yamashita, and Y. Nakao
*Luminescence of the Pb^{2+}-Ion Dimer Center in CaS and CaSe Phosphors
(b) Vol. **89**, No. 2, p. 663—673 (1978).
See also *13.4*, p. *181*; 22.4.1, p. 466; 22.4.2, p. 469.

H. Böttger, V. V. Bryksin, and P. Kleinert
Theory of Resonant Raman Scattering in Systems with Localized Electron States
(b) Vol. **89**, No 2., p. 675—685 (1978).
See also *6*, p. *76*; *13*, p. *148*; 22, p. 418.

I. K. Kristensen
*Decay of Photoluminescence in GaSe
(b) Vol. **89**, No. 2, p. K155—K157 (1978).
See also 13.4, p. 182; 22.3, p. 459.

B. Baranov, L. Däweritz, V. B. Gutan, G. Jungk, H. Neumann, and H. Raidt
Growth and Properties of $Al_xGa_{1-x}N$ Epitaxial Layers
(a) Vol. **49**, No. 2, p. 629—636 (1978).
See also *1.5*, p. *39*; 13.4, p. 182; 14.3.1, p. 228; 20.1, p. 335; 22.2.4, p. 456.

R. Mach, W. Gericke, H. Treptow, and W. Ludwig
*Direct Impact Excitation of Luminescent Centres in ZnSe Single Crystals
(a) Vol. **49**, No. 2, p. 667—674 (1978).
See also 13.4, p. 182; *14.3.3*, p. *237*; 22.4.2, p. 469.

S. Bhushan, B. Rao Kaza, and A. N. Pandey
*Photo- and Electroluminescence of Rare Earth (Pr, Gd, and Er) Doped ZnO Phosphors
(a) Vol. **49**, No. 2, p. K167—K171 (1978).
See also 22.4, p. 462.

A. S. Filipchenko, T. B. Kurenkeev, and L. P. Bolshakov
*Velocity of the Spontaneous Radiative Recombination of Current Carriers in Degenerate Semiconductors
(a) Vol. **49**, No. 2, p. K189—K192 (1978).
See also 22.2.1, p. 448; 22.2.3, p. 454.

E. Walentynowicz and S. Szuba
*The Excitation Mechanism of Mn Luminescence Centres in ZnS:Mn Thin Films
(a) Vol. **49**, No. 2, p. K201—K203 (1978).
See also 22.4.1, p. 466.

H. Schrey and C. Klingshirn
*Biexcitons, Impurity Transitions, and Electron–Hole Plasma in Highly Excited II–VI Compounds?
(b) Vol. **90**, No. 1, p. 67—76 (1978).
See also 13.4, p. 183; *13.5.1*, p. *195*; 22.4.1, p. 466.

E. Schuberth and M. Creuzburg
*Luminescence from Rare-Gas-Doped Solid Neon
(b) Vol. **90**, No. 1, p. 189—196 (1978).
See also *13.5.1*, p. *196*; 23, p. 531.

R. Schwabe, F. Thuselt, T. Hänsel, and F. Lévy
*Electron Beam Excited Free Exciton Luminescence of GaSe
(b) Vol. **90**, No. 1, p. 233—239 (1978).
See also *13.5.1*, p. *196*; 22.3, p. 459.

Yu. D. Arepev, V. P. Oleinik, E. A. Salkov, and K. B. Tolpygo
*The Anomalous Stark Shift of the Ground State Energy of a Donor–Acceptor Pair in an Electric Field
(b) Vol. **90**, No. 1, p. 355—360 (1978).
See also *13.4*, p. *183*; 22.4.1, p. 466.

S. B. S. Sastry and K. Balasubramanyam
*Thermoluminescence of Irradiated RbCl and RbCl:Sn Crystals
(b) Vol. **90**, No. 1, p. 375—383 (1978).
See also *10.2*, p. *120*; 22.5.2, p. 485.

V. L. Broude, N. A. Vidmont, and V. V. Korshunov
*Singlet Exciton Annihilation Kinetics in Anthracene Crystals
(b) Vol. **90**, No. 1, p. K53—K58 (1978).
See also *13.5.1*, p. *196*; 22.9, p. 528.

I. A. Maltseva, A. Mamedov, Yu. V. Rud, and Yu. K. Undalov
*Luminescence of In-Doped $CdGeP_2$ Crystals
(a) Vol. **50**, No. 1, p. 139–146 (1978).
See also *10.2*, p. *120*; 22.8, p. 508.

P. Iacconi, D. Lapraz, and R. Caruba
*Traps and Emission Centres in Thermoluminescent ZrO_2
(a) Vol. **50**, No. 1, p. 275–283 (1978).
See also 10.2, p. 121; 13.4, p. 184; 22.6, p. 497.

V. K. Jain and S. P. Kathuria
*Z_3 Centre Thermoluminescence in LiF TLD Phosphor
(a) Vol. **50**, No. 1, p. 329–333 (1978).
See also *10.2*, p. *121*; 20.1, p. 336; 22.5.2, p. 485.

K. Atobe
*The Role of F-Centers in Thermoluminescence Processes in Reactor Irradiated SrF_2 Crystals
(a) Vol. **50**, No. 1, p. K77–K80 (1978).
See also *10.2*, p. *121*; 11, p. 130; 22.5.3, p. 488.

E. Grilli, P. Cappelletti, and M. Guzzi
*Photoluminescence of $CdIn_2S_4$ Single Crystals
(a) Vol. **50**, No. 1, p. K93–K96 (1978).
See also 22.8, p. 508.

R. A. Noack
*Inversion of the Γ and L Conduction Bands of GaSb under Uniaxial Stress
(b) Vol. **90**, No. 2, p. 615–620 (1978).
See also 12.2, p. 141; *13.1*, p. *161*; 22.2.3, p. 454.

S. Benci, R. Capelletti, F. Fermi, M. Manfredi, J. Z. Damm, and E. Mugeński
*Spectral and Time Decay Analysis of Photoluminescence Excited in the A Band Region in KCl:Pb
(b) Vol. **90**, No. 2, p. 657–666 (1978).
See also 13.4, p. 185; *20.1*, p. *337*; 22.5.2, p. 486.

V. B. Sharan, D. N. S. Srivastwa, and S. C. Sen
Absorption, Excitation, and Fluorescence Spectra of Thallium-Activated Ammonium Bromide
(b) Vol. **90**, No. 2, p. 671–678 (1978).
See also *13.4*, p. *185*; 20.1, p. 337; 22.5, p. 476.

C. Lawther and J. Woods
*Blue Light Emission in Forward-Biased ZnS MIS Diodes
(a) Vol. **50**, No. 2, p. 491–502 (1978).
See also *14.3.3*, p. *238*; 22.4.1, p. 467.

L. N. Skuja, A. R. Silin, and J. Mareš
*Decay Time and Polarization Properties of Luminescence Centers in Vitreous Silica
(a) Vol. **50**, No. 2, p. K149–K152 (1978).
See also *2*, p. *53*; 22.6, p. 498.

H.-J. Kos and S. Mieke
*Effect of Deformation on the Thermoluminescence of Titanium Doped Lithium Fluoride
(a) Vol. **50**, No. 2, p. K165–K168 (1978).
See also 11, p. 131; *12.2*, p. *141*; 22.5.2, p. 486.

A. S. Filipchenko, T. B. Kurenkeev, and T. Seisenbaev
*Photoluminescence of Doped n-Type Indium Arsenide and Indium Antimonide Crystals
(a) Vol. **50**, No. 2, p. K251–K254 (1978).
See also 13.4, p. 186; 22.2.1, p. 449; 22.2.3, p. 455.

21. Metals. Alloys

Review Article

V. S. Postnikov, V. V. Postnikov, and V. M. Fedorov
Metastability and Superconductivity in Thin Films
(b) Vol. **87**, No. 1, p. 11–23 (1978).
See also 1.2, p. 15; *1.6*, p. *40*; *14.2*, p. *213*; 22, p. 408.

Original Papers and Short Notes

V. I. Grebennikov, O. B. Sokolov, and E. A. Turov
Effects of Correlation Interaction of Electrons with Deep Holes in X-Ray Spectra of Metals
(b) Vol. **85**, No. 1, p. 127–134 (1978).
See also *13*, p. *142*; 20, p. 315.

C.-O. Almbladh and P. Minnhagen
On the Thermal Broadening of Core Levels in Metals Due to Many-Electron Effects
(b) Vol. **85**, No. 1, p. 135–143 (1978).
See also *13*, p. *142*; 17, p. 269; 20, p. 315.

E. Krüger
Superconductivity Originating from Quasi-Orbital Electrons (I): The Model of Quasi-Orbital Conduction Electrons
(b) Vol. **85**, No. 1, p. 261–270 (1978).
See also 13.1, p. 150; *14.2*, p. *213*.

G. S. Agarwal and H. D. Vollmer
Surface Polariton Effects in Spontaneous Emission (II): Effects of Spatial Dispersion
(b) Vol. **85**, No. 1, p. 301–310 (1978).
See also *13.3*, p. *164*; 13.5.2, p. 198; 20.1, p. 320; 22.4.1, p. 463.

C. L. Briant and J. J. Burton
Icosahedral Microclusters. A Possible Structural Unit in Amorphous Metals
(b) Vol. **85**, No. 1, p. 393–402 (1978).
See also *2*, p. *42*.

G. Lautenschläger and E. Mrosan
Change in Electronic Charge Density around 3d-Transition Metal Impurities in Al and Cu
(b) Vol. **85**, No. 1, p. K19–K21 (1978).
See also *13.4*, p. *169*; 21.1, p. 368.

R. Kamel, A. R. Ali, and Z. Farid
Annealing Spectrum of Precipitating Al–Li Alloys
(a) Vol. **45**, No. 1, p. 47–52 (1978).
See also *10.1*, p. *99*; 12.1, p. 134; 21.2, p. 392.

R. Z. Valiev and O. A. Kaibyshev
Mechanism of Superplastic Deformation in a Magnesium Alloy (III): The Deformation Model
(a) Vol. **45**, No. 1, p. 77–82 (1978).
See also *10.1*, p. *99*; 21.3, p. 394.

W. ZINGG and H. WARLIMONT
On Imaging of Aluminium by Field Ion Microscopy
(a) Vol. **45**, No. 1, p. 117—122 (1978).
See also *17.1*, p. *272*.

A. P. SUTTON and R. C. POND
Electron Microscope Images of Coincidence Boundaries
(a) Vol. **45**, No. 1, p. 149—155 (1978).
See also *1.4*, p. *29*.

J.-M. DEROCHETTE
Adsorption of Nitrogen on Iridium. Influence of Steps on the Surface Potential Studied by Field Emission
(a) Vol. **45**, No. 1, p. 163—169 (1978).
See also *1.6*, p. *40*; *17.1*, p. *272*.

T. HIRATA
Solute Atom Clustering during Ageing an Aluminium–Magnesium–Silicon Solid Solution under Stress
(a) Vol. **45**, No. 1, p. 201—205 (1978).
See also *10.1*, p. *99*; 21.3, p. **394**.

R. MATTHEIS
Electrical Resistivity of PdH_x Films
(a) Vol. **45**, No. 1, p. 219—227 (1978).
See also *14.1*, p. *205*; *14.2*, p. *213*.

G. FUCHS
Scaling Law in Superconducting NbTi Alloys
(a) Vol. **45**, No. 1, p. 239—247 (1978).
See also *14.2*, p. *213*.

C. E. LACIANA, A. J. PEDRAZA, and E. J. SAVINO
Lattice Distortion and Migration Energy of Oxygen in Vanadium and Hydrogen in Niobium, Tantalum, and Vanadium
(a) Vol. **45**, No. 1, p. 315—326 (1978).
See also *10.1*, p. *100*; 21.1, p. 369.

A. A. GALKIN, V. V. TOKII, and B. P. FILATOV
Dislocation Multiplication (II): The Effect of the Initial Defect Structure and Deformation Conditions on the Dislocation Multiplication
(a) Vol. **45**, No. 1, p. 327—332 (1978).
See also *10.1*, p. *100*.

A. N. ALESHIN, V. YU. ARISTOV, B. S. BOKSHTEIN, and L. S. SHVINDLERMAN
Kinetic Properties of ⟨111⟩ Tilt Boundaries in Aluminium
(a) Vol. **45**, No. 1, p. 359—366 (1978).
See also *1.4*, p. *29*; **9**, p. 91.

P. SVOBODA
Size Effect Correction to the Longitudinal Magnetoresistance of Thin Wires
(a) Vol. **45**, No. 1, p. K43—K46 (1978).
See also *14.1*, p. *205*; 21.6, p. 401.

R. ROSEI, E. COLAVITA, S. MODESTI, and N. NISTICÒ
Electronic Structure of Dilute α-Phase Ag–Cd Alloys from Thermoreflectance and XPS Measurements
(b) Vol. **85**, No. 2, p. 445—452 (1978).
See also *13.1*, p. *151*; 20.1, p. 321; 21.6, p. 401.

E. KRÜGER
Superconductivity Originating from Quasi-Orbital Electrons (II): The Superconducting Ground State of Quasi-Orbital Conductivity Electrons
(b) Vol. **85**, No. 2, p. 493—504 (1978).
See also 13.1, p. 151; *14.2*, p. *213*.

M. BRAUWERS and A. K. BHATTACHARYA
Temperature Dependence of the Electrical Resistivity and Hall Coefficient in a Model of Concentrated Alloy
(b) Vol. **85**, No. 2, p. 519—524 (1978).
See also 13.1, p. 151; *14.1*, p. *206*.

K. FLACHBART, A. FEHER, Š. JÁNOŠ, Z. MÁLEK, and A. RYSKA
Thermal Conductivity of Nb–Ti Alloy in the Low-Temperature Range
(b) Vol. **85**, No. 2, p. 545—551 (1978).
See also *8*, p. *83*; 14.2, p. 213.

F. GOEDSCHE and G. VOJTA
Spin-Disorder Resistivity of Ternary Alloys in the CPA Weak Scattering Limit
(b) Vol. **85**, No. 2, p. K81—K84 (1978).
See also *14.1*, p. *206*; 18.2, p. 282; 21.1, p. 370.

E. MROSAN and G. LEHMANN
Charge Density Perturbation around Impurities in Metals
(b) Vol. **85**, No. 2, p. K85—K88 (1978).
See also *13.4*, p. *171*.

F. FUJIMOTO, N. SUMIDA, H. FUJITA, and Y. UCHIDA
Electron Channeling in ⟨111⟩ Direction of Tungsten Crystals (II): Energy Range of 1000 to 2000 keV
(a) Vol. **45**, No. 2, p. 411—418 (1978).
See also *11*, p. *124*; 13, p. 143.

R. KAMEL, A. R. ALI, and Z. FARID
Annealing Spectrum of Cold-Worked Dilute Al–Li Alloys
(a) Vol. **45**, No. 2, p. 419—422 (1978).
See also *10.1*, p. *100*; 12.1, p. 134.

H. D. CARSTANJEN, J. DÜNSTL, G. LÖBL, and R. SIZMANN
Lattice Location and Determination of Thermal Amplitudes of Deuterium in α-$PdD_{0.007}$ by Channeling
(a) Vol. **45**, No. 2, p. 529—536 (1978).
See also *1.1*, p. *11*; 11, p. 124.

R. A. VARIN, J. W. WYRZYKOWSKI, W. ŁOJKOWSKI, and M. W. GRABSKI
Spreading of Extrinsic Grain Boundary Dislocations in Plastically Deformed Aluminium
(a) Vol. **45**, No. 2, p. 565—569 (1978).
See also *1.4*, p. *29*; 10.1, p. 100.

R. G. Pillay and P. N. Tandon
Hyperfine Magnetic Field at Sn in Heusler Alloys Rh_2MnSn and Rh_2NiSn
(a) Vol. **45**, No. 2, p. K109—K112 (1978).
See also *5*, p. *60*; *18.2*, p. *283*.

R. Feder and J. Kirschner
Surface Structure Determination via LEED Rotation Diagrams and the Relaxation of W (001)
(a) Vol. **45**, No. 2, p. K117—K119 (1978).
See also *1.6*, p. *40*.

D. Gignoux and H. Asmat
Magnetic Properties and Structures of TbAlGa and HoAlGa
(a) Vol. **45**, No. 2, p. K149—K153 (1978).
See also *1.1*, p. *11*; *18.4*, p. *298*; 21.4, p. 396.

J. Shy-Yih Wang and M. Rasolt
Effect of Nonlocal Potential on the Fermi Surface of Metals and the Phase Shift Parametrization Scheme
(b) Vol. **86**, No. 1, p. 37—40 (1978).
See also *13.2*, p. *162*.

H. Gollisch and L. Fritsche
Relativistic One-Particle Equation for Electron States of Heavy Metals
(b) Vol. **86**, No. 1, p. 145—150 (1978).
See also *13*, p. *143*.

F. Goedsche, R. Richter, and G. Vojta
Coherent Potential Approximation for the Electrical Resistivity of Partially Ordered Binary Magnetic Alloys
(b) Vol. **86**, No. 1, p. 177—182 (1978).
See also *14.1*, p. *206*; 18.2, p. 283.

V. Christoph and W. Schiller
Force-Correlation Function Method Applied to the Electrical Resistivity of Many-Band Metals
(b) Vol. **86**, No. 1, p. 231—237 (1978).
See also *14.1*, p. *206*.

E. Mrosan, Ch. Schober, and G. Lehmann
Multiple Scattering Theory for Magnetoresistance of Localized Impurities in Metals
(b) Vol. **86**, No. 1, p. 249—256 (1978).
See also 13.4, p. 172; *14.1*, p. *206*; 18, p. 273.

E. Kolley and W. Kolley
Singularities in the Paramagnetic Susceptibility of Random Hubbard Alloys
(b) Vol. **86**, No. 1, p. 397—408 (1978).
See also *13*, p. *143*; *18.1*, p. *277*; 18.2, p. 284.

K. Míšek
Anomalous Attenuation of Ultrasound in a Niobium Single Crystal at Low Temperatures
(b) Vol. **86**, No. 1, p. K31—K33 (1978).
See also *7*, p. *80*; 14.2, p. 213.

B. K. Chakrabarti, B. P. Barua, and S. K. Sinha
Coupling of the Martensitic Transformation with the Superconducting Transition
(b) Vol. **86**, No. 1, p. K55—K58 (1978).
See also *1.2*, p. *17*; *14.2*, p. *214*.

Th. Eifrig, E. Kolley, and W. Kolley
Electron Correlations and Ferromagnetism in Disordered Alloys
(b) Vol. **86**, No. 1, p. K79—K83 (1978).
See also *13*, p. *143*; 18.2, p. 284.

V. Drchal and J. Kudrnovský
The T-Matrix Approximation for the Anderson Model: A Self-Consistent Solution
(b) Vol. **86**, No. 1, p. K89—K92 (1978).
See also *13.4*, p. *172*.

J. Wortmann
Optically Stimulated Exoelectron Emission of Tribomechanically Activated Aluminium
(a) Vol. **46**, No. 1, p. 69—76 (1978).
See also *13.3*, p. *164*; *17*, p. *269*.

D. R. Krahn, R. L. Herny, D. B. Tanner, and P. E. Wigen
The Critical Fields of Superconducting Palladium Hydride
(a) Vol. **46**, No. 1, p. 209—212 (1978).
See also *14.2*, p. *214*.

Jong K. Lee and William C. Johnson
Calculation of the Elastic Strain Field of a Cuboidal Precipitate in an Anisotropic Matrix
(a) Vol. **46**, No. 1, p. 267—272 (1978).
See also *10.1*, p. *101*; *12.1*, p. *134*.

V. Raman and P. Mukhopadhyay
Mechanical Equation of State in Zirconium Alloys
(a) Vol. **46**, No. 1, p. K5—K8 (1978).
See also *10.1*, p. *101*.

A. Ślebarski
Anomalous Change of the Curie Temperature of $GdAl_2$
(a) Vol. **46**, No. 1, p. K29—K34 (1978).
See also *8*, p. *84*; *18.2*, p. *284*; 21.4, p. 397.

G. Ferron and M. Quintard
Evidence of an Internal Friction Phase Transformation Peak in Niobium–Hydrogen Alloys
(a) Vol. **46**, No. 1, p. K43—K46 (1978).
See also *10.1*, p. *101*; *12.1*, p. *134*.

G. C. Vezzoli
Optical and Electrical Effects during the Solidification of Mercury
(a) Vol. **46**, No. 1, p. K49—K53 (1978).
See also *1.2*, p. *17*; 14.1, p. 206; 20.1, p. 324.

A. Murasik, H. Ptasiewicz-Bąk, and A. Zygmunt
Magnetic Structure and Crystal-Field Effects in CeZn
(a) Vol. **46**, No. 1, p. K75—K78 (1978).
See also *18.2*, p. *284*; *18.4*, p. *299*; 21.4, p. 397.

P. R. Antoniewicz
The Depolarization Interaction of Adsorbed Non-Polar Molecules
(b) Vol. **86**, No. 2, p. 645—652 (1978).
See also *13.3*, p. *165*.

N. A. Tyapunina, E. K. Naimi, S. V. Gasparyan, and G. M. Zinenkova
The Orientation Dependence of Dislocation Internal Friction in Real Crystals (II): Hexagonal Crystals
(a) Vol. **46**, No. 2, p. 411—420 (1978).
See also *10.1*, p. *101*; *12.1*, p. *135*; 21.1, p. 372; 21.3, p. 394.

L. M. Caspers, R. H. J. Fastenau, A. van Veen, and W. F. W. M. van Heugten
Mutation of Vacancies to Divacancies by Helium Trapping in Molybdenum. Effect on the Onset of Percolation
(a) Vol. **46**, No. 2, p. 541—546 (1978).
See also *10.1*, p. *102*.

E. Seitz, W. Schmatz, G. S. Bauer, and W. Just
A Trigonal Distortion Field for Substitutionally Dissolved Bi Atoms in Pb
(a) Vol. **46**, No. 2, p. 557—565 (1978).
See also 1.3, p. 25; *10.1*, p. *102*.

R. Kamel, A. R. Ali, and Z. Farid
On the Recovery of Quenched and Irradiated Dilute Al–Li Alloys
(a) Vol. **46**, No. 2, p. 697—701 (1978).
See also *10.1*, p. *102*; 12.1, p. 135; 14.1, p. 207.

R. A. Varin and J. W. Wyrzykowski
"In Situ" Observation of Dislocation Generation from Grain Boundary in Aluminium
(a) Vol. **46**, No. 2, p. K79—K81 (1978).
See also *10.1*, p. *102*.

A. E. Karkin, B. N. Goshchitskii, V. E. Arkhipov, E. E. Valiev, and S. K. Sidorov
Electrical Resistivity and Transition Temperature of Highly Disordered Nb_3Sn
(a) Vol. **46**, No. 2, p. K87—K90 (1978).
See also 11, p. 125; *14.2*, p. *214*.

W. F. Huang and Y. C. Leung
On the Analysis of the Angular Correlation Data from Positron Annihilation in Metals
(a) Vol. **46**, No. 2, p. K107—K109 (1978).
See also *10.1*, p. *102*.

P. Ziesche, H. Wonn, Ch. Müller, V. V. Nemoshkalenko, and V. P. Krivitskii
Theoretical and Experimental Investigation of the Electronic Structure of the B2-Compound FeAl
(b) Vol. **87**, No. 1, p. 129—134 (1978).
See also *13.1*, p. *154*; 20, p. 316; 21.1.1, p. 386.

P. Pawlicki
Heat Magnetization in the Singlet–Triplet Model
(b) Vol. **87**, No. 1, p. K17—K20 (1978).
See also *18.2*, p. *286*.

E. Mrosan and G. Lehmann
Electronic States of 3d-Transition Metal Impurities in Aluminium
(b) Vol. 87, No. 1, p. K21—K24 (1978).
See also *13.4*, p. *175*.

J. Cornelis, J. Nihoul, F. Brosens, and J. T. Devreese
Interatomic Potential for Molybdenum
(b) Vol. 87, No. 1, p. K61—K64 (1978).
See also 6, p. 71; *10.1*, p. *102*.

S. Roth and J. Sonntag
Temperature Dependence of the Electrical Resistivity of Semiconducting CrAl Solid Solutions at High Temperatures
(b) Vol. 87, No. 1, p. K69—K72 (1978).
See also *14.3*, p. *220*; *18.4*, p. *300*; 21.1, p. 372.

V. M. Chernov and M. M. Savin
Elastic Interaction of 60° and Screw Dislocations with Point Defects in F.C.C. Crystals
(a) Vol. 47, No. 1, p. 45—55 (1978).
See also *10.1*, p. *103*.

I. P. Jones
Systematic Electron Diffraction by H.C.P. Metals (I): Theory and Main Beam Transmission
(a) Vol. 47, No. 1, p. 163—178 (1978).
See also *1.1*, p. *12*.

C. Esnouf and G. Fantozzi
Theoretical Aspects of the Bordoni Relaxation
(a) Vol. 47, No. 1, p. 201—211 (1978).
See also *10.1*, p. *103*; *12.1*, p. *135*.

M. Geddo, A. Stella, and G. Soncini
Photoelectron–Grain Boundary Scattering in Al Films of MOS Structures
(a) Vol. 47, No. 1, p. 223—227 (1978).
See also *16*, p. *264*.

D. Varchon, C. Oytana, and P. Delobelle
Application of a Method for Analysing Inelastic Relaxation-Extension to Creep at High Temperature
(a) Vol. 47, No. 1, p. 259—269 (1978).
See also *10.1*, p. *103*.

P. Erlenwein
On a Relationship between the Symmetries of Faceted Cones on Ion Bombarded Single-Crystal Faces and Ejection Patterns
(a) Vol. 47, No. 1, p. K9—K10 (1978).
See also *1.4*, p. *30*; 11, p. 126; 21.6, p. 402.

K. Melidis
On the Temperature Dependence of the Hall Effect and Electrical Resistivity of Cd, Zn, and Mg Single Crystals
(a) Vol. 47, No. 1, p. K27—K30 (1978).
See also *14.1*, p. *207*; 21.1, p. 373; 21.3, p. 394.

E. A. Vasilev and A. E. Gelyasin
Magnetic Properties of the $Mn_{1+x}Sb_{1-x}Sn_x$ System
(a) Vol. **47**, No. 1, p. K55—K57 (1978).
See also *18.2*, p. *286*; *18.3*, p. *295*.

B. Huber and G. Sicking
Tritium Diffusion in Palladium Alloys with Pt and Cu
(a) Vol. **47**, No. 1, p. K85—K89 (1978).
See also *9*, p. *92*.

I. G. Ratishvili
Stepped Ordering in Binary Solid Solutions
(b) Vol. **87**, No. 2, p. 461—471 (1978).
See also *1.2*, p. *19*; 1.3, p. 26.

F. Perrot
Phenomenological Study of Vacancies in Aluminium within Density-Functional and Second-Order Perturbation Formalisms
(b) Vol. **87**, No. 2, p. 707—718 (1978).
See also *10.1*, p. *103*; 14.1, p. 208.

I. P. Jones
Systematic Electron Diffraction by H.C.P. Metals (II): The Critical Voltage Effect
(a) Vol. **47**, No. 2, p. 385—395 (1978).
See also *1.1*, p. *13*.

C. E. Laciana, A. J. Pedraza, and E. J. Savino
Lattice Distortion Due to Oxygen and Nitrogen Di-Interstitial Clusters in Niobium and Vanadium
(a) Vol. **47**, No. 2, p. 425—436 (1978).
See also *10.1*, p. *103*; 21.1, p. 374.

E. Stracke and Ch. Herzig
Electromigration, Thermomigration, and Solubility of Copper in Lead
(a) Vol. **47**, No. 2, p. 513—521 (1978).
See also *9*, p. *93*; 14.1, p. 208.

R. H. J. Fastenau, A. van Veen, P. Penning, and L. M. Caspers
Monte Carlo Simulation of the Interaction of Interstitial Helium, Selfinterstitials, and Vacancies with Vacancies or Vacancy Clusters
(a) Vol. **47**, No. 2, p. 577—587 (1978).
See also *10.1*, p. *104*.

D. Kuhlmann-Wilsdorf
Deviations from Schmid's Law in Uni-Directional Strain
(a) Vol. **47**, No. 2, p. 639—649 (1978).
See also *10.1*, p. *104*; *12.1*, p. *135*.

Y. Kimura and M. Umeda
Temperature Dependence of the Electrical Resistivity of Laves Phase Compounds V_2(Hf, Zr)
(a) Vol. **47**, No. 2, p. K111—K113 (1978).
See also *1.2*, p. *20*; *14.1*, p. *208*.

K. Balcerek and T. Tyc
Heat Flux Rectification in Tin–α-Brass System
(a) Vol. **47**, No. 2, p. K125—K128 (1978).
See also *8*, p. *86*.

H. Misiorek, T. Zakrzewski, and J. Rafałowicz
Influence of Plastic Deformation on the Thermal Conductivity Maximum of Copper and Aluminium in the Temperature Range 4.2 to 70 K
(a) Vol. **47**, No. 2, p. K137—K140 (1978).
Erratum
(a) Vol. **50**, No. 2, p. K261 (1978).
See also *8*, p. *87*; *10.1*, p. *104*; 21.1, p. 375.

E. A. Pushkarev, N. S. Petrenko, and V. A. Finkel
Thermal Expansion of the Superconducting Compound HfV_2 at Low Temperatures
(a) Vol. **47**, No. 2, p. K145—K148 (1978).
See also *8*, p. *87*; *14.2*, p. *214*.

G. Dharmadurai and B. A. Ratnam
A Non-Ohmic Potential Transverse to a Long Superconducting Weak-Link
(a) Vol. **47**, No. 2, p. K187—K190 (1978).
See also *14.2*, p. *214*.

K. Yonemitsu, K. Takano, and T. Matsuda
Hall Coefficient of Dilute Aluminium Alloys (I): Calculation by Four-OPW Approximation
(b) Vol. **88**, No. 1, p. 273—282 (1978).
See also 13.2, p. 162; *14.1*, p. *209*.

B. K. Agarwal and R. K. Johri
X-Ray K-Absorption Structure of Zirconium Metal and Its Compounds
(b) Vol. **88**, No. 1, p. 309—313 (1978).
See also *13.1*, p. *156*; *20*, p. *317*.

J. Cornelis, J. Nihoul, F. Brosens, and J. T. Devreese
Interatomic Potentials from Experimental Phonon Spectra. Application to Au, Cu, K, Ni, Pd, Pt, and Rb
(b) Vol. **88**, No. 2, p. K93—K96 (1978).
See also *4*, p. *58*; 6, p. 74.

J. Mucha and J. Rafałowicz
Thermal Conductivity Minimum of Aluminium
(a) Vol. **48**, No. 1, p. 221—224 (1978).
See also *8*, p. *87*.

O. A. Troitskii
Fractionation of Stable Isotopes of Metals by Combined Electric Migration and Zone Melting
(a) Vol. **48**, No. 1, p. 229—234 (1978).
See also *3*, p. *55*; 9, p. 93.

V. A. Streltsov
Shear Stresses Induced by Hydrostatic Pressure near the Surface of an Inclusion
(a) Vol. **48**, No. 1, p. K23—K25 (1978).
See also *1.4*, p. *32*; *12.1*, p. *135*.

P. Vostrý
On the Influence of Silver on the Recovery Spectrum of Queched Cadmium Foils
(a) Vol. 48, No. 1, p. K39—K42 (1978).
See also *10.1*, p. *104*; *14.1*, p. *209*.

A. F. Akkerman
Reflection of Slow Hydrogen and Helium Ions from Solid Surfaces
(a) Vol. 48, No. 1, p. K47—K50 (1978).
See also *11*, p. *128*; 21.1.1, p. 388.

O. Simmich and H. Löffler
A Simple Model to Describe Isothermic Phase Transformations in Binary Alloys
(a) Vol. 48, No. 1, p. K67—K70 (1978).
See also *1.2*, p. *21*; 10.1, p. 104.

S. Panchanadeeswaran, R. W. Ure, Jr., and J. G. Byrne
Positron Trapping at Precipitates in Aluminium-4wt% Copper Single Crystals
(a) Vol. 48, No. 1, p. K83—K87 (1978).
See also *1.3*, p. *27*; *10.1*, p. *105*.

C. Koenig
On the Positron Localization in Ordered and Disordered Metallic Alloys
(b) Vol. 88, No. 2, p. 569—579 (1978).
See also *1.3*, p. *27*; 10.1, p. 105.

H. Szymczak
On Magnetostriction in Amorphous Thin Films
(b) Vol. 88, No. 2, p. K97—K101 (1978).
See also *2*, p. *47*; *18*, p. *275*.

H. Bossac, H.-G. Fabian, and H. Löffler
A Quantitative Interpretation of the Decomposition Kinetics of Some Al-Zn-Mg Alloys
(a) Vol. 48, No. 2, p. 369—375 (1978).
See also *1.3*, p. *27*; *10.1*, p. *105*; 14.1, p. 209; 21.1, p. 376; 21.3, p. 395.

M. van Sande, R. de Ridder, G. van Tendeloo, J. van Landuyt, and S. Amelinckx
High Resolution Study of One-Dimensional Long Period Superstructures in Cu_3Sn with Additions of Zinc and Nickel
(a) Vol. 48, No. 2, p. 383—394 (1978).
See also *1.3*, p. *27*; 1.4, p. 32; 21.1, p. 376.

D. Rouby and P. Fleischmann
Spectral Analysis of Acoustic Emission from Aluminium Single Crystals Undergoing Plastic Deformation
(a) Vol. 48, No. 2, p. 439—445 (1978).
See also *7*, p. *81*; 10.1, p. 105.

T. Hirata
Structural Changes during Annealing Vacuum-Evaporated Thin Aluminium Films at Temperatures near the Thermodynamic Melting Point
(a) Vol. 48, No. 2, p. 451—458 (1978).
See also *1.4*, p. *32*; *1.5*, p. *37*; 2, p. 48.

R. Gauthier, M. Babout, and C. Guittard
Laser Induced Desorption Studied with Mirror Electron Microscopy
(a) Vol. 48, No. 2, p. 459—464 (1978).
See also *17*, p. *270*.

W. Richter and R. Diller
Flat Thin Film SQUIDs with Current Transformers
(a) Vol. **48**, No. 2, p. 513—521 (1978).
See also *14.2*, p. *215*.

M. L. Johnson, S. Saterlie, D. Boice, and J. G. Byrne
The Removal of Defects from Solids as Observed with Positron Annihilation
(a) Vol. **48**, No. 2, p. 551—554 (1978).
See also *1.3*, p. *27*; *10.1*, p. *105*; 21.1, p. 376.

E. Vlachá, B. Sprušil, and P. Vostrý
Dynamic Defect Annealing during Plastic Deformation of Al
(a) Vol. **48**, No. 2, p. K113—K114 (1978).
See also *10.1*, p. *105*.

M. Ohta, T. Kanadani, A. Sakakibara, and M. Yamada
Aging and Reversion in Low-Concentrated Al-Zn Alloys
(a) Vol. **48**, No. 2, p. K141—K144 (1978).
See also *14.1*, p. *209*.

R. Richter, F. Goedsche, and G. Vojta
Statistical Theory of Electrical Resistivity of Magnetic Ternary Alloys with Partial Long-Range Order
(b) Vol. **89**, No. 1, p. 77—83 (1978).
See also 13, p. 147; *14.1*, p. *209*; 18.4, p. 301.

K. Saermark
Azbel-Kaner Cyclotron Resonance as a Long-Wavelength Limit of an Exact Expression for the Surface Impedance of a Metal in a Magnetic Field
(b) Vol. **89**, No. 1, p. 117—121 (1978).
See also *13.2*, p. *163*; 14.1, p. 210.

J. Richter and W. Schiller
Residual Resistivity of Disordered Systems
(b) Vol. **89**, No. 1, p. 143—150 (1978).
See also *2*, p. *48*; *14.1*, p. *210*.

N. V. Volkenshtein, V. E. Startsev, and V. I. Cherepanov
Low Temperature Anomalies of the Hall Effect in Tungsten
(b) Vol. **89**, No. 1, p. K53—K56 (1978).
See also *14.1*, p. *210*.

D. R. Mašović and S. Zeković
Model Pseudopotential for Aluminium
(b) Vol. **89**, No. 1, p. K57—K60 (1978).
See also *13.1*, p. *158*.

K. Müller and K. Heinz
Reconstructed Surfaces
(a) Vol. **49**, No. 1, p. 39—49 (1978).
See also *1.6*, p. *41*; 21.6, p. 404; 22.1.2, p. 436.

H. Löffler, O. Kabisch, and R. Kroggel
About a Relation between Microhardness and the Average Radius of the Guinier-Preston Zones in Al–Zn Alloys
(a) Vol. **49**, No. 1, p. 299–303 (1978).
See also *10.1*, p. *106*; 21.1, p. 377.

W. Andrä, H. Danan, R. Mattheis, and U. Zeimer
On the Electrical Resistivity of PdH_x
(a) Vol. **49**, No. 1, p. K97–K100 (1978).
See also *14.1*, p. *210*.

L. Fritsche, H. Gollisch, and J. Noffke
Dipole Barriers at Structured Metal Surfaces
(b) Vol. **89**, No. 2, p. 367–373 (1978).
See also *13.3*, p. *167*.

J. Reichelt and K. Winzer
Magnetoresistance and de Haas-van Alphen Effect on $LaAl_2$
(b) Vol. **89**, No. 2, p. 489–494 (1978).
See also *13.2*, p. *163*.

K. Elk, J. Richter, and V. Christoph
Electrical Conductivity of Disordered Alloys with Strong Hybridization
(b) Vol. **89**, No. 2, p. 537–540 (1978).
See also 13.1, p. 159; *14.1*, p. *211*.

G. Seifert E. Mrosan and H. Müller
Electronic Structure of 4d and 5d Transition Metal Clusters
(b) Vol. **89**, No. 2, p. 553–560 (1978).
See also *13*, p. *148*.

H. Sato, T. Babauchi, and K. Yonemitsu
Hall Coefficient of Dilute Aluminium Alloys (II): Application
(b) Vol. **89**, No. 2, p. 571–576 (1978).
See also *14.1*, p. *211*.

P. S. Grover
Computation of Van Der Waals Forces across Metal Surfaces
(b) Vol. **89**, No. 2, p. K163–K166 (1978).
See also *4*, p. *59*.

G. Seifert, E. Mrosan, H. Müller, and P. Ziesche
SW-Xα Calculations of Small Niobium Clusters
(b) Vol. **89**, No. 2, p. K175–K178 (1978).
See also *13.1*, p. *159*.

J. Gastaldi and C. Jourdan
New Possibilities for Recrystallization Study by X-Ray Synchrotron Radiation Topography
(a) Vol. **49**, No. 2, p. 529–537 (1978).
See also 1.4, p. 33; *3*, p. *56*.

A. Dunlop, N. Lorenzelli, and J. C. Jousset
Dechanneling of α Particles by Carbon Dissolved in Palladium
(a) Vol. **49**, No. 2, p. 643–650 (1978).
See also *11*, p. *130*; 14.1, p. 211.

M. IGNAT, R. BONNET, D. CAILLARD, and J. L. MARTIN
Creep of the Lamellar Al–$CuAl_2$ Composite (I): Microstructural Observations
(a) Vol. **49**, No. 2, p. 675–684 (1978).
See also *10.1*, p. *106*; 21.1, p. 379.

E. M. SAVITSKII, P. V. GELD, V. E. ZINOVEV, N. B. GORINA, V. I. SPERELUP, V. P. POLYAKOVA, and A. L. SOKOLOV
Transport Properties of Osmium Single Crystals at High Temperatures
(a) Vol. **49**, No. 2, p. K117–K120 (1978).
See also *8*, p. *88*; *14.1*, p. *211*.

B. B. SAHAY and P. I. SRIVASTAVA
Isothermal Compressibility for Various Liquid Metals by Harmonic Model Potential
(b) Vol. **90**, No. 1, p. 93–97 (1978).
See also 2, p. 51; *12.1*, p. *136*; 21.1, p. 379; 21.2, p. 393.

A. D. MOKRUSHIN and R. KH. SABIROV
Positron Thermalization as Quantum Statistical Process
(b) Vol. **90**, No. 1, p. 99–104 (1978).
See also 6, p. 77; *13*, p. *148*.

L. DUMOULIN, P. NÉDELLEC, C. ARZOUMANIAN, and J. P. BURGER
Optical Phonons in PdH_x and PdD_x by Superconducting Tunneling Measurements
(b) Vol. **90**, No. 1, p. 207–210 (1978).
See also *6*, p. *77*; *14.2*, p. *215*.

A. FREUDENHAMMER
Consequences of the Linear Part of the Specific Heat of Dilute Magnetic Alloys
(b) Vol. **90**, No. 1, p. 269–275 (1978).
See also *8*, p. *89*; 13, p. 148.

H. P. AUBAUER
Residual Electrical Resistivity of Alloys
(b) Vol. **90**, No. 1, p. 345–354 (1978).
See also *14.1*, p. *211*.

N. N. SINHA and P. L. SRIVASTAVA
Resistivity and Thermo-Electric Power of Monovalent, Divalent, and Trivalent Liquid Metals
(b) Vol. **90**, No. 1, p. 369–374 (1978).
See also 2, p. 51; *14.1*, p. *211*; *15*, p. *261*; 21.1, p. 380; 21.2, p. 393.

A. N. CHERKASOV, A. A. GALKIN, and L. T. TSYMBAL
W and Mo RF Surface Impendance. New Type of Oscillations
(b) Vol. **90**, No. 1, p. K17–K20 (1978).
See also *13.3*, p. *168*; *14.1*, p. *211*; 19, p. 313.

W. F. W. M. VAN HEUGTEN, A. VAN VEEN, and L. M. CASPERS
Molybdenium–Helium and Helium–Helium Interaction Potential from Helium Desorption Experiments
(b) Vol. **90**, No. 1, p. K89–K94 (1978).
See also 6, p. 78; *10.1*, p. *107*.

R. SCHÖPKE and E. MROSAN
Influence of Multiple Scattering Effects on the Residual Resistivity of 3d-Transition Metal Impurities in Aluminium
(b) Vol. **90**, No. 1, p. K95—K98 (1978).
See also *13.4*, p. *184*; *14.1*, p. *212*.

E. KOLLEY and W. KOLLEY
Electron Correlation Effects on the Spin Wave Stiffness Constant in Itinerant Ferromagnets
(b) Vol. **90**, No. 1, p. K103—K107 (1978).
See also *13.5.4*, p. *202*; 18.2, p. 291.

P. DUVAL and A. HAUET
Interpretation of Diffuse Intensity Lines Observed in Electron Diffraction in Cu_3Ga Alloy
(a) Vol. **50**, No. 1, p. 67—72 (1978).
See also *1.4*, p. *34*; 21.1, p. 380.

R. YAMAMOTO, T. MIHARA, and M. DOYAMA
Comments on the Glass Forming Region of the Metal–Metalloid Systems Predicted by the Polk Model
(a) Vol. **50**, No. 1, p. 165—169 (1978).
See also *2*, p. *52*.

H. U. HABERMEIER, R. AOKI, and H. KRONMÜLLER
The Quantitative Determination of Flux-Density Gradients in Type-II Superconductors by Means of the Magneto-Optical Faraday Effect
(a) Vol. **50**, No. 1, p. 187—194 (1978).
See also *14.2*, p. *215*; 18, p. 275; 20, p. 318.

I. P. GREBENNIK and S. P. YURCHENKO
The Use of Two-Dimensional Electron Diffraction Effects for the Initial Stage Research of γ-Phase Growth in the Metal–Melt Contact Zone
(a) Vol. **50**, No. 1, p. K1—K3 (1978).
See also *1.5*, p. *39*; 21.6, p. 405.

S. K. PABI
Mechanism of the Dissolution of Si in Al
(a) Vol. **50**, No. 1, p. K9—K12 (1978).
See also *9*, p. *94*; 14.3.4, p. 246; 22.1.2, p. 438.

L. ROTHKIRCH, R. HERRMANN, H. DWELK, and U. PREPPERNAU
Surface Impedance Investigations of Superconducting Niobium
(b) Vol. **90**, No. 2, p. 517—524 (1978).
See also 13.3, p. 168; *14.2*, p. *216*.

W. BRAUN, M. NEUMANN, M. IWAN, and E. E. KOCH
UPS Measurements of CO Chemisorbed and Condensed on Rh (111) Using Synchrotron Radiation
(b) Vol. **90**, No. 2, p. 525—533 (1978).
See also *13.1*, p. *160*; *17*, p. *270*.

W. F. W. M. VAN HEUGTEN, F. VAN DEN BERG, L. M. CASPERS, A. VAN VEEN, and J. TH. M. DE HOSSON
The Influence of Interatomic Potentials on the Interaction of He with a ½ ⟨111⟩ {110} Edge Dislocation in Molybdenum
(b) Vol. **90**, No. 2, p. 643—648 (1978).
See also *10.1*, p. *107*.

E. Krüger

Superconductivity Originating from Quasi-Orbital Electrons (III): Quasi-Orbital Conduction Electrons in Non-Adiabatic Systems
(b) Vol. **90**, No. 2, p. 719—731 (1978).
See also 13.1, p. 161; *14.2*, p. *216*.

L. Rothkirch

Magnetic Surface States in Superconductors
(b) Vol. **90**, No. 2, p. K115—K119 (1978).
See also *13.3*, p. *168*; *14.2*, p. *216*.

T. Soma

Equation of State under Pressure for Metallic β-Sn
(b) Vol. **90**, No. 2, p. K129—K131 (1978).
See also 1.2, p. 24; *4*, p. *59*; *12.1*, p. *136*.

K. Masuda

On the Electronic Structure of Ordered Adlayers
(b) Vol. **90**, No. 2, p. K179—K183 (1978).
See also *13.3*, p. *168*.

R. I. Kuznetsova and N. N. Zhukov

Structural Peculiarities of the Superplastic Behaviour of the Alloy Al–4 wt%Ge Due to Porosity
(a) Vol. **50**, No. 2, p. 385—391 (1978).
See also 1.4, p. 34; *10.1*, p. *108*.

Y. Motohashi and S. Ohtake

The Effect of Self-Energy of Dislocations on the Choice of Slip Systems in Crystals
(a) Vol. **50**, No. 2, p. 449—458 (1978).
See also *10.1*, p. *108*; *12.1*, p. *136*.

H. Düsterhöft

Dependence of Secondary Ion Emission on the Mass of the Bombarding Ions (Z_1-Dependence)
(a) Vol. **50**, No. 2, p. 503—507 (1978).
See also *17*, p. *271*.

J. Wortmann

Upon the Photostimulated Electron Emission from Tribomechanically Activated Aluminium
(a) Vol. **50**, No. 2, p. 509—516 (1978).
See also *13.3*, p. *168*; *17*, p. *271*.

M. van Sande, R. de Ridder, J. van Landuyt, and S. Amelinckx

A Study by Means of Electron Microscopy and Electron Diffraction of Vacancy Ordering in Ternary Alloys of the System AlCuNi
(a) Vol. **50**, No. 2, p. 587—599 (1978).
See also *1.3*, p. *28*; *1.4*, p. *34*; 10.1, p. 108; 21.1, p. 382.

N. P. KOBELEV and YA. M. SOIFER
Electronic Damping of Dislocations in Lead
(a) Vol. **50**, No. 2, p. K185—K188 (1978).
See also 7, p. 82; *10.1*, p. *108*; *14.2*, p. *216*.

YU. M. VYSOTSKII, E. A. VASILEV, A. E. GELYASIN, and A. V. SHABLOVSKII
Nuclear Zeeman Effect in the $Mn_{1+x}Sb_{1-x}Sn_x$ System
(a) Vol. **50**, No. 2, p. K205—K208 (1978).
See also 5, p. 63; *18.2*, p. *292*; *18.3*, p. *297*.

21.1 Transition Metals (Group Sc to Zn) and Their Alloys

Review Articles

K. Bärner
Exchange Interactions and the Coupled Spin–Lattice Phase Transitions in Manganese Pnictides
(b) Vol. 88, No. 1, p. 13–23 (1978).
See also 18.1, p. 277; *18.2*, p. *281*; 18.3, p. 295; *18.4*, p. *298*.

M. Tomášek and Š. Pick
Projected Surface Energy Bands and Shockley Surface States of Transition Metals. General Considerations and the Example of the B.C.C. Lattice
(b) Vol. 89, No. 1, p. 11–26 (1978).
See also 13.1, p. 150; *13.3*, p. *164*; 21.1.1, p. 384.

Original Papers and Short Notes

G. C. Aers, A. D. Boardman, and P. Clark
A Computational Study of the Correlation of Attenuated Total Reflection Minima with Surface Polariton Modes
(b) Vol. 85, No. 1, p. 171–178 (1978).
See also 13.5, p. 187; *18*, p. *273*; *20.1*, p. *319*.

H. Eckseler and Ch. Herzig
On the Mass Dependence of the Energy Factor ΔK of the Isotope Effect for Impurity Diffusion in Copper
(b) Vol. 85, No. 1, p. 185–193 (1978).
See also *9*, p. *91*; 10.1, p. 99.

C. Norris and G. P. Williams
Photoelectron Measurements of α-Phase Alloys CuZn, CuGa, and CuGe
(b) Vol. 85, No. 1, p. 325–330 (1978).
See also *13.4*, p. *169*; *17*, p. *269*.

G. Lautenschläger and E. Mrosan
Change in Electronic Charge Density around 3d-Transition Metal Impurities in Al and Cu
(b) Vol. 85, No. 1, p. K19–K21 (1978).
See also *13.4*, p. *169*; 21, p. 352.

B. G. Mulimani and M. A. Angadi
A Note on the Analysis of the dHvA Effect in Zinc Based Kondo Alloys with Magnetic Breakdown
(b) Vol. 85, No. 1, p. K33–K38 (1978).
See also *13.2*, p. *162*.

A. V. Deryagin, N. V. Kudrevatykh, and V. N. Moskalev
Magnetic Anisotropy and Intersublattice Exchange Interaction in $Er_2(Co_{1-x}M_x)_{17}$ Intermetallic Compounds
(a) Vol. 45, No. 1, p. 71–76 (1978).
See also *18.2*, p. *281*; 21.4, p. 396.

H. Neuhäuser and G. Schulz
Dislocation Path Lengths in Cu Single Crystals of Low Dislocation Density after Pulse Loading
(a) Vol. **45**, No. 1, p. 89—97 (1978).
See also *10.1*, p. *99*.

S. Takahashi
Compression-Induced Magnetic Anisotropy and Superlattice Dislocations in Ni_3Fe (III): The Annealing Effect
(a) Vol. **45**, No. 1, p. 133—138 (1978).
See also *10.1*, p. *99*; *18.2*, p. *281*; 21.1.1, p. 384.

P. Turchi, Y. Calvayrac, and F. Plicque
Effects of Configurational Order in a Ni_3Fe Single Crystal on Debye Temperature Deduced from Elastic Constants and X-Ray Measurements
(a) Vol. **45**, No. 1, p. 229—238 (1978).
See also *1.3*, p. *25*; 6, p. 65; 12.1, p. 134; 21.1.1, p. 384.

S. S. Sheinin and J. M. Corbett
Electron Microscope Images of Twin Boundaries (II): Contrast Mechanisms
(a) Vol. **45**, No. 1, p. 289—295 (1978).
See also *1.4*, p. *29*.

C. E. Laciana, A. J. Pedraza, and E. J. Savino
Lattice Distortion and Migration Energy of Oxygen in Vanadium and Hydrogen in Niobium, Tantalum, and Vanadium
(a) Vol. **45**, No. 1, p. 315—326 (1978).
See also *10.1*, p. *100*; 21, p. 353.

I. M. Belii, F. F. Komarov, V. S. Tishkov, and V. M. Yankovskii
Formation of Chemical Compounds by Ion Bombardment of Thin Transition Metal Films
(a) Vol. **45**, No. 1, p. 343—352 (1978).
See also *1.2*, p. *15*; *1.5*, p. *36*; 11, p. 123.

S. A. Mahmoud, M. A. Semary, and F. Mohsen
Reduction of the Magnetization Curves for Ni5%Mn Alloy
(a) Vol. **45**, No. 1, p. K5—K7 (1978).
See also *18.2*, p. *281*.

J. Schneider, A. Handstein, and R. Hesske
Mictomagnetic to Ferromagnetic Transition in Amorphous Ni–Fe Based Alloys
(a) Vol. **45**, No. 1, p. K47—K50 (1978).
See also *2*, p. *42*; *18.2*, p. *282*; 21.1.1, p. 384.

A. Bessis, Y. Bismuth, and J. Bigot
Study of Damage in Chronium by Neutron Irradiation at 24.5 K
(a) Vol. **45**, No. 1, p. K71—K76 (1978).
See also *11*, p. *123*; *14.1*, p. *205*.

M. V. Medvedev and E. L. Rumyantsev
On the Existence of a Spin-Glass State in an Ising Alloy Model with Mixed Nearest-Neighbour Exchange Interactions
(b) Vol. **85**, No. 2, p. 427—438 (1978).
See also *18.2*, p. *282*; *18.4*, p. *298*; 21.1.1, p. 384.

F. Goedsche and G. Vojta
Spin-Disorder Resistivity of Ternary Alloys in the CPA Weak Scattering Limit
(b) Vol. **85**, No. 2, p. K81—K84 (1978).
See also *14.1*, p. *206*; 18.2, p. 282; 21, p. 354.

A. G. Khachaturyan and A. F. Rumynina
Strain Energy Controlled Shape of Crystals Forming Due to a Martensite Transformation or Decomposition
(a) Vol. **45**, No. 2, p. 393—402 (1978).
See also *10.1*, p. *100*; 21.1.1, p. 385.

J. Roggen, J. Nihoul, L. Stals, and J. Cornelis
α-Particle Irradiation Damage and Stage I Recovery in Zinc
(a) Vol. **45**, No. 2, p. 439—443 (1978).
See also *10.1*, p. *100*; *11*, p. *124*.

H. Maeda
Magnetocrystalline Anisotropy of Pr $(Co_{1-x}Cu_x)_5$ Compounds. Experiment and Theoretical Analysis
(a) Vol. **45**, No. 2, p. 445—452 (1978).
See also *18.2*, p. *283*; 21.4, p. 396.

M. van Sande, G. van Tendeloo, J. van Landuyt, and S. Amelinckx
A Study by Transmission Electron Microscopy and Diffraction of the Ternary Alloy System Cu_2AuPd
(a) Vol. **45**, No. 2, p. 553—558 (1978).
See also *1.2*, p. *16*; *1.4*, p. *29*.

R. C. O'Handley and D. S. Boudreaux
Magnetic Properties of Transition Metal–Metalloid Glasses. A Charge Transfer Model
(a) Vol. **45**, No. 2, p. 607—615 (1978).
See also *2*, p. *43*; *18.2*, p. *283*.

S. K. Sen, M. B. Dutt, and A. K. Barua
The Diffusion of Iron in Copper and of Nickel in Silver
(a) Vol. **45**, No. 2, p. 657—663 (1978).
See also *9*, p. *91*; 21.1.1, p. 385; 21.6, p. 401.

G. Dlubek, J. Pielaszek, and O. Brümmer
Comparison of Positron Annihilation and X-Ray Estimations of the Dislocation Density and Depth of Dislocation Profiles in Nickel Transformed into Hydride and Decomposed
(a) Vol. **45**, No. 2, p. K87—K90 (1978).
See also *10.1*, p. *100*.

J. Barton and E. Wieser
Investigation of Magnetic Preferred Directions in an Fe–Mn Base Alloy
(a) Vol. **45**, No. 2, p. K105—K108 (1978).
See also *5*, p. *60*; *18.2*, p. *283*; 21.1.1, p. 385.

E. Burzo, D. P. Lazar, and L. Palit
Paramagnetic Behaviour of $Gd(Fe_xCo_{1-x})_2$ Compounds
(a) Vol. **45**, No. 2, p. K145—K148 (1978).
See also *18.1*, p. *277*; 18.3, p. 295; 21.1.1, p. 385; 21.4, p. 396.

F. Sacchetti, P. Bosi, F. Dupré, G. Frollani, F. Menzinger, and M. C. Spinelli
Observation of Precursory Phonons of an F.C.C.–H.C.P. Transition
(b) Vol. **86**, No. 1, p. 77—81 (1978).
See also *1.2*, p. *16*; *6*, p. *67*.

N. I. Kulikov, V. N. Borzunov, and A. D. Zvonkov
The Electronic Band Structure and Interatomic Bond in Nickel and Titanium Hydrides
(b) Vol. **86**, No. 1, p. 83—91 (1978).
See also 4, p. 57; *13.1*, p. *152*.

K. M. Miller and P. D. Bristowe
A Computer Simulation of Point Defects and Planar Defects in Nickel
(b) Vol. **86**, No. 1, p. 93—102 (1978).
See also *10.1*, p. *101*; 12.1, p. 134.

W. F. W. M. van Heugten
An Interatomic Potential for Copper from the Phonon Spectra
(b) Vol. **86**, No. 1, p. 277—281 (1978).
See also 6, p. 67; *10.1*, p. *101*.

H. Wolf and H. Solbrig
Model Calculations on Surface-Resonance Fine Structures in Low-Energy Electron Diffraction
(b) Vol. **86**, No. 1, p. K15—K19 (1978).
See also *1.6*, p. *40*.

D. Werneth
Mechanical and Magnetic Relaxion Measurements on Pure and Doped Nickel after Plastic Deformation and Electron Irradiation
(a) Vol. **46**, No. 1, p. 99—106 (1978).
See also *10.1*, p. *101*; 12.1, p. 134; 18.2, p. 284.

H. Maeda
Anistropy of Saturation Magnetization in $Pr(Co_{1-x}Cu_x)_5$ Compounds
(a) Vol. **46**, No. 1, p. 233—238 (1978).
See also *18.2*, p. *284*; 21.4, p. 396.

G. van Tendeloo and S. Amelinckx
Erratum to: Long-Period Antiphase Boundary Structures in the Ordered Alloy Au_4Zn (I)
(a) Vol. **46**, No. 1, p. 399 (1978).
See also *1.3*, p. *25*; *1.4*, p. *30*; 21.6, p. 401.

O. P. Sharma
A Pseudopotential Study of the Electronic Properties of Noble Metals
(b) Vol. **86**, No. 2, p. 483—492 (1978).
See also 4, p. 57; *6*, p. *68*; 14.1, p. 207; 21.6, p. 401.

W. John, D. Hamann, and P. Urwank
Partial Densities of States and Electron–Phonon Interaction in Transition Metals
(b) Vol. **86**, No. 2, p. 569—577 (1978).
See also *13.1*, p. *154*; 14.2, p. 214; 21.1.1, p. 386.

A. Freudenhammer
On the RKKY Interaction in Spin Glass Systems
(b) Vol. **86**, No. 2, p. 579—584 (1978).
See also *13*, p. *144*; *18*, p. *274*.

F. M. Galperin
Atomic Magnetic Moments of Fe, Co, Ni and Hyperfine Fields at Impurities in Ferromagnetic Iron
(b) Vol. **86**, No. 2, p. 679–684 (1978).
See also *13*, p. *144*; *18.2*, p. *285*; 21.1.1, p. 386.

Anand P. Pathak
Stopping Power of Solids in Planar Channelling
(b) Vol. **86**, No. 2, p. 751–758 (1978).
See also *11*, p. *125*; 13, p. 144; 21.6, p. 402.

K. Berndt, O. Brümmer, and U. Marx
Study of the Fe 2p States by Means of Photoelectron Spectroscopy in Iron and Iron Alloys
(b) Vol. **86**, No. 2, p. K93–K95 (1978).
See also *13*, p. *144*; 20, p. 316; 21.1.1, p. 386.

N. A. Tyapunina, E. K. Naimi, S. V. Gasparyan, and G. M. Zinenkova
The Orientation Dependence of Dislocation Internal Friction in Real Crystals (II): Hexagonal Crystals
(a) Vol. **46**, No. 2, p. 411–420 (1978).
See also *10.1*, p. *101*; *12.1*, p. *135*; 21, p. 357; 21.3, p. 394.

P. K. Madden and J. E. Harbottle
The Imaging of Slip Bands in Nickel
(a) Vol. **46**, No. 2, p. 639–652 (1978).
See also *1.5*, p. *36*; 11, p. 125.

M. Kopcewicz
Mössbauer Study of the Fast Magnetization Reversal Forced in Permalloy and Invar by an External RF Magnetic Field
(a) Vol. **46**, No. 2, p. 675–685 (1978).
See also *5*, p. *61*; *18.2*, p. *285*; 21.1.1, p. 386.

H. J. Blythe and L. H. Dunleavy
Point-Defect Relaxations in Dilute Iron–Nickel Alloys after Low-Temperature Neutron Irradiation
(a) Vol. **46**, No. 2, p. K95–K99 (1978).
See also *10.1*, p. *102*; 11, p. 125; *18.2*, p. *285*; 21.1.1, p. 386.

H.-R. Müller and R. Perthel
Model Calculations for the Anisotropy Formation by Resputtering Processes during the Growth of Amorphous Gd–Co Films
(b) Vol. **87**, No. 1, p. 203–211 (1978).
See also *2*, p. *44*; 3, p. 54; *18.2.1*, p. *293*; 21.4, p. 397.

A. Eroglu, S. Arajs, C. A. Moyer, and K. V. Rao
Electrical Resistivity of Chromium–Gold Alloys between 77 and 700 K
(b) Vol. **87**, No. 1, p. 287–291 (1978).
See also *14.1*, p. *207*; 21.6, p. 402.

S. Roth and J. Sonntag
Temperature Dependence of the Electrical Resistivity of Semiconducting CrAl Solid Solutions at High Temperatures
(b) Vol. **87**, No. 1, p. K69–K72 (1978).
See also *14.3*, p. *220*; *18.4*, p. *300*; 21, p. 358.

G. VAN TENDELOO, R. WOLF, D. VAN DYCK, and S. AMELINCKX
The Study of Ordering in the Au_5Mn_2 Alloy by Means of Structure Images
(a) Vol. **47**, No. 1, p. 105–122 (1978).
See also *1.3*, p. *25*; *1.4*, p. *30*; 21.6, p. 402.

U. RINDELHARDT, S. MAGER, W. BLAU, and E. HEGENBARTH
Recovery of Low Temperature Specific Heat, Resistivity, and Coercive Force in Fe–Cr–Al Alloys
(a) Vol. **47**, No. 1, p. 157–161 (1978).
See also 8, p. 85; *10.1*, p. *103*; 14.1, p. 207; 18.2, p. 286; 21.1.1, p. 387.

R. WOLF, G. VAN TENDELOO, J. VAN LANDUYT, and S. AMELINCKX
Evidence for the Occurrence of the AABB Structure in the Gold–Manganese System
(a) Vol. **47**, No. 1, p. 241–244 (1978).
See also *1.2*, p. *18*; 1.3, p. 25; 1.4, p. 30; 21.6, p. 402.

A. J. MORTON
Faulted Defects in γ-Brass-Type Alloys
(a) Vol. **47**, No. 1, p. 293–304 (1978).
See also *1.3*, p. *26*; *1.4*, p. *30*; 10.1, p. 103.

A. M. MOISIN
Neutron Scattering Study of Local Ordering in Fe–Ni and Fe–Al Alloys
(a) Vol. **47**, No. 1, p. 305–309 (1978).
See also *18*, p. *274*; 21.1.1, p. 387.

C. L. BAUER and A. G. JORDAN
Study of Interfacial Reactions in Bimetallic Thin-Film Couples by Contact Resistance Measurements
(a) Vol. **47**, No. 1, p. 321–328 (1978).
See also *1.5*, p. *37*; 14.1, p. 207.

K. MELIDIS
On the Temperature Dependence of the Hall Effect and Electrical Resistivity of Cd, Zn, and Mg Single Crystals
(a) Vol. **47**, No. 1, p. K27–K30 (1978).
See also *14.1*, p. *207*; 21, p. 358; 21.3, p. 394.

M. MASZKIEWICZ
Specific Heat of High Purity Nickel Single Crystal near the Critical Temperature
(a) Vol. **47**, No. 1, p. K77–K80 (1978).
See also *8*, p. *86*.

K. MASUDA
Chemisorption on a Model Transition Metal. Coverage Dependence of the Adsorption Behaviour
(b) Vol. **87**, No. 2, p. 739–749 (1978).
See also *13.3*, p. *165*.

Y. WASEDA and H. S. CHEN
Calculation of Electrical Resistivity and Its Temperature Coefficient of Amorphous Cu_{60}–Zr_{40} Alloy Using *t*-Matrix
(b) Vol. **87**, No. 2, p. 777–782 (1978).
See also *2*, p. *45*; *14.1*, p. *208*.

CHANG PENG-HSIANG and G. SCHROEDER
Domain Patterns and Dislocations in Cyclically Deformed Ni Single Crystals and Their Behaviour during Annealing
(a) Vol. **47**, No. 2, p. 397—406 (1978).
See also *10.1*, p. *103*; *18.2*, p. *287*.

J. A. ASHBY, H. M. FLOWER, and R. D. RAWLINGS
Correlation of Electrical Resistivity with Microstructure in an Fe–Co–2% V Alloy
(a) Vol. **47**, No. 2, p. 407—414 (1978).
See also *1.4*, p. *31*; *14.1*, p. *208*; 21.1.1, p. 387.

C. E. LACIANA, A. J. PEDRAZA, and E. J. SAVINO
Lattice Distortion Due to Oxygen and Nitrogen Di-Interstitial Clusters in Niobium and Vanadium
(a) Vol. **47**, No. 2, p. 425—436 (1978).
See also *10.1*, p. *103*; 21, p. 359.

H. J. BAUER, M. BECKER. H. PRETSCH, and M. ZWICK
Magnetic Study of the Activation Energy of the Phase Transition Kinetics between Nickel and Nickel Hydride
(a) Vol. **47**, No. 2, p. 445—450 (1978).
See also *1.2*, p. *19*; 18.2, p. 287.

J. F. DINHUT, J. P. RIVIÈRE, and J. C. DESOYER
Point Defects Produced by Plastic Deformation and Ordering Phenomena in Fe–Co and Fe–Co–2% V Alloys
(a) Vol. **47**, No. 2, p. 469—478 (1978).
See also *10.1*, p. *104*; 14.1, p. 208; 21.1.1, p. 387.

J. KNAPPE, H.-R. MÜLLER, and B. SPRINGMANN
Homogeneity of Amorphous Gd–Co Films Prepared by Bias Sputtering
(a) Vol. **47**, No. 2, p. 523—532 (1978).
See also *1.5*, p. *37*; *2*, p. *45*; 18.3, p. 296; 21.4, p. 398.

G. VAN TENDELOO, R. WOLF, J. VAN LANDUYT, and S. AMELINCKX
Electron Diffraction and High Resolution Electron Microscopic Study of Ordering in the Gold–Manganese System (I): The Orthorhombic and Monoclinic I Structure in $Au_{11}Mn_4$
(a) Vol. **47**, No. 2, p. 539—554 (1978).
See also *1.2*, p. *19*; 1.3, p. 26; 1.4, p. 31; 21.6, p. 403.

G. VAN TENDELOO and S. AMELINCKX
Direct Observation by High Resolution Electron Microscopy of Coincidence Sites in Domain Boundaries in Ordered Gold–Manganese Alloys
(a) Vol. **47**, No. 2, p. 555—564 (1978).
See also *1.3*, p. *26*; *1.4*, p. *31*; 21.6, p. 403.

P. L. ROSSITER and M. E. HOUGHTON
Magnetic Properties and Microstructure of an Fe–27.5 Cr–17.5 Co–0.5 Al Alloy
(a) Vol. **47**, No. 2, p. 597—608 (1978).
See also *1.4*, p. *31*; *18.2*, p. *287*; 21.1.1, p. 387.

H. METZGER, H. JO, S. C. MOSS, and D. G. WESTLAKE
Single Crystal X-Ray Study of the Superstructure Modulation and Long-Range Order in V_2D
(a) Vol. **47**, No. 2, p. 631—638 (1978).
See also *1.2*, p. *19*; 1.3, p. 26.

H. Misiorek, T. Zakrzewski, and J. Rafałowicz
Influence of Plastic Deformation on the Thermal Conductivity Maximum of Copper and Aluminium in the Temperature Range 4.2 to 70 K
(a) Vol. **47**, No. 2, p. K137—K140 (1978).
Erratum
(a) Vol. **50**, No. 2, p. K261 (1978).
See also *8*, p. *87*; *10.1*, p. *104*; 21, p. 360.

R. Enderlein, R. E. Hummel, J. B. Andrews, R. J. Nastasi-Andrews, and C. W. Shanley
Interpretation of Compositional Modulation Spectra to Determine Some Optical Properties of Alloys
(b) Vol. **88**, No. 1, p. 173—178 (1978).
See also *13.1*, p. *156*; *20.1*, p. *329*.

R. Wolf, G. van Tendeloo, J. van Landuyt, and S. Amelinckx
Electron Diffraction and High Resolution Electron Microscopic Study of Ordering in the Gold–Manganese System (II): The Monoclinic II Structure in $Au_{11}Mn_4$ and Its Relation to Other LPAPB Phases
(a) Vol. **48**, No. 1, p. 39—52 (1978).
See also *1.2*, p. *20*; 1.3, p. 27; 1.4, p. 31; 21.6, p. 403.

M. E. Houghton and P. L. Rossiter
Induced Anistropy in an Fe–27.5 Cr–17.5 Co–0.5 Al Alloy
(a) Vol. **48**, No. 1, p. 71—77 (1978).
See also *1.4*, p. *31*; *18.2*, p. *289*; 21.1.1, p. 388.

S. Besse, M. Quintard, Ph. Mazot, and J. de Fouquet
Influence of Hydrogen on the Internal Friction of Titanium
(a) Vol. **48**, No. 1, p. 99—103 (1978).
See also *10.1*, p. *104*; *12.1*, p. *135*.

W. I. Khan and D. Melville
Magnetic Anistropy Associated with Iron Substitution in $Y_2(Co_xFe_{1-x})_{17}$
(a) Vol. **48**, No. 1, p. 209—214 (1978).
See also *18.2*, p. *289*.

O. K. Alekseeva, V. N. Bykov, V. A. Levdik, N. F. Miron, and V. P. Shantarovich
Positron Study of Electron-Irradiated Vanadium
(a) Vol. **48**, No. 1, p. K97—K99 (1978).
See also *11*, p. *128*.

J. Thiemann
Host Nuclear Magnetic Resonance in Copper-Band Magnetic Alloys
(b) Vol. 88, No. 2, p. 397—406 (1978).
See also 10.1, p. 105; *19*, p. *310*.

Farid A. Khwaja, A. A. Katsnelson, and V. M. Silonov
Many-Electron Effects on the Characteristics of Short-Range Order in the Pseudopotential Approximation
(b) Vol. 88, No. 2, p. 477—483 (1978).
See also *1.3*, p. *27*; *13*, p. *146*.

R. Richter and F. Goedsche
Complex Band Structure and Density of States for a One-Dimensional Model of Transition Metal Alloys
(b) Vol. 88, No. 2, p. 485—491 (1978).
See also *13.1*, p. *157*.

B. K. Ponomarev and V. G. Thiessen
Band Ferromagnetism in Nickel–Palladium Alloys
(b) Vol. 88, No. 2, p. K139—K142 (1978).
See also *18.2*, p. *289*; 18.4, p. 301.

I. Pop, A. Giurgiu, E. Pop, and V. Iuşan
Coexistence of Itinerant Antiferromagnetism with Paramagnetism of Localized Magnetic Moments in Low Concentrated Cr–Fe Alloys
(b) Vol. 88, No. 2, p. K181—K185 (1978).
See also *18.1*, p. *278*; *18.4*, p. *301*; 21.1.1, p. 388.

H. Bossac, H.-G. Fabian, and H. Löffler
A Quantitative Interpretation of the Decomposition Kinetics of Some Al–Zn–Mg Alloys
(a) Vol. 48, No. 2, p. 369—375 (1978).
See also *1.3*, p. *27*; *10.1*, p. *105*; 14.1, p. 209; 21, p. 361; 21.3, p. 395.

M. van Sande, R. de Ridder, G. van Tendeloo, J. van Landuyt, and S. Amelinckx
High Resolution Study of One-Dimensional Long Period Superstructures in Cu_3Sn with Additions of Zinc and Nickel
(a) Vol. 48, No. 2, p. 383—394 (1978).
See also *1.3*, p. *27*; 1.4, p. 32; 21, p. 361.

G. Hilscher and E. Gratz
An Estimation of the Density of States from Magnetization and Resistivity Measurements in $Ti(Fe_{1-x}Co_x)$ and $Zr(Fe_{1-x}Co_x)_2$
(a) Vol. 48, No. 2, p. 473—480 (1978).
See also *13*, p. *147*; 14.1, p. 209; 18.1, p. 279; 18.2, p. 289; 21.1.1, p. 388.

A. Bostanjoglo and H. P. Gemünd
Extraordinary Viscous Magnetic Wall Motion in a Wide Range of Velocities
(a) Vol. 48, No. 2, p. 481—488 (1978).
See also *18.2.1*, p. *293*; 21.1.1, p. 388.

W. H. M. Alsem, P. M. Bronsveld, and E. W. van Royen
Superlattice Dislocations in Cu_2NiZn
(a) Vol. 48, No. 2, p. 497—502 (1978).
See also *1.3*, p. *27*; *10.1*, p. *105*.

M. L. Johnson, S. Saterlie, D. Boice, and J. G. Byrne
The Removal of Defects from Solids as Observed with Positron Annihilation
(a) Vol. 48, No. 2, p. 551—554 (1978).
See also *1.3*, p. *27*; *10.1*, p. *105*; 21, p. 362.

C. Bansal
Concentration Dependence of the Average Magnetic Moment for F.C.C. Co–Fe Alloys
(a) Vol. 48, No. 2, p. K119—K121 (1978).
See also *18.2*, p. *289*; 21.1.1, p. 389.

T. Skośkiewicz
Preparation of Superconducting Hydrides of Palladium–Nickel Alloys under High Hydrogen Pressures
(a) Vol. **48**, No. 2, p. K165—K168 (1978).
See also *14.2*, p. *215*; 21.6, p. 404.

S. Y. Chuang, P. K. Tseng, G. J. Jan, and H. S. Chen
An Experimental Study of the Angular Correlation of Positron Annihilation Radiations in Electron Irradiated Pd–Ni–Si Metallic Glass
(a) Vol. **48**, No. 2, p. K181—K183 (1978).
See also *1.4*, p. *32*; *2*, p. *48*; 21.6, p. 404.

V. G. Thiessen, V. E. Antonov, I. T. Belash, B. K. Ponomarev, and E. G. Ponkatovskii
Magnetization of Co–H Solid Solutions
(a) Vol. **48**, No. 2, p. K185—K187 (1978).
See also *18.2*, p. *290*.

P. Blaudeck and R. Lenk
Multiple Scattering in Condensed Materials (II): Calculation of the Density of States for Non-Crystalline Copper and Iron
(b) Vol. **89**, No. 1, p. 199—202 (1978).
See also *2*, p. *49*; 13, p. 147; 21.1.1, p. 389.

H.-J. Nowak, H. Leitz, and W. Buckel
On the Structure of Amorphous Metallic Layers of the System Germanium–Copper
(a) Vol. **49**, No. 1, p. 73—82 (1978).
See also *1.5*, p. *38*; *2*, p. *49*; 22.1.1, p. 427.

H. Gärtner and W. Auer
The Influence of Nickel Clusters on the Specific Resistance of Paramagnetic Cu–Ni Alloys
(a) Vol. **49**, No. 1, p. 149—152 (1978).
See also *10.1*, p. *105*; *14.1*, p. *210*.

I. Pop, M. Coldea, and V. U. S. Rao
NMR and Magnetic Susceptibility of Intermetallic Compounds ZrV_2, $Zr_{0.99}Pt_{0.01}V_2$, and $Zr_{0.9}Fe_{0.1}V_2$
(a) Vol. **49**, No. 1, p. 207—210 (1978).
See also *18.1*, p. *279*; *19*, p. *311*.

H. Löffler, O. Kabisch, and R. Kroggel
About a Relation between Microhardness and the Average Radius of the Guinier-Preston Zones in Al–Zn Alloys
(a) Vol. **49**, No. 1, p. 299—303 (1978).
See also *10.1*, p. *106*; 21, p. 363.

H.-J. Ullrich, U. Reinhold, S. Däbritz, P. Paufler, K. Kleinstück, and B. Pietrass
On the Lattice Transformation in Plastically Deformed V_3Si Single Crystals Studied by the Kossel Technique
(a) Vol. **49**, No. 1, p. 323—330 (1978).
See also *1.2*, p. *22*; *10.1*, p. *106*; 14.2, p. 215.

G. van Tendeloo and S. Amelinckx
A High Resolution Study of Ordering in Au_4Mn (I): The Microstructure of Fully Ordered Au_4Mn
a) Vol. **49**, No. 1, p. 337—346 (1978).
See also *1.1*, p. *13*; *1.4*, p. *33*; 21.6, p. 404.

Y. Waseda and H. S. Chen
A Structural Study of Metallic Glasses Containing Boron (Fe–B, Co–B, and Ni–B)
(a) Vol. **49**, No. 1, p. 387–392 (1978).
See also *2*, p. *49*.

Ya. M. Pogosyan, S. D. Gevorkyan, and R. Straubel
Submicroscopical Magnetization Structures in Permalloy Films Induced by Fields Normal to the Film Plane
(a) Vol. **49**, No. 1, p. K3–K7 (1978).
See also *18.2.1*, p. *294*; 21.1.1, p. 389.

K. Fukamichi
Magnetic Phase Diagram of the Dilute Cr–Sb Alloy System
(a) Vol. **49**, No. 1, p. K37–K40 (1978).
See also 8, p. 88; 14.1, p. 210; *18.4*, p. *301*.

H.-R. Müller
Influence of Anisotropic Short-Range Order on Magnetic Properties of the Amorphous Ferrimagnet Gd–Co
(a) Vol. **49**, No. 1, p. K63–K66 (1978).
See also *2*, p. *50*; *18.3*, p. *296*; 21.4, p. 398.

R. Kosicik, P. Duda, W. Keilig, and R. Mattheis
The Influence of Ion Bombardement on the Magnetic Anisotropy in Evaporated Amorphous Gd–Co Films
(a) Vol. **49**, No. 1, p. K67–K69 (1978).
See also *2*, p. *50*; 11, p. 129; *18.3*, p. *296*; 21.4, p. 398.

J. Čermák, V. Tvarožek, R. Harman, and J. Kaczér
Contribution to the Technology of Gd–Co Films
(a) Vol. **49**, No. 1, p. K81–K83 (1978).
See also 2, p. 50; *18.2.1*, p. *294*; 21.4, p. 398.

J. R. Nicholls and R. D. Rawlings
Activation Analysis of Plastic Deformation of Ni_3Al–10 at%Fe
(a) Vol. **49**, No. 1, p. K85–K89 (1978).
See also *10.1*, p. *106*; 21.1.1, p. 389.

Y. D. Yao and S. Arajs
Determination of the Curie Temperature of Ferromagnetic Alloys Using the Deviation from Matthiessen's Rule
(b) Vol. **89**, No. 2, p. K201–K205 (1978).
See also *14.1*, p. *211*; 18.2, p. 290.

R. Rodloff and H. Neuhäuser
Mean Velocity of Edge and Screw Dislocation Groups in Neutron-Irradiated Copper Single Crystals
(a) Vol. **49**, No. 2, p. 445–454 (1978).
See also *10.1*, p. *106*; 11, p. 130.

M. Mayer, O. Vöhringer, and E. Macherauch
On the Quantitative Characterization of the Portevin-Le Chatelier Effect in Homogeneous CuNi and CuNiZn Alloys
(a) Vol. **49**, No. 2, p. 473–482 (1978).
See also *10.1*, p. *106*.

K. Bärner and H. Berg
Elastic Anomalies at the Second-Order Phase Transition in MnAs
(a) Vol. **49**, No. 2, p. 545—554 (1978).
See also 6, p. 76; *12.1*, p. *136*; 18.1, p. 279; 18.2, p. 290.

G. van Tendeloo, R. de Ridder, and S. Amelinckx
A High Resolution Study of Ordering in Au_4Mn (II): The Evolution of Ordering in Au_4Mn
(a) Vol. **49**, No. 2, p. 655—666 (1978).
See also *1.1*, p. *14*; *1.4*, p. *33*; 21.6, p. 405.

M. Ignat, R. Bonnet, D. Caillard, and J. L. Martin
Creep of the Lamellar Al–$CuAl_2$ Composite (I): Microstructural Observations
(a) Vol. **49**, No. 2, p. 675—684 (1978).
See also *10.1*, p. *106*; 21, p. 364.

N. Rihon
Energy Distribution Study of Field-Emitted Electrons from (100) and (111) Nickel Planes
(a) Vol. **49**, No. 2, p. 697—703 (1978).
See also *17.1*, p. *272*.

M. Maček, V. Kraševec, and V. Marinković
On the Relief Transformation Strains in γ-Phase Cr_5Al_8
(a) Vol. **49**, No. 2, p. 767—771 (1978).
See also *1.2*, p. *23*; *1.4*, p. *34*.

M. I. Zakharova, S. A. Kirov, and A. G. Khundzhua
Formation of Metastable and Equilibrium Phases in the Decomposition of β Solid Solution in Zr Alloys
(a) Vol. **49**, No. 2, p. 803—810 (1978).
See also *1.2*, p. *23*; 1.3, p. 28.

S. Martelli, G. Matteucci, and M. V. Antisari
Shadow and Interference Electron Microscopy of Magnetic Domain Walls
(a) Vol. **49**, No. 2, p. K103—K107 (1978).
See also *18.2.1*, p. *294*.

G. Krautheim, A. Neidhardt, and U. Reinhold
Impurity Diffusion of ^{110m}Ag in Cu
(a) Vol. **49**, No. 2, p. K125—K128 (1978).
See also *9*, p. *94*.

Y. Andersson, S. Rundqvist, O. Beckman, L. Lundgren, and P. Nordblad
Properties of Fe_2P Crystals Prepared from a Liquid Copper Medium
(a) Vol. **49**, No. 2, p. K153—K156 (1978).
See also *3*, p. *56*; 18.2, p. 291; 21.1.1, p. 389.

J. D. Browne, N. J. Hance, G. B. Johnston, and C. F. Sampson
The Magnetic Behaviour of Mn_3In
(a) Vol. **49**, No. 2, p. K177—K179 (1978).
See also *18.3*, p. *296*; 18.4, p. 301.

B. B. Sahay and P. L. Srivastava
Isothermal Compressibility for Various Liquid Metals by Harmonic Model Potential
(b) Vol. **90**, No. 1, p. 93—97 (1978).
See also 2, p. 51; *12.1*, p. *136*; 21, p. 364; 21.2, p. 393.

E. MARSCHALL and H. BROSS
Ferromagnetic Nickel: Self-Consistent Calculation of Energy Bands
(b) Vol. **90**, No. 1, p. 241–250 (1978).
See also *13.1*, p. *160*; 18.1, p. 280; 18.2, p. 291.

N. N. SINHA and P. L. SRIVASTAVA
Resistivity and Thermo-Electric Power of Monovalent, Divalent, and Trivalent Liquid Metals
(b) Vol. **90**, No. 1, p. 369–374 (1978).
See also 2, p. 51; *14.1*, p. *211*; *15*, p. *261*; 21, p. 364; 21.2, p. 393.

A. ČÍŽEK, M. ŠOB, and I. YA. DEKHTYAR
Investigation of the Surface of Vanadium Single Crystals by Positron Annihilation
(b) Vol. **90**, No. 1, p. K1–K3 (1978).
See also *13.3*, p. *168*.

S. TANIGAWA, S. NANAO, K. KURIBAYASHI, and M. DOYAMA
Positron Annihilation in Dezincified α-Brass
(b) Vol. **90**, No. 1, p. K31–K34 (1978).
See also *10.1*, p. *107*.

O. HUNDERI and P. O. NILSSON
The Electronic Band Structure of Zinc: A Theroetical Note
(b) Vol. **90**, No. 1, p. K69–K72 (1978).
See also *13.1*, p. *160*.

B. MRYGOŃ
Scaling Hypothesis for Non-Ideal System
(b) Vol. **90**, No. 1, p. K99–K101 (1978).
See also *1.2*, p. *24*; *8*, p. *89*.

J. M. ROLAND, X. QUILLARD, A. MOREAU, and R. GAY
Analysis of Electron Diffraction Micrographs of Cu–Al Alloys by the Patterson Function
(a) Vol. **50**, No. 1, p. 41–52 (1978).
See also *1.3*, p. *28*; 14.1, p. 212.

G. VAN TENDELOO and S. AMELINCKX
Long-Period Antiphase Boundary Structures in the Ordered Alloy Au_4Zn(II): Growth Mechanism of the LPAPB Structure
(a) Vol. **50**, No. 1, p. 53–66 (1978).
See also *1.3*, p. *28*; *1.4*, p. *34*; 21.6, p. 405.

P. DUVAL and A. HAUET
Interpretation of Diffuse Intensity Lines Observed in Electron Diffraction in Cu_3Ga Alloy
(a) Vol. **50**, No. 1, p. 67–72 (1978).
See also *1.4*, p. *34*; 21, p. 365.

J. FIDLER and P. SKALICKY
TEM Investigation of Grown-in Dislocations and Stacking Faults in $SmCo_5$ Single Crystals
(a) Vol. **50**, No. 1, p. 73–81 (1978).
See also *1.4*, p. *34*; 10.1, p. 107; 21.4, p. 398.

E. M. SCHULSON and R. B. TURNER
The Electrical Resistivity of Ordered Zr_3Al
(a) Vol. **50**, No. 1, p. 83–86 (1978).
See also *14.1*, p. *212*.

P. Hendy and E. W. Lee
A Powder Neutron Diffraction Study of Some Rare-Earth–Co_2 Compounds
(a) Vol. **50**, No. 1, p. 101—107 (1978).
See also *18.3*, p. *296*; 21.4, p. 398.

C. Radomsky and H. Löffler
On the Relation between X-Ray Small-Angle Scattering and Electrical Resistivity
(a) Vol. **50**, No. 1, p. 123—130 (1978).
See also *10.1*, p. *107*; 14.1, p. 212.

G. Pompe, J. Sonntag, A. Biedermann, and S. Schieffer
Abnormal Thermal Conductivity of Chromium–Aluminium Alloys in the Temperature Range 5 to 100 K
(a) Vol. **50**, No. 1, p. 147—152 (1978).
See also *8*, p. *89*; 14.1, p. 212.

G. Hausch and E. Török
Elastic, Magnetoelastic, and Thermal Properties of Some Ferromagnetic Metallic Glasses
(a) Vol. **50**, No. 1, p. 159—164 (1978).
See also *2*, p. *51*; 8, p. 89; 12.1, p. 136; *18.2*, p. *291*; 21.1.1, p. 390.

M. I. Darby, M. N. Read, and K. N. R. Taylor
The Energy of Interstitial Hydrogen in α-Zirconium
(a) Vol. **50**, No. 1, p. 203—211 (1978).
See also *10.1*, p. *107*; *13.4*, p. *184*.

Y. Aoki and K. Fukamichi
Ferromagnetic Behaviour of Co-Rich Co–V Alloys Annealead at 600 °C
(a) Vol. **50**, No. 1, p. 263—267 (1978).
See also *18.2*, p. *291*.

G. Ya. Akimov, V. I. Zaitsev, and V. A. Streltsov
Stress Relaxation in Hydrostatically Compressed Zinc Single Crystals
(a) Vol. **50**, No. 1, p. K33—K36 (1978).
See also *10.1*, p. *107*; *12.1*, p. *136*.

A. G. Shishkov, E. N. Ilicheva, N. G. Kanavina, O. S. Kolotov, and R. V. Telesnin
Bloch-Line Mobility in Permalloy Films in a Wide Thickness Range
(a) Vol. **50**, No. 1, p. K101—K104 (1978).
See also *18.2.1*, p. *294*; 21.1.1, p. 390.

D. Deutges, A. Freudenhammer, and G. Heber
Spin Glass Transition. A Dynamical Instability
(b) Vol. **90**, No. 2, p. 475—485 (1978).
See also *18*, p. *276*.

K. Berndt, U. Marx, and O. Brümmer
Electronic Structure of FeNi Alloys Studied by Means of X-Ray Photoelectron Spectroscopy (I): Valence Bands
(b) Vol. **90**, No. 2, p. 487—496 (1978).
See also *13.1*, p. *160*; *20*, p. *318*; 21.1.1, p. 390.

H. Eschrig and I. Bergert
An Optimized LCAO Version for Band Structure Calculations. Application to Copper
(b) Vol. **90**, No. 2, p. 621—628 (1978).
See also *13.1*, p. *161*.

D. E. G. Williams, H. W. G. Spraget, and R. Perrin
The Magnetic Moment of $Cr_{0.6}.Mn_{0.4}$
(b) Vol. **90**, No. 2, p. 667—670 (1978).
See also *18.4*, p. *302*.

N. N. Delyagin, V. I. Krylov, and V. I. Nesterov
Temperature Anomalies of the Magnetic Hyperfine Field for Sn in Metallic Ferromagnets and the Thermal Expansion of the Crystal Lattice
(b) Vol. **90**, No. 2, p. 741—749 (1978).
See also *8*, p. *90*; *18.2*, p. *292*.

H. Puslat
On the Temperature Dependence of the Hall Coefficient in Cr–Al Alloys
(b) Vol. **90**, No. 2, p. K121—K123 (1978).
See also *14.1*, p. *212*.

S. S. Hsu, J. H. Chen, and P. C. Clapp
Acoustic Absorption and Emission Associated with the Martensitic Transformation in an FeNi Alloy
(a) Vol. **50**, No. 2, p. 393—398 (1978).
See also 1.3, p. 28; *7*, p. *82*; 21.1.1, p. 390.

G. Frommeyer
High-Frequency Conductivity, Classical and Anomalous Skin Effect of Eutectic Silver–Copper Composite Wires
(a) Vol. **50**, No. 2, p. 517—528 (1978).
See also *14.1*, p. *212*; 21.6, p. 405.

G. J. den Otter and R. Vetter
A Model Comprising Work Hardening and Solid Solution Hardening
(a) Vol. **50**, No. 2, p. 529—536 (1978).
See also *10.1*, p. *108*; *12.1*, p. *137*.

H.-R. Müller, W. Keilig, R. Kosicik, P. Rosemann, and Z. Frait
Significance and Interpretation of Magnetic Anisotropy Measurements in Amorphous Gd–Co Films
(a) Vol. **50**, No. 2, p. 537—542 (1978).
See also *18.3*, p. *297*; 21.4, p. 399.

M. van Sande, R. de Ridder, J. van Landuyt, and S. Amelinckx
A Study by Means of Electron Microscopy and Electron Diffraction of Vacancy Ordering in Ternary Alloys of the System AlCuNi
(a) Vol. **50**, No. 2, p. 587—599 (1978).
See also *1.3*, p. *28*; *1.4*, p. *34*; 10.1, p. 108; 21, p. 366.

V. Lantto
Contribution of High-Angle Deflections to the Fast-Electron Transmission
(a) Vol. **50**, No. 2, p. 611—616 (1978).
See also *11*, p. *131*.

V. L. Sedov and L. V. Solomatina
Structural Inhomogeneity in Invar Alloys Studied by Positron Annihilation
(a) Vol. **50**, No. 2, p. 757—765 (1978).
See also *10.1*, p. *108*; 21.1.1, p. 391.

V. I. Kolomytsev, V. A. Lobodyuk, and L. G. Khandros
Structural Changes at Heating of the β-Phase in NiTi Alloys
(a) Vol. **50**, No. 2, p. K127–K129 (1978).
See also *1.2*, p. *24*.

N. N. Evtikhiev, N. A. Ekonomov, A. R. Krebs, N. A. Zamyatina, A. S. Komalov, V. G. Pinko, and L. V. Ivaeva
FePt and CoPt Alloy Films for Magneto-Optic Storage
(a) Vol. **50**, No. 2, p. K153–K155 (1978).
See also *18.2.1*, p. *294*; *20*, p. *318*; 21.1.1, p. 391.

21.1.1 Fe and Fe Alloys

Review Article

M. Tomášek and Š. Pick
Projected Surface Energy Bands and Shockley Surface States of Transition Metals. General Considerations and the Example of the B.C.C. Lattice
(b) Vol. **89**, No. 1, p. 11–26 (1978).
See also 13.1, p. 150; *13.3*, p. *164*; 21.1, p. 368.

Original Papers and Short Notes

S. Takahashi
Compression-Induced Magnetic Anisotropy and Superlattice Dislocations in Ni_3Fe (III): The Annealing Effect
(a) Vol. **45**, No. 1, p. 133–138 (1978).
See also *10.1*, p. *99*; *18.2*, p. *281*; 21.1, p. 369.

J. P. Rivière
Electrical Resistivity of B_2-Type Fe–Al Ordered Alloys
(a) Vol. **45**, No. 1, p. 157–162 (1978).
See also 10.1, p. 99; *14.1*, p. *205*.

W. Kümmerle and U. Gradmann
Magnetic Properties of Ferromagnetic γ-Fe Films on Cu (111)
(a) Vol. **45**, No. 1, p. 171–180 (1978).
See also *18.2.1*, p. *293*.

P. Turchi, Y. Calvayrac, and F. Plicque
Effects of Configurational Order in a Ni_3Fe Single Crystal on Debye Temperature Deduced from Elastic Constants and X-Ray Measurements
(a) Vol. **45**, No. 1, p. 229–238 (1978).
See also *1.3*, p. *25*; 6, p. 65; 12.1, p. 134; 21.1, p. 369.

R. Yamamoto, H. Matsuoka, and M. Doyama
Structural Relaxation of the Dense Random Packing Model for Amorphous Iron
(a) Vol. **45**, No. 1, p. 305–314 (1978).
See also *2*, p. *42*.

R. Huster and E. Huster
Hydrogen Absorption and Magnetic Behaviour of Palladium–Iron Alloys
(a) Vol. **45**, No. 1, p. K23 (1978).
See also *18*, p. *273*.

J. Schneider, A. Handstein, and R. Hesske
Mictomagnetic to Ferromagnetic Transition in Amorphous Ni–Fe Based Alloys
(a) Vol. **45**, No. 1, p. K47–K50 (1978).
See also *2*, p. *42*; *18.2*, p. *282*; 21.1, p. 369.

M. V. Medvedev and E. L. Rumyantsev
On the Existence of a Spin-Glass State in an Ising Alloy Model with Mixed Nearest-Neighbour Exchange Interactions
(b) Vol. **85**, No. 2, p. 427–438 (1978).
See also *18.2*, p. *282*; *18.4*, p. *298*; 21.1, p. 369.

R. CISZEWSKI, A. JABŁONKA, A. SOBASZEK, and D. SIKORSKA
Temperature Shift of the Maximum of the Neutron Critical Scattering in an Iron Monocrystal
(b) Vol. **85**, No. 2, p. K73—K75 (1978).
See also *1*, p. *9*; 18.2, p. 282.

R. CISZEWSKI, A. JABŁONKA, and A. SOBASZEK
Experimental Anisotropy of the Neutron Critical Scattering from a Ferromagnet Monocrystal
(b) Vol. **85**, No. 2, p. K77—K79 (1978).
See also *1*, p. *9*; 18.2, p. 282.

A. G. KHACHATURYAN and A. F. RUMYNINA
Strain Energy Controlled Shape of Crystals Forming Due to a Martensite Transformation or Decomposition
(a) Vol. **45**, No. 2, p. 393—402 (1978).
See also *10.1*, p. *100*; 21.1, p. 370.

J. DEGAUQUE, B. ASTIE, and L. P. KUBIN
Evidence for the Interaction between Magnetic Domain Walls and Dislocations in High-Purity Iron from Magnetomechanical Damping Experiments
(a) Vol. **45**, No. 2, p. 493—501 (1978).
See also *10.1*, p. *100*; *18.2*, p. *283*.

S. K. SEN, M. B. DUTT, and A. K. BARUA
The Diffusion of Iron in Copper and of Nickel in Silver
(a) Vol. **45**, No. 2, p. 657—663 (1978).
See also *9*, p. *91*; 21.1, p. 370; 21.6, p. 401.

J. BARTON and E. WIESER
Investigation of Magnetic Preferred Directions in an Fe–Mn Base Alloy
(a) Vol. **45**, No. 2, p. K105—K108 (1978).
See also *5*, p. *60*; *18.2*, p. *283*; 21.1, p. 370.

E. BURZO, D. P. LAZĂR, and L. PALII
Paramagnetic Behaviour of $Gd(Fe_xCo_{1-x})_2$ Compounds
(a) Vol. **45**, No. 2, p. K145—K148 (1978).
See also *18.1*, p. *277*; 18.3, p. 295; 21.1, p. 370; 21.4, p. 396.

O. K. ALEKSEEVA, U. V. MOISH, V. I. SARRAK, G. A. FILIPPOV, and V. P. SHANTAROVICH
Positron Annihilation for Martensite Defect Identification
(a) Vol. **45**, No. 2, p. K159—K162 (1978).
See also *10.1*, p. *101*.

V. HIVERT, P. GROH, W. FRANK, I. RITCHIE, and P. MOSER
Internal Friction Peaks Due to Dislocation Relaxations in Plastically Deformed and/or Irradiated High-Purity α-Iron (II): Interpretation
(a) Vol. **46**, No. 1, p. 89—98 (1978).
See also *10.1*, p. *101*; 11, p. 125.

A. TARI
The Quenching Effect of Iron Substitution on the Holmium Moment in the $Ho_{0.01}Gd_{0.99}(Rh_{1-x}Fe_x)_2$ Compounds
(a) Vol. **46**, No. 1, p. 173—177 (1978).
See also *18.2*, p. *284*; 19, p. 305; 21.4, p. 396.

J. P. Jakubovics and D. J. Fathers
Image Profils and Resolution of Magnetic Domains in the Scanning Electron Microscope
(a) Vol. **46**, No. 1, p. 291—303 (1978).
See also *18.2*, p. *284*; 21.4, p. 397.

P. Weber, W. Richter, and A. Feuerpfeil
Improved Sensitivity of Thin Film SQUIDs with Permalloy Films
(a) Vol. **46**, No. 1, p. K67—K69 (1978).
See also *14.2*, p. *214*; 18.2.1, p. 293.

W. John, D. Hamann, and P. Urwank
Partial Densities of States and Electron–Phonon Interaction in Transition Metals
(b) Vol. **86**, No. 2, p. 569—577 (1978).
See also *13.1*, p. *154*; 14.2, p. 214; 21.1, p. 371.

F. M. Galperin
Atomic Magnetic Moments of Fe, Co, Ni and Hyperfine Fields at Impurities in Ferromagnetic Iron
(b) Vol. **86**, No. 2, p. 679—684 (1978).
See also *13*, p. *144*; *18.2*, p. *285*; 21.1, p. 372.

K. Berndt, O. Brümmer, and U. Marx
Study of the Fe 2p States by Means of Photoelectron Spectroscopy in Iron and Iron Alloys
(b) Vol. **86**, No. 2, p. K93—K95 (1978).
See also *13*, p. *144*; 20, p. 316; 21.1, p. 372.

R. Yamamoto, T. Hoshiai, H. Matsuoka, K. Taira, and M. Doyama
The Calculations of the Heat of Crystallization and Bulk Modulus of Glassy Iron
(a) Vol. **46**, No. 2, p. 671—674 (1978).
See also *2*, p. *44*; *12.1*, p. *135*.

M. Kopcewicz
Mössbauer Study of the Fast Magnetization Reversal Forced in Permalloy and Invar by an External RF Magnetic Field
(a) Vol. **46**, No. 2, p. 675—685 (1978).
See also *5*, p. *61*; *18.2*, p. *285*; 21.1, p. 372.

H. J. Blythe and L. H. Dunleavy
Point-Defect Relaxations in Dilute Iron–Nickel Alloys after Low-Temperature Neutron Irradiation
(a) Vol. **46**, No. 2, p. K95—K99 (1978).
See also *10.1*, p. *102*; 11, p. 125; *18.2*, p. *285*; 21.1, p. 372.

C. P. van E. Oosthuizen and P. de V. du Plessis
Uniaxial Magnetic Anisotropy Induced in a Compression Deformed Iron Single Crystal
(a) Vol. **46**, No. 2, p. K111—K113 (1978).
See also *18.2*, p. *285*.

P. Ziesche, H. Wonn, Ch. Müller, V. V. Nemoshkalenko, and V. P. Krivitskii
Theoretical and Experimental Investigation of the Electronic Structure of the B2-Compound FeAl
(b) Vol. **87**, No. 1, p. 129—134 (1978).
See also *13.1*, p. *154*; 20, p. 316; 21, p. 357.

J. TH. M. DE HOSSON
On the Vibrational Entropy of a ½ ⟨111⟩ {110} Edge Dislocation in B.C.C. Iron
(b) Vol. **87**, No. 1, p. 151–161 (1978).
See also *10.1*, p. *102*.

G. SCHROEDER
Investigation of the Dislocation Arrangement in Plastically Deformed Iron Single Crystals by Means of Ferromagnetic Techniques
(a) Vol. **47**, No. 1, p. 145–156 (1978).
See also *10.1*, p. *103*; 18.2, p. 286.

U. RINDELHARDT, S. MAGER, W. BLAU, and E. HEGENBARTH
Recovery of Low Temperature Specific Heat, Resistivity, and Coercive Force in Fe–Cr–Al Alloys
(a) Vol. **47**, No. 1, p. 157–161 (1978).
See also 8, p. 85; *10.1*, p. *103*; 14.1, p. 207; 18.2, p. 286; 21.1, p. 373.

A. M. MOISIN
Neutron Scattering Study of Local Ordering in Fe–Ni and Fe–Al Alloys
(a) Vol. **47**, No. 1, p. 305–309 (1978).
See also *18*, p. *274*; 21.1, p. 373.

S. V. SEMENOVSKAYA
Effect of Correlation on the X-Ray Diffuse Scattering
(b) Vol. **87**, No. 2, p. 733–737 (1978).
See also *1.3*, p. *26*.

N. K. GOBRAN and N. S. GHALI
Ultrasonic Anomalies in Iron at Elevated Temperatures
(a) Vol. **47**, No. 2, p. 365–370 (1978).
See also *7*, p. *81*; 18.2, p. 287.

J. A. ASHBY, H. M. FLOWER, and R. D. RAWLINGS
Correlation of Electrical Resistivity with Microstructure in an Fe–Co–2% V Alloy
(a) Vol. **47**, No. 2, p. 407–414 (1978).
See also *1.4*, p. *31*; *14.1*, p. *208*; 21.1, p. 374.

J. F. DINHUT, J. P. RIVIÈRE, and J. C. DESOYER
Point Defects Produced by Plastic Deformation and Ordering Phenomena in Fe–Co and Fe–Co–2% V Alloys
(a) Vol. **47**, No. 2, p. 469–478 (1978).
See also *10.1*, p. *104*; 14.1, p. 208; 21.1, p. 374.

P. L. ROSSITER and M. E. HOUGHTON
Magnetic Properties and Microstructure of an Fe–27.5 Cr–17.5 Co–0.5 Al Alloy
(a) Vol. **47**, No. 2, p. 597–608 (1978).
See also *1.4*, p. *31*; *18.2*, p. *287*; 21.1, p. 374.

Y. SHIMADA and H. KOJIMA
Thermal Stability of Amorphous Ferromagnetic Thin Films Produced by Sputtering
(a) Vol. **47**, No. 2, p. K119–K121 (1978).
See also *2*, p. *46*; *18.2.1*, p. *293*.

W. Gustaw, T. Goworek, and J. Wawryszczuk
Influence of Fatigue on Positron Annihilation in Steel
(a) Vol. **47**, No. 2, p. K141—K144 (1978).
See also *12.1*, p. *135*.

M. E. Houghton and P. L. Rossiter
Induced Anisotropy in an Fe–27.5 Cr–17.5 Co–0.5 Al Alloy
(a) Vol. **48**, No. 1, p. 71—77 (1978).
See also *1.4*, p. *31*; *18.2*, p. *289*; 21.1, p. 375.

M. Kikuchi, K. Fukamichi, T. Masumoto, T. Jagielinski, K. I. Arai, and N. Tsuya
Giant ΔE Effect and Elinvar Characteristics in Amorphous Fe–B Binary Alloys
(a) Vol. **48**, No. 1, p. 175—181 (1978).
See also 2, p. 47; *18.2*, p. *289*.

W. Raczyński
Permeability, Diffusivity, and Solubility of Hydrogen and Deuterium in Pure Iron at 10 to 60 °C
(a) Vol. **48**, No. 1, p. K27—K30 (1978).
See also *9*, p. *93*.

A. F. Akkerman
Reflection of Slow Hydrogen and Helium Ions from Solid Surfaces
(a) Vol. **48**, No. 1, p. K47—K50 (1978).
See also *11*, p. *128*; 21, p. 361.

J. Barton, E. Wieser, and M. Müller
Investigation of the α–γ Phase Transformation by Annealing of a Cold-Worked Fe–Mn Base Alloy
(a) Vol. **48**, No. 1, p. K51—K53 (1978).
See also *1.2*, p. *21*; 5, p. 62.

S. D. Zvonkov, V. D. Sedykh, Yu. S. Stark, and A. Sh. Chavchanidze
Electronic Structure and Hyperfine Interactions in α-Fe-Based Dilute Solid Solutions
(b) Vol. **88**, No. 2, p. K103—K107 (1978).
See also *13.4*, p. *179*.

I. Pop, A. Giurgiu, E. Pop, and V. Iuşan
Coexistence of Itinerant Antiferromagnetism with Paramagnetism of Localized Magnetic Moments in Low Concentrated Cr–Fe Alloys
(b) Vol. **88**, No. 2, p. K181—K185 (1978).
See also *18.1*, p. *278*; *18.4*, p. *301*; 21.1, p. 376.

G. Hilscher and E. Gratz
An Estimation of the Density of States from Magnetization and Resistivity Measurements in $Ti(Fe_{1-x}Co_x)$ and $Zr(Fe_{1-x}Co_x)_2$
(a) Vol. **48**, No. 2, p. 473—480 (1978).
See also *13*, p. *147*; 14.1, p. 209; 18.1, p. 279; 18.2, p. 289; 21.1, p. 376.

A. Bostanjoglo and H P. Gemünd
Extraordinary Viscous Magnetic Wall Motion in a Wide Range of Velocities
(a) Vol. **48**, No. 2, p. 481—488 (1978).
See also *18.2.1*, p. *293*; 21.1, p. 376.

C. Bansal
Concentration Dependence of the Average Magnetic Moment for F.C.C. Co–Fe Alloys
(a) Vol. **48**, No. 2, p. K119—K121 (1978).
See also *18.2*, p. *289*; **21.1**, p. **376**.

P. Blaudeck and R. Lenk
Multiple Scattering in Condensed Materials (II): Calculation of the Density of States for Non-Crystalline Copper and Iron
(b) Vol. **89**, No. 1, p. 199—202 (1978).
See also *2*, p. *49*; 13, p. 147; **21.1**, p. **377**.

Ya. M. Pogosyan, S. D. Gevorkyan, and R. Straubel
Submicroscopical Magnetization Structures in Permalloy Films Induced by Fields Normal to the Film Plane
(a) Vol. **49**, No. 1, p. K3—K7 (1978).
See also *18.2.1*, p. *294*; **21.1**, p. **378**.

H. Bittel, W. Grosse-Nobis, and Th. Jannemann
On the Magnetostriction of Grain-Oriented Ferromagnetic Material
(a) Vol. **49**, No. 1, p. K21—K25 (1978).
See also *18.2*, p. *290*.

J. R. Nicholls and R. D. Rawlings
Activation Analysis of Plastic Deformation of Ni_3Al–10 at%Fe
(a) Vol. **49**, No. 1, p. K85—K89 (1978).
See also *10.1*, p. *106*; **21.1**, p. **378**.

P. Raj, S. K. Kulshreshtha, and J. V. Yakhmi
Mössbauer and X-Ray Studies of $Fe_{1.67}Ge$
(a) Vol. **49**, No. 1, p. K91—K95 (1978).
See also *1.3*, p. *28*; *5*, p. *62*; 18.2, p. 290.

W. A. Theiner and H. P. Geserich
Anderson Transition in $FeSi_2$ Films
(b) Vol. **89**, No. 2, p. 441—447 (1978).
See also *1.2*, p. *22*; *1.5*, p. *39*; 2, p. 50; 14.1, p. 210; 20.1, p. 334.

V. M. Aliabev and V. A. Pavlov
Creep and Fracture of α-Iron between 77 and 673 K
(a) Vol. **49**, No. 2, p. 509—515 (1978).
See also *10.1*, p. *106*.

H. O. K. Kirchner, H. Mughrabi, F. Prinz, and H. Schmid
Dislocation Loops in α-Iron
(a) Vol. **49**, No. 2, p. 539—543 (1978).
See also *10.1*, p. *106*; 12.1, p. 136.

H. J. Blythe and L. H. Dunleavy
Low-Temperature Magnetic Relaxation in Neutron-Irradiated α-Iron
(a) Vol. **49**, No. 2, p. 609—619 (1978).
See also 10.1, p. 106; *11*, p. *130*; *18.2*, p. *290*.

Y. Andersson, S. Rundqvist, O. Beckman, L. Lundgren, and P. Nordblad
Properties of Fe_2P Crystals Prepared from a Liquid Copper Medium
(a) Vol. **49**, No. 2, p. K153—K156 (1978).
See also *3*, p. *56*; 18.2, p. 291; 21.1, p. 379.

M. MITERA, M. NAKA, T. MASUMOTO, N. KAZAMA, and K. WATANABE
Effects of Metalloids on the Magnetic Properties of Iron Based Amorphous Alloys
(a) Vol. **49**, No. 2, p. K163—K166 (1978).
See also *2*, p. *51*; *18.2*, p. *291*.

J. D. PATTERSON
The Low Temperature Lorenz Number of Spin Glasses
(b) Vol. **90**, No. 1, p. K85—K88 (1978).
See also *2*, p. *51*; *8*, p. *89*; 14.1, p. 212.

G. HAUSCH and E. TÖRÖK
Elastic, Magnetoelastic, and Thermal Properties of Some Ferromagnetic Metallic Glasses
(a) Vol. **50**, No. 1, p. 159—164 (1978).
See also *2*, p. *51*; 8, p. 89; 12.1, p. 136; *18.2*, p. *291*; 21.1, p. 381.

E. A. DOROFEEVA, V. P. MAKAROV, B. V. MOLOTILOV, and L. P. BABICHEV
On the Nature of Additional Losses in Fe–3%Si Single Crystals
(a) Vol. **50**, No. 1, p. K5—K7 (1978).
See also *18.2*, p. *292*.

T. JAGIELIŃSKI, K. I. ARAI, N. TSUYA, and K. FUKAMICHI
Magnetoelastic Effects in Invar-Type Amorphous Fe–B Alloys
(a) Vol. **50**, No. 1, p. K25—K28 (1978).
See also *2*, p. *52*; *18.2*, p. *292*.

M. URNER-WILLE, T. S. TE VELDE, and P. G. VAN ENGEN
Kerr-Effect Enhancement in Amorphous GdFe Films
(a) Vol. **50**, No. 1, p. K29—K31 (1978).
See also 2, p. 52; *18.3*, p. *297*; *20*, p. *318*; 21.4, p. 398.

E. JAPA and K. KROP
Mössbauer Studies of Ordering in Fe–Si Alloys
(a) Vol. **50**, No. 1, p. K89—K91 (1978).
See also 1.3, p. 28; *5*, p. *62*.

A. G. SHISHKOV, E. N. ILICHEVA, N. G. KANAVINA, O. S. KOLOTOV, and R. V. TELESNIN
Bloch-Line Mobility in Permalloy Films in a Wide Thickness Range
(a) Vol. **50**, No. 1, p. K101—K104 (1978).
See also *18.2.1*, p. *294*; 21.1, p. 381.

K. BERNDT, U. MARX, and O. BRÜMMER
Electronic Structure of FeNi Alloys Studied by Means of X-Ray Photoelctron Spectroscopy (I): Valence Bands
(b) Vol. **90**, No. 2, p. 487—496 (1978).
See also *13.1*, p. *160*; *20*, p. *318*; 21.1, p. 381.

S. S. HSU, J. H. CHEN, and P. C. CLAPP
Acoustic Absorption and Emission Associated with the Martensitic Transformation in an FeNi Alloy
(a) Vol. **50**, No. 2, p. 393—398 (1978).
See also 1.3, p. 28; *7*, p. *82*; 21.1, p. 382.

G. SCHROEDER, R. SCHÄFER, and H. KRONMÜLLER
Magneto-Optical Investigation of the Domain Structure of Amorphous $Fe_{80}B_{20}$ Alloys
(a) Vol. **50**, No. 2, p. 475—481 (1978).
See also 2, p. 52; *18.2*, p. *292*; *20*, p. *318*.

G. Tomasi, G. Frohberg, Th. Hehenkamp, and H. Wever
Hall-Field Electrotransport of Carbon in Iron
(a) Vol. **50**, No. 2, p. 563–572 (1978).
See also *9*, p. *95*; 14.1, p. 212.

I. G. Ritchie, J. F. Dufresne, and P. Moser
Internal Friction in Deformed Pure Iron
(a) Vol. **50**, No. 2, p. 617–626 (1978).
See also *10.1*, p. *108*; *12.1*, p. *137*.

V. L. Sedov and L. V. Solomatina
Structural Inhomogeneity in Invar Alloys Studied by Positron Annihilation
(a) Vol. **50**, No. 2, p. 757–765 (1978).
See also *10.1*, p. *108*; 21.1, p. 382.

N. N. Evtikhiev, N. A. Ekonomov, A. R. Krebs, N. A. Zamyatina, A. S. Komalov, V. G. Pinko, and L. V. Ivaeva
FePt and CoPt Alloy Films for Magneto-Optic Storage
(a) Vol. **50**, No. 2, p. K153–K155 (1978).
See also *18.2.1*, p. *294*; *20*, p. *318*; 21.1, p. 383.

A. Kalz, F. Storbeck, and G. Blasek
Absorption of H_2O on α-Fe(111) under the Influence of 2.5 keV Electrons
(a) Vol. **50**, No. 2, p. K247–K250 (1978).
See also *1.6*, p. *41*; 11, p. 132.

21.2 Alkali Metals

Original Papers and Short Notes

V. G. Vaks, E. V. Zarochentsev, S. P. Kravchuk, V. P. Safronov, and A. V. Trefilov
Lattice Heat Capacity and Thermal Vibrations in Alkali Metals
(b) Vol. **85**, No. 1, p. 63—74 (1978).
See also *6*, p. *64*; *8*, p. *83*.

R. Kamel, A. R. Ali, and Z. Farid
Annealing Spectrum of Precipitating Al–Li Alloys
(a) Vol. **45**, No. 1, p. 47—52 (1978).
See also *10.1*, p. *99*; **12.1**, p. **134**; **21**, p. **352**.

V. G. Vaks, E. V. Zarochentsev, S. P. Kravchuk, V. P. Safronov, and A. V. Trefilov
Thermal Expansion and Grüneisen Parameters in Alkali Metals
(b) Vol. **85**, No. 2, p. 749—759 (1978).
See also *6*, p. *66*; *8*, p. *83*.

V. Ramamurthy and K. K. Singh
On the Crossing of Dispersion Curves of Alkali Metals in [ζ00] Direction
(b) Vol. **85**, No. 2, p. 761—768 (1978).
See also *6*, p. *66*; **12.1**, p. **134**.

R. S. Sorbello
Electromigration in Liquid Metal Alloys
(b) Vol. **86**, No. 2, p. 671—678 (1978).
See also *2*, p. *44*; *14.1*, p. *207*.

K. N. Khanna a nd P. K. Sharma
Susceptib ility of Alkali Metals
(b) Vol. **86**, No. 2, p. K167—K170 (1978).
See also *18.1*, p. *278*.

W. H. Tanttila
Self-Diffusion of Solid Sodium
(b) Vol. **86**, No. 2, p. K175—K179 (1978).
Erratum
(b) Vol. **89**, No. 1, p. K109 (1978).
See also *9*, p. *92*.

M. A. Ball and M. M. Islam
Effects of Core and Correlation Corrections on the Electrical Resistivity of Lithium
(b) Vol. **87**, No. 1, p. 145—150 (1978).
See also *13*, p. *145*; *14.1*, p. *207*.

B. Prasad and R. S. Srivastava
Thermal Properties and Grüneisen Parameters of Alkali Metals Employing the Model Potential of Krasko and Gurskii
(b) Vol. **87**, No. 2, p. 771—776 (1978).
See also *6*, p. *72*; 8, p. 86.

E. N. Mgbenu and A. Maduemezia
X-Ray Emission Spectrum of Metallic Lithium
(b) Vol. **88**, No. 2, p. 805–808 (1978).
See also *13.1*, p. *157*; 13.2, p. 162; *20*, p. *317*.

K. N. Khanna and P. K. Sharma
A Model Potential and Its Calculations
(b) Vol. **88**, No. 2, p. 809–814 (1978).
See also *12.1*, p. *136*; 13.2, p. 162; 18.1, p. 278.

R. N. Singh and P. L. Srivastava
Model Potential Calculation of Binding Energy and Compressibility of Alkali Metals
(b) Vol. **89**, No. 1, p. K13–K16 (1978).
See also *4*, p. *59*; *13*, p. *147*.

H. Humberg and H. Merz
Polarization Dependence of the Photoelectron Energy Distribution Curves of Cesium
(b) Vol. **89**, No. 2, p. 397–402 (1978).
See also *17*, p. *270*.

P. Steiner, F. J. Reiter, H. Höchst, and S. Hüfner
The KLL Auger Spectra of Na and Mg Metal and Their Plasmon Structure
(b) Vol. **90**, No. 1, p. 45–51 (1978).
See also *13.1*, p. *159*; *13.5.2*, p. *199*; 21.3, p. 395.

B. B. Sahay and P. L. Srivastava
Isothermal Compressibility for Various Liquid Metals by Harmonic Model Potential
(b) Vol. **90**, No. 1, p. 93–97 (1978).
See also 2, p. 51; *12.1*, p. *136*; 21, p. 364; 21.1, p. 379.

V. Ramamurthy and M. Satishkumar
Analysis of Phonon Frequencies of Alkali Metals
(b) Vol. **90**, No. 1, p. 117–122 (1978).
See also *6*, p. *77*; 12.1, p. 136.

W. F. W. M. van Heugten and J. Th. M. de Hosson
Interatomic Potentials for Alkali Metals. A Comparative Study
(b) Vol. **90**, No. 1, p. 225–232 (1978).
See also *4*, p. *59*; 6, p. 77; 10.1, p. 107.

N. N. Sinha and P. L. Srivastava
Resistivity and Thermo-Electric Power of Monovalent, Divalent, and Trivalent Liquid Metals
(b) Vol. **90**, No. 1, p. 369–374 (1978).
See also 2, p. 51; *14.1*, p. *211*; *15*, p. *261*; 21, p. 364; 21.1, p. 380.

F. A. Zhivopistsev and F. C. Hernández
Fast Electron Scattering in Metals with the Excitation of a Core Electron
(b) Vol. **90**, No. 2, p. 441–447 (1978).
See also *13*, p. *149*; 13.5.2, p. 199.

21.3 Alkali-Earth Metals

Original Papers and Short Notes

R. Z. Valiev and O. A. Kaibyshev
Mechanism of Superplastic Deformation in a Magnesium Alloy (III): The Deformation Model
(a) Vol. **45**, No. 1, p. 77–82 (1978).
See also *10.1*, p. *99*; 21, p. 352.

T. Hirata
Solute Atom Clustering during Ageing an Aluminium–Magnesium–Silicon Solid Solution under Stress
(a) Vol. **45**, No. 1, p. 201–205 (1978).
See also *10.1*, p. *99*; 21, p. 353.

A. Ranga Rao and C. Suryanarayana
Solute–Vacancy Binding Energies in Magnesium Alloys
(a) Vol. **45**, No. 2, p. K131–K133 (1978).
See also *10.1*, p. *100*; 12.1, p. 134.

P. Meriel, A. Herr, and A. Meyer
Neutron Diffraction Study of the Magnetic Structure of $TbBe_{13}$ at 4.2 K
(a) Vol. **46**, No. 1, p. K35–K38 (1978).
See also *18.4*, p. *299*; 21.4, p. 397.

N. A. Tyapunina, E. K. Naimi, S. V. Gasparyan, and G. M. Zinenkova
The Orientation Dependence of Dislocation Internal Friction in Real Crystals (II): Hexagonal Crystal
(a) Vol. **46**, No. 2, p. 411–420 (1978).
See also *10.1*, p. *101*; *12.1*, p. *135*; 21, p. 357; 21.1, p. 372.

P. Rennert, Th. Dörre, and U. Gläser
Theoretical Investigation of the Anisotropy of Beryllium by Polarized K-Emission, APS, and Compton Profiles
(b) Vol. **87**, No. 1, p. 221–226 (1978).
See also *13.1*, p. *154*; 20, p. 316.

K. Melidis
On the Temperature Dependence of the Hall Effect and Electrical Resistivity of Cd, Zn, and Mg Single Crystals
(a) Vol. **47**, No. 1, p. K27–K30 (1978).
See also *14.1*, p. *207*; 21, p. 358; 21.1, p. 373.

P. Airo
On the Superstructure of the Alloy Au–18 at%Mg
(a) Vol. **47**, No. 2, p. K107–K110 (1978).
See also *1.2*, p. *20*; *1.3*, p. *26*; 21.6, p. 403.

H. Bossac, H.-G. Fabian, and H. Löffler
A Quantitative Interpretation of the Decomposition Kinetics of Some Al–Zn–Mg Alloys
(a) Vol. **48**, No. 2, p. 369–375 (1978).
See also *1.3*, p. *27*; *10.1*, p. *105*; 14.1, p. 209; 21, p. 361; 21.1, p. 376.

P. Steiner, F. J. Reiter, H. Höchst, and S. Hüfner
The KLL Auger Spectra of Na and Mg Metal and Their Plasmon Structure
(b) Vol. **90**, No. 1, p. 45–51 (1978).
See also *13.1*, p. *159*; *13.5.2*, p. *199*; 21.2, p. 393.

21.4 Rare-Earth Metals

Original Papers and Short Notes

Š. JÁNOŠ, A. FEHER, and H. U. ÅSTRÖM
Thermal Conductivity of Neodymium from 0.5 to 8 K
(b) Vol. **85**, No. 1, p. K31—K32 (1978).
See also *8*, p. *83*.

A. V. DERYAGIN, N. V. KUDREVATYKH, and V. N. MOSKALEV
Magnetic Anisotropy and Intersublattice Exchange Interaction in $Er_2(Co_{1-x}M_x)_{17}$ Intermetallic Compounds
(a) Vol. **45**, No. 1, p. 71—76 (1978).
See also *18.2*, p. *281*; 21.1, p. 368.

H. DEXPERT and P. CARO
Reaction Processes in the Solid Phase of Some Rare Earth Compounds
(a) Vol. **45**, No. 1, p. 297—304 (1978).
See also *1.2*, p. *15*.

H. MAEDA
Magnetocrystalline Anisotropy of $Pr(Co_{1-x}Cu_x)_5$ Compounds. Experiment and Theoretical Analysis
(a) Vol. **45**, No. 2, p. 445—452 (1978).
See also *18.2*, p. *283*; 21.1, p. 370.

E. BURZO, D. P. LAZĂR, and L. PALII
Paramagnetic Behaviour of $Gd(Fe_xCo_{1-x})_2$ Compounds
(a) Vol. **45**, No. 2, p. K145—K148 (1978).
See also *18.1*, p. *277*; 18.3, p. 295; 21.1, p. 370; 21.1.1, p. 385.

D. GIGNOUX and H. ASMAT
Magnetic Properties and Structures of TbAlGa and HoAlGa
(a) Vol. **45**, No. 2, p. K149—K153 (1978).
See also *1.1*, p. *11*; *18.4*, p. *298*; 21, p. 355.

H. KRIZEK, K. N. R. TAYLOR, and W. D. CORNER
Resistivity of Some Light–Heavy Rare-Earth Alloys (II): Magnetic Contributions to ϱ_{res}, ϱ_{s-f} and T_n
(a) Vol. **46**, No. 1, p. 39—43 (1978).
See also *14.1*, p. *206*.

A. TARI
The Quenching Effect of Iron Substitution on the Holmium Moment in the $Ho_{0.01}Gd_{0.99}(Rh_{1-x}Fe_x)_2$ Compounds
(a) Vol. **46**, No. 1, p. 173—177 (1978).
See also *18.2*, p. *284*; 19, p. 305; 21.1.1, p. 385.

H. MAEDA
Anisotropy of Saturation Magnetization in $Pr(Co_{1-x}Cu_x)_5$ Compounds
(a) Vol. **46**, No. 1, p. 233—238 (1978).
See also *18.2*, p. *284*; 21.1, p. 371.

J. P. Jakubovics and D. J. Fathers
Image Profils and Resolution of Magnetic Domains in the Scanning Electron Microscope
(a) Vol. **46**, No. 1, p. 291–303 (1978).
See also *18.2*, p. *284*; 21.1.1, p. 386.

A. Ślebarski
Anomalous Change of the Curie Temperature of $GdAl_2$
(a) Vol. **46**, No. 1, p. K29–K34 (1978).
See also *8*, p. *84*; *18.2*, p. *284*; 21, p. 356.

P. Meriel, A. Herr, and A. Meyer
Neutron Diffraction Study of the Magnetic Structure of $TbBe_{13}$ at 4.2K
(a) Vol. **46**, No. 1, p. K35–K38 (1978).
See also *18.4*, p. *299*; 21.3, p. 394.

M. Drulis and Z. Biegański
Low Temperature Phase Transition in Cerium Trihydride. Specific Heat Measurements
(a) Vol. **46**, No. 1, p. K39–K41 (1978).
See also *8*, p. *84*; *18*, p. *274*.

A. Murasik, H. Ptasiewicz-Bąk, and A. Zygmunt
Magnetic Structure and Crystal-Field Effects in CeZn
(a) Vol. **46**, No. 1, p. K75–K78 (1978).
See also *18.2*, p. *284*; *18.4*, p. *299*; 21, p. 356.

H.-R. Müller and R. Perthel
Model Calculations for the Anisotropy Formation by Resputtering Processes during the Growth of Amorphous Gd–Co Films
(b) Vol. **87**, No. 1, p. 203–211 (1978).
See also *2*, p. *44*; 3, p. 54; *18.2.1*, p. *293*; 21.1, p. 372.

D. M. Eagles
Widths of 5d Energy Bands in the Heavy Rare-Earth Metals
(b) Vol. **87**, No. 1, p. 301–307 (1978).
See also *13.1*, p. *154*.

I. O. Bashkin, E. G. Ponyatovskii, and M. E. Kost
Phase Transformations in Hydrides of Rare-Earth Metals under Pressure (II): The Sc–H and Yb–H Systems under High Hydrogen Pressure
(b) Vol. **87**, No. 1, p. 369–372 (1978).
See also *1.2*, p. *18*; 12.1, p. 135.

Ya. A. Kraftmakher and T. Yu. Pinegina
Anomalies of Resistivity and Thermopower of Samarium at High Temperatures
(a) Vol. **47**, No. 1, p. K81–K83 (1978).
See also 1.2, p. 18; *14.1*, p. *208*; *15*, p. *260*.

A. Andreeff, L. P. Kaun, B. Lippold, W. Matz, N. I. Moreva, and K. Walther
Crystal Field Levels of Pr^{3+} in $PrAl_3$ Studied by Inelastic Neutron Scattering
(b) Vol. **87**, No. 2, p. 535–541 (1978).
See also *13.4*, p. *176*.

R. Ramji Rao and A. Ramanand
Lattice Dynamics, Specific Heat, and Bulk Modulus of Scandium
(b) Vol. **87**, No. 2, p. 751–756 (1978).
See also *6*, p. *72*; 8, p. 86; 12.1, p. 135.

R. Herz and H. Kronmüller
Field-Induced Phase Transitions in the Helical State of Dysprosium
(a) Vol. **47**, No. 2, p. 451—458 (1978).
See also *18.2*, p. *287*.

B. Knappe, H.-R. Müller, and B. Springmann
Homogeneity of Amorphous Gd–Co Films Prepared by Bias Sputtering
(a) Vol. **47**, No. 2, p. 523—532 (1978).
See also *1.5*, p. *37*; *2*, p. *45*; 18.3, p. 296; 21.1, p. 374.

V. V. Nemoshkalenko, Vl. N. Antonov, V. N. Antonov, P. Ziesche, and H. Wonn
On the Influence of Relativistic Effects on the Electronic Structure of Luthecium
(b) Vol. **88**, No. 1, p. K83—K86 (1978).
See also *13.1*, p. *157*.

R. Glocker and L. Fritsche
Pressure Dependence of the Electron–Phonon Coupling in Lanthanum
(b) Vol. **88**, No. 2, p. 639—644 (1978).
See also *6*, p. *74*; 12.1, p. 136; *13.1*, p. *157*; 14.2, p. 215.

H.-R. Müller
Influence of Anisotropic Short-Range Order on Magnetic Properties of the Amorphous Ferrimagnet Gd–Co
(a) Vol. **49**, No. 1, p. K63—K66 (1978).
See also *2*, p. *50*; *18.3*, p. *296*; 21.1, p. 378.

R. Kosicik, P. Duda, W. Keilig, and R. Mattheis
The Influence of Ion Bombardment on the Magnetic Anisotropy in Evaporated Amorphous Gd–Co Films
(a) Vol. **49**, No. 1, p. K67—K69 (1978).
See also *2*, p. *50*; 11, p. 129; *18.3*, p. *296*; 21.1, p. 378.

J. Čermák, V. Tvarožek, R. Harman, and J. Kaczér
Contribution to the Technology of Gd–Co Films
(a) Vol. **49**, No. 1, p. K81—K83 (1978).
See also 2, p. 50; *18.2.1*, p. *294*; 21.1, p. 378.

J. Fidler and P. Skalicky
TEM Investigation of Grown-in Dislocations and Stacking Faults in $SmCo_5$ Single Crystals
(a) Vol. **50**, No. 1, p. 73—81 (1978).
See also *1.4*, p. *34*; 10.1, p. 107; 21.1, p. 380.

P. Hendy and E. W. Lee
A Powder Neutron Diffraction Study of Some Rare-Earth–Co_2 Compounds
(a) Vol. **50**, No. 1, p. 101—107 (1978).
See also *18.3*, p. *296*; 21.1, p. 381.

M. Urner-Wille, T. S. te Velde, and P. G. van Engen
Kerr-Effect Enhancement in Amorphous GdFe Films
(a) Vol. **50**, No. 1, p. K29—K31 (1978).
See also 2, p. 52; *18.3*, p. *297*; *20*, p. *318*; 21.1.1, p. 390.

L. M. Noskova, L. A. Ugodnikova, and S. G. Chuchman
The Néel Points of Yttrium–Rare Earth Metal Alloys
(a) Vol. **50**, No. 1, p. K47—K49 (1978).
See also *18.1*, p. *280*; *18.4*, p. *301*.

H.-R. Müller, W. Keilig, R. Kosicik, P. Rosemann, and Z. Frait
Significance and Interpretation of Magnetic Anisotropy Measurements in Amorphous Gd–Co Films
(a) Vol. **50**, No. 2, p. 537—542 (1978).
See also *18.3*, p. *297*; 21.1, p. 382.

21.5 Actinides

Short Note

A. Čížek, M. Šob, E. Dębowska, and B. Rozenfeld
Positron Annihilation in Some Uranium and Thorium Compounds
(b) Vol. **89**, No. 2, p. K159—K161 (1978).
See also *13*, p. *148*.

21.6 Noble Metals

Original Papers and Short Notes

G. Rein, H. Mehrer, and K. Maier
Diffusion of ^{197}Pt and ^{199}Au in Platinum at Low Temperatures
(a) Vol. **45**, No. 1, p. 253–261 (1978).
See also *9*, p. *91*; 10.1, p. 99.

P. Svoboda
Size Effect Correction to the Longitudinal Magnetoresistance of Thin Wires
(a) Vol. **45**, No. 1, p. K43–K46 (1978).
See also *14.1*, p. *205*; 21, p. 353.

R. Rosei, E. Colavita, S. Modesti, and N. Nisticò
Electronic Structure of Dilute α-Phase Ag–Cd Alloys from Thermoreflectance and XPS Measurements
(b) Vol. **85**, No. 2, p. 445–452 (1978).
See also *13.1*, p. *151*; 20.1, p. 321; 21, p. 354.

B. Prasad and R. S. Srivastava
Thermal Properties and Grüneisen Parameters of Gold Using Toya's First Principle Approach
(b) Vol. **85**, No. 2, p. 789–793 (1978).
See also *6*, p. *66*; *8*, p. *83*.

V. S. Postnikov, V. V. Postnikov, and V. M. Fedorov
Instability and Superconductivity in Pd–Ag–D and Pd–H Systems
(b) Vol. **85**, No. 2, p. K115–K117 (1978).
See also *14.2*, p. *213*.

S. K. Sen, M. B. Dutt, and A. K. Barua
The Diffusion of Iron in Copper and of Nickel in Silver
(a) Vol. **45**, No. 2, p. 657–663 (1978).
See also *9*, p. *91*; 21.1, p. 370; 21.1.1, p. 385.

J. Takahashi, K. Ozawa, S. Yamaguchi, Y. Fujino, O. Yoshinari, and M. Hirabayashi
Temperature Dependence of the Thermal Vibrations of Deuterium in a $Pb_{0.8}Au_{0.2}$ Alloy Studied by Ion Channelling
(a) Vol. **46**, No. 1, p. 217–223 (1978).
See also *1.3*, p. *25*; 6, p. 68; *11*, p. *125*.

G. van Tendeloo and S. Amelinckx
Erratum to: Long-Period Antiphase Boundary Structures in the Ordered Alloy Au_4Zn (I)
(a) Vol. **46**, No. 1, p. 399 (1978).
See also *1.3*, p. *25*; *1.4*, p. *30*; 21.1, p. 371.

O. P. Sharma
A Pseudopotential Study of the Electronic Properties of Noble Metals
(b) Vol. **86**, No. 2, p. 483–492 (1978).
See also 4, p. 57; *6*, p. *68*; 14.1, p. 207; 21.1, p. 371.

ANAND P. PATHAK
Stopping Power of Solids in Planar Channelling
(b) Vol. **86**, No. 2, p. 751—758 (1978).
See also *11*, p. *125*; 13, p. 144; 21.1, p. 372.

R. KIRSCH
An Automated FSA Ellipsometer System for in Situ Investigations
(a) Vol. **46**, No. 2, p. 459—466 (1978).
See also 14.4, p. 248; *20.1*, p. *325*.

P. JUNG
Subthreshold Irradiation Effects in Quenched Platinum
(a) Vol. **46**, No. 2, p. 527—531 (1978).
See also *10.1*, p. *102*; 11, p. 125; 14.1, p. 207.

J. G. ERLINGS and F. W. SCHAPINK
Dislocation Networks in Twin Boundaries
(a) Vol. **46**, No. 2, p. 653—657 (1978).
See also *10.1*, p. *102*.

A. EROGLU, S. ARAJS, C. A. MOYER, and K. V. RAO
Electrical Resistivity of Chromium–Gold Alloys between 77 and 700 K
(b) Vol. **87**, No. 1, p. 287—291 (1978).
See also *14.1*, p. *207*; 21.1, p. 372.

Y. THOMAS
On the Variation of the Debye Temperature at High Temperatures
(b) Vol. **87**, No. 1, p. K13—K16 (1978).
See also *6*, p. *70*; *8*, p. *85*.

G. VAN TENDELOO, R. WOLF, D. VAN DYCK, and S. AMELINCKX
The Study of Ordering in the Au_5Mn_2 Alloy by Means of Structure Images
(a) Vol. **47**, No. 1, p. 105—122 (1978).
See also *1.3*, p. *25*; *1.4*, p. *30*; 21.1, p. 373.

R. WOLF, G. VAN TENDELOO, J. VAN LANDUYT, and S. AMELINCKX
Evidence for the Occurrence of the AABB Structure in the Gold–Manganese System
(a) Vol. **47**, No. 1, p. 241—244 (1978).
See also *1.2*, p. *18*; 1.3, p. 25; 1.4, p. 30; 21.1, p. 373.

P. ERLENWEIN
On a Relationship between the Symmetries of Faceted Cones on Ion Bombarded Single-Crystal Faces and Ejection Patterns
(a) Vol. **47**, No. 1, p. K9—K10 (1978).
See also *1.4*, p. *30*; 11, p. 126; 21, p. 358.

W. BRAUNE, J. LEBECH, and K. SAERMARK
Propagation of High-Frequency Electromagnetic Waves in the Voigt Configuration through Samples of Bi at 45, 70, and 300 GHz
(b) Vol. **87**, No. 2, p. 527—533 (1978).
See also *13.2*, p. *162*; *14.1*, p. *208*.

S. HORIGUCHI, K. KOYAMA, and Y. H. OHTSUKI
Auger Neutralization of Slow Protons at Solid Surfaces
(b) Vol. **87**, No. 2, p. 757—763 (1978).
See also *13.3*, p. *166*.

CH. HERZIG and H. E. PELTNER
Electromigration and Residual Resistivity of Cobalt in Gold
(a) Vol. **47**, No. 2, p. 497–506 (1978).
See also *9*, p. *93*; 14.1, p. 208.

G. VAN TENDELOO, R. WOLF, J. VAN LANDUYT, and S. AMELINCKX
Electron Diffraction and High Resolution Electron Microscopic Study of Ordering in the Gold–Manganese System (I): The Orthorhombic and Monoclinic I Structure in $Au_{11}Mn_4$
(a) Vol. **47**, No. 2, p. 539–554 (1978).
See also *1.2*, p. *19*; 1.3, p. 26; 1.4, p. 31; 21.1, p. 374.

G. VAN TENDELOO and S. AMELINCKX
Direct Observation by High Resolution Electron Microscopy of Coincidence Sites in Domain Boundaries in Ordered Gold–Manganese Alloys
(a) Vol. **47**, No. 2, p. 555–564 (1978).
See also *1.3*, p. *26*; *1.4*, p. *31*; 21.1, p. 374.

P. AIRO
On the Superstructure of the Alloy Au–18 at%Mg
(a) Vol. **47**, No. 2, p. K107–K110 (1978).
See also *1.2*, p. *20*; *1.3*, p. *26*; 21.3, p. 394.

S. B. SHIRVASTAVA and H. P. BONDE
Positron Annihilation Rates in the Metals of Ag and Au Series
(b) Vol. **88**, No. 1, p. 269–272 (1978).
See also *13.1*, p. *156*.

R. WOLF, G. VAN TENDELOO, J. VAN LANDUYT, and S. AMELINCKX
Electron Diffraction and High Resolution Electron Microscopic Study of Ordering in the Gold–Manganese System (II): The Monoclinic II Structure in $Au_{11}Mn_4$ and Its Relation to Other LPAPB Phases
(a) Vol. **48**, No. 1, p. 39–52 (1978).
See also *1.2*, p. *20*; 1.3, p. 27; 1.4, p. 31; 21.1, p. 375.

K. NARAYANDAS, M. RADHAKRISHNAN, and C. BALASUBRAMANIAN
*Annealing Study of the Electrical Resistivity and Defect Density in Silver Films
(a) Vol. **48**, No. 1, p. K71–K74 (1978).
See also 10.1, p. 104; *14.1*, p. *209*.

H. E. PELTNER and CH. HERZIG
New Application of the Knudsen Method for a Direct Determination of Thermodynamic Activities
(a) Vol. **48**, No. 1, p. K79–K81 (1978).
See also *3*, p. *55*.

D. HEITMANN
Theory and Experiment of the Light Emission of Surface Plasmons Induced by Fast Electrons on Modulated Surfaces
(b) Vol. **88**, No. 2, p. 493–506 (1978).
See also *13.3*, p. *166*; *13.5.2*, p. *198*; 20.3, p. 346.

F. COSANDEY, Y. KOMEM, and C. L. BAUER
Characterization of [001] Tilt Boundaries Produced by Epitaxial Growth on Bicrystalline Substrates of NaCl
(a) Vol. **48**, No. 2, p. 555–563 (1978).
See also *1.5*, p. *38*.

T. Skośkiewicz
Preparation of Superconducting Hydrides of Palladium–Nickel Alloys under High Hydrogen Pressures
(a) Vol. **48**, No. 2, p. K165–K168 (1978).
See also *14.2*, p. *215*; 21.1, p. 377.

S. Y. Chuang, P. K. Tseng, G. J. Jan, and H. S. Chen
An Experimental Study of the Angular Correlation of Positron Annihilation Radiations in Electron Irradiated Pb–Ni–Si Metallic Glass
(a) Vol. **48**, No. 2, p. K181–K183 (1978).
See also *1.4*, p. *32*; *2*, p. *48*; 21.1, p. 377.

C. Mande and A. Pimpale
Evidence for Surface Plasmons in X-Ray Emission Bands
(b) Vol. **89**, No. 1, p. K21–K24 (1978).
See also *13.3*, p. *167*; *13.5.2*, p. *198*; 20, p. 317.

R. Kutner, I. Sosnowska, and W. Kley
Next-Neighbour Jumps of Hydrogen in Quasi-Elastic Neutron Scattering from α-PdH_x
(b) Vol. **89**, No. 1, p. K29–K33 (1978).
See also *9*, p. *93*.

K. Müller and K. Heinz
Reconstructed Surfaces
(a) Vol. **49**, No. 1, p. 39–49 (1978).
See also *1.6*, p. *41*; 21, p. 362; 22.1.2, p. 436.

D. Möckel, W. Rössler, and F. Baumann
Electrical Resistivity of Amorphous Sb Films and Sb–Au Films
(a) Vol. **49**, No. 1, p. 103–110 (1978).
See also 2, p. 49; *14.1*, p. *210*; *14.3.1*, p. *228*; 21.7, p. 407.

K. Abdelmoula, B. Pardo, and D. Renard
Study by Electron Microscopy of Some Twinned Gold Crystals with Variations in Thickness
(a) Vol. **49**, No. 1, p. 267–278 (1978).
See also *1.4*, p. *33*; 1.5, p. 38.

T. J. Magee and J. Peng
Investigation of Solid–Solid Reactions of Au Films on Silicon
(a) Vol. **49**, No. 1, p. 313–322 (1978).
See also *1.3*, p. *28*; 1.5, p. 38; *14.3.2*, p. *231*; 22.1.2, p. 437.

G. van Tendeloo and S. Amelinckx
A High Resolution Study of Ordering in Au_4Mn (I): The Microstructure of Fully Ordered Au_4Mn
(a) Vol. **49**, No. 1, p. 337–346 (1978).
See also *1.1*, p. *13*; *1.4*, p. *33*; 21.1, p. 377.

V. M. Ievlev, A. V. Bugakov, and V. A. Ammer
Interphase Coincidence Boundaries in Au–Si and Au–CaF_2 Systems
(a) Vol. **49**, No. 1, p. 413–420 (1978).
See also *1.4*, p. *33*; *1.5*, p. *38*; 3, p. 56.

Y. Waseda and W. A. Miller
Structural Relaxation in Metallic Glasses
(a) Vol. **49**, No. 1, p. K31—K36 (1978).
See also *2*, p. *49*.

C. R. Brooks
The Heat Capacity at Constant Volume of Platinum
(b) Vol. **89**, No. 2, p. K123—K125 (1978).
See also *8*, p. *88*.

G. van Tendeloo, R. de Ridder, and S. Amelinckx
A High Resolution Study of Ordering in Au_4Mn (II): The Evolution of Ordering in Au_4Mn
(a) Vol. **49**, No. 2, p. 655—666 (1978).
See also *1.1*, p. *14*; *1.4*, p. *33*; 21.1, p. 379.

K. Narayandas, M. Radhakrishnan, and C. Balasubramanian
Annealing Effect of Silver–Tellurium Films
(a) Vol. **49**, No. 2, p. K213—K215 (1978).
See also 1.2, p. 23; 2, p. 51; *14.1*, p. *211*; 22.1.3, p. 442.

G. van Tendeloo and S. Amelinckx
Long-Period Antiphase Boundary Structures in the Ordered Alloy Au_4Zn (II): Growth Mechanism of the LPAPB Structure
(a) Vol. **50**, No. 1, p. 53—66 (1978).
See also *1.3*, p. *28*; *1.4*, p. *34*; 21.1, p. 380.

J. Bihr, H. Mehrer, and K. Maier
A Comparision between Microsectioning Studies of Low-Temperature Self Diffusion in Silver
(a) Vol. **50**, No. 1, p. 171—178 (1978).
See also *9*, p. *94*.

I. P. Grebennik and S. P. Yurchenko
The Use of Two-Dimensional Electron Diffraction Effects for the Initial Stage Research of γ-Phase Growth in the Metal–Melt Contact Zone
(a) Vol. **50**, No. 1, p. K1—K3 (1978).
See also *1.5*, p. *39*; 21, p. 365.

G. Frommeyer
High-Frequency Conductivity, Classical and Anomalous Skin Effect of Eutectic Silver–Copper Composite Wires
(a) Vol. **50**, No. 2, p. 517—528 (1978).
See also *14.1*, p. *212*; 21.1, p. 382.

K. Maier, H. Mehrer, and J. Bihr
A Novel Method for the Direct Determination of the Activation Enthalpy in Tracer Diffusion Studies with Applications to "Low"-Temperature Self-Diffusion in Silver
(a) Vol. **50**, No. 2, p. 543—549 (1978).
See also *9*, p. *95*.

21.7 Semi-Metals (Semiconducting Alloys See 22.7)

Original Papers and Short Notes

S. Matsuo, T. Nishimoto, S. Hayashi, and S. Noguchi
Magnetothermal Oscillation and Spin Splitting of Electrons in Dilute Bismuth–Antimony Alloys
(b) Vol. **86**, No. 1, p. 169–176 (1978).
See also *13.1*, p. *152*; 14.1, p. 206.

D. V. Gitsu, F. M. Muntyanu, M. I. Onu, and A. I. Russu
Peculiarities of the Thermomagnetic Power in $Bi_{0.91}Sb_{0.09}$ Alloy in Strong Magnetic Field
(b) Vol. **86**, No. 1, p. K71–K73 (1978).
See also *15*, p. *259*.

L. G. Khvostantsev and V. A. Sidorov
Influence of Pressure up to 90 kbar on the Thermoelectric Properties of Antimony
(a) Vol. **46**, No. 1, p. 305–309 (1978).
See also *15*, p. *259*.

A. Delunas and F. Aymerich
On Electrolytic Amorphous Antimony
(a) Vol. **46**, No. 2, p. 703–706 (1978).
See also *1.2*, p. *18*; *2*, p. *44*.

T. Soma
Lattice Dynamics of Group IV Elements with the Local Heine-Abarenkov Model Potential
(b) Vol. **87**, No. 1, p. 345–350 (1978).
See also *6*, p. *70*; 12.2, p. 139; 22.1.1, p. 425; 22.1.2, p. 432.

A. R. Patel, K. V. Rao, and G. K. Shivakumar
Initial Stages of Bismuth Deposition
(a) Vol. **47**, No. 1, p. K1–K3 (1978).
See also *1.5*, p. *37*.

B. Borie
On the Observation of Forbidden Bragg Maxima for White Tin
(a) Vol. **47**, No. 1, p. K39–K42 (1978).
See also *1.1*, p. *12*; 6, p. 71.

D. W. Field
Comment on "On the Observation of Forbidden Bragg Maxima for White Tin"
(a) Vol. **47**, No. 1, p. K43 (1978).
See also *1.1*, p. *12*; 6, p. 71.

D. W. Field
Anharmonicity in X-Ray Diffraction Intensities in White Tin
(a) Vol. **47**, No. 2, p. 651–656 (1978).
See also *6*, p. *72*.

D. Segers, M. Dorikens, and L. Dorikens-Vanpraet
Temperature Dependence of Positron Annihilation Parameters in Bismuth
(a) Vol. **48**, No. 1, p. 133–136 (1978).
See also *10.1*, p. *104*.

K. Bartkowski, D. Włosewicz, and J. Rafałowicz
Electrical Resistivity Anisotropy of Tin Monocrystals of Different Concentration of Chemical Impurities in the Temperature Range 5 to 70 K
(a) Vol. **48**, No. 1, p. 215—219 (1978).
See also *14.1*, p. *209*.

G. C. Vezzoli
Peaked Electrical Conductivity in Liquid Bismuth and Associated Optical Scattering
(a) Vol. **48**, No. 1, p. K75—K77 (1978).
See also *2*, p. *47*; *14.1*, p. *209*; 20.1, p. 330.

G. Oelgart, R. Stegmann, and R. Herrmann
Propagation of Electromagnetic Waves in a Direction Parallel to the Magnetic Field in Semiconductives Bi–Sb Alloys
(b) Vol. **88**, No. 2, p. 507—516 (1978).
See also *13.2*, p. *162*; 14.3, p. 223; 19, p. 310.

A. Bittar and D. Lesueur
Quenching and Annealing of Defects in Bismuth and Its Dilute Alloys
(a) Vol. **48**, No. 2, p. K123—K126 (1978).
See also *10.1*, p. *105*; *14.1*, p. *209*.

W. Braune, G.Kuka, H.-J. Gollnest, and R. Herrmann
Microwave Spectroscopy in Semimetallic Bismuth–Antimony Alloys
(b) Vol. **89**, No. 1, p. 95—101 (1978).
See also *13.1*, p. *158*; *19*, p. *311*.

D. Möckel, W. Rössler, and F. Baumann
Electrical Resistivity of Amorphous Sb Films and Sb–Au Films
(a) Vol. **49**, No. 1, p. 103—110 (1978).
See also 2, p. 49; *14.1*, p. *210*; *14.3.1*, p. *228*; 21.6, p. 404.

G. Kuka, W. Kraak, H.-J. Gollnest, and R. Herrmann
Temperature Dependence of the Resistivity in Semimetals of the Bismuth Type
(b) Vol. **89**, No. 2, p. 547—551 (1978).
See also *14.1*, p. *211*.

Y. Motohashi and S. Ohtake
Elastic Energy, Stability, and Interaction of Slip Dislocations and the Choice of Slip Systems in Rhombohedral Crystals, Bi, Sb, As, and Hg
(a) Vol. **50**, No. 1, p. 335—347 (1978).
See also *10.1*, p. *107*.

22. Semiconductors and Ionic Crystals

Review Articles

J. A. Wilson
Modelling the Contrasting Semimetallic Characters of TiS_2 and $TiSe_2$
(b) Vol. **86**, No. 1, p. 11–36 (1978).
See also 6, p. 64; 13.1, p. 150; *14.1*, p. *205*.

A. S. Borukhovich
Kinetic Properties of IVa and Va Subgroup Transition Metal Monocarbides
(a) Vol. **46**, No. 1, p. 11–37 (1978).
See also 8, p. 83; 13.1, p. 150; *14.3*, p. *217*; 15, p. 259; 22.3, p. 457; 22.3.1, p. 460.

V. S. Postnikov, V. V. Postnikov, and V. M. Fedorov
Metastability and Superconductivity in Thin Films
(b) Vol. **87**, No. 1, p. 11–23 (1978).
See also 1.2, p. 15; *1.6*, p. *40*; *14.2*, p. *213*; 21, p. 352.

V. I. Ivanov-Omskii, L. I. Korovin, and E. M. Shereghii
Phonon-Assisted Cyclotron Resonance in Semiconductors
(b) Vol. **90**, No. 1, p. 11–32 (1978).
See also *6*, p. *64*; 13, p. 142; *19*, p. *303*; 22.2.3, p. 452; 22.4.1, p. 463.

Original Papers and Short Notes

V. V. Hizhnyakov rand A. V. Sherman
On the Theoy of Resonant Secondary Radiation of Excitons Weakly Interacting with Phonons
(b) Vol. **85**, No. 1, p. 51–61 (1978).
See also *6*, p. *64*; *13.5.1*, p. *189*; 20.1, p. 319; 20.3, p. 341.

E. L. Ivchenko, I. G. Lang, and S. T. Pavlov
Resonant Secondary Radiation in Polar Semiconductors
(b) Vol. **85**, No. 1, p. 81–85 (1978).
See also *6*, p. *64*; *20.3*, p. *341*.

I. P. Zvyagin
Comments on the Theory of Hopping Conduction
(b) Vol. **85**, No. 1, p. 105–109 (1978).
See also 13.4, p. 169; *14.3*, p. *217*.

R. Keiper and R. Schuchardt
On the Theory of Photo-Induced DC Hopping Conductivity in Disordered Semiconductors
(b) Vol. **85**, No. 1, p. 155–162 (1978).
See also *16*, p. *262*.

R. Pässler
Temperature Dependences of the Nonradiative Multiphonon Carrier Capture and Ejection Properties of Deep Traps in Semiconductors (I): Theoretical Results
(b) Vol. **85**, No. 1, p. 203–215 (1978).
See also *13.4*, p. *169*.

P. Kleinert and F. Bechstedt
Influence of Mass Defect Disorder on Resonance Raman Scattering
(b) Vol. **85**, No. 1, p. 253—260 (1978).
See also *6*, p. *64*; *20.1*, p. *319*.

V. M. Fomin and E. P. Pokatilov
Non-Linear Optical Properties of the Band Charge Carrier Gas
(b) Vol. **85**, No. 1, p. 403—415 (1978).
See also *14.3.4*, p. *240*; *20.2*, p. *339*.

V. Čápek and J. Přívratská
Spin Polarons in the s–d Model with Off-Diagonal Exchange
(b) Vol. **85**, No. 1, p. K1—K4 (1978).
See also *13.5.3*, p. *200*; *18.2*, p. *281*.

M. Kitao, H. Ikeda, H. Hasegawa, and S. Yamada
Photoconductivity of Amorphous As_2Se_3
(a) Vol. **45**, No. 1, p. 107—111 (1978).
See also 2, p. 42; *16*, p. *262*.

G. H. Döhler
Analytical Solution for the Junction Behaviour of Relaxation Semiconductors
(a) Vol. **45**, No. 1, p. 113—116 (1978).
See also *14.3.3*, p. *233*; *14.3.4*, p. *240*.

C. J. Rossouw and M. J. Whelan
Thickness Determination of Thin Graphite and Aluminium Nitride Foils by Transmission Electron Microscopy Using Diffraction and Energy Loss Analysis
(a) Vol. **45**, No. 1, p. 277—288 (1978).
See also *1.5*, p. *36*; 13.5.2, p. 198; 22.1, p. 422.

H. Haug and D. B. Tran Thoai
Dynamical Screening of Excitons by Free Carriers
(b) Vol. **85**, No. 2, p. 561—568 (1978).
See also *13.5.1*, p. *190*; 13.5.2, p. 198.

R. M. M. Fonville, W. Geertsma, and C. Haas
On the Nature of the Top of the Valence Band in Layered Mo and W Dichalcogenides
(b) Vol. **85**, No. 2, p. 621—627 (1978).
See also *13.1*, p. *151*; 14.3, p. 217; 18.1, p. 277; 19, p. 304.

W. I. Friesen, D. B. Litvin, and B. Bergersen
Theory of the X-Ray and Optical Properties of Polysulphur Nitride
(b) Vol. **85**, No. 2, p. 715—722 (1978).
See also *13*, p. *143*; *20*, p. *315*.

S. C. Varshney and A. A. Gundjian
Internal Strain and Piezoelectricity in Zincblende Crystals
(b) Vol. **85**, No. 2, p. 733—742 (1978).
See also 6, p. 66; *12.2*, p. *138*; *14.3*, p. *217*.

P. J. Lin-Chung
Many Special Point Scheme for Noncubic Lattices
(b) Vol. **85**, No. 2, p. 743—748 (1978).
See also *13.1*, p. *151*.

V. T. Bublik
The Mean Square Atomic Displacements and Enthalpies of Vacancy Formation in Some Semiconductors
(a) Vol. **45**, No. 2, p. 543–548 (1978).
See also *10.2*, p. *110*; 22.2, p. **443**; 22.4, p. 461.

A. N. Zyuganov, Yu. G. Pismennyi, S. V. Svechnikov, and P. S. Smertenko
Regime of Contact Emission Limited Current with Weak Harmonic Distortion of Steady Bias
(a) Vol. **45**, No. 2, p. 631–638 (1978).
See also 13.4, p. 171; *14.3.4*, p. *240*.

Ch. Schnittler
Some New Aspects of the SCLC Characteristic of Homogeneous Insulators with Traps
(a) Vol. **45**, No. 2, p. K179–K182 (1978).
See also 13.4, p. 171; *14.3.4*, p. *241*.

P. Konsin
Structural Phase Transitions of Antiferroelectric and Displacive Modulated Types Caused by Electron–Phonon Interaction
(b) Vol. **86**, No. 1, p. 57–66 (1978).
See also *1.2*, p. *16*; 6, p. 67; 14.4.1, p. 254.

S. H. El-Mahalawy and B. L. Evans
Pressure Dependence of the Electrical Conductivity in 2H–MoS_2 and 2H–WSe_2
(b) Vol. **86**, No. 1, p. 151–157 (1978).
See also 12.2, p. 138; *14.3*, p. *218*.

E. Majerníková and Š. Barta
Magnetic Freeze-Out Effect in Noncrystalline Semiconductors
(b) Vol. **86**, No. 1, p. 183–194 (1978).
See also 2, p. 43; *13.1*, p. *152*; *13.5.1*, p. *190*; 18, p. 273.

V. P. Oleinik and V. A. Sinyak
To the Theory of the Tunnel Effect in a Semiconductor in Constant Electric and Magnetic Fields
(b) Vol. **86**, No. 1, p. 319–327 (1978).
See also *13.1*, p. *153*; 14.3, p. 218.

M. Combescot
Thermodynamics of an Electron–Hole System in Semiconductors
(b) Vol. **86**, No. 1, p. 349–358 (1978).
See also *13.5.1*, p. *190*.

D. Genzow, K. H. Herrmann, H. Kostial, I. Rechenberg, and A. E. Yunovich
Interband Absorption Edge in $Pb_{1-x}Sn_xTe$ Epitaxial Layers
(b) Vol. **86**, No. 1, p. K21–K25 (1978).
See also *20.1*, p. *323*.

R. Pässler
Intermediate-Temperature Dependences of the Nonradiative Multiphonon Carrier-Capture Cross-Section of Deep Traps in Semiconductors
(b) Vol. **86**, No. 1, p. K39–K44 (1978).
See also *13.4*, p. *172*; 22.2.1, p. 445.

R. Pässler
A Simple Approximation for the Temperature Dependences of the Nonradiative Multiphonon Carrier–Capture Cross-Section of Deep Traps in Semiconductors
(b) Vol. **86**, No. 1, p. K45—K47 (1978).
See also *13.4*, p. *172*.

A. P. Dmitriev, A. E. Stefanovich, and L. D. Tsendin
Discontinuities in Drift Solutions Due to Reversal in Ambipolar Drift. Injection and Accumulation of Hot Carriers
(a) Vol. **46**, No. 1, p. 45—53 (1978).
See also *13*, p. *144*; *14.3.4*, p. *241*.

S. Cristoloveanu, A. Chovet, and G. Kamarinos
Some Comments on "The Effect of Effective Lifetime Modulation in Semiconductors"
(a) Vol. **46**, No. 1, p. K13—K16 (1978).
See also *13.3*, p. *165*; 14.3.3, p. 234.

I. Munteanu
On the Effective Lifetime Modulation in Semiconductors
(a) Vol. **46**, No. 1, p. K17—K20 (1978).
See also *13.3*, p. *165*; 14.3.3, p. 234.

I. M. Kotina, S. R. Novikov, and T. I. Pirozhkova
On the Determination of the Parameter of Impurity Centres from the Thermally Stimulated Current of p–n Junctions
(a) Vol. **46**, No. 1, p. K21—K23 (1978).
See also *13.4*, p. *173*; *14.3.3*, p. *234*.

J. Richter and P. Seidel
An Analytic Expression for the Zero-Bias Offset in the Conductance–Voltage Characteristic of Metal–Insulator–Metal Tunnel Junctions
(a) Vol. **46**, No. 1, p. K25—K28 (1978).
See also *14.3.3*, p. *234*.

R. Car, G. Ciucci, and L. Quartapelle
Electronic Band Structure of SnSe
(b) Vol. **86**, No. 2, p. 471—478 (1978).
See also *13.1*, p. *153*.

E. Gerlach and M. Rautenberg
Ionized Impurity Scattering in Semiconductors
(b) Vol. **86**, No. 2, p. 479—482 (1978).
See also *13.4*, p. *174*; *14.3*, p. *219*.

Vo Hong Anh and Nguyne Nhu Dat
Parametric Excitation in Kane's Semiconductors
(b) Vol. **86**, No. 2, p. 585—591 (1978).
See also *6*, p. *69*; *13.5.2*, p. *198*; 20, p. 316.

R. Ramani, K. K. Mani†, and R. P. Singh
Long Optical Vibrations and Elastic Constants of Hexagonal Boron Nitride
(b) Vol. **86**, No. 2, p. 759—763 (1978).
See also *6*, p. *69*; 12.2, p. 138.

I. M. Dykman and P. M. Tomchuk
Electromagnetic Waves in Inhomogeneous Semiconductors with Non-Equilibrium Carriers
(b) Vol. **86**, No. 2, p. 773–780 (1978).
See also *13*, p. *144*; *14.3*, p. *220*; 20, p. 316.

B. V. Egorov and M. A. Krivoglaz
The Theory of Autolocalized Electron States in Ferromagnetic Semiconductors at Arbitrary Relation between the Band Width and the Spin–Electron Interaction
(b) Vol. **86**, No. 2, p. 781–794 (1978).
See also *13*, p. *144*; 13.5.3, p. 200; 18.2, p. 285.

V. G. Plekhanov and A. A. O'Connel-Bronin
Polariton Emission in LiH, LiD, and NaI Single Crystals
(b) Vol. **86**, No. 2, p. K123–K125 (1978).
See also *13.5*, p. *187*; 20.3, p. 343; 22.5.2, p. 480.

D. Genzow
Anisotropy of Photoconductivity in Lead Chalcogenides
(b) Vol. **86**, No. 2, p. K143–K146 (1978).
See also *16*, p. *263*.

T. Taki
^{31}P Multiple-Pulse Nuclear Magnetic Resonance Experiments in Inorganic Solids
(b) Vol. **86**, No. 2, p. K147–K150 (1978).
See also *19*, p. *306*.

G. J. Russell and J. Woods
An Electron Diffraction Study of Phases in the Cu_xS System
(a) Vol. **46**, No. 2, p. 433–444 (1978).
See also *1.6*, p. *41*; 22.4.1, p. 464.

J. van Landuyt, G. A. Wiegers, and S. Amelinckx
A New Type of Deformation Modulated Superstructure in 1T-VSe_2 and Its Relation with Other Superstructures in Transition Metal Dichalcogenides
(a) Vol. **46**, No. 2, p. 479–492 (1978).
See also *1.1*, p. *12*; 1.2, p. 17.

G. Weinhausen
Theoretical Study of the Depletion-Layer Capacitance in Abrupt p–n Semiconductor Junctions
(a) Vol. **46**, No. 2, p. 517–525 (1978).
See also *14.3.3*, p. *234*; 14.3.4, p. 241.

M. Hou, P. Delavignette, and A. Art
Detection of Implanted Ions by X-Ray Emission Analysis in TEM
(a) Vol. **46**, No. 2, p. K167–K169 (1978).
See also *11*, p. *126*.

V. M. Shatalov, V. A. Telezhkin, and K. B. Tolpygo
Inadequacy of the Direct Interband Transition Scheme Applied to Valence Semiconductor Fundamental Absorption Spectra
(b) Vol. **87**, No. 1, p. 79–86 (1978).
See also *13.1*, p. *154*; 13.5.1, p. 191; 20.1, p. 325.

H. L. ALTHAUS, G. WEISER, and S. NAGEL
Optical Spectra and Band Structure of As_2Se_3
(b) Vol. 87, No. 1, p. 117–128 (1978).
See also *13.1*, p. *154*; *20.1*, p. *326*.

S. GUHA and S. GHOSH
Hybrid Mode Interaction in Inhomogeneous Semiconductor Plasmas
(b) Vol. 87, No. 1, p. K9–K12 (1978).
See also *14.3*, p. *220*; *20*, p. *316*.

R. ERLANDSSON, U. BIRKHOLZ, and S.-E. KARLSSON
Study of Si_2Te_3 Surface Reactions with Auger Electron Spectroscopy
(a) Vol. 47, No. 1, p. 85–90 (1978).
See also *1.6*, p. *41*.

G. I. MAKOVETSKII and G. M. SHAKHLEVICH
Magnetic Properties of the $CrS_{1-x}Se_x$ System
(a) Vol. 47, No. 1, p. 219–222 (1978).
See also *18.4*, p. *300*.

V. CHÁB, L. ŠTOURAČ, and B. VELICKÝ
The Composition Dependence of Activation Energy in the As_x–Te_{100-x} and $As_2Te_3+Tl_x$ Glassy Systems
(a) Vol. 47, No. 1, p. 271–276 (1978).
See also *2*, p. *45*; *14.3*, p. *221*; 15, p. 260.

J. M. DUSSEAU, L. ENSUQUE, IM-SAROEUN, and T. P. LEPETRE
High Electric Field Effect in Silicon–Boron Compounds such as $B_{14}Si$
(a) Vol. 47, No. 1, p. K11–K15 (1978).
See also *14.3.4*, p. *242*.

C. ALBERS, CH. WALKOWIAK, P. SCHÄFER, and I. RECHENBERG
Time Dependent Negative Weak-Field Magnetoresistance in n-Type Lead Telluride Epitaxial Films
(a) Vol. 47, No. 1, p. K31–K34 (1978).
See also *14.3.1*, p. *227*.

O. PAVLOV and Z. BONTCHEVA-MLADENOVA
Measurements of the Ag_2Te–Ag_2S System at 300 K
(a) Vol. 47, No. 1, p. K59–K61 (1978).
See also *1.3*, p. *26*; 8, p. 86; *14.3*, p. *221*; 15, p. 260.

I. M. CHAPNIK
Atomic Volume and Semiconducting Properties in Sulfides of Non-Magnetic Transition Metals
(a) Vol. 47, No. 1, p. K71–K75 (1978).
See also *1.3*, p. *26*; 14.1, p. 207; *14.3*, p. *221*.

C. RAZZETTI and P. P. LOTTICI
Resonant Raman Scattering in As_2S_3 Amorphous Thin Films
(b) Vol. 87, No. 2, p. 479–483 (1978).
See also *2*, p. *45*; *20.1*, p. *327*.

V. L. Vinetskii and T. A. Kudykina
Localized Bipolarons in Ionic Crystals
(b) Vol. **87**, No. 2, p. 507—511 (1978).
See also *13.5.3*, p. *200*.

V. L. Volkov
The Electron-Phonon Interaction in $Pb_{1-x}Sn_xTe$
(b) Vol. **87**, No. 2, p. K93—K96 (1978).
See also *6*, p. *72*; *13.1*, p. *155*.

K. O. Kechechyan and A. A. Kirakossyan
Impurity Centers in Semiconductors Containing Screw Dislocations
(b) Vol. **87**, No. 2, p. K123—K126 (1978).
See also *10.2*, p. *113*; *13.4*, p. *176*.

C. R. Whitehouse and A. A. Balchin
Non-Stoichiometry in ZrS_2 and $ZrSe_2$
(a) Vol. **47**, No. 2, p. K173—K176 (1978).
See also *1.2*, p. *20*.

E. Mlitzke
On the Original Method of Balarin and Zetzsche
(a) Vol. **47**, No. 2, p. K181—K182 (1978).
See also *13.4*, p. *177*; *20.3*, p. *345*.

Gh. Ilonca
Electrical Resistivity and Magnetoresistivity of Amorphous $Cr_xPd_{82-x}Ge_{18}$ Alloys
(b) Vol. **88**, No. 1, p. 37—41 (1978).
See also *2*, p. *46*; *14.3*, p. *221*.

V. M. Shatalov
Quasi-Molecular Model Interpretation of the Resonances in the Fundamental Absorption Region of Valence Semiconductors
(b) Vol. **88**, No. 1, p. 129—133 (1978).
See also *13.1*, p. *155*; 13.5.1, p. 192; 20.1, p. 328.

M. Matlak and W. Nolting
Influence of Finite Carrier Concentrations on the Electronic Spectrum of a Ferromagnetic Semiconductor
(b) Vol. **88**, No. 1, p. 231—240 (1978).
See also *13.5.4*, p. *201*; 18.2, p. 288.

M. Baleva
Thermoreflectance of PbTe
(b) Vol. **88**, No. 1, p. 335—339 (1978).
See also *13.1*, p. *156*; *20.1*, p. *329*.

S. Ueda and T. Shimizu
Nuclear Magnetic Resonance of ^{77}Se in Amorphous $Cu_x(As_{0.4}Se_{0.6})_{1-x}$
(b) Vol. **88**, No. 1, p. K1—K3 (1978).
See also *2*, p. *46*; *19*, p. *309*.

G. Röpke, K. Kilimann, D. Kremp, W. D. Kraeft, and R. Zimmermann
The Influence of Dynamical Effects on the Two-Particle States (Excitons) in the Electron-Hole Plasma
(b) Vol. 88, No. 1, p. K59—K63 (1978).
See also *13.5*, p. *187*; *13.5.1*, p. *193*.

V. A. Gritsenko, E. E. Meerson, and S. P. Sinitsa
Unsteady Silicon Nitride Conductivity in High Electric Fields
(a) Vol. 48, No. 1, p. 31—37 (1978).
See also 2, p. 47; *14.3.2*, p. *230*; *14.3.4*, p. *243*.

R. Sh. Malkovich and V. A. Pokoeva
Diffusion of an Ionized Impurity in a Semi-Infinite Semiconductor
(a) Vol. 48, No. 1, p. 241—248 (1978).
See also *9*, p. *93*.

T. Cornelissens, G. van Tendeloo, J. van Landuyt, and S. Amelinckx
Peierls Distortion. Chain Polytypism, and Dislocation Coupling in NbS_3
(a) Vol. 48, No. 1, p. K5—K9 (1978).
Erratum
(a) Vol. 49, No. 2, p. K217 (1978).
See also *1.4*, p. *31*; 10.2, p. 115.

J. Kočka
Influence of Contacts on the AC Conductivity of Glasses
(a) Vol. 48, No. 1, p. K59—K62 (1978).
See also *2*, p. *47*; *14.3*, p. *222*; 14.3.4, p. 243.

P. Doubrava, V. L. Averyanov, and J. Škácha
Spectral Characteristics of Photodarkening of a-$As_{60}Se_{40}$ Films
(a) Vol. 48, No. 1, p. K93—K96 (1978).
See also *2*, p. *47*; 16, p. 265; *20.1*, p. *330*.

I. M. Dykman, V. M. Rosenbaum, and F. T. Vasko
Hot Electrons in Semiconductors with Quasi-Relativistic Band Structure
(b) Vol. 88, No. 2, p. 385—395 (1978).
See also *13*, p. *146*; *14.3.4*, p. *243*; 16, p. 265.

K. Saiki, M. Yoshimi, and S. Tanaka
Modulation Spectroscopy on the Group IV and VI Transition-Metal Dichalcogenides
(b) Vol. 88, No. 2, p. 607—614 (1978).
See also 13.1, p. 157; *13.5.1*, p. *193*; *20.1*, p. *330*.

F. Galeskii, A. E. Yunovich, K. H. Herrmann, H. Kostial, I. Rechenberg, and P. Schäfer
Stimulated Emission, Absorption Spectra, and Recombination in Epitaxial PbTe
(b) Vol. 88, No. 2, p. 675—681 (1978).
See also 13.4, p. 178; 20.1, p. 330; *20.2*, p. *339*.

C. G. Patil and B. S. Krishnamurthy
Magnon Drag Thermoelectric Power in Ferromagnetic Semiconductors
(b) Vol. 88, No. 2, p. K117—K122 (1978).
See also *13.5.4*, p. *202*; *15*, p. *260*; 18.2, p. 289.

S. Jandl, G. Perluzzo, and P. E. Girard
Mobility of Electrons in SnSSe Wingle Crystals
(b) Vol. 88, No. 2, p. K135—K137 (1978).
See also *14.3*, p. *223*.

R. H. Williams
The Electronic Structure of Tin Diselenide Containing Cyclopropylamine
(b) Vol. 88, No. 2, p. K157—K159 (1978).
See also *13.1*, p. *158*; 17, p. 270.

N. A. Poklonskii, V. F. Stelmakh, V. D. Tkachev, and S. V. Voitikov
Screening of Electrostatic Fields in Semiconductors by Multicharged Defects
(b) Vol. 88, No. 2, p. K165—K168 (1978).
See also *13.4*, p. *179*; *14.3*, p. *223*.

Ch. Schnittler
A Simplified Approach to the SCLC Characteristic of Homogeneous Solids with Traps Distributed in Energy
(a) Vol. 48, No. 2, p. 357—368 (1978).
See also 13.4, p. 179; *14.3.4*, p. *244*.

M. H. El-Fouly and J. T. Edmond
Electrical Conductivity of Some Chalcogenide Glasses as a Function of Electric Field and Temperature
(a) Vol. 48, No. 2, p. 395—405 (1978).
See also *2*, p. *48*; *14.3.4*, p. *244*.

W. R. Fahrner, D. Bräunig, R. Ferretti, and B. Müller
A Rapid Destruction-Free Determination of Breakdown Voltage of Thin Insulating Layers
(a) Vol. 48, No. 2, p. 533—541 (1978).
See also *14.3.1*, p. *227*; *14.3.4*, p. *244*.

A. Buroff, E. Nebauer, and P. Süptitz
Influence of an Electric DC Field on the Migration of Silver in Amorphous As_2S_3
(a) Vol. 48, No. 2, p. K109—K112 (1978).
See also *2*, p. *48*; *9*, p. *93*; 14.3.4, p. 244.

O. K. Alekseeva, V. I. Mikhailov, and V. P. Shantarovich
Positron Annihilation in Point Defects of the Glassy As–Se System
(a) Vol. 48, No. 2, p. K169—K173 (1978).
See also *2*, p. *48*; *10.2*, p. *117*.

M. Moldovanova and R. Assenov
A New Interpretation of the Temperature Dependence of the Hall Coefficient in "Pure" and Doped SnTe
(a) Vol. 48, No. 2, p. K193—K197 (1978).
See also *14.3*, p. *223*; 15, p. 261.

Y. Lépine, G. Bader, and D. Matz
Ground State Energy of Bound Polarons
(b) Vol. 89, No. 1, p. 53—59 (1978).
See also *13.5.3*, p. *200*.

A. PHILIPP and K. SEEGER
Influence of Halogens on the DC Conductivity and Magnetoresistance of $(SN)_x$
(b) Vol. **89**, No. 1, p. 187—192 (1978).
See also *14.1*, p. *210*.

M. WAUTELET and L. D. LAUDE
Electronic Structure of Solids under Laser Irradiation
(b) Vol. **89**, No. 1,p. 275—279 (1978).
See also 11, p. 129; *13.1*, p. *158*; 20.2, p. 339.

A. R. FOROUHI and I. BLOOMER
A Quantum Mechanical Approach to the Velocity of Dislocations in Ice
(b) Vol. **89**, No. 1, p. 309—312 (1978).
See also *10.2*, p. *117*.

H. D. DIMITROV
On the Surface Electrical Resistivity of Thin Semiconducting and Semimetallic Films
(b) Vol. **89**, No. 1, p. 319—328 (1978).
See also *13.3*, p. *166*; *14.3.1*, p. *228*; 14.3.2, p. 231.

V. K. BASHENOV, D. I. MARVAKOV, and A. G. PETUKHOV
Average Structure Factors and Stability of Binary Crystals
(b) Vol. **89**, No. 1, p. K5—K8 (1978).
See also *4*, p. *59*; *13.1*, p. *158*.

D. GENZOW
On the Interband Absorption in Lead Chalcogenides under Linearly Polarized Light Excitation
(b) Vol. **89**, No. 1, p. K9—K11 (1978).
See also *20.1*, p. *332*.

K. W. BÖER
The Physics of the Photovoltaic Effect
(a) Vol. **49**, No. 1, p. 13—26 (1978).
See also *14.3.3*, p. *236*; *16*, p. *265*; 22.4.1, p. 465.

H. P. BAUER and U. BIRKHOLZ
Electrical Conductivity of Passivated Si_2Te_3 Single Crystals
(a) Vol. **49**, No. 1, p. 127—131 (1978).
See also 14.3.2, p. 231; *14.3.4*, p. *244*.

V. M. EFIMOV, V. A. KOLOSANOV, and S. P. SINITSA
Electron and Hole Conduction in Silicon Nitride at Moderate Electric Fields
(a) Vol. **49**, No. 1, p. 217—223 (1978).
See also *14.4*, p. *250*.

A. V. MOSKALONOV, and J. R. ZAKIS
X-Ray Induced Paramagnetic States and Luminescence in As_2S_3 and As_2Se_3
(a) Vol. **49**, No. 1, p. 231—235 (1978).
See also 2, p. 49; *13.4*, p. *181*; 19, p. 312; *20.3*, p. *347*.

J.-M. MORET and R. LACROIX
Paramagnetic Resonance of Mn^{2+} in Na_2S Single Crystals
(a) Vol. **49**, No. 1, p. K55—K56 (1978).
See also *19*, p. *312*.

H. Atzmüller and U. Schröder
Theoretical Investigations of Haynes' Rule
(b) Vol. **89**, No. 2, p. 349–356 (1978).
See also *13.4*, p. *181*; *13.5.1*, p. *194*.

P. Kacman, J. Blinowski, and M. Grynberg
Dynamic Dielelectric Function in the Three-Band Model for Zero-Gap Semiconductors in a Magnetic Field
(b) Vol. **89**, No. 2, p. 521–530 (1978).
See also *13.1*, p. *159*; 14.4, p. 251; 18, p. 275.

R. Enderlein
On the Theory of Light Propagation through Disordered Solids. Scattering from Configurational Fluctuations
(b) Vol. **89**, No. 2, p. 625–632 (1978).
See also 2, p. 50; *6*, p. *76*; *20.1*, p. *334*.

B. Esser
On the Theory of Resonance Raman Scattering in Disordered Semiconductors
(b) Vol. **89**, No. 2, p. 633–643 (1978).
See also 2, p. 50; *6*, p. *76*; 13.1, p. 159; *20.1*, p. *334*.

V. M. Kenkre
Model for Trapping Rates for Sensitized Fluorescence in Molecular Crystals
(b) Vol. **89**, No. 2, p. 651–654 (1978).
See also *20.3*, p. *348*.

H. Böttger, V. V. Bryksin, and P. Kleinert
Theory of Resonant Raman Scattering in Systems with Localized Electron States
(b) Vol. **89**, No. 2, p. 675–685 (1978).
See also *6*, p. *76*; *13*, p. *148*; 20.3, p. 348.

N. F. Berk, L. Yarmus, W. Bizzaro, and J. Rosenthal
The Source of "Residual" Linewidth in EPR of Triplet Excitons in Molecular Crystals
(b) Vol. **89**, No. 2, p. K167–K170 (1978).
See also *13.5.1*, p. *195*; *19*, p. *313*.

S. Guha and S. Ghosh
Drift Wave Instability in Solid-State Plasmas
(b) Vol. **89**, No. 2, p. K171–K174 (1978).
See also *14.3*, p. *224*; *20*, p. *317*.

C. Clemen, X. I. Saldaña, P. Munz, and E. Bucher
Photovoltaic Properties of Some Semiconducting Layer Structures
(a) Vol. **49**, No. 2, p. 437–443 (1978).
See also *16*, p. *266*.

K. W. Böer
Heterojunction Interface Double Layer and Consequences for Photovoltaic Cells, Specifically $Cd_zZn_{1-z}S/Cu_2S$
(a) Vol. **49**, No. 2, p. 455–462 (1978).
See also *14.3.3*, p. *237*; *16*, p. *266*; 22.4.1, p. 466.

J. Baars, D. Bassett, and M. Schulz
Metal–Semiconductor Barrier Studies of PbTe
(a) Vol. **49**, No. 2, p. 483—488 (1978).
See also *14.3.4*, p. *245*.

R. Durný
EPR of Mn^{2+} in Ge–S Glasses
(a) Vol. **49**, No. 2, p. K113—K116 (1978).
See also *2*, p. *51*; *19*, p. *313*.

G. Rudlof, J. Becherer, and H. Glaefeke
Behaviour of the Fractional Glow Technique with First-Order Detrapping Processes, Traps Distributed in Energy or Frequency Factor
(a) Vol. **49**, No. 2, p. K121—K124 (1978).
See also *13.4*, p. *182*; *17*, p. *270*.

Z. S. Kachlishvili
On a Certain Feature of Nonlinear Galvanomagnetic Characteristics
(a) Vol. **49**, No. 2, p. K135—K137 (1978).
See also *13.4*, p. *182*; *14.3.4*, p. *245*.

S. Nešpůrek and J. Sworakowski
A Differential Method of Analysis of Steaty-State Space-Charge-Limited Currents: An Extension
(a) Vol. **49**, No. 2, p. K149—K152 (1978).
See also 13.4, p. 183; *14.3.4*, p. *245*.

R. Zimmermann, K. Kilimann, W. D. Kraeft, D. Kremp, and G. Röpke
Dynamical Screening and Self-Energy of Excitons in the Electron–Hole Plasma
(b) Vol. **90**, No. 1, p. 175—187 (1978).
See also *13.5*, p. *188*; *13.5.1*, p. *196*.

O. Ziep, M. Mocker, D. Genzow, and K. H. Herrmann
Auger Recombination in PbSnTe-Like Semiconductors
(b) Vol. **90**, No. 1, p. 197—205 (1978).
See also *13.1*, p. *159*.

H. Van Cong
Influence of Band-Tail Parameters on Electronic Specific Heat and Transport Coefficients in Heavily Doped Semiconductors
(b) Vol. **90**, No. 1, p. 401—407 (1978).
See also *8*, p. *89*; 13.4, p. 184; *14.3*, p. *224*.

A. T. Fromhold, Jr. and A.-B. Chen
Hopon Mechanism for Ion Transport in Ionic Compounds
(b) Vol. **90**, No. 1, p. K21—K25 (1978).
See also *9*, p. *94*; *10.2*, p. *120*.

H. Flietner
Schottky-Barrier Theory and the Complex Band Structure
(b) Vol. **90**, No. 1, p. K27—K30 (1978).
See also *13.3*, p. *168*; *14.3.2*, p. *231*.

S. A. Boiko, D. I. Bletskan, and S. F. Terekhova
Exciton Absorption of Germanium Diselenide
(b) Vol. **90**, No. 1, p. K49—K52 (1978).
See also *13.5.1*, p. *196*; *20.1*, p. *336*.

A. V. Goltsev, I. G. Lang, and S. T. Pavlov
Oscillations of Secondary Radiation Intensity in Polar Semiconductors in a Magnetic Field
(b) Vol. **90**, No. 1, p. K63—K67 (1978).
See also *6*, p. *78*; *20.1*, p. *336*.

D. I. Tsiulyanu, A. M. Andriesh, and E. P. Kolomeyko
Arsenic Trisulphide in Isotype Amorphous-Crystalline Heterojunctions
(a) Vol. **50**, No. 1, p. 195—202 (1978).
See also *2*, p. *52*; *14.3.3*, p. *237*; 14.3.4, p. 246.

R. de Ridder, D. van Dyck, and S. Amelinckx
Diffuse Intensity Contours Observed in Electron Diffraction Patterns of Transition Metal Tri- and Di-Chalcogenides: A Structural Interpretation
(a) Vol. **50**, No. 1, p. 349—362 (1978).
See also *1.1*, p. *14*; *1.4*, p. *34*.

N. G. Nilsson
Empirical Approximations Applied to the Generalized Einstein Relation for Degenerate Semiconductors
(a) Vol. **50**, No. 1, p. K43—K45 (1978).
See also *14.3*, p. *224*.

P. Macko and J. Doupovec
Thermal and Electrical Properties of Neutron Irradiated GeS_{11} Glass
(a) Vol. **50**, No. 1, p. K51—K54 (1978).
See also 2, p. 52; *8*, p. *89*; 11, p. 130; *14.3*, p. *225*.

S. D. Shutov and M. S. Iovu
Dispersive Charge Transport in Vitreous 0.55 $As_2S_3 \cdot 0.45\ Sb_2S_3$
(a) Vol. **50**, No. 1, p. K59—K61 (1978).
See also *2*, p. *52*; *14.3.1*, p. *229*.

R. de Ridder, D. van Dyck, and S. Amelinckx
A Simple Geometrical Interpretation of the Superstructures $\sqrt{p} \times \sqrt{p}$ ($p = 3, 7, 19, 31, 49, \ldots$) Observed on Surfaces with Hexagonal Symmetry by Means of LEED and RHEED
(a) Vol. **50**, No. 1, p. K67—K71 (1978).
See also *1.6*, p. *41*.

Ľ. Hrivnák
On the Switching Effect in Crystalline Semiconductors
(a) Vol. **50**, No. 1, p. K97—K99 (1978).
See also *13.4*, p. *184*; *14.3.4*, p. *246*.

D. Genzow, A. G. Mironov, and O. Ziep
On the Interband Absorption in Lead Chalcogenides
(b) Vol. **90**, No. 2, p. 535—542 (1978).
See also *13.1*, p. *160*; *20.1*, p. *337*.

A. M. Altshuler, Yu. Kh. Vekilov, and G. R. Umarov
Characteristic Features of the Electronic Spectrum and the Stability of Ternary Diamond-Like Semiconductors
(b) Vol. **90**, No. 2, p. 733—740 (1978).
See also 1, p. 10; *13.1*, p. *161*.

N. A. Kulagin and V. I. Tutlis
Change of the Praseodymium Spectrum under the Free State-to-Crystal Transition
(b) Vol. **90**, No. 2, p. K109—K113 (1978).
See also *13.4*, p. *185*; 20.1, p. 337.

E. S. Alekseev, R. G. Arkhipov, and S. V. Popova
Band Structure of the Metastable F.C.C. Tantalum Nitride
(b) Vol. **90**, No. 2, p. K133—K136 (1978).
See also *13.1*, p. *161*.

K.-A. Chao and A. Ferreira da Silva
Matrix Elements in the Matsubara-Toyozawa Theory of Impurity Band
(b) Vol. **90**, No. 2, p. K153—K155 (1978).
See also *13.4*, p. *185*.

F. L. Edelman, V. Z. Latuta, B. N. Zaitsev, and A. A. Khoromenko
The Structure of Silicon Nitride Films (I): Stoichiometric Silicon Nitride
(a) Vol. **50**, No. 2, p. 573—578 (1978).
See also *1.2*, p. *24*; *1.5*, p. *39*.

C. Riekel, M. Thomas, and R. Schöllhorn
Structure Refinement on Stoichoimetric $TiTe_2$ by Neutron Diffraction
(a) Vol. **50**, No. 2, p. K231—K234 (1978).
See also *1.1*, p. *14*.

22.1 Elements

Original Papers and Short Notes

C. J. Rossouw and M. J. Whelan
Thickness Determination of Thin Graphite and Aluminium Nitride Foils by Transmission Electron Microscopy Using Diffraction and Energy Loss Analysis
(a) Vol. **45**, No. 1, p. 277–288 (1978).
See also *1.5*, p. *36*; 13.5.2, p. 198; 22, p. 409.

R. Enderlein, B. Seifert, and F. Bechstedt
Electronic Elementary Excitations in the Far UV-Spectral Region. Core Level Exciton and Core Hole Polarization
(b) Vol. **85**, No. 2, p. 569–576 (1978).
See also *13.1*, p. *151*; *13.5.1*, p. *190*; 20, p. 315.

A. Zaręba
Piezoconductivity of β-Rhombohedral Boron
(a) Vol. **45**, No. 2, p. K183–K187 (1978).
See also *14.3*, p. *218*.

T. Soma
Local Heine-Abarenkov Model Potential for Covalent Crystals
(b) Vol. **86**, No. 1, p. 263–268 (1978).
See also *4*, p. *57*; 13.1, p. 153; 22.1.1, p. 424; 22.1.2, p. 430.

V. K. Bashenov, D. I. Marvakov, and A. M. Mutal
Internal Strain Parameter in Diamond-Type Crystals
(b) Vol. **86**, No. 1, p. K7–K10 (1978).
See also *4*, p. *57*; 12.2, p. 138; 22.1.1, p. 424; 22.1.2, p. 431.

V. K. Bashenov, Y. K. Markolenko, and V. V. Timofeenko
Semiconductor–Metal Transition of Diamond-Type Crystals
(b) Vol. **87**, No. 2, p. K73–K75 (1978).
See also *1.2*, p. *19*; *13.1*, p. *155*.

V. K. Bashenov, D. I. Marvakov, and A. G. Petukhov
Lattice Dynamics of Diamond-Type Crystals from Keating's Valence-Force Field
(b) Vol. **88**, No. 2, p. K161–K164 (1978).
See also *6*, p. *74*; 22.1.1, p. 427; 22.1.2, p. 435.

K. Ellmer and J. Sturm
Energy–Angle Distributions of ^{14}N and ^{20}Ne Ions after Passage through Thin Carbon Foils
(b) Vol. **89**, No. 2, p. 611–618 (1978).
See also *11*, p. *129*.

N. Terao
Experimental Study of Crack Propagation in Brittle Amorphous Solids: Cracks Formed in Glassy Carbon
(a) Vol. **49**, No. 2, p. K101–K102 (1978).
See also *2*, p. *51*; *12.2*, p. *141*.

M. Zanini, D. Grubisic, and J. E. Fischer
Optical Anisotropy of Highly Oriented Pyrolytic Graphite. Effect of Surface Preparation
(b) Vol. **90**, No. 1, p. 151–156 (1978).
See also *20.1*, p. *335*.

Y. H. Lee, P. R. Brosious, and J. W. Corbett
High-Temperature Ion Implantation in Diamond
(a) Vol. **50**, No. 1, p. 237–242 (1978).
See also 10.2, p. 120; *11*, p. *130*; 19, p. 314.

P. T. Landsberg, D. J. Robbins, and E. Schöll
Threshold Switching as a Generation–Recombination Induced Non-Equilibrium Phase Transition
(a) Vol. **50**, No. 2, p. 423–426 (1978).
See also 2, p. 52; 13.4, p. 185; *14.3.4*, p. *247*.

L. G. Aravin, R. A. Salukvadze, V. P. Shantarovich, and A. A. Khomenko
Positron Annihilation in Carbon Structures
(a) Vol. **50**, No. 2, p. K213–K217 (1978).
See also *1.4*, p. *35*; *10.2*, p. *122*.

22.1.1 Germanium

Original Papers and Short Notes

B. G. Helme and P. J. King
The Phonon Viscosity Tensor of Si, Ge, GaAs, and InSb
(a) Vol. **45**, No. 1, p. K33—K37 (1978).
See also 6, p. 65; *7*, p. *80*; 22.1.2, p. 429; 22.2.1, p. 445; 22.2.3, p. 452.

S. Marklund
Electron States Associated with the Core Region of the 60° Dislocation in Silicon and Germanium
(b) Vol. **85**, No. 2, p. 673—681 (1978).
See also *13.4*, p. *170*; 22.1.2, p. 429.

M. Nishida
Cluster Model Calculation for Threefold Coordinated Site Chemisorption of Cl on the Ge(111) Surface by the Extended Hückel Theory
(b) Vol. **85**, No. 2, p. K69—K72 (1978).
See also *13.3*, p. *164*.

S. Winter
A Comment on the Optical Absorption in Germanium with 60°-Dislocations
(b) Vol. **85**, No. 2, p. K95—K98 (1978).
See also 10.2, p. 110; *13.4*, p. *171*; *20.1*, p. *321*.

W. Lochmann
Phonon-Assisted Auger Recombination in Indirect Gap Semiconductors
(a) Vol. **45**, No. 2, p. 423—432 (1978).
See also *6*, p. *66*; *13.1*, p. *152*; 20, p. 315; 22.1.2, p. 430.

H. Siethoff
On the Creep of Germanium and Silicon
(a) Vol. **45**, No. 2, p. K135—K137 (1978).
See also *9*, p. *92*; *10.2*, p. *110*; 22.1.2, p. 430.

T. Soma
Local Heine-Abarenkov Model Potential for Covalent Crystals
(b) Vol. **86**, No. 1, p. 263—268 (1978).
See also *4*, p. *57*; 13.1, p. 153; 22.1, p. 422; 22.1.2, p. 430.

J. Humlíček
On the Nature of E_1-Transitions in Germanium
(b) Vol. **86**, No. 1, p. 303—310 (1978).
See also 13.1, p. 153; *14.3.4*, p. *241*; *20.1*, p. *322*.

V. K. Bashenov, D. I. Marvakov, and A. W. Mutal
Internal Strain Parameter in Diamond-Type Crystals
(b) Vol. **86**, No. 1, p. K7—K10 (1978).
See also *4*, p. *57*; 12.2, p. 138; 22.1, p. 422; 22.1.2, p. 431.

J. Dumas
Analysis of the Intensity of the ESR Line in Amorphous Si–Ge Films
(b) Vol. **86**, No. 1, p. K75—K77 (1978).
See also *2*, p. *43*; *19*, p. *305*; 22.1.2, p. 431.

R. S. Nakhmanson and L. K. Popov
Measurements of Small-Signal Photo-EMF of Si and Ge MIS Structures Using Scanning Light Probe
(a) Vol. **46**, No. 1, p. 59—68 (1978).
See also *14.3.2*, p. *230*; *14.3.3*, p. *233*; 16, p. 263; 22.1.2, p. 431.

P. Thomas and D. Wuertz
Low Temperature Polaronic Hopping Conduction
(b) Vol. **86**, No. 2, p. 541—548 (1978).
See also *2*, p. *43*; *14.3*, p. *219*; 15, p. 260; 22.1.2, p. 432.

M. Singh
Phonon Scattering and Effect of Dislocations on the Low-Temperature Phonon Conductivity in Ge
(b) Vol. **86**, No. 2, p. 691—696 (1978).
See also *6*, p. *69*; 8, p. 84; 10.2, p. 111.

H. A. Kassim and K. S. Dubey
Lattice Thermal Resistivity Due to the Presence of Electrons
(b) Vol. **86**, No. 2, p. K107—K111 (1978).
See also *6*, p. *69*; *8*, p. *84*.

J. F. Graczyk
Defect Structures in Amorphous Germanium Films
(a) Vol. **46**, No. 2, p. 611—617 (1978).
See also *1.5*, p. *36*; *2*, p. *44*.

T. Soma
Lattice Dynamics of Group IV Elements with the Local Heine-Abarenkov Model Potential
(b) Vol. **87**, No. 1, p. 345—350 (1978).
See also *6*, p. *70*; 12.2, p. 139; 21.7, p. 406; 22.1.2, p. 432.

M. Leppihalme and T. Tuomi
Phonon Energies and Barrier Height from Photovoltage Spectra of Au–Intrinsic Ge Diodes
(a) Vol. **47**, No. 1, p. 245—250 (1978).
See also 6, p. 71; 13.5.1, p. 191; *14.3.3*, p. *235*; *16*, p. *264*.

K. O. Lee
The Meyer-Neldel Rule and the DC Conductivity in Amorphous Ge Films
(a) Vol. **47**, No. 1, p. K47—K49 (1978).
See also *2*, p. *45*; *14.3.1*, p. *227*.

M. David, G. van Tendeloo, J. van Landuyt, and R. Gevers
The Critical Voltage Effect in Transmission Electron Microscopy (X): Experimental Obervations in the Presence of Non-Systematic Reflections
(b) Vol. **87**, No. 2, p. 419—432 (1978).
See also *1.1*, p. *13*; 22.6, p. 494.

K. S. Dubey
Heat Transfer in a Doped Crystal Lattice at Low Temperatures: Application to Sb-Doped Ge
(b) Vol. **87**, No. 2, p. K133—K136 (1978).
See also *6*, p. *72*; *8*, p. *86*.

N. N. Sirota, V. I. Lukashova, and L. A. Kurochkin
Irradiation Effect on Cyclotron Resonance in Germanium
(a) Vol. **47**, No. 2, p. 677–682 (1978).
See also *11*, p. *127*; 19, p. 309.

Z. Vučić, D. Subašić, and Z. Ogorelec
The Determination of the Crystal–Liquid Interfacial Tension in Germanium
(a) Vol. **47**, No. 2, p. 703–710 (1978).
See also *2*, p. *46*.

S.-L. Chang
On the 'Solution' Dispersion Surfaces of Two-Beam X-Ray Dynamical Bragg Reflection for Infinitely Thick Crystals
(a) Vol. **47**, No. 2, p. 717–722 (1978).
See also *1.1*, p. *13*.

B. Pohorylles and R. Piotrzkowski
Effect of Hydrostatic Pressure on the Photoconductivity of Plastically Deformed n-Type Germanium
(a) Vol. **47**, No. 2, p. K115–K117 (1978).
See also 13.4, p. 177; *16*, p. *264*.

E. N. Mgbenu
Room Temperature Au–Ge Interdiffusion
(a) Vol. **47**, No. 2, p. K153–K155 (1978).
See also *9*, p. *93*; 14.3.1, p. 227.

D. P. Bhattacharya and Z. S. Kachlishvili
Faraday Rotation in Presence of Attractive Traps
(b) Vol. **88**, No. 1, p. 283–287 (1978).
See also *13.4*, p. *177*; *14.3.4*, p. *243*; 18, p. 274.

Mahiradhwaj Singh
Phonon Attenuation in n-Type Semiconductors with ZB Structure in Presence of Internal Strains and Magnetic Field
(b) Vol. **88**, No. 1, p. 295–300 (1978).
See also *6*, p. *73*; 7, p. 81; 12.2, p. 140; 18, p. 274; 22.2.3, p. 453.

K. I. Papunashvili
H-Linear "Planar" Hall Effects of Hot Electrons in n-Ge
(b) Vol. **88**, No. 1, p. K27–K30 (1978).
See also *13*, p. *146*; *14.3.4*, p. *243*.

T. Soma
Equation of State under Pressure for Si and Ge
(b) Vol. **88**, No. 1, p. K69–K72 (1978).
See also *4*, p. *58*; 22.1.2, p. 434.

G. Lautz and K. Sachse
Noise Spectra of Hot Carriers in n-Germanium
(a) Vol. **48**, No. 1, p. 121–131 (1978).
See also *13*, p. *146*; *14.3.4*, p. *243*.

V. DUTTA, P. NATH, and K. L. CHOPRA
´ Sructural and Electrical Properties of Polycrystalline Ge Films
(a) Vol. **48**, No. 1, p. 257—262 (1978).
See also *1.5*, p. *37*; *14.3.1*, p. *227*.

H. H. BOGHOSIAN and K. S. DUBEY
Role of Electron–Phonon Interaction and Peripheral Phonons in the Lattice Thermal Conductivity of a Doped Semiconductor at Low Temperatures: Application to Phosphorous-Doped Germanium
(b) Vol. **88**, No. 2, p. 417—427 (1978).
See also *6*, p. *74*; *8*, p. *87*.

NGUYEN VAN DONG and TRAN QUOC HAI
Electronic Transport in Doped a-Ge and a-Si Prepared by DC Cathodic Sputtering
(b) Vol. **88**, No. 2, p. 555—561 (1978).
See also *2*, p. *47*; *14.3.1*, p. *227*; 22.1.2, p. 435.

V. K. BASHENOV, D. I. MARVAKOV, and A. G. PETUKHOV
Lattice Dynamics of Diamond-Type Crystals from Keating's Valence-Force Field
(b) Vol. **88**, No. 2, p. K161—K164 (1978).
See also *6*, p. *74*; 22.1, p. 422; 22.1.2, p. 435.

J. VON WIENSKOWSKI
Carbon Monoxide Adsorption on Cleaved Si (111) and Ge (111) Surfaces
(b) Vol. **88**, No. 2, p. K169—K172 (1978).
See also *13.3*, p. *166*; 22.1.2, p. 435.

F. R. KESSLER, J. METZDORF, and W. ZUREK
Interband Faraday Ellipticity in Germanium
(b) Vol. **89**, No. 1, p. 203—211 (1978).
See also 13.1, p. 158; 13.5.1, p. 194; *18*, p. *275*; *20*, p. *317*.

G. LANGOUCHE, I. DEZSI, M. VAN ROSSUM, J. DE BRUYN, and R. COUSSEMENT
On the Existence of a Quadrupole Interaction at ^{57}Fe Implanted in Si and Ge
(b) Vol. **89**, No. 1, p. K17—K19 (1978).
See also *5*, p. *62*; 22.1.2, p. 436.

H.-J. NOWAK, H. LEITZ, and W. BUCKEL
On the Structure of Amorphous Metallic Layers of the System Germanium–Copper
(a) Vol. **49**, No. 1, p. 73—82 (1978).
See also *1.5*, p. *38*; *2*, p. *49*; 21.1, p. 377.

K. DETTMER, H. MÜLLER, and F. R. KESSLER
Surface Influence of Glass Substrates on Lattice Parameters and Phase Transition Temperature of Evaporated Ge Thin Films
(a) Vol. **49**, No. 1, p 255—260 (1978).
See also *1.5*, p. *38*; *2*, p. *49*.

V. DUTTA, P. NATH, V. D. VANKAR, and K. L. CHOPRA
Structural Properties of Polycrystalline Ge–Metal Films
(a) Vol. **49**, No. 1, p. 379—385 (1978).
See also 1.4, p. 33; *1.5*, p. *38*; 2, p. 49.

F. R. Kessler and W. Olsowski
Exciton Contribution to Oscillatory Magnetophotoconductivity in Ge
(b) Vol. **89**, No. 2, p. 389–396 (1978).
See also *13.5.1*, p. *194*; *16*, p. *266*.

S. Winter
Electron States below the Conduction Band in Germanium Originating from Dissociated 60°-Dislocations
(b) Vol. **90**, No. 1, p. 289–293 (1978).
See also *13.4*, p. *183*.

R. M. Westervelt, J. L. Staehli, and E. E. Haller
Nucleation Phenomena in Electron–Hole Drop Formation in Ge and Si (III): Experimental Results for Ge
(b) Vol. **90**, No. 2, p. 557–569 (1978).
See also *13.5.1*, p. *196*.

R. S. Markiewicz
Acoustic Phonon Scattering by Electron–Hole Droplets in Ge (II): Total Scattering Rate and Anisotopy of Scattering, as Reflected in the Shape of a Droplet Cloud
(b) Vol. **90**, No. 2, p. 585–596 (1978).
See also *13.5.1*, p. *197*; 14.3, p. 225.

A. Milchev
Frequency- and Wave-Vector-Dependent Dielectric Function of a Model Semiconductor
(b) Vol. **90**, No. 2, p. 679–688 (1978).
See also *13.1*, p. *161*; *14.4*, p. *252*; 22.1.2, p. 439; 22.2.1, p. 449; 22.4.2, p. 469.

G. P. Srivastava
Role of Optical Phonons in the High-Temperature Lattice Thermal Conductivity of Semiconductors
(b) Vol. **90**, No. 2, p. K125–K128 (1978).
See also *6*, p. *78*; *8*, p. *90*; 22.1.2, p. 439; 22.2.1, p. 449; 22.2.3, p. 455.

V. F. Kiselev, V. A. Matveev, and R. V. Prudnikov
The Proton Mechanism of Neutralization of Surface Recombination and Fast Capture Centres on Germanium
(a) Vol. **50**, No. 2, p. 739–744 (1978).
See also *13.3*, p. *168*.

22.1.2 Silicon

Review Article

V. V. TITOV
Ion Implantation — Problems and Perspectives
(a) Vol. **48**, No. 1, p. 13—22 (1978).
See also *3*, p. *54*; *11*, p. *123*.

Original Papers and Short Notes

H.-J. WÜNSCHE, V. E. KHARTSIEV, and K. HENNEBERGER
Binding-Energy Calculations for Multiple Bound Excitons in Silicon
(b) Vol. **85**, No. 1, p. K53—K56 (1978).
See also *13.5.1*, p. *189*.

B. JOSEPH
Influence of Insulator Surface Conduction on the Transient Response of MIS Capacitors
(a) Vol. **45**, No. 1, p. K25—K28 (1978).
See also *14.3.2*, p. *230*; 14.3.3, p. 233.

B. G. HELME and P. J. KING
The Phonon Viscosity Tensor of Si, Ge, GaAs, and InSb
(a) Vol. **45**, No. 1, p. K33—K37 (1978).
See also 6, p. 65; *7*, p. *80*; 22.1.1, p. 424; 22.2.1, p. 445; 22.2.3, p. 452.

Z. JANÁČEK and J. KUBĚNA
Integrated Intensity of the Laue Diffraction from an Elastically Bent Crystal
(a) Vol. **45**, No. 1, p. K55—K57 (1978).
See also *1.1*, p. *11*.

M. HÖHNE
EPR of Lithium in Mechanically Affected Silicon
(b) Vol. **85**, No. 2, p. 525—534 (1978).
See also 1.6, p. 40; 13.4, p. 170; *19*, p. *304*.

S. MARKLUND
Electron States Associated with the Core Region of the 60° Dislocation in Silicon and Germanium
(b) Vol. **85**, No. 2, p. 673—681 (1978).
See also *13.4*, p. *170*; 22.1.1, p. 424.

G. P. SRIVASTAVA
Electronic and Optical Properties of Crystalline and Amorphous Si
(b) Vol. **85**, No. 2, p. K121—K126 (1978).
See also 2, p. 43; *13.1*, p. *152*; *20.1*, p. *321*.

V. V. KALININ, N. N. GERASIMENKO, and S. I. STENIN
The Effect of Implantation Temperature on the Mechanism of Misfit Dislocation Formation
(a) Vol. **45**, No. 2, p. 377—385 (1978).
See also 1.4, p. 29; *10.2*, p. *110*; *11*, p. *124*.

W. Lochmann
Phonon-Assisted Auger Recombination in Indirect Gap Semiconductors
(a) Vol. **45**, No. 2, p. 423—432 (1978).
See also *6*, p. *66*; *13.1*, p. *152*; 20, p. 315; 22.1.1, p. 424.

H. Runge and H. Wörl
Autoradiographic Detection of Getter Effect of Argon-Implanted Layers in Silicon
(a) Vol. **45**, No. 2, p. 509—512 (1978).
See also *9*, p. *91*; *11*, p. *124*; 14.3, p. 218.

L. N. Aleksandrov, R. N. Lovyagin, P. A. Simonov, and I. S. Bzinkovskaya
p–n Junctions in the Surface Region of Silicon Obtained by Evaporation of Silicon in Ultrahigh Vacuum
(a) Vol. **45**, No. 2, p. 521—527 (1978).
See also 9, p. 91; *14.3.3*, p. *233*.

Y. Ochiai and E. Matsuura
Spin–Lattice Relaxation at High Temperatures in Heavily Doped n-Type Silicon
(a) Vol. **45**, No. 2, p. K101—K104 (1978).
See also *19*, p. *305*.

G. Mende
A Comparison of Calculated Arsenic Implantation Profiles in Silicon with Experimental Results
(a) Vol. **45**, No. 2, p. K121—K123 (1978).
See also *11*, p. *124*.

D. Landgraf-Dietz
A Simple Method for the Determination of Parameters of Ion Implanted Doping Profiles by Means of Threshold Voltage Measurements of MOSFET's
(a) Vol. **45**, No. 2, p. K125—K129 (1978).
See also *11*, p. *124*; *14.3.3*, p. *233*.

H. Siethoff
On the Creep of Germanium and Silicon
(a) Vol. **45**, No. 2, p. K135—K137 (1978).
See also *9*, p. *92*; *10.2*, p. *110*; 22.1.1, p. 424.

M. Höhne
Nonrandom Strain in "Crushed" Silicon. EPR of Thermally Excited Lithium Donors
(b) Vol. **86**, No. 1, p. 119—126 (1978).
See also 1.6, p. 40; 13.4, p. 172; *19*, p. *305*.

T. Soma
Local Heine-Abarenkov Model Potential for Covalent Crystals
(b) Vol. **86**, No. 1, p. 263—268 (1978).
See also *4*, p. *57*; 13.1, p. 153; 22.1, p. 422; 22.1.1, p. 424.

H. Neubrand
ESR from Boron in Silicon at Zero and Small External Stress (I): Line Positions and Line Structure
(b) Vol. **86**, No. 1, p. 269—275 (1978).
See also 13.4, p. 172; *19*, p. *305*.

V. K. Bashenov, D. I. Marvakov, and A. M. Mutal
Internal Strain Parameter in Diamond-Type Crystals
(b) Vol. **86**, No. 1, p. K7—K10 (1978).
See also *4*, p. *57*; 12.2, p. 138; 22.1, p. 422; 22.1.1, p. 424.

V. V. Kaplin, D. E. Popov, and S. A. Vorobev
Quantum-Mechanical Treatment of Anomalous Rutherford Scattering of Swift Electrons in Crystals
(b) Vol. **86**, No. 1, p. K35—K38 (1978).
See also *11*, p. *124*; 13, p. 143.

J. Dumas
Analysis of the Intensity of the ESR Line in Amorphous Si–Ge Films
(b) Vol. **86**, No. 1, p. K75—K77 (1978).
See also *2*, p. *43*; *19*, p. *305*; 22.1.1, p. 424.

R. S. Nakhmanson and L. K. Popov
Measurements of Small-Signal Photo-EMF of Si and Ge MIS Structures Using Scanning Light Probe
(a) Vol. **46**, No. 1, p. 59—68 (1978).
See also *14.3.2*, p. *230*; *14.3.3*, p. *233*; 16, p. 263; 22.1.1, p. 425.

P. Gaworzewski and H. Riemann
On the Effect of Heat Treatment upon the Charge Carrier Concentration of Dislocation-Free Floating-Zoned Silicon Crystals
(a) Vol. **46**, No. 1, p. 117—124 (1978).
See also 13.4, p. 173; *14.3*, p. *219*.

P. L. F. Hemment and R. W. Bicknell
1.5 MeV H^+ and He^+ Irradiations of Arsenic Diffused Layers in ⟨100⟩ Silicon
(a) Vol. **46**, No. 1, p. 225—231 (1978).
See also *11*, p. *125*; 13.4, p. 173; 14.3, p. 219.

S. Kopta, E. Maydell-Ondrusz, and B. Rajchel
Backscattering Measurements of Postimplantation Damage Profiles in Silicon Single Crystals
(a) Vol. **46**, No. 1, p. 285—290 (1978).
See also *11*, p. *125*.

M. V. Kruglov, V. N. Shchemelev, and G. G. Kareva
Detection of Atomic Plane Displacements in Near-Surface Layers of Crystals Using the Angular Dependence of Photoemission during Bragg Diffraction of X-Rays
(a) Vol. **46**, No. 1, p. 343—350 (1978).
See also *1.1*, p. *11*; 1.6, p. 40; 17, p. 269.

V. L. Indenbom, V. I. Nikitenko, E. V. Suvorov, and V. M. Kaganer
Section Topography as Single-Crystal Interferometry
(a) Vol. **46**, No. 1, p. 379—386 (1978).
See also *10.2*, p. *111*.

H.-J. Wünsche, K. Henneberger, and V. E. Khartsiev
Binding-Energy Calculations for Multiple Bound Excitons
(b) Vol. **86**, No. 2, p. 505—515 (1978).
See also *13.5.1*, p. *190*.

P. THOMAS and D. WUERTZ
Low Temperature Polaronic Hopping Conduction
(b) Vol. **86**, No. 2, p. 541–548 (1978).
See also *2*, p. *43*; *14.3*, p. *219*; 15, p. 260; 22.1.1, p. 425.

A. BOHG
Measurement of Stresses in Thin Films on Single Crystalline Substrates
(a) Vol. **46**, No. 2, p. 445–450 (1978).
See also *1.5*, p. *36*; 10.2, p. 112.

A. E. GERSHINSKII, L. V. MIRONOVA, and E. I. CHEREPOV
On the Conductivity of the Si–Thermal SiO_2–Electrolyte System in Superhigh Electric Fields
(a) Vol. **46**, No. 2, p. 467–473 (1978).
See also *14.3.1*, p. *226*; *14.3.4*, p. *241*; 22.6, p. 492.

F. M. LIVINGSTONE, R. M. DE LA RUE, and W. DUNCAN
pSi–nCdS Heterojunction Photodetectors
(a) Vol. **46**, No. 2, p. 595–601 (1978).
See also *14.3.3*, p. *234*; 16, p. 263; 22.4.1, p. 464.

O. BRÜMMER, H.-R. HÖCHE, and J. NIEBER
Some Remarks on Extremely Asymmetric X-Ray Diffraction in the Laue Case of Grazing Incidence
(a) Vol. **46**, No. 2, p. K131–K133 (1978).
See also *1.1*, p. *12*.

L. P. KHIZNICHENKO, E. OTENIYAZOV, and N. S. TULYAGANOVA
Internal Friction in Neutron-Irradiated Silicon
(a) Vol. **46**, No. 2, p. K147–K149 (1978).
See also *10.2*, p. *112*; 11, p. 126; *12.2*, p. *139*.

G. C. JAIN, A. PRASAD, and B. C. CHAKRAVARTY
Diffusion of Ga into Si through SiO_2
(a) Vol. **46**, No. 2, p. K151–K155 (1978).
See also *9*, p. *92*.

B. I. RESER and V. V. DYAKIN
Instability of the OPW Method in Relation to the Error in Core Eigenvalues
(b) Vol. **87**, No. 1, p. 41–47 (1978).
See also *13.1*, p. *154*.

T. SOMA
Lattice Dynamics of Group IV Elements with the Local Heine-Abarenkov Model Potential
(b) Vol. **87**, No. 1, p. 345–350 (1978).
See also *6*, p. *70*; 12.2, p. 139; 21.7, p. 406; 22.1.1, p. 425.

C. WAGNER and F. BURKHARDT
The Annealing Characteristics of Phosphorus-Implanted Silicon Investigated at Low Temperatures
(a) Vol. **47**, No. 1, p. 131–138 (1978).
See also *11*, p. *126*; *14.3*, p. *220*.

F.-G. Kirscht, J. Doerschel, and C. Hansch
Some Observations Concerning the Initial Stage of Plastic Flow in Silicon
(a) Vol. **47**, No. 1, p. K17—K20 (1978).
See also *10.2*, p. *113*.

N. N. Gerasimenko, P. P. Dobrovolskii, N. P. Kibalina, and V. F. Stas
Investigation of a High-Resistivity Surface Layer in Electron Irradiated Silicon by Schottky Capacitance Measurements
(a) Vol. **47**, No. 1, p. K91—K94 (1978).
See also 11, p. 127; *14.3.2*, p. *230*.

R. Wedell
Radiative Damping Effect for Ultrarelativistic Channelled Particles
(b) Vol. **87**, No. 2, p. 491—496 (1978).
See also *11*, p. *127*; 13, p. 145.

L. Pasemann
On the Screening of the Ions in Solids. Band Structure Calculations of Si with Several Screening Variants
(b) Vol. **87**, No. 2, p. 553—564 (1978).
See also *13.1*, p. *155*.

V. A. Tyagai, O. V. Snitko, A. M. Evstigneev, and A. N. Krasiko
Surface Phenomena in the Polarization-Dependent Electroreflectance Spectra of Si
(b) Vol. **87**, No. 2, p. 613—618 (1978).
See also 13.3, p. 165; *14.3.4*, p. *242*; *20.1*, p. *327*.

C. K. Kim and A. B. Kunz
Electronic Structure of Silicon with Non-Spherical Potentials for the Two Basis Atoms
(b) Vol. **87**, No. 2, p. 765—769 (1978).
See also *13.1*, p. *155*.

W. Ecke, A. A. Lebedev, and A. T. Mamadalimov
Deep Defect States in Evaporated Heteroepitaxial Silicon Films
(a) Vol. **47**, No. 2, p. 507—512 (1978).
See also *13.4*, p. *176*; *14.3.1*, p. *227*.

G. Foti, E. Rimini, and S. U. Campisano
Laser Annealing of Pb–Implanted Silicon
(a) Vol. **47**, No. 2, p. 533—538 (1978).
See also *11*, p. *127*.

H. F. Kappert, K. F. Heidemann, B. Grabe, and E. te Kaat
Range and Damage Profiling after Heavy Ion Implantation in the MeV Region
(a) Vol. **47**, No. 2, p. 751—762 (1978).
See also *11*, p. *127*.

K. P. Arefiev, A. A. Tsoi, and S. A. Vorobiev
Influence of High-Temperature Annealing on Positron Annihilation in Electron-Irradiated Silicon
(a) Vol. **47**, No. 2, p. K149—K152 (1978).
See also 10.2, p. 114; *11*, p. *127*.

W. Eberhardt, G. Kalkoffen, C. Kunz, D. Aspnes, and M. Cardona
Photoemission Studies of 2p Core Levels of Pure and Heavily Doped Silicon
(b) Vol. 88, No. 1, p. 135–143 (1978).
See also *13.1*, p. *156*; 17, p. 270; 20.1, p. 328.

T. Soma
Equation of State under Pressure for Si and Ge
(b) Vol. 88, No. 1, p. K69–K72 (1978).
See also *4*, p. *58*; 22.1.1, p. 426.

D. Wuertz and P. Thomas
The Pre-Exponential in DC-Hopping Conduction
(b) Vol. 88, No. 1, p. K73–K77 (1978).
See also *2*, p. *46*; *14.3*, p. 222.

E. Kuroda, M. Matsuda, and M. Maki
Growth and Characterization of Silicon Ribbon Crystals Grown with Wetting and Non-Wetting Dies
(a) Vol. 48, No. 1, p. 105–111 (1978).
See also *3*, p. *55*; 14.3, p. 222.

C. S. G. Cousins, L. Gerward, and J. Staun Olsen
Multiple Diffraction in Crystals Studied by an X-Ray Energy-Dispersive Method
(a) Vol. 48, No. 1, p. 113–119 (1978).
See also *1.1*, p. *13*.

P. Viktorovitch, G. Kamarinos, P. Even, and E. Fabre
Correlation between Interface Surface and MIS Silicon Solar Cell Performances
(a) Vol. 48, No. 1, p. 137–145 (1978).
See also 13.3, p. 166; *14.3.3*, p. *235*; *16*, p. *264*.

J. Rużyłło
Model of Structural Transformations on the Chemically Etched Silicon Surface during Thermal Oxidation
(a) Vol. 48, No. 1, p. 199–202 (1978).
See also *1.2*, p. *21*; 1.6, p. 41; *3*, p. *55*; 22.6, p. 494.

A. Wolkenberg
Capacitance Measurements of the Silicon–Silicon Dioxide–Potassium Chloride Water Solution Interface as a Method to Determine the Fixed Charge Value and Its Centroid
(a) Vol. 48, No. 1, p. 203–208 (1978).
See also *14.3.2*, p. *230*; *14.3.3*, p. *235*.

J. Borkowicz, J. Korec, and E. Nossarzewska-Orłowska
Optimum Growth Conditions in Silicon Vapour Epitaxy
(a) Vol. 48, No. 1, p. 225–228 (1978).
See also *3*, p. *55*.

D. E. Ioannou and S. M. Davidson
SEM Observation of Dislocations in Boron Implanted Silicon Using Schottky Barrier EBIC Technique
(a) Vol. 48, No. 1, p. K1–K4 (1978).
See also *10.2*, p. *115*.

F.-G. Kirscht, C. Hansch, and J. Doerschel
Dislocation Reactions in Plastically Deformed Silicon Observed by X-Ray Topography
(a) Vol. 48, No. 1, p. K11–K14 (1978).
See also *10.2*, p. *115*.

P. M. Mooney, R. T. Young, J. Karins, Y. H. Lee, and J. W. Corbett
Defects in Laser Damaged Silicon Observed by DLTS
(a) Vol. 48, No. 1, p. K31–K34 (1978).
See also *11*, p. *128*; *13.4*, p. *178*.

H. Mell, B. Movaghar, and L. Schweitzer
Magnetic Field Dependence of the Photoconductivity in Amorphous Silicon
(b) Vol. 88, No. 2, p. 531–535 (1978).
See also *2*, p. *47*; *16*, p. *265*.

Nguyen van Dong and Tran Quoc Hai
Electronic Transport in Doped a-Ge and a-Si Prepared by DC Cathodic Sputtering
(b) Vol. 88, No. 2, p. 555–561 (1978).
See also *2*, p. *47*; *14.3.1*, p. *227*; 22.1.1, p. 427.

V. K. Bashenov, D. I. Marvakov, and A. G. Petukhov
Lattice Dynamics of Diamond-Type Crystals from Keating's Valence-Force Field
(b) Vol. 88, No. 2, p. K161–K164 (1978).
See also *6*, p. *74*; 22.1, p. 422; 22.1.1, p. 427.

J. von Wienskowski
Carbon Monoxide Adsorption on Cleaved Si (111) and Ge (111) Surfaces
(b) Vol. 88, No. 2, p. K169–K172 (1978).
See also *13.3*, p. *166*; 22.1.1, p. 427.

U. Köster
Crystallization of Amorphous Silicon Films
(a) Vol. 48, No. 2, p. 313–321 (1978).
See also *1.5*, p. *37*; 2, p. 48; *3*, p. *55*.

L. F. Kurtenok, E. A. Movchan, O. G. Sarbey, V. V. Mitin, and M. Asche
Static High Field Domains in n-Si
(a) Vol. 48, No. 2, p. 323–328 (1978).
See also *14.3.4*, p. *243*.

L. Baldi, G. F. Cerofolini, G. Ferla, and G. Frigerio
Gold Solubility in Silicon and Gettering by Phosphorus
(a) Vol. 48, No. 2, p. 523–532 (1978).
See also *10.2*, p. *116*; 14.3.3, p. 235.

M. Lefeld-Sosnowska
X-Ray Diffraction Contrast of a 60° Dislocation Investigated by Means of Section Topography
(a) Vol. 48, No. 2, p. 565–569 (1978).
See also *1.4*, p. *32*; *10.2*, p. *116*.

G. Weidner
Application of Hillock Etching of Microdefects to the Investigation of Czochralski Silicon by Transmission Electron Microscopy
(a) Vol. 48, No. 2, p. K105–K107 (1978).
See also *1.4*, p. *32*.

Yu. V. Bogatyrev and F. P. Korshunov
Investigation of Properties of the Si–SiO_2 Interface in MOS Structures Irradiated by Fast Electrons
(a) Vol. **48**, No. 2, p. K127–K130 (1978).
See also *11*, p. *128*; *14.3.2*, p. *231*; 22.6, p. 495.

M. Manzel and W. Ecke
The Influence of Surface Recombination on the Double-Injection Switching
(a) Vol. **48**, No. 2, p. K161–K163 (1978).
See also *14.3.2*, p. *231*; *14.3.4*, p. *244*.

W. Fuhs, U. Holzhauer, S. Mantl, F. W. Richter, and R. Sturm
Annihilation of Positrons in Electron-Irradiated Silicon Crystals
(b) Vol. **89**, No. 1, p. 69–75 (1978).
See also *10.2*, p. *117*; 11, p. 128.

C. A. J. Ammerlaan and J. C. Wolfrat
Extended-Hückel-Theory Calculations for the Positive Divacancy in Silicon
(b) Vol. **89**, No. 1, p. 85–94 (1978).
See also *13.4*, p. *179*.

D. Wruck
Vibrational Properties of the ⟨001⟩ Split Interstitial in Silicon
(b) Vol. **89**, No. 1, p. 177–186 (1978).
See also *6*, p. *75*; 20.1, p. 332.

G. Langouche, I. Dezsi, M. van Rossum, J. de Bruyn, and R. Coussement
On the Existence of a Quadrupole Interaction at ^{57}Fe Implanted in Si and Ge
(b) Vol. **89**, No. 1, p. K17–K19 (1978).
See also *5*, p. *62*; 22.1.1, p. 427.

K. Müller and K. Heinz
Reconstructed Surfaces
(a) Vol. **49**, No. 1, p. 39–49 (1978).
See also *1.6*, p. *41*; 21, p. 362; 21.6, p. 404.

K. Graff and H. Pieper
Degradation of Carrier Lifetime in Silicon Crystals at Room Temperature
(a) Vol. **49**, No. 1, p. 137–144 (1978).
See also 10.2, p. 118; *14.3*, p. *224*.

I. Dudeck and R. Kassing
Determination of the Temperature Dependence of the Capture Cross-Sections of the Gold Acceptor Level and of the Temperature of Current Filaments in Silicon pin Diodes
(a) Vol. **49**, No. 1, p. 153–161 (1978).
See also *13.4*, p. *180*; 14.3.3, p. 236; *14.3.4*, p. *244*.

T. Puritis
Plasmon Mechanism for Secondary Breakdown
(a) Vol. **49**, No. 1, p. 225–230 (1978).
See also *13.5.2*, p. *199*; *14.3.4*, p. *244*.

M. T. Pham
A Very Simple Method for Profiling the Ion-Implanted Si-Surface
(a) Vol. **49**, No. 1, p. 261–265 (1978).
See also *11*, p. *129*; *14.3.2*, p. *231*.

T. J. Magee and J. Peng
Investigation of Solid–Solid Reactions of Au Films on Silicon
(a) Vol. **49**, No. 1, p. 313—322 (1978).
See also *1.3*, p. *28*; 1.5, p. 38; *14.3.2*, p. *231*; 21.6, p. 404.

G. Battaglin, G. Della Mea, A. V. Drigo, G. Foti, G. G. Bentini, and M. Servidori
Two-Stage Laser Annealing of Lattice Disorder in Phosphorus Implanted Silicon
(a) Vol. **49**, No. 1, p. 347—352 (1978).
See also *11*, p. *129*.

V. S. Lysenko and A. N. Nazarov
Effect of Laser Beam Irradiation on the Electrophysical Properties of the Interface in Boron Ion Implanted MIS Structures
(a) Vol. **49**, No. 1, p. 405—411 (1978).
See also 11, p. 129; 13.3, p. 167; *14.3.3*, p. *236*.

G. Stetter, H. Coufal, and E. Lüscher
EPR Study of the Annealing Behaviour of Neutron-Transmutation-Doped Silicon
(a) Vol. **49**, No. 1, p. K17—K20 (1978).
See also *11*, p. *129*; 14.3, p. 224; *19*, p. *312*.

W. Fuhs, M. Milleville, and J. Stuke
Drift Mobility and Photoconductivity in Amorphous Silicon
(b) Vol. **89**, No. 2, p. 495—502 (1978).
See also 2, p. 50; *14.3.1*, p. *228*; *16*, p. *266*.

D. Jäger, E. Kähler, and R. Heidemann
Negative Conductance in SCLC Diodes with Traps
(a) Vol. **49**, No. 2, p. 431—435 (1978).
See also 13.4, p. 182; *14.3.3*, p. *236*; *14.3.4*, p. *245*.

W. Bergholz and W. Schröter
Precipitation of Cobalt in Silicon Studied by Mössbauer Spectroscopy
(a) Vol. **49**, No. 2, p. 489—498 (1978).
See also *5*, p. *62*; 9, p. 94; *10.2*, p. *119*.

A. Siems
Quantitative Study of Kikuchi Structures (I): Energy Analysis of Intensity Profiles
(a) Vol. **49**, No. 2, p. 621—628 (1978).
See also *1.1*, p. *14*; 6, p. 77; 13.5.2, p. 199.

V. M. Harutyunyan and F. V. Gasparyan
Impedance of p^+–n–n^+ Structures Based on Semiconductors Compensated by Deep Doubly Ionized Centres
(a) Vol. **49**, No. 2, p. 797—801 (1978).
See also *13.4*, p. *182*; *14.3.3*, p. *237*.

H. Neubrand
ESR from Boron in Silicon at Zero and Small External Stress (II): Linewidth and Crystal Defects
(b) Vol. **90**, No. 1, p. 301—308 (1978).
See also 13.4, p. 183; *19*, p. *313*.

J. von Wienskowski
Oxidation of Cleaved Si(111) Surfaces
(b) Vol. **90**, No. 1, p. K9—K12 (1978).
See also *13.3*, p. *168*.

W. Szkiełko
Spin-Dependent Recombination at Dislocations in Silicon Revealed at Reverse Current in a p–n Junction
(b) Vol. **90**, No. 1, p. K81—K83 (1978).
See also *10.2*, p. *120*; *14.3.3*, p. *237*; 16, p. 267; 19, p. 314.

J. Reichel
The Influence of White Light on the Annealing Process of Self-Interstitials in p-Type Silicon
(a) Vol. **50**, No. 1, p. 87—93 (1978).
See also 9, p. 94; *10.2*, p. *120*.

S. Ristić and V. Cvekić
Some Effects of Localized Stress on Silicon Planar Transistors
(a) Vol. **50**, No. 1, p. 153—157 (1978).
See also *14.3.3*, p. *237*.

E. V. Suvorov, V. I. Nikitenko, and K. Yu. Mukhin
On the Possibilities of X-Ray Dynamic Focussing to Study Dislocation Microstress Fields
(a) Vol. **50**, No. 1, p. 213—219 (1978).
See also *10.2*, p. *120*.

K. H. Yang, H. F. Kappert, and G. H. Schwuttke
Minority Carrier Lifetime in Annealed Silicon Crystals Containing Oxygen
(a) Vol. **50**, No. 1, p. 221—235 (1978).
See also *10.2*, p. *120*; *14.3.3*, p. *238*.

A. Wolkenberg
Energetic Barrier Height Hypothesis on the Silicon–Electrolyte and Gallium Arsenide–Electrolyte Interfaces
(a) Vol. **50**, No. 1, p. 249—256 (1978).
See also 13.3, p. 168; *14.3.2*, p. *232*; *14.3.3*, p. *238*; 22.2.1, p. 449.

Z. Janáček, J. Kuběna, and V. Holý
X-Ray Laue Diffraction from a Bent Crystal. Integrated Intensity
(a) Vol. **50**, No. 1, p. 285—291 (1978).
See also *1.1*, p. *14*; 10.2, p. 121.

S. K. Pabi
Mechanism of the Dissolution of Si in Al
(a) Vol. **50**, No. 1, p. K9—K12 (1978).
See also *9*, p. *94*; 14.3.4, p. 246; 21, p. 365.

S. Mantovani, E. Mazzega, S. Valeri, and U. del Pennino
Edge Dislocation Energy Level in Silicon
(a) Vol. **50**, No. 1, p. K123—K126 (1978).
See also *10.2*, p. *121*; *13.4*, p. *184*.

H. STÖHR and H. BROSS
Solid Silicon Selfconsistently with X_α Potentials
(b) Vol. **90**, No. 2, p. 497—507 (1978).
See also *13.1*, p. *160*.

I. R. ENTIN
Theoretical and Experimental Study of X-Ray Acoustic Resonance in Perfect Silicon Crystals
(b) Vol. **90**, No. 2, p. 575—584 (1978).
See also *20*, p. *318*.

A. MILCHEV
Frequency- and Wave-Vector-Dependent Dielectric Function of a Model Semiconductor
(b) Vol. **90**, No. 2, p. 679—688 (1978).
See also *13.1*, p. *161*; *14.4*, p. *252*; 22.1.1, p. 428; 22.2.1, p. 449; 22.4.2, p. 469.

N. N. GERASIMENKO, M. ROLLÉ, LI-JEN CHENG, Y. H. LEE, J. C. CORELLI, and J. W. CORBETT
Infrared Absorption of Silicon Irradiated by Protons
(b) Vol. **90**, No. 2, p. 689—695 (1978).
See also *6*, p. *78*; 11, p. 131; *20.1*, p. *337*.

G. P. SRIVASTAVA
Role of Optical Phonons in the High-Temperature Lattice Thermal Conductivity of Semiconductors
(b) Vol. **90**, No. 2, p. K125—K128 (1978).
See also *6*, p. *78*; *8*, p. *90*; 22.1.1, p. 428; 22.2.1, p. 449; 22.2.3, p. 455.

A. ROIZÈS and R. SCHUTTLER
Galvanomagnetic Effects in p-Type Silicon. Application to Electrically Active Impurity Contents
(a) Vol. **50**, No. 2, p. 399—409 (1978).
See also 13.4, p. 185; *14.3*, p. *225*.

T. TUOMI, M. TILLI, V. KELHÄ, and J. D. STEPHENSON
Pendellösung Fringes in Synchrotron X-Ray Topographs of a Wedge-Shaped Silicon Crystal
(a) Vol. **50**, No. 2, p. 427—431 (1978).
See also *1.1*, p. *14*.

W. R. FAHRNER and E. KLAUSMANN
Characterization of MOS Structures with Buried Layers
(a) Vol. **50**, No. 2, p. 433—444 (1978).
See also 11, p. 131; *14.3.2*, p. *232*; *14.3.3*, p. *238*.

I. YONENAGA and K. SUMINO
Dislocation Dynamics in the Plastic Deformation of Silicon Crystals (I): Experiments
(a) Vol. **50**, No. 2, p. 685—693 (1978).
See also *10.2*, p. *122*.

A. P. DOLGOLENKO and I. I. FISHCHUK
Defect Clusters and Simple Defect Build-Up Kinetics in Fast-Neutron Irradiated n-Si
(a) Vol. **50**, No. 2, p. 751—755 (1978).
See also 10.2, p. 122; *11*, p. *131*.

A. K. Semenyuk and M. D. Varentsov
Anisotropy of Galvanomagnetic Effects in Irradiated Silicon with Layer Inhomogeneity
(a) Vol. **50**, No. 2, p. K169–K173 (1978).
See also 11, p. 131; *14.3*, p. *225*.

G. Götz, W. Karthe, B. Schnabel, and N. Sobolev
ESR Line Shape of Amorphous Centres in Ion Implanted Silicon
(a) Vol. **50**, No. 2, p. K209–K212 (1978).
See also 11, p. 132; *19*, p. *314*.

22.1.3 Group VI Elements

Review Article

P. A. THROWER and R. M. MAYER
Point Defects and Self-Diffusion in Graphite
(a) Vol. **47**, No. 1, p. 11—37 (1978).
See also *9*, p. *91*; *10.2*, p. *109*; 13.4, p. 169.

Original Papers and Short Notes

W. EICHLER
On the Analysis of Galvano- and Thermomagnetic Effects in Pure Perfect Tellurium Crystals
(a) Vol. **45**, No. 1, p. 195—199 (1978).
See also *14.3*, p. *217*; *15*, p. *259*.

R. M. MEHRA, A. K. KATHURIA, RADHEY SHYAM, A. L. DAWAR, and P. C. MATHUR
Determination of the Density of Localized States in a-Se Films
(a) Vol. **45**, No. 1, p. K59—K62 (1978).
See also *2*, p. *43*; *13.4*, p. *170*; 14.3.1, p. 226.

M. PÖHLER and G. STAUPENDAHL
Investigation of the Simultaneous Absorption of Three CO_2-Laser Photons by Tellurium
(b) Vol. **86**, No. 1, p. 225—230 (1978).
See also *13.5.1*, p. *190*; *20.1*, p. *322*.

J. A. SAWICKI and B. D. SAWICKA
Correlation of ^{57}Fe Isomer Shift, Quadrupole Coupling, and Interatomic Distance in Group-IV Elements
(b) Vol. **86**, No. 2, p. K159—K161 (1978).
See also *5*, p. *61*.

V. M. BONDAR, V. S. RADCHENKO, and L. S. SOLONCHUK
Conductivity of Warm Holes in Tellurium
(b) Vol. **87**, No. 2, p. K97—K100 (1978).
See also *14.3.4*, p. *242*.

N. P. GAVALESHKO, P. N. GORLEY, P. V. KAVYUK, L. S. SOLONCHUK, I. I. TARASYUK, and V. A. SHENDEROVSKII
Transverse and Longitudinal Nernst-Ettingshausen Effect in p-Te
(b) Vol. **88**, No. 1, p. 179—188 (1978).
See also *13.4*, p. *177*; 14.3, p. 222; *15*, p. *260*.

G. JUŠKA, S. VENGRIS, and K. ARLAUSKAS
Electron Lifetime in Amorphous Selenium
(a) Vol. **48**, No. 2, p. K149—K151 (1978).
See also *2*, p. *48*; *14.3*, p. *223*.

G. STAUPENDAHL and K. SCHINDLER
Self-Defocusing of CO_2-Laser Radiation in Tellurium
(a) Vol. **48**, No. 2, p. K199—K201 (1978).
See also *20.2*, p. *339*.

M. Meissner, A. Tausend, and D. Wobig
Heat Capacities and Thermodynamic Properties of α-Monoclinic Selenium in the Range 3 to 300 K
(a) Vol. **49**, No. 1, p. 59–66 (1978).
See also *8*, p. *88*.

J. P. Estienne and J. L. Farvacque
Occurence of an "***a***" Screw Dislocation Substructure in Tellurium by a Torsion Experiment
(a) Vol. **49**, No. 2, p. K109–K111 (1978).
See also *10.2*, p. *119*.

K. Narayandas, M. Radhakrishnan, and C. Balasubramanian
Annealing Effect of Silver–Tellurium Films
(a) Vol. **49**, No. 2, p. K213–K215 (1978).
See also 1.2, p. 23; 2, p. 51; *14.1*, p. *211*; 21.6, p. 405.

22.2 III–V Compounds

Review Article

P. C. Mathur, R. Shyam, and Sushil Jain
Transport Phenomena in III–V Compound Semiconductors
(a) Vol. **50**, No. 1, p. 11–40 (1978).
See also 6, p. 64; *13.1*, p. *150*; 13.4, p. 169; *14.3*, p. *217*; 20.1, p. 319.

Original Papers and Short Notes

E. M. Baskin, M. D. Blokh, M. V. Entin, and L. I. Magarill
Erratum to: Current Quadratic in Field and Photogalvanic Effect in Crystals without Inversion Centre
(b) Vol. **85**, No. 2, p. K135 (1978).
See also *14.3.4*, p. *240*; 16, p. 262.

V. T. Bublik
The Mean Square Atomic Displacements and Enthalpies of Vacancy Formation in Some Semiconductors
(a) Vol. **45**, No. 2, p. 543–548 (1978).
See also *10.2*, p. *110*; 22, p. 410; 22.4, p. 461.

M. Lindén, E. Ejder, and R. Hellborg
Channeling Studies of As-Grown GaN
(a) Vol. **45**, No. 2, p. K167–K170 (1978).
See also *11*, p. *124*.

A. Milchev and K. Unger
Third-Order Contribution to the Effective Ion–Ion Interaction and Formation of Angular Forces in Crystalline Solids (I): General Formulation of the Expression for the Total Energy
(b) Vol. **87**, No. 1, p. 227–237 (1978).
See also *4*, p. *58*; 13, p. 145.

V. K. Garg, D. S. Puri, and M. P. Verma
On the Second- and Third-Order Elastic Constants of ZnS Structure Solids
(b) Vol. **87**, No. 2, p. 401–407 (1978).
See also *6*, p. *70*; *12.2*, p. *139*; 22.4, p. 461; 22.5, p. 475.

A. Milchev and K. Unger
Third-Order Contribution to the Effective Ion–Ion Interaction and Formation of Angular Forces in Crystalline Solids (II): Application to Tetrahedral Solids
(b) Vol. **87**, No. 2, p. 473–478 (1978).
See also *4*, p. *58*; 13, p. 145.

F. Flores, C. Tejedor, and A. Martín-Rodero
Surface States in the (111) and $(\bar{1}\bar{1}\bar{1})$ Faces of Zincblende Compounds
(b) Vol. **88**, No. 2, p. 591–597 (1978).
See also *13.3*, p. *166*.

V. A. Savastenko and A. U. Sheleg
Study of the Elastic Properties of Gallium Nitride
(a) Vol. **48**, No. 2, p. K135–K139 (1978).
See also *12.2*, p. *140*.

MAHIRADHWAJ SINGH
The Influence of Internal Strain on Bound Hole–Phonon Interaction in p-Type Semiconductors with Zincblende Structure
(b) Vol. **89**, No. 1, p. 299—307 (1978).
See also *6*, p. *75*; 8, p. 88.

A. HAUG, D. KERKHOFF, and W. LOCHMANN
Calculation of Auger Coefficients for III–V Semiconductors with Emphasis on GaSb
(b) Vol. **89**, No. 2, p. 357—365 (1978).
See also *13.1*, p. *159*; 20, p. 317; 22.2.3, p. 454.

B. PERTZSCH and U. RÖSSLER
Stress Dependence of Shallow Acceptor States in Semiconductors with Zincblende Symmetry
(b) Vol. **89**, No. 2, p. 475—487 (1978).
See also 12.2, p. 140; *13.4*, p. *181*.

22.2.1 Arsenides

Original Papers and Short Notes

H. Gottschalk, G. Patzer, and H. Alexander
Stacking Fault Energy and Ionicity of Cubic III–V Compounds
(a) Vol. **45**, No. 1, p. 207–217 (1978).
See also 4, p. 57; *10.2*, p. *109*; 22.2.2, p. 450; 22.2.3, p. 452.

I. A. Kurova and N. N. Ormont
IR Luminescence of Acoustoelectric Domains in n-GaAs
(a) Vol. **45**, No. 1, p. 333–342 (1978).
See also *7*, p. *80*; *20.3*, p. *341*.

B. G. Helme and P. J. King
The Phonon Viscosity Tensor of Si, Ge, GaAs, and InSb
(a) Vol. **45**, No. 1, p. K33–K37 (1978).
See also 6, p. 65; *7*, p. *80*; 22.1.1, p. 424; 22.1.2, p. 429; 22.2.3, p. 452.

V. C. Kieu
Current Controlled Negative Resistance in GaAs at Low Temperatures
(a) Vol. **45**, No. 2, p. 571–578 (1978).
See also *13.4*, p. *171*; *14.3.4*, p. *240*.

P. M. Batev, M. D. Ivanovitch, E. I. Kafediiska, and S. S. Simeonov
Schottky Barrier on W–GaAs Contact
(a) Vol. **45**, No. 2, p. 671–675 (1978).
See also *14.3.3*, p. *233*; 14.3.4, p. 240.

K. D. Glinchuk and A. V. Prokhorovich
The Role of Different Local Centres in the Determination of the Concentration Dependence of the Intrinsic Emission Intensity in n-Type GaAs
(a) Vol. **45**, No. 2, p. K91–K94 (1978).
See also *13.4*, p. *171*; *20.3*, p. *342*.

B. Jensen
The Infrared Absorption Spectrum of n-Type GaAs
(b) Vol. **86**, No. 1, p. 291–301 (1978).
See also *14.3*, p. *218*; *20.1*, p. *322*.

R. Pässler
Intermediate-Temperature Dependences of the Nonradiative Multiphonon Carrier-Capture Cross-Section of Deep Traps in Semiconductors
(b) Vol. **86**, No. 1, p. K39–K44 (1978).
See also *13.4*, p. *172*; 22, p. 410.

R. Zimmermann
Final State Interactions in the Gain and Absorption Spectra of Electron–Hole Liquids
(b) Vol. **86**, No. 1, p. K63–K66 (1978).
See also *13.5*, p. *187*; *20.1*, p. *323*.

P. C. Mathur, Vijender Sharma, and Praveen Saxena
A Small Signal Noise Analysis for Realistic GaAs IMPATT Diode
(a) Vol. **46**, No. 1, p. 321–326 (1978).
See also *14.3.3*, p. *234*.

V. T. Bublik and V. N. Leikin
Calculation of the Pseudobinary Alloy Semiconductor Phase Diagrams
(a) Vol. **46**, No. 1, p. 365–372 (1978).
See also *1.2*, p. *17*; 22.2.2, p. 450; 22.2.3, p. 452.

M. C. Al-Edani and K. S. Dubey
Phonon Conductivity of InSb and GaAs
(b) Vol. **86**, No. 2, p. 741–750 (1978).
See also *6*, p. *69*; 8, p. 84; 22.2.3, p. 452.

P. I. Baranskii, L. S. Demidenko, S. L. Korolyuk, and A. G. Samoilovich
Impurity Scattering of Electrons in Quantizing Magnetic Fields
(b) Vol. **86**, No. 2, p. 765–772 (1978).
See also 13, p. 144; *14.3*, p. *219*.

B. Hughes and G. H. Narayanan
On Annealing-Induced Prismatic Dislocation Loops and Electrical Changes in Heavily Te-Doped GaAs
(a) Vol. **46**, No. 2, p. 627–637 (1978).
See also *1.4*, p. *30*; *10.2*, p. *112*; 14.3, p. 220.

R. Mansfield and G. Azhdarov
Negative Magnetoresistance in Indium Arsenide in the Extreme Quantum Limit
(b) Vol. **87**, No. 1, p. 163–167 (1978).
See also *13.4*, p. *174*; *14.3.4*, p. *242*.

Mahiradhwaj Singh
Theory of Resonance Scattering for Electron–Phonon Interaction in Doped Semiconductors with Zincblende Structure
(b) Vol. **87**, No. 1, p. 333–344 (1978).
See also *6*, p. *70*; 8, p. 85; 15, p. 260; 22.2.3, p. 453.

A. S. Filipchenko, L. P. Bolshakov, and A. Naurizbaev
The Moss-Burstein Effect in n-Type Indium Arsenide Crystals Doped with Different Donor Impurities
(b) Vol. **87**, No. 1, p. K1–K3 (1978).
See also 13.1, p. 155; *20.1*, p. *326*.

M. C. Al-Edani and K. S. Dubey
Boundary Scattering and Phonon Conductivity: Application to GaAs
(b) Vol. **87**, No. 1, p. K47–K52 (1978).
See also *6*, p. *70*; 8, p. 85.

W. Thielemann, P. Walther, and H. Neumann
Evidence for Different Absorption Tails in the Surface Depletion Layer and in the Bulk of n-Type GaAs
(a) Vol. **47**, No. 1, p. K35–K38 (1978).
See also *20.1*, p. *327*.

R. Ruppin
Validity Range of the Maxwell-Garnett Theory
(b) Vol. **87**, No. 2, p. 619–624 (1978).
See also 6, p. 71; *20.1*, p. *328*; 22.6, p. 494.

H. Reichl, D. Huber, and J. Müller
Photoluminescence Investigations of Diffused GaAs Ligh Emitting Diodes
(a) Vol. **47**, No. 2, p. 489–495 (1978).
See also *14.3.3*, p. *235*; *20.3*, p. *345*.

I. Mojzes
Formation of AuGe Contacts to n-GaAs
(a) Vol. **47**, No. 2, p. K183–K185 (1978).
See also *14.3.4*, p. *243*.

A. I. Bobrysheva and V. I. Vybornov
The Biexciton in Polar Crystals
(b) Vol. **88**, No. 1, p. 315–319 (1978).
See also *13.5.1*, p. *193*.

Z. M. Alekseeva, A. P. Vyatkin, G. F. Karavaev, and N. P. Krivorotov
Effect of Pressure on Hole Effective Masses in GaAs
(b) Vol. **88**, No. 1, p. 321–326 (1978).
See also *12.2*, p. *140*; *13.1*, p. *156*; 14.3, p. 222.

M. Godlewski and A. M. Hennel
EPR Measurements on GaAs:Co
(b) Vol. **88**, No. 1, p. K11–K13 (1978).
See also *19*, p. *309*.

I. Ya. Dekhtyar, V. T. Adonkin, S. P. Likhtorovich, and Sh. A. Rustamov
Positronium-Like States in GaAs at Low Temperature
(b) Vol. **88**, No. 1, p. K55–K58 (1978).
See also *13.1*, p. *157*.

T. I. Kolchenko and V. M. Lomako
Removal and Scattering of Charge Carriers by Defect Clusters in Semiconductors
(a) Vol. **48**, No. 1, p. 263–269 (1978).
See also *11*, p. *128*; 13.4, p. 178; 14.3, p. 222.

O. Yu. Borkovskaya, N. L. Dmitruk, R. V. Konakova, and Yu. A. Tkhorik
Investigation of Radiation Defects in GaAs by Means of Schottky Diode Characteristics
(a) Vol. **48**, No. 1, p. K55–K58 (1978).
See also *11*, p. *128*; 14.3.3, p. 235.

E. O. Goebel and O. Hildebrand
Thermalization of the Electron–Hole Plasma in GaAs. The Case of Resonant Excitation
(b) Vol. **88**, No. 2, p. 645–652 (1978).
See also *13.5*, p. *187*; *20.3*, p. *346*.

B. Stébé and G. Munschy
Binding of an Exciton to a Neutral Acceptor
(b) Vol. **88**, No. 2, p. 713–725 (1978).
See also *13.4*, p. *178*; *13.5.1*, p. *193*; 14.3, p. 223.

V. P. Dolgopolskii, R. P. Seisyan, and A. L. Efros, Jr.
The Ground State of a Wannier-Mott Exciton in Diamond-Like Semiconductors in "Intermediate" Magnetic Fields
(b) Vol. **88**, No. 2, p. K177–K180 (1978).
See also *13.1*, p. *158*; *13.5.1*, p. *194*.

K. D. GLINCHUK, A. V. PROKHOROVICH, V. E. RODIONOV, and V. I. VOVNENKO
Study of Non-Linear Extrinsic Luminescence in GaAs
(a) Vol. **48**, No. 2, p. 593–602 (1978).
See also *13.4*, p. *179*; *20.3*, p. *346*.

P. KOCEVER and E. FITZ
The Influence of Acoustic Phonon Disturbances on the Conductivity of Piezo-Electric Semiconductors
(b) Vol. **89**, No. 1, p. 225–233 (1978).
See also *6*, p. *75*; 14.3, p. 224.

H. STEINHARDT and P. HAASEN
Creep and Dislocation Velocities in Gallium Arsenide
(a) Vol. **49**, No. 1, p. 93–101 (1978).
See also *10.2*, p. *118*.

W. HOYER, G. LAUTZ, and M. HEYEN
Low-Frequency Current Noise of GaAs Epitaxial Layers between 4.4 and 300 K
(a) Vol. **49**, No. 1, p. 111–117 (1978).
Erratum
(a) Vol. **50**, No. 2, p. 771 (1978).
See also 13.4, p. 180; *14.3.1*, p. *228*; *14.3.3*, p. *236*.

K. METTLER
Effect of Reabsorbed Recombination Radiation on Photoluminescence and Photoconductivity in a Semi-Infinite Direct-Gap Semiconductor
(a) Vol. **49**, No. 1, p. 163–168 (1978).
See also *16*, p. *266*; *20.3*, p. *347*.

A. S. FILIPCHENKO, L. P. BOLSHAKOV, and A. NAURIZBAEV
Absorption in Indium Antimonide and Arsenide Crystals Doped with Various Admixtures
(a) Vol. **49**, No. 1, p. 183–188 (1978).
See also *13.4*, p. *180*; *20.1*, p. *333*; 22.2.3, p. 454.

S. ÖBERG
Electronic States Associated with 60° Dislocations in Gallium Arsenide
(b) Vol. **89**, No. 2, p. K127–K130 (1978).
See also *13.4*, p. *181*.

ATSUO KASUYA, S. TOGASHI, T. GOTO, and Y. NISHINA
Two-Photon Photoemission in PbI_2 and GaAs
(b) Vol. **89**, No. 2, p. K145–K147 (1978).
See also *17*, p. *270*; 22.5, p. 476.

A. S. FILIPCHENKO, T. B. KURENKEEV, and L. P. BOLSHAKOV
Velocity of the Spontaneous Radiative Recombination of Current Carriers in Degenerate Semiconductors
(a) Vol. **49**, No. 2, p. K189–K192 (1978).
See also *20.3*, p. *349*; 22.2.3, p. 454.

E. VIGIL, J. A. RODRÍGUEZ, and R. PÉREZ ALVAREZ
Optical Constants of n- and p-Type GaAs between 2.5 and 3.5 eV
(b) Vol. **90**, No. 1, p. 409–414 (1978).
See also *20.1*, p. *335*.

A. Wolkenberg
Energetic Barrier Height Hypothesis on the Silicon–Electrolyte and Gallium Arsenide–Electrolyte Interfaces
(a) Vol. **50**, No. 1, p. 249–256 (1978).
See also 13.3, p. 168; *14.3.2*, p. *232*; *14.3.3*, p. *238*; 22.1.2, p. 438.

A. le Bloa, P. N. Favennec, and Y. Colin
Determination of Complementary Transition Energy Threshold by Photocapacitance Measurements. Application to Oxygen Implanted GaAs
(a) Vol. **50**, No. 1, p. 311–322 (1978).
See also *13.4*, p. *184*; 14.3.3, p. 238.

A. Milchev
Frequency- and Wave-Vector-Dependent Dielectric Function of a Model Semiconductor
(b) Vol. **90**, No. 2, p. 679–688 (1978).
See also *13.1*, p. *161*; *14.4*, p. *252*; 22.1.1, p. 428; 22.1.2, p. 439; 22.4.2, p. 469.

G. P. Srivastava
Role of Optical Phonons in the High-Temperature Lattice Thermal Conductivity of Semiconductors
(b) Vol. **90**, No. 2, p. K125–K128 (1978).
See also *6*, p. *78*; *8*, p. *90*; 22.1.1, p. 428; 22.1.2, p. 439; 22.2.3, p. 455.

J. P. Leburton and R. Evrard
Oscillatory Electron Drift Velocity in Polar Semiconductor in Non-Ohmic Regime at Low Temperature
(b) Vol. **90**, No. 2, p. K137–K139 (1978).
See also *13*, p. *149*; *14.3.4*, p. *246*; 22.2.3, p. 455.

A. S. Filipchenko, T. B. Kurenkeev, and T. Seisenbaev
Photoluminescence of Doped n-Type Indium Arsenide and Indium Antimonide Crystals
(a) Vol. **50**, No. 2, p. K251–K254 (1978).
See also 13.4, p. 186; *20.3*, p. *351*; 22.2.3, p. 455.

22.2.2 Phosphides

Original Papers and Short Notes

H. GOTTSCHALK, G. PATZER, and H. ALEXANDER
Stacking Fault Energy and Ionicity of Cubic III–V Compounds
(a) Vol. **45**, No. 1, p. 207–217 (1978).
See also 4, p. 57; *10.2*, p. *109*; 22.2.1, p. 445; 22.2.3, p. 452.

P. R. JAY and B. TUCK
Mercury Probe *C–U* Measurements on GaP
(a) Vol. **46**, No. 1, p. 199–208 (1978).
See also *9*, p. *92*; *14.3*, p. *219*.

V. T. BUBLIK and V. N. LEIKIN
Calculation of the Pseudobinary Alloy Semiconductor Phase Diagrams
(a) Vol. **46**, No. 1, p. 365–372 (1978).
See also *1.2*, p. *17*; 22.2.1, p. 446; 22.2.3, p. 452.

J. M. HESS, P. H. NGUYEN, B. LEPLEY, and S. RAVELET
Barrier Height Study on Au–InP Schottky Diodes
(a) Vol. **46**, No. 1, p. K55–K59 (1978).
See also *14.3.3*, p. *234*; 14.3.4, p. 241.

M. SCHNEIDER and E. NEBAUER
Diffusivity and Solubility of Copper in LEC-GaP
(a) Vol. **46**, No. 1, p. K71–K74 (1978).
See also *9*, p. *92*.

B. ULRICI and E. JAHNE
Two-Phonon Infrared Absorption in $In_{1-x}Ga_xP$ Mixed Crystals
(b) Vol. **86**, No. 2, p. 517–526 (1978).
See also *6*, p. *68*; *20.1*, p. *324*; 22.2.4, p. 456.

M. SINYUKOV, R. TROMMER, and M. CARDONA
Resonant Raman Scattering in InP
(b) Vol. **86**, No. 2, p. 563–568 (1978).
See also *6*, p. *68*; *20.1*, p. *324*.

R. BINDEMANN, R. SCHWABE, and T. HÄNSEL
The "Quasidirect" Radiative Recombination of Free Holes at Neutral Shallow Donors in GaP
(b) Vol. **87**, No. 1, p. 169–177 (1978).
See also *13.4*, p. *174*; *20.3*, p. *344*.

P. KRISPIN and J. MAEGE
Capacitance–Frequency Dispersion and Electroluminescence Efficiency of GaP p–n Junctions
(a) Vol. **47**, No. 1, p. 91–97 (1978).
See also 13.4, p. 175; *14.3.3*, p. *235*; *20.3*, p. *344*.

H. MAAREF, J. BARRAU, M. BROUSSEAU, J. COLLET, J. MAZZASCHI, and M. PUGNET
First-Order Mott Transition of the Exciton Gas in GaP
(b) Vol. **88**, No. 1, p. 261–268 (1978).
See also *13.5.1*, p. *192*; 20.3, p. 345.

H. MENNIGER, G. VOIGT, H. RAIDT, H. FABIG, and J. MAEGE
Scanning Electron Microscopic Investigations of GaP
(a) Vol. **48**, No. 2, p. 407–415 (1978).
See also *1.4*, p. *32*; 14.3.2, p. 230; 20.3, p. 346.

T. D. DZHAFAROV, M. N. KOLOTOV, and A. A. LITVIN
The Influence of γ-Irradiation on the Degradation of Red-Emitting GaP LED's
(a) Vol. **49**, No. 1, p. 211–215 (1978).
See also 11, p. 129; *14.3.3*, p. *236*; *20.3*, p. *347*.

E. NEBAUER
A Comparative Study of Zinc Tracer Diffusion Profiles in LEC, SSD, and CVD GaP
(a) Vol. **49**, No. 1, p. 249–254 (1978).
See also *9*, p. *94*; 10.2, p. 118.

R. BINDEMANN, H. FISCHER, and K. KREHER
The Spectral Dependence of the Photoneutralization Cross-Section of Oxygen in GaP Near the Threshold
(a) Vol. **49**, No. 1, p. 331–336 (1978).
See also *13.4*, p. *181*; 13.5.1, p. 194; *20.3*, p. *347*.

J. LLOPIS and J. PIQUERAS
Cathodoluminescence from Grain Boundaries in MgO and GaP
(a) Vol. **49**, No. 1, p. K9–K12 (1978).
See also *10.2*, p. *118*; *20.3*, p. *348*; 22.6, p. 496.

W. SENSKE, M. SONDERGELD, and H. J. QUEISSER
Impurity Interaction on Excited Acceptor States in GaP
(b) Vol. **89**, No. 2, p. 467–474 (1978).
See also *13.4*, p. *181*; *20.3*, p. *348*.

R. SCHWABE, F. THUSELT, H. WEINERT, R. BINDEMANN, and K. UNGER
Luminescence from Electron–Hole Drops in Undoped GaP
(b) Vol. **89**, No. 2, p. 561–570 (1978).
See also *13.5.1*, p. *195*; *20.3*, p. *348*.

H. NEUMANN and B. JACOBS
Electrical Properties of Liquid Phase Epitaxial InP Grown from Tin Solution
(a) Vol. **49**, No. 2, p. K139–K144 (1978).
See also *14.3.3*, p. *237*.

G. WEIMANN and W. SCHLAPP
Ohmic Contacts on Indium Phosphide
(a) Vol. **50**, No. 2, p. K219–K223 (1978).
See also *14.3.4*, p. *247*.

22.2.3 Antimonides

Review Article

V. I. Ivanov-Omskii, L. I. Korovin, and E. M. Shereghii
Phonon-Assisted Cyclotron Resonance in Semiconductors
(b) Vol. **90**, No. 1, p. 11—32 (1978).
See also *6*, p. *64*; 13, p. 142; *19*, p. *303*; 22, p. 408; 22.4.1, p. 463.

Original Papers and Short Notes

H. Gottschalk, G. Patzer, and H. Alexander
Stacking Fault Energy and Ionicity of Cubic III–V Compounds
(a) Vol. **45**, No. 1, p. 207—217 (1978).
See also 4, p. 57; *10.2*, p. *109*; 22.2.1, p. 445; 22.2.2, p. 450.

B. G. Helme and P. J. King
The Phonon Viscosity Tensor of Si, Ge, GaAs, and InSb
(a) Vol. **45**, No. 1, p. K33—K37 (1978).
See also 6, p. 65; *7*, p. *80*; 22.1.1, p. 424; 22.1.2, p. 429; 22.2.1, p. 445.

Yu. V. Gulyaev, S. N. Ivanov, and A. G. Kozorezov
On the Electronic Absorption and Amplification of Sound in Semiconductors at "Intermediate" Frequencies
(b) Vol. **85**, No. 2, p. 487—492 (1978).
See also 6, p. 66; *7*, p. *80*.

A. Dargys, R. Sedrakyan, and J. Poželа
Measurement of the Electron Drift Velocity in InSb up to Fields of 800 V/cm in the Presence of Impact Ionization
(a) Vol. **45**, No. 2, p. 387—392 (1978).
See also *14.3.4*, p. *240*.

W. Müller, M. Overhamm, G. Bauer, E. Gornik, P. Grosse, H. W. Pötzl, and M. Rautenberg
On the Influence of Background Radiation on Landau Emission Spectra
(b) Vol. **86**, No. 1, p. 205—213 (1978).
See also 13.1, p. 152; *14.3.4*, p. *241*; 20.1, p. 322; *20.3*, p. *343*.

V. T. Bublik and V. N. Leikin
Calculation of the Pseudobinary Alloy Semiconductor Phase Diagrams
(a) Vol. **46**, No. 1, p. 365—372 (1978).
See also *1.2*, p. *17*; 22.2.1, p. 446; 21.2.2, p. 450.

G. Appold, H. Pascher, R. Ebert, U. Steigenberger, and M. von Ortenberg
Determination of the Effective *g*-Value of Electrons in n-Type InSb by Spin-Flip Raman Scattering and Electric-Dipole-Excited Electron Spin Resonance
(b) Vol. **86**, No. 2, p. 557—561 (1978).
See also *13.4*, p. *174*; 19, p. 306; 20.1, p. 324.

M. C. Al-Edani and K. S. Dubey
Phonon Conductivity of InSb and GaAs
(b) Vol. **86**, No. 2, p. 741—750 (1978).
See also *6*, p. *69*; 8, p. 84; 22.2.1, p. 446

Mahiradhwaj Singh
Theory of Resonance Scattering for Electron–Phonon Interaction in Doped Semiconductors with Zincblende Structure
(b) Vol. **87**, No. 1, p. 333–344 (1978).
See also *6*, p. *70*; 8, p. 85; 15, p. 260; 22.2.1, p. 446.

N. Y. Chernysheva, G. A. Kachurin, and V. A. Bogatyriov
Amorphisation and Recrystallisation of InSb Ion Implanted Layers
(a) Vol. **47**, No. 1, p. K5–K8 (1978).
See also 1.5, p. 37; *11*, p. *126*.

A. L. Dawar, K. V. Krishna, O. P. Taneja, and P. C. Mathur
Electrical Transport Properties of Polycrystalline p-Type InSb Films
(a) Vol. **47**, No. 2, p. 375–378 (1978).
See also 13.4, p. 176; *14.3.1*, p. *227*.

D. Vajda
Dependence of the Threshold Electric Field of the Acoustoelectric Current Instabilities in Monocrystalline n-InSb on the Length of a Sample
(a) Vol. **47**, No. 2, p. K123–K124 (1978).
See also *7*, p. *81*; *14.3.4*, p. *243*.

P. Pfeffer and W. Zawadzki
Inelastic Scattering of Electrons by Optic Phonons in InSb-Type Semiconductors
(b) Vol. **88**, No. 1, p. 247–260 (1978).
See also 6, p. 73; 13.1, p. 156; *14.3*, p. *222*.

Mahiradhwaj Singh
Phonon Attenuation in n-Type Semiconductors with ZB Structure in Presence of Internal Strains and Magnetic Field
(b) Vol. **88**, No. 1, p. 295–300 (1978).
See also *6*, p. *73*; 7, p. 81; 12.2, p. 140; 18, p. 274; 22.1.1, p. 426.

C. K. Sarkar and D. Chattopadhyay
Effect of the Electron Spin on the Magneto-Seebeck Coefficient of Polar Semiconductors at Large Magnetic Fields
(b) Vol. **88**, No. 2, p. K109–K112 (1978).
See also 13.1, p. 157; *15*, p. *260*.

A. S. Filipchenko, T. B. Kurenkeev, L. P. Bolshakov, and E. I. Chaikina
Photoluminescence of Heavily Doped n-Type Gallium Antimonide
(a) Vol. **48**, No. 2, p. 281–285 (1978).
See also 13.4, p. 179; *20.3*, p. *346*.

A. S. Filipchenko, L. P. Bolshakov, A. Naurizbaev, A. G. Braginskaya, and A. N. Popkov
Moss-Burstein Effect in n-Type Gallium Antimonide Crystals
(a) Vol. **48**, No. 2, p. K115–K117 (1978).
See also *13.4*, p. *179*; 20.1, p. 331.

S. Lehr and H. Pagnia
Work Function of Irradiated Real and Cleaned Indium Antimonide Surfaces
(a) Vol. **49**, No. 1, p. 83–91 (1978).
See also *13.3*, p. *167*; 16, p. 265; 17, p. 270.

A. S. Filipchenko, L. P. Bolshakov, and A. Naurizbaev
Absorption on Indium Antimonide and Arsenide Crystals Doped with Various Admixtures
(a) Vol. **49**, No. 1, p. 183—188 (1978).
See also *13.4*, p. *180*; *20.1*, p. *333*; 22.2.1, p. 448.

S. Myhra
Defect Annealing and Transport Properties in High-Purity InSb Irradiated at 80 K
(a) Vol. **49**, No. 1, p. 285—292 (1978).
See also *10.2*, p. *118*; *14.3*, p. *224*; 15, p. 261.

S. Guha and S. Ghosh
Effect of a Density Gradient on the Electromagnetic Wave Instability in Semiconductor Plasmas
(a) Vol. **49**, No. 1, p. 363—370 (1978).
See also *14.3.4*, p. *245*.

A. Soliman and N. Saleh
Photovoltaic Operation of Nonuniformly Doped Intrinsic Photoconductors
(a) Vol. **49**, No. 1, p. 393—403 (1978).
See also *14.3.3*, p. *236*; *16*, p. *266*.

A. Haug, D. Kerkhoff, and W. Lochmann
Calculation of Auger Coefficients for III–V Semiconductors with Emphasis on GaSb
(b) Vol. **89**, No. 2, p. 357—365 (1978).
See also *13.1*, p. *159*; 20, p. 317; 22.2, p. 444.

D. Ferré and J. L. Farvacque
Asymmetrical Behaviour of "60°" Dislocation Glide in Indium Antimonide
(a) Vol. **49**, No. 2, p. 737—745 (1978).
See also *10.2*, p. *119*.

A. S. Filipchenko, T. B. Kurenkeev, and L. P. Bolshakov
Velocity of the Spontaneous Radiative Recombination of Current Carriers in Degenerate Semiconductors
(a) Vol. **49**, No. 2, p. K189—K192 (1978).
See also *20.3*, p. *349*; 22.2.1, p. 448.

J. T. Devreese, R. Evrard, and E. Kartheuser
Note on the Solution of the Boltzmann Equation for Electron–LO Phonon Scattering
(b) Vol. **90**, No. 1, p. K73—K76 (1978).
See also *6*, p. *78*; *14.3.4*, p. *246*.

Z. Dobrovolskis, A. Krotkus, and J. Požela
Negative Differential Conductivity in Hydrostatically Pressed Indium Antimonide at Room Temperature
(a) Vol. **50**, No. 1, p. K73—K75 (1978).
See also 12.2, p. 141; *14.3.4*, p. *246*.

R. A. Noack
Inversion of the Γ and L Conduction Bands of GaSb under Uniaxial Stress
(b) Vol. **90**, No. 2, p. 615—620 (1978).
See also 12.2, p. 141; *13.1*, p. *161*; *20.3*, p. *350*.

G. P. Srivastava
Role of Optical Phonons in the High-Temperature Lattice Thermal Conductivity of Semiconductors
(b) Vol. **90**, No. 2, p. K125—K128 (1978).
See also *6*, p. *78*; *8*, p. *90*; 22.1.1, p. 428; 22.1.2, p. 439; 22.2.1, p. 449.

J. P. Leburton and R. Evrard
Oscillatory Electron Drift Velocity in Polar Semiconductor in Non-Ohmic Regime at Low Temperature
(b) Vol. **90**, No. 2, p. K137—K139 (1978).
See also *13*, p. *149*; *14.3.4*, p. *246*; 22.2.1, p. 449.

G. E. Alberga and J. T. Devreese
Mobility of Electrons in n-Type InSb at 77 K in the Avalanche Regime: Theory and Experiment
(b) Vol. **90**, No. 2, p. K149—K152 (1978).
See also *6*, p. *78*; *14.3.4*, p. *247*.

V. K. Malyutenko, S. S. Bolgov, and Yu. M. Malozovsky
Magnetoconcentration Effect at Nonlinear Recombination of Current Carriers
(a) Vol. **50**, No. 2, p. 723—732 (1978).
See also *14.3*, p. *225*.

A. F. Plotnikov, V. A. Tolokonnikov, and V. E. Shubin
Investigation of the Energy Band Diagram of the Anodic Oxide-Indium Antimonide System by Internal Photoemission
(a) Vol. **50**, No. 2, p. 733—738 (1978).
See also *13.1*, p. *161*; *14.3.3*, p. *239*; 16, p. 268.

A. S. Filipchenko, T. B. Kurenkeev, and T. Seisenbaev
Photoluminescence of Doped n-Type Indium Arsenide and Indium Antimonide Crystals
(a) Vol. **50**, No. 2, p. K251—K254 (1978).
See also 13.4, p. 186; *20.3*, p. *351*; 22.2.1, p. 449.

22.2.4 Mixed Crystals

Original Papers and Short Notes

L. LEIBLER
Theory of Electron Spin Resonance in Graded Mixed Semiconductors
(b) Vol. **85**, No. 2, p. 611—619 (1978).
See also 13.1, p. 151; *19*, p. *304*; 22.4.4, p. 472.

A. FREY
Composition Dependence of Nitrogen and Tellurium Bound State Energies in $In_{1-x}Ga_xP$
(b) Vol. **85**, No. 2, p. K107—K110 (1978).
See also *13.4*, p. *171*; *20.3*, p. *342*.

R. STREHLOW
On the No-Phonon Structure in Mixed Crystals with Indirect Band Structure
(b) Vol. **86**, No. 1, p. 311—318 (1978).
See also 6, p. 67; 13.1, p. 153; 13.5.1, p. 190; *20.1*, p. *322*.

B. ULRICI and E. JAHNE
Two-Phonon Infrared Absorption in $In_{1-x}Ga_xP$ Mixed Crystals
(b) Vol. **86**, No. 2, p. 517—526 (1978).
See also *6*, p. *68*; *20.1*, p. *324*; 22.2.2, p. 450.

V. K. BASHENOV, A. M. MUTAL, and V. V. TIMOFEENKO
Interaction Parameters in III–V Semiconductor Solid Solutions
(b) Vol. **86**, No. 2, p. K139—K142 (1978).
See also *1.1*, p. *11*; 1.2, p. 17.

G. MÜLLER, M. HAUBOLD, R. SCHIMKO, M. TRAPP, and G. SCHWARZ
Investigations on the Diffusion on Implanted Zinc in $GaAs_{1-x}P_x$ by Ion Microprobe
(a) Vol. **49**, No. 1, p. 279—284 (1978).
See also *9*, p. *94*; *11*, p. *129*.

B. BARANOV, L. DÄWERITZ, V. B. GUTAN, G. JUNGK, H. NEUMANN, and H. RAIDT
Growth and Properties of $Al_xGa_{1-x}N$ Epitaxial Layers
(a) Vol. **49**, No. 2, p. 629—636 (1978).
See also *1.5*, p. *39*; 13.4, p. 182; 14.3.1, p. 228; 20.1, p. 335; 20.3, p. 348.

R. STREHLOW, W. KÜHN, H. FRIEDRICH, and H. LANGE
On the Quantitative Characterization of Chemical Disorder of $In_{1-x}Ga_xP$ fromWavelength-Modulated No-Phonon Absorption
(b) Vol. **90**, No. 1, p. 167—173 (1978).
See also 6, p. 77; 13.1, p. 159; 13.5.1, p. 195; *20.1*, p. *335*.

22.3 IV–IV, II–V, and III–VI Compounds

Review Article

A. S. Borukhovich
Kinetic Properties of IVa and Va Subgroup Transition Metal Monocarbides
(a) Vol. **46**, No. 1, p. 11–37 (1978).
See also 8, p. 83; 13.1, p. 150; *14.3*, p. *217*; 15, p. 259; 22, p. 408; 22.3.1, p. 460.

Original Papers and Short Notes

F. E. Faradev, N. M. Gasanly, B. N. Mavrin, and N. N. Melnik
Raman Scattering on Some III–VI Layer Single Crystals
(b) Vol. **85**, No. 1, p. 381–386 (1978).
See also *6*, p. *65*; *20.1*, p. *320*.

G. L. Belenkii and M. O. Godzhaev
Excitons in Gallium Sulphide
(b) Vol. **85**, No. 2, p. 453–458 (1978).
See also *13.5.1*, p. *189*; *20.3*, p. *341*.

Y. Yamada
Band Structure Calculation of ZnSb
(b) Vol. **85**, No. 2, p. 723–732 (1978).
See also *13.1*, p. *151*.

J. Kočka and J. Krištofik
AC Conductivity of $CdAs_2$-Based Glasses
(a) Vol. **45**, No. 2, p. 559–564 (1978).
See also *2*, p. *43*; *14.3*, p. *218*.

L. Sodomka
Mechanoluminescent Spectra of Silicon Carbide Powders
(a) Vol. **45**, No. 2, p. K171–K174 (1978).
See also *20.3*, p. *342*.

T. Ikari and K. Hashimoto
Electrical Properties of TlTe Single Crystals
(b) Vol. **86**, No. 1, p. 239–248 (1978).
Erratum
(b) Vol. **90**, No. 2, p. 751 (1978).
See also 13.1, p. 153; *14.3*, p. *218*; *15*, p. *259*.

I. S. Gorban, V. A. Gorinya, R. A. Dashkovskaya, V. I. Lugovoi, A. P. Makovetskaya, and I. I. Tichina
One- and Two-Phonon States in Tetragonal ZnP_2 and CdP_2 Crystals
(b) Vol. **86**, No. 1, p. 419–423 (1978).
See also *6*, p. *67*; *20.1*, p. *323*.

L. P. Mokhracheva, V. A. Tskhai, and P. V. Geld
Charge Distribution and Ionicity of the Chemical Bonds in TiC, VC, and TiN
(b) Vol. **87**, No. 1, p. 49–52 (1978).
See also *4*, p. *58*; *13*, p. *145*.

B. Dowgiałło-Plenkiewicz and P. Plenkiewicz
Spin–Orbit and Crystal Field Splittings of the Valence Band in II_3V_2 Semiconducting Compounds
(b) Vol. **87**, No. 1, p. 309–315 (1978).
See also *13.1*, p. *154*.

V. Riede, Hoang Xuan Nguyen, H. Neumann, H. Sobotta, and F. Levy
Refractive Index of GaS in the Far Infrared
(b) Vol. **87**, No. 1, p. K65–K67 (1978).
See also *20.1*, p. *326*.

L. Gouskov, A. Gouskov, and B. Girault
Electrooptical Properties of SnO_2 Spray–GaSe and SnO_2 Spray–GaTe Heterojunctions
(a) Vol. **47**, No. 1, p. 195–199 (1978).
See also *14.3.3*, p. *235*; 14.3.4, p. 242; 16, p. 264; 20.3, p. 344.

K. R. Allakhverdiev, E. A. Vinogradov, N. N. Melnik, M. A. Nizametdinova, E. Yu. Salaev, and R. M. Sardarly
Vibrational Properties of TlSe
(b) Vol. **87**, No. 2, p. K115–K117 (1978).
See also *6*, p. *72*; *20.1*, p. *328*.

D. Bartzokas, C. Manolikas, and J. Spyridelis
Electron Microscopic Study of the Destabilization of Stabilized γ-Phase of Indium Sesquisulphide
(a) Vol. **47**, No. 2, p. 459–467 (1978).
See also *1.2*, p. *19*; 1.4, p. 31.

J. M. Pawlikowski, J. Misiewicz, N. Mirowska, and P. Becla
Absorption Edge of Zn_3P_2 at 5, 80, and 300 K
(b) Vol. **88**, No. 2, p. K153–K156 (1978).
See also *13.1*, p. *158*; *20.1*, p. *331*.

I. S. Gorban, G. A. Grishchenko, A. P. Sakalas, A. S. Sodeika, I. I. Tychina, and A. K. Tkachenko
Electrical Properties of ZnP_2 Single Crystals of Tetragonal and Monoclinic Modifications
(a) Vol. **48**, No. 2, p. 329–334 (1978).
See also *13.4*, p. *179*; *14.3*, p. *223*.

K. H. Yang and G. H. Schwuttke
Analysis of Vapor-Deposited Silicon Carbide Films on Silicon Ribbon Surfaces
(a) Vol. **48**, No. 2, p. 335–343 (1978).
See also *1.5*, p. *37*; 3, p. 55.

A. V. Gaisanyuk, K. I. Kugel, V. A. Petrov, and V. V. Sokolov
Resistivity of the ZrC–C System: An Example of Percolation Behaviour
(a) Vol. **48**, No. 2, p. K131–K133 (1978).
See also *1.2*, p. *22*; *14*, p. *204*.

N. M. Gasanly, B. M. Yavadov, V. I. Tagirov, and E. A. Vinogradov
Infrared and Raman Spectra of Layer InSe Single Crystals
(b) Vol. **89**, No. 1, p. K43–K48 (1978).
See also *6*, p. *75*; *20.1*, p. *332*.

V. K. Bashenov, I. Baumann, and D. I. Marvakov
Lattice Dynamics of Gallium Selenide in the One-Layer Approach
(b) Vol. **89**, No. 2, p. K115–K118 (1978).
See also *6*, p. *76*.

I. K. Kristensen
Decay of Photoluminescence in GaSe
(b) Vol. **89**, No. 2, p. K155–K157 (1978).
See also 13.4, p. 182; *20.3*, p. *348*.

R. Schwabe, F. Thuselt, T. Hänsel, and F. Lévy
Electron Beam Excited Free Exciton Luminescence of GaSe
(b) Vol. **90**, No. 1, p. 233–239 (1978).
See also *13.5.1*, p. *196*; *20.3*, p. *349*.

G. B. Dubrovskii, S. T. Pavlov, and M. S. Vershinin
Interband Optical Absorption in Semiconductors with One-Dimensional Electronic Superlattice
(b) Vol. **90**, No. 1, p. 429–432 (1978).
See also *13.1*, p. *160*; *20.1*, p. *336*.

V. K. Bashenov, D. I. Marvakov, and A. G. Petukhov
Bond-Orbital Model of Layer Gallium Selenide
(b) Vol. **90**, No. 1, p. K5–K8 (1978).
See also *4*, p. *59*; *13.1*, p. *160*.

V. A. Gadzhiev, S. G. Abdullaeva, F. K. Aliev, F. K. Allakhverdiev, and A. Z. Badalov
Electroabsorption in GaTe
(b) Vol. **90**, No. 1, p. K77–K79 (1978).
See also 13.5.1, p. 196; *14.3.4*, p. *246*; *20.1*, p. *336*.

V. T. Cherepin, A. A. Kosyachkov, and M. A. Vasilev
Secondary Ion Emission of Transition Metal Carbides
(a) Vol. **50**, No. 1, p. K113–K116 (1978).
See also *17*, p. *270*.

G. L. Bleris, Th. Karakostas, N. A. Economou, and R. de Ridder
The Ordered State of In_2Te_3 and Its Relation to the Transition State
(a) Vol. **50**, No. 2, p. 579–586 (1978).
See also *1.2*, p. *24*; *1.3*, p. *28*; 10.2, p. 122.

22.3.1 Mixed Crystals

Review Article

A. S. Borukhovich
Kinetic Properties of IV a and V a Subgroup Transition Metal Monocarbides
(a) Vol. **46**, No. 1, p. 11—37 (1978).
See also 8, p. 83; 13.1, p. 150; *14.3*, p. *217*; 15, p. 259; 22, p. 408; 22.3, p. 457.

Original Papers and Short Notes

B. Etlinger, R. Trojko, and U. V. Desnica
Erratum to: Thermal Decomposition of $Al_2In_{12}S_{21}$
(a) Vol. **46**, No. 2, p. K173 (1978).
See also *3*, p. *54*.

V. I. Tagirov, N. M. Gasanly, B. M. Yavadov, M. A. Sobeikh, and V. M. Salmanov
Photoluminescence of Layered $Ga_{1-x}In_xSe$ Crystals at Two-Phonon Optical Excitation
(a) Vol. **47**, No. 2, p. K157—K160 (1978).
See also *20.3*, p. *345*.

C. Manfredotti, A. M. Mancini, A. Rizzo, R. Murri, and L. Vasanelli
Electrical Properties of $GaS_xSe_{(1-x)}$ Solid Solutions
(a) Vol. **48**, No. 2, p. 293—296 (1978).
See also *13.4*, p. *179*; *14.3*, p. *223*.

C. Paorici and L. Zecchina
Thermal Conductivity in In–Se–Te Alloys
(a) Vol. **50**, No. 2, p. K235—K238 (1978).
See also *8*, p. *90*.

22.4 II–VI Compounds

Original Papers and Short Notes

K. Hümmer and P. Gebhardt
Angular Dependence of the Reflection Spectra and Directional Dispersion of the Anisotropic Exciton Polaritons in ZnO
(b) Vol. **85**, No. 1, p. 271–282 (1978).
See also *13.5.1*, p. *189*; *20.1*, p. *320*.

M. Hamera
Calculation of the *g*-Shift and Superhyperfine Structure Tensor of Mn^{2+} Ion in $A_{II}B_{IV}$ Tetrahedral Semiconductors
(b) Vol. **85**, No. 2, p. 645–655 (1978).
See also *13.4*, p. *170*; 19, p. 304.

V. T. Bublik
The Mean Square Atomic Displacements and Enthalpies of Vacancy Formation in Some Semiconductors
(a) Vol. **45**, No. 2, p. 543–548 (1978).
See also *10.2*, p. *110*; 22, p. 410; 22.2, p. 443.

V. Gomathy, C. Basu, and U. S. Ghosh
Configuration Interaction and Intensity of Crystal Field Spectra of the Tetrahedral Ni^{2+} Ion
(b) Vol. **86**, No. 1, p. 379–384 (1978).
See also *13.4*, p. *172*; 20.1, p. 322.

K. Hümmer, R. Helbig, and M. Baumgärtner
Manifestation of *k*-Linear Term Effects on the Zeeman Splitting of the A1-Exciton Emission Lines in ZnO
(b) Vol. **86**, No. 2, p. 527–533 (1978).
See also *13.5.1*, p. *191*; *20.3*, p. *343*.

S. Bhushan, A. N. Pandey, and Balakrishna Rao Kaza
Brightness Waves of Electroluminescence in ZnO: La Phosphors
(a) Vol. **46**, No. 2, p. K123–K126 (1978).
See also *20.3*, p. *343*.

P. K. C. Pillai, R. G. Mendiratta, C. K. Pillai, and R. Nath
Photo-Induced AC Impedance Measurements in Single Crystal Photoconductors
(a) Vol. **46**, No. 2, p. K127–K130 (1978).
See also *14.4*, p. *248*; *16*, p. *264*; 22.5, p. 475.

V. K. Garg, D. S. Puri, and M. P. Verma
On the Second- and Third-Order Elastic Constants of ZnS Structure Solids
(b) Vol. **87**, No. 1, p. 401–407 (1978).
See also *6*, p. *70*; *12.2*, p. *139*; 22.2, p. 443; 22.5, p. 475.

I. Balslev
The Excitonic Surface Potential in Semiconductors
(b) Vol. **88**, No. 1, p. 155–161 (1978).
See also *13.5.1*, p. *192*; 20.1, p. 329.

G. HEILAND and D. KOHL
Interpretation of Surface Phenomena on ZnO by the Compensation Model
(a) Vol. **49**, No. 1, p. 27–37 (1978).
See also 13.3, p. 167; 13.4, p. 180; *14.3.2*, p. *231*; 14.3.4, p. 244; *16*, p. *265*.

M. ALTWEIN, H. FINKENRATH, and D. GERHARDT
Temperature Dependence of the Magnetoresistance in CdO
(a) Vol. **49**, No. 1, p. 67–71 (1978).
See also *14.3.1*, p. *228*.

F. G. GÄRTNER and E. MOLLWO
IR Absorption of OH and OD Centres and OH/OH, OD/OD and OH/OD Complexes in Cu-Doped ZnO Single Crystals (I): Experimental Results
(b) Vol. **89**, No. 2, p. 381–388 (1978).
See also 13.4, p. 181; *20.1*, p. *333*.

S. W. KOCH, H. HAUG, G. SCHMIEDER, W. BOHNERT, and C. KLINGSHIRN
Stimulated Intrinsic Recombination Processes in II–VI Compounds
(b) Vol. **89**, No. 2, p. 431–440 (1978).
See also 13.5.1, p. 195; *20.2*, p. *339*; *20.3*, p. *348*; 22.4.1, p. 465; 22.4.3, p. 471.

S. BHUSHAN, B. RAO KAZA, and A. N. PANDEY
Photo- and Electroluminescence of Rare Earth (Pr, Gd, and Er) Doped ZnO Phosphors
(a) Vol. **49**, No. 2, p. K167–K171 (1978).
See also *20.3*, p. *349*.

F. G. GÄRTNER and E. MOLLWO
IR-Absorption of OH and OD Centres and OH/OH, OD/OD, and OH/OD Complexes in Cu-Doped ZnO Single Crystals (II): Discussion and Quantitative Interpretation of Models
(b) Vol. **90**, No. 1, p. 33–44 (1978).
See also 13.4, p. 183; *20.1*, p. *335*.

22.4.1 Sulphides

Review Article

V. I. Ivanov-Omskii, L. I. Korovin, and E. M. Shereghii
Phonon-Assisted Cyclotron Resonance in Semiconductors
(b) Vol. **90**, No. 1, p. 11–32 (1978).
See also *6*, p. *64*; 13, p. 142; *19*, p. *303*; 22, p. 408; 22.2.3, p. 452.

Original Papers and Short Notes

G. A. Boiko, V. S. Dneprovskii, M. V. Kraevskii, K. Marinova, S. M. Oak, E. K. Silina, and V. S. Fokin
A Study of the Kinetics of Recombination Radiation of CdS and CdSe Crystals
(b) Vol. **85**, No. 1, p. 111–119 (1978).
See also *13.5.1*, p. *189*; *20.3*, p. *341*; 22.4.2, p. 468.

D. Suisky, R. Boyn, R. Enderlein, and B. Biering
First-Harmonic Electromodulation in Uniaxial Crystals
(b) Vol. **85**, No. 1, p. 217–224 (1978).
See also 13.4, p. 169; *14.3.4*, p. *240*; *20.1*, p. *319*.

O. Goede, H.-J. Krug, and A. Horst
Deformation Effects on Exciton Excitation Spectra in CdS
(b) Vol. **85**, No. 1, p. 239–245 (1978).
See also *13.5.1*, p. *189*; *20.1*, p. *319*.

G. S. Agarwal and H. D. Vollmer
Surface Polariton Effects in Spontaneous Emission (II): Effects of Spatial Dispersion
(b) Vol. **85**, No. 1, p. 301–310 (1978).
See also *13.3*, p. *164*; 13.5.2, p. 198; 20.1, p. 320; 21, p. 352.

H. J. Eichler, Ch. Hartig, and J. Knof
Laser Induced Gratings in CdS
(a) Vol. **45**, No. 2, p. 433–438 (1978).
See also 13.4, p. 171; *20.2*, p. *339*.

G. Jacobsen and W. Maenhout-van der Vorst
Photoconductivity and the Negative Differential Photo-Effect in Sulfur-Annealed ZnS Single Crystals
(a) Vol. **45**, No. 2, p. 463–468 (1978).
See also *16*, p. *262*.

M. K. Roy
Coupling of Surface Acoustic Waves Propagating in Two Separated Piezoelectric Media
(a) Vol. **45**, No. 2, p. 549–552 (1978).
See also *7*, p. *80*; *13.3*, p. *164*; 14.3.2, p. 230; 22.6.1, p. 499.

P. Pipinys and A. Rimeika
Current Pulses in ZnS:Cu Crystals Stimulated by Electric Field at Low Temperatures
(a) Vol. **45**, No. 2, p. K113–K116 (1978).
See also *6*, p. *66*; 13.4, p. 171; *14.3.4*, p. *240*.

V. A. Korneichuk, T. I. Mitrofanova, P. E. Mozol, and G. S. Pekar
Photoconductivity of CdS:Cu under Ruby Laser Excitation
(b) Vol. 86, No. 2, p. K155—K158 (1978).
See also 13.5.1, p. 191; *16*, p. *263*.

G. J. Russell and J. Woods
An Electron Diffraction Study of Phases in the Cu_xS System
(a) Vol. 46, No. 2, p. 433—444 (1978).
See also *1.6*, p. *41*; 22, p. 412.

F. M. Livingstone, R. M. de La Rue, and W. Duncan
pSi–nCdS Heterojunction Photodetectors
(a) Vol. 46, No. 2, p. 595—601 (1978).
See also *14.3.3*, p. *234*; 16, p. 263; 22.1.2, p. 432.

R. Kümmel, H. Rauh, and E. Bangert
Theory of Stark-Ladder Currents
(b) Vol. 87, No. 1, p. 99—110 (1978).
See also 6, p. 70; *14.3.4*, p. *242*.

A. Król, M. Kozielski, and W. Nazarewicz
Localized Vibrational Modes of Al–Cu Complexes in Zinc Sulphide
(b) Vol. 87, No. 1, p. K25—K28 (1978).
See also *6*, p. *70*; *20.1*, p. *326*.

S. W. Biernacki
Jahn-Teller Coupling of Cr^{2+} Ion in ZnS, ZnSe, and ZnTe
(b) Vol. 87, No. 2, p. 607—612 (1978).
See also *13.4*, p. *176*; *20.1*, p. *327*; 22.4.2, p. 468; 22.4.3, p. 470.

I. Rückmann, J. Puls, and J. Voigt
Induced Absorption of CdS Single Crystals in the Energy Ranges of Exciton and Exciton–Biexciton Transitions
(b) Vol. 87, No. 2, p. K111—K114 (1978).
See also *13.5.1*, p. *192*; *20.1*, p. *328*.

A. A. Klyukanov, E. P. Pokatilov, V. N. Gladilin, and Tr. H. Wu
Pinning Structure of Cyclotron Resonance Lines
(b) Vol. 87, No. 2, p. K159—K162 (1978).
See also *19*, p. *308*.

M. Kobayashi, K. Takesawa, and I. Yokota
Carrier Concentration Dependence of Ultrasonic Attenuation in Piezoelectric Semiconductors
(a) Vol. 47, No. 2, p. 657—660 (1978).
See also *7*, p. *81*.

L. Risch
The Influence of Surface Layers on Photoreflectance in the Exciton Region of CdS
(b) Vol. 88, No. 1, p. 111—116 (1978).
See also *13.3*, p. *166*; 13.5.1, p. 192; 20.1, p. 328.

F. Bechstedt and D. Haus
Theory of Second-Order Resonance Raman Scattering in the Case of Strong Excitonic Effects
(b) Vol. **88**, No. 1, p. 163–171 (1978).
See also *6*, p. *73*; 13.5.1, p. 192; *20.1*, p. *329*.

M. Egée
Effect of a Direct Electric Field on the Photoluminescence Processes and Model of the Luminogen Centre in ZnS:Cu
(a) Vol. **48**, No. 1, p. 89–98 (1978).
See also 13.4, p. 177; 14.3, p. 222; *20.3*, p. *345*.

A. V. Bobyl, G. S. Pekar, and M. K. Sheinkman
Dark Conductivity and Photoconductivity of Nickel-Doped CdS Single Crystals
(a) Vol. **48**, No. 1, p. 249–256 (1978).
See also *13.4*, p. *178*; *16*, p. *264*.

A. V. Sachenko, V. A. Tyagai, and A. G. Kundzich
Exciton Luminescence in Semiconductors. Surface Recombination and Space-Charge-Layer Effects
(b) Vol. **88**, No. 2, p. 797–804 (1978).
See also 13.3, p. 166; *13.5.1*, p. *193*; *20.3*, p. *346*.

K. W. Böer
The Physics of the Photovoltaic Effect
(a) Vol. **49**, No. 1, p. 13–26 (1978).
See also *14.3.3*, p. *236*; *16*, p. *265*; 22, p. 417.

R. Butendeich and W. Ruppel
Photovoltage at the Metal–CdS Schottky Contact
(a) Vol. **49**, No. 1, p. 169–175 (1978).
See also *16*, p. *266*.

L. Risch
Investigation of Surface Properties of CdS by Means of Photoreflectance in the Exciton Region
(a) Vol. **49**, No. 1, p. 177–182 (1978).
See also *13.3*, p. *167*; 16, p. 266; 20.1, p. 333.

H. Nelkowski and K. Petermann
Preparation and Investigation of ZnS–$CuGaS_2$ Heterojunctions
(a) Vol. **49**, No. 1, p. K13–K16 (1978).
See also 1.5, p. 39; 3, p. 56; *14.3.3*, p. *236*; *20.3*, p. *348*; 22.8, p. 507.

W. Stössel and H. J. Wagner
Comparison of Additional Boundary Conditions Based on Thermoreflectance Spectra of the $A_{n=1}$-Exciton in CdS Crystals at Oblique Incidence
(b) Vol. **89**, No. 2, p. 403–417 (1978).
See also *13.5.1*, p. *194*; *20.1*, p. *333*.

S. W. Koch, H. Haug, G. Schmieder, W. Bohnert, and C. Klingshirn
Stimulated Intrinsic Recombination Processes in II–VI Compounds
(b) Vol. **89**, No. 2, p. 431–440 (1978).
See also 13.5.1, p. 195; *20.2*, p. *339*; *20.3*, p. *348*; 22.4, p. 462; 22.4.3, p. 471.

S. Asano, N. Yamashita, and Y. Nakao
Luminescence of the Pb^{2+}-Ion Dimer Center in CaS and CaSe Phosphors
(b) Vol. **89**, No. 2, p. 663—673 (1978).
See also *13.4*, p. *181*; *20.3*, p. *348*; 22.4.2, p. 469.

P. Franzosi
Exciton–Phonon Interaction in CdS
(b) Vol. **89**, No. 2, p. K119—K122 (1978).
See also *6*, p. *76*; *13.5.1*, p. *195*; 20.1, p. 334.

K. W. Böer
Heterojunction Interface Double Layer and Consequences for Photovoltaic Cells, Specifically $Cd_zZn_{1-z}S/Cu_2S$
(a) Vol. **49**, No. 2, p. 455—462 (1978).
See also *14.3.3*, p. *237*; *16*, p. *266*; 22, p. 418.

K. G. Germanova
On Some Stationary and Non-Stationary Temperature-Electric Instabilities in Photoconducting CdS Crystals
(a) Vol. **49**, No. 2, p. 499—507 (1978).
See also *14.3.4*, p. *245*; 16, p. 266.

E. Walentynowicz and S. Szuba
The Excitation Mechanism of Mn Luminescence Centres in ZnS:Mn Thin Films
(a) Vol. **49**, No. 2, p. K201—K203 (1978).
See also *20.3*, p. *349*.

H. Schrey and C. Klingshirn
Biexcitons, Impurity Transitions, and Electron–Hole Plasma in Highly Excited II–VI Compounds?
(b) Vol. **90**, No. 1, p. 67—76 (1978).
See also 13.4, p. 183; *13.5.1*, p. *195*; *20.3*, p. *349*.

I. Broser, M. Rosenzweig, R. Broser, M. Richard, and E. Birkicht
A Quantitative Study of Excitonic Polariton Reflectance in CdS
(b) Vol. **90**, No. 1, p. 77—91 (1978).
See also *13.5.1*, p. *195*; *20.1*, p. *335*.

Yu. D. Arepev, V. P. Oleinik, E. A. Salkov, and K. B. Tolpygo
The Anomalous Stark Shift of the Ground State Energy of a Donor–Acceptor Pair in an Electric Field
(b) Vol. **90**, No. 1, p. 355—360 (1978).
See also *13.4*, p. *183*; *20.3*, p. *349*.

R. Baubinas, B.-P. Kietis, R. Rėksnys, and A. Sakalas
Bipolar Photoconductivity and Acoustoelectric Current in CdTe, CdS, and CdSe
(a) Vol. **50**, No. 1, p. K63—K65 (1978).
See also *7*, p. *82*; *16*, p. *267*; 22.4.2, p. 469; 22.4.3, p. 471.

A. Król, M. J. Kozielski, and W. Nazarewicz
Infrared Studies of Al Complexes in Zinc Sulphide
(b) Vol. **90**, No. 2, p. 649—656 (1978).
See also *6*, p. *78*; 10.2, p. 121; *20.1*, p. *337*.

C. LAWTHER and J. WOODS
Blue Light Emission in Forward-Biased ZnS MIS Diodes
(a) Vol. **50**, No. 2, p. 491–502 (1978).
See also *14.3.3*, p. *238*; *20.3*, p. *350*.

N. E. KORSUNSKAYA, I. V. MARKEVICH, T. V. TORCHINSKAYA, and M. K. SHEINKMAN
Some Peculiarities of Thermostimulated Conductivity and Optical Quenching of the Photocurrent in Crystals with the Auger Excitation of Recombination Centres
(a) Vol. **50**, No. 2, p. 767–770 (1978).
See also *13.4*, p. *186*; *14.3*, p. *225*; 16, p. 268.

22.4.2 Selenides

Original Papers and Short Notes

G. A. Boiko, V. S. Dneprovskii, M. V. Kraevskii, K. Marinova, S. M. Oak, E. K. Silina, and V. S. Fokin
A Study of the Kinetics of Recombination Radiation of CdS and CdSe Crystals
(b) Vol. **85**, No. 1, p. 111–119 (1978).
See also *13.5.1*, p. *189*; *20.3*, p. *341*; 22.4.1, p. 463.

A. P. Radlinski
On the Properties of Cobalt Impurities in Zinc Selenide Crystals (II): The Optical Properties in Near-Infrared. $^4A_2(^4F) \leftrightarrow {}^4T_2(^4F)$ Transitions
(b) Vol. **86**, No. 1, p. 41–46 (1978).
See also *13.4*, p. *172*; 20.1, p. 321; 20.3, p. 342.

C. Lawther and J. Woods
The Calculation of Donor Densities from *C–U* Measurements on MIS ZnSe Diodes
(a) Vol. **46**, No. 1, p. 245–250 (1978).
See also *13.4*, p. *173*; *14.3.1*, p. *226*.

B. R. Sethi and P. C. Mathur
Effect of Heavy Doping of ZnSe Crystals with Indium in Creating Compensating Acceptors
(a) Vol. **46**, No. 2, p. 717–721 (1978).
See also *13.4*, p. *174*; *14.3*, p. *220*.

A. Martinaitis and A. Sakalas
High Temperature Investigation of the Hall Effect in p-CdSe in Selenium Vapour Atmosphere
(a) Vol. **46**, No. 2, p. K141–K144 (1978).
See also 10.2, p. 112; *14.3*, p. *220*.

A. V. Martinaitis, A. P. Sakalas, Z. V. Januškevičius, and J. K. Viščakas
Defect Structure of CdSe in Selenium Vapour
(a) Vol. **47**, No. 1, p. 187–194 (1978).
See also *10.2*, p. *113*; *14.3*, p. *220*.

S. W. Biernacki
Jahn-Teller Coupling of Cr^{2+} Ion in ZnS, ZnSe, and ZnTe
(b) Vol. **87**, No. 2, p. 607–612 (1978).
See also *13.4*, p. *176*; *20.1*, p. *327*; 22.4.1, p. 464; 22.4.3, p. 470.

W. Szuszkiewicz, A. M. Witowski, and M. Grynberg
The Dynamic Dielectric Function in HgSe and HgTe
(b) Vol. **87**, No. 2, p. 637–645 (1978).
See also *13.1*, p. *155*; 14.4, p. 249; 20.1, p. 328; 22.4.3, p. 470.

B. R. Sethi, P. L. Talwar, O. P. Sharma, and P. C. Mathur
Effect of Indium Doping and Zinc Annealing on the Electrical Transport Properties of Zinc Selenide Single Crystals
(a) Vol. **47**, No. 2, p. 699–702 (1978).
See also *13.4*, p. *176*; *14.3*, p. *221*

S. Asano, N. Yamashita, and Y. Nakao
Luminescence of the Pb^{2+}-Ion Dimer Center in CaS and CaSe Phosphors
(b) Vol. **89**, No. 2, p. 663–673 (1978).
See also *13.4*, p. *181*; *20.3*, p. *348*; 22.4.1, p. 466.

R. Mach, W. Gericke, H. Treptow, and W. Ludwig
Direct Impact Excitation of Luminescent Centres in ZnSe Single Crystals
(a) Vol. **49**, No. 2, p. 667–674 (1978).
See also 13.4, p. 182; *14.3.3*, p. *237*; *20.3*, p. *349*.

A. Manabe, H. Noguchi, and A. Mitsuishi
Dielectric Function in HgSe at 2, 95, and 300 K
(b) Vol. **90**, No. 1, p. 157–166 (1978).
See also 6, p. 77; 13.1, p. 159; 13.5.2, p. 199; *20.1*, p. *335*.

V. I. Sokolov, N. A. Moskvina, and B. V. Melnikov
Discovery of Exciton–Impurity Complexes in Zinc Selenide by Electroabsorption
(b) Vol. **90**, No. 1, p. K39–K42 (1978).
See also *13.4*, p. *184*; *13.5.1*, p. *196*; 14.3.4, p. 246; 20.1, p. 336.

R. Baubinas, B.-P. Kietis, R. Rėksnys, and A. Sakalas
Bipolar Photoconductivity and Acoustoelectric Current in CdTe, CdS, and CdSe
(a) Vol. **50**, No. 1, p. K63–K65 (1978).
See also *7*, p. *82*; *16*, p. *267*; 22.4.1, p. 466; 22.4.3, p. 471.

A. Milchev
Frequency- and Wave-Vector-Dependent Dielectric Function of a Model Semiconductor
(b) Vol. **90**, No. 2, p. 679–688 (1978).
See also *13.1*, p. *161*; *14.4*, p. *252*; 22.1.1, p. 428; 22.1.2, p. 439; 22.2.1, p. 449.

V. V. Serdyuk, A. P. Chebanenko und G. G. Chemeresyuk
Current Switching Mechanism in Inhomogeneous Zinc Selenide Crystals
(a) Vol. **50**, No. 2, p. 717–722 (1978).
See also *13.4*, p. *186*; *14.3.4*, p. *247*; 16, p. 267.

22.4.3 Tellurides

Original Papers and Short Notes

V. G. Golubev, E. I. Georgitse, and V. I. Ivanov-Omskii
Photoconductivity of HgTe in Magnetic Field at 337 μm
(b) Vol. **85**, No. 1, p. K39—K41 (1978).
See also *16*, p. *262*; 18, p. 273; *20.1*, p. *320*.

D. Boulanger and M.-S. Martin
Crystal Growth and EPR Study of Cr^+-Doped MgTe Single Crystals
(b) Vol. **85**, No. 2, p. 597—601 (1978).
See also 3, p. 54; 13.4, p. 170; *19*, p. *304*.

J. Nahum and F. J. Bryant
Exciton Emission of Ytterbium-Implanted Cadmium Telluride
(b) Vol. **85**, No. 2, p. 629—637 (1978).
See also *13.5.1*, p. *190*; *20.3*, p. *342*.

B. L. Gelmont, V. I. Ivanov-Omskii, and E. N. Ukraintsev
The Frequency Dependent Dielectric Function of Gapless Semiconductors in a Magnetic Field
(b) Vol. **87**, No. 1, p. 381—383 (1978).
See also *13.5*, p. *187*; 14.4, p. 248.

T. Takebe, J. Saraie, and T. Tanaka
Study of the Surface Barrier of the Metal–n-CdTe Contact
(a) Vol. **47**, No. 1, p. 123—130 (1978).
See also 13.3, p. 165; *14.3.2*, p. *230*; 14.3.4, p. 242.

S. W. Biernacki
Jahn-Teller Coupling of Cr^{2+} Ion in ZnS, ZnSe, and ZnTe
(b) Vol. **87**, No. 2, p. 607—612 (1978).
See also *13.4*, p. *176*; *20.1*, p. *327*; 22.4.1, p. 464; 22.4.2, p. 468.

W. Szuszkiewicz, A. M. Witowski, and M. Grynberg
The Dynamic Dieletric Function in HgSe and HgTe
(b) Vol. **87**, No. 2, p. 637—645 (1978).
See also *13.1*, p. *155*; 14.4, p. 249; 20.1, p. 328; 22.4.2, p. 468.

N. Romeo, G. Sberveglieri, L. Tarricone, J. Vidal, and A. Wojtowicz
Electrical Properties of Sb-Doped ZnTe Thin Films
(a) Vol. **47**, No. 2, p. 371—374 (1978).
See also *14.3.1*, p. *227*.

K. Germanova, K. Marinova, and S. Stoyanov
Effect of Surface Layers on the Free Exciton Reflectance of ZnTe
(b) Vol. **88**, No. 1, p. K19—K22 (1978).
See also *13.5.1*, p. *193*; 20.1, p. 329.

P. Höschl, P. Moravec, A. Martinaitis, and A. Sakalas
Study of Photo-Hall Effect on Single Crystals of Cadmium Telluride Doped with Chlorine
(a) Vol. **48**, No. 1, p. K43—K46 (1978).
See also *13.4*, p. *178*; *16*, p. *265*.

K. Maeda, K. Nakagawa, and S. Takeuchi
Activation Analysis of Plastic Deformation of CdTe below Room Temperature
(a) Vol. **48**, No. 2, p. 587–591 (1978).
See also *10.2*, p. *116*; *12.2*, p. *140*.

P. Świątek, A. M. Witowski, and M. Grynberg
Experimental and Theoretical Study of the Dynamic Dielectric Function in p-HgTe
(b) Vol. **89**, No. 1, p. K1–K4 (1978).
See also *13.1*, p. *158*; 14.4, p. 250; 20.1, p. 332.

Z. Liliental and G. Kästner
Voltage Dependence on Precipitation Contrast
(a) Vol. **49**, No. 1, p. K1–K2 (1978).
See also *10.2*, p. *118*.

S. W. Koch, H. Haug, G. Schmieder, W. Bohnert, and C. Klingshirn
Stimulated Intrinsic Recombination Processes in II–VI Compounds
(b) Vol. **89**, No. 2, p. 431–440 (1978).
See also 13.5.1, p. 195; *20.2*, p. *339*; *20.3*, p. *348*; 22.4, p. 462; 22.4.1, p. 465.

H. Friedrich, Yu. V. Ozerov, R. Strehlow, and A. N. Georgobiani
On the Electron–Phonon and Exciton–Phonon Interaction in ZnTe
(b) Vol. **89**, No. 2, p. 603–609 (1978).
See also *13.5.1*, p. *195*; *20.1*, p. *334*.

B. Rabin, H. Tabatabai, and P. Siffert
Capacitance of Cadmium Telluride Schottky Diodes
(a) Vol. **49**, No. 2, p. 577–584 (1978).
See also *13.4*, p. *182*; *14.3.3*, p. *237*.

D. Shaw and E. Watson
The Shift of Optical Absorption Edge in CdTe with Temperature and Doping Level
(a) Vol. **49**, No. 2, p. 593–602 (1978).
See also 10.2, p. 119; *20.1*, p. *335*.

R. Baubinas, B.-P. Kietis, R. Rėksnys, and A. Sakalas
Bipolar Photoconductivity and Acoustoelectric Current in CdTe, CdS, and CdSe
(a) Vol. **50**, No. 1, p. K63–K65 (1978).
See also *7*, p. *82*; *16*, p. *267*; 22.4.1, p. 466; 22.4.2, p. 469.

C. Lhermitte, D. Carles, C. Viger, and C. Vautier
Transient Photoconductivity Effects on n-Type Cadmium Telluride Films in Presence of a Uniform Distribution of Traps. Theoretical Model and Experimental Results
(a) Vol. **50**, No. 2, p. 651–659 (1978).
See also 13.4, p. 186; *16*, p. *267*.

22.4.4 Mixed Crystals

Original Papers and Short Notes

L. LEIBLER
Theory of Electron Spin Resonance in Graded Mixed Semiconductors
(b) Vol. **85**, No. 2, p. 611—619 (1978).
See also 13.1, p. 151; *19*, p. *304*; 22.2.4, p. 456.

J. J. DUBOWSKI
Disorder Scattering in $Cd_xHg_{1-x}Te$ Mixed Crystals
(b) Vol. **85**, No. 2, p. 663—672 (1978).
See also *13.1*, p. *151*; *14.3*, p. *217*; 15, p. 259.

V. D. VANKAR, S. R. DAS, PREM NATH, and K. L. CHOPRA
Structure of Vacuum Evaporated $Cd_xZn_{1-x}S$ Thin Films
(a) Vol. **45**, No. 2, p. 665—669 (1978).
See also *1.5*, p. *36*.

V. I. IVANOV-OMSKII, V. A. MALTSEVA, A. D. BRITOV, and S. D. SIVACHENKO
Photo- and Cathodoluminescence of $Cd_{0.3}Hg_{0.7}Te$ Alloys
(a) Vol. **46**, No. 1, p. 77—80 (1978).
See also *13.4*, p. *173*; *20.3*, p. *343*.

E. POPKO and J. M. PAWLIKOWSKI
Contribution of the Lattice Dilatation to the Temperature Shift of the Energy Gap of $Cd_xHg_{1-x}Te$
(a) Vol. **46**, No. 1, p. K9—K12 (1978).
See also *8*, p. *84*; *13.1*, p. *153*.

J. KOSSUT
The Disorder Scattering in Zincblende Narrow-Gap Semiconducting Mixed Crystals
(b) Vol. **86**, No. 2, p. 593—601 (1978).
See also *13.1*, p. *154*; *13.4*, p. *174*; 14.3, p. 219.

C. G. PATIL and B. S. KRISHNAMURTHY
Carrier Mobility in Ferromagnetic Semiconductors
(b) Vol. **86**, No. 2, p. 725—732 (1978).
See also 13.5.4, p. 201; *14.3*, p. *219*; 18.2, p. 285.

W. GĘBICKI and W. NAZAREWICZ
Far Infrared Reflection Spectra of $Mn_xCd_{1-x}Te$ Mixed Crystals
(b) Vol. **86**, No. 2, p. K135—K138 (1978).
See also *6*, p. *69*; *20.1*, p. *325*.

A. BANERJEE, PREM NATH, V. D. VANKAR, and K. L. CHOPRA
Properties of $Zn_xCd_{1-x}S$ Films Prepared by Solution Spray Technique
(a) Vol. **46**, No. 2, p. 723—728 (1978).
See also *1.5*, p. *36*; *14.3.1*, p. *226*; 16, p. 263.

A. KUROWSKA
Theory of Free Carrier Absorption in Mixed Semiconductors with Composition Fluctuations
(b) Vol. **87**, No. 1, p. 69—78 (1978).
See also *13.1*, p. *154*; *20.1*, p. *325*.

M. A. MEKHTIEV and S. M. SEID-RZAEVA
Quasi-Local Acceptor Levels and Electron Mobility in $Hg_{1-x}Cd_xTe$
(b) Vol. **87**, No. 2, p. 447–454 (1978).
See also *13.4*, p. *175*; *14.3*, p. *221*.

M. JACZYŃSKI, J. KOSSUT, and R. R. GAŁĄZKA
Influence of Exchange Interaction on the Quantum Transport Phenomena in $Hg_{1-x}Mn_xTe$
(b) Vol. **88**, No. 1, p. 73–85 (1978).
See also *13.2*, p. *162*; *14.3*, p. *221*; 15, p. 260.

S. D. BENESLAVSKII and O. ZIEP
Hot Electrons in Zero-Gap Semiconductors with Non-Parabolic Electronic Band Structure
(b) Vol. **88**, No. 1, p. 221–230 (1978).
See also *13.1*, p. *156*; *14.3.4*, p. *243*; 14.4, p. 249.

A. KOZANECKI, Z. WERNER, H. RZEWUSKI, and J. J. LOFERSKI
A Study of Excitonic Spectra in CdS_xSe_{1-x} Mixed Crystals
(b) Vol. **89**, No. 1, p. 313–317 (1978).
See also *13.5.1*, p. *194*; *20.3*, p. *347*.

O. GOEDE, L. JOHN, and D. HENNIG
Compositional Disorder-Induced Broadening for Free Excitons in II–VI Semiconducting Mixed Crystals
(b) Vol. **89**, No. 2, p. K183–186 (1978).
See also *13.5.1*, p. *195*; *20.1*, p. *334*.

P. J. FLATAU
Kinetics of the Intraband Absorption and Magnetoabsorption Coefficients in Mixed Semiconductors with Composition Fluctuations
(b) Vol. **90**, No. 1, p. 251–260 (1978).
See also *1.3*, p. *28*; *20.1*, p. *335*.

22.5 Halides

Review Article

P. P. Seregin, F. S. Nasredinov, and L. N. Vasilev
A Study of Radiation Defects in Solids by Means of Mössbauer Spectroscopy
(a) Vol. **45**, No. 1, p. 11–45 (1978).
See also *5*, p. *60*; *11*, p. *123*; 22.8, p. 503; 22.9, p. 522.

Original Papers and Short Notes

H. Spiering, S. Hösl, and H. Vogel
Polarization Effects in Mössbauer Absorption of a $FeCl_2 \cdot 4\,H_2O$ Single Crystal in an External Magnetic Field
(b) Vol. **85**, No. 1, p. 87–93 (1978).
See also *5*, p. *60*.

V. Valvoda and J. Ječný
X-Ray Diffraction Study of Anharmonic Thermal Vibrations in CuCl
(a) Vol. **45**, No. 1, p. 269–275 (1978).
See also *6*, p. *65*.

Č. Konák, L. Budík, and L. Novák
Electro-Optical Properties of NH_4Cl Single Crystals near Their Order–Disorder Phase Transition
(a) Vol. **45**, No. 1, p. K63–K66 (1978).
See also *1.2*, p. *16*; 14, p. 203; 20, p. 315.

J. C. Castro and M. de Souza
The OCN^- Vibrational Spectrum and the Phase Transitions in KCN_xKCl_{1-x} Single Crystals
(b) Vol. **86**, No. 1, p. 137–144 (1978).
See also *1.2*, p. *16*; *6*, p. *67*; 12.2, p. 138; 20.1, p. 321; 22.5.4, p. 489.

Vu Duy Phach, A. Bivas, B. Hönerlage, and J. B. Grun
Two-Photon Resonant Raman Scattering via Biexcitons
(b) Vol. **86**, No. 1, p. 159–168 (1978).
See also *13.5.1*, p. *190*; 20.1, p. 321; 20.3, p. 342.

K. Fukushi, M. Nippus, and R. Claus
The Origin of the Anomalous TO-Doublet in Cubic CuCl Studied by Light Scattering
(b) Vol. **86**, No. 1, p. 257–262 (1978).
See also *6*, p. *67*; *20.1*, p. *322*.

W. Ekardt and M. I. Sheboul
Many-Exciton Effects in the Theory of Biexcitons
(b) Vol. **86**, No. 2, p. 535–540 (1978).
See also *13.5.1*, p. *191*.

M. B. Haxton
The Phonon Spectra of γ- and β-CuBr
(b) Vol. **86**, No. 2, p. 661–670 (1978).
See also *6*, p. *69*.

P. K. C. PILLAI, R. G. MENDIRATTA, C. K. PILLAI, and R. NATH
Photo-Induced AC Impedance Measurements in Single Crystal Photoconductors
(a) Vol. **46**, No. 2, p. K127—K130 (1978).
See also *14.4*, p. *248*; *16*, p. *264*; 22.4, p. 461.

V. K. AGRAWAL
Rotation Twinning in Cadmium Iodide Crystals
(a) Vol. **46**, No. 2, p. K145—K146 (1978).
See also *10.2*, p. *112*.

A. R. HASSAN
Exciton Molecule Creation by Pure and Phonon-Assisted Two-Photon Transitions in Solids
(b) Vol. **87**, No. 1, p. 31—40 (1978).
See also *13.5.1*, p. *191*; 20.1, p. 325.

H. J. L. VAN DER VALK, P. MEERTENS, and C. HAAS
Electron Paramagnetic Resonance of Divalent V, Mn, and Co in Single Crystals of CdI_2 and PbI_2
(b) Vol. **87**, No. 1, p. 135—144 (1978).
See also *19*, p. *307*.

V. K. GARG, D. S. PURI, and M. P. VERMA
On the Second- and Third-Order Elastic Constants of ZnS Structure Solids
(b) Vol. **87**, No. 1, p. 401—407 (1978).
See also *6*, p. *70*; *12.2*, p. *139*; 22.2, p. 443; 22.4, p. 461.

M. KALM and CH. UIHLEIN
Investigations on the Temperature-Dependent TPA Linewidth of the Z_3-Exciton in CuCl
(b) Vol. **87**, No. 2, p. 575—580 (1978).
See also 6, p. 71; *13.5.1*, p. *192*; *20.1*, p. *327*.

E. WISZT
A Contribution to the Phenomenological Theory of the Anomalies of the Elastic Constants near a Structural Phase Transition
(b) Vol. **87**, No. 2, p. 675—688 (1978).
See also *1.2*, p. *19*; 12.2, p. 139.

T. IWASAKI, T. GOTO, and Y. NISHINA
Oscillator Strength of Excitons in Highly Excited PbI_2
(b) Vol. **88**, No. 1, p. 289—293 (1978).
See also *13.5.1*, p. *193*; 20.1, p. 329; 20.3, p. 345.

B. PREVOT, C. SCHWAB, and B. DORNER
Phonon Dispersion in Red HgI_2
(b) Vol. **88**, No. 1, p. 327—333 (1978).
See also *6*, p. *73*; 7, p. 81; *20.1*, p. *329*.

L. GRAF
Electronic Transitions in the Raman Spectra of $FeCl_2 \cdot 2\,H_2O$ and $FeCl_2 \cdot 2\,D_2O$
(b) Vol. **88**, No. 2, p. 429—437 (1978).
See also *6*, p. *74*; *13.5.4*, p. *201*; 20.1, p. 330.

B. HÖNERLAGE, VU DUY PHACH, and J. B. GRUN
Polarization Properties of Biexciton Luminescence and Two-Photon Raman Emission in CuCl
(b) Vol. **88**, No. 2, p. 545—553 (1978).
See also *13.5.1*, p. *193*; *20.3*, p. *346*.

Atsuo Kasuya, S. Togashi, T. Goto, and Y. Nishina
Two-Photon Photoemission in PbI_2 and GaAs
(b) Vol. **89**, No. 2, p. K145—K147 (1978).
See also *17*, p. *270*; 22.2.1, p. **448**.

G. Behnke and H. Büttner
The Exciton Dispersion in Polar Semiconductors
(b) Vol. **90**, No. 1, p. 53—59 (1978).
See also *13.5.1*, p. *195*; 13.5.3, p. 200.

A. Lachter, J. Salardenne, and A. S. Barrière
Optical Absorption Study of Iron Triflouride Thin Films
(b) Vol. **90**, No. 1, p. 147—150 (1978).
See also *20.1*, p. *335*.

S. A. Seryshev, A. G. Lundin, and M. L. Afanasev
NMR in Tungsten Hexafluoride at High Hydrostatic Pressure
(b) Vol. **90**, No. 1, p. K35—K37 (1978).
See also *19*, p. *314*.

F. Adduci, L. Baldassarre, G. Maggipinto, A. Minafra, and F. Lévy
Photoelectromagnetic Investigation of PbI_2
(a) Vol. **50**, No. 1, p. 257—262 (1978).
See also *16*, p. *267*.

V. B. Sharan, D. N. S. Srivastwa, and S. C. Sen
Absorption, Excitation, and Fluorescence Spectra of Thallium-Activated Ammonium Bromide
(b) Vol. **90**, No. 2, p. 671—678 (1978).
See also *13.4*, p. *185*; 20.1, p. 337; 20.3, p. 350.

22.5.1 Silver Halides

Original Papers and Short Notes

CH. DIAINE, J. L. GISCLON, and J. DUPUY
Colloid Centres in AgCl Induced by Heavy Ion Implantation
(b) Vol. **87**, No. 1, p. 293–300 (1978).
See also *10.2*, p. *113*; *11*, p. *126*; 20.1, p. 326.

K. D. BECKER
Nuclear Spin Relaxation by Quadrupolar Interactions. The Diffusion of Substitutional Impurities in Ionic Crystals
(b) Vol. **87**, No. 2, p. 589–597 (1978).
See also *9*, p. *92*; *19*, p. *308*.

V. PLATIKANOVA and J. MALINOWSKI
Neutralization of Trapped Photoholes in AgBr
(a) Vol. **47**, No. 2, p. 683–690 (1978).
See also *10.2*, p. *114*; 13.4, p. 176.

R. S. EACHUS, R. E. GRAVES, and M. T. OLM
Electron Spin Resonance of Mobile Electrons in AgCl
(b) Vol. **88**, No. 2, p. 705–711 (1978).
See also *10.2*, p. *116*; 16, p. 265; *19*, p. *310*.

M. B. O'REILLY
A Phenomenological Theory for Ionic Conductivity in Solid Electrolytes
(a) Vol. **48**, No. 2, p. 489–496 (1978).
See also *10.2*, p. *116*; *14.4*, p. *250*.

J. KALUS, K. FUNKE, and A. HEIDEMANN
Search for Quasielastic Neutron Scattering by Silver in β-AgI
(a) Vol. **48**, No. 2, p. K179–K180 (1978).
See also *9*, p. *93*.

P. A. VAROTSOS
An Estimate of the Pressure Dependence of the Dielectric Constant in Alkali Halides
(b) Vol. **90**, No. 1, p. 339–343 (1978).
See also 4, p. 59; 6, p. 77; 12.2, p. 141; *14.4*, p. *251*; 22.5.2, p. 485.

M. E. VAN HULLE and W. MAENHOUT-VAN DER VORST
Ionic Conductivity of β-AgI Microcrystals
(a) Vol. **50**, No. 2, p. K175–K180 (1978).
See also *10.2*, p. *122*; *14.4*, p. *252*.

22.5.2 Alkali Halides

Original Papers and Short Notes

V. K. Garg and M. P. Verma
Erratum to: Evaluation of Some Important Lattice Sums on NaCl and CsCl Structures
(b) Vol. **85**, No. 1, p. 417 (1978).
See also *1.1*, p. *11*; 12.2, p. 138.

S. B. S. Sastry, A. Scharmann, and D. Schwabe
Luminescence of RbI:Pb^{2+} under B-Band Excitation
(b) Vol. **85**, No. 1, p. K15—K17 (1978).
Erratum
(b) Vol. **87**, No. 2, p. K163 (1978).
See also *20.3*, p. *341*.

L. Bosi and M. Nimis
New Lifetime Data for the Excited State of the M-Centre in Alkali Halides
(b) Vol. **85**, No. 1, p. K27—K29 (1978).
Erratum
(b) Vol. **89**, No. 1, p. K107 (1978).
See also *10.2*, p. *109*; *13.4*, p. *170*.

K. Hoffmann and D. Pöss
Electron Spin Resonance Investigations on Lithium Fluoride Doped with Titanium
(a) Vol. **45**, No. 1, p. 263—268 (1978).
See also 13.4, p. 170; *19*, p. *303*.

P. Varotsos
Diffusion of Monovalent Ions in an Alkali Halide Crystal
(a) Vol. **45**, No. 1, p. K9—K12 (1978).
See also *9*, p. *91*.

L. Mărculescu, N. Vlahovici, and T. Velula
On the Decay of the Luminescence Emission of Pb^{2+} Centres in Alkali Halide Crystals
(b) Vol. **85**, No. 2, p. 439—443 (1978).
Erratum
(b) Vol. **89**, No. 1, p. 335 (1978).
See also *13.4*, p. *170*; *20.3*, p. *341*.

S. C. Goyal, R. Prakash, and S. P. Tripathi
Many-Body Lattice Theory of Elastic Dielectric and Its Application to Solids with NaCl Structure
(b) Vol. **85**, No. 2, p. 477—486 (1978).
Erratum
(b) Vol. **89**, No. 1, p. 335 (1978).
See also *6*, p. *65*; 12.2, p. 138.

P. Schrey, R. Balzer, and H. Peisl
Release of Stored Energy of X-Irradiated KCl and CsBr
(b) Vol. **85**, No. 2, p. 553—559 (1978).
See also *10.2*, p. *109*; *11*, p. *123*.

M. Vanhaelst, P. Matthys, and E. Boesman
The Phonon Induced Zero Field Splitting in Vanadium Doped Alkali Chlorides
(b) Vol. **85**, No. 2, p. 639–644 (1978).
See also *19*, p. *304*.

F. Lüty and A. Diaz-Gongora
Electro-Birefringence from Paraelectric Defects in Cubic Crystals
(b) Vol. **85**, No. 2, p. 693–706 (1978).
See also 10.2, p. 109; *14*, p. *203*; *20*, p. *315*.

V. I. Altukhov and V. G. Kvachadze
Phonon Scattering on Small-Radius Colloids in KCl Crystals
(b) Vol. **85**, No. 2, p. 769–775 (1978).
See also 6, p. 66; 8, p. 83; *10.2*, p. *109*.

R. A. Evarestov and A. N. Ermoshkin
Large Unit Cell Calculations of the Band Structure of Ionic Crystals Using the Mulliken-Ruedenberg Approximation
(b) Vol. **86**, No. 1, p. 47–55 (1978).
See also *13.1*, p. *152*.

A. Diaz-Gongora and F. Lüty
Kerr Effect and Birefringence from Paraelectric CN^- Defects in Nine Alkali Halides
(b) Vol. **86**, No. 1, p. 127–136 (1978).
See also 10.2, p. 110; *14*, p. *203*; *20*, p. *316*.

P. G. Baranov, R. A. Zhitnikov, and V. A. Khramtsov
EPR and Optical Absorption of Zn^+ Ions in NaCl Crystals
(b) Vol. **86**, No. 1, p. K67–K70 (1978).
See also *19*, p. *305*; *20.1*, p. *323*.

R. Steinbrech, W. Skrotzki, and P. Haasen
Latent Hardening of Doped Alkali Halide Single Crystals
(a) Vol. **46**, No. 1, p. 107–115 (1978).
See also *10.2*, p. *110*.

P. Hennl
Optical Properties of Chalcogen–Vacancy Centres in Alkali Halides
(a) Vol. **46**, No. 1, p. 147–152 (1978).
See also *10.2*, p. *111*; *20.1*, p. *323*; 20.3, p. 343.

G. A. Andreev, M. Hartmanová, and V. A. Klimov
Precipitation of Pb Impurity from the Lattic of NaCl Crystals
(a) Vol. **46**, No. 1, p. 259–265 (1978).
See also *10.2*, p. *111*; 20.1, p. 324.

V. N. Kadchenko and M. Elango
Diffusion Parameters of Hot Holes Created by Ionizing Radiation in NaCl(Ag)
(a) Vol. **46**, No. 1, p. 315–319 (1978).
See also *13.4*, p. *173*; *20.3*, p. *343*.

N. A. Tyapunina, G. M. Zinenkova, and E. V. Shtrom
Dislocation Multiplication in Alkali Halide Crystals Exposed to Ultrasonic Waves. The Original Stage
(a) Vol. **46**, No. 1, p. 327–336 (1978).
See also *10.2*, p. *111*.

N. A. Tyapunina, E. K. Naimi, and S. V. Gasparyan
The Orientation Dependence of Dislocation Internal Friction in Real Crystals (I): Alkali Halide Crystals
(a) Vol. **46**, No. 1, p. 351–356 (1978).
See also *10.2*, p. *111*; *12.2*, p. *138*.

T. M. Bulatova and E. M. Nadgornyi
Internal Stresses and Dislocation Mobility in NaCl Single Crystals
(a) Vol. **46**, No. 1, p. 357–364 (1978).
See also *10.2*, p. *111*.

V. N. Kadchenko and M. Elango
Diffusion Approach to the Photostimulated Electron Emission of Ionic Crystals after Photoexcitation
(a) Vol. **46**, No. 1, p. K61–K65 (1978).
See also *17*, p. *269*.

S. Lefrant, J. P. Buisson, M. Ghomi, and L. Taurel
First-Order Raman Scattering of CsF Doped with F Centres
(b) Vol. **86**, No. 2, p. 493–496 (1978).
See also *6*, p. *68*; *20.1*, p. *324*.

G. G. Nilsson, U. Karlsson, and G. Raunio
Experimental Phonon Frequencies of RbBr at 80, 290, and 370 K
(b) Vol. **86**, No. 2, p. 653–660 (1978).
See also *6*, p. *69*; *8*, p. *84*.

K. D. Nierzewski, T. Todorov, and M. Georgiev
Room Temperature F–F′ Conversion in KCl Crystals of Various Purity and Method of Coloration
(b) Vol. **86**, No. 2, p. 697–704 (1978).
See also *10.2*, p. *112*; 20.1, p. 324.

V. Hizhnyakov and S. Zazubovich
Vacancy-Induced Splitting of Excited States and the Structure of $Sn^{2+}v_c^-$ Centres in Alkali Halides
(b) Vol. **86**, No. 2, p. 733–739 (1978).
See also *13.4*, p. *174*; *20.1*, p. *324*.

V. G. Plekhanov and A. A. O'Connel-Bronin
Polariton Emission in LiH, LiD, and NaI Single Crystals
(b) Vol. **86**, No. 2, p. K123–K125 (1978).
See also *13.5*, p. *187*; 20.3, p. 343; 22, p. 412.

M. Adachi and N. Takeuchi
ESR Study of Copper(II) Ions in NaCl X-Irradiated at Room Temperature
(b) Vol. **86**, No. 2, p. K151–K153 (1978).
See also *19*, p. *306*.

G. A. Andreev and V. A. Klimov
Flotation Study of Ca^{2+} and Sr^{2+} Precipitation in KCl Crystals
(a) Vol. **46**, No. 2, p. 421–426 (1978).
See also *1.4*, p. *30*; *10.2*, p. *112*.

C. Pedrini
Cu$^+$ Luminescent Centre in LiCl Single Crystals
(b) Vol. **87**, No. 1, p. 273—286 (1978).
See also *13.4*, p. *175*; *20.3*, p. *344*.

S. L. A. de Queiroz, B. Koiller, B. Maffeo, and H. S. Brandi
A Cluster-Bethe Lattice Treatment for the F-Center in Alkali Halides
(b) Vol. **87**, No. 1, p. 351—359 (1978).
See also *13.4*, p. *175*; *20.1*, p. *326*.

L. Nikolova and T. Todorov
Photoinduced Reorientation of F_A Centres in KCl:Na near Room Temperature
(a) Vol. **47**, No. 1, p. 213—217 (1978).
See also *10.2*, p. *113*; *20.1*, p. *326*.

M. C. Wintersgill and P. D. Townsend
The Effectiveness of the Implantation of Oxygen in Increasing the Luminescence from LiF:Mg:Ti
(a) Vol. **47**, No. 1, p. K67—K69 (1978).
See also 13.4, p. 175; *20.3*, p. *344*.

C. Jaccard and M. Ecabert
Optical Detection of Nuclear Magnetic Resonance with Distant F-Centre Pairs in KCl
(b) Vol. **87**, No. 2, p. 497—505 (1978).
See also 13.4, p. 175; *19*, p. *308*; *20.3*, p. *344*.

S. K. Sarkar and S. Sengupta
Lundqvist Three-Body Interaction and Lattice Dynamics of Ionic Crystals
(b) Vol. **87**, No. 2, p. 517—525 (1978).
See also *4*, p. *58*; 12.2, p. 139.

G. C. Kuczynski, P. E. Dietz, W. H. Hamill, and J. H. Jischke
H_2O Related Centers in Potassium Chloride
(b) Vol. **87**, No. 2, p. 581—588 (1978).
See also *10.2*, p. *113*; *20.1*, p. *327*.

K. C. Bleijenberg and C. W. M. Timmermans
Luminescence and Electrical Conductivity of Uranium-Activated Sodium Fluoride Crystals
(a) Vol. **47**, No. 2, p. 589—595 (1978).
See also 10.2, p. 114; *14.4*, p. *249*; *20.3*, p. *345*.

D. E. Aboltin, V. J. Grabovskis, A. R. Kangro, Ch. Lushchik, A. A. O'Konnel-Bronin, I. K. Vitol, and V. E. Zirap
Thermally Stimulated and Tunneling Luminescence and Frenkel Defect Recombination in KCl and KBr at 4.2 to 77 K
(a) Vol. **47**, No. 2, p. 667—675 (1978).
See also *10.2*, p. *114*; *20.3*, p. *345*.

S. B. S. Sastry and K. Balasubramanyam
Studies on Tin Centres in RbCl Crystals
(a) Vol. **47**, No. 2, p. 711—716 (1978).
See also *10.2*, p. *114*; 11, p. 127; *20.1*, p. *328*.

P. A. Varotsos and K. Alexopoulos
Calculation of the Migration Volume of Vacancies in Ionic Solids from Macroscopic Parameters
(a) Vol. **47**, No. 2, p. K133—K136 (1978).
See also *10.2*, p. *114*.

R. K. Gartia and B. S. Acharya
On the Production of Z_1 Centres
(a) Vol. **47**, No. 2, p. K165—K168 (1978).
See also *10.2*, p. *114*; 20.3, p. 345.

P. R. Sarode
Correlation between the Chemical Shifts in X-Ray Absorption Spectra of Ionic Solids and Contribution to the Chemical Bond
(b) Vol. **88**, No. 1, p. K35—K38 (1978).
See also *4*, p. *58*; *20*, p. *317*.

A. Watterich
Z_4 Like Centre in NaCl Doped by Divalent Metals: Mg, Mn, Ni, Cd, or Pb
(b) Vol. **88**, No. 1, p. K51—K54 (1978).
See also *10.2*, p. *115*; 20.1, p. 330.

J. R. Hanscomb and P. Chantikul
The Effect of Strontium Chloride as Impurity on the Thermoelectric Power of Potassium Chloride
(a) Vol. **48**, No. 1, p. 79—87 (1978).
See also 10.2, p. 115; *15*, p. *260*.

W. H. Robinson, A. J. Glover, and A. Wolfenden
Electrical–Mechanical Coupling of Dislocations in KCl, NaCl, LiF, and CaF_2
(a) Vol. **48**, No. 1, p. 155—163 (1978).
See also *10.2*, p. *115*; 22.5.3, p. 488.

R. K. Gartia, B. S. Acharya, and V. V. Ratnam
Effect of Re-Irradiation on Z_1 Centres
(a) Vol. **48**, No. 1, p. 235—239 (1978).
See also *10.2*, p. *115*; *20.3*, p. *345*.

M. Wagner
On the Non-Adiabatic Properties of Tunneling Centres
(b) Vol. **88**, No. 2, p. 517—530 (1978).
See also *6*, p. *74*; *10.2*, p. *115*.

Ch. Hoentzsch and J. M. Spaeth
Localized Vibrations of Substitutional Atomic Hydrogen in KCl and RbCl Studied by ENDOR
(b) Vol. **88**, No. 2, p. 581—589 (1978).
See also *13.4*, p. *178*; *19*, p. *310*.

E. Goovaerts and D. Schoemaker
Inelastic Light Scattering of the V_K Centers in the Alkali Halides
(b) Vol. **88**, No. 2, p. 615—621 (1978).
See also *10.2*, p. *116*; *20.1*, p. *330*.

E. Doni, R. Girlanda, and G. Pastori Parravicini
A Note on the Polarization Dependence of the Two-Photon Absorption Coefficient in Crystals. Extension to Circularly Polarized Light
(b) Vol. 88, No. 2, p. 773—786 (1978).
See also *20.1*, p. *331*.

R. Mehrotra and M. M. Pant
Long Wave Phonon Modes and Infrared Absorption of a NaI Slab
(b) Vol. 88, No. 2, p. K131—K133 (1978).
See also *6*, p. *74*; 20.1, p. 331.

E. V. Shvedov and V. M. Ievlev
Nucleus Saturation Density and Epitaxy in the Condensation of Metals on Alkali-Halide Crystals (II): Oriented Growth of Islands
(a) Vol. 48, No. 2, p. 603—608 (1978).
See also *3*, p. *55*; *10.2*, p. *116*.

S. V. Plotnikov, E. I. Rozum, V. A. Kuzminikh, and S. A. Vorobiev
Channeling Effect of Fast Electrons on the Luminescence Yield of NaCl:Tl Crystals
(a) Vol. 48, No. 2, p. K175—K178 (1978).
See also 11, p. 128; *20.3*, p. *346*.

S. B. S. Sastry and S. Sapru
Thermoluminescence of γ-Irradiated Rubidium Chloride Doped with Europium
(a) Vol. 48, No. 2, p. K189—K191 (1978).
See also *10.2*, p. *117*; *20.3*, p. *347*.

C. S. Lam and Y. P. Varshni
Binding Energies of Positron to F and F′ Colour Centres in Alkali Halides
(b) Vol. 89, No. 1, p. 103—108 (1978).
See also *10.2*, p. *117*; *13.4*, p. *179*.

F. Lohse, G. Reuter, and J. M. Spaeth
Optical Investigation of Atomic Hydrogen Centres in Alkali Chlorides Doped with I^- or Br^-
(b) Vol. 89, No. 1, p. 109—116 (1978).
See also *10.2*, p. *117*; 13.4, p. 180; *20.1*, p. *332*.

M. Vanhaelst, P. Matthys, and E. Boesman
Electron Spin Transfer in Alkali Chlorides and Alkali Bromides Doped with Divalent Vanadium
(b) Vol. 89, No. 1, p. 151—157 (1978).
See also *19*, p. *311*; *20.1*, p. *332*.

F. Carlier and G. Jacobs
Transient Infrared Absorption of Bound Polarons in NaCl
(b) Vol. 89, No. 1, p. 193—198 (1978).
See also *13.5.3*, p. *200*; *20.1*, p. *332*.

V. C. Jain and J. Shanker
The Exchange Charge Shell Model Analysis of Strain Derivatives of Static Polarizability and Effective Charge Parameter of Alkali Halides
(b) Vol. 89, No. 1, p. 213—219 (1978).
See also *4*, p. *59*; 6, p. 75; 14.4, p. 250.

L. Bosi, A. Longoni, and M. Nimis
Lifetime of Color Centers in NaBr
(b) Vol. **89**, No. 1, p. 221—224 (1978).
See also *10.2*, p. *117*; 13.4, p. 180; *20.3*, p. *347*.

I. Földvári, R. Voszka, and Z. Morlin
The Properties of Ni Ions in NaCl Single Crystals (I): Vacuum Ultraviolet, Ionic Conductivity, and X-Ray Diffraction Studies
(b) Vol. **89**, No. 1, p. 235—240 (1978).
See also 1.2, p. 22; *14.4*, p. *250*; *20.1*, p. *332*; 22.5.4, p. 489.

F. Meseguer and C. Sánchez
Piezo-Optical Properties of Alkali Halides and Their Temperature Dependence
(b) Vol. **89**, No. 1, p. 281—288 (1978).
See also *20.1*, p. *332*.

M. Vanhaelst, J. Broeckx, P. Matthys, and E. Boesman
Spin–Lattic Relaxation of the V^{2+} Centre in NaCl, KCl, and RbCl
(b) Vol. **89**, No. 1, p. K25—K28 (1978).
See also *19*, p. *311*.

Debjani Bose and A. N. Basu
Second-Neighbour Interaction Potential in Alkali Halide Crystals
(b) Vol. **89**, No. 1, p. K35—K38 (1978).
See also *4*, p. *59*; *12.2*, p. *140*.

U. Holland and F. Lüty
Hydrostatic Pressure Effects on the Absorption and Emission of F_A(Li) Centers in KCl
(b) Vol. **89**, No. 1, p. K73—K77 (1978).
See also 12.2, p. 140; *13.4*, p. *180*; *20.3*, p. *347*.

G. Guillot and A. Nouailhat
Influence of the Purity of Crystals on the Stability of F-Centres and Interstitials in Alkali Halides
(b) Vol. **89**, No. 1, p. K85—K89 (1978).
See also *10.2*, p. *117*; *20.1*, p. *333*.

I. A. Parfianovich, E. E. Penzina, V. M. Khulugurov, V. M. Metzik, and L. M. Sobolev
Luminescence and Absorption of Colour Centres on RbCl: Eu and KCl: Eu Crystals
(b) Vol. **89**, No. 1, p. K91—K94 (1978).
See also *10.2*, p. *117*; 20.1, p. **333**; *20.3*, p. *347*.

F. Carlier and G. Jakobs
Transient Infrared Absorption of Bound Polarons in KBr
(b) Vol. **89**, No. 1, p. K95—K98 (1978).
See also *10.2*, p. *118*; 13.5.3, p. 200; *20.1*, p. *333*.

A. L. Guerrero, S. C. Jain, and P. L. Pratt
The Role of Dimers and Trimers in the Aggregation of Impurity–Vacancy Dipoles in Alkali Halides
(a) Vol. **49**, No. 1, p. 353—362 (1978).
See also *10.2*, p. *118*; *14.4*, p. *250*.

A. Ioan, V. Topa, and M. Giurgea
New Colour Centres in KCl:($Tl^+ + Ca^{2+}$) and KCl:($Tl^+ + Sr^{2+}$) Crystals
(b) Vol. **89**, No. 2, p. 513–520 (1978).
See also *10.2*, p. *118*; 20.1, p. **334**.

I. K. Amanis, J. G. Kliava and J. J. Purāns
New Paramagnetic Centres in Mixed Alkali Halides: Ag_c^0 (Br_1) in KCl:KBr
(b) Vol. **89**, No. 2, p. 531–535 (1978).
See also *10.2*, p. *119*; 19, p. **312**.

J. Poźniak
Dipolar Relaxation Inducted by an AC Field on Various Crystallographic Directions of Me^{2+}-Doped Alkali Halide Crystals (I): Nearest-Neighbour Approach
(b) Vol. **89**, No. 2, p. 577–582 (1978).
See also *10.2*, p. *119*; 14.4, p. **251**.

M. Dubiel, G. Berg, and F. Fröhlich
Aggregation of Point Defects in Highly Doped NaCl:Ca^{++} Crystals
(b) Vol. **89**, No. 2, p. 595–602 (1978).
See also *10.2*, p. *119*; 19, p. **312**.

A. Watterich and K. Polák
ESR Investigation of Isolated V^{2+} Ions in NaCl
(b) Vol. **89**, No. 2, p. K187–K190 (1978).
See also 11, p. 130; *19*, p. *313*.

J. Poźniak
Dipolar Relaxation Induced by an AC Field in Various Crystallographic Directions of Me^{2+}-Doped Alkali Halide Crystals (II): Next-Nearest Neighbour Approach
(b) Vol. **90**, No. 1, p. 261–267 (1978).
See also *10.2*, p. *119*; 14.4, p. **251**.

P. A. Varotsos
An Estimate of the Pressure Dependence of the Dielectric Constant in Alkali Halides
(b) Vol. **90**, No. 1, p. 339–343 (1978).
See also 4, p. 59; 6, p. 77; 12.2, p. 141; *14.4*, p. *251*; 22.5.1, p. **477**.

S. B. S. Sastry and K. Balasubramanyam
Thermoluminescence of Irradiated RbCl and RbCl:Sn Crystals
(b) Vol. **90**, No. 1, p. 375–383 (1978).
See also *10.2*, p. *120*; *20.3*, p. *349*.

Mahiradhwaj Singh
Resonance Scattering of Phonons in the Thermal Conductivity of CN Doped KCl
(b) Vol. **90**, No. 1, p. K43–K47 (1978).
See also *6*, p. *78*; *8*, p. *89*.

V. K. Jain and S. P. Kathuria
Z_3 Centre Thermoluminescence in LiF TLD Phosphor
(a) Vol. **50**, No. 1, p. 329–333 (1978).
See also *10.2*, p. *121*; 20.1, p. **336**; *20.3*, p. *350*.

W. Raupach
Atomistic Calculation of the Core Configuration of Screw Dislocations in Rock Salt Crystals
(a) Vol. **50**, No. 1, p. K81–K84 (1978).
See also *10.2*, p. *121*.

B. Rambabu, C. Ramasastry, and B. V. R. Chowdari
Optical Absorption of Off-Centre Cu^+ in CsBr
(b) Vol. **90**, No. 2, p. 465–473 (1978).
See also *10.2*, p. *121*; *20.1*, p. *337*.

S. Benci, R. Capelletti, F. Fermi, M. Manfredi, J. Z. Damm, and E. Mugeński
Spectral and Time Decay Analysis of Photoluminescence Excited in the A Band Region in KCl:Pb
(b) Vol. **90**, No. 2, p. 657–666 (1978).
See also 13.4, p. 185; *20.1*, p. *337*; *20.3*, p. *350*.

J.-J. Pilloud and C. Jaccard
ESR of (X_2^-) Centres in CsBr:Tl^+ and CsCl:Tl^+
(b) Vol. **90**, No. 2, p. K161–K164 (1978).
See also *19*, p. *314*.

L. C. Jen and J. F. Merklin
Optically Induced Dichroism of Z_2'-Centers in LiF:Mg, Ti (TLD-100)
(a) Vol. **50**, No. 2, p. 469–474 (1978).
See also *10.2*, p. *121*; *20.1*, p. *338*.

J. A. A. Khan and J. Leteurtre
An Experimental Investigation of the Stress Oriented Dislocation Loop Nucleation in Whiskers
(a) Vol. **50**, No. 2, p. K161–K164 (1978).
See also *10.2*, p. *122*; 11, p. 131.

H.-J. Kos and S. Mieke
Effect of Deformation on the Thermoluminescence of Titanium Doped Lithium Fluoride
(a) Vol. **50**, No. 2, p. K165–K168 (1978).
See also 11, p. 131; *12.2*, p. *141*; *20.3*, p. *351*.

22.5.3 Alkali-Earth Halides

Original Papers and Short Notes

R. Alcalá, V. M. Orera, and J. Beamonte
A New Type of F_3-Centre in Additively Coloured CaF_2
(b) Vol. **85**, No. 1, p. 283–290 (1978).
See also *10.2*, p. *109*; *20.1*, p. *320*.

P. J. Alonso and R. Alcalá
Optical Absorption of X-Irradiated CaF_2:Co
(b) Vol. **85**, No. 1, p. K23–K26 (1978).
See also *11*, p. *123*; *20.1*, p. *320*.

W. Bollmann
Association Parameters of [Me^{3+}–F_i^-] Dipoles in CaF_2 Crystals
(a) Vol. **45**, No. 1, p. 187–193 (1978).
See also *10.2*, p. *109*; 14.4, p. 248.

L. E. Oliveira, P. Murilo Oliveira, and B. Maffeo
F Centers in Alkaline-Earth Fluorides. Inadequacy of the Muffin-Tin Approximation
(b) Vol. **87**, No. 1, p. 25–30 (1978).
See also *10.2*, p. *112*; *20.1*, p. *325*.

Ch. S. Kuo and D. Schroeer
Mössbauer Study of Dy^{3+} in CaF_2
(b) Vol. **87**, No. 1, p. 325–331 (1978).
See also *5*, p. *61*; 10.2, p. 113.

R. Droste, R. Gmeiner, and R. Geick
Electronic Absorption of $SrCl_2$:Fe^{2+} and $SrCl_2$:Co^{2+}
(b) Vol. **87**, No. 2, p. 543–552 (1978).
See also *13.4*, p. *176*; *20.1*, p. *327*.

N. Guskos, J. Kuriata, and T. Rewaj
Spin Lattice Coefficients of Gd^{3+} in SrF_2 Single Crystals
(b) Vol. **87**, No. 2, p. K155–K158 (1978).
See also *19*, p. *308*.

R. Hultzsch
Passive Q-Switching of the Ruby Laser by Means of Colour Centres in SrF_2
(a) Vol. **47**, No. 2, p. 415–423 (1978).
See also 10.2, p. 114; *20.2*, p. *339*.

R. Capelletti, E. Okuno, G. E. Matthews, and J. H. Crawford, Jr.
ITC Spectra of Impurity Aggregate in CaF_2 Crystals Doped with Trivalent Rare-Earth Ions
(a) Vol. **47**, No. 2, p. 617–624 (1978).
See also *8*, p. *86*; 10.2, p. 114.

F. Krok, W. Bogusz, B. Jarosz, and W. Jakubowski
Electrode Polarization in CdF_2 Crystals
(a) Vol. **47**, No. 2, p. K103–K105 (1978).
See also *14.4*, p. *249*.

N. B. YUNUSOV and V. P. ZENTSOV
Electronic Structure and ESR of Ti^{2+} in Fluorite Type Crystals
(b) Vol. **88**, No. 1, p. 87–97 (1978).
See also *13.4*, p. *177*; 19, p. 309.

T. REWAJ and M. KRUPSKI
The Effect of Hydrostatic Stress and Temperature on the EPR Spectrum of Gd^{3+} Ion in CdF_2: $-b_4^0$ as a Function of Interionic Distances
(b) Vol. **88**, No. 1, p. K65–K68 (1978).
See also *19*, p. *309*.

W. H. ROBINSON, A. J. GLOVER, and A. WOLFENDEN
Electrical–Mechanical Coupling of Dislocations in KCl, NaCl, LiF, and CaF_2
(a) Vol. **48**, No. 1, p. 155–163 (1978).
See also *10.2*, p. *115*; 22.5.2, p. 482.

K. THÖRMER and F. LÜTY
A New Off-Center Defect System, RbCl:Li^+, Studied with Dielectric Techniques under Hydrostatic Pressure
(b) Vol. **90**, No. 1, p. 277–287 (1978).
See also 10.2, p. 120; *14.4*, p. *251*.

E. KAMINSKA and A. PIOTROWSKA
Thermal Stability of CdF_2 in Vacuum
(a) Vol. **50**, No. 1, p. 243–247 (1978).
See also *14.3.3*, p. *238*.

K. ATOBE
The Role of F-Centers in Thermoluminescence Processes in Reactor Irradiated SrF_2 Crystals
(a) Vol. **50**, No. 1, p. K77–K80 (1978).
See also *10.2*, p. *121*; 11, p. 130; *20.3*, p. *350*.

22.5.4 Mixed Crystals

Original Papers and Short Notes

R. Ethiraj, V. G. Krishna Murty, and K. G. Bansigir
Piezo-Optic Birefringence in Mixed Crystals of KCl–KBr
(a) Vol. **45**, No. 1, p. 83—87 (1978).
See also 12.2, p. 138; *20*, p. *315*.

M. V. Eremin and Yu. V. Rakitin
Channel Model in Isotropic Exchange Theory (III)
(b) Vol. **85**, No. 2, p. 783—788 (1978).
See also 1, p. 9; *13*, p. *143*.

T. Gotô, T. Fujimura, and K. Kamiyoshi
Ultrasonic Investigation of the Phase Transitions in $NH_4Cl_{1-x}Br_x$ Mixed Crystals
(a) Vol. **45**, No. 2, p. 513—519 (1978).
See also *1.2*, p. *16*; *7*, p. *80*; 14.4, p. 248.

J. C. Castro and M. de Souza
The OCN^- Vibrational Spectrum and the Phase Transitions in KCN_xKCl_{1-x} Single Crystals
(b) Vol. **86**, No. 1, p. 137—144 (1978).
See also *1.2*, p. *16*; *6*, p. *67*; 12.2, p. 138; 20.1, p. 321; 22.5, p. 474.

A. E. Kalnin and M. F. Trinkler
Spectroscopy of Mixed Alkali Halide Crystals Activated by Thallium
(b) Vol. **88**, No. 2, p. 445—455 (1978).
See also *13.4*, p. *178*; 20.1, p. 330; *20.3*, p. *346*.

A. G. Coutts and P. Day
Charge Transfer Spectra of Transition Metal Ions in Cadmium Halide Lattices
(b) Vol. **88** No. 2 p. 767—772 (1978).
See also *20.1*, p. *331*.

C. J. de Pater
Dielectric Constant and Ferroelectric Hysteresis of Rb_2ZnBr_4
(a) Vol. **48**, No. 2, p. 503—512 (1978).
See also *14.4.1*, p. *256*.

I. Földvári, R. Voszka, and Z. Morlin
The Properties of Ni Ions in NaCl Single Crystals (I): Vacuum Ultraviolet, Ionic Conductivity, and X-Ray Diffraction Studies
(b) Vol. **89**, No. 1, p. 235—240 (1978).
See also 1.2, p. 22; *14.4*, p. *250*; *20.1*, p. *332*; 22.5.2, p. 484.

H. Bill
Parametrization of the Measured Temperature Dependence of the Crystal Field Splitting in Axial ^8S State Ion Centers
(b) Vol. **89**, No. 1, p. K49—K52 (1978).
See also *13.4*, p. *180*; 19, p. 311.

V. C. Sahni and M. L. Bansal
Raman Study of the Mixed System $[(NH_4)_{1-x}K_x]_2CuCl_4 \cdot H_2O$
(b) Vol. **90**, No. 1, p. 415–422 (1978).
See also *6*, p. *78*; *20.1*, p. *336*.

V. B. Sharan, D. N. S. Srivastwa, and S. C. Sen
Diffuse Absorption Spectrum of NH_4Cl + TlCl Mixed Phosphors
(a) Vol. **50**, No. 1, p. 323–327 (1978).
See also *13.4*, p. *184*; *20.1*, p. *336*.

22.6 Simple Oxides (CdO, ZnO See 22.4)

Original Papers and Short Notes

V. K. Yarmarkin, S. P. Teslenko, and A. I. Knyazev
Mössbauer Study of Sb and Fe Impurities in Semiconducting and Insulating TiO_2
(a) Vol. **45**, No. 1, p. 63–69 (1978).
See also *5*, p. *60*; 9, p. 91; 10.2, p. 109.

E. I. Terukov, W. Reichelt, and H. Oppermann
On the Influence of Doping on the Phase Transition of V_3O_5
(a) Vol. **45**, No. 1, p. K77–K79 (1978).
See also *14.1*, p. *205*; *14.3.1*, p. *226*.

H. H. Boghossian and K. S. Dubey
Heat Transfer in a Lattice at Low Temperatures: Application to a Doped Sample of Al_2O_3
(b) Vol. **85**, No. 2, p. K99–K102 (1978).
See also *6*, p. *66*; *8*, p. *84*.

V. Cristea and V. Babes
The Seebeck Effect in Nb-Doped TiO_2 Rutile
(a) Vol. **45**, No. 2, p. 617–624 (1978).
See also 14.3, p. 218; *15*, p. *259*.

J. Narayan
a ⟨100⟩ Dislocation Loops in Magnesium Oxide
(a) Vol. **45**, No. 2, p. 625–630 (1978).
See also *10.2*, p. *110*.

C. Boulesteix, G. Nihoul-Boutang, and B. Yangui
Influence of Upper Laue Areas on Fine Structure of Bend Contours near a High Symmetry Axis
(a) Vol. **45**, No. 2, p. 639–645 (1978).
See also *1.1*, p. *11*; 1.4, p. 29.

V. Dolocan
Some Electrical Properties of Bi_2O_3 Thin Films
(a) Vol. **45**, No. 2, p. K155–K157 (1978).
See also *14.3.1*, p. *226*; 14.3.4, p. 241.

M. Hirose, A. Fischer, and K. Ploog
Growth of Al_2O_3 Layer on MBE GaAs
(a) Vol. **45**, No. 2, p. K175–K177 (1978).
See also *3*, p. *54*; *14.3.1*, p. *226*.

A. N. Trukhin
Investigation of the Photoelectric and Photoluminescent Properties of Crystalline Quartz and Vitreous Silica in the Fundamental Absorption Region. A Model for Electronic Structure and Migration of Energy in SiO_2
(b) Vol. **86**, No. 1, p. 67–75 (1978).
See also 2, p. 43; *13.1*, p. *152*; 16, p. 263; 20.3, p. 342.

R. Gründler, K. Breuer, and W. Tews

Optical Properties of Amorphous and Polycrystalline BeO Thin Films from Electron Energy Loss Measurements
(b) Vol. **86**, No. 1, p. 329—338 (1978).
See also 2, p. **43**; 13.1, p. 153; *14.4*, p. *248*; *20.1*, p. *322*.

J. Szńajd

Cross-Over from Discontinuous to Continuous Phase Transition under an External Magnetic Field
(b) Vol. **86**, No. 1, p. K11—K14 (1978).
See also *1.2*, p. *16*; 18.1, p. 277; 18.4, p. 298.

I. J. Dayawansa and C. F. Bohren

The Effect of Substrate and Aggregation on Infrared Extinction Spectra of MgO Particles
(b) Vol. **86**, No. 1, p. K27—K30 (1978).
See also *20.1*, p. *323*.

R. W. A. Hillhouse and J. Woods

Plasma Growth of Rutile Crystals and Their Photoelectronic Properties
(a) Vol. **46**, No. 1, p. 163—172 (1978).
See also *3*, p. *54*; 13.4, p. 173; 16, p. 263.

J. Narayan and S. M. Ohr

Residual Image Contrast of Dislocation Loops
(a) Vol. **46**, No. 1, p. 281—284 (1978).
See also *1.4*, p. *29*; *10.1*, p. *101*.

R. J. Gaboriaud, M. F. Denanot, M. Boisson and J. Grilhé

Study of Dislocations by Electron Microscopy in Plastically Deformed Single-Crystalline Yttrium Oxide (Y_2O_3)
(a) Vol. **46**, No. 1, p. 387—396 (1978).
See also *10.2*, p. *111*.

K. Hübner and A. Lehmann

Chemical Bond and Related Properties of SiO_2 (IV): Structure of Amorphous SiO_2
(a) Vol. **46**, No. 2, p. 451—457 (1978).
See also 2, p. 44; *4*, p. *57*.

A. E. Gershinskii, L. V. Mironova, and E. I. Cherepov

On the Conductivity of the Si–Thermal SiO_2–Electrolyte System in Superhigh Electric Fields
(a) Vol. **46**, No. 2, p. 467—473 (1978).
See also *14.3.1*, p. *226*; *14.3.4*, p. *241*; 22.1.2, p. 432.

A. D. Buonaquisti and R. A. Collins

Energy Distribution of Electrons Emited from MIM Thin-Film Sandwich Cathodes
(a) Vol. **46**, No. 2, p. K91—K93 (1978).
See also 14.3.3, p. 235; *17*, p. *269*.

G. Smith, E. R. Vance, Z. Hasan, A. Edgar, and W. A. Runciman

A Charge Transfer Mechanism for the Colour of Rose Quartz
(a) Vol. **46**, No. 2, p. K135—K140 (1978).
See also *20.1*, p. *325*.

A. GOLTZENE and C. SCHWAB
EPR Investigation of the Photomemory Effect of Cu_2O
(a) Vol. **46**, No. 2, p. K157—K161 (1978).
See also *19*, p. *307*.

E. BURZO and L. STĂNESCU
On the Charge Compensation of Iron Ions in the V_2O_5 Lattice
(a) Vol. **46**, No. 2, p. K163—K166 (1978).
See also *5*, p. *61*.

S. NIKITINE, S. G. ELKOMOSS, and C. SCHWAB
Study of Linewidth and Exciton–Impurity Scattering in Cu_2O
(b) Vol. **87**, No. 1, p. 261—272 (1978).
See also *13.5.1*, p. *191*; *20.1*, p. *326*.

R. S. DE BIASI
Forbidden $\Delta m_s = 2$ Transitions in the EPR Spectrum of Fe^{3+} in MgO
(b) Vol. **87**, No. 1, p. K29—K31 (1978).
See also *19*, p. *307*.

S. V. KOPTSIK, V. A. POVITSKII, M. YU. MESSINEV, YU. V. BALDOKHIN, and N. F. SHCHUKIN
Mössbauer Study on Shocked Hematite
(a) Vol. **47**, No. 1, p. 39—43 (1978).
See also *5*, p. *61*; 18.4, p. 300.

A. JEŻOWSKI and J. RAFAŁOWICZ
Heat Flow Asymmetry on a Junction of Quartz with Graphite
(a) Vol. **47**, No. 1, p. 229—232 (1978).
See also *8*, p. *85*.

B. BALLAND and P. PINARD
Interfacial Tunnel Interaction in MOS Elements in Depletion
(a) Vol. **47**, No. 1, p. 251—258 (1978).
See also *13.4*, p. *175*; *14.3.2*, p. *230*.

B. YANGUI, M. REKIK, G. NIHOUL-BOUTANG, C. JOUANIN, C. BOULESTEIX, and D. RENARD
Theroretical and Experimental Study of the Transmitted Intensity as a Function of Thickness when the Incident Beam Is Parallel to a Crystallographic Axis
(a) Vol. **47**, No. 1, p. 311—320 (1978).
See also *1.1*, p. *12*; 1.4, p. 30.

C. E. WICKERSHAM and J. E. GREENE
The Effect of Substrate Bias on the Electrical and Optical Properties of In_2O_3 Films Grown by RF Sputtering
(a) Vol. **47**, No. 1, p. 329—337 (1978).
See also 13.4, p. 175; *14.3.1*, p. *226*; *20.1*, p. *327*.

W. BRÜCKNER and W. MOLDENHAUER
Resistivity–Temperature Characteristic and Thermal Switching
(a) Vol. **47**, No. 1, p. K99—K102 (1978).
See also *8*, p. *86*; *14.3.4*, p. *242*.

M. DAVID, G. VAN TENDELOO, J. VAN LANDUYT, and R. GEVERS
The Critical Voltage Effect in Transmission Electron Microscopy (X): Experimental Observations in the Presence of Non-Systematic Reflections
(b) Vol. **87**, No. 2, p. 419–432 (1978).
See also *1.1*, p. *13*; 22.1.1, p. 425.

R. RUPPIN
Validity Range of the Maxwell-Garnett Theory
(b) Vol. **87**, No. 2, p. 619–624 (1978).
See also 6, p. 71; *20.1*, p. *328*; 22.2.1, p. 446.

M. VANHAELST and P. CLAUWS
EPR Spectrum of the Oxygen Vacancy in Single Crystals V_2O_5
(b) Vol. **87**, No. 2, p. 719–723 (1978).
See also 10.2, p. 113; *19*, p. *308*.

R. P. GOYAL and S. C. GOYAL
Dielectric Behaviour of Transition Metal Oxides
(b) Vol. **87**, No. 2, p. K119–K122 (1978).
See also *14.4*, p. *249*.

R. PRAKASH, L. K. BANSAL, and S. C. GOYAL
Photoelastic Constants of the Alkaline Earth Oxides
(b) Vol. **87**, No. 2, p. K127–K131 (1978).
See also *4*, p. *58*; *12.2*, p. *139*.

S. NEOV, I. GERASSIMOVA, K. KREZHOV, B. SYDZHIMOV, and V. KOZHUKHAROV
Atomic Arrangement in Tellurite Glasses Studied by Neutron Diffraction
(a) Vol. **47**, No. 2, p. 743–750 (1978).
See also *2*, p. *46*; 4, p. 58.

W. BRÜCKNER, W. MOLDENHAUER, and H. OPPERMANN
Conductivity Transitions in (Ga + Mo)-Doped VO_2
(a) Vol. **47**, No. 2, p. K129–K132 (1978).
See also *1.2*, p. *20*; *14.3*, p. *221*.

V. LEMOS, F. CERDEIRA, and M. CARDONA
Atomic Motions Corresponding to Zone Center Phonons in Paratellurite
(b) Vol. **88**, No. 1, p. 199–206 (1978).
See also *6*, p. *73*; *20.1*, p. *329*.

K. HÜBNER
Chemical Bond and Related Properties of SiO_2 (V): Pseudopotential Study of Bonding and Structure
(a) Vol. **48**, No. 1, p. 147–154 (1978).
See also 2, p. 47; *4*, p. *59*; *13.1*, p. *157*; 14.4, p. 249.

J. RUZYŁŁO
Model of Structural Transformations on the Chemically Etched Silicon Surface during Thermal Oxidation
(a) Vol. **48**, No. 1, p. 199–202 (1978).
See also *1.2*, p. *21*; 1.6, p. 41; *3*, p. *55*; 22.1.2, p. 434.

V. N. Abramov, B. G. Ivanov, A. I. Kuznetsov, I. A. Meriloo, and M. I. Musatov
On the Vacuum Ultraviolet Transparency of Sapphire
(a) Vol. **48**, No. 2, p. 287–292 (1978).
See also *20.1*, p. *331*.

P. Nath, J.-E. Sundgren, and H. T. G. Nilsson
Structural and Mechanical Properties of Composite Films of Iron Particles in an Oxide Matrix
(a) Vol. **48**, No. 2, p. 345–355 (1978).
See also *1.5*, p. *37*; 10.2, p. 116.

E. I. Terukov, W. Reichelt, M. Wolf, H. Hemschik, and H. Oppermann
On the Influence of the Substisution of O^{16} by O^{18} in a Few Vanadium Oxides on Their Semiconductor–Metal Transition
(a) Vol. **48**, No. 2, p. 377–381 (1978).
See also *1.2*, p. *21*; 6, p. 75; 14.3, p. 223.

J. P. Huvenne, G. Vergoten, B. Boniface, G. Fleury, and P. Legrand
Low-Frequency Vibrations of Lead Oxide Pb_3O_4. Relationship between the Development of Spectra and Ferroelectric and Ferroelastic Phase Transitions
(a) Vol. **48**, No. 2, p. 417–423 (1978).
See also *6*, p. *75*; *20.1*, p. *331*.

M. Treilleux, P. Thévenard, G. Chassagne, and L. W. Hobbs
Observation of Implanted Potassium Aggregates in MgO Single Crystals
(a) Vol. **48**, No. 2, p. 425–430 (1978).
See also *1.4*, p. *32*; 11, p. 128.

A. Haydar and A. Coret
Excitons and Structural Defects in Cu_2O at Low Temperature
(a) Vol. **48**, No. 2, p. 465–471 (1978).
See also *13.5.1*, p. *194*; *16*, p. *265*.

K. I. Kirov, S. S. Georgiev, E. V. Gerova, and S. P. Aleksandrova
Investigation of SiO_2 Layers Deposited by Plasma Decomposition of Tetra-Ethoxy Silane in a Planar Reactor
(a) Vol. **48**, No. 2, p. 609–614 (1978).
See also *1.5*, p. *38*; *14.3.1*, p. *228*.

M. Jedlička and V. Valvoda
The Macroscopic Structure of PbO Layers
(a) Vol. **48**, No. 2, p. K101–K104 (1978).
See also *1.5*, p. *38*.

Yu. V. Bogatyrev and F. P. Korshunov
Investigation of Properties of the Si–SiO_2 Interface in MOS Structures Irradiated by Fast Electrons
(a) Vol. **48**, No. 2, p. K127–K130 (1978).
See also *11*, p. *128*; *14.3.2*, p. *231*; 22.1.2, p. 436.

V. N. Andreev, F. A. Chudnovskii, A. V. Petrov, and E. I. Terukov
Thermal Conductivity of VO_2, V_3O_5, and V_2O_3
(a) Vol. **48**, No. 2, p. K153–K156 (1978).
See also *8*, p. *88*.

A. M. Kapitonov, E. M. Smokotin, A. P. Korenko, and G. N. Stepanov
Ultrasonic-Wave Velocities in Cr_2O_3 near the Magnetic Phase Transition
(a) Vol. **48**, No. 2, p. K203—K206 (1978).
See also *1.2*, p. *22*; 7, p. 82; *18*, p. *275*.

G. A. Puchkovskaya, V. L. Strizhevskii, Yu. A. Frolkov, N. M. Chepilko, and Yu. N. Yashkir
Crystal Optics of Surface Polaritons in Anisotropic Crystals
(b) Vol. **89**, No. 1, p. 27—36 (1978).
See also *13.5*, p. *188*; 20.1, p. 331.

H. H. Boghossian and K. S. Dubey
Heat Transfer in Doped Al_2O_3 at Low Temperatures
(b) Vol. **89**, No. 1, p. K65—K67 (1978).
See also *6*, p. *76*; *8*, p. *88*.

F. Dettmann
Properties of Quantum Interference Devices with V–VO_x–Pb Tunnel Junctions
(a) Vol. **49**, No. 1, p. 293—298 (1978).
See also *14.2*, p. *215*.

U. Saxena and O. N. Srivastava
Effect of Nonstoichiometry on the Electrical Conductivity of Erbium Oxide Films
(a) Vol. **49**, No. 1, p. 371—378 (1978).
See also *1.5*, p. *38*; 2, p. 49; 13.1, p. 158; *14.3.1*, p. *228*.

J. Llopis and J. Piqueras
Cathodoluminescence from Grain Boundaries in MgO and GaP
(a) Vol. **49**, No. 1, p. K9—K12 (1978).
See also *10.2*, p. *118*; *20.3*, p. *348*; 22.2.2, p. 451.

E. Wimmer, P. Weinberger, A. Kosakov, and G. Leonhardt
On the Electronic Structure of As_2O_3, Sb_2O_3, and As_2S_3
(b) Vol. **89**, No. 2, p. 619—624 (1978).
See also 2, p. 50; *13*, p. *148*; *20*, p. *317*.

P. P. Tirodkar and R. Pratap
Equivalent Rectangular Shape of a Potential Barrier in a Tunnel Junction
(a) Vol. **49**, No. 2, p. 637—641 (1978).
See also *14.3.3*, p. *237*; 14.4, p. 251.

L. Trépied and J.-C. Doukhan
Dissociated **a**-Dislocations in Quartz. Influence on Plastic Deformation
(a) Vol. **49**, No. 2, p. 713—724 (1978).
See also *1.4*, p. *33*; *10.2*, p. *119*.

A. Mansingh, R. Singh, and M. Sayer
Dielectric Behaviour of Vanadium Dioxide
(a) Vol. **49**, No. 2, p. 773—779 (1978).
See also *14.4*, p. *251*.

V. A. Izvozchikov, G. A. Bordovskii, V. T. Avanesyan and V. A. Bordovskii
On the Problem of Ferroelectricity in Photoconducting Lead Oxides
(a) Vol. **49**, No. 2, p. K173—K175 (1978).
See also *14.4.1*, p. *256*.

H. J. de Bruin and S. P. S. Badwal
Faradaic Reaction Kinetics in Solid Electrolytes by Impedance Dispersion Analysis
(a) Vol. **49**, No. 2, p. K181—K184 (1978).
See also *13.3*, p. *167*; *14.3.4*, p. *246*.

G. Gobsch, H. Haberlandt, H.-J. Weckner, and J. Reinhold
Calculation of the g-Tensor and ^{29}Si Hyperfine Tensors of the E_1' Centre in Silicon Dioxide
(b) Vol. **90**, No. 1, p. 309—317 (1978).
See also *13.4*, p. *183*; *19*, p. *313*.

P. Iacconi, D. Lapraz, and R. Caruba
Traps and Emission Centres in Thermoluminescent ZrO_2
(a) Vol. **50**, No. 1, p. 275—283 (1978).
See also 10.2, p. 121; 13.4, p. 184; *20.3*, p. *350*.

K. I. Kirov, E. D. Atanasova, and N. A. Ivanov
Investigation of Al_2O_3 Layers Obtainied by DC Planar-Magnetron Sputtering
(a) Vol. **50**, No. 1, p. 363—368 (1978).
See also *1.5*, p. *39*; *14.3.3*, p. *238*; 14.4, p. 252.

L. Trepied and J.-C. Doukhan
Some Precisions on the Brazil Twins in Quartz
(a) Vol. **50**, No. 1, p. K37—K41 (1978).
See also *1.4*, p. *34*.

J. K. Wysocki and C. M. Vodenicharov
Approximate Formula for the Emission Current Density in the Second Intermediate Region of Christov
(a) Vol. **50**, No. 2, p. 411—415 (1978).
See also *14.3.3*, p. *238*; *17*, p. *271*.

R. E. Blaha and W. R. Fahrner
Technology and Reliability of Thermal SiO_2 Layers as Used for Passivating Silicon Power Devices
(a) Vol. **50**, No. 2, p. 551—562 (1978).
See also 11, p. 131; *14.3.1*, p. *229*; *14.3.3*, p. *238*.

A. A. Konova and M. G. Michailov
Crystalline Anodic Silicon Dioxide on Silicon
(a) Vol. **50**, No. 2, p. 627—633 (1978).
See also *1.5*, p. *39*.

S. Kawamura and B. S. H. Royce
Thermally Stimulated Current Studies of Electron and Hole Traps in Single Crystal Al_2O_3
(a) Vol. **50**, No. 2, p. 669—677 (1978).
See also *13.4*, p. *186*; *14.3*, p. *225*.

L. N. Skuja, A. R. Silin, and J. Mareš
Decay Time and Polarization Properties of Luminescence Centers in Vitreous Silica
(a) Vol. **50**, No. 2, p. K149—K152 (1978).
See also *2*, p. *53*; *20.3*, p. *350*.

S. E. Ormrod and D. L. Kirk
The Effect of Path Tortuosity and Anisotropy upon the Diffusion of Ions through Polycrystalline β Aluminas
(a) Vol. **50**, No. 2, p. K193—K196 (1978).
See also *9*, p. *95*.

22.6.1 Mixed Crystals

Original Papers and Short Notes

E. WIESER, W. BRÜCKNER, B. THUSS, and U. GERLACH
Mössbauer Effect Studies on Modified VO_2. $V_{1-x}Fe_xO_{2-y}$ and $V_{1-x-z}Fe_xMo_zO_2$
(a) Vol. **45**, No. 1, p. 123–131 (1978).
See also *1.2*, p. *15*; *5*, p. *60*; 18.1, p. 277.

A. CZAPLA, M. JACHIMOWSKI, E. KUSIOR, and J. SZCZYRBOWSKI
Optical Properties of $Pb_xSn_{1-x}O_2$ Films
(a) Vol. **45**, No. 2, p. 537–542 (1978).
See also *20.1*, p. *321*.

M. K. ROY
Coupling of Surface Acoustic Waves Propagating in Two Separated Piezoelectric Media
(a) Vol. **45**, No. 2, p. 549–552 (1978).
See also *7*, p. *80*; *13.3*, p. *164*; 14.3.2, p. 230; 22.4.1, p. 463.

D. GOURIER, D. VIVIEN, J. THERY, J. LIVAGE, and R. COLLONGUES
ESR of Copper-Doped β-Alumina. Localization of Cu^{2+} and Linewidth Effect
(a) Vol. **45**, No. 2, p. 599–606 (1978).
See also *19*, p. *305*.

R. NAGARAJAN and J. K. SRIVASTAVA
Anomalous Mössbauer Lineshapes in a Magnetically Ordered Ga_2O_3–Fe_2O_3 System in Presence of an External Electric Field Applied during the Solid Solution Formation
(b) Vol. **89**, No. 1, p. K79–K83 (1978).
See also *5*, p. *62*; *18.1*, p. *279*.

B. LORENZ
On the Magnetic Susceptibility of $(Ti_{1-x}V_x)_4O_7$
(b) Vol. **89**, No. 1, p. K99–K101 (1978).
See also *18.1*, p. *279*.

J. K. SRIVASTAVA and B. W. DALE
Cross Relaxation Studies in the Cr^{3+}-Doped Magnetically Dilute Paramagnetic Al_2O_3–Fe_2O_3 System (I): Effect of Cr^{3+} Concentration
(b) Vol. **90**, No. 1, p. 391–399 (1978).
See also *5*, p. *62*; 13.4, p. 184.

J. K. SRIVASTAVA and B. W. DALE
Cross Relaxation Studies in the Cr^{3+}-Doped Magnetically Dilute Paramagnetic Al_2O_3–Fe_2O_3 System (II): Effect of Temperature and External Magnetic Field
(b) Vol. **90**, No. 2, p. 571–574 (1978).
See also *5*, p. *62*; 13.4, p. 185.

J. de Oliveira, O. de O. Damasceno, J. R. F. Ferreira, A. L. de Oliveira, J. D. Fabris, P. Boyer, and A. Baudry

Electric Field Gradient Measurements at ^{181}Ta Nuclei in HfO_2–CaO and HfO_2–MgO Systems
(b) Vol. **90**, No. 2, p. K169—K174 (1978).
See also *10.2*, p. *121*; *13*, p. *149*; 14.4, p. 252.

J. K. Srivastava and S. Muralidhara Rao

Search for Reclustering of Fe^{3+} Ions in the Incompletely Annealed Cr_2O_3–Fe_2O_3 Solid Solutions
(b) Vol. **90**, No. 2, p. K175—K177 (1978).
See also *5*, p. *63*.

22.7 Semiconducting Intermetallic Compounds

Review Article

R. Suryanarayanan
Valence Related Optical and Other Studies of Sm and Tm Chalcogenides
(b) Vol. **85**, No. 1, p. 9–43 (1978).
See also 13.1, p. 150; 14.1, p. 205; 14.3, p. 217; *20.1*, p. *319*.

Original Papers and Short Notes

V. Wagner, G. Dolling, B. M. Powell, and G. Landwehr
Lattice Vibrations of Bi_2Te_3
(b) Vol. **85**, No. 1, p. 311–317 (1978).
See also *6*, p. *65*; *20.1*, p. *320*.

H. Neumann and A. Kosakov
Tight-Binding Calculation of the Density of Valence States of SnS, SnSe, and SnTe
(b) Vol. **85**, No. 1, p. K11–K14 (1978).
See also *13.1*, p. *150*; 17, p. 269.

J. Gazsó and J. Hajtó
Self-Controlled Laser Beam Chopping Effect in $GeSe_2$ Thick Films
(a) Vol. **45**, No. 1, p. 181–186 (1978).
See also 2, p. 42; 11, p. 123; *20.1*, p. *320*.

K. V. S. Rama Rao, T. Rajasekharan, V. Seshubai, and C. Ramasastry
Ferromagnetic Resonance Study in Thin Films of $Mn_{1-x}Cu_xSb$
(a) Vol. **45**, No. 1, p. K51–K53 (1978).
See also *19*, p. *303*.

A. Aladgadgyan and A. Toneva
Study of the Valence Band Structure of $Pb_{1-x}Sn_xTe$
(b) Vol. **85**, No. 2, p. K131–K134 (1978).
See also *13.1*, p. *152*; 14.3, p. 218.

P. A. Alekseev, I. P. Sadikov, I. A. Markova, and O. D. Chistyakov
On the Effects of the Crystalline Electric Field in $PrSn_3$
(b) Vol. **86**, No. 1, p. K59–K62 (1978).
See also *6*, p. *68*; *18*, p. *273*.

W. Eichler and G. Simon
Galvanomagnetic Transport Properties of Sb_2Te_3
(b) Vol. **86**, No. 1, p. K85–K88 (1978).
See also *14.3*, p. *218*.

N. N. Koren, V. V. Krasnova, and E. E. Matyas
Investigation of Alloys of the System GeS–GeSe
(a) Vol. **46**, No. 1, p. K1–K3 (1978).
See also *1.2*, p. *17*; *1.3*, p. *25*.

L. Vescan, H. Stötzel, T. Stoica, and A. Kottwitz
Photoconductivity in a-GeSe
(b) Vol. **87**, No. 1, p. 239–246 (1978).
See also 2, p. 44; *16*, p. *264*.

R. Stegmann and G. Oelgart
Hybrid Resonance in a Quantizing Magnetic Field in BiSb
(b) Vol. **87**, No. 2, p. 485—489 (1978).
See also *13.2*, p. *162*; *19*, p. *308*.

V. K. Bashenov, A. M. Mutal, and V. V. Timofeenko
Valence-Band Density of States for Mg_2Si from Pseudopotential Calculation
(b) Vol. **87**, No. 2, p. K77—K79 (1978).
See also *13.1*, p. *155*.

T. Grochulski, K. Leibler, and A. Sienkiewicz
Magnetic Resonance in MnTe at High Hydrostatic Pressures
(a) Vol. **47**, No. 2, p. K169—K171 (1978).
See also *19*, p. *309*.

W. Kraak, G. Oelgart, G. Schneider, and R. Herrmann
The Semiconductor–Semimetal Transition in $Bi_{1-x}Sb_x$ Alloys with $x \geqq 0.22$
(b) Vol. **88**, No. 1, p. 105—110 (1978).
See also *13.4*, p. *177*; 14.1, p. 208; *14.3*, p. *221*.

Mahiradhwaj Singh
Electron–Phonon Interaction and Phonon Conductivity of the n-Type Semiconductors Mg_2Ge and Mg_2Sn at Low Temperatures
(b) Vol. **90**, No. 1, p. 385—389 (1978).
See also *6*, p. *77*; 8, p. 89.

H. Köhler and W. Ossau
Magneto-Seebeck and Shubnikov-de Haas Effect in n-Type $(Bi_{1-x}Sb_x)_2Te_3$
(b) Vol. **90**, No. 2, p. 509—515 (1978).
See also *13.1*, p. *160*; 14.3, p. 225; *15*, p. *261*.

22.8 Tertiary and Higher Compounds

Review Article

P. P. Seregin, F. S. Nasredinov, and L. N. Vasilev
A Study of Radiation Defects in Solids by Means of Mössbauer Spectroscopy
(a) Vol. **45**, No. 1, p. 11–45 (1978).
See also *5*, p. *60*; *11*, p. *123*; 22.5, p. 474; 22.9, p. 522.

Original Papers and Short Notes

W. Hörig, H. Neumann, B. Schumann, and G. Kühn
Optical Properties of $CuGaSe_2$ near and above the Fundamental Absorption Edge
(b) Vol. **85**, No. 1, p. K57–K61 (1978).
See also *13.1*, p. *150*; *20.1*, p. *320*.

S. Grande, H.-D. Mecke, and St. Idziak
High-Resolution Proton Magnetic Resonance in Single Crystals of Potassium Trihydrogen Selenite
(a) Vol. **45**, No. 1, p. K29–K32 (1978).
See also *4*, p. *57*; *19*, p. *303*.

C. J. Schinkel, Z. Żołnierek, and R. Troć
High Field Magnetization of U_2N_2Te
(a) Vol. **45**, No. 1, p. K39–K41 (1978).
See also *18.2*, p. *282*.

M. C. Bonnet, E. Antic-Fidancev, P. Porcher, M. Lemaitre-Blaise, and P. Caro
The Optical Spectra of Rare-Earth Cobalticyanides
(b) Vol. **85**, No. 2, p. 587–596 (1978).
See also *20.1*, p. *321*; *20.3*, p. *342*.

F. W. Boswell, A. Prodan, W. R. Vaughan, and J. M. Corbett
On the Ordering of Fe Atoms in Fe_xNbS_2
(a) Vol. **45**, No. 2, p. 469–481 (1978).
See also *1.2*, p. *16*; *1.3*, p. *25*.

M. Krupski and J. Stankowski
New High-Pressure Phase in $[Ni(NH_3)_6]\,(ClO_4)_2$ Detected by EPR
(a) Vol. **45**, No. 2, p. 585–589 (1978).
See also *1.2*, p. *16*; 19, p. 304.

S. Niziol, H. Bińczycka, A. Szytuła, J. Todorović, R. Fruchart, J. P. Senateur and D. Fruchart
Magnetic Structure of MnCoSi
(a) Vol. **45**, No. 2, p. 591–597 (1978).
See also *18.2*, p. *283*; *18.4*, p. *298*.

G. A. Medvedkin, Yu. V. Rud, Yu. A. Valov, and V. I. Sokolova
Photoconductivity of p-Type $CdSnP_2$ Single Crystals
(a) Vol. **45**, No. 2, p. K95–K99 (1978).
See also *16*, p. *262*.

N. M. Gasanly, B. N. Mavrin, Kh. E. Sterin, V. I. Tagirov, and Z. D. Khalafov
Raman Study of Layer $TlGaS_2$, β-$TlInS_2$, and $TlGaSe_2$ Crystals
(b) Vol. **86**, No. 1, p. K49—K53 (1978).
See also *6*, p. *68*; *20.1*, p. *323*.

W. K. Unger, H. Meuth, J. C. Irwin, and H. Pink
Photoluminescence and Resonant Raman Scattering in $CdIn_2S_4$ and $ZnIn_2S_4$
(a) Vol. **46**, No. 1, p. 81—87 (1978).
See also *20.1*, p. *323*; *20.3*, p. *343*.

S. Unterricker and J. Hausbrand
PAC Examination of Radiation Damaged $CdSiP_2$
(a) Vol. **46**, No. 1, p. 125—130 (1978).
See also 10.2, p. 110; *11*, p. *125*.

E. A. Tsvetkov
Calculation of the Optical Fluorescence Intensities and Crystal Field Parameters of Nd^{3+} in Calcium Fluorophosphate
(a) Vol. **46**, No. 1, p. 311—314 (1978).
See also *13.4*, p. *173*; 20.2, p. 339; 20.3, p. 343.

M.-F. Charlot, J.-J. Girerd, and O. Kahn
Molecular Orbital Approach of Antiferromagnetic Coupling in Linear Chains $AMCl_3$ with $M = V$, Mn, and Ni and $A = Cs$ or $N(CH_3)_4$
(b) Vol. **86**, No. 2, p. 497—504 (1978).
See also *18.4*, p. *299*.

I. V. Bodnar, A. G. Karoza, and G. F. Smirnova
Infrared Reflectivity Spectra of $CuGa_xIn_{1-x}S_2$ Solid Solutions
(b) Vol. **86**, No. 2, p. K171—K174 (1978).
See also *6*, p. *70*; *20.1*, p. *325*.

R. M. Henson, R. R. Zeyfang, and E. Linhart
Pyroelectric Properties of Pb(Ti, Zr, $[Fe_{1/2}Ta_{1/2}])O_3$ Polycrystalline Solid Solutions
(a) Vol. **46**, No. 2, p. 511—515 (1978).
See also *14.4.1*, p. *254*.

S. Grande, H.-D. Mecke, and L. A. Shuvalov
Deuterium Resonance of $KD_3(SeO_3)_2$ Single Crystals above and below Phase Transition
(a) Vol. **46**, No. 2, p. 547—555 (1978).
See also *1.2*, p. *17*; *4*, p. *57*; 19, p. 306.

L. Baldassarre, V. Capozzi, G. Maggipinto, and A. Minafra
Space Charge Limited Current Measurements in $ZnIn_2S_4$
(a) Vol. **46**, No. 2, p. 589—593 (1978).
See also *14.3.4*, p. *241*.

H. H. Heikens, R. S. Kuindersma, C. F. van Bruggen, and C. Haas
Magnetic Properties of $MnEr_2S_4$, $MgEr_2S_4$, and $Mn_xMg_{1-x}Y_2S_4$
(a) Vol. **46**, No. 2, p. 687—695 (1978).
See also *18.1*, p. *278*; *18.4*, p. *299*.

S. Siek, A. Szytuła, and J. Leciejewicz
The Crystal and Magnetic Structure of $CeMn_2Si_2$
(a) Vol. **46**, No. 2, p. K101—K105 (1978).
See also *1.1*, p. *12*; *18.4*, p. *299*.

R. S. Saraswat and G. C. Upreti
Estimation of T_1 of Fast Relaxing Co^{2+} Ions from the Observed EPR Linewidth of Mn^{2+} and Vo^{2+} Impurities in Some Cobalt Salt Single Crystals
(b) Vol. 87, No. 1, p. K39—K42 (1978).
See also *19*, p. *307*.

A. Migdał-Mikuli, E. Mikuli, M. Rachwalska, and S. Hodorowicz
Investigation of the Phase Situation in Polycrystalline $[Mg(NH_3)_6](ClO_4)_2$ by Adiabatic Calorimetry and X-Ray Diffraction Methods
(a) Vol. 47, No. 1, p. 57—64 (1978).
See also *1.2*, p. *18*; 8, p. 85.

J. Stasz
ESR Line Intensity near the Curie Point in $CdCr_2Se_4$, $HgCr_2Se_4$, and $CdCr_2S_4$ Single Crystals
(a) Vol. 47, No. 1, p. K23—K26 (1978).
See also *18.2*, p. *286*; *19*, p. *308*.

A. Kobayashi and K. Suzuki
High Field Transport Properties of a Magnetic Semiconductor
(a) Vol. 47, No. 1, p. K51—K53 (1978).
See also *14.3.4*, p. *242*; *18.2*, p. *286*.

B. I. Swanson
Displacive Phase Transformations in K_2SnBr_6
(a) Vol. 47, No. 1, p. K95—K98 (1978).
See also *1.2*, p. *19*; 6, p. 71; 20.1, p. 327.

M. A. B. Whitaker
The Motional Effect in Nuclear Double-Resonance
(b) Vol. 87, No. 2, p. 689—695 (1978).
See also 6, p. 72; *19*, p. *308*.

K. J. Michalski
Magnetic Symmetry of Ordered Phases in Mn_3GaC
(b) Vol. 87, No. 2, p. K85—K87 (1978).
See also *18.2*, p. *287*; 18.4, p. 300.

A. I. Merkulov, S. I. Radautsan, and V. E. Tezlevan
The Effect of the Degree of Doping on Magneto-Electrical Properties of n-$CdCr_2Se_4$ Monocrystals
(b) Vol. 87, No. 2, p. K141—K143 (1978).
See also *14.3*, p. *221*.

H. Ptasiewicz-Bąk, J. Leciejewicz, and A. Zygmunt
Magnetic Ordering in UGeY (Y = S, Se, Te) by Neutron Diffraction
(a) Vol. 47, No. 2, p. 349—356 (1978).
See also *18.2*, p. *287*; *18.4*, p. *300*.

N. Terao
Experimental Study of Crack Propagation in Brittle Amorphous Solids, Bifurcation of a Crack in Glass
(a) Vol. 47, No. 2, p. 565—575 (1978).
See also *2*, p. *45*; *12.2*, p. *140*.

V. M. Ryzhkovskii
On Structural and Magnetic Properties of the Mn_2As–Mn_2Sb Quasibinary System
(a) Vol. 47, No. 2, p. K177—K180 (1978).
See also 1.2, p. 20; *18.3*, p. *296*; *18.4*, p. *300*.

K. Allakhverdiev, R. Sardarly, F. Wondre, and J. F. Ryan
Raman and Infrared Spectra of $TlGaSe_2$
(b) Vol. 88, No. 1, p. K5—K9 (1978).
See also *6*, p. *73*; *20.1*, p. *329*.

M. A. Arriandiaga, M. J. Tello, J. Fernandez, H. Arend, and J. Roos
Calorimetric Study of Order–Disorder Phase Transitions in Long-Chain $(C_nH_{2n+1}NH_3)_2CuCl_4$ Compounds
(a) Vol. 48, No. 1, p. 53—57 (1978).
See also *1.2*, p. *21*.

F. Holuj and A. Jesmanowicz
EPR of Fe^{3+} in Andalusite and Kyanite at V-Band and the Pair Spectra
(a) Vol. 48, No. 1, p. 191—198 (1978).
See also 13.4, p. 177; *19*, p. *310*.

R. Florian, J. Pelzl, M. Rosenberg, H. Vargas, and R. Wernhardt
Photoacoustic Detection of Phase Transitions
(a) Vol. 48, No. 1, p. K35—K38 (1978).
See also *1.2*, p. *21*; *7*, p. *81*.

E. Ziegler, W. Siegel, G. Kühnel, and E. Buhrig
Incorporation of Gallium in $ZnSiP_2$
(a) Vol. 48, No. 1, p. K63—K65 (1978).
See also 13.4, p. 178; *14.3*, p. *222*.

V. E. Petrashen, Yu. V. Yablokov, and R. L. Davidovich
Electron Paramagnetic Resonance Study of Cu^{2+} in $K_2Zn(ZrF_6)_2 \cdot 6\,H_2O$
(b) Vol. 88, No. 2, p. 439—443 (1978).
See also *13.4*, p. *178*; *19*, p. *310*.

R. Chandra, R. P. Tripathi, and S. Lokanathan
Search for Anisotropy of Mössbauer Absorption in Biotite
(b) Vol. 88, No. 2, p. 633—637 (1978).
See also *5*, p. *62*.

M. Rachwalska and T. Stanek
An Adiabatic Calorimetric Study of the Phase Situation in $[Ni(H_2O)_6](ClO_4)_2$
(a) Vol. 48, No. 2, p. 297—300 (1978).
See also *1.2*, p. *21*; 8, p. 87.

F. Studer and B. Raveau
Electron Microscopic Study of Intergrowth Phenomena in Barium Siliconiobates (I): The Compounds $(Ba_3Nb_6Si_4O_{26})_n \cdot Ba_3Nb_4Ti_4O_{21}$ with n between 3/2 and 29/2
(a) Vol. 48, No. 2, p. 301—312 (1978).
See also *1.1*, p. *13*; *1.4*, p. *32*.

J. H. von Bardeleben, A. Goltzene, B. Meyer, and C. Schwab
Effects of Iron Content and Stoichiometry on the Coloration of $CuGaS_2$
(a) Vol. 48, No. 2, p. K145—K148 (1978).
See also *10.2*, p. *116*; 19, p. 311; 20.1, p. 331.

F. Studer and B. Raveau
Electron Microscopic Study of Intergrowth Phenomena in Barium Siliconiobates (II): Multiple Intergrowths and Microdomains
(a) Vol. **49**, No. 1, p. 189—205 (1978).
See also *1.1*, p. *13*; *1.4*, p. *33*.

V. Kobzarenko, F. Donika, and S. Radautsan
Thin Films of the One-Packed Polytype of $ZnIn_2S_4$(I)
(a) Vol. **49**, No. 1, p. K5—K7 (1978).
See also *1.5*, p. *39*.

H. Nelkowski and K. Petermann
Preparation and Investigation of ZnS–$CuGaS_2$ Heterojunctions
(a) Vol. **49**, No. 1, p. K13—K16 (1978).
See also 1.5, p. 39; 3, p. 56; *14.3.3*, p. *236*; *20.3*, p. *348*; 22.4.1, p. 465.

J. A. Gaj, J. Ginter, and R. R. Gałązka
Exchange Interaction of Manganese $3d^5$ States with Band Electrons in $Cd_{1-x}Mn_xTe$
(b) Vol. **89**, No. 2, p. 655—662 (1978).
See also *13.1*, p. *159*; 18.2, p. 290.

L. C. Scavarda do Carmo and F. Lüty
Librational and Phonon Sidebands of the Infrared CN^- Stretching Mode in NaCN and KCN Crystals
(b) Vol. **89**, No. 2, p. K149—K153 (1978).
See also *6*, p. *76*; *20.1*, p. *334*.

V. Kumar and R. Kumar
On the *g* and Hyperfine Interaction Parameters of Er^{3+} in Three Hydroxides
(b) Vol. **89**, No. 2, p. K195—K197 (1978).
See also *13.4*, p. *182*; 18.1, p. 279.

E. Johnson, J. Ferrer, and L. T. Chadderton
Radiolytic Radiation Damage of Sodalite
(a) Vol. **49**, No. 2, p. 585—591 (1978).
See also 10.2, p. 119; *11*, p. *130*.

G. Kühnel, W. Siegel, and E. Ziegler
Metal-n-$ZnSiP_2$ Schottky Contacts
(a) Vol. **49**, No. 2, p. 705—711 (1978).
See also 13.3, p. 167; *14.3.4*, p. *245*.

V. N. Brudnyi, M. A. Kirov, A. I. Potapov, I. K. Polushina, V. D. Prochukhan, and Yu. V. Rud
Change of Electrical Properties in Electron Irradiated $CdGeAs_2$ Crystals
(a) Vol. **49**, No. 2, p. 761—765 (1978).
See also *11*, p. *130*; 13.4, p. 182; *14.3*, p. *224*.

E. Ziegler, W. Siegel, and G. Kühnel
Potential Barrier Scattering in n-Type Single Crystals of $ZnSiP_2$
(a) Vol. **49**, No. 2, p. K205—K208 (1978).
See also 13.4, p. 183; *14.3*, p. *224*.

M. Baran, H. Szymczak, W. Wardzyński, and B. M. Wanklyn
EPR and Optical Absorption Spectra of $K_2Cr_5F_{17}$ Monocrystals
(b) Vol. **90**, No. 1, p. K59—K61 (1978).
See also *19*, p. *314*; *20.1*, p. *336*.

I. A. Maltseva, A. Mamedov, Yu. V. Rud, and Yu. K. Undalov
Luminescence of In-Doped $CdGeP_2$ Crystals
(a) Vol. **50**, No. 1, p. 139–146 (1978).
See also *10.2*, p. *120*; *20.3*, p. *350*.

T. Rajasekharan and K. V. S. Rama Rao
Nuclear Magnetic Resonance Study of Ferromagnetic $Mn_{1-x}Co_xSb$
(a) Vol. **50**, No. 1, p. 303–310 (1978).
See also *18.2*, p. *291*; *19*, p. *314*.

N. A. Zakharov, V. S. Krikorov, E. F. Kustov, and S. Yu. Stefanovich
New Non-Linear Crystals in the $A_2B_2O_7$ Series
(a) Vol. **50**, No. 1, p. K13–K16 (1978).
See also 3, p. 56; *14.4.1*, p. *257*; *20*, p. *318*.

E. Grilli, P. Cappelletti, and M. Guzzi
Photoluminescence of $CdIn_2S_4$ Single Crystals
(a) Vol. **50**, No. 1, p. K93–K96 (1978).
See also *20.3*, p. *350*.

J. Maetz, M. Müllner, and H. Jex
The Cubic to Tetragonal Phase Transition in $RbCaF_3$ Studied with Neutron Diffraction
(a) Vol. **50**, No. 1, p. K117–K121 (1978).
See also *1.2*, p. *24*.

V. N. Brudnyi, D. L. Budnitskii, M. A. Krivov, R. V. Masagutova, V. D. Prochukhan, and Yu. V. Rud
The Electrical and Optical Properties of 2.0 MeV Electron-Irradiated $ZnGeP_2$
(a) Vol. **50**, No. 2, p. 379–384 (1978).
See also 11, p. 131; 13.4, p. 185; *14.3*, p. *225*; *20.1*, p. *337*.

I. A. Maltseva, A. Mamedov, Yu. V. Rud, and Yu. K. Undalov
The Photosensitivity of Diodes Based on $CdGeP_2$:In Crystals
(a) Vol. **50**, No. 2, p. 445–448 (1978).
See also 14.3, p. 225; *16*, p. *267*.

G. Luther, H. E. Müser, and A. Levstik
Dielectric Behaviour of Ferroelectric $Na(H_{1-x}D_x)_3(SeO_3)_2$ at RF and Microwave Frequencies
(a) Vol. **50**, No. 2, p. 459–468 (1978).
See also *14.4.1*, p. *257*.

A. Anedda, L. Garbato, F. Raga, and A. Serpi
Photoconductivity and Trap Distribution in $CdIn_2S_4$
(a) Vol. **50**, No. 2, p. 643–650 (1978).
See also 13.4, p. 186; *16*, p. *267*.

R. Jakubas, Z. Czapla, L. Sobczyk, and S. Olejnik
Ferroelectric Properties of Selenate Alums
(a) Vol. **50**, No. 2, p. K137–K140 (1978).
See also *14.4.1*, p. *257*.

Sh. M. Efendiev, V. E. Bagiev, A. Ch. Zeinally, and V. M. Skorikov
Optical Transitions in $Bi_{12}SiO_{20}$
(a) Vol. **50**, No. 2, p. K141–K143 (1978).
See also 6, p. 79; 13.1, p. 161; *20.1*, p. *338*.

A. K. Moskalev, I. A. Belobrova, I. P. Aleksandrova, Sh. Sawada, and Y. Shiroishi
Investigation of the Incommensurate Phase of Rb_2ZnCl_4 by the NQR Method
(a) Vol. **50**, No. 2, p. K157–K159 (1978).
See also *14.4.1*, p. *257*; *19*, p. *314*.

F. Smutný
Specific Heat Anomaly at the $F\bar{4}3c \rightarrow Pca2_1$ Phase Transition in Nickel Chlorine Boracite, $Ni_3B_7O_{13}Cl$
(a) Vol. **50**, No. 2, p. K201–K204 (1978).
See also *8*, p. *90*; *14.4.1*, p. *257*.

22.8.1 Salts of Oxy-Acids (Carbonates, Phosphates, Silicates Including Mica etc.)

Original Papers and Short Notes

S. I. Fărcaş, A. Darabont, and A. Nicula
Theoretical Study of the Forbidden ESR Transitions of Cr^{3+} and Fe^{3+} in Alums
(b) Vol. **85**, No. 1, p. 75–79 (1978).
See also *19*, p. *303*.

G. Neumann and H. Vogt
Rayleigh Wing Scattering in Disordered Sodium Nitrate
(b) Vol. **85**, No. 1, p. 179–184 (1978).
See also *1.2*, p. *15*; 2, p. 42; *6*, p. *64*; 20.1, p. 319.

A. V. Shendrik and D. M. Yudin
A New Concept of the Model for the Paramagnetic E'-Centre in the SiO_{2-x} System
(b) Vol. **85**, No. 1, p. 343–349 (1978).
See also 2, p. 42; *13.4*, p. *169*; *19*, p. *303*.

I. N. Geifman
Electron Paramagnetic Resonance of Ti^{3+} in $KTaO_3$
(b) Vol. **85**, No. 1, p. K5–K6 (1978).
See also *19*, p. *303*.

K. S. Aleksandrov, I. P. Aleksandrova, V. F. Shabanov, V. I. Yuzvak, Y. Z. Nozik, and L. I. Fikin
Mechanism of Ferroelectric Phase Transition in $NaNH_4SeO_4 \cdot 2H_2O$
(a) Vol. **45**, No. 1, p. 53–62 (1978).
See also 1.2, p. 15; *14.4.1*, p. *253*; 19, p. 303; 20, p. 315.

V. I. Kovalechiv, L. A. Shuvalov, and T. R. Volk
Spontaneous Polarization Reversal and Photorefractive Effect in Single-Domain Iron-Doped Lithium Niobate Crystals
(a) Vol. **45**, No. 1, p. 249–252 (1978).
See also *14.4.1*, p. *253*.

Hj. Bernhardt
Studies of the Colour of Lead Molybdate Crystals
(a) Vol. **45**, No. 1, p. 353–357 (1978).
See also 13.4, p. 170; *20.1*, p. *321*; *20.3*, p. *341*.

A. S. Bhalla, L. E. Cross, and L. Tongson
Passivation of the *c* Surfaces of Single Crystal Gadolinium Molybdate ($Gd_2(MoO_4)_3$) against Attack by Hydrofluoric Acid by Inert Ion Beam Irradiation
(a) Vol. **45**, No. 1, p. K1–K4 (1978).
See also *11*, p. *123*; 14.4.1, p. 253.

I. B. Barkan, S. I. Marennikov, and M. V. Entin
Dynamic Optical Storage in $LiNbO_3$ Crystals
(a) Vol. **45**, No. 1, p. K17–K22 (1978).
See also *13.4*, p. *170*; 14.4.1, p. 253; 16, p. 262; *20.1*, p. *321*.

Z. Wróbel
Effect of Proportional Compositions of $Pb(Zr_xTi_{1-x})O_3$ Solid Solutions on Their Electric Properties with Particular Regard to the Morphotropic Phase Boundary
(a) Vol. **45**, No. 1, p. K67—K69 (1978).
See also *14.4.1*, p. *253*.

J. J. Bara and B. F. Bogacz
Addendum to: Investigations of Thickness Dependences of Mössbauer Absorption and Scattering Lines
(a) Vol. **45**, No. 1, p. K81 (1978).
See also *5*, p. *60*.

S. R. Shukla and S. D. Pandey
EPR of Mn^{2+} in Cobalt Potassium Selenate Hexahydrate Tutton Salt
(b) Vol. **85**, No. 2, p. K103—K105 (1978).
See also *19*, p. *304*.

P. Braillon, L. Kubin, and J. Serughetti
Plastic Deformation of Calcite Single Crystals Deformed in Compression Parallel to [111]
(a) Vol. **45**, No. 2, p. 453—462 (1978).
See also *10.2*, p. *110*.

B. Strocka
Imaging of Photorefractive Effects in $LiNbO_3$ by Double Crystal X-Ray Reflection Topography
(a) Vol. **45**, No. 2, p. K83—K86 (1978).
See also *10.2*, p. *110*; 14.4.1, p. 253; *16*, p. *262*.

M. Schwab
Phase Transition in the Mixed Crystals $Tb_pY_{1-p}AsO_4$
(b) Vol. **86**, No. 1, p. 195—203 (1978).
See also *1.2*, p. *16*; 13.4, p. 172; 18.4, p. 298; 20.1, p. 322.

J. Bergmann, K. Kneipp, and H. E. Ponath
Raman-Induced Kerr Effect with Elliptical Polarization in Optically Active and Anisotropic Crystals
(b) Vol. **86**, No. 1, p. 215—223 (1978).
See also *6*, p. *67*; *20.1*, p. *322*.

C. A. S. Lima, A. Scalabrin, L. C. M. Miranda, H. Vargas, and S. P. S. Porto
Temperature Behaviour of the Dielectric Constant of Tetragonal $BaTiO_3$
(b) Vol. **86**, No. 1, p. 373—378 (1978).
See also *14.4.1*, p. *254*.

M. Scrocco and L. Mattioli
The Structure of $Sr(NO_3)_2$: A Single Crystal Raman Spectra Study
(b) Vol. **86**, No. 1, p. K1—K5 (1978).
See also *6*, p. *68*; *20.1*, p. *323*.

C. R. A. Catlow
Ion Distribution Functions for Complex Solids and Their Application to the Conductivity of Glasses
(a) Vol. **46**, No. 1, p. 191—198 (1978).
See also *14.4*, p. *248*.

L. Benguigui
Disordered Ferroelectrics: $Ba_xSr_{1-x}TiO_3$ Single Crystals
(a) Vol. **46**, No. 1, p. 337—342 (1978).
See also *14.4.1*, p. *254*.

S. Nadolski, A. M. Balbashov, and A. Ya. Chervonenkis
Influence of Photomagnetic Effects on the ^{57}Fe NMR in $YFeO_3$
(a) Vol. **46**, No. 1, p. K47—K48 (1978).
See also *18.4*, p. *299*; *19*, p. *306*.

H. Vogt and G. Neumann
Hyper-Rayleigh Scattering in Disordered Sodium Nitrate
(b) Vol. **86**, No. 2, p. 615—620 (1978).
See also 1.2, p. 17; 6, p. 69; *20.1*, p. *324*.

F. Milia, M. Voudouris, A. Kandušer, A. Levstik, and R. Blinc
^{75}As Nuclear Quadrupole Resonance Study of the Phase Transition in Na_2HAsO_4
(b) Vol. **86**, No. 2, p. K119—K121 (1978).
See also *1.2*, p. *17*; 14.4.1, p. 254; *19*, p. *306*.

V. Chandra Mouli and G. Sivarama Sastry
ESR Studies of Cu^{2+} Doped $Na_2HAsO_4 \cdot 7\,H_2O$ Single Crystals
(b) Vol. **86**, No. 2, p. K127—K130 (1978).
See also *19*, p. *306*.

D. Samaras, A. Collomb, and J. C. Joubert
Study of Magnetic Properties of the Compound $SrNd_2Fe_2O_7$ — Determination of Magnetic Structures by Neutron Diffraction
(a) Vol. **46**, No. 2, p. 493—498 (1978).
See also *18.4*, p. *299*.

A. I. Otko, N. M. Nesterenko, and L. V. Povstyanyi
Phenomenological Approach to Structural Phase Transitions in Trigonal Double Molybdates and Tungstates
(a) Vol. **46**, No. 2, p. 577—587 (1978).
See also *1.2*, p. *18*.

V. M. Fridkin and B. N. Popov
The Anomalous Photovoltaic Effect and Photoconductivity in Ferroelectrics
(a) Vol. **46**, No. 2, p. 729—733 (1978).
See also *14.4.1*, p. *254*; *16*, p. *263*.

B. Pura and J. Przedmojski
X-Ray Critical Scattering in $NaNO_2$
(a) Vol. **46**, No. 2, p. K115—K117 (1978).
See also *1.2*, p. *18*.

H. Kalbfleisch and K. Melchior
Susceptibility of $DyPO_4$ near the Magnetic Phase Boundaries
(a) Vol. **46**, No. 2, p. K119—K121 (1978).
See also *18.4*, p. *299*.

U. Roy, K. Petrov, I. Tsolovski, and P. Peshev
Erratum to: On the Structural Phase Transition and the Thermal Expansion of Spinel Lithium Titanate
(a) Vol. **46**, No. 2, p. K171 (1978).
See also *1.2*, p. *18*; 8, p. 85.

F. Prokert and R. Schalge
Neutron and Light Scattering from Ferroelectric Fluctuations and Domain Walls of Strontium Barium Niobate
(b) Vol. **87**, No. 1, p. 179–184 (1978).
See also *14.4.1*, p. *254*.

C. V. Manjunath and R. Srinivasan
ESR Study of a Cr^{5+} Centre in Ferroelectric Ammonium Sulphate
(b) Vol. **87**, No. 1, p. 395–400 (1978).
See also *14.4.1*, p. *254*; *19*, p. *307*.

E. Beyer, R. Biederbick, A. Hofstaetter, R. Oeder, A. Scharmann, D. Schwabe, and B. Vitt
Some Features of the Creation of Intrinsic Hole Centres in Scheelite Type Crystals
(b) Vol. **87**, No. 1, p. K35–K38 (1978).
See also *10.2*, p. *113*; 11, p. 126; *19*, p. *307*.

V. S. Postnikov, S. A. Gridnev, B. N. Prasolov, and S. K. Turkov
Specific Features of Low-Frequency Sound Attenuation in $LiNbO_3$ Poly- and Single-Domain Crystals
(a) Vol. **47**, No. 1, p. 65–69 (1978).
See also *7*, p. *80*; *12.2*, p. *139*; 14.4.1, p. 255.

R. Wernicke
The Influence of Kinetic Processes on the Electrical Conductivity of Donor-Doped $BaTiO_3$ Ceramics
(a) Vol. **47**, No. 1, p. 139–144 (1978).
See also 10.2, p. 113; *14.3*, p. *220*.

J. Koshy
Effect of Neutron Bombardment on the Structure of Barium Sulphate Single Crystals
(a) Vol. **47**, No. 1, p. K21–K22 (1978).
See also *11*, p. *126*.

K. A. Mirza, P. D. Townsend, and G. L. Destefanis
Light Emission from the Surface of $LiNbO_3$
(a) Vol. **47**, No. 1, p. K63–K66 (1978).
See also 13.3, p. 165; 14.4.1, p. 255; *20.3*, p. *344*.

H. K. Diemer, E. J. Bijvank, and H. W. den Hartog
EPR Investigations of ADP:Cr^{3+} above and below T_c
(b) Vol. **87**, No. 2, p. 697–706 (1978).
See also *18.1*, p. *278*; *19*, p. *308*.

E. Burzo and I. Ardelean
EPR Study of Fe^{3+} Ions in Lead Borate Glasses
(b) Vol. **87**, No. 2, p. K137–K140 (1978).
See also *19*, p. *308*.

S. A. GRIDNEV and B. M. DARINSKII
Attenuation of Low-Frequency Elastic Oscillations in KH_2PO_4-Type Ferroelectric Crystals
(a) Vol. 47, No. 2, p. 379—384 (1978).
See also *12.2*, p. *139*; 14.4.1, p. 255.

H. IHRIG
The Positive Temperature Coefficient Resistivity of $BaTiO_3$ Ceramics as a Function of the Amount of Titanium-Rich Second Phase
(a) Vol. 47, No. 2, p. 437—444 (1978).
See also *14.4.1*, p. *255*.

I. L. CHISTYI, L. CSILLAG, V. F. KITAEVA, N. KROÓ, and N. N. SOBOLEV
Investigation of Molecular Light Scattering in $PbMoO_4$
(a) Vol. 47, No. 2, p. 609—615 (1978).
See also *7*, p. *81*; *20.1*, p. *328*.

S. RADHAKRISHNA and T. BHASKAR RAO
Optical Absorption and ESR Studies of Nickel Ions in Sodium Zinc Sulphate Tetrahydrate
(a) Vol. 47, No. 2, p. 723—729 (1978).
See also 13.4, p. 176; *19*, p. *309*; *20.1*, p. *328*.

R. R. LEVITSKII, N. A. KORYNEVSKII, and I. V. STASYUK
Distribution Functions and Thermodynamical Properties of KD_2PO_4 and $ND_4D_2PO_4$ Type Crystals
(b) Vol. 88, No. 1, p. 51—63 (1978).
See also *1.2*, p. *20*; *14.4.1*, p. *255*.

J. GROSSMANN, D. MÜLLER, and J. PETERSSON
Study of the Ultraslow Critical Dynamics of the Order Parameter of the Ferroelectric $AgNa(NO_2)_2$ by the Spin Relaxation Times T_1, $T_{1\varrho}$, T_2 of the ^{23}Na Nucleus
(b) Vol. 88, No. 1, p. 99—103 (1978).
See also *14.4.1*, p. *255*; *19*, p. *309*.

S. FAKIOĞLU
Susceptibility Calculations on Spin-Doped Crystals
(b) Vol. 88, No. 1, p. 301—308 (1978).
See also 13.4, p. 177; *18.1*, p. *278*.

M. CHABIN, F. GILLETTA, and C. RIDOU
Thermal Properties of $TlCdF_3$ and $RbCaF_3$ near Their Phase Transitions
(a) Vol. 48, No. 1, p. 67—70 (1978).
See also *1.2*, p. *21*; *8*, p. *87*.

V. V. GAGULIN, N. V. FADEEVA, A. G. BELOUS, A. V. TITOV, K. P. MITROFANOV, M. V. PLOTNIKOVA, S. P. SOLOVIEV, and YU. N. VENEVTSEV
New Seignette-Magnets with Hexagonal Barium Titanate Structure
(a) Vol. 48, No. 1, p. 183—189 (1978).
See also *1*, p. *9*; 5, p. 61; *14.4.1*, p. *255*; 18.2, p. 289; 18.4, p. 301; 19, p. 310.

C. S. SUNANDANA
ESR Saturation of SO_3^- in X-Irradiated $KHSO_4$
(a) Vol. 48, No. 1, p. K19—K21 (1978).
See also 11, p. 128; *19*, p. *310*.

H. Konwent and J. Lorenc
A Thermodynamic Theory of the Improper Antiferroelectric Phase Transition in ADP
(b) Vol. **88**, No. 2, p. 747–755 (1978).
See also 1.2, p. 21; *12.2*, p. *140*; *14.4.1*, p. *256*.

V. K. Jain and P. Venkateswarlu
EPR of Mn^{2+} in $MgSeO_4 \cdot 6\, H_2O$ Single Crystals
(b) Vol. **88**, No. 2, p. K173–K175 (1978).
See also *19*, p. *311*.

L. E. Cross and A. Bhalla
Domain "Decoration" in $Gd_2(MoO_4)_3$
(a) Vol. **48**, No. 2, p. 431–437 (1978).
See also *1.6*, p. *41*; *14.4.1*, p. *256*.

C. Meyer, Y. Gros, B. Bochu, A. Collomb, J. Chenavas, J. C. Joubert, and M. Marezio
Synthesis, Crystal Structure, and Mössbauer Study of a Series of Perovskite-Like Compounds $[ACu_3](M, Fe)_4O_{12}$
(a) Vol. **48**, No. 2, p. 581–586 (1978).
See also *1.1*, p. *13*; *5*, p. *62*.

B. Milsch and W. Windsch
Electron Paramagnetic Resonance of Mn^{2+} in Guanidinium Aluminium Sulfate Hexahydrate
(b) Vol. **89**, No. 1, p. 241–251 (1978).
See also 13.4, p. 180; *14.4.1*, p. *256*; *19*, p. *311*.

S. Fakioğlu and Ş. Erkoç
Magnetic and Thermal Properties of Nd^{3+} in Scheelite-Type Crystals
(b) Vol. **89**, No. 1, p. 289–298 (1978).
See also *8*, p. *88*; *18.1*, p. *279*.

H. E. Müser, W. Kuhn, and J. Albers
Optical Investigation of the Cubic–Tetragonal Phase Transition in $BaTiO_3$ Single Crystals
(a) Vol. **49**, No. 1, p. 51–58 (1978).
See also *1.2*, p. *22*; *14.4.1*, p. *256*; 20, p. **317**.

W. Eigermann, G. Müller-Vogt, and W. Wendl
Solubility Curves in High-Temperature Melts for the Growth of Single Crystals of Rare Earth Vanadates and Phosphates
(a) Vol. **49**, No. 1, p. 145–148 (1978).
See also 1.3, p. 27; *3*, p. *55*.

J. G. Kliava and J. J. Purāns
Analysis of Mn^{2+} EPR Spectra in Polycrystals Using Bir's Method to Account for the Transition Probability
(a) Vol. **49**, No. 1, p. K43–K47 (1978).
See also *19*, p. *312*.

A. Hofstaetter, R. Oeder, A. Scharmann, D. Schwabe, and B. Vitt
Paramagnetic Resonance and Thermoluminescence of the $PbWO_4/PbMoO_4$ Mixed Crystal System
(b) Vol. **89**, No. 2, p. 375–380 (1978).
See also *19*, p. *312*; 20.3, p. **348**.

R. von Baltz
Theory of the Anomalous Bulk Photovoltaic Effect in Ferroelectrics
(b) Vol. **89**, No. 2, p. 419–429 (1978).
See also *14.4.1*, p. *256*; *16*, p. *266*.

R. Biederbick, A. Hofstaetter, and A. Scharmann
Temperature-Dependent Mn^{2+}-^6S State Splitting in Scheelites
(b) Vol. **89**, No. 2, p. 449–458 (1978).
See also *13.4*, p. *181*; *19*, p. *312*.

M. Razeghi and B. Houlier
EPR Study of Gd^{3+} around the Ferroelastic Transition Point of $Pb_3(PO_4)_2$
(b) Vol. **89**, No. 2, p. K135–K137 (1978).
See also *1.2*, p. *23*; *19*, p. *313*.

J. A. Groenink and D. A. van Wezep
Photoconductivity of $PbMoO_4$ and $PbWO_4$ Single Crystals
(a) Vol. **49**, No. 2, p. 651–654 (1978).
See also *16*, p. *267*.

T. I. Chekmasova and I. P. Aleksandrova
NMR Investigation of the High-Pressure Phase in $LiNH_4SO_4$
(a) Vol. **49**, No. 2, p. K185–K188 (1978).
See also *1.2*, p. *23*; *14.4.1*, p. *256*; **19**, p. **313**.

G. Schmidt and I. Boczek
Pseudoelasticity and Shape Memory of PLZT Ceramic
(a) Vol. **50**, No. 1, p. K109–K111 (1978).
See also 12.2, p. **141**; *14.4.1*, p. *257*.

M. A. F. Scarparo, R. S. Katiyar, R. Srivastava, and S. P. S. Porto
Evidence of Non-Linear Behaviour of Soft Modes in Crystals with KDP Structure
(b) Vol. **90**, No. 2, p. 543–556 (1978).
See also *14.4.1*, p. *257*; 20.1, p. **337**.

A. Collomb, D. Samaras, and J. C. Joubert
Neutron Diffraction Determination of Magnetic Structures of $SrTCrO_4$ Compounds (T=La, Nd)
(a) Vol. **50**, No. 2, p. 635–642 (1978).
See also *18.4*, p. *302*.

S. Tsunekawa and H. Takei
Twinning Structure of Ferroelastic $LaNbO_4$ and $NdNbO_4$ Crystals
(a) Vol. **50**, No. 2, p. 695–702 (1978).
See also *1*, p. *10*; 10.2, p. 122; 14.4.1, p. 257.

A. A. Bukharaev and N. R. Yafaev
Laser and Thermal Bleaching of Colour Centres in Sodium Borate Glasses
(a) Vol. **50**, No. 2, p. 711–716 (1978).
See also 2, p. 53; *13.4*, p. *186*; *20.1*, p. *338*.

A. A. Kaminskii, S. E. Sarkisov, Tran Ngoc, B. I. Denker, V. V. Osiko, and A. M. Prokhorov
Stimulated Emission Spectroscopy of Concentrated Lithium Neodymium Phosphate Glasses in $^4F_{3/2}\rightarrow{}^4I_{11/2}$ and $^4F_{3/2}\rightarrow{}^4I_{13/2}$ Transitions
(a) Vol. **50**, No. 2, p. 745–750 (1978).
See also 13.4, p. 186; *20.1*, p. *338*; *20.2*, p. *340*.

Z. Jirák, S. Vratislav, and J. Zajíček
The Magnetic Structure of the Tetragonal $Mn_{2.5}Cr_{0.5}O_4$
(a) Vol. **50**, No. 2, p. K131–K135 (1978).
See also *18.3*, p. *297*.

Z. Surowiak
The Effect of a Stable Electric Field on the Properties of $BaTiO_3$ Films
(a) Vol. **50**, No. 2, p. K145–K148 (1978).
See also *14.4.1*, p. *257*.

H. Hesse and S. Kapphan
Doping of $SrTiO_3$ with Hydrogen and Deuterium
(a) Vol. **50**, No. 2, p. K243–K245 (1978).
See also 10.2, p. 122; *14.4.1*, p. *257*; *20.1*, p. *338*.

V. M. Fridkin, E. I. Boikova, B. N. Popov, and G. I. Rosenman
The Influence of the Anomalous Photovoltaic Effect on the Exoelectron Emission in $LiNbO_3$
(a) Vol. **50**, No. 2, p. K255–K259 (1978).
See also 14.4.1, p. 258; *16*, p. *268*; *17*, p. *271*.

22.8.2 Spinels, Garnets, Ferrites

Original Papers and Short Notes

T. Merceron, A. Marais, M. Porte, and L. Brossard
Study of a Series of Manganese Ferrites with Manganese Excess by Directional Ordering. Cation Valency and Repartition of Ions on Sites A and B
(a) Vol. **45**, No. 2, p. 503–507 (1978).
See also *18.3*, p. *295*.

J. V. de Menezes and F. de S. Barros
On the Mössbauer Spectrum of $FeC_2O_4 \cdot 2\ H_2O$
(a) Vol. **45**, No. 2, p. K139–K143 (1978).
See also *5*, p. *60*.

M. W. Muller
Site Ordering and Magnetostriction in Garnet (II): Iron
(b) Vol. **86**, No. 1, p. 345–348 (1978).
See also *1*, p. *9*; *18.3*, p. *295*.

V. A. Antonov, P. A. Arsenev, and D. S. Petrova
Analysis of Spectral and Structural Regularities of Neodymium Ions in Garnet Monocrystals
(a) Vol. **46**, No. 1, p. 55–58 (1978).
See also *1.4*, p. *29*; 20, p. 316.

M. Wendt and A. Schmidt
Improved Reproductivity of Energy-Dispersive X-Ray Microanalysis by Normalization to the Background
(a) Vol. **46**, No. 1, p. 179–183 (1978).
See also *1.3*, p. *25*.

S. Prasad and N. C. Goel
DC Resistivity of Doped Mn–Zn Polycrystalline Ferrites
(a) Vol. **46**, No. 1, p. 373–377 (1978).
See also *14.3*, p. *219*.

S. Ligenza
Exchange Splitting of Ionic S-State in $MnFe_2O_4$ Studied by Neutron Spectroscopy
(b) Vol. **86**, No. 2, p. 635–643 (1978).
See also 13.5.4, p. 201; *18.3*, p. *295*.

M. N. Iliev, E. Anastassakis, and T. Arai
Raman Scattering of $HgCr_2Se_4$
(b) Vol. **86**, No. 2, p. 717–723 (1978).
See also *6*, p. *69*; 18.2, p. 285; *20.1*, p. *324*.

R. Hergt and P. Görnert
Growth Rate Anisotropy and Kinetic Coefficients of Vicinal Faces at LPE Garnet Films (I): Growth Rate Anisotropy
(a) Vol. **46**, No. 2, p. 427–432 (1978).
See also *3*, p. *54*.

I. RATTAN and N. C. SRIVASTAVA
Magnetoelastic Interaction Energy of a Cubic Ferromagnet in an Arbitrary Coordinate System
(a) Vol. **46**, No. 2, p. 475–478 (1978).
See also *12.2*, p. *139*; *18.2*, p. *285*.

I. S. EDELMAN and V. D. BAURIN
Influence of Cr^{3+} Ions on the Magnetooptical Properties of Cobalt Ferrite
(a) Vol. **46**, No. 2, p. K83–K86 (1978).
See also *18.2.1*, p. *293*; *20.1*, p. *325*.

HJ. BERNHARDT
Bound Polarons in YAG Crystals
(b) Vol. **87**, No. 1, p. 213–219 (1978).
See also *10.2*, p. *112*; *13.5.3*, p. *200*.

V. NEKVASIL
The Crystal Field for Nd^{3+} in Garnets
(b) Vol. **87**, No. 1, p. 317–323 (1978).
See also *13.4*, p. *175*; *20.1*, p. *326*.

P. GÖRNERT and R. HERGT
Growth Rate Anisotropy and Kinetic Coefficients of Vicinal Faces at LPE Garnet Films (II): Kinetic Coefficients
(a) Vol. **47**, No. 1, p. 99–104 (1978).
See also *3*, p. *54*.

M. MARYŠKO
The Influence of Electromagnetic Propagation on Ferromagnetic Resonance in an Infinite Cylinder
(a) Vol. **47**, No. 1, p. 277–286 (1978).
See also *18.2*, p. *286*; 18.3, p. 295; *19*, p. *307*.

V. A. POVITSKII, A. N. SALUGIN, YU. V. BALDOKHIN, E. F. MAKAROV, and N. V. ELISTRATOV
Spin Reorientation in Hematite (α-Fe_2O_3) with Al Impurity
(b) Vol. **87**, No. 2, p. K101–K104 (1978).
See also *5*, p. *61*; 18.3, p. 295.

S. A. MARSHALL, T. MARSHALL, and R. A. SERWAY
Electron Spin Resonance Absorption Spectrum of Trivalent Gadolinium in the Oxide YAlG
(a) Vol. **48**, No. 1, p. 165–173 (1978).
See also *19*, p. *310*.

W. K. UNGER
Search for Light-Induced Magnetization in Magnetic Semiconductors
(a) Vol. **48**, No. 1, p. K89–K91 (1978).
See also *18.4*, p. *301*.

YA. L. SHAIOVICH and R. A. ZVINCHUK
Conditions of Magnetite Precipitation during Decomposition of Wüstite $Fe_{1-x}O$
(a) Vol. **48**, No. 2, p. 543–550 (1978).
See also *1.2*, p. *21*; 10.2, p. 116.

R. A. SERWAY, F. H. YANG, and S. A. MARSHALL
Line Broadening in the ESR Absorption Spectra of Fe^{3+}, Cr^{3+}, and Gd^{3+}in the Oxide YAlG
(b) Vol. **89**, No. 1, p. 267–273 (1978).
See also 13.4, p. 180; *19*, p. *311*.

A. A. KAMINSKII, V. V. OSIKO, S. E. SARKISOV, M. T. TIMOSHECHKIN, E. V. ZHARIKOV, J. BOHM, P. REICHE, and D. SCHULTZE
Growth, Spectroscopic Investigations, and Some New Stimulated Emission Data of $Gd_3Ga_5O_{12}$:Nd^{3+} Single Crystals
(a) Vol. **49**, No. 1, p. 305–311 (1978).
See also 3, p. 56; 20.1, p. 333; *20.2*, p. *339*; *20.3*, p. *347*.

M. MARYŠKO
Note on FMR Absorption in a Low Resistivity Ferrite Cylinder
(a) Vol. **49**, No. 1, p. K71–K73 (1978).
See also 18.3, p. 296; *19*, p. *312*.

S. NADOLSKI, E. JEDRYKA, K. PIOTROWSKI, and B. WANKLYN
Cr^{53} Nuclear Magnetic Resonance in Domain Walls of Yttrium Orthochromite
(a) Vol. **49**, No. 2, p. K193–K196 (1978).
See also *18.4*, p. *301*; *19*, p. *313*.

K. G. NIKIFOROV, S. I. RADAUTSAN, and V. E. TEZLEVAN
Negative Magnetoresistance in the Ga-Doped Magnetic Semiconductor $CdCr_2S_4$
(a) Vol. **49**, No. 2, p. K197–K200 (1978).
See also *14.3.4*, p. *246*.

S. LIGENZA and H. PTASIEWICZ-BĄK
Magneto-Electric Splitting of the Crystal Field F_2-Level of Cr^{3+} Ions in $MgCr_2O_4$ Studied by Neutron Spectroscopy
(b) Vol. **90**, No. 1, p. 319–326 (1978).
See also *13.4*, p. *183*; *18*, p. *275*.

B. GILLOT, F. BOUTON, F. CHASSAGNEUX, and A. ROUSSET
Electrical Conductivity of $(Fe^{2+}Al^{3+}_{2-x}Cr^{3+}_x)O^{2-}_4$ Spinels and Defect Phases $\gamma(Fe^{3+}_{1/3}Al^{3+}_{2/3-y}Cr^{3+}_y)_2O^{2-}_3$
(a) Vol. **50**, No. 1, p. 109–116 (1978).
See also *14.3*, p. *224*; 18.3, p. 296.

E. M. SMOKOTIN, G. F. GUSYATSKII, L. M. PROTOPOPOVA, and A. M. KAPITONOV
The Effect of Inhomogeneous Stress on the FMR in Ferrites
(a) Vol. **50**, No. 1, p. 269–274 (1978).
See also *18.3*, p. *296*; *19*, p. *314*.

A. BOMBIK, B. LEŚNIEWSKA, and A. OLEŚ
Atomic and Magnetic Structures of $HoVO_3$
(a) Vol. **50**, No. 1, p. K17–K20 (1978).
See also *1.1*, p. *14*; *18.3*, p. *296*.

Z. JIRÁK, S. VRATISLAV, and P. NOVÁK
Study of Cubic and Tetragonal Structures in the System $Mn_xCr_{3-x}O_4$
(a) Vol. **50**, No. 1, p. K21–K24 (1978).
See also *1.1*, p. *14*.

R. E. Vandenberghe
Comment on Neutron Diffraction Studies of the Spinel Oxide $CuMn_2O_4$
(a) Vol. **50**, No. 1, p. K85—K88 (1978).
See also *1*, p. *10*.

Y. Watanabe, K. Urade, and S. Saito
Contribution of the Fe^{2+} Ion to the Magnetic Anisotropy Constant in Ferrites
(b) Vol. **90**, No. 2, p. 697—702 (1978).
See also *18.3*, p. *297*.

J. Jelonek and M. Szczygielski
Dispersion of Initial Magnetic Permeability for $HgCr_2Se_4$
(a) Vol. **50**, No. 2, p. 679—683 (1978).
See also *18.2*, p. *292*.

M.-L. Craus and N. Rezlescu
The Influence of the Composition and Thermal Treatment on the Cation Redistribution in the $Mg_xFe_{3-x}O_4$ System
(a) Vol. **50**, No. 2, p. K225—K229 (1978).
See also *18.3*, p. *297*.

22.9 Organic Semiconductors

Review Article

P. P. Seregin, F. S. Nasredinov, and L. N. Vasilev
A Study of Radiation Defects in Solids by Means of Mössbauer Spectroscopy
(a) Vol. 45, No. 1, p. 11–45 (1978).
See also *5*, p. *60*; *11*, p. *123*; 22.5, p. 474; 22.8, p. 503.

Original Papers and Short Notes

M. E. Michel-Beyerle, R. Haberkorn, J. Kinder, and H. Seidlitz
Direct Evidence for the Singlet–Triplet Exciton Annihilation in Anthracene Crystals
(b) Vol. 85, No. 1, p. 45–49 (1978).
See also *13.5.1*, p. *189*; 20.3, p. 341.

B. Reimer and H. Bässler
Transient Photoconduction in a Polydiacetylene Single Crystal. Determination of Transit Time, Free Lifetime, and Recombination Time of Charge Carriers
(b) Vol. 85, No. 1, p. 145–153 (1978).
See also *16*, p. *262*.

Th. Nattermann
Static and Dynamic Critical Behaviour of Uniaxial Ferroelectrics and the Phase Transition in TGS
(b) Vol. 85, No. 1, p. 291–300 (1978).
See also 1.2, p. 15; *14.4.1*, p. *253*.

S. Franck, A. Kühnel, and S. Wendt
On the Microscopic Mechanism of the Ferroelectric Phase Transition of TSCC: Mn^{2+}
(b) Vol. 85, No. 1, p. 319–323 (1978).
See also *14.4.1*, p. *253*; 19, p. 303.

E. L. Bokhenkov, E. M. Rodina, E. F. Sheka, and I. Natkaniec
Inelastic Incoherent Neutron Scattering Spectra at Different Temperatures and Computer Experiment for External Phonon Modes of Naphthalene Crystals
(b) Vol. 85, No. 1, p. 331–342 (1978).
See also *6*, p. *65*.

G. J. Ashwell
A Comment upon the Metallic and Semiconducting Phases of $(DEPE)^{2+}(TCNQ)_4^{2-}$
(b) Vol. 85, No. 1, p. K7–K9 (1978).
See also *14.1*, p. *205*; *14.3*, p. *217*.

A. I. Belkind, A. M. Brodskii, and V. V. Grechov
Theory of Photoemission from Molecular Crystals
(b) Vol. 85, No. 2, p. 465–472 (1978).
See also 13.1, p. 151; *17*, p. *269*.

M. E. Michel-Beyerle and R. Haberkorn
Temperature-Independent Intersystem Crossing in Anthracene Crystals
(b) Vol. 85, No. 2, p. 473–475 (1978).
See also *13.5.1*, p. *189*; *20.3*, p. *342*.

N. Graber and I. Zschokke-Gränacher
Delayed Fluorescence Excitation Spectra in Doped Anthracene Crystals
(b) Vol. **85**, No. 2, p. 505–510 (1978).
See also *20.3*, p. *342*.

H. Jex and C. J. Maetz
Lattice Dynamics of the Molecular Crystals KCN and NaCN in Their Disordered Cubic Phase
(b) Vol. **85**, No. 2, p. 511–517 (1978).
See also *6*, p. *66*; 8, p. 83.

J. Alizon, G. Berthet, J. P. Blanc, J. Gallice, H. Robert, J. M. Fabre, and L. Giral
Dynamic Nuclear Polarization in TCNQ Salts
(b) Vol. **85**, No. 2, p. 603–610 (1978).
See also *18*, p. *273*; 19, p. **304**.

G. C. Stevens and D. Bloor
ESR Studies of a Diacetylene Polymer (I): Partially Crystalline Polymer Extracts
(a) Vol. **45**, No. 2, p. 483–491 (1978).
See also 13.4, p. 171; *19*, p. *304*.

J. L. Jacquemin and G. Bordure
Numerical Simulation of the AC Electrical Properties of Random Inhomogeneous Systems. Application to Heat-Treated Polymers
(a) Vol. **45**, No. 2, p. 579–584 (1978).
See also *14*, p. *203*; 14.4, p. 248.

S. D. Phadke and R. N. Karekar
Photocurrents through Polythiourea and Polymalononitrile Thin Films
(a) Vol. **45**, No. 2, p. K163–K165 (1978).
See also 14.3.1, p. 226; *16*, p. *263*.

N. N. Tyutyulkov, J. I. Petkov, und P. Schuster
Band Structure of Bis (Catena-Poly(Methylarsen))
(b) Vol. **86**, No. 1, p. 385–395 (1978).
See also *13.1*, p. *153*.

G. M. Georgiev and J. S. Nikolov
Spontaneous Parametric Scattering of Light in Ammonium Bioxalate Monohydrate
(b) Vol. **86**, No. 1, p. 431–437 (1978).
See also *6*, p. *68*; *20.1*, p. *323*.

G. C. Stevens and D. Bloor
ESR Studies of a Diacetylene Polymer (II): Spin Concentrations and g-Values for Single Crystal Samples
(a) Vol. **46**, No. 1, p. 141–146 (1978).
See also 13.4, p. 173; *19*, p. *305*.

S. Lahiry, J. Sharma, G. D. Sootha, and H. O. Gupta
Cu^{2+} in Pseudotetrahedral Symmetry:Magnetic Anisotropy, Susceptibility, and EPR of Bis(Trimethyl- and Triethyl Benzyl Ammonium) Tetrachlorocuprate(II) from 300 to 77 K
(a) Vol. **46**, No. 1, p. 153–162 (1978).
See also *18.1*, p. *277*; *19*, p. *305*.

M. Punkkinen and L. P. Ingman
Molecular Motion and Spin–Lattice Relaxation in Polyethylene
(a) Vol. **46**, No. 1, p. 213–216 (1978).
See also *19*, p. *306*.

E. Hadjoudis, F. Milia, J. Seliger, and R. Blinc
Temperature Dependence of the ^{14}N Quadrupole Resonance in Some Thermochromic and Photochromic N-Salicylideneanilines
(a) Vol. **46**, No. 1, p. 239–244 (1978).
See also *10.2*, p. *111*; *19*, p. *306*.

M. Samoć and Z. Zboiński
Trapping of Charge Carriers in the Deformed Surface Region of Anthracene Single Crystals
(a) Vol. **46**, No. 1, p. 251–257 (1978).
See also *13.3*, p. *165*; 14.3, p. 219; 16, p. 263.

S. Nešpurek, J. Obrda, and J. Sworakowski
A Study of Traps for Current Carriers in Organic Solids. N,N′-diphenyl-p-phenylenediamine
(a) Vol. **46**, No. 1, p. 273–280 (1978).
See also 13.4, p. 173; *14.3.4*, p. *241*.

V. A. Benderskii, V. Kh. Brikenshtein, A. G. Lavrushko, and P. G. Filippov
Non-Linear Fluorescence Quenching in Molecular Crystals (I): Recombination of Localized Excitons
(b) Vol. **86**, No. 2, p. 449–458 (1978).
See also *13.5.1*, p. *190*; *20.3*, p. *343*.

G. J. Ashwell
Organic Metals: Chain Stabilisation in Bipyridinium–TCNQ Salts
(b) Vol. **86**, No. 2, p. 705–715 (1978).
See also *1.1*, p. *11*; 14, p. 203.

H. Arndt, G. Schmidt, N. R. Ivanov, and L. A. Shuvalov
Electrooptic Switching Anomalies in Ferroelectric TGS
(a) Vol. **46**, No. 2, p. 533–540 (1978).
See also *14.4.1*, p. *254*; *20.1*, p. *325*.

G. C. Stevens and D. Bloor
ERS Studies of a Diacetylene Polymer (III): Thermal History Effects and Temperature Dependence of Polycrystalline Samples
(a) Vol. **46**, No. 2, p. 619–626 (1978).
See also 13.4, p. 174; *19*, p. *306*.

A. K. Kostin, V. V. Savelev, and A. V. Vannikov
Free Charges and Excitons in Stilbene Single Crystals Excited by Electron Pulses
(b) Vol. **87**, No. 1, p. 255–259 (1978).
See also *13.5.1*, p. *191*; 20.1, p. 326; 20.3, p. 344.

E. L. Frankevich, M. M. Tribel, I. A. Sokolik, and A. I. Pristupa
Magnetic-Resonant Modulation of Photoconductivity of Crystalline Charge Transfer Complexes. Anthracene–Tetracyanbenzene
(b) Vol. **87**, No. 1, p. 373–379 (1978).
See also 13.5.1, p. 191; *16*, p. *264*; 20.3, p. 344.

E. Heinze and S. Grande
Nuclear Spin–Lattice Relaxation in the Solid State of Mesogenic p-Alkoxybenzylidene-Alkylanilines
(b) Vol. **87**, No. 1, p. K43–K45 (1978).
See also *19*, p. *307*.

R. Müller and G. Scheler
Reorientation of Water Molecules in Monocrystals of Oxalic Acid Dihydrate ($(COOH)_2 \cdot 2\ H_2O$) Investigated by High Resolution NMR in Solids
(b) Vol. **87**, No. 1, p. K53–K55 (1978).
See also *19*, p. *307*.

L. Benguigui and D. Cabib
Determination of the Tilt Angle in the Smectic C Phase of Bis-(Heptyloxy)-Azoxybenzene by Means of Dielectric Measurements
(a) Vol. **47**, No. 1, p. 71–78 (1978).
See also 1.4, p. 30; *14.4*, p. *249*.

R. Kind, S. Plesko, and J. Roos
Structural Phase Transitions in the Perovskite-Type Layer Compound $NH_3(CH_2)_3NH_3MnCl_4$
(a) Vol. **47**, No. 1, p. 233–240 (1978).
See also *1.2*, p. *18*; 19, p. 307.

W. Wanarski and Z. Zagórski
Influence of γ-Irradiation on the Electric Permittivity of Thiourea Single Crystal
(a) Vol. **47**, No. 1, p. K45–K46 (1978).
See also *11*, p. *126*; *14.4.1*, p. *255*.

V. Šubrta
Exciton Propagation and Memory Effect in Molecular Crystals at Low Temperatures
(b) Vol. **87**, No. 2, p. 647–652 (1978).
See also *13.5.1*, p. *192*.

M. Z. Zgierski and W. H. Henneker
Vibronic Spectra of Molecular Systems
(b) Vol. **87**, No. 2, p. 653–661 (1978).
See also *6*, p. *72*; 13.5.1, p. 192.

Z. Zboiński
A Kinetic Model of Triplet Exciton Dynamics in Organic Crystals with Discrete Trapping Levels
(b) Vol. **87**, No. 2, p. 663–674 (1978).
See also *13.5.1*, p. *192*; 20.3, p. 344.

H. Pykacz
Pyroelectric Properties of TGS with Internal Bias Field
(a) Vol. **47**, No. 2, p. 625–630 (1978).
See also *14.4.1*, p. *255*.

M. Godlewska and M. Rachwalska
An Adiabatic Calorimetric Study of the Phase Situation in 2,2-Dinitropropane
(a) Vol. **47**, No. 2, p. 661–665 (1978).
See also *1.2*, p. *20*; 8, p. 86.

K. S. Bhatki, S. D. Pradhan, and B. V. Thosar
Further Studies on Positron Lifetimes in γ-Irradiated Teflon in Low Dose Range
(a) Vol. **47**, No. 2, p. 691–698 (1978).
See also *10.2*, p. *114*; 11, p. 127; 14.4, p. 249; 19, p. 309.

E. Heinze and S. Grande
Nuclear Spin Relaxation in the Dipolar System in Solid p-Alkoxybenzylidene-p-n-Alkylanilines
(b) Vol. **88**, No. 1, p. K23–K25 (1978).
See also 1.2, p. 20; 6, p. 74; *19*, p. *309*.

P. Schlotter and H. Bässler
Sensitized Injection into Anthracene Crystals from Surface Oxidation Products
(a) Vol. **48**, No. 1, p. K15–K18 (1978).
See also *13.3*, p. *166*; *16*, p. *264*.

A. A. Gogolin and V. I. Melnikov
Frequency Dependences of Conductivity and Dielectric Permeability of a One-Dimensional Disordered Conductor
(b) Vol. **88**, No. 2, p. 377–383 (1978).
See also *14*, p. *204*.

H. Asami, T. Nakayama, T. Chong, and N. Itoh
Reaction Rate Constants of Triplet Ex citons in Naphthalene Single Crystals
(b) Vol. **88**, No. 2, p. 623–631 (1978).
See also *13.5.1*, p. *193*; 20.3, p. 346.

K. Lochner, H. Bässler, B. Tieke, and G. Wegner
Photoconduction in Polydiacetylene Multilayer Structures and Single Crystals. Evidence for Band-to-Band Excitation
(b) Vol. **88**, No. 2, p. 653–661 (1978).
See also *13.1*, p. *157*; *16*, p. *265*.

C. Papatriantafillou
NMP-TCNQ: A Realization of the Disordered Hubbard Model
(b) Vol. **88**, No. 2, p. 663–674 (1978).
See also *13*, p. *146*; 14.4, p. 249.

P. Petelenz
Effect of the Mixing between Frenkel Excitons and Charge Transfer States on Charge Carrier Transfer in Molecular Crystals (II): Physical Consequences
(b) Vol. **88**, No. 2, p. 689–697 (1978).
See also *13.5.1*, p. *193*.

Th. Eifrig
On the Electronic Properties of Disordered Quasi-One-Dimensional Systems with Electron Correlation
(b) Vol. **89**, No. 1, p. 171–175 (1978).
See also *2*, p. *48*; *14.1*, p. *210*; 18.1, p. 279.

T. Goworek, C. Rybka, and J. Wawryszczuk
Positronium in Mixed Molecular Crystals
(b) Vol. **89**, No. 1, p. 253–259 (1978).
See also *10.2*, p. *117*.

W. G. ALBRECHT, H. COUFAL, R. HABERKORN, and M. E. MICHEL-BEYERLE
Excitation Spectra of Exciton Fission in Organic Crystals
(b) Vol. **89**, No. 1, p. 261–265 (1978).
See also *13.5.1*, p. *194*; 20.1, p. 332.

P. P. REDDY and Y. P. REDDY
Electronic Absorption Spectrum of Co^{2+} in Strontium Tartrate Tetrahydrate Single Crystals
(b) Vol. **89**, No. 1, p. K61–K63 (1978).
See also *20.1*, p. *333*.

H. KIESS and R. CLARKE
Pyroelectric Effect on PTS
(a) Vol. **49**, No. 1, p. 133–136 (1978).
See also *1*, p. *9*; *14.4*, p. *250*; 15, p. 261.

L. LEBIODA, S. HODOROWICZ, and K. LEWINSKI
Low-Temperature Phase-Transition in Urea
(a) Vol. **49**, No. 1, p. K27–K30 (1978).
See also *1.2*, p. *22*.

W. MYCIELSKI and A. LIPIŃSKI
SCL Currents and Trap Distribution in Polycrystalline Tetracene
(a) Vol. **49**, No. 1, p. K41–K42 (1978).
See also *13.4*, p. *181*; *14.3.4*, p. *245*.

M. VODENICHAROVA
Thermionic-Field Emission from a Metal into Organic Polymer Semiconductor Films
(a) Vol. **49**, No. 1, p. K49–K53 (1978).
See also *14.3.3*, p. *236*; *14.3.4*, p. *245*.

S. EWERT, G. HEILAND, H. LÜTH, U. ROLL, and W. SANDER
Molecular Vibrations in a Polymethine Dye Layer Studied by Inelastic Electron Tunneling Spectroscopy
(b) Vol. **89**, No. 2, p. 459–465 (1978).
See also 6, p. 76; *14.3.1*, p. *228*; *20.1*, p. *334*.

H.-D. MAIER, D. MÜLLER, and J. PETERSSON
Temperature and Frequency Dependence of the Dielectric Constant and Proton NMR near the Structural Phase Transition of Squaric Acid
(b) Vol. **89**, No. 2, p. 587–593 (1978).
See also *14.4*, p. *251*; *19*, p. *312*.

H. KUZMANY
Instability of the Raman Spectra of Tetrathiofulvalene-Tetracyanoquinodimethan (TTF-TCNQ) and Tetrathiofulvalene (TTF)
(b) Vol. **89**, No. 2, p. K139–K144 (1978).
See also 11, p. 130; *20.1*, p. *334*.

H. BERTAGNOLLI and H. G. HERTZ
Preservation and Loss of Structural Features of Solid Acetic Acid during the Melting Process
(a) Vol. **49**, No. 2, p. 463–472 (1978).
See also *1*, p. *10*; *2*, p. *50*.

W. Auch, W. Steudle, J. U. von Schütz, and H. C. Wolf
The Phase Transition in the 1:1 CT-Complex Anthracene/TCNB at 206 K. A NMR Study
(a) Vol. **49**, No. 2, p. 563—567 (1978).
See also 1.2, p. 23; 6, p. 77; *19*, p. *313*.

V. P. Arora and V. K. Agarwal
Dielectric Behaviour of Nematic Phase OHMBBA at Radio Frequencies
(a) Vol. **49**, No. 2, p. 757—760 (1978).
See also *2*, p. *51*; *14.4*, p. *251*.

W. Rehwald and A. Vonlanthen
Ultrasonic Studies of the Structural Phase Transition in Squaric Acid
(b) Vol. **90**, No. 1, p. 61—66 (1978).
See also *1.2*, p. *23*; 7, p. 82; 12.2, p. 141.

V. Šubrta and V. Čápek
Fading Memory as Insufficient Condition for the Exciton Diffusion
(b) Vol. **90**, No. 1, p. 295—299 (1978).
See also *13.5.1*, p. *196*.

V. I. Lesin
Analytical Form of the Magnetic Field Dependences of Rate Constants of Triplet Exciton Annihilation and Triplet Exciton Quenching by Free Radicals in Molecular Crystals
(b) Vol. **90**, No. 1, p. K13—K16 (1978).
See also *13.5.1*, p. *196*.

V. L. Broude, N. A. Vidmont, and V. V. Korshunov
Singlet Exciton Annihilation Kinetics in Anthracene Crystals
(b) Vol. **90**, No. 1, p. K53—K58 (1978).
See also *13.5.1*, p. *196*; *20.3*, p. *349*.

K. Lochner, Th. Hinrichsen, W. Hofberger, and H. Bässler
Polymerization of a Polydiacetylene Single Crystal under Hydrostatic Pressure
(a) Vol. **50**, No. 1, p. 95—100 (1978).
See also *1*, p. *10*.

H. Möhwald and A. Thaer
Raman Study of the Phase Transition in the CT-Crystal Anthracene–TCNB
(a) Vol. **50**, No. 1, p. 131—138 (1978).
See also *1.2*, p. *24*; 6, p. 78; 20.1, p. 336.

M. Przybylski, A. Graja, A. Rajchel, M. Gawron, and T. Borowiak
DC Conductivity of Complex Salts of TCNQ with N-Methyl-Pyridines
(a) Vol. **50**, No. 1, p. K105—K107 (1978).
See also *14*, p. *204*.

A. S. Davydov
The Role of Proteins in Electron Transport at Large Distances
(b) Vol. **90**, No. 2, p. 457—464 (1978).
See also *14*, p. *204*.

A. S. Bhalla, L. E. Cross, and D. A. Payne
Thermal, Optical, and Elastic Properties of Ferroelastic Isopropyl Ammonium Chloroplatinate (IPACP) and Chlorostannate (IPACS)
(a) Vol. **50**, No. 2, p. 661—667 (1978).
See also *1.2*, p. *24*; 8, p. 90; 12.2, p. 141; 20.1, p. 338.

H. Kresse, D. Demus, and A. Wiegeleben
Dielectric Relaxation of Liquid Crystals with Nematic, Smectic A, and Smectic C Phases
(a) Vol. **50**, No. 2, p. K181—K183 (1978).
See also *2*, p. *53*; *14.4*, p. *252*.

C. Hamann, H.-J. Höhne, F. Kersten, M. Müller, F. Przyborowski, and M. Starke
Switching Effects on Polycrystalline Films of Lead Phthalocyanine (PbPc)
(a) Vol. **50**, No. 2, p. K189—K192 (1978).
See also *14.3.3*, p. *239*.

I. V. Krivoshei, V. P. Babiyczuk, I. M. Guella, I. F. Golovkina, N. V. Mansia, V. A. Starodub, and S. A. Chueva
A New Conducting Compound: Tetrafluoroborate of Dibenztetrathiofulvalene: Electrical and Magnetic Properties
(a) Vol. **50**, No. 2, p. K197—K200 (1978).
See also *14*, p. *204*; *18*, p. *276*.

L. Brehmer and M. Pinnow
Estimation of the Charge Carrier Mobility in Low Density Polyethylene by Means of Current Oscillations
(a) Vol. **50**, No. 2, p. K239—K242 (1978).
See also *14.3.4*, p. *247*.

23. Solidified Gases

Original Papers and Short Notes

A. T. Macrander and R. K. Crawford
Erratum to: Density of Solid Argon at Melting
(a) Vol. **46**, No. 1, p. 397 (1978).
See also *1.2*, p. *17*.

R. Resta
Quantum Defect Theory of Excitons. Solid Neon
(b) Vol. **86**, No. 2, p. 627—633 (1978).
See also *13.5.1*, p. *191*.

M. S. Kushwaha and S. S. Kushwaha
Unpaired Elastic Force Model in the Lattice Dynamical Study of Rare Gas Solids
(b) Vol. **87**, No. 1, p. 247—253 (1978).
See also *6*, p. *70*; 12.2, p. 139.

V. I. Zubov
Higher Anharmonicities in the Improved Unsymmetrized Self-Consistent Field Approximation for a Crystal (I): General Consideration
(b) Vol. **87**, No. 1, p. 385—394 (1978).
See also *6*, p. *70*; *8*, p. *85*.

I. J. Gupta and S. K. Trikha
The Reduced Conductivity of Dieletrics when Both Boundary and Dislocation Scatterings are Important at Low Temperatures
(b) Vol. **87**, No. 2, p. 513—516 (1978).
See also *6*, p. *71*; *8*, p. *86*.

V. I. Zubov
Higher Anharmonicities in the Improved Unsymmetrized Self-Consistent Field Approximation for a Crystal (II): Comparison of Calculations with Experiment
(b) Vol. **88**, No. 1, p. 43—50 (1978).
See also *6*, p. *73*; *8*, p. *87*.

I. J. Gupta and S. K. Trikha
Lattice Thermal Conductivity of Solid Neon in the Temperature Range 0.5 to 10 K
(b) Vol. **88**, No. 2, p. 815—818 (1978).
See also *6*, p. *74*; *8*, p. *87*.

A. T. Macrander
Density of Solid Krypton at Melting and Isochoric Equation of State of Solid Krypton and Solid Argon
(a) Vol. **48**, No. 2, p. 571—579 (1978).
See also *1.2*, p. *22*; 8, p. 87.

M. S. Kushwaha and S. S. Kushwaha
Frequency Spectra and Thermal Properties of Rare Gas Solids (Argon and Krypton)
(b) Vol. **89**, No. 2, p. 583—586 (1978).
See also *6*, p. *76*; 8, p. 88.

E. Schuberth and M. Creuzburg
Luminescence from Rare-Gas-Doped Solid Neon
(b) Vol. **90**, No. 1, p. 189–196 (1978).
See also *13.5.1*, p. *196*; *20.3*, p. *349*.

I. A. Oxton and O. Knop
Infrared Spectra of $SiHD_3$ in Crystalline Germane
(a) Vol. **50**, No. 1, p. K55–K58 (1978).
See also *1.2*, p. *24*; *20.1*, p. *337*.

Author Index

K. Abdelmoula	(a) **49**, 267 (1978)	*1.4*, p. *33*; 1.5, p. 38; 21.6, p. 404
S. G. Abdullaeva	(b) **90**, K77 (1978)	13.5.1, p. 196; *14.3.4*, p. *246*; *20.1*, p. *336*; 22.3, p. 459
D. E. Aboltin	(a) **47**, 667 (1978)	*10.2*, p. *114*; *20.3*, p. *345*; 22.5.2, p. 481
V. N. Abramov	(a) **48**, 287 (1978)	*20.1*, p. *331*; 22.6, p. 495
B. S. Acharya	(a) **47**, K165 (1978)	*10.2*, p. *114*; 20.3, p. 345; 22.5.2, p. 482
	(a) **48**, 235 (1978)	*10.2*, p. *115*; *20.3*, p. *345*; 22.5.2, p. 482
M. Adachi	(b) **86**, K151 (1978)	*19*, p. *306*; 22.5.2, p. 480
F. Adduci	(a) **50**, 257 (1978)	*16*, p. *267*; 22.5, p. 476
V. T. Adonkin	(b) **88**, K55 (1978)	*13.1*, p. *157*; 22.2.1, p. 447
G. C. Aers	(b) **85**, 171 (1978)	13.5, p. 187; *18*, p. *273*; *20.1*, p. *319*; 22.1, p. 368
M. L. Afanasev	(b) **90**, K35 (1978)	*19*, p. *314*; 22.5, p. 476
B. K. Agarwal	(b) **88**, 309 (1978)	*13.1*, p. *156*; *20*, p. *317*; 21, p. 360
G. S. Agarwal	(b) **85**, 301 (1978)	*13.3*, p. *164*; 13.5.2, p. 198; 20.1, p. 320; 21, p. 352; 22.4.1, p. 463
V. K. Agarwal	(a) **49**, 757 (1978)	*2*, p. *51*; *14.4*, p. *251*; 22.9, p. 528
V. K. Agrawal	(a) **46**, K145 (1978)	*10.2*, p. *112*; 22.5, p. 475
P. Airo	(a) **47**, K107 (1978)	*1.2*, p. *20*; *1.3*, p. *26*; 21.3, p. 394; 21.6, p. 403
G. Ya. Akimov	(a) **50**, K33 (1978)	*10.1*, p. *107*; *12.1*, p. *136*; 21.1, p. 381
A. F. Akkerman	(a) **48**, K47 (1988)	*11*, p. *128*; 21, p. 361; 21.1.1, p. 388
	(b) **89**, 329 (1978)	*11*, p. *129*; 13, p. 147
V. L. Aksionov	(b) **88**, K43 (1978)	*13*, p. *146*; *18.2*, p. *288*
M. C. Al-Edani	(b) **86**, 741 (1978)	*6*, p. *69*; 8, p. 84; 22.2.1, p. 446; 22.2.3, p. 452
	(b) **87**, K47 (1978)	*6*, p. *70*; 8, p. 85; 22.2.1, p. 446
A. Aladgadgyan	(b) **85**, K131 (1978)	*13.1*, p. *152*; 14.3, p. 218; 22.7, p. 501
G. E. Alberga	(b) **90**, K149 (1978)	*6*, p. *78*; *14.3.4*, p. *247*; 22.2.3, p. 455
C. Albers	(a) **47**, K31 (1978)	*14.3.1*, p. *227*; 22, p. 413
J. Albers	(a) **49**, 51 (1978)	*1.2*, p. *22*; *14.4.1*, p. *256*; 20, p. 317; 22.8.1, p. 515
G. Albrecht	(a) **49**, 569 (1978)	*14.2*, p. *215*
W. G. Albrecht	(b) **89**, 261 (1978)	*13.5.1*, p. *194*; 20.1, p. 332; 22.9, p. 527
R. Alcalá	(b) **85**, 283 (1978)	*10.2*, p. *109*; *20.1*, p. *320*; 22.5.3, p. 487
	(b) **85**, K23 (1978)	*11*, p. *123*; *20.1*, p. *320*; 22.5.3, p. 487
K. S. Aleksandrov	(a) **45**, 53 (1978)	1.2, p. 15; *14.4.1*, p. *253*; 19, p. 303; 20, p. 315; 22.8.1, p. 510
L. N. Aleksandrov	(a) **45**, 521 (1978)	9, p. 91; *14.3.3*, p. *233*; 22.1.2, p. 430
I. P. Aleksandrova	(a) **45**, 53 (1978)	1.2, p. 15; *14.4.1*, p. *253*; 19, p. 303; 20, p. 315; 22.8.1, p. 510
	(a) **49**, K185 (1978)	*1.2*, p. *23*; *14.4.1*, p. *256*; 19, p. 313; 22.8.1, p. 516
	(a) **50**, K157 (1978)	*14.4.1*, p. *257*; *19*, p. *314*; 22.8, p. 509

S. P. Aleksandrova	(a) **48**, 609 (1978)	*1.5*, p. *38*; *14.3.1*, p. *228*; 22.6, p. 495
E. S. Alekseev	(b) **90**, K113 (1978)	*13.1*, p. *161*; 22, p. 421
P. A. Alekseev	(b) **86**, K59 (1978)	*6*, p. *68*; *18*, p. *273*; 22.7, p. 501
O. K. Alekseeva	(a) **45**, K159 (1978)	*10.1*, p. *101*; 21.1.1, p. 385
	(a) **48**, K97 (1978)	*11*, p. *128*; 21.1, p. 375
	(a) **48**, K169 (1978)	*2*, p. *48*; *10.2*, p. *117*; 22, p. 416
Z. M. Alekseeva	(b) **88**, 321 (1978)	*12.2*, p. *140*; *13.1*, p. *156*; 14.3, p. 222; 22.2.1, p. 447
A. N. Aleshin	(a) **45**, 359 (1978)	*1.4*, p. *29*; 9, p. 91; 21, p. 353
H. Alexander	(a) **45**, 207 (1978)	4, p. 57; *10.2*, p. *109*; 22.2.1, p. 445; 22.2.2, p. 450; 22.2.3, p. 452
S. P. Alexandrova	(a) **49**, 781 (1978)	*13.3*, p. *167*; *14.3.2*, p. *231*
K. Alexopoulos	(a) **47**, K133 (1978)	*10.2*, p. *114*; 22.5.2, p. 482
A. R. Ali	(a) **45**, 47 (1978)	*10.1*, p. *99*; 12.1, p. 134; 21, p. 352; 21.2, p. 392
	(a) **45**, 419 (1978)	*10.1*, p. *100*; 12.1, p. 134; 21, p. 354
	(a) **46**, 697 (1978)	*10.1*, p. *102*; 12.1, p. 135; 14.1, p. 207; 21, p. 357
V. M. Aliabev	(a) **49**, 509 (1978)	*10.1*, p. *106*; 21.1.1, p. 389
F. K. Aliev	(b) **90**, K77 (1978)	13.5.1, p. 196; *14.3.4*, p. *246*; *20.1*, p. *336*; 22.3, p. 459
J. Alizon	(b) **85**, 603 (1978)	*18*, p. *273*; 19, p. 304; 22.9, p. 523
K. R. Allakhverdiev	(b) **87**, K115 (1978)	*6*, p. *72*; *20.1*, p. *328*; 22.3, p. 458
	(b) **88**, K5 (1978)	*6*, p. *73*; *20.1*, p. *329*; 22.8, p. 506
	(b) **90**, K77 (1978)	13.5.1, p. 196; *14.3.4*, p. *246*; *20.1*, p. *336*; 22.3, p. 459
C.-O. Almbladh	(b) **85**, 135 (1978)	*13*, p. *142*; 17, p. 269; 20, p. 315; 21, p. 352
P. J. Alonso	(b) **85**, K23 (1978)	*11*, p. *123*; *20.1*, p. *320*; 22.5.3, p. 487
W. H. M. Alsem	(a) **48**, 497 (1978)	*1.3*, p. *27*; *10.1*, p. *105*; 21.1, p. 376
H. L. Althaus	(b) **87**, 117 (1978)	*13.1*, p. *154*; *20.1*, p. *326*; 22, p. 413
A. M. Altshuler	(b) **90**, 733 (1978)	1, p. 10; *13.1*, p. *161*; 22, p. 421
V. I. Altukhov	(b) **85**, 769 (1978)	6, p. 66; 8, p. 83; *10.2*, p. *109*; 22.5.2, p. 479
M. Altwein	(a) **49**, 67 (1978)	*14.3.1*, p. *228*; 22.4, p. 462
R. Pérez Alvarez	(b) **90**, 409 (1978)	*20.1*, p. *335*; 22.2.1, p. 448
I. K. Amanis	(b) **89**, 531 (1978)	*10.2*, p. *119*; 19, p. 312; 22.5.2, p. 485
S. Amelinckx	(a) **45**, 553 (1978)	*1.2*, p. *16*; *1.4*, p. *29*; 21.1, p. 370
	(a) **46**, 399 (1978)	*1.3*, p. *25*; *1.4*, p. *30*; 21.1, p. 371; 21.6, p. 401
	(a) **46**, 479 (1978)	*1.1*, p. *12*; 1.2, p. 17; 22, p. 412
	(a) **47**, 105 (1978)	*1.3*, p. *25*; *1.4*, p. *30*; 21.1, p. 373; 21.6, p. 402
	(a) **47**, 241 (1978)	*1.2*, p. *18*; 1.3, p. 25; 1.4, p. 30; 21.1, p. 373; 21.6, p. 402
	(a) **47**, 539 (1978)	*1.2*, p. *19*; 1.3, p. 26; 1.4, p. 31; 21.1, p. 374; 21.6, p. 403
	(a) **47**, 555 (1978)	*1.3*, p. *26*; *1.4*, p. *31*; *21.1*, p. *374*; 21.6, p. 403
	(a) **48**, 383 (1978)	*1.3*, p. *27*; 1.4, p. 32; 21, p. 361; 21.1, p. 376
	(a) **48**, 39 (1978)	*1.2*, p. *20*; 1.3, p. 27; 1.4, p. 31; 21.1, p. 375; 21.6, p. 403
	(a) **48**, K5 (1978)	*1.4*, p. *31*; 10.2, p. 115; 22, p. 415
	(a) **49**, 337 (1978)	*1.1*, p. *13*; *1.4*, p. *33*; 21.1, p. 377; 21.6, p. 404
	(a) **49**, 655 (1978)	*1.1*, p. *14*; *1.4*, p. *33*; 21.1, p. 379; 21.6, p. 405
	(a) **49**, K217 (1978)	*1.4*, p. *31*; 10.2, p. 115; 22, p. 415
	(a) **50**, 53 (1978)	*1.3*, p. *28*; *1.4*, p. *34*; 21.1, p. 380; 21.6, p. 405
	(a) **50**, 349 (1978)	*1.1*, p. *14*; *1.4*, p. *34*; 22, p. 420
	(a) **50**, K67 (1978)	*1.6*, p. *41*; 22, p. 420
	(a) **50**, 587 (1978)	*1.3*, p. *28*; *1.4*, p. *34*; 10.1, p. 108; 21, p. 366; 21.1, p. 382
V. A. Ammer	(a) **49**, 413 (1978)	*1.4*, p. *33*; *1.5*, p. *38*; 3, p. 56; 21.6, p. 404

C. A. J. Ammerlaan	(b) **89**, 85 (1978)	*13.4*, p. *179*; 22.1.2, p. 436
	(b) **89**, 541 (1978)	*18.1*, p. *279*
E. Anastassakis	(b) **86**, 717 (1978)	*6*, p. *69*; 18.2, p. 285; *20.1*, p. *324*; 22.8.2, p. 518
Y. Andersson	(a) **49**, K153 (1978)	*3*, p. *56*; 18.2, p. 291; 21.1, p. 379; 21.1.1, p. 389
W. Andrä	(a) **49**, K97 (1978)	*14.1*, p. *210*; 21, p. 363
A. Andreeff	(b) **87**, 535 (1978)	*13.4*, p. *176*; 21.4, p. 397
G. A. Andreev	(a) **46**, 259 (1978)	*10.2*, p. *111*; 20.1, p. 324; 22.5.2, p. 479
	(a) **46**, 421 (1978)	*1.4*, p. *30*; *10.2*, p. *112*; 22.5.2, p. 480
V. N. Andreev	(a) **48**, K153 (1978)	*8*, p. *88*; 22.6, p. 495
J. B. Andrews	(b) **88**, 173 (1978)	*13.1*, p. *156*; *20.1*, p. *329*; 21.1, p. 375
A. M. Andriesh	(a) **50**, 195 (1978)	*2*, p. *52*; *14.3.3*, p. *237*; 14.3.4, p. 246; 22, p. 420
A. Anedda	(a) **50**, 643 (1978)	13.4, p. 186; *16*, p. *267*; 22.8, p. 508
M. A. Angadi	(b) **85**, K33 (1978)	*13.2*, p. *162*; 21.1, p. 368
Vo Hong Anh	(b) **86**, 585 (1978)	*6*, p. *69*; *13.5.2*, p. *198*; 20, p. 316; 22, p. 411
E. Antic-Fidancev	(b) **85**, 587 (1978)	*20.1*, p. *321*; *20.3*, p. *342*; 22.8, p. 503
M. V. Antisari	(a) **49**, K103 (1978)	*18.2.1*, p. *294*; 21.1, p. 379
P. R. Antoniewicz	(b) **86**, 339 (1978)	*13.3*, p. *164*
	(b) **86**, 645 (1978)	*13.3*, p. *165*; 21, p. 357
V. A. Antonov	(a) **46**, 55 (1978)	*1.4*, p. *29*; 20, p. 316; 22.8.2, p. 518
V. E. Antonov	(a) **48**, K185 (1978)	*18.2*, p. *290*; 21.1, p. 377
Vl. N. Antonov	(b) **88**, K83 (1978)	*13.1*, p. *157*; 21.4, p. 398
V. N. Antonov	(b) **88**, K83 (1978)	*13.1*, p. *157*; 21.4, p. 398
R. Aoki	(a) **50**, 187 (1978)	*14.2*, p. *215*; 18, p. 275; 20, p. 318; 21, p. 365
Y. Aoki	(a) **50**, 263 (1978)	*18.2*, p. *291*; 21.1, p. 381
G. Appold	(b) **86**, 557 (1978)	*13.4*, p. *174*; 19, p. 306; 20.1, p. 324; 22.2.3, p. 452
K. I. Arai	(a) **48**, 175 (1978)	2, p. 47; *18.2*, p. *289*; 21.1.1, p. 388
	(a) **50**, K25 (1978)	*2*, p. *52*; *18.2*, p. *292*; 21.1.1, p. 390
T. Arai	(b) **86**, 717 (1978)	*6*, p. *69*; 18.2, p. 285; *20.1*, p. *324*; 22.8.2, p. 518
S. Arajs	(b) **87**, 287 (1978)	*14.1*, p. *207*; 21.1, p. 372; 21.6, p. 402
	(b) **89**, K201 (1978)	*14.1*, p. *211*; 18.2, p. 290; 21.1, p. 378
M. M. Arakelyan	(b) **85**, 459 (1978)	*6*, p. *65*; *13*, p. *142*
	(b) **89**, 37 (1978)	*6*, p. *75*; 7, p. 82
L. G. Aravin	(a) **50**, K213 (1978)	*1.4*, p. *35*; *10.2*, p. *122*; 22.1, p. 423
I. Ardelean	(b) **87**, K137 (1978)	*19*, p. *308*; 22.8.1, p. 513
K. P. Arefiev	(a) **47**, K149 (1978)	10.2, p. 114; *11*, p. *127*; 22.1.2, p. 433
H. Arend	(a) **48**, 53 (1978)	*1.2*, p. *21*; 22.8, p. 506
Yu. D. Arepev	(b) **90**, 355 (1978)	*13.4*, p. *183*; *20.3*, p. *349*; 22.4.1, p. 466
V. Yu. Aristov	(a) **45**, 359 (1978)	*1.4*, p. *29*; 9, p. 91; 21, p. 353
K. A. Aristova	(b) **85**, 351 (1978)	*6*, p. *65*; *13.5.1*, p. *189*; 20.1, p. 320
G. R. Arkhipov	(b) **90**, K133 (1968)	*13.1*, p. *161*; 22, p. 421
V. E. Arkhipov	(a) **46**, K87 (1978)	11, p. 125; *14.2*, p. *214*; 21, p. 357
K. Arlauskas	(a) **48**, K149 (1978)	*2*, p. *48*; *14.3*, p. *223*; 22.1.3, p. 441
H. Arndt	(a) **46**, 533 (1978)	*14.4.1*, p. *254*; *20.1*, p. *325*; 22.9, p. 524
V. P. Arora	(a) **49**, 757 (1978)	*2*, p. *51*; *14.4*, p. *251*; 22.9, p. 528
M. A. Arriandiaga	(a) **48**, 53 (1978)	*1.2*, p. *21*; 22.8, p. 506
P. A. Arsenev	(a) **46**, 55 (1978)	*1.4*, p. *29*; 20, p. 316; 22.8.2, p. 518
A. Art	(a) **46**, K167 (1978)	*11*, p. *126*; 22, p. 412
C. Arzoumanian	(b) **90**, 207 (1978)	*6*, p. *77*; *14.2*, p. *215*; 21, p. 364
H. Asami	(b) **88**, 623 (1978)	*13.5.1*, p. *193*; 20.3, p. 346; 22.9, p. 526
S. Asano	(b) **89**, 663 (1978)	*13.4*, p. *181*; *20.3*, p. *348*; 22.4.1, p. 466; 22.4.2, p. 469
M. Asche	(a) **48**, 323 (1978)	*14.3.4*, p. *243*; 22.1.2, p. 435
J. A. Ashby	(a) **47**, 407 (1978)	*1.4*, p. *31*; *14.1*, p. *208*; 21.1, p. 374; 21.1.1, p. 387

G. J. Ashwell	(b) **85**,	K7 (1978)	*14.1*, p. *205*; *14.3*, p. *217*; 22.9, p. 522
	(b) **86**,	705 (1978)	*1.1*, p. *11*; 14, p. 203; 22.9, p. 524
B. M. Askerov	(b) **86**,	103 (1978)	14.3.1, p. 226; *15*, p. *259*
H. Asmat	(a) **45**,	K149 (1978)	*1.1*, p. *11*; *18.4*, p. *298*; 21, p. 355; 21.4, p. 396
D. Aspnes	(b) **88**,	135 (1978)	*13.1*, p. *156*; 17, p. 270; 20.1, p. 328; 22.1.2, p. 434
R. Assenov	(a) **48**,	K193 (1978)	*14.3*, p. *223*; 15, p. 261; 22, p. 416
B. Astie	(a) **45**,	493 (1978)	*10.1*, p. *100*; *18.2*, p. *283*; 21.1.1, p. 385
H. U. Åström	(b) **85**,	K31 (1978)	*8*, p. *83*; 21.4, p. 396
E. D. Atanasova	(a) **50**,	363 (1978)	*1.5*, p. *39*; *14.3.3*, p. *238*; 14.4, p. 252; 22.6, p. 497
K. Atobe	(a) **50**,	K77 (1978)	*10.2*, p. *121*; 11, p. 130; *20.3*, p. *350*; 22.5.3, p. 488
H. Atzmüller	(b) **89**,	349 (1978)	*13.4*, p. *181*; *13.5.1*, p. *194*; 22, p. 418
H. P. Aubauer	(b) **90**,	345 (1978)	*14.1*, p. *211*; 21, p. 364
W. Auch	(a) **49**,	563 (1978)	1.2, p. 23; 6, p. 77; *19*, p. *313*; 22.9, p. 528
W. Auer	(a) **49**,	149 (1978)	*10.1*, p. *105*; *14.1*, p. *210*; 21.1, p. 377
V. T. Avanesyan	(a) **49**,	K173 (1978)	*14.4.1*, p. *256*; 22.6, p. 497
V. L. Averyanov	(a) **48**,	K93 (1978)	*2*, p. *47*; 16, p. 265; *20.1*, p. *330*; 22, p. 415
F. Aymerich	(a) **46**,	703 (1978)	*1.2*, p. *18*; *2*, p. *44*; 21.7, p. 406
G. Azhdarov	(b) **87**,	163 (1978)	*13.4*, p. *174*; *14.3.4*, p. *242*; 22.2.1, p. 446
J. Baars	(a) **49**,	483 (1978)	*14.3.4*, p. *245*; 22, p. 419
T. Babauchi	(b) **89**,	571 (1978)	*14.1*, p. *211*; 21, p. 363
V. Babeş	(a) **45**,	617 (1978)	14.3, p. 218; *15*, p. *259*; 22.6, p. 491
L. P. Babichev	(a) **50**,	K5 (1978)	*18.2*, p. *292*; 21.1.1, p. 390
V. P. Babiyczuk	(a) **50**,	K197 (1978)	*14*, p. *204*; *18*, p. *276*; 22.9, p. 529
M. Babout	(a) **48**,	459 (1978)	*17*, p. *270*; 21, p. 361
A. Z. Badalov	(b) **90**,	K77 (1978)	13.5.1, p. 196; *14.3.4*, p. *246*; *20.1*, p. *336*; 22.3, p. 459
G. Bader	(b) **89**,	53 (1978)	*13.5.3*, p. *200*; 22, p. 416
S. P. S. Badwal	(a) **49**,	K181 (1978)	*13.3*, p. *167*; *14.3.4*, p. *246*; 22.6, p. 497
V. E. Bagiev	(a) **50**,	K141 (1978)	6, p. 79; 13.1, p. 161; *20.1*, p. *338*; 22.8, p. 508
V. M. Baginskii	(b) **90**,	423 (1978)	*1.2*, p. *23*; *13*, p. *148*; 14, p. 204; 18, p. 275
C. Balasubramanian	(a) **48**,	K71 (1978)	10.1, p. 104; *14.1*, p. *209*; 21.6, p. 403
	(a) **49**,	K213 (1978)	1.2, p. 23; 2, p. 51; *14.1*, p. *211*; 21.6, p. 405; 22.1.3, p. 442
K. Balasubramanyam	(a) **47**,	711 (1978)	*10.2*, p. *114*; 11, p. 127; *20.1*, p. *328*; 22.5.2, p. 481
	(b) **90**,	375 (1978)	*10.2*, p. *120*; *20.3*, p. *349*; 22.5.2, p. 485
A. M. Balbashov	(a) **46**,	K47 (1978)	*18.4*, p. *299*; *19*, p. *306*; 22.8.1, p. 512
K. Balcerek	(a) **47**,	K125 (1978)	*8*, p. *86*; 21, p. 360
A. A. Balchin	(a) **47**,	K173 (1967)	*1.2*, p. *20*; 22, p. 414
L. Baldassarre	(a) **46**,	589 (1978)	*14.3.4*, p. *241*; 22.8, p. 504
	(a) **50**,	257 (1978)	*16*, p. *267*; 22.5, p. 476
L. Baldi	(a) **48**,	523 (1978)	*10.2*, p. *116*; 14.3.3, p. 235; 22.1.2, p. 435
Yu. V. Baldokhin	(b) **87**,	K101 (1978)	*5*, p. *61*; 18.3, p. 295; 22.8.2, p. 519
	(a) **47**,	39 (1978)	*5*, p. *61*; 18.4, p. 300; 22.6, p. 493
M. Baleva	(b) **88**,	335 (1978)	*13.1*, p. *156*; *20.1*, p. *329*; 22, p. 414
C. A. B. Ball	(a) **46**,	707 (1978)	*1.5*, p. *36*; *10.2*, p. *112*
M. A. Ball	(b) **87**,	145 (1978)	*13*, p. *145*; *14.1*, p. *207*; 21.2, p. 392
B. Balland	(a) **47**,	251 (1978)	*13.4*, p. *175*; *14.3.2*, p. *230*; 22.6, p. 493
I. Balslev	(b) **88**,	155 (1978)	*13.5.1*, p. *192*; 20.1, p. 329; 22.4, p. 461
R. von Baltz	(b) **89**,	419 (1978)	*14.4.1*, p. *256*; *16*, p. *266*; 22.8.1, p. 516
R. Balzer	(b) **85**,	553 (1978)	*10.2*, p. *109*; *11*, p. *123*; 22.5.2, p. 478
A. Banerjee	(a) **46**,	723 (1978)	*1.5*, p. *36*; *14.3.1*, p. *226*; 16, p. 263; 22.4.4, p. 472

E. Bangert	(b) **87**, 99 (1978)	6, p. 70; *14.3.4*, p. *242*; 22.4.1, p. 464
C. Bansal	(a) **48**, K119 (1978)	*18.2*, p. *289*; 21.1, p. 376; 21.1.1, p. 389
L. K. Bansal	(b) **87**, K127 (1978)	*4*, p. *58*; *12.2*, p. *139*; 22.6, p. 494
M. L. Bansal	(b) **90**, 415 (1978)	*6*, p. *78*; *20.1*, p. *336*; 22.5.4, p. 490
K. G. Bansigir	(a) **45**, 83 (1978)	12.2, p. 138; *20*, p. *315*; 22.5.4, p. 489
J. J. Bara	(a) **45**, K81 (1978)	*5*, p. *60*; 22.8.1, p. 511
M. Baran	(b) **90**, K59 (1978)	*19*, p. *314*; *20.1*, p. *336*; 22.8, p. 507
B. Baranov	(a) **49**, 629 (1978)	*1.5*, p. *39*; 13.4, p. 182; 14.3.1, p. 228; 20.1, p. 335; 20.3, p. 348; 22.2.4, p. 456
P. G. Baranov	(b) **86**, K67 (1978)	*19*, p. *305*; *20.1*, p. *323*; 22.5.2, p. 479
P. I. Baranskii	(b) **86**, 765 (1978)	13, p. 144; *14.3*, p. *219*; 22.2.1, p. 446
T. W. Barbee, Jr.	(a) **47**, 357 (1978)	*14.2*, p. *214*
H. J. von Bardeleben	(a) **48**. K145 (1978)	*10.2*, p. *116*; 19, p. 311; 20.1, p. 331; 22.8, p. 506
I. B. Barkan	(a) **45**, K17 (1978)	*13.4*, p. *170*; 14.4.1, p. 253; 16, p. 262; *20.1*, *321*; 22.8.1, p. 510
K. Bärner	(b) **88**, 13 (1978)	18.1, p. 277; *18.2*, p. *281*; 18.3, p. 295; *18.4*, p. *298*; 21.1, p. 368
	(a) **49**, 545 (1978)	6, p. 76; *12.1*, p. *136*; 18.1, p. 279; 18.2, p. 290; 21.1, p. 379
J. Barrau	(b) **88**, 261 (1978)	*13.5.1*, p. *192*; 20.3, p. 345; 22.2.2, p. 450
F. C. de Sa Barreto	(b) **87**, K105 (1978)	*1.2*, p. *19*; *2*, p. *45*; 14.4, p. 249
A. S. Barrière	(b) **90**, 147 (1978)	*20.1*, p. *335*; 22.5, p. 476
F. de S. Barros	(a) **45**, K139 (1978)	*5*, p. *60*; 22.8.2, p. 518
Š. Barta	(b) **86**, 183 (1978)	2, p. 43; *13.1*, p. *152*; *13.5.1*, p. *190*; 18, p. 273; 22, p. 410
K. Bartkowski	(a) **48**, 215 (1978)	*14.1*, p. *209*; 21.7, p. 407
J. Barton	(a) **45**, K105 (1978)	*5*, p. *60*; *18.2*, p. *283*; 21.1, p. 370; 21.1.1, p. 385
	(a) **48**, K51 (1978)	*1.2*, p. *21*; 5, p. 62; 21.1.1, p. 388
I. Bartoš	(b) **87**, 111 (1978)	*13.3*, p. *165*
J. Bartoš	(b) **85**, K127 (1978)	*13.3*, p. *164*
D. Bartzokas	(a) **47**, 459 (1978)	*1.2*, p. *19*; 1.4, p. 31; 22.3, p. 458
A. K. Barua	(a) **45**, 657 (1978)	*9*, p. *91*; 21.1, p. 370; 21.1.1, p. 385; 21.6, p. 401
B. P. Barua	(b) **86**, K55 (1978)	*1.2*, p. *17*; *14.2*, p. *214*; 21, p. 356
V. G. Baryshevskii	(b) **88**, 351 (1978)	*1.1*, p. *13*; *11*, p. *127*
V. K. Bashenov	(b) **86**, K7 (1978)	*4*, p. *57*; 12.2, p. 138; 22.1, p. 422; 22.1.1, p. 424; 22.1.2, p. 431
	(b) **86**, K139 (1978)	*1.1*, p. *11*; 1.2, p. 17; 22.2.4, p. 456
	(b) **87**, K73 (1978)	*1.2*, p. *19*; *13.1*, p. *155*; 22.1, p. 422
	(b) **87**, K77 (1978)	*13.1*, p. *155*; 22.7, p. 502
	(b) **88**, K161 (1978)	*6*, p. *74*; 22.1, p. 422; 22.1.1, p. 427; 22.1.2, p. 435
	(b) **89**, K5 (1978)	*4*, p. *59*; *13.1*, p. *158*; 22, p. 417
	(b) **89**, K115 (1978)	*6*, p. *76*; 22.3, p. 459
	(b) **90**, K5 (1978)	*4*, p. *59*; *13.1*, p. *160*; 22.3, p. 459
I. O. Bashkin	(b) **87**, 369 (1978)	*1.2*, p. *18*; 12.1, p. 135; 21.4, p. 397
E. M. Baskin	(b) **85**, K135 (1978)	*14.3.4*, p. *240*; 16, p. 262; 22.2, p. 443
D. Bassett	(a) **49**, 483 (1978)	*14.3.4*, p. *245*; 22, p. 419
H. Bässler	(b) **85**, 145 (1978)	*16*, p. *262*; 22.9, p. 522
	(b) **88**, 653 (1978)	*13.1*, p. *157*; *16*, p. *265*; 22.9, p. 526
	(a) **48**, K15 (1978)	*13.3*, p. *166*; *16*, p. *264*; 22.9, p. 526
	(a) **50**, 95 (1978)	*1*, p. *10*; 22.9, p. 528
J. H. Basson	(a) **46**, 707 (1978)	*1.5*, p. *36*; *10.2*, p. *112*
A. N. Basu	(b) **89**, K35 (1978)	*4*, p. *59*; *12.2*, p. *140*; 22.5.2, p. 484
C. Basu	(b) **86**, 379 (1978)	*13.4*, p. *172*; 20.1, p. 322; 22.4, p. 461
P. M. Batev	(a) **45**, 671 (1978)	*14.3.3*, p. *233*; 14.3.4, p. 240; 22.2.1, p. 445
G. Battaglin	(a) **49**, 347 (1978)	*11*, p. *129*; 22.1.2, p. 437

E. G. Batyev	(b) **90**, 449 (1978)	*1.2*, p. *24*; *13.5.3*, p. *200*
R. Baubinas	(a) **50**, K63 (1978)	*7*, p. *82*; *16*, p. *267*; 22.4.1, p. 466; 22.4.2, p. 469; 22.4.3, p. 471
A. Baudry	(b) **90**, K169 (1978)	*10.2*, p. *121*; 13, p. 149; 14.4, p. 252; 22.6.1, p. 500
C. L. Bauer	(a) **47**, 321 (1978)	*1.5*, p. *37*; 14.1, p. 207; 21.1, p. 373
	(a) **48**, 555 (1978)	*1.5*, p. *38*; 21.6, p. 403
G. Bauer	(b) **86**, 205 (1978)	13.1, p. 152; *14.3.4*, p. *241*; 20.1, p. 322; *20.3*, p. *343*; 22.2.3, p. 452
G. S. Bauer	(a) **46**, 557 (1978)	1.3, p. 25; *10.1*, p. *102*; 21, p. 357
H. J. Bauer	(a) **47**, 445 (1978)	*1.2*, p. *19*; 18.2, p. 287; 21.1, p. 374
H. P. Bauer	(a) **49**, 127 (1978)	14.3.2, p. 231; *14.3.4*, p. *244*; 22, p. 417
F. Baumann	(a) **49**, 103 (1978)	2, p. 49; *14.1*, p. *210*; *14.3.1*, p. *228*; 21.6, p. 404; 21.7, p. 407
I. Baumann	(b) **89**, K115 (1978)	*6*, p. *76*; 22.3, p. 459
M. Baumgärtner	(b) **86**, 527 (1978)	*13.5.1*, p. *191*; *20.3*, p. *343*; 22.4, p. 461
V. D. Baurin	(a) **46**, K83 (1978)	*18.2.1*, p. *293*; *20.1*, p. *325*; 22.8.2, p. 519
J. Beamonte	(b) **85**, 283 (1978)	*10.2*, p. *109*; *20.1*, p. *320*; 22.5.3, p. 487
J. Beauvillain	(a) **46**, 659 (1978)	*1.1*, p. *12*
J. Becherer	(a) **49**, K121 (1978)	*13.4*, p. *182*; *17*, p. *270*; 22, p. 419
F. Bechstedt	(b) **85**, 253 (1978)	*6*, p. *64*; *20.1*, p. *319*; 22, p. 409
	(b) **85**, 569 (1978)	*13.1*, p. *151*; *13.5.1*, p. *190*; 20, p. 315; 22.1, p. 422
	(b) **88**, 163 (1978)	*6*, p. *73*; 13.5.1, p. 192; *20.1*, p. *329*; 22.4.1, p. 465
K. D. Becker	(b) **87**, 589 (1978)	*9*, p. *92*; *19*, p. *308*; 22.5.1, p. 477
M. Becker	(a) **47**, 445 (1978)	*1.2*, p. *19*; 18.2, p. 287; 21.1, p. 374
O. Beckman	(a) **49**, K153 (1978)	*3*, p. *56*; 18.2, p. 291; 21.1, p. 379; 21.1.1, p. 389
P. Becla	(b) **88**, K153 (1978)	*13.1*, p. *158*; *20.1*, p. *331*; 22.3, p. 458
U. Behn	(b) **88**, 699 (1978)	*13*, p. *147*; 14, p. 204
G. Behnke	(b) **90**, 53 (1978)	*13.5.1*, p. *195*; 13.5.3, p. 200; 22.5, p. 476
I. T. Belash	(a) **48**, K185 (1978)	*18.2*, p. *290*; 21.1, p. 377
G. L. Belenkii	(b) **85**, 453 (1978)	*13.5.1*, p. *189*; *20.3*, p. *341*; 22.3, p. 457
I. M. Belii	(a) **45**, 343 (1978)	*1.2*, p. *15*; *1.5*, p. *36*; 11, p. 123; 21.1, p. 369
A. I. Belkind	(b) **85**, 465 (1978)	13.1, p. 151; *17*, p. *269*; 22.9, p. 522
I. A. Belobrova	(a) **50**, K157 (1978)	*14.4.1*, p. *257*; *19*, p. *314*; 22.8, p. 509
A. G. Belous	(a) **48**, 183 (1978)	*1*, p. *9*; 5, p. 61; *14.4.1*, p. *255*; 18.2, p. 289; 18.4, p. 301; 19, p. 310; 22.8.1, p. 514
G. N. Belozerskii	(a) **46**, 131 (1978)	*5*, p. *61*; *18.2*, p. *284*
S. Benci	(b) **90**, 657 (1978)	13.4, p. 185; *20.1*, p. *337*; *20.3*, p. *350*; 22.5.2, p. 486
V. A. Benderskii	(b) **86**, 449 (1978)	*13.5.1*, p. *190*; *20.3*, p. *343*; 22.9, p. 524
S. D. Beneslavskii	(b) **88**, 221 (1978)	*13.1*, p. *156*; *14.3.4*, p. *243*; 14.4, p. 249; 22.4.4, p. 473
L. Benguigui	(a) **46**, 337 (1978)	*14.4.1*, p. *254*; 22.8.1, p. 512
	(a) **47**, 71 (1978)	1.4, p. 30; *14.4*, p. *249*; 22.9, p. 525
G. G. Bentini	(a) **49**, 347 (1978)	*11*, p. *129*; 22.1.2, p. 437
A. M. Berezhkovskii	(b) **85**, 121 (1978)	*13*, p. *142*
F. van den Berg	(b) **90**, 643 (1978)	*10.1*, p. *107*; 21, p. 365
G. Berg	(b) **89**, 595 (1978)	*10.2*, p. *119*; 19, p. 312; 22.5.2, p. 485
H. Berg	(a) **49**, 545 (1978)	6, p. 76; *12.1*, p. *136*; 18.1, p. 279; 18.2, p. 290; 21.1, p. 379
B. Bergersen	(b) **85**, 715 (1978)	*13*, p. *143*; *20*, p. *315*; 22, p. 409
I. Bergert	(b) **90**, 621 (1978)	*13.1*, p. *161*; 21.1, p. 381
W. Bergholz	(a) **49**, 489 (1978)	*5*, p. *62*; 9, p. 94; *10.2*, p. *119*; 22.1.2, p. 437
J. Bergmann	(b) **86**, 215 (1978)	*6*, p. *67*; *20.1*, p. *322*; 22.8.1, p. 511
N. F. Berk	(b) **89**, K167 (1978)	*13.5.1*, p. *195*; *19*, p. *313*; 22, p. 418

K. Berndt	(b) **86**, K93 (1978)	*13*, p. *144*; 20, p. 316; 21.1, p. 372; 21.1.1, p. 386
	(b) **90**, 487 (1978)	*13.1*, p. *160*; *20*, p. *318*; 21.1, p. 381; 21.1.1, p. 390
Hj. Bernhardt	(a) **45**, 353 (1978)	13.4, p. 170; *20.1*, p. *321*; *20.3*, p. *341*; 22.8.1, p. 510
	(b) **87**, 213 (1978)	*10.2*, p. *112*; *13.5.3*, p. *200*; 22.8.2, p. 519
H. Bertagnolli	(a) **49**, 463 (1978)	*1*, p. *10*; *2*, p. *50*; 22.9, p. 527
G. Berthet	(b) **85**, 603 (1978)	*18*, p. *273*; 19, p. 304; 22.9, p. 523
S. Besse	(a) **48**, 99 (1978)	*10.1*, p. *104*; 12.1, p. 135; 21.1, p. 375
A. Bessis	(a) **45**, K71 (1978)	*11*, p. *123*; *14.1*, p. *205*; 21.1, p. 369
E. Beyer	(b) **87**, K35 (1978)	*10.2*, p. *113*; 11, p. 126; *19*, p. *307*; 22.8.1, p. 513
A. S. Bhalla	(a) **45**, K1 (1978)	*11*, p. *123*; 14.4.1, p. 253; 22.8.1, p. 510
	(a) **48**, 431 (1978)	*1.6*, p. *41*; *14.4.1*, p. *256*; 22.8.1, p. 515
	(a) **50**, 661 (1978)	*1.2*, p. *24*; 8, p. 90; 12.2, p. 141; 20.1, p. 338; 22.9, p. 528
T. Bhaskar Rao	see under letter "R"	
K. S. Bhatki	(a) **47**, 691 (1978)	*10.2*, p. *114*; 11, p. 127; 14.4, p. 249; 19, p. 309; 22.9, p. 526
A. K. Bhattacharya	(b) **85**, 519 (1978)	13.1, p. 151; *14.1*, p. *206*; 21, p. 354
D. P. Bhattacharya	(b) **88**, 283 (1978)	*13.4*, p. *177*; *14.3.4*, p. *243*; 18, p. 274; 22.1.1, p. 426
S. Bhushan	(a) **46**, K123 (1978)	*20.3*, p. *343*; 22.4, p. 461
	(a) **49**, K167 (1978)	*20.3*, p. *349*; 22.4, p. 462
R. S. de Biasi	(b) **87**, K29 (1978)	*19*, p. *307*; 22.6, p. 493
R. W. Bicknell	(a) **46**, 225 (1978)	*11*, p. *125*; 13.4, p. 173; 14.3, p. 219; 22.1.2, p. 431
R. Biederbick	(b) **87**, K35 (1978)	*10.2*, p. *113*; 11, p. 126; *19*, p. *307*; 22.8.1, p. 513
	(b) **89**, 449 (1978)	*13.4*, p. *181*; *19*, p. *312*; 22.8.1, p. 516
A. Biedermann	(a) **50**, 147 (1978)	*8*, p. *89*; 14.1, p. 212; 21.1, p. 381
Z. Biegański	(a) **46**, K39 (1978)	*8*, p. *84*; *18*, p. *274*; 21.4, p. 397
B. Biering	(b) **85**, 217 (1978)	13.4, p. 169; *14.3.4*, p. *240*; *20.1*, p. *319*; 22.4.1, p. 463
S. W. Biernacki	(b) **87**, 607 (1978)	*13.4*, p. *176*; *20.1*, p. *327*; 22.4.1, p. 464; 22.4.2, p. 468; 22.4.3, p. 470
J. Bigot	(a) **45**, K71 (1978)	*11*, p. *123*; *14.1*, p. *205*; 21.1, p. 369
J. Bihr	(a) **50**, 171 (1978)	*9*, p. *94*; 21.6, p. 405
	(a) **50**, 543 (1798)	*9*, p. *95*; 21.6, p. 405
E. J. Bijvank	(b) **87**, 697 (1978)	*18.1*, p. *278*; *19*, p. *308*; 22.8.1, p. 513
O. Bílek	(b) **85**, 225 (1978)	*13*, p. *142*
H. Bill	(b) **89**, K49 (1978)	*13.4*, p. *180*; 19, p. 311; 22.5.4, p. 489
H. Bińczycka	(a) **45**, 591 (1978)	*18.2*, p. *283*; *18.4*, p. *298*; 22.8, p. 503
R. Bindemann	(b) **87**, 169 (1978)	*13.4*, p. *174*; *20.3*, p. *344*; 22.2.2, p. 450
	(a) **49**, 331 (1978)	*13.4*, p. *181*; 13.5.1, p. 194; *20.3*, p. *347*; 22.2.2, p. 451
	(b) **89**, 561 (1978)	*13.5.1*, p. *195*; *20.3*, p. *348*; 22.2.2, p. 451
U. Birkholz	(a) **47**, 85 (1978)	*1.6*, p. *41*; 22, p. 413
	(a) **49**, 127 (1978)	14.3.2, p. 231; *14.3.4*, p. *244*; 22, p. 417
E. Birkicht	(b) **90**, 77 (1978)	*13.5.1*, p. *195*; *20.1*, p. *335*; 22.4.1, p. 466
Y. Bismuth	(a) **45**, K71 (1978)	*11*, p. *123*; *14.1*, p. *205*; 21.1, p. 369
A. Bittar	(a) **48**, K123 (1978)	*10.1*, p. *105*; *14.1*, p. *209*; 21.7, p. 407
H. Bittel	(a) **49**, K21 (1978)	*18.2*, p. *290*; 21.1.1, p. 389
A. Bivas	(b) **86**, 159 (1978)	*13.5.1*, p. *190*; 20.1, p. 321; 20.3, p. 342; 22.5, p. 474
W. Bizzaro	(b) **89**, K167 (1978)	*13.5.1*, p. *195*; *19*, p. *313*; 22, p. 418

R. E. Blaha	(a) **50**, 551 (1978)	11, p. 131; *14.3.1*, p. *229*; *14.3.3*, p. *238*; 22.6, p. 497
J. P. Blanc	(b) **85**, 603 (1978)	*18*, p. *273*; 19, p. 304; 22.9, p. 523
G. Blasek	(a) **50**, K247 (1978)	*1.6*, p. *41*; 11, p. 132; 21.1.1, p. 391
W. Blau	(a) **47**, 157 (1978)	8, p. 85; *10.1*, p. *103*; 14.1, p. 207; 18.2, p. 286; 21.1, p. 373; 21.1.1, p. 387
P. Blaudeck	(b) **89**, 199 (1978)	*2*, p. *49*; 13, p. 147; 21.1, p. 377; 21.1.1, p. 389
K. C. Bleijenberg	(a) **47**, 589 (1978)	10.2, p. 114; *14.4*, p. *249*; *20.3*, p. *345*; 22.5.2, p. 481
G. L. Bleris	(a) **50**, 579 (1978)	*1.2*, p. *24*; *1.3*, p. *28*; 10.2, p. 122; 22.3, p. 459
	(a) **50**, 703 (1978)	*1.1*, p. *14*; *1.4*, p. *35*
D. I. Bletskan	(b) **90**, K49 (1978)	*13.5.1*, p. *196*; *20.1*, p. *336*; 22, p. 420
R. Blinc	(a) **46**, 239 (1978)	*10.2*, p. *111*; *19*, p. *306*; 22.9, p. 524
	(b) **86**, K119 (1978)	*1.2*, p. *17*; 14.4.1, p. 254; *19*, p. *306*; 22.8.1, p. 512
	(b) **87**, K105 (1978)	*1.2*, p. *19*; *2*, p. *45*; 14.4, p. 249
J. Blinowski	(b) **89**, 521 (1978)	*13.1*, p. *159*; 14.4, p. 251; 18, p. 275; 22, p. 418
A. le Bloa	(a) **50**, 311 (1978)	*13.4*, p. *184*; 14.3.3, p. 238; 22.2.1, p. 449
M. D. Blokh	(b) **85**, K135 (1978)	*14.3.4*, p. *240*; 16, p. 262; 22.2, p. 443
I. Bloomer	(b) **89**, 309 (1978)	*10.2*, p. *117*; 22, p. 417
D. Bloor	(a) **45**, 483 (1978)	13.4, p. 171; *19*, p. *304*; 22.9, p. 523
	(a) **46**, 141 (1978)	13.4, p. 173; *19*, p. *305*; 22.9, p. 523
	(a) **46**, 619 (1978)	13.4, p. 174; *19*, p. *306*; 22.9, p. 524
K. Blüthner	(a) **49**, 569 (1978)	*14.2*, p. *215*
H. J. Blythe	(a) **46**, K95 (1978)	*10.1*, p. *102*; 11, p. 125; *18.2*, p. *285*; 21.1, p. 372; 21.1.1, p. 386
	(a) **49**, 609 (1978)	10.1, p. 106; *11*, p. *130*; *18.2*, p. *290*; 21.1.1, p. 389
A. D. Boardman	(b) **85**, 171 (1978)	13.5, p. 187; *18*, p. *273*; *20.1*, p. *319*; 21.1, p. 368
A. I. Bobrysheva	(b) **88**, 315 (1978)	*13.5.1*, p. *193*; 22.2.1, p. 447
A. V. Bobyl	(a) **48**, 249 (1978)	*13.4*, p. *178*; *16*, p. *264*; 22.4.1, p. 465
B. Bochu	(a) **48**, 581 (1978)	*1.1*, p. *13*; *5*, p. *62*; 22.8.1, p. 515
I. Boczek	(a) **50**, K109 (1978)	12.2, p. 141; *14.4.1*, p. *257*; 22.8.1, p. 516
I. V. Bodnar	(b) **86**, K171 (1978)	*6*, p. *70*; *20.1*, p. *325*; 22.8, p. 504
K. W. Böer	(a) **49**, 13 (1978)	*14.3.3*, p. *236*; *16*, p. *265*; 22, p. 417; 22.4.1, p. 465
	(a) **49**, 455 (1978)	*14.3.3*, p. *237*; *16*, p. *266*; 22, p. 418; 22.4.1, p. 466
E. Boesman	(b) **85**, 639 (1978)	*19*, p. *304*; 22.5.2, p. 479
	(b) **89**, 151 (1978)	*19*, p. *311*; *20.1*, p. *332*; 22.5.2, p. 483
	(b) **89**, K25 (1978)	*19*, p. *311*; 22.5.2, p. 484
B. F. Bogacz	(a) **45**, K81 (1978)	*5*, p. *60*; 22.8.1, p. 511
Yu. V. Bogatyrev	(a) **48**, K127 (1978)	*11*, p. *128*; *14.3.2*, p. *231*; 22.1.2, p. 436; 22.6, p. 495
V. A. Bogatyriov	(a) **47**, K5 (1978)	1.5, p. 37; *11*, p. *126*; 22.2.3, p. 453
H. H. Boghossian	(b) **85**, K99 (1978)	*6*, p. *66*; *8*, p. *84*; 22.6, p. 491
	(b) **88**, 417 (1978)	*6*, p. *74*; *8*, p. *87*; 22.1.1, p. 427
	(b) **89**, K65 (1978)	*6*, p. *76*; *8*, p. *88*; 22.6, p. 496
W. Bogusz	(a) **47**, K103 (1978)	*14.4*, p. *249*; 22.5.3, p. 487
A. Bohg	(a) **46**, 445 (1978)	*1.5*, p. *36*; 10.2, p. 112; 22.1.2, p. 432
J. Bohm	(a) **49**, 305 (1978)	3, p. 56; 20.1, p. 333; *20.2*, p. *339*; *20.3*, p. *347*; 22.8.2, p. 520
W. Bohnert	(b) **89**, 431 (1978)	13.5.1, p. 195; *20.2*, p. *339*; *20.3*, p. *348*; 22.4, p. 462; 22.4.1, p. 465; 22.4.3, p. 471
C. F. Bohren	(b) **86**, K27 (1978)	*20.1*, p. *323*; 22.6, p. 492

D. Boice	(a) **48**, 551 (1978)	*1.3*, p. *27*; *10.1*, p. *105*; 21, p. 362; 21.1, p. 376
G. A. Boiko	(b) **85**, 111 (1978)	*13.5.1*, p. *189*; *20.3*, p. *341*; 22.4.1, p. 463; 22.4.2, p. 468
S. A. Boiko	(b) **90**, K49 (1978)	*13.5.1*, p. *196*; *20.1*, p. *336*; 22, p. 420
E. I. Boikova	(a) **50**, K255 (1978)	14.4.1, p. 258; *16*, p. *268*; *17*, p. *271*; 22.8.1, p. 517
M. Boisson	(a) **46**, 387 (1978)	*10.2*, p. *111*; 22.6, p. 492
E. L. Bokhenkov	(b) **85**, 331 (1978)	*6*, p. *65*; 22.9, p. 522
B. S. Bokshtein	(a) **45**, 359 (1978)	*1.4*, p. *29*; 9, p. 91; 21, p. 353
S. S. Bolgov	(a) **50**, 723 (1978)	*14.3*, p. *225*; 22.2.3, p. 455
W. Bollmann	(a) **45**, 187 (1978)	*10.2*, p. *109*; 14.4, p. 248; 22.5.3, p. 487
L. P. Bolshakov	(b) **87**, K1 (1978)	13.1, p. 155; *20.1*, p. *326*; 22.2.1, p. 446
	(a) **48**, 281 (1978)	13.4, p. 179; *20.3*, p. *346*; 22.2.3, p. 453
	(a) **48**, K115 (1978)	*13.4*, p. *179*; 20.1, p. 331; 22.2.3, p. 453
	(a) **49**, 183 (1978)	*13.4*, p. *180*; *20.1*, p. *333*; 22.2.1, p. 448; 22.2.3, p. 454
	(a) **49**, K189 (1978)	*20.3*, p. *349*; 22.2.1, p. 448; 22.2.3, p. 454
A. Bombik	(a) **50**, K17 (1978)	*1.1*, p. *14*; *18.3*, p. *296*; 22.8.2, p. 520
V. M. Bondar	(b) **87**, K97 (1978)	*14.3.4*, p. *242*; 22.1.3, p. 441
H. P. Bonde	(b) **88**, 269 (1978)	*13.1*, p. *156*; 21.6, p. 403
B. Boniface	(a) **48**, 417 (1978)	*6*, p. *75*; *20.1*, p. *331*; 22.6, p. 495
M. C. Bonnet	(b) **85**, 587 (1978)	*20.1*, p. *321*; *20.3*, p. *342*; 22.8, p. 503
R. Bonnet	(a) **49**, 675 (1978)	*10.1*, p. *106*; 21, p. 364; 21.1, p. 379
Z. Bontcheva-Mladenova	(a) **47**, K59 (1978)	*1.3*, p. *26*; 8, p. 86; *14.3*, p. *221*; 15, p. 260; 22, p. 413
G. A. Bordovskii	(a) **49**, K173 (1978)	*14.4.1*, p. *256*; 22.6, p. 497
V. A. Bordovskii	(a) **49**, K173 (1978)	*14.4.1*, p. *256*; 22.6, p. 497
G. Bordure	(a) **45**, 579 (1978)	*14*, p. *203*; 14.4, p. 248; 22.9, p. 523
W. Borgieł	(b) **88**, K47 (1978)	*13.5.4*, p. *201*; 18.1, p. 278; 18.2, p. 288
B. Borie	(a) **47**, K39 (1978)	*1.1*, p. *12*; 6, p. 71; 21.7, p. 406
O. Yu. Borkovskaya	(a) **48**, K55 (1978)	*11*, p. *128*; 14.3.3, p. 235; 22.2.1, p. 447
J. Borkowicz	(a) **48**, 225 (1978)	*3*, p. *55*; 22.1.2, p. 434
T. Borowiak	(a) **50**, K105 (1978)	*14*, p. *204*; 22.9, p. 528
G. Borstel	(b) **88**, K123 (1978)	*13.5*, p. *187*
A. S. Borukhovich	(a) **46**, 11 (1978)	8, p. 83; 13.1, p. 150; *14.3*, p. *217*; 15, p. 259; 22, p. 408; 22.3, p. 457; 22.3.1, p. 460
V. N. Borzunov	(b) **86**, 83 (1978)	*4*, p. 57; *13.1*, p. *152*; 21.1, p. 371
Debjani Bose	(b) **89**, K35 (1978)	*4*, p. *59*; *12.2*, p. *140*; 22.5.2, p. 484
L. Bosi	(b) **85**, K27 (1978)	*10.2*, p. *109*; *13.4*, p. *170*; 22.5.2, p. 478
	(b) **89**, 221 (1978)	*10.2*, p. *117*; 13.4, p. 180; *20.3*, p. *347*; 22.5.2, p. 484
	(b) **89**, K107 (1978)	*10.2*, p. *109*; *13.4*, p. *170*; 22.5.2, p. 478
P. Bosi	(b) **86**, 77 (1978)	*1.2*, p. *16*; *6*, p. *67*; 21.1, p. 371
H. Bossac	(a) **48**, 369 (1978)	*1.3*, p. *27*; *10.1*, p. *105*; 14.1, p. 209; 21, p. 361; 21.1, p. 376; 21.3, p. 395
A. Bostanjoglo	(a) **48**, 481 (1978)	*18.2.1*, p. *293*; 21.1, p. 376; 21.1.1, p. 388
F. W. Boswell	(a) **45**, 469 (1978)	*1.2*, p. *16*; *1.3*, p. *25*; 22.8, p. 503
H. Böttger	(b) **89**, 675 (1978)	*6*, p. *76*; *13*, p. *148*; 20.3, p. 348; 22, p. 418
D. S. Boudreaux	(a) **45**, 607 (1978)	*2*, p. *43*; *18.2*, p. *283*; 21.1, p. 370
D. Boulanger	(b) **85**, 597 (1978)	3, p. 54; 13.4, p. 170; *19*, p. *304*; 22.4.3, p. 470
C. Boulesteix	(a) **45**, 639 (1978)	*1.1*, p. *11*; 1.4, p. 29; 22.6, p. 491
	(a) **47**, 311 (1978)	*1.1*, p. *12*; 1.4, p. 30; 22.6, p. 493
F. Bouton	(a) **50**, 109 (1978)	*14.3*, p. *224*; 18.3, p. 296; 22.8.2, p. 520
P. Boyer	(b) **90**, K169 (1978)	*10.2*, p. *121*; *13*, p. *149*; 14.4, p. 252; 22.6.1, p. 500

R. Boyn	(b) **85**, 217 (1978)	13.4, p. 169; *14.3.4*, p. *240*; *20.1*, p. *319*; 22.4.1, p. 463
A. G. Braginskaya	(a) **48**, K115 (1978)	*13.4*, p. *179*; 20.1, p. 331; 22.2.3, p. 453
P. Braillon	(a) **45**, 453 (1978)	*10.2*, p. *110*; 22.8.1, p. 511
H. S. Brandi	(b) **87**, 351 (1978)	*13.4*, p. *175*; *20.1*, p. *326*; 22.5.2, p. 481
W. Braun	(b) **90**, 525 (1978)	*13.1*, p. *160*; *17*, p. *270*; 21, p. 365
W. Braune	(b) **87**, 527 (1978)	*13.2*, p. *162*; *14.1*, p. *208*; 21.6, p. 402
	(b) **89**, 95 (1978)	*13.1*, p. *158*; *19*, p. *311*; 21.7, p. 407
D. Bräunig	(a) **48**, 533 (1978)	*14.3.1*, p. *227*; *14.3.4*, p. *244*; 22, p. 416
M. Brauwers	(b) **85**, 519 (1978)	13.1, p. 151; *14.1*, p. *206*; 21, p. 354
L. Brehmer	(a) **50**, K239 (1978)	*14.3.4*, p. *247*; 22.9, p. 529
K. Breuer	(b) **86**, 329 (1978)	2, p. 43; 13.1, p. 153; *14.4*, p. *248*; *20.1*, p. *322*; 22.6, p. 492
C. L. Briant	(b) **85**, 393 (1978)	*2*, p. *42*; 21, p. 352
V. Kh. Brikenshtein	(b) **86**, 449 (1978)	*13.5.1*, p. *190*; *20.3*, p. *343*; 22.9, p. 524
P. D. Bristowe	(b) **86**, 93 (1978)	*10.1*, p. *101*; 12.1, p. 134; 21.1, p. 371
A. D. Britov	(a) **46**, 77 (1978)	*13.4*, p. *173*; *20.3*, p. *343*; 22.4.4, p. 472
W. Brodkorb	(a) **48**, K157 (1978)	*14.3.2*, p. *231*; *14.3.4*, p. *244*
A. M. Brodskii	(b) **85**, 465 (1978)	13.1, p. 151; *17*, p. *269*; 22.9, p. 522
J. Broeckx	(b) **89**, K25 (1978)	*19*, p. *311*; 22.5.2, p. 484
P. M. van den Broek	(b) **90**, 211 (1978)	*1*, p. *10*; *13*, p. *148*
P. M. Bronsveld	(a) **48**, 497 (1978)	*1.3*, p. *27*; *10.1*, p. *105*; 21.1, p. 376
C. R. Brooks	(b) **89**, K123 (1978)	*8*, p. *88*; 21.6, p. 405
F. Brosens	(b) **87**, K61 (1978)	6, p. 71; *10.1*, p. *102*; 21, p. 358
	(b) **88**, K93 (1978)	*4*, p. *58*; 6, p. 74; 21, p. 360
I. Broser	(b) **90**, 77 (1978)	*13.5.1*, p. *195*; *20.1*, p. *335*; 22.4.1, p. 466
R. Broser	(b) **90**, 77 (1978)	*13.5.1*, p. *195*; *20.1*, p. *335*; 22.4.1, p. 466
P. R. Brosious	(a) **50**, 237 (1978)	10.2, p. 120; *11*, p. *130*; 19, p. 314; 22.1, p. 423
H. Bross	(b) **90**, 241 (1978)	*13.1*, p. *160*; 18.1, p. 280; 18.2, p. 291; 21.1, p. 380
	(b) **90**, 497 (1978)	*13.1*, p. *160*; 22.1.2, p. 439
L. Brossard	(a) **45**, 503 (1978)	*18.3*, p. *295*; 22.8.2, p. 518
V. L. Broude	(b) **90**, K53 (1978)	*13.5.1*, p. *196*; *20.3*, p. *349*; 22.9, p. 528
M. Brousseau	(b) **88**, 261 (1978)	*13.5.1*, p. *192*; 20.3, p. 345; 22.2.2, p. 450
H. A. Brown	(b) **88**, K31 (1978)	*2*, p. *46*; *18.2*, p. *288*
	(b) **88**, K39 (1978)	*18.2*, p. *288*
	(b) **90**, K165 (1978)	*2*, p. *52*; *18.2*, p. *292*
J. D. Browne	(a) **49**, K177 (1978)	*18.3*, p. *296*; 18.4, p. 301; 21.1, p. 379
W. Brückner	(a) **45**, 123 (1978)	*1.2*, p. *15*; *5*, p. *60*; 18.1, p. 277; 22.6.1, p. 499
	(a) **47**, K99 (1978)	*8*, p. *86*; *14.3.4*, p. *242*; 22.6, p. 493
	(a) **47**, K129 (1978)	*1.2*, p. *20*; *14.3*, p. *221*; 22.6, p. 494
V. N. Brudnyi	(a) **49**, 761 (1978)	*11*, p. *130*; 13.4, p. 182; *14.3*, p. *224*; 22.8, p. 507
	(a) **50**, 379 (1978)	11, p. 131; 13.4, p. 185; *14.3*, p. *225*; *20.1*, p. *337*; 22.8, p. 508
C. F. van Bruggen	(a) **46**, 687 (1978)	*18.1*, p. *278*; *18.4*, p. *299*; 22.8, p. 504
H. J. de Bruin	(a) **49**, K181 (1978)	*13.3*, p. *167*; *14.3.4*, p. *246*; 22.6, p. 497
O. Brümmer	(a) **45**, K87 (1978)	*10.1*, p. *100*; 21.1, p. 370
	(b) **86**, K93 (1978)	*13*, p. *144*; 20, p. 316; 21.1, p. 372; 21.1.1, p. 386
	(a) **46**, K131 (1978)	*1.1*, p. *12*; 22.1.2, p. 432
	(b) **90**, 487 (1978)	*13.1*, p. *160*; *20*, p. *318*; 21.1, p. 381; 21.1.1, p. 390
J. de Bruyn	(b) **89**, K17 (1978)	*5*, p. *62*; 22.1.1, p. 427; 22.1.2, p. 436
F. J. Bryant	(b) **85**, 629 (1978)	*13.5.1*, p. *190*; *20.3*, p. *342*; 22.4.3, p. 470
V. V. Bryksin	(b) **89**, 675 (1978)	*6*, p. *76*; *13*, p. *148*; 20.3, p. 348; 22, p. 418

V. T. Bublik	(a) **45**, 543 (1978)	*10.2*, p. *110*; 22, p. 410; 22.2, p. 443; 22.4, p. 461
	(a) **46**, 365 (1978)	*1.2*, p. *17*; 22.2.1, p. 446; 22.2.2, p. 450; 22.2.3, p. 452
E. Bucher	(a) **49**, 437 (1978)	*16*, p. *266*; 22, p. 418
W. Buckel	(a) **49**, 73 (1978)	*1.5*, p. *38*; *2*, p. *49*; 21.1, p. 377; 22.1.1, p. 427
L. Budík	(a) **45**, K63 (1978)	*1.2*, p. *16*; 14, p. 203; 20, p. 315; 22.5, p. 474
D. L. Budnitskii	(a) **50**, 379 (1978)	11, p. 131; 13.4, p. 185; *14.3*, p. *225*; *20.1*, p. *337*; 22.8, p. 508
A. V. Bugakov	(a) **49**, 413 (1978)	*1.4*, p. *33*; *1.5*, p. *38*; 3, p. 56; 21.6, p. 404
E. Buhrig	(a) **48**, K63 (1978)	13.4, p. 178; *14.3*, p. *222*; 22.8, p. 506
J. P. Buisson	(b) **86**, 493 (1978)	*6*, p. *68*; *20.1*, p. *324*; 22.5.2, p. 480
A. A. Bukharaev	(a) **50**, 711 (1978)	2, p. 53; *13.4*, p. *186*; *20.1*, p. *338*; 22.8.1, p. 516
T. M. Bulatova	(a) **46**, 357 (1978)	*10.2*, p. *111*; 22.5.2, p. 480
A. D. Buonaquisti	(a) **46**, K91 (1978)	14.3.3, p. 235; *17*, p. *269*; 22.6, p. 492
J. P. Burger	(b) **90**, 207 (1978)	*6*, p. *77*; *14.2*, p. *215*; 21, p. 364
F. Burkhardt	(a) **47**, 131 (1978)	*11*, p. *126*; *14.3*, p. *220*; 22.1.2, p. 432
A. Buroff	(a) **48**, K109 (1978)	*2*, p. *48*; *9*, p. *93*; 14.3.4, p. 244; 22, p. 416
J. J. Burton	(b) **85**, 393 (1978)	*2*, p. *42*; 21, p. 352
E. Burzo	(a) **45**, K145 (1978)	*18.1*, p. *277*; 18.3, p. 295; 21.1, p. 370; 21.1.1, p. 385; 21.4, p. 396
	(a) **46**, K163 (1978)	*5*, p. *61*; 22.6, p. 493
	(b) **87**, K137 (1978)	*19*, p. *308*; 22.8.1, p. 513
R. Butendeich	(a) **49**, 169 (1978)	*16*, p. *266*; 22.4.1, p. 465
H. Büttner	(b) **90**, 53 (1978)	*13.5.1*, p. *195*; 13.5.3, p. 200; 22.5, p. 476
J. Buxo	(a) **46**, 185 (1978)	*13.3*, p. *164*; *14.3.3*, p. *234*
	(a) **49**, 603 (1978)	*14.3.3*, p. *237*; *16*, p. *267*
V. N. Bykov	(a) **48**, K97 (1978)	*11*, p. *128*; 21.1, p. 375
J. G. Byrne	(a) **48**, 551 (1978)	*1.3*, p. *27*; *10.1*, p. *105*; 21, p. 362; 21.1, p. 376
	(a) **48**, K83 (1978)	*1.3*, p. *27*; *10.1*, p. *105*; 21, p. 361
I. S. Bzinkovskaya	(a) **45**, 521 (1978)	9, p. 91; *14.3.3*, p. *233*; 22.1.2, p. 430
D. Cabib	(a) **47**, 71 (1978)	1.4, p. 30; *14.4*, p. *249*; 22.9, p. 525
D. Caillard	(a) **49**, 675 (1978)	*10.1*, p. *106*; 21, p. 364; 21.1, p. 379
Y. Calvayrac	(a) **45**, 229 (1978)	*1.3*, p. *25*; 6, p. 65; 12.1, p. 134; 21.1, p. 369; 21.1.1, p. 384
S. U. Campisano	(a) **47**, 533 (1978)	*11*, p. *127*; 22.1.2, p. 433
V. Čápek	(b) **85**, K1 (1978)	*13.5.3*, p. *200*; *18.2*, p. *281*; 22, p. 409
	(b) **90**, 295 (1978)	*13.5.1*, p. *196*; 22.9, p. 528
R. Capelletti	(a) **47**, 617 (1978)	*8*, p. *86*; 10.2, p. 114; 22.5.3, p. 487
	(b) **90**, 657 (1978)	13.4, p. 185; *20.1*, p. *337*; *20.3*, p. *350*; 22.5.2, p. 486
V. Capozzi	(a) **46**, 589 (1978)	*14.3.4*, p. *241*; 22.8, p. 504
P. Cappelletti	(a) **50**, K93 (1978)	*20.3*, p. *350*; 22.8, p. 508
R. Car	(b) **86**, 471 (1978)	*13.1*, p. *153*; 22, p. 411
M. Cardona	(b) **86**, 563 (1978)	*6*, p. *68*; *20.1*, p. *324*; 22.2.2, p. 450
	(b) **88**, 135 (1978)	*13.1*, p. *156*; 17, p. 270; 20.1, p. 328; 22.1.2, p. 434
	(b) **88**, 199 (1978)	6, p. 73; *20.1*, p. *329*; 22.6, p. 494
D. Carles	(a) **50**, 651 (1978)	13.4, p. 186; *16*, p. *267*; 22.4.3, p. 471
F. Carlier	(b) **89**, 193 (1978)	*13.5.3*, p. *200*; *20.1*, p. *332*; 22.5.2, p. 483
	(b) **89**, K95 (1978)	*10.2*, p. *118*; 13.5.3, p. 200; *20.1*, p. *333*; 22.5.2, p. 484
L. C. Scavarda do Carmo	(b) **89**, K149 (1978)	6, p. *76*; *20.1*, p. *334*; 22.8, p. 507
P. Caro	(a) **45**, 297 (1978)	*1.2*, p. *15*; 21.4, p. 396
	(b) **85**, 587 (1978)	*20.1*, p. *321*; *20.3*, p. *342*; 22.8, p. 503

H. D. Carstanjen	(a) **45**, 529 (1978)	*1.1*, p. *11*; 11, p. 124; 21, p. 354
R. Caruba	(a) **50**, 275 (1978)	10.2, p. 121; 13.4, p. 184; *20.3*, p. *350*; 22.6, p. 497
L. M. Caspers	(a) **46**, 541 (1978)	*10.1*, p. *102*; 21, p. 357
	(a) **47**, 577 (1978)	*10.1*, p. *104*; 21, p. 359
	(b) **90**, K89 (1978)	6, p. 78; *10.1*, p. *107*; 21, p. 364
	(b) **90**, 643 (1978)	*10.1*, p. *107*; 21, p. 365
J. C. Castro	(b) **86**, 137 (1978)	*1.2*, p. *16*; *6*, p. *67*; 12.2, p. 138; 20.1, p. 321; 22.5, p. 474; 22.5.4, p. 489
C. R. A. Catlow	(a) **46**, 191 (1978)	*14.4*, p. *248*; 22.8.1, p. 511
F. Cerdeira	(b) **88**, 199 (1978)	*6*, p. *73*; *20.1*, p. *329*; 22.6, p. 494
J. Čermák	(a) **49**, K81 (1978)	2, p. 50; *18.2.1*, p. *294*; 21.1, p. 378; 21.4, p. 398
G. F. Cerofolini	(a) **48**, 523 (1978)	*10.2*, p. *116*; 14.3.3, p. 235; 22.1.2, p. 435
V. Cháb	(a) **47**, 271 (1978)	*2*, p. *45*; *14.3*, p. *221*; 15, p. 260; 22, p. 413
M. Chabin	(a) **48**, 67 (1978)	*1.2*, p. *21*; *8*, p. *87*; 22.8.1, p. 514
L. T. Chadderton	(a) **49**, 585 (1978)	10.2, p. 119; *11*, p. *130*; 22.8, p. 507
G. Chaddha	(b) **85**, 777 (1978)	*18.2*, p. *282*
	(b) **90**, K145 (1978)	*18.2*, p. *292*
E. I. Chaikina	(a) **48**, 281 (1978)	13.4, p. 179; *20.3*, p. *346*; 22.2.3, p. 453
B. K. Chakrabarti	(b) **86**, K55 (1978)	*1.2*, p. *17*; *14.2*, p. *214*; 21, p. 356
B. C. Chakravarty	(a) **46**, K151 (1978)	*9*, p. *92*; 22.1.2, p. 432
R. Chandra	(b) **88**, 633 (1978)	*5*, p. *62*; 22.8, p. 506
V. Chandra Mouli	see under letter "M"	
Chang Peng-Hsiang	see under letter "P"	
S.-L. Chang	(a) **47**, 717 (1978)	*1.1*, p. *13*; 22.1.1, p. 426
P. Chantikul	(a) **48**, 79 (1978)	10.2, p. 115; *15*, p. *260*; 22.5.2, p. 482
K. A. Chao	(b) **87**, 625 (1978)	13.1, p. 155; *13.4*, p. *176*; *18.4*, p. *300*
	(b) **90**, K153 (1978)	*13.4*, p. *185*; 22, p. 421
I. M. Chapnik	(a) **47**, K71 (1978)	*1.3*, p. *26*; 14.1, p. 307; *14.3*, p. *221*; 22, p. 413
M.-F. Charlot	(b) **86**, 497 (1978)	*18.4*, p. *299*; 22.8, p. 504
G. Chassagne	(a) **48**, 425 (1978)	*1.4*, p. *32*; 11, p. 128; 22.6, p. 495
F. Chassagneux	(a) **50**, 109 (1978)	*14.3*, p. *224*; 18.3, p. 296; 22.8.2, p. 520
P. Chatterjee	(b) **90**, 609 (1978)	*13.1*, p. *161*; *14.2*, p. *216*
S. Chatterjee	(b) **90**, 609 (1978)	*13.1*, p. *161*; *14.2*, p. *216*
D. Chattopadhyay	(b) **88**, K109 (1978)	13.1, p. 157; *15*, p. *260*; 22.2.3, p. 453
P. Chaudhari	(b) **87**, K5 (1978)	*1.5*, p. *36*; *2*, p. *44*
A. Sh. Chavchanidze	(b) **88**, K103 (1978)	*13.4*, p. *179*; 21.1.1, p. 388
A. P. Chebanenko	(a) **50**, 717 (1978)	*13.4*, p. *186*; *14.3.4*, p. *247*; 16, p. 267; 22.4.2, p. 469
T. I. Chekmasova	(a) **49**, K185 (1978)	*1.2*, p. *23*; *14.4.1*, p. *256*; 19, p. 313; 22.8.1, p. 516
G. G. Chemeresyuk	(a) **50**, 717 (1978)	*13.4*, p. *186*; *14.3.4*, p. *247*; 16, p. 267; 22.4.2, p. 469
A.-B. Chen	(b) **90**, K21 (1978)	*9*, p. *94*; *10.2*, p. *120*; 22, p. 419
H. S. Chen	(b) **87**, 777 (1978)	*2*, p. *45*; *14.1*, p. *208*; 21.1, p. 373
	(a) **48**, K181 (1978)	*1.4*, p. *32*; *2*, p. *48*; 21.1, p. 377; 21.6, p. 404
	(a) **49**, 387 (1978)	*2*, p. *49*; 21.1, p. 378
J. H. Chen	(a) **50**, 393 (1978)	1.3, p. 28; *7*, p. *82*; 21.1, p. 382; 21.1.1, p. 390
J. Chenavas	(a) **48**, 581 (1978)	*1.1*, p. *13*; *5*, p. *62*; 22.8.1, p. 515
Li-Jen Cheng	(b) **90**, 689 (1978)	*6*, p. *78*; 11, p. 131; *20.1*, p. *337*; 22.1.2, p. 439
N. M. Chepilko	(b) **89**, 27 (1978)	*13.5*, p. *188*; 20.1, p. 331; 22.6, p. 496
V. I. Cherepanov	(b) **89**, K53 (1978)	*14.1*, p. *210*; 21, p. 362
V. T. Cherepin	(a) **50**, K113 (1978)	*17*, p. *270*; 22.3, p. 459
E. I. Cherepov	(a) **46**, 467 (1978)	*14.3.1*, p. *226*; *14.3.4*, p. *241*; 22.1.2, p. 432; 22.6, p. 492
A. N. Cherkasov	(b) **90**, K17 (1978)	*13.3*, p. *168*; *14.1*, p. *211*; 19, p. 313; 21, p. 364

G. Ya. Chernov	(b) **89**, 329 (1978)	*11*, p. *129*; 13, p. 147
V. M. Chernov	(a) **47**, 45 (1978)	*10.1*, p. *103*; 21, p. 358
N. Y. Chernysheva	(a) **47**, K5 (1978)	1.5, p. 37; *11*, p. *126*; 22.2.3, p. 453
A. Ya. Chervonenkis	(a) **46**, K47 (1978)	*18.4*, p. *299*; *19*, p. *306*; 22.8.1, p. 512
O. D. Chistyakov	(b) **86**, K59 (1978)	*6*, p. *68*; *18*, p. *273*; 22.7, p. 501
I. L. Chistyi	(a) **47**, 609 (1978)	*7*, p. *81*; *20.1*, p. *328*; 22.8.1, p. 514
T. Chong	(b) **88**, 623 (1978)	*13.5.1*, p. *193*; 20.3, p. 346; 22.9, p. 526
V. M. Choodnovskii	(b) **87**, K145 (1978)	*2*, p. *45*; *6*, p. *72*; 7, p. 81
K. L. Chopra	(a) **45**, 665 (1978)	*1.5*, p. *36*; 22.4.4, p. 472
	(a) **46**, 723 (1978)	*1.5*, p. *36*; *14.3.1*, p. *226*; 16, p. 263; 22.4.4, p. 472
	(a) **48**, 257 (1978)	*1.5*, p. *37*; *14.3.1*, p. *227*; 22.1.1, p. 427
	(a) **49**, 379 (1978)	1.4, p. 33; *1.5*, p. *38*; 2, p. 49; 22.1.1, p. 427
A. Chovet	(a) **46**, K13 (1968)	*13.3*, p. *165*; 14.3.3, p. 234; 22, p. 411
B. V. R. Chowdari	(b) **90**, 465 (1978)	*10.2*, p. *121*; *20.1*, p. *337*; 22.5.2, p. 486
V. Christoph	(b) **86**, 231 (1978)	*14.1*, p. *206*; 21, p. 355
	(b) **89**, 537 (1978)	13.1, p. 159; *14.1*, p. *211*; 21, p. 363
S. Y. Chuang	(a) **48**, K181 (1978)	*1.4*, p. *32*; *2*, p. *48*; 21.1, p. 377; 21.6, p. 404
S. G. Chuchman	(a) **50**, K47 (1978)	*18.1*, p. *280*; *18.4*, p. *301*; 21.4, p. 399
F. A. Chudnovskii	(a) **48**, K153 (1978)	*8*, p. *88*; 22.6, p. 495
S. A. Chueva	(a) **50**, K197 (1978)	*14*, p. *204*; *18*, p. *276*; 22.9, p. 529
D. Y. Chung	(a) **46**, 603 (1978)	*12*, p. *133*
J. Cirák	(a) **47**, 287 (1978)	5, p. 61; *14.2*, p. *214*
R. Ciszewski	(b) **85**, K73 (1978)	*1*, p. *9*; 18.2, p. 282; 21.1.1, p. 385
	(b) **85**, K77 (1978)	*1*, p. *9*; 18.2, p. 282; 21.1.1, p. 385
G. Ciucci	(b) **86**, 471 (1978)	*13.1*, p. *153*; 22, p. 411
A. Čížek	(b) **89**, K159 (1978)	*13*, p. *148*; 21.5, p. 400
	(b) **90**, K1 (1978)	*13.3*, p. *168*; 21.1, p. 380
P. C. Clapp	(a) **50**, 393 (1978)	1.3, p. 28; *7*, p. *82*; 21.1, p. 382; 21.1.1, p. 390
P. Clark	(b) **85**, 171 (1978)	13.5, p. 187; *18*, p. *273*; *20.1*, p. *319*; 21.1, p. 368
R. Clarke	(a) **49**, 133 (1978)	*1*, p. *9*; *14.4*, p. *250*; 15, p. 261; 22.9, p. 527
F. Claro	(b) **88**, K147 (1978)	*13.1*, p. *158*
R. Claus	(b) **86**, 257 (1978)	*6*, p. *67*; *20.1*, p. *322*; 22.5, p. 474
	(b) **88**, 683 (1978)	*13.5*, p. *187*; *14.4*, p. *250*
	(b) **88**, K123 (1978)	*13.5*, p. *187*
	(b) **89**, 159 (1978)	*13.5*, p. *188*; 14.4, p. 250
P. Clauws	(b) **87**, 719 (1978)	10.2, p. 113; *19*, p. *308*; 22.6, p. 494
C. Clemen	(a) **49**, 437 (1978)	*16*, p. *266*; 22, p. 418
E. Colavita	(b) **85**, 445 (1978)	*13.1*, p. *151*; 20.1, p. 321; 21, p. 354; 21.6, 401
M. Coldea	(a) **49**, 207 (1978)	*18.1*, p. *279*; *19*, p. *311*; 21.1, p. 377
Y. Colin	(a) **50**, 311 (1978)	*13.4*, p. *184*; 14.3.3, p. 238; 22.2.1, p. 449
J. Collet	(b) **88**, 261 (1978)	*13.5.1*, p. *192*; 20.3, p. 345; 22.2.2, p. 450
R. A. Collins	(a) **46**, K91 (1978)	14.3.3, p. 235; *17*, p. *269*; 22.6, p. 492
A. Collomb	(a) **46**, 493 (1978)	*18.4*, p. *299*; 22.8.1, p. 512
	(a) **48**, 581 (1978)	*1.1*, p. *13*; *5*, p. *62*; 22.8.1, p. 515
	(a) **50**, 635 (1978)	*18.4*, p. *302*; 22.8.1, p. 516
R. Collongues	(a) **45**, 599 (1978)	*19*, p. *305*; 22.6.1, p. 499
M. Combescot	(b) **86**, 349 (1978)	*13.5.1*, p. *190*; 22, p. 410
H. van Cong	(b) **90**, 401 (1978)	*8*, p. *89*; 13.4, p. 184; *14.3*, p. *224*; 22, p. 419
J. M. Corbett	(a) **45**, 289 (1978)	*1.4*, p. *29*; 21.1, p. 369
	(a) **45**, 469 (1978)	*1.2*, p. *16*; *1.3*, p. *25*; 22.8, p. 503
J. W. Corbett	(a) **48**, K31 (1978)	*11*, p. *128*; *13.4*, p. *178*; 22.1.2, p. 435
	(a) **50**, 237 (1978)	10.2, p. 120; *11*, p. *130*; 19, p. 314; 22.1, p. 423
	(b) **90**, 689 (1978)	*6*, p. *78*; 11, p. 131; *20.1*, p. *337*; 22.1.2, p. 439
A. Corciovei	(b) **87**, 61 (1978)	*18.2.1*, p. *293*

J. C. Corelli	(b) **90**, 689 (1978)	*6*, p. *78*; 11, p. 131; *20.1*, p. *337*; 22.1.2, p. 439
A. Coret	(a) **48**, 465 (1978)	*13.5.1*, p. *194*; *16*, p. *265*; 22.6, p. 495
J. Cornelis	(a) **45**, 439 (1978)	*10.1*, p. *100*; *11*, p. *124*; 21.1, p. 370
	(b) **87**, K61 (1978)	6, p. 71; *10.1*, p. *102*; 21, p. 358
	(b) **88**, K93 (1978)	*4*, p. *58*; 6, p. 74; 21, p. 360
T. Cornelissens	(a) **48**, K5 (1978)	*1.4*, p. *31*; 10.2, p. 115; 22, p. 415
	(a) **49**, K217 (1978)	*1.4*, p. *31*; 10.2, p. 115; 22, p. 415
W. D. Corner	(a) **46**, 39 (1978)	*14.1*, p. *204*; 21.4, p. 396
J. F. Cornwell	(b) **90**, 211 (1978)	*1*, p. *10*; *13*, p. *148*
F. Cosandey	(a) **48**, 555 (1978)	*1.5*, p. *38*; 21.6, p. 403
A. A. Cottey	(b) **88**, 207 (1978)	*13.1*, p. *156*; 13.3, p. 166; *14.3.1*, p. *227*
H. Coufal	(b) **89**, 261 (1978)	*13.5.1*, p. *194*; 20.1, p. 332; 22.9, p. 527
	(a) **49**, K17 (1978)	*11*, p. *129*; 14.3, p. 224; *19*, p. *312*; 22.1.2, p. 437
C. S. G. Cousins	(a) **48**, 113 (1978)	*1.1*, p. *13*; 22.1.2, p. 434
R. Coussement	(b) **89**, K17 (1978)	*5*, p. *62*; 22.1.1, p. 427; 22.1.2, p. 436
A. G. Coutts	(b) **88**, 767 (1978)	*20.1*, p. *331*; 22.5.4, p. 489
M.-L. Craus	(a) **50**, K225 (1978)	*18.3*, p. *297*; 22.8.2, p. 521
J. H. Crawford, Jr.	(a) **47**, 617 (1978)	*8*, p. *86*; 10.2, p. 114; 22.5.3, p. 487
R. K. Crawford	(a) **46**, 397 (1978)	*1.2*, p. *17*; 23, p. 530
M. Creuzburg	(b) **90**, 189 (1978)	*13.5.1*, p. *196*; *20.3*, p. *349*; 23, p. 531
M. Crişan	(b) **87**, K151 (1978)	*13*, p. *145*
V. Cristea	(a) **45**, 617 (1978)	14.3, p. 218; *15*, p. *259*; 22.6, p. 491
S. Cristoloveanu	(a) **46**, K13 (1978)	*13.3*, p. *165*; 14.3.3, p. 234; 22, p. 411
L. E. Cross	(a) **45**, K1 (1978)	*11*, p. *123*; 14.4.1, p. 253; 22.8.1, p. 510
	(a) **48**, 431 (1978)	*1.6*, p. *41*; *14.4.1*, p. *256*; 22.8.1, p. 515
	(a) **50**, 661 (1978)	*1.2*, p. *24*; 8, p. 90; 12.2, p. 141; 20.1, p. 338; 22.9, p. 528
L. Csillag	(a) **47**, 609 (1978)	*7*, p. *81*; *20.1*, p. *328*; 22.8.1, p. 514
V. Cvekić	(a) **50**, 153 (1978)	*14.3.3*, p. *237*; 22.1.2, p. 438
B. Cvikl	(b) **88**, 25 (1978)	*2*, p. *46*; *6*, p. *73*
A. Czapla	(a) **45**, 537 (1978)	*20.1*, p. *321*; 22.6.1, p. 499
Z. Czapla	(a) **50**, K137 (1978)	*14.4.1*, p. *257*; 22.8, p. 508
S. Däbritz	(a) **49**, 323 (1978)	*1.2*, p. *22*; *10.1*, p. *106*; 14.2, p. 215; 21.1, p. 377
D. Dădârlat	(b) **87**, K161 (1978)	*13*, p. *145*
B. W. Dale	(b) **90**, 391 (1978)	*5*, p. *62*; 13.4, p. 184; 22.6.1, p. 499
	(b) **90**, 571 (1978)	*5*, p. *62*; 13.4, p. 185; 22.6.1, p. 499
O. de O. Damasceno	(b) **90**, K169 (1978)	*10.2*, p. *121*; *13*, p. *149*; 14.4, p. 252; 22.6.1, p. 500
J. Z. Damm	(b) **90**, 657 (1978)	13.4, p. 185; *20.1*, p. *337*; *20.3*, p. *350*; 22.5.2, p. 486
H. Danan	(a) **49**, K97 (1978)	*14.1*, p. *210*; 21, p. 363
A. Darabont	(b) **85**, 75 (1978)	*19*, p. *303*; 22.8.1, p. 510
M. I. Darby	(b) **85**, K63 (1978)	*2*, p. *42*
	(a) **50**, 203 (1978)	*10.1*, p. *107*; *13.4*, p. *184*; 21.1, p. 381
A. Dargys	(a) **45**, 387 (1978)	*14.3.4*, p. *240*; 22.2.3, p. 452
B. M. Darinskii	(a) **47**, 379 (1078)	*12.2*, p. *139*; 14.4.1, p. 255; 22.8.1, p. 514
S. R. Das	(a) **45**, 665 (1978)	*1.5*, p. *36*; 22.4.4, p. 472
R. A. Dashkovskaya	(b) **86**, 419 (1978)	*6*, p. *67*; *20.1*, p. *323*; 22.3, p. 457
Nguyen Nhu Dat	(b) **86**, 585 (1978)	*6*, p. *69*; *13.5.2*, p. *198*; 20, p. 316; 22, p. 411
M. David	(b) **87**, 419 (1978)	*1.1*, p. *13*; 22.1.1, p. 425; 22.6, p. 494
R. L. Davidovich	(b) **88**, 439 (1978)	*13.4*, p. *178*; *19*, p. *310*; 22.8, p. 506
S. M. Davidson	(a) **48**, K1 (1978)	*10.2*, p. *115*; 22.1.2, p. 434
A. S. Davydov	(b) **90**, 457 (1978)	*14*, p. *204*; 22.9, p. 528

A. L. Dawar	(a) **45**, K59 (1978)	*2*, p. *43*; *13.4*, p. *170*; 14.3.1, p. 226; 22.1.3, p. 441
	(a) **47**, 375 (1978)	13.4, p. 176; *14.3.1*, p. *227*; 22.2.3, p. 453
L. Däweritz	(a) **49**, 629 (1978)	*1.5*, p. *39*; 13.4, p. 182; 14.3.1, p. 228; 20.1, p. 335; 20.3, p. 348; 22.2.4, p. 456
P. Day	(b) **88**, 767 (1978)	*20.1*, p. *331*; 22.5.4, p. 489
I. J. Dayawansa	(b) **86**, K27 (1978)	*20.1*, p. *323*; 22.6, p. 492
E. Dębowska	(b) **89**, K159 (1978)	*13*, p. *148*; 21.5, p. 400
J. Degauque	(a) **45**, 493 (1978)	*10.1*, p. *100*; *18.2*, p. *283*; 21.1.1, p. 385
I. Ya. Dekhtyar	(b) **88**, K55 (1978)	*13.1*, p. *157*; 22.2.1, p. 447
	(b) **90**, K1 (1978)	*13.3*, p. *168*; 21.1, p. 380
P. Delavignette	(a) **46**, K167 (1978)	*11*, p. *126*; 22, p. 412
	(a) **50**, 703 (1978)	*1.1*, p. *14*; *1.4*, p. *35*
T. H. O'Dell	see under letter "O"	
G. Della Mea	(a) **49**, 347 (1978)	*11*, p. *129*; 22.1.2, p. 437
P. Delobelle	(a) **47**, 259 (1978)	*10.1*, p. *103*; 21, p. 358
A. Delunas	(a) **46**, 703 (1978)	*1.2*, p. *18*; *2*, p. *44*; 21.7, p. 406
N. N. Delyagin	(b) **90**, 741 (1978)	*8*, p. *90*; *18.2*, p. *292*; 21.1, p. 382
L. S. Demidenko	(b) **86**, 765 (1978)	13, p. 144; *14.3*, p. *219*; 22.2.1, p. 446
D. Demus	(a) **50**, K181 (1978)	*2*, p. *53*; *14.4*, p. *252*; 22.9, p. 529
M. F. Denanot	(a) **46**, 387 (1978)	*10.2*, p. *111*; 22.6, p. 492
B. I. Denker	(a) **50**, 745 (1978)	13.4, p. 186; *20.1*, p. *338*; *20.2*, p. *340*; 22.8.1, p. 516
J.-M. Derochette	(a) **45**, 163 (1978)	*1.6*, p. *40*; *17.1*, p. *272*; 21, p. 353
A. V. Deryagin	(a) **45**, 71 (1978)	*18.2*, p. *281*; 21.1, p. 368; 21.4, p. 396
U. V. Desnica	(a) **46**, K173 (1978)	*3*, p. *54*; 22.3.1, p. 460
J. C. Desoyer	(a) **47**, 469 (1978)	*10.1*, p. *104*; 14.1, p. 208; 21.1, p. 374; 21.1.1, p. 387
G. L. Destefanis	(a) **47**, K63 (1978)	13.3, p. 165; 14.4.1, p. 255; *20.3*, p. *344*; 22.8.1, p. 513
V. Detcheva	(b) **86**, 425 (1978)	*13.3*, p. *164*; 14.3.3, p. 233
F. Dettmann	(a) **49**, 293 (1978)	*14.2*, p. *215*; 22.6, p. 496
	(a) **49**, 573 (1978)	*14.2*, p. *215*
	(a) **49**, K209 (1978)	*14.2*, p. *215*
	(a) **50**, 417 (1978)	*14.2*, p. *216*
K. Dettmer	(a) **49**, 255 (1978)	*1.5*, p. *38*; *2*, p. *49*; 22.1.1, p. 427
D. Deutges	(b) **90**, 475 (1978)	*18*, p. *276*; 21.1, p. 381
J. T. Devreese	(b) **87**, K61 (1978)	6, p. 71; *10.1*, p. *102*; 21, p. 358
	(b) **88**, K93 (1978)	*4*, p. *58*; 6, p. 74; 21, p. 360
	(b) **90**, K73 (1978)	*6*, p. *78*; *14.3.4*, p. *246*; 22.2.3, p. 454
	(b) **90**, K149 (1978)	*6*, p. *78*; *14.3.4*, p. *247*; 22.2.3, p. 455
H. Dexpert	(a) **45**, 297 (1978)	*1.2*, p. *15*; 21.4, p. 396
I. Dezsi	(b) **89**, K17 (1978)	*5*, p. *62*; 22.1.1, p. 427; 22.1.2, p. 436
G. Dharmadurai	(a) **47**, K187 (1978)	*14.2*, p. *214*; 21, p. 360
Ch. Diaine	(b) **87**, 293 (1978)	*10.2*, p. *113*; *11*, p. *126*; 20.1, p. 326; 22.5.1, p. 477
A. Diaz-Gongora	(b) **85**, 693 (1978)	10.2, p. 109; *14*, p. *203*; *20*, p. *315*; 22.5.2, p. 479
	(b) **86**, 127 (1978)	10.2, p. 110; *14*, p. *203*; *20*, p. *316*; 22.5.2, p. 479
H. K. Diemer	(b) **87**, 697 (1978)	*18.1*, p. *278*; *19*, p. *308*; 22.8.1, p. 513
P. E. Dietz	(b) **87**, 581 (1978)	*10.2*, p. *113*; *20.1*, p. *327*; 22.5.2, p. 481
Diep-The-Hung	(b) **86**, K181 (1978)	1.6, p. 40; *13.5.4*, p. *201*; 18.4, p. 299
R. Diller	(a) **48**, 513 (1978)	*14.2*, p. *215*; 21, p. 362
H. D. Dimitrov	(b) **89**, 319 (1978)	*13.3*, p. *166*; *14.3.1*, p. *228*; 14.3.2, p. 231; 22, p. 417

J. F. Dinhut	(a) **47**, 469 (1978)	*10.1*, p. *104*; 14.1, p. 208; 21.1, p. 374; 21.1.1, p. 387
G. Dlubek	(a) **45**, K87 (1978)	*10.1*, p. *100*; 21.1, p. 370
A. P. Dmitriev	(a) **46**, 45 (1978)	*13*, p. *144*; *14.3.4*, p. *241*; 22, p. 411
N. L. Dmitruk	(a) **48**, K55 (1978)	*11*, p. *128*; 14.3.3, p. 235; 22.2.1, p. 447
V. S. Dneprovskii	(b) **85**, 111 (1978)	*13.5.1*, p. *189*; *20.3*, p. *341*; 22.4.1, p. 463; 22.4.2, p. 468
P. P. Dobrovolskii	(a) **47**, K91 (1978)	11, p. 127; *14.3.2*, p. *230*; 22.1.2, p. 433
Z. Dobrovolskis	(a) **50**, K73 (1978)	12.2, p. 141; *14.3.4*, p. *246*; 22.2.3, p. 454
J. Doerschel	(a) **47**, K17 (1978)	*10.2*, p. *113*; 22.1.2, p. 433
	(a) **48**, K11 (1978)	*10.2*, p. *115*; 22.1.2, p. 435
G. H. Döhler	(a) **45**, 113 (1978)	*14.3.3*, p. *233*; *14.3.4*, p. *240*; 22, p. 409
V. P. Dolgopolskii	(b) **88**, K177 (1978)	*13.1*, p. *158*; *13.5.1*, p. *194*; 22.2.1, p. 447
G. Dolling	(b) **85**, 311 (1978)	*6*, p. *65*; *20.1*, p. *320*; 22.7, p. 501
V. Dolocan	(a) **45**, K155 (1978)	*14.3.1*, p. *226*; 14.3.4, p. 241; 22.6, p. 491
Nguyen van Dong	(b) **88**, 555 (1978)	*2*, p. *47*; *14.3.1*, p. *227*; 22.1.1, p. 427; 22.1.2, p. 435
E. Doni	(b) **88**, 773 (1978)	*20.1*, p. *331*; 22.5.2, p. 483
F. Donika	(a) **49**, K5 (1978)	*1.5*, p. *39*; 22.8, p. 507
A. P. Dolgolenko	(a) **50**, 751 (1978)	10.2, p. 122; *11*, p. *131*; 22.1.2, p. 439
M. Dorikens	(a) **48**, 133 (1978)	*10.1*, p. *104*; 21.7, p. 406
L. Dorikens-Vanpraet	(a) **48**, 133 (1978)	*10.1*, p. *104*; 21.7, p. 406
B. Dorner	(b) **88**, 327 (1978)	*6*, p. *73*; 7, p. 81; *20.1*, p. *329*; 22.5, p. 475
E. A. Dorofeeva	(a) **50**, K5 (1978)	*18.2*, p. *292*; 21.1.1, p. 390
Th. Dörre	(b) **87**, 221 (1978)	*13.1*, p. *154*; 20, p. 316; 21.3, p. 394
P. Doubrava	(a) **48**, K93 (1978)	*2*, p. *47*; 16, p. 265; *20.1*, p. *330*; 22, p. 415
J.-D. Doukhan	(a) **49**, 713 (1978)	*1.4*, p. *33*; *10.2*, p. *119*; 22.6, p. 496
	(a) **50**, K37 (1978)	*1.4*, p. *34*; 22.6, p. 497
J. Doupovec	(a) **50**, K51 (1978)	2, p. 52; *8*, p. *89*; 11, p. 130; *14.3*, p. *225*; 22, p. 420
B. Dowgiałło-Plenkiewicz	(b) **87**, 309 (1978)	*13.1*, p. *154*; 22.3, p. 458
M. Doyama	(a) **45**, 305 (1978)	*2*, p. *42*; 21.1.1, p. 384
	(a) **46**, 671 (1978)	*2*, p. *44*; *12.1*, p. *135*; 21.1.1, p. 386
	(b) **90**, K31 (1978)	*10.1*, p. *107*; 21.1, p. 380
	(b) **50**, 165 (1978)	*2*, p. *52*; 21, p. 365
V. Drchal	(a) **86**, K89 (1978)	*13.4*, p. *172*; 21, p. 356
A. V. Drigo	(a) **49**, 347 (1978)	*11*, p. *129*; 22.1.2, p. 437
R. Droste	(b) **87**, 543 (1978)	*13.4*, p. *176*; *20.1*, p. *327*; 22.5.3, p. 487
M. Drulis	(a) **46**, K39 (1978)	*8*, p. *84*; *18*, p. *274*; 21.4, p. 397
B. M. Dshavadov	(a) **47**, K157 (1978)	*20.3*, p. *345*; 22.3.1, p. 460
	(b) **89**, K43 (1978)	*6*, p. *75*; *20.1*, p. *332*; 22.3, p. 458
K. S. Dubey	(b) **85**, K99 (1978)	*6*, p. *66*; *8*, p. *84*; 22.6, p. 491
	(b) **86**, 741 (1978)	*6*, p. *69*; 8, p. 84; 22.2.1, p. 446; 22.2.3, p. 452
	(b) **86**, K107 (1978)	*6*, p. *69*; *8*, p. *84*; 22.1.1, p. 425
	(b) **87**, K47 (1978)	*6*, p. *70*; 8, p. 85; 22.2.1, p. 446
	(b) **87**, K133 (1978)	*6*, p. *72*; *8*, p. *86*; 22.1.1, p. 425
	(b) **88**, 417 (1978)	*6*, p. *74*; *8*, p. *87*; 22.1.1, p. 427
	(b) **89**, K65 (1978)	*6*, p. *76*; *8*, p. *88*; 22.6, p. 496
M. Dubiel	(b) **89**, 595 (1978)	*10.2*, p. *119*; 19, p. 312; 22.5.2, p. 485
I. Ya. Dubovskaya	(b) **88**, 351 (1978)	*1.1*, p. *13*; *11*, p. *127*
J. J. Dubowski	(b) **85**, 663 (1978)	*13.1*, p. *151*; *14.3*, p. *217*; 15, p. 259; 22.4.4, p. 472
G. B. Dubrovskii	(b) **90**, 429 (1978)	*13.1*, p. *160*; *20.1*, p. *336*; 22.3, p. 459
A. Duda	(b) **88**, 735 (1978)	*18.2.1*, p. *293*
P. Duda	(a) **49**, K67 (1978)	*2*, p. *50*; 11, p. 129; *18.3*, p. *296*; 21.1, p. 378; 21.4, p. 398
I. Dudeck	(a) **49**, 153 (1978)	*13.4*, p. *180*; 14.3.3, p. 236; *14.3.4*, p. *244*; 22.1.2, p. 436
J. F. Dufresne	(a) **50**, 617 (1978)	*10.1*, p. *108*; *12.1*, p. *137*; 21.1.1, p. 391

J. Dumas	(b) **86**, K75 (1978)	*2*, p. *43*; *19*, p. *305*; 22.1.1, p. 424; 22.1.2, p. 431
L. Dumoulin	(b) **90**, 207 (1978)	*6*, p. *77*; *14.2*, p. *215*; 21, p. 364
W. Duncan	(a) **46**, 595 (1978)	*14.3.3*, p. *234*; 16, p. 263; 22.1.2, p. 432; 22.4.1, p. 464
L. H. Dunleavy	(a) **46**, K95 (1978)	*10.1*, p. *102*; 11, p. 125; *18.2*, p. *285*; 21.1, p. 372; 21.1.1, p. 386
	(a) **49**, 609 (1978)	10.1, p. 106; *11*, p. *130*; *18.2*, p. *290*; 21.1.1, p. 389
A. Dunlop	(a) **49**, 643 (1978)	*11*, p. *130*; 14.1, p. 211; 21, p. 363
J. Dünstl	(a) **45**, 529 (1978)	*1.1*, p. *11*; 11, p. 124; 21, p. 354
F. Dupré	(b) **86**, 77 (1978)	*1.2*, p. *16*; *6*, p. *67*; 21.1, p. 371
J. Dupuy	(b) **87**, 293 (1978)	*10.2*, p. *113*; *11*, p. *126*; 20.1, p. 326; 22.5.1, p. 477
R. Durný	(a) **49**, K113 (1978)	*2*, p. *51*; *19*, p. *313*; 22, p. 419
J. M. Dusseau	(a) **47**, K11 (1978)	*14.3.4*, p. *242*; 22, p. 413
H. Düsterhöft	(a) **50**, 503 (1978)	*17*, p. *271*; 21, p. 366
M. B. Dutt	(a) **45**, 657 (1978)	*9*, p. *91*; 21.1, p. 370; 21.1.1, p. 385; 21.6, p. 401
V. Dutta	(a) **48**, 257 (1978)	*1.5*, p. *37*; *14.3.1*, p. *227*; 22.1.1, p. 427
	(a) **49**, 379 (1978)	1.4, p. 33; *1.5*, p. *38*; 2, p. 49; 22.1.1, p. 427
P. Duval	(a) **50**, 67 (1978)	*1.4*, p. *34*; 21, p. 365; 21.1, p. 380
Vu Duy Phach	see under letter "P"	
H. Dwelk	(b) **90**, 517 (1978)	13.3, p. 168; *14.2*, p. *216*; 21, p. 365
V. V. Dyakin	(b) **87**, 41 (1978)	*13.1*, p. *154*; 22.1.2, p. 432
D. van Dyck	(a) **47**, 105 (1978)	*1.3*, p. *25*; *1.4*, p. *30*; 21.1, p. 373; 21.6, p. 402
	(a) **50**, 349 (1978)	*1.1*, p. *14*; *1.4*, p. *34*; 22, p. 420
	(a) **50**, K67 (1978)	*1.6*, p. *41*; 22, p. 420
I. M. Dykman	(b) **86**, 773 (1978)	*13*, p. *144*; *14.3*, p. *220*; 20, p. 316; 22, p. 412
	(b) **88**, 385 (1978)	*13*, p. *146*; *14.3.4*, p. *243*; 16, p. 265; 22, p. 415
M. I. Dykman	(b) **88**, 463 (1978)	*6*, p. *74*; *13.1*, p. *157*; 20.1, p. 330
T. D. Dzhafarov	(a) **49**, 211 (1978)	11, p. 129; *14.3.3*, p. *236*; *20.3*, p. *347*; 22.2.2, p. 451
R. S. Eachus	(b) **88**, 705 (1978)	*10.2*, p. *116*; 16, p. 265; *19*, p. *310*; 22.5.1, p. 477
D. M. Eagles	(b) **87**, 301 (1978)	*13.1*, p. *154*; 21.4, p. 397
W. Eberhardt	(b) **88**, 135 (1978)	*13.1*, p. *156*; 17, p. 270; 20.1, p. 328; 22.1.2, p. 434
R. Ebert	(b) **86**, 557 (1978)	*13.4*, p. *174*; 19, p. 306; 20.1, p. 324; 22.2.3, p. 452
M. Ecabert	(b) **87**, 497 (1978)	13.4, p. 175; *19*, p. *308*; *20.3*, p. *344*; 22.5.2, p. 481
W. Ecke	(a) **47**, 507 (1978)	*13.4*, p. *176*; *14.3.1*, p. *227*; 22.1.2, p. 433
	(a) **48**, K161 (1978)	*14.3.2*, p. *231*; *14.3.4*, p. *244*; 22.1.2, p. 436
H. Eckseler	(b) **85**, 185 (1978)	*9*, p. *91*; 10.1, p. 99; 21.1, p. 368
N. A. Economou	(a) **50**, 579 (1978)	*1.2*, p. *24*; *1.3*, p. *28*; 10.2, p. 122; 22.3, p. 459
M. C. Al-Edani see under letter "A"		
F. L. Edelman	(a) **50**, 573 (1978)	*1.2*, p. *24*; *1.5*, p. *39*; 22, p. 421
I. S. Edelman	(a) **46**, K83 (1978)	*18.2.1*, p. *293*; *20.1*, p. *325*; 22.8.2, p. 519
A. Edgar	(a) **46**, K135 (1978)	*20.1*, p. *325*; 22.6, p. 492
J. T. Edmond	(a) **48**, 395 (1978)	*2*, p. *48*; *14.3.4*, p. *244*; 22, p. 416
A. V. Efanov	(b) **90**, 449 (1978)	*1.2*, p. *24*; *13.5.3*, p. *200*
Sh. M. Efendiev	(a) **50**, K141 (1978)	6, p. 79; 13.1, p. 161; *20.1*, p. *338*; 22.8, p. 508
V. M. Efimov	(a) **49**, 217 (1978)	*14.4*, p. *250*; 22, p. 417
A. L. Efros, Jr.	(b) **88**, K177 (1978)	*13.1*, p. *158*; *13.5.1*, p. *194*; 22.2.1, p. 447
M. Egée	(a) **48**, 89 (1978)	13.4, p. 177; 14.3, p. 222; *20.3*, p. *345*; 22.4.1, p. 465

B. V. Egorov	(b) **86**, 781 (1978)	*13*, p. *144*; 13.5.3, p. 200; 18.2, p. 285; 22, p. 412
H. J. Eichler	(a) **45**, 433 (1978)	13.4, p. 171; *20.2*, p. *339*; 22.4.1, p. 463
W. Eichler	(a) **45**, 195 (1978)	*14.3*, p. *217*; *15*, p. *259*; 22.1.3, p. 441
	(b) **86**, K85 (1978)	*14.3*, p. *218*; 22.7, p. 501
Th. Eifrig	(b) **86**, K79 (1978)	13, p. 143; 18.2, p. 284; 21, p. 356
	(b) **89**, 171 (1978)	*2*, p. *48*; *14.1*, p. *210*; 18.1, p. 279; 22.9, p. 526
W. Eigermann	(a) **49**, 145 (1978)	1.3, p. 27; *3*, p. *55*; 22.8.1, p. 515
E. Ejder	(a) **45**, K167 (1978)	*11*, p. *124*; 22.2, p. 443
W. Ekardt	(b) **86**, 535 (1978)	*13.5.1*, p. *191*; 22.5, p. 474
N. A. Ekonomov	(a) **50**, K153 (1978)	*18.2.1*, p. *294*; *20*, p. *318*; 21.1, p. 383; 21.1.1, p. 391
M. H. El-Fouly	(a) **48**, 395 (1978)	*2*, p. *48*; *14.3.4*, p. *244*; 22, p. 416
S. H. El-Mahalawy	(b) **86**, 151 (1978)	12.2, p. 138; *14.3*, p. *218*; 22, p. 410
M. Elango	(a) **46**, 315 (1978)	*13.4*, p. *173*; *20.3*, p. *343*; 22.5.2, p. 479
	(a) **46**, K61 (1978)	*17*, p. *269*; 22.5.2, p. 480
N. V. Elistratov	(b) **87**, K101 (1978)	*5*, p. *61*; 18.3, p. 295; 22.8.2, p. 519
K. Elk	(b) **89**, 537 (1978)	13.1, p. 159; *14.1*, p. *211*; 21, p. 363
S. G. Elkomoss	(b) **87**, 261 (1978)	*13.5.1*, p. *191*; *20.1*, p. *326*; 22.6, p. 493
K. Ellmer	(b) **85**, K67 (1978)	*11*, p. *123*
	(b) **89**, 611 (1978)	*11*, p. *129*; 22.1, p. 422
R. Enderlein	(b) **85**, 217 (1978)	13.4, p. 169; *14.3.4*, p. *240*; *20.1*, p. *319*; 22.4.1, p. 463
	(b) **85**, 569 (1978)	*13.1*, p. *151*; *13.5.1*, p. *190*; 20, p. 315; 22.1, p. 422
	(b) **88**, 173 (1978)	*13.1*, p. *156*; *20.1*, p. *329*; 21.1, p. 375
	(b) **89**, 625 (1978)	2, p. 50; *6*, p. *76*; *20.1*, p. *334*; 22, p. 418
P. G. van Engen	(a) **50**, K29 (1978)	2, p. 52; *18.3*, p. *297*; 20, p. 318; 21.1.1, p. 390; 21.4, p. 398
L. Ensuque	(a) **47**, K11 (1978)	*14.3.4*, p. *242*; 22, p. 413
I. R. Entin	(b) **90**, 575 (1978)	*20*, p. *318*; 22.1.2, p. 439
M. V. Entin	(a) **45**, K17 (1978)	*13.4*, p. *170*; 14.4.1, p. 253; 16, p. 262; *20.1*, p. *321*; 22.8.1, p. 510
	(b) **85**, K135 (1978)	*14.3.4*, p. *240*; 16, p. 262; 22.2, p. 443
M. V. Eremin	(b) **85**, 783 (1978)	1, p. 9; *13*, p. *143*; 22.5.4, p. 489
	(b) **89**, 503 (1978)	1, p. 10; *13*, p. *147*
	(b) **90**, 123 (1978)	1, p. 10; *13*, p. *148*
Ş. Erkoç	(b) **89**, 289 (1978)	*8*, p. *88*; *18.1*, p. *279*; 22.8.1, p. 515
R. Erlandsson	(a) **47**, 85 (1978)	*1.6*, p. *41*; 22, p. 413
P. Erlenwein	(a) **47**, K9 (1978)	*1.4*, p. *30*; 11, p. 126; 21, p. 358; 21.6, p. 402
J. G. Erlings	(a) **46**, 653 (1978)	*10.1*, p. *102*; 21.6, p. 402
A. N. Ermoshkin	(b) **86**, 47 (1978)	*13.1*, p. *152*; 22.5.2, p. 479
A. Eroglu	(b) **87**, 287 (1978)	*14.1*, p. *207*; 21.1, p. 372; 21.6, p. 402
H. Eschrig	(b) **90**, 621 (1978)	*13.1*, p. *161*; 21.1, p. 381
C. Esnouf	(a) **47**, 201 (1978)	*10.1*, p. *103*; *12.1*, p. *135*; 21, p. 358
B. Esser	(b) **89**, 633 (1978)	2, p. 50; *6*, p. *76*; 13.1, p. 159; *20.1*, p. *334*; 22, p. 418
D. Esteve	(a) **46**, 185 (1978)	*13.3*, p. *164*; *14.3.3*, p. *234*
J. P. Estienne	(a) **49**, K109 (1978)	*10.2*, p. *119*; 22.1.3, p. 442
R. Ethiraj	(a) **45**, 83 (1978)	12.2, p. 138; *20*, p. *315*; 22.5.4, p. 489
B. Etlinger	(a) **46**, K173 (1978)	*3*, p. *54*; 22.3.1, p. 460
B. L. Evans	(b) **86**, 151 (1978)	12.2, p. 138; *14.3*, p. *218*; 22, p. 410
G. R. Evans	(b) **85**, K63 (1978)	*2*, p. *42*
R. A. Evarestov	(b) **86**, 47 (1978)	*13.1*, p. *152*; 22.5.2, p. 479
P. Even	(a) **48**, 137 (1978)	13.3, p. 166; *14.3.3*, p. *235*; *16*, p. *264*; 22.1.2, p. 434

R. Evrard	(b) **90**, K73 (1978)	*6*, p. *78*; *14.3.4*, p. *246*; 22.2.3, p. 454
	(b) **90**, K137 (1978)	*13*, p. *149*; *14.3.4*, p. *246*; 22.2.1, p. 449; 22.2.3, p. 455
A. M. Evstigneev	(b) **87**, 613 (1978)	13.3, p. 165; *14.3.4*, p. *242*; *20.1*, p. *327*; 22.1.2, p. 433
N. N. Evtikhiev	(a) **50**, K153 (1978)	*18.2.1*, p. *294*; *20*, p. *318*; 21.1, p. 383; 21.1.1, p. 391
S. Ewert	(b) **89**, 459 (1978)	6, p. 76; *14.3.1*, p. *228*; *20.1*, p. *334*; 22.9, p. 527
H.-G. Fabian	(a) **48**, 369 (1978)	*1.3*, p. *27*; *10.1*, p. *105*; 14.1, p. 209; 21, p. 361; 21.1, p. 376; 21.3, p. 395
H. Fabig	(a) **48**, 407 (1978)	*1.4*, p. *32*; 14.3.2, p. 230; 20.3, p. 346; 22.2.2, p. 451
E. Fabre	(a) **48**, 137 (1978)	13.3, p. 166; *14.3.3*, p. *235*; *16*, p. *264*; 22.1.2, p. 434
J. M. Fabre	(b) **85**, 603 (1978)	*18*, p. *273*; 19, p. 304; 22.9, p. 523
J. D. Fabris	(b) **90**, K169 (1978)	*10.2*, p. *121*; *13*, p. *149*; 14.4, p. 252; 22.6.1, p. 500
N. V. Fadeeva	(a) **48**, 183 (1978)	*1*, p. *9*; 5, p. 61; *14.4.1*, p. *255*; 18.2, p. 289; 18.4, p. 301; 19, p. 310; 22.8.1, p. 514
W. R. Fahrner	(a) **48**, 533 (1978)	*14.3.1*, p. *227*; *14.3.4*, p. *244*; 22, p. 416
	(a) **50**, 433 (1978)	11, p. 131; *14.3.2*, p. *232*; *14.3.3*, p. *238*; 22.1.2, p. 439
	(a) **50**, 551 (1978)	11, p. 131; *14.3.1*, p. *229*; *14.3.3*, p. *238*; 22.6, p. 497
Ş. Fakioğlu	(b) **88**, 301 (1978)	13.4, p. 177; *18.1*, p. *278*; 22.8.1, p. 514
	(b) **89**, 289 (1978)	*8*, p. *88*; *18.1*, p. *279*; 22.8.1, p. 515
G. Fantozzi	(a) **47**, 201 (1978)	*10.1*, p. *103*; *12.1*, p. *135*; 21, p. 358
F. E. Faradev	(b) **85**, 381 (1978)	*6*, p. *65*; *20.1*, p. *320*; 22.3, p. 457
S. I. Fărcaş	(b) **85**, 75 (1978)	*19*, p. *303*; 22.8.1, p. 510
Z. Farid	(a) **45**, 47 (1978)	*10.1*, p. *99*; 12.1, p. 134; 21, p. 352; 21.2, p. 392
	(a) **45**, 419 (1978)	*10.1*, p. *100*; 12.1, p. 134; 21, p. 354
	(a) **46**, 697 (1978)	*10.1*, p. *102*; 12.1, p. 135; 14.1, p. 207; 21, p. 357
J. L. Farvacque	(a) **49**, 737 (1978)	*10.2*, p. *119*; 22.2.3, p. 454
	(a) **49**, K109 (1978)	*10.2*, p. *119*; 22.1.3, p. 442
R. H. J. Fastenau	(a) **46**, 541 (1978)	*10.1*, p. *102*; 21, p. 357
	(a) **47**, 577 (1978)	*10.1*, p. *104*; 21, p. 359
D. J. Fathers	(a) **46**, 291 (1978)	*18.2*, p. *284*; 21.1.1, p. 386; 21.4, p. 397
P. N. Favennec	(a) **50**, 311 (1978)	*13.4*, p. *184*; 14.3.3, p. 238; 22.2.1, p. 449
R. Feder	(a) **45**, K117 (1978)	*1.6*, p. *40*; 21, p. 355
V. M. Fedorov	(b) **85**, K115 (1978)	*14.2*, p. *213*; 21.6, p. 401
	(b) **87**, 11 (1978)	1.2, p. 15; *1.6*, p. *40*; *14.2*, p. *213*; 21, p. 352; 22, p. 408
V. K. Fedyanin	(b) **89**, 123 (1978)	*13.5*, p. *188*; 18.4, p. 301
A. Feher	(b) **85**, K31 (1978)	*8*, p. *83*; 21.4, p. 396
	(b) **85**, 545 (1978)	*8*, p. *83*; 14.2, p. 213; 21, p. 354
V. A. Feigin	(b) **87**, K33 (1978)	*18.2*, p. *286*
G. Ferla	(a) **48**, 523 (1978)	*10.2*, p. *116*; 14.3.3, p. 235; 22.1.2, p. 435
F. Fermi	(b) **90**, 657 (1978)	13.4, p. 185; *20.1*, p. *337*; *20.3*, p. *350*; 22.5.2, p. 486
J. Fernandez	(a) **48**, 53 (1978)	*1.2*, p. *21*; 22.8, p. 506
D. Ferré	(a) **49**, 737 (1978)	*10.2*, p. *119*; 22.2.3, p. 454
J. R. F. Ferreira	(b) **90**, K169 (1978)	*10.2*, p. *121*; *13*, p. *149*; 14.4, p. 252; 22.6.1, p. 500

A. Ferreira da Silva	(b) **90**, K153 (1978)	*13.4*, p. *185*; 22, p. 421
J. Ferrer	(a) **49**, 585 (1978)	10.2, p. 119; *11*, p. *130*; 22.8, p. 507
R. Ferretti	(a) **48**, 533 (1978)	*14.3.1*, p. *227*; *14.3.4*, p. *244*; 22, p. 416
G. Ferron	(a) **46**, K43 (1978)	*10.1*, p. *101*; *12.1*, p. *134*; 21, p. 356
A. Feuerpfeil	(a) **46**, K67 (1978)	*14.2*, p. *214*; 18.2.1, p. 293; 21.1.1, p. 386
J. Fidler	(a) **50**, 73 (1978)	*1.4*, p. *34*; 10.1, p. 107; 21.1, p. 380; 21.4, p. 398
D. W. Field	(a) **47**, K43 (1978)	*1.1*, p. *12*; 6, p. 71; 21.7, p. 406
	(a) **47**, 651 (1978)	*6*, p. *72*; 21.7, p. 406
S. R. Figarova	(b) **86**, 103 (1978)	14.3.1, p. 226; *15*, p. *259*
L. I. Fikin	(a) **45**, 53 (1978)	1.2, p. 15; *14.4.1*, p. *253*; 19, p. 303; 20, p. 315; 22.8.1, p. 510
B. P. Filatov	(a) **45**, 327 (1978)	*10.1*, p. *100*; 21, p. 353
A. S. Filipchenko	(b) **87**, K1 (1978)	13.1, p. 155; *20.1*, p. *326*; 22.2.1, p. 446
	(a) **48**, 281 (1978)	13.4, p. 179; *20.3*, p. *346*; 22.2.3, p. 453
	(a) **48**, K115 (1978)	*13.4*, p. *179*; 20.1, p. 331; 22.2.3, p. 453
	(a) **49**, 183 (1978)	*13.4*, p. *180*; *20.1*, p. *333*; 22.2.1, p. 448; 22.2.3, p. 454
	(a) **49**, K189 (1978)	*20.3*, p. *349*; 22.2.1, p. 448; 22.2.3, p. 454
	(a) **50**, K251 (1978)	13.4, p. 186; *20.3*, p. *351*; 22.2.1, p. 449; 22.2.3, p. 455
G. A. Filippov	(a) **45**, K159 (1978)	*10.1*, p. *101*; 21.1.1, p. 385
P. G. Filippov	(b) **86**, 449 (1978)	*13.5.1*, p. *190*; *20.3*, p. *343*; 22.9, p. 524
V. A. Finkel	(a) **47**, K145 (1978)	*8*, p. *87*; *14.2*, p. *214*; 21, p. 360
H. Finkenrath	(a) **49**, 67 (1978)	*14.3.1*, p. *228*; 22.4, p. 462
A. Fischer	(a) **45**, K175 (1978)	*3*, p. *54*; *14.3.1*, p. *226*; 22.6, p. 491
H. Fischer	(a) **49**, 331 (1978)	*13.4*, p. *181*; 13.5.1, p. 194; *20.3*, p. *347*; 22.2.2, p. 451
J. E. Fischer	(b) **90**, 151 (1978)	*20.1*, p. *335*; 22.1, p. 423
I. I. Fishchuk	(b) **89**, 61 (1978)	*2*, p. *48*; *14*, p. *204*
	(a) **50**, 751 (1978)	10.2, p. 122; *11*, p. *131*; 22.1.2, p. 439
E. Fitz	(b) **89**, 225 (1978)	*6*, p. *75*; 14.3, p. 224; 22.2.1, p. 448
K. Flachbart	(b) **85**, 545 (1978)	*8*, p. *83*; 14.2, p. 213; 21, p. 354
P. J. Flatau	(b) **90**, 251 (1978)	*1.3*, p. *28*; *20.1*, p. *335*; 22.4.4, p. 473
P. Fleischmann	(a) **48**, 439 (1978)	*7*, p. *81*; 10.1, p. 105; 21, p. 361
G. Fleury	(a) **48**, 417 (1978)	*6*, p. *75*; *20.1*, p. *331*; 22.6, p. 495
H. Flietner	(b) **90**, K27 (1978)	*13.3*, p. *168*; *14.3.2*, p. *231*; 22, p. 419
F. Flores	(b) **88**, 591 (1978)	*13.3*, p. *166*; 22.2, p. 443
R. Florian	(a) **48**, K53 (1978)	*1.2*, p. *21*; *7*, p. *81*; 22.8, p. 506
H. M. Flower	(a) **47**, 407 (1978)	*1.4*, p. *31*; *14.1*, p. *208*; 21.1, p. 374; 21.1.1, p. 387
M. E. Foglio	(b) **86**, 459 (1978)	*13.4*, p. *174*
	(b) **87**, 87 (1978)	*13.4*, p. *174*
V. S. Fokin	(b) **85**, 111 (1978)	*13.5.1*, p. *189*; *20.3*, p. *341*; 22.4.1, p. 463; 22.4.2, p. 468
I. Földvári	(b) **89**, 235 (1978)	1.2, p. 22; *14.4*, p. *250*; *20.1*, p. *332*; 22.5.2, p. 484; 22.5.4, p. 489
H. Föll	(a) **49**, 555 (1978)	*1.4*, p. *33*; *10*, p. *97*
V. M. Fomin	(b) **85**, 403 (1978)	*14.3.4*, p. *240*; *20.2*, p. *339*; 22, p. 409
	(b) **87**, 725 (1978)	6, p. 72; 14.3.4, p. 242; *20*, p. *316*
R. M. M. Fonville	(b) **85**, 621 (1978)	*13.1*, p. *151*; 14.3, p. 217; 18.1, p. 277; 19, p. 304; 22, p. 409
A. R. Forouhi	(b) **89**, 309 (1978)	*10.2*, p. *117*; 22, p. 417
G. Foti	(a) **47**, 533 (1978)	*11*, p. *127*; 22.1.2, p. 433
	(a) **49**, 347 (1978)	*11*, p. *129*; 22.1.2, p. 437
M. H. El-Fouly	see under letter "E"	
J. de Fouquet	(a) **48**, 99 (1978)	*10.1*, p. *104*; *12.1*, p. *135*; 21.1, p. 375

Z. Frait	(a) **50,**	537 (1978)	*18.3*, p. *297*; 21.1, p. 382; 21.4, p. 399
S. Franck	(b) **85,**	319 (1978)	*14.4.1*, p. *253*; 19, p. 303; 22.9, p. 522
W. Frank	(a) **46,**	89 (1978)	*10.1*, p. *101*; 11, p. 125; 21.1.1, p. 385
E. L. Frankevich	(b) **87,**	373 (1978)	13.5.1, p. 191; *16*, p. *164*; 20.3, p. 344; 22.9, p. 524
P. Franzosi	(b) **89,**	K119 (1978)	*6*, p. *76*; *13.5.1*, p. *195*; 20.1, p. 334; 22.4.1, p. 466
Th. Frauenheim	(b) **88,**	457 (1978)	*13*, p. *146*; *18*, p. *274*
A. Freudenhammer	(b) **86,**	579 (1978)	*13*, p. *144*; *18*, p. *274*; 21.1, p. 371
	(b) **90,**	269 (1978)	*8*, p. *89*; 13, p. 148; 21, p. 364
	(b) **90,**	475 (1978)	*18*, p. *276*; 21.1, p. 381
A. Frey	(b) **85,**	K107 (1978)	*13.4*, p. *171*; *20.3*, p. *342*; 22.2.4, p. 456
V. M. Fridkin	(a) **46,**	729 (1978)	*14.4.1*, p. *254*; *16*, p. *263*; 22.8.1, p. 512
	(a) **50,**	K255 (1978)	14.4.1, p. 258; *16*, p. *268*; *17*, p. *271*; 22.8.1, p. 517
H. Friedrich	(b) **89,**	603 (1978)	*13.5.1*, p. *195*; *20.1*, p. *334*; 22.4.3, p. 471
	(b) **90,**	167 (1978)	6, p. 77; 13.1, p. 159; 13.5.1, p. 195; *20.1*, p. *335*; 22.2.4, p. 456
W. I. Friesen	(b) **85,**	715 (1978)	*13*, p. *143*; *20*, p. *315*; 22, p. 409
G. Frigerio	(a) **48,**	523 (1978)	*10.2*, p. *116*; 14.3.3, p. 235; 22.1.2, p. 435
L. Fritsche	(b) **86,**	145 (1978)	*13*, p. *143*; 21, p. 355
	(b) **88,**	639 (1978)	*6*, p. *74*; 12.1, p. 136; *13.1*, p. *157*; 14.2, p. 215; 21.4, p. 398
	(b) **89,**	367 (1978)	*13.3*, p. *167*; 21, p. 363
G. Frohberg	(a) **50,**	563 (1978)	*9*, p. *95*; 14.1, p. 212; 21.1.1, p. 391
F. Fröhlich	(b) **89,**	595 (1978)	*10.2*, p. *119*; 19, p. 312; 22.5.2, p. 485
Yu. A. Frolkov	(b) **89,**	27 (1978)	*13.5*, p. *188*; 20.1, p. 331; 22.6, p. 496
G. Frollani	(b) **86,**	77 (1978)	*1.2*, p. *16*; *6*, p. *67*; 21.1, p. 371
A. T. Fromhold, Jr.	(b) **90,**	K21 (1978)	*9*, p. *94*; *10.2*, p. *120*; 22, p. 419
G. Frommeyer	(a) **50,**	517 (1978)	*14.1*, p. *212*; 21.1, p. 382; 21.6, p. 405
D. Fruchart	(a) **45,**	591 (1978)	*18.2*, p. *283*; *18.4*, p. *298*; 22.8, p. 503
R. Fruchart	(a) **45,**	591 (1978)	*18.2*, p. *283*; *18.4*, p. *298*; 22.8, p. 503
G. Fuchs	(a) **45,**	239 (1978)	*14.2*, p. *213*; 21, p. 353
W. Fuchs	(b) **89,**	69 (1978)	*10.2*, p. *117*; 11, p. 128; 22.1.2, p. 436
	(b) **89,**	495 (1978)	2, p. 50; *14.3.1*, p. *228*; *16*, p. *266*; 22.1.2, p. 437
F. Fujimoto	(a) **45,**	99 (1978)	*1.1*, p. *11*
	(a) **45,**	411 (1978)	*11*, p. *124*; 13, p. 143; 21, p. 354
T. Fujimura	(a) **45,**	513 (1978)	*1.2*, p. *16*; *7*, p. *80*; 14.4, p. 248; 22.5.4, p. 489
Y. Fujino	(a) **46,**	217 (1978)	*1.3*, p. *25*; 6, p. 68; *11*, p. *125*; 21.6, p. 401
H. Fujita	(a) **45,**	411 (1978)	*11*, p. *124*; 13, p. 143; 21, p. 354
K. Fukamichi	(a) **48,**	175 (1978)	2, p. 47; *18.2*, p. *289*; 21.1.1, p. 388
	(a) **49,**	K37 (1978)	8, p. 88; 14.1, p. 210; *18.4*, p. *301*; 21.1, p. 378
	(a) **50,**	263 (1978)	*18.2*, p. *291*; 21.1, p. 381
	(a) **50,**	K25 (1978)	*2*, p. *52*; *18.2*, p. *292*; 21.1.1, p. 390
K. Fukushi	(b) **86,**	257 (1978)	*6*, p. *67*; *20.1*, p. *322*; 22.5, p. 474
K. Funke	(a) **48,**	K179 (1978)	*9*, p. *93*; 22.5.1, p. 477
R. J. Gaboriaud	(a) **46,**	387 (1978)	*10.2*, p. *111*; 22.6, p. 492
V. A. Gadzhiev	(b) **90,**	K77 (1978)	13.5.1, p. 196; *14.3.4*, p. *246*; *20.1*, p. *336*; 22.3, p. 459
V. V. Gagulin	(a) **48,**	183 (1978)	*1*, p. *9*; 5, p. 61; *14.4.1*, p. *255*; 18.2, p. 289; 18.4, p. 301; 19, p. 310; 22.8.1, p. 514
B. K. D. Gairola	(b) **85,**	577 (1978)	*10*, p. *96*
A. V. Gaisanyuk	(a) **48,**	K131 (1978)	*1.2*, p. *22*; *14*, p. *204*; 22.3, p. 458
J. A. Gaj	(b) **89,**	655 (1978)	*13.1*, p. *159*; 18.2, p. 290; 22.8, p. 507

R. R. Gałązka	(b) **88**, 73 (1978)	*13.2*, p. *162*; *14.3*, p. *221*; 15, p. 260; 22.4.4, p. 473
	(b) **89**, 655 (1978)	*13.1*, p. *159*; 18.2, p. 290; 22.8, p. 507
F. Galeskii	(b) **88**, 675 (1978)	13.4, p. 178; 20.1, p. 330; *20.2*, p. *339*; 22, p. 415
A. A. Galkin	(a) **45**, 327 (1978)	*10.1*, p. *100*; 21, p. 353
	(b) **90**, K17 (1978)	*13.3*, p. *168*; *14.1*, p. *211*; 19, p. 313; 21, p. 364
J. Gallice	(b) **85**, 603 (1978)	*18*, p. *273*; 19, p. 304; 22.9, p. 523
J.-P. Gallinar	(b) **88**, K143 (1978)	*13*, p. *147*; *14*, p. *204*
F. M. Galperin	(b) **86**, 679 (1978)	*13*, p. *144*; *18.2*, p. *285*; 21.1, p. 372; 21.1.1, p. 386
L. Garbato	(a) **50**, 643 (1978)	13.4, p. 186; *16*, p. *267*; 22.8, p. 508
V. K. Garg	(b) **85**, 417 (1978)	*1.1*, p. *11*; 12.2, p. 138; 22.5.2, p. 478
	(b) **87**, 401 (1978)	*6*, p. *70*; *12.2*, p. *139*; 22.2, p. 443; 22.4, p. 461; 22.5, p. 475
R. K. Gartia	(a) **47**, K165 (1978)	*10.2*, p. *114*; 20.3, p. 345; 22.5.2, p. 482
	(a) **48**, 235 (1978)	*10.2*, p. *115*; *20.3*, p. *345*; 22.5.2, p. 482
F. G. Gärtner	(b) **89**, 381 (1978)	13.4, p. 181; *20.1*, p. *333*; 22.4, p. 462
	(b) **90**, 33 (1978)	13.4, p. 183; *20.1*, p. *335*; 22.4, p. 462
H. Gärtner	(a) **49**, 149 (1978)	*10.1*, p. *105*; *14.1*, p. *210*; 21.1, p. 377
N. M. Gasanly	(b) **85**, 381 (1978)	*6*, p. *65*; *20.1*, p. *320*; 22.3, p. 457
	(b) **86**, K49 (1978)	*6*, p. *68*; *20.1*, p. *323*; 22.8, p. 504
	(a) **47**, K157 (1978)	*20.3*, p. *345*; 22.3.1, p. 460
	(b) **89**, K43 (1978)	*6*, p. *75*; *20.1*, p. *332*; 22.3, p. 458
F. V. Gasparyan	(a) **49**, 797 (1978)	*13.4*, p. *182*; *14.3.3*, p. *237*; 22.1.2, p. 437
S. V. Gasparyan	(a) **46**, 351 (1978)	*10.2*, p. *111*; *12.2*, p. *138*; 22.5.2, p. 480
	(a) **46**, 411 (1978)	*10.1*, p. *101*; *12.1*, p. *135*; 21, p. 357; 21.1, p. 372; 21.3, p. 394
W. Gasser	(b) **87**, 185 (1978)	*13.5.2*, p. *198*; *14*, p. *203*; 18, p. 274
	(b) **89**, 165 (1978)	*13.5.2*, p. *198*; 14, p. 204; 18.1, p. 279
J. Gastaldi	(a) **49**, 529 (1978)	1.4, p. 33; *3*, p. *56*; 21, p. 363
R. Gauthier	(a) **48**, 459 (1978)	*17*, p. *270*; 21, p. 361
N. P. Gavaleshko	(b) **88**, 179 (1978)	*13.4*, p. *177*; 14.3, p. 222; *15*, p. *260*; 22.1.3, p. 441
P. Gaworzewski	(a) **46**, 117 (1978)	13.4, p. 173; *14.3*, p. *219*; 22.1.2, p. 431
M. Gawron	(a) **50**, K105 (1978)	*14*, p. *204*; 22.9, p. 528
R. Gay	(a) **50**, 41 (1978)	*1.3*, p. *28*; 14.1, p. 212; 21.1, p. 380
J. Gazsó	(a) **45**, 181 (1978)	2, p. 42; 11, p. 123; *20.1*, p. *320*; 22.7, p. 501
P. Gebhardt	(b) **85**, 271 (1978)	*13.5.1*, p. *189*; *20.1*, p. *320*; 22.4, p. 461
W. Gębicki	(b) **86**, K135 (1978)	*6*, p. *69*; *20.1*, p. *325*; 22.4.4, p. 472
M. Geddo	(a) **47**, 223 (1978)	*16*, p. *264*; 21, p. 358
W. Geertsma	(b) **85**, 621 (1978)	*13.1*, p. *151*; 14.3, p. 217; 18.1, p. 277; 19, p. 304; 22, p. 409
W. Gehlhoff	(b) **88**, 599 (1978)	*19*, p. *310*
R. Geick	(b) **87**, 543 (1978)	*13.4*, p. *176*; *20.1*, p. *327*; 22.5.3, p. 487
I. N. Geifman	(b) **85**, K5 (1978)	*19*, p. *303*; 22.8.1, p. 510
P. V. Geld	(b) **87**, 49 (1978)	*4*, p. *58*; *13*, p. *145*; 22.3, p. 457
	(a) **49**, K117 (1978)	*8*, p. *88*; *14.1*, p. *211*; 21, p. 364
B. L. Gelmont	(b) **87**, 381 (1978)	*13.5*, p. *187*; 14.4, p. 248; 22.4.3, p. 470
A. E. Gelyasin	(a) **47**, K55 (1978)	*18.2*, p. *286*; *18.3*, p. *295*; 21, p. 359
	(a) **50**, K205 (1978)	5, p. 63; *18.2*, p. *292*; *18.3*, p. *297*; 21, p. 367
H. P. Gemünd	(a) **48**, 481 (1978)	*18.2.1*, p. *293*; 21.1, p. 376; 21.1.1, p. 388
D. Genzow	(b) **86**, K21 (1978)	*20.1*, p. *323*; 22, p. 410
	(b) **86**, K143 (1978)	*16*, p. *263*; 22, p. 412
	(b) **89**, K9 (1978)	*20.1*, p. *332*; 22, p. 417
	(b) **90**, 197 (1978)	*13.1*, p. *159*; 22, p. 419
	(b) **90**, 535 (1978)	*13.1*, p. *160*; *20.1*, p. *337*; 22, p. 420

G. Georgiev	(b) **85**, 657 (1978)	*13.5.4*, p. *201*; *18.2*, p. *282*
G. M. Georgiev	(b) **86**, 431 (1978)	*6*, p. *68*; *20.1*, p. *323*; 22.9, p. 523
M. Georgiev	(b) **86**, 697 (1978)	*10.2*, p. *112*; 20.1, p. 324; 22.5.2, p. 480
S. S. Georgiev	(a) **48**, 609 (1978)	*1.5*, p. *38*; *14.3.1*, p. *228*; 22.6, p. 495
E. I. Georgitse	(b) **85**, K39 (1978)	*16*, p. *262*; 18, p. 273; *20.1*, p. *320*; 22.4.3, p. 470
A. N. Georgobiani	(b) **89**, 603 (1978)	*13.5.1*, p. *195*; *20.1*, p. *334*; 22.4.3, p. 471
N. N. Gerasimenko	(a) **45**, 377 (1978)	1.4, p. 29; *10.2*, p. *110*; *11*, p. *124*; 22.1.2, p. 429
	(a) **47**, K91 (1978)	11, p. 127; *14.3.2*, p. *230*; 22.1.2, p. 433
	(b) **90**, 689 (1978)	*6*, p. *78*; 11, p. 131; *20.1*, p. *337*; 22.1.2, p. 439
I. Gerassimova	(a) **47**, 743 (1978)	*2*, p. *46*; 4, p. 58; 22.6, p. 494
D. Gerhardt	(a) **49**, 67 (1978)	*14.3.1*, p. *228*; 22.4, p. 462
W. Gericke	(a) **49**, 667 (1978)	13.4, p. 182; *14.3.3*, p. *237*; *20.3*, p. *349*; 22.4.2, p. 469
E. Gerlach	(b) **86**, 479 (1978)	*13.4*, p. *174*; *14.3*, p. *219*; 22, p. 411
U. Gerlach	(a) **45**, 123 (1978)	*1.2*, p. *15*; *5*, p. *60*; 18.1, p. 277; 22.6.1, p. 499
K. G. Germanova	(b) **88**, K19 (1978)	*13.5.1*, p. *193*; 20.1, p. 329; 22.4.3, p. 470
	(a) **49**, 499 (1978)	*14.3.4*, p. *245*; 16, p. 266; 22.4.1, p. 466
E. V. Gerova	(a) **48**, 609 (1978)	*1.5*, p. *38*; *14.3.1*, p. *228*; 22.6, p. 495
A. E. Gershinskii	(a) **46**, 467 (1978)	*14.3.1*, p. *226*; *14.3.4*, p. *241*; 22.1.2, p. 432; 22.6, p. 492
L. Gerward	(a) **48**, 113 (1978)	*1.1*, p. *13*; 22.1.2, p. 434
H. P. Geserich	(b) **89**, 441 (1978)	*1.2*, p. *22*; *1.5*, p. *39*; 2, p. 50; 14.1, p. 210; 20.1, p. 334; 21.1.1, p. 389
R. Gevers	(b) **87**, 419 (1978)	*1.1*, p. *13*; 22.1.1, p. 425; 22.6, p. 494
S. D. Gevorkyan	(a) **49**, K3 (1978)	*18.2.1*, p. *294*; 21.1, p. 378; 21.1.1, p. 389
N. S. Ghali	(a) **47**, 365 (1978)	*7*, p. *81*; 18.2, p. 287; 21.1.1, p. 387
M. Ghomi	(b) **86**, 493 (1978)	*6*, p. *68*; *20.1*, p. *324*; 22.5.2, p. 480
S. Ghosh	(b) **87**, K9 (1978)	*14.3*, p. *220*; *20*, p. *316*; 22, p. 413
	(b) **89**, K171 (1978)	*14.3*, p. *224*; *20*, p. *317*; 22, p. 418
	(a) **49**, 363 (1978)	*14.3.4*, p. *245*; 22.2.3, p. 454
U. S. Ghosh	(b) **86**, 379 (1978)	*13.4*, p. *172*; 20.1, p. 322; 22.4, p. 461
D. Gignoux	(a) **45**, K149 (1978)	*1.1*, p. *11*; *18.4*, p. *298*; 21, p. 355; 21.4, p. 396
F. Gilletta	(a) **48**, 67 (1978)	*1.2*, p. *21*; *8*, p. *87*; 22.8.1, p. 514
B. Gillot	(a) **50**, 109 (1978)	*14.3*, p. *224*; 18.3, p. 296; 22.8.2, p. 520
C. Trallero Giner	see under letter "T"	
J. Ginter	(b) **89**, 655 (1978)	*13.1*, p. *159*; 18.2, p. 290; 22.8, p. 507
L. Giral	(b) **85**, 603 (1978)	*18*, p. *273*; 19, p. 304; 22.9, p. 523
P. E. Girard	(b) **88**, K135 (1978)	*14.3*, p. *223*; 22, p. 416
B. Girault	(a) **47**, 195 (1978)	*14.3.3*, p. *235*; 14.3.4, p. 242; 16, p. 264; 20.3, p. 344; 22.3, p. 458
J.-J. Girerd	(b) **86**, 497 (1978)	*18.4*, p. *299*; 22.8, p. 504
R. Girlanda	(b) **88**, 773 (1978)	*20.1*, p. *331*; 22.5.2, p. 483
J. L. Gisclon	(b) **87**, 293 (1978)	*10.2*, p. *113*; *11*, p. *126*; 20.1, p. 326; 22.5.1, p. 477
D. V. Gitsu	(b) **86**, K71 (1978)	*15*, p. *259*; 21.7, p. 406
M. Giurgea	(b) **89**, 513 (1978)	*10.2*, p. *118*; 20.1, p. 334; 22.5.2, p. 485
A. Giurgiu	(b) **88**, K181 (1978)	*18.1*, p. *278*; *18.4*, p. *301*; 21.1, p. 376; 21.1.1, p. 388
V. N. Gladilin	(b) **87**, K159 (1978)	*19*, p. *308*; 22.4.1, p. 464
H. Glaefeke	(a) **49**, K121 (1978)	*13.4*, p. *182*; *17*, p. *270*; 22, p. 419
U. Gläser	(b) **87**, 221 (1978)	*13.1*, p. *154*; 20, p. 316; 21.3, p. 394
K. D. Glinchuk	(a) **45**, K91 (1978)	*13.4*, p. *171*; *20.3*, p. *342*; 22.2.1, p. 445
	(a) **48**, 593 (1978)	*13.4*, p. *179*; *20.3*, p. *346*; 22.2.1, p. 448
R. Glocker	(b) **88**, 639 (1978)	*6*, p. *74*; 12.1, p. 136; *13.1*, p. *157*; 14.2, p. 215; 21.4, p. 398

A. J. Glover	(a) **48**, 155 (1978)	*10.2*, p. *115*; 22.5.2, p. 482; 22.5.3, p. 488
K. Głowacki	(b) **85**, K119 (1978)	*18.2*, p. *283*
R. Gmeiner	(b) **87**, 543 (1978)	*13.4*, p. *176*; *20.1*, p. *327*; 22.5.3, p. 487
N. K. Gobran	(a) **47**, 365 (1978)	*7*, p. *81*; 18.2, p. 287; 21.1.1, p. 387
G. Gobsch	(b) **90**, 309 (1978)	*13.4*, p. *183*; *19*, p. *313*; 22.6, p. 497
I. Gochev	(b) **85**, 657 (1978)	*13.5.4*, p. *201*; *18.2*, p. *282*
M. Godlewska	(a) **47**, 661 (1978)	*1.2*, p. *20*; 8, p. 86; 22.9, p. 525
M. Godlewski	(b) **88**, K11 (1978)	*19*, p. *309*; 22.2.1, p. 447
M. O. Godzhaev	(b) **85**, 453 (1978)	*13.5.1*, p. *189*; *20.3*, p. *341*; 22.3, p. 457
E. O. Goebel	(b) **88**, 645 (1978)	*13.5*, p. *187*; *20.3*, p. *346*; 22.2.1, p. 447
O. Goede	(b) **85**, 239 (1978)	*13.5.1*, p. *189*; *20.1*, p. *319*; 22.4.1, p. 463
	(b) **89**, K183 (1978)	*13.5.1*, p. *195*; *20.1*, p. *334*; 22.4.4, p. 473
F. Goedsche	(b) **85**, K81 (1978)	*14.1*, p. *206*; 18.2, p. 282; 21, p. 354; 21.1, p. 370
	(b) **86**, 177 (1978)	*14.1*, p. *206*; 18.2, p. 283; 21, p. 355
	(b) **88**, 189 (1978)	*13*, p. *146*; *18*, p. *274*; 18.1, p. 278; 18.2, p. 288; 18.4, p. 300
	(b) **88**, 485 (1978)	*13.1*, p. *157*; 21.1, p. 376
	(b) **89**, 77 (1978)	13, p. 147; *14.1*, p. *209*; 18.4, p. 301; 21, p. 362
N. C. Goel	(a) **46**, 373 (1978)	*14.3*, p. *219*; 22.8.2, p. 518
A. A. Gogolin	(b) **88**, 377 (1978)	*14*, p. *204*; 22.9, p. 526
H. Gollisch	(b) **86**, 145 (1978)	*13*, p. *143*; 21, p. 355
	(b) **89**, 367 (1978)	13.3, p. 167; 21, p. 363
H.-J. Gollnest	(b) **89**, 95 (1978)	*13.1*, p. *158*; *19*, p. *311*; 21.7, p. 407
	(b) **89**, 547 (1978)	*14.1*, p. *211*; 21.7, p. 407
I. F. Golovkina	(a) **50**, K197 (1978)	*14*, p. *204*; *18*, p. *276*; 22.9, p. 529
A. V. Goltsev	(b) **90**, K63 (1978)	*6*, p. *78*; *20.1*, p. *336*; 22, p. 420
A. Goltzene	(a) **46**, K157 (1978)	*19*, p. *307*; 22.6, p. 493
	(a) **48**, K145 (1978)	*10.2*, p. *116*; 19, p. 311; 20.1, p. 331; 22.8, p. 506
V. G. Golubev	(b) **85**, K39 (1978)	*16*, p. *262*; 18, p. 273; *20.1*, p. *320*; 22.4.3, p. 470
V. Gomathy	(b) **86**, 379 (1978)	*13.4*, p. *172*; 20.1, p. 322; 22.4, p. 461
E. Goovaerts	(b) **88**, 615 (1978)	*10.2*, p. *116*; *20.1*, p. *330*; 22.5.2, p. 482
I. S. Gorban	(b) **86**, 419 (1978)	*6*, p. *67*; *20.1*, p. *323*; 22.3, p. 457
	(a) **48**, 329 (1978)	*13.4*, p. *179*; *14.3*, p. *223*; 22.3, p. 458
N. K. Gorban	(a) **47**, 365 (1978)	*7*, p. *81*; 18.2, p. 287; 21.1.1, p. 387
A. Ya. Gordon	(b) **87**, K57 (1978)	1.2, p. 18; 6, p. 71; 8, p. 85; *14.4.1*, p. *254*
	(b) **89**, K69 (1978)	*1.2*, p. *22*; *14.4.1*, p. *256*
N. B. Gorina	(a) **49**, K117 (1978)	*8*, p. *88*; *14.1*, p. *211*; 21, p. 364
V. A. Gorinya	(b) **86**, 419 (1978)	*6*, p. *67*; *20.1*, p. *323*; 22.3, p. 457
P. N. Gorley	(b) **88**, 179 (1978)	*13.4*, p. *177*; 14.3, p. 222; *15*, p. *260*; 22.1.3, p. 441
P. Görnert	(a) **46**, 427 (1978)	*3*, p. *54*; 22.8.2, p. 518
	(a) **47**, 99 (1978)	*3*, p. *54*; 22.8.2, p. 519
E. Gornik	(b) **86**, 205 (1978)	13.1, p. 152; *14.3.4*, p. *241*; 20.1, p. 322; *20.3*, p. *343*; 22.2.3, p. 452
Yu. N. Gornostirev	(a) **47**, 731 (1978)	*10*, p. *96*
	(a) **49**, 517 (1978)	*10*, p. *97*
B. N. Goshchitskii	(a) **46**, K87 (1978)	11, p. 125; *14.2*, p. *214*; 21, p. 357
T. Gotô	(a) **45**, 513 (1978)	*1.2*, p. *16*; *7*, p. *80*; 14.4, p. 248; 22.5.4, p. 489
	(b) **88**, 289 (1978)	*13.5.1*, p. *193*; 20.1, p. 329; 20.3, p. 345; 22.5, p. 475
	(b) **89**, K145 (1978)	*17*, p. *270*; 22.2.1, p. 448; 22.5, p. 476
H. Gottschalk	(a) **45**, 207 (1978)	4, p. 57; *10.2*, p. *109*; 22.2.1, p. 445; 22.2.2, p. 450; 22.2.3, p. 452
G. Götz	(a) **50**, K209 (1978)	11, p. 132; *19*, p. *314*; 22.1.2, p. 440

D. Gourier	(a) **45**, 599 (1978)	*19*, p. *305*; 22.6.1, p. 499
A. Gouskov	(a) **47**, 195 (1978)	*14.3.3*, p. *235*; 14.3.4, p. 242; 16, p. 264; 20.3, p. 344; 22.3, p. 458
L. Gouskov	(a) **47**, 195 (1978)	*14.3.3*, p. *235*; 14.3.4, p. 242; 16, p. 264; 20.3, p. 344; 22.3, p. 458
T. Goworek	(a) **47**, K141 (1978)	*12.1*, p. *135*; 21.1.1, p. 388
	(b) **89**, 253 (1978)	*10.2*, p. *117*; 22.9, p. 526
R. P. Goyal	(b) **87**, K119 (1978)	*14.4*, p. *249*; 22.6, p. 494
S. C. Goyal	(b) **85**, 477 (1978)	*6*, p. *65*; 12.2, p. 138; 22.5.2, p. 478
	(b) **87**, K119 (1978)	*14.4*, p. *249*; 22.6, p. 494
	(b) **87**, K127 (1978)	*4*, p. *58*; *12.2*, p. *139*; 22.6, p. 494
	(b) **89**, 335 (1978)	*6*, p. *65*; 12.2, p. 138; 22.5.2, p. 478
B. Grabe	(a) **47**, 751 (1978)	*11*, p. *127*; 22.1.2, p. 433
N. Graber	(b) **85**, 505 (1978)	*20.3*, p. *342*; 22.9, p. 523
V. J. Grabovskis	(a) **47**, 667 (1978)	*10.2*, p. *114*; *20.3*, p. *345*; 22.5.2, p. 481
M. W. Grabski	(a) **45**, 565 (1978)	*1.4*, p. *29*; 10.1, p. 100; 21, p. 354
J. F. Graczyk	(a) **46**, 611 (1978)	*1.5*, p. *36*; *2*, p. *44*; 22.1.1, p. 425
	(b) **87**, K5 (1978)	*1.5*, p. *36*; *2*, p. *44*
U. Gradmann	(a) **45**, 171 (1978)	*18.2.1*, p. *293*; 21.1.1, p. 384
L. Graf	(b) **88**, 429 (1978)	*6*, p. *74*; *13.5.4*, p. *201*; 20.1, p. 330; 22.5, p. 475
K. Graff	(a) **49**, 137 (1978)	10.2, p. 118; *14.3*, p. *224*; 22.1.2, p. 436
A. Graja	(a) **50**, K105 (1978)	*14*, p. *204*; 22.9, p. 528
S. Grande	(a) **45**, K29 (1978)	*4*, p. *57*; *19*, p. *303*; 22.8, p. 503
	(a) **46**, 547 (1978)	*1.2*, p. *17*; *4*, p. *57*; 19, p. 306; 22.8, p. 504
	(b) **87**, K43 (1978)	*19*, p. *307*; 22.9, p. 525
	(b) **88**, K23 (1978)	1.2, p. 20; 6, p. 74; *19*, p. *309*; 22.9, p. 526
E. Gratz	(a) **48**, 473 (1978)	*13*, p. *147*; 14.1, p. 209; 18.1, p. 279; 18.2, p. 289; 21.1, p. 376; 21.1.1, p. 388
R. E. Graves	(b) **88**, 705 (1978)	*10.2*, p. *116*; 16, p. 265; *19*, p. *310*; 22.5.1, p. 477
I. P. Grebennik	(a) **50**, K1 (1978)	*1.5*, p. *39*; 21, p. 365; 21.6, p. 405
V. I. Grebennikov	(b) **85**, 127 (1978)	*13*, p. *142*; 20, p. 315; 21, p. 352
V. V. Grechov	(b) **85**, 465 (1978)	13.1, p. 151; *17*, p. *269*; 22.9, p. 522
B. A. Greenberg	(a) **45**, 403 (1978)	*10*, p. *96*
	(a) **47**, 731 (1978)	*10*, p. *96*
	(a) **49**, 517 (1978)	*10*, p. *97*
J. E. Greene	(a) **47**, 329 (1978)	13.4, p. 175; *14.3.1*, p. *226*; *20.1*, p. *327*; 22.6, p. 493
D. Grensing	(b) **86**, K131 (1978)	*13*, p. *145*
S. A. Gridnev	(a) **47**, 65 (1978)	*7*, p. *80*; *12.2*, p. *139*; 14.4.1, p. 255; 22.8.1, p. 513
	(a) **47**, 379 (1978)	*12.2*, p. *139*; 14.4.1, p. 255; 22.8.1, p. 514
J. Grilhé	(a) **46**, 387 (1978)	*10.2*, p. *111*; 22.6, p. 492
E. Grilli	(a) **50**, K93 (1978)	*20.3*, p. *350*; 22.8, p. 508
G. A. Grishchenko	(a) **48**, 329 (1978)	*13.4*, p. *179*; *14.3*, p. *223*; 22.3, p. 458
V. A. Gritsenko	(a) **48**, 31 (1978)	2, p. 47; *14.3.2*, p. *230*; *14.3.4*, p. *243*; 22, p. 415
T. Grochulski	(a) **47**, K169 (1978)	*19*, p. *309*; 22.7, p. 502
J. A. Groenink	(a) **49**, 651 (1978)	*16*, p. *267*; 22.8.1, p. 516
P. Groh	(a) **46**, 89 (1978)	*10.1*, p. *101*; 11, p. 125; 21.1.1, p. 385
Y. Gros	(a) **48**, 581 (1978)	*1.1*, p. *13*; *5*, p. *62*; 22.8.1, p. 515
P. Grosse	(b) **86**, 205 (1978)	13.1, p. 152; *14.3.4*, p. *241*; 20.1, p. 322; *20.3*, p. *343*; 22.2.3, p. 452
W. Grosse-Nobis	(a) **49**, K21 (1978)	*18.2*, p. *290*; 21.1.1, p. 389
J. Grossmann	(b) **88**, 99 (1978)	*14.4.1*, p. *255*; *19*, p. *309*; 22.8.1, p. 514
P. S. Grover	(b) **89**, K163 (1978)	*4*, p. *59*; 21, p. 363

A. O. GRUBICH	(b) **88**, 351 (1978)	*1.1*, p. *13*; *11*, p. *127*
D. GRUBISIC	(b) **90**, 151 (1978)	*20.1*, p. *335*; 22.1, p. 423
J. B. GRUN	(b) **86**, 159 (1978)	*13.5.1*, p. *190*; 20.1, p. 321; 20.3, p. 342; 22.5, p. 474
	(b) **88**, 545 (1978)	*13.5.1*, p. *193*; *20.3*, p. *346*; 22.5, p. 475
R. GRÜNDLER	(b) **86**, 329 (1978)	2, p. 43; 13.1, p. 153; *14.4*, p. *248*; *20.1*, p. *322*; 22.6, p. 492
G. R. GRUZALSKI	(b) **88**, K127 (1978)	*2*, p. *47*; *18.2*, p. *289*
M. GRYNBERG	(b) **87**, 637 (1978)	*13.1*, p. *155*; 14.4, p. 249; 20.1, p. 328; 22.4.2, p. 468; 22.4.3, p. 470
	(b) **89**, K1 (1978)	*13.1*, p. *158*; 14.4, p. 250; 20.1, p. 332; 22.4.3, p. 471
	(b) **89**, 521 (1978)	*13.1*, p. *159*; 14.4, p. 251; 18, p. 275; 22, p. 418
I. M. GUELLA	(a) **50**, K197 (1978)	*14*, p. *204*; *18*, p. *276*; 22.9, p. 529
A. L. GUERRERO	(a) **49**, 353 (1978)	*10.2*, p. *118*; *14.4*, p. *250*; 22.5.2, p. 484
S. GUHA	(b) **87**, K9 (1978)	*14.3*, p. *220*; *20*, p. *316*; 22, p. 413
	(b) **89**, K171 (1978)	*14.3*, p. *224*; *20*, p. *317*; 22, p. 418
	(a) **49**, 363 (1978)	14.3.4, p. 245; 22.2.3, p. 454
G. GUILLOT	(b) **89**, K85 (1978)	*10.2*, p. *117*; *20.1*, p. *333*; 22.5.2, p. 484
C. GUITTARD	(a) **48**, 459 (1978)	*17*, p. *270*; 21, p. 361
YU. V. GULYAEV	(b) **85**, 487 (1978)	6, p. 66; *7*, p. *80*; 22.2.3, p. 452
A. A. GUNDJIAN	(b) **85**, 733 (1978)	6, p. 66; *12.2*, p. *138*; *14.3*, p. *217*; 22, p. 409
H. O. GUPTA	(a) **46**, 153 (1978)	*18.1*, p. *277*; *19*, p. *305*; 22.9, p. 523
I. J. GUPTA	(b) **87**, 513 (1978)	*6*, p. *71*; *8*, p. *86*; 23, p. 530
	(b) **88**, 815 (1978)	*6*, p. *74*; *8*, p. *87*; 23, p. 530
N. GUSKOS	(b) **87**, K155 (1978)	*19*, p. *308*; 22.5.3, p. 487
W. GUSTAW	(a) **47**, K141 (1978)	*12.1*, p. *135*; 21.1.1, p. 388
G. F. GUSYATSKII	(a) **50**, 269 (1978)	*18.3*, p. *296*; *19*, p. *314*; 22.8.2, p. 520
V. B. GUTAN	(a) **49**, 629 (1978)	*1.5*, p. *39*; 13.4, p. 182; 14.3.1, p. 228; 20.1, p. 335; 20.3, p. 348; 22.2.4, p. 456
M. GUZZI	(a) **50**, K93 (1978)	*20.3*, p. *350*; 22.8, p. 508
C. HAAS	(b) **85**, 621 (1978)	*13.1*, p. *151*; 14.3, p. 217; 18.1, p. 277; 19, p. 304; 22, p. 409
	(a) **46**, 687 (1978)	*18.1*, p. *278*; *18.4*, p. *299*; 22.8, p. 504
	(b) **87**, 135 (1978)	*19*, p. *307*; 22.5, p. 475
P. HAASEN	(a) **46**, 107 (1978)	*10.2*, p. *110*; 22.5.2, p. 479
	(a) **49**, 93 (1978)	*10.2*, p. *118*; 22.2.1, p. 448
R. HABERKORN	(b) **85**, 45 (1978)	*13.5.1*, p. *189*; 20.3, p. 341; 22.9, p. 522
	(b) **85**, 473 (1978)	*13.5.1*, p. *189*; *20.3*, p. *342*; 22.9, p. 522
	(b) **89**, 261 (1978)	*13.5.1*, p. *194*; 20.1, p. 332; 22.9, p. 527
W. HABERKORN	(a) **47**, K161 (1978)	*14.2*, p. *214*
H. HABERLANDT	(b) **90**, 309 (1978)	*13.4*, p. *183*; *19*, p. *313*; 22.6, p. 497
H. U. HABERMEIER	(a) **50**, 187 (1978)	*14.2*, p. *215*; 18, p. 275; 20, p. 318; 21, p. 365
E. HADJOUDIS	(a) **46**, 239 (1978)	*10.2*, p. *111*; *19*, p. *306*; 22.9, p. 524
S. HAGEGE	(a) **50**, 703 (1978)	*1.1*, p. *14*; *1.4*, p. *35*
TRAN QUOC HAI	(b) **88**, 555 (1978)	*2*, p. *47*; *14.3.1*, p. *227*; 22.1.1, p. 427; 22.1.2, p. 435
J. HAJTÓ	(a) **45**, 181 (1978)	2, p. 42; 11, p. 123; *20.1*, p. *320*; 22.7, p. 501
E. E. HALLER	(b) **90**, 557 (1978)	*13.5.1*, p. *196*; 22.1.1, p. 428
C. HAMANN	(a) **50**, K189 (1978)	*14.3.3*, p. *239*; 22.9, p. 529
D. HAMANN	(b) **86**, 569 (1978)	*13.1*, p. *154*; 14.2, p. 214; 21.1, p. 371; 21.1.1, p. 386
M. HAMERA	(b) **85**, 645 (1978)	*13.4*, p. *170*; 19, p. 304; 22.4, p. 461
W. H. HAMILL	(b) **87**, 581 (1978)	*10.2*, p. *113*; *20.1*, p. *327*; 22.5.2, p. 481
N. J. HANCE	(a) **49**, K177 (1978)	*18.3*, p. *296*; 18.4, p. 301; 21.1, p. 379

R. C. O'Handley	see under letter "O"	
A. Handstein	(a) **45**, K47 (1978)	*2*, p. *42*; *18.2*, p. *282*; 21.1, p. 369; 21.1.1, p. 384
C. Hansch	(a) **47**, K17 (1978)	*10.2*, p. *113*; 22.1.2, p. 433
	(a) **48**, K11 (1978)	*10.2*, p. *115*; 22.1.2, p. 435
J. R. Hanscomb	(a) **48**, 79 (1978)	10.2, p. 115; *15*, p. *260*; 22.5.2, p. 482
T. Hänsel	(b) **87**, 169 (1978)	*13.4*, p. *174*; *20.3*, p. *344*; 22.2.2, p. 450
	(b) **90**, 233 (1978)	*13.5.1*, p. *196*; *20.3*, p. *349*; 22.3, p. 459
J. E. Harbottle	(a) **46**, 639 (1978)	*1.5*, p. *36*; 11, p. 125; 21.1, p. 372
R. Harman	(a) **45**, K81 (1978)	2, p. 50; *18.2.1*, p. *294*; 21.1, p. 378; 21.4, p. 398
Ch. Hartig	(a) **45**, 433 (1978)	13.4, p. 171; *20.2*, p. *339*; 22.4.1, p. 463
M. Hartmanová	(a) **46**, 259 (1978)	*10.2*, p. *111*; 20.1, p. 324; 22.5.2, p. 479
H. W. den Hartog	(b) **87**, 697 (1978)	*18.1*, p. *278*; *19*, p. *308*; 22.8.1, p. 513
V. M. Harutyunyan	(a) **49**, 797 (1978)	*13.4*, p. *182*; *14.3.3*, p. *237*; 22.1.2, p. 437
Z. Hasan	(a) **46**, K135 (1978)	*20.1*, p. *325*; 22.6, p. 492
H. Hasegawa	(a) **45**, 107 (1978)	2, p. 42; *16*, p. *262*; 22, p. 409
K. Hashimoto	(b) **86**, 239 (1978)	13.1, p. 153; *14.3*, p. *218*; *15*, p. *259*; 22.3, p. 457
	(b) **90**, 751 (1978)	13.1, p. 153; *14.3*, p. *218*; *15*, p. *259*; 22.3, p. 457
A. R. Hassan	(b) **87**, 31 (1978)	*13.5.1*, p. *191*; 20.1, p. 325; 22.5, p. 475
M. Haubold	(a) **49**, 279 (1978)	*9*, p. *94*; *11*, p. *129*; 22.2.4, p. 456
A. Hauet	(a) **50**, 67 (1978)	*1.4*, p. *34*; 21, p. 365; 21.1, p. 380
A. Haug	(b) **89**, 357 (1978)	*13.1*, p. *159*; 20, p. 317; 22.2, p. 444; 22.2.3, p. 454
H. Haug	(b) **85**, 561 (1978)	*13.5.1*, p. *190*; 13.5.2, p. 198; 22, p. 409
	(b) **89**, 431 (1978)	13.5.1, p. 195; *20.2*, p. *339*; *20.3*, p. *348*; 22.4, p. 462; 22.4.1, p. 465; 22.4.3, p. 471
D. Haus	(b) **88**, 163 (1978)	*6*, p. *73*; 13.5.1, p. 192; *20.1*, p. *329*; 22.4.1, p. 465
J. Hausbrand	(a) **46**, 125 (1978)	10.2, p. 110; *11*, p. *125*; 22.8, p. 504
G. Hausch	(a) **50**, 159 (1978)	*2*, p. *51*; 8, p. 89; 12.1, p. 136; *18.2*, p. *291*; 21.1, p. 381; 21.1.1, p. 390
M. B. Haxton	(b) **86**, 661 (1978)	*6*, p. *69*; 22.5, p. 474
S. Hayashi	(b) **86**, 169 (1978)	*13.1*, p. *152*; 14.1, p. 206; 21.7, p. 406
A. Haydar	(a) **48**, 465 (1978)	*13.5.1*, p. *194*; *16*, p. *265*; 22.6, p. 495
G. Heber	(b) **90**, 475 (1978)	*18*, p. *276*; 21.1, p. 381
E. Hegenbarth	(a) **47**, 157 (1978)	8, p. 85; *10.1*, p. *103*; 14.1, p. 207; 18.2, p. 286; 21.1, p. 373; 21.1.1, p. 387
Th. Hehenkamp	(a) **50**, 563 (1978)	*9*, p. *95*; 14.1, p. 212; 21.1.1, p. 391
A. Heidemann	(a) **48**, K179 (1978)	*9*, p. *93*; 22.5.1, p. 477
K. F. Heidemann	(a) **47**, 751 (1978)	*11*, p. *127*; 22.1.2, p. 433
R. Heidemann	(a) **49**, 431 (1978)	13.4, p. 182; *14.3.3*, p. *236*; *14.3.4*, p. *245*; 22.1.2, p. 437
H. H. Heikens	(a) **46**, 687 (1978)	*18.1*, p. *278*; *18.4*, p. *299*; 22.8, p. 504
G. Heiland	(a) **49**, 27 (1978)	13.3, p. 167; 13.4, p. 180; *14.3.2*, p. *231*; 14.3.4, p. 244; *16*, p. *265*; 22.4, p. 462
	(b) **89**, 459 (1978)	6, p. 76; *14.3.1*, p. *228*; *20.1*, p. *334*; 22.9, p. 527
L. von Heimendahl	(b) **86**, 549 (1978)	*2*, p. *43*; *6*, p. *68*
J. Heinrichs	(b) **85**, 163 (1978)	*13*, p. *142*
K. Heinz	(a) **49**, 39 (1978)	*1.6*, p. *41*; 21, p. 362; 21.6, p. 404; 22.1.2, p. 436
E. Heinze	(b) **87**, K43 (1978)	*19*, p. *307*; 22.9, p. 525
	(b) **88**, K23 (1978)	1.2, p. 20; 6, p. 74; *19*, p. *309*; 22.9, p. 526
M. Heise	(b) **85**, 535 (1978)	*13*, p. *143*; *18.2*, p. *282*

D. Heitmann	(b) **88**, 493 (1978)	*13.3*, p. *166*; *13.5.2*, p. *198*; 20.3, p. 346; 21.6, p. 403
R. Helbig	(b) **86**, 527 (1978)	*13.5.1*, p. *191*; *20.3*, p. *343*; 22.4, p. 461
R. Hellborg	(a) **45**, K167 (1978)	*11*, p. *124*; 22.2, p. 443
B. G. Helme	(a) **45**, K33 (1978)	6, p. 65; *7*, p. *80*; 22.1.1, p. 424; 22.1.2, p. 429; 22.2.1, p. 445; 22.2.3, p. 452
P. L. F. Hemment	(a) **46**, 225 (1978)	*11*, p. *125*; 13.4, p. 173; 14.3, p. 219; 22.1.2, p. 431
H. Hemschik	(a) **48**, 377 (1978)	*1.2*, p. *21*; 6, p. 75; 14.3, p. 223; 22.6, p. 495
P. Hendy	(a) **50**, 101 (1978)	*18.3*, p. *296*; 21.1, p. 381; 21.4, p. 398
K. Henneberger	(b) **85**, K53 (1978)	*13.5.1*, p. *189*; 22.1.2, p. 429
	(b) **86**, 505 (1978)	*13.5.1*, p. *190*; 22.1.2, p. 431
W. H. Henneker	(b) **87**, 653 (1978)	*6*, p. *72*; 13.5.1, p. 192; 22.9, p. 525
A. M. Hennel	(b) **88**, K11 (1978)	*19*, p. *309*; 22.2.1, p. 447
D. Hennig	(b) **89**, K183 (1978)	*13.5.1*, p. *195*; *20.1*, p. *334*; 22.4.4, p. 473
P. Hennl	(a) **46**, 147 (1978)	*10.2*, p. *111*; *20.1*, p. *323*; 20.3, p. 343; 22.5.2, p. 479
R. L. Henry	(a) **46**, 209 (1978)	*14.2*, p. *214*; 21, p. 356
R. M. Henson	(a) **46**, 511 (1978)	*14.4.1*, p. *254*; 22.8, p. 504
R. Hergt	(a) **46**, 427 (1978)	*3*, p. *54*; 22.8.2, p. 518
	(a) **47**, 99 (1978)	*3*, p. *54*; 22.8.2, p. 519
F. C. Hernández	(b) **90**, 441 (1978)	*13*, p. *149*; 13.5.2, p. 199; 21.2, p. 393
A. Herr	(a) **46**, K35 (1978)	*18.4*, p. *299*; 21.3, p. 394; 21.4, p. 397
K. H. Herrmann	(b) **86**, K21 (1978)	*20.1*, p. *323*; 22, p. 410
	(b) **88**, 675 (1978)	13.4, p. 178; 20.1, p. 330; *20.2*, p. *339*; 22, p. 415
	(b) **90**, 197 (1978)	*13.1*, p. *159*; 22, p. 419
R. Herrmann	(b) **88**, 105 (1978)	*13.4*, p. *177*; 14.1, p. 208; *14.3*, p. *221*; 22.7, p. 502
	(b) **88**, 507 (1978)	*13.2*, p. *162*; 14.3, p. 223; 19, p. 310; 21.7, p. 407
	(b) **89**, 95 (1978)	*13.1*, p. *158*; *19*, p. *311*; 21.7, p. 407
	(b) **89**, 547 (1978)	*14.1*, p. *211*; 21.7, p. 407
	(b) **90**, 517 (1978)	13.3, p. 168; *14.2*, p. *216*; 21, p. 365
H. G. Hertz	(a) **49**, 463 (1978)	*1*, p. *10*; *2*, p. *50*; 22.9, p. 527
R. Herz	(a) **47**, 451 (1978)	*18.2*, p. *287*; 21.4, p. 398
Ch. Herzig	(b) **85**, 185 (1978)	*9*, p. *91*; 10.1, p. 99; 21.1, p. 368
	(a) **47**, 497 (1978)	*9*, p. *93*; 14.1, p. 208; 21.6, p. 403
	(a) **47**, 513 (1978)	*9*, p. *93*; 14.1, p. 208; 21, p. 359
	(a) **48**, K79 (1978)	*3*, p. *55*; 21.6, p. 403
J. M. Hess	(a) **46**, K55 (1978)	*14.3.3*, p. *234*; 14.3.4, p. 241; 22.2.2, p. 450
H. Hesse	(a) **50**, K243 (1978)	10.2, p. 122; *14.4.1*, p. *257*; *20.1*, p. *338*; 22.8.1, p. 517
R. Hesske	(a) **45**, K47 (1978)	*2*, p. *42*; *18.2*, p. *282*; 21.1, p. 369; 21.1.1, p. 384
W. F. W. M. van Heugten		
	(b) **86**, 277 (1978)	6, p. 67; *10.1*, p. *101*; 21.1, p. 371
	(a) **46**, 541 (1978)	*10.1*, p. *102*; 21, p. 357
	(b) **90**, 225 (1978)	*4*, p. *59*; 6, p. 77; 10.1, p. 107; 21.2, p. 393
	(b) **90**, K89 (1978)	6, p. 78; *10.1*, p. *107*; 21, p. 364
	(b) **90**, 643 (1978)	*10.1*, p. *107*; 21, p. 365
M. Heyen	(a) **49**, 111 (1978)	13.4, p. 180; *14.3.1*, p. *228*; *14.3.3*, p. *236*; 22.2.1, p. 448
	(a) **50**, 771 (1978)	13.4, p. 180; *14.3.1*, p. *228*; *14.3.3*, p. *236*; 22.2.1, p. 448
M. Hietschold	(b) **87**, 111 (1978)	*13.3*, p. *165*
W. Hild	(a) **50**, 417 (1978)	*14.2*, p. *216*

O. Hildebrand	(b) **88**, 645 (1978)	*13.5*, p. *187*; *20.3*, p. *346*; 22.2.1, p. 447
R. W. A. Hillhouse	(a) **46**, 163 (1978)	*3*, p. *54*; 13.4, p. 173; 16, p. 263; 22.6, p. 492
G. Hilscher	(a) **48**, 473 (1978)	*13*, p. *147*; 14.1, p. 209; 18.1, p. 279; 18.2, p. 289; 21.1, p. 376; 21.1.1, p. 388
Th. Hinrichsen	(a) **50**, 95 (1978)	*1*, p. *10*; 22.9, p. 528
W. Hinz	(a) **47**, 79 (1978)	*2*, p. *45*; *11*, p. *126*
M. Hirabayashi	(a) **46**, 217 (1978)	*1.3*, p. *25*; 6, p. 68; *11*, p. *125*; 21.6, p. 401
T. Hirata	(a) **45**, 201 (1978)	*10.1*, p. *99*; 21, p. 353; 21.3, p. 394
	(a) **48**, 451 (1978)	*1.4*, p. *32*; *1.5*, p. *37*; 2, p. 48; 21, p. 361
M. Hirose	(a) **45**, K75 (1978)	*3*, p. *54*; *14.3.1*, p. *226*; 22.6, p. 491
V. Hivert	(a) **46**, 89 (1978)	*10.1*, p. *101*; 11, p. 125; 21.1.1, p. 385
V. Hizhnyakov	(b) **86**, 733 (1978)	*13.4*, p. *174*; *20.1*, p. *324*; 22.5.2, p. 480
V. V. Hizhnyakov	(b) **85**, 51 (1978)	*6*, p. *64*; *13.5.1*, p. *189*; 20.1, p. 319; 20.3, p. 341; 22, p. 408
Hoang Xuan Nguyen	see under letter "N"	
L. W. Hobbs	(a) **48**, 425 (1978)	*1.4*, p. *32*; 11, p. 128; 22.6, p. 495
H.-R. Höche	(a) **46**, K131 (1978)	*1.1*, p. *12*; 22.1.2, p. 432
H. Höchst	(b) **90**, 45 (1978)	*13.1*, p. *159*; *13.5.2*, p. *199*; 21.2, p. 393; 21.3, p. 395
S. Hodorowicz	(a) **47**, 57 (1978)	*1.2*, p. *18*; 8, p. 85; 22.8, p. 505
	(a) **49**, K27 (1978)	*1.2*, p. *22*; 22.9, p. 527
Ch. Hoentzsch	(b) **88**, 581 (1978)	*13.4*, p. *178*; *19*, p. *310*; 22.5.2, p. 482
W. Hofberger	(a) **50**, 95 (1978)	*1*, p. *10*; 22.9, p. 528
K. Hoffmann	(a) **45**, 263 (1978)	13.4, p. 170; *19*, p. *303*; 22.5.2, p. 478
A. Hofstaetter	(b) **87**, K35 (1978)	*10.2*, p. *113*; 11, p. 126; *19*, p. *307*; 22.8.1, p. 513
	(b) **89**, 375 (1978)	*19*, p. *312*; 20.3, p. 348; 22.8.1, p. 515
	(b) **89**, 449 (1978)	*13.4*, p. *181*; *19*, p. *312*; 22.8.1, p. 516
H.-J. Höhne	(a) **50**, K189 (1978)	*14.3.3*, p. *239*; 22.9, p. 529
M. Höhne	(b) **85**, 525 (1978)	1.6, p. 40; 13.4, p. 170; *19*, p. *304*; 22.1.2, p. 429
	(b) **86**, 119 (1978)	1.6, p. 40; 13.4, p. 172; *19*, p. *305*; 22.1.2, p. 430
U. Holland	(b) **89**, K73 (1978)	12.2, p. 140; *13.4*, p. *180*; *20.3*, p. *347*; 22.5.2, p. 484
F. Holuj	(a) **48**, 191 (1978)	13.4, p. 177; *19*, p. *310*; 22.8, p. 506
V. Holý	(a) **50**, 285 (1978)	*1.1*, p. *14*; 10.2, p. 121; 22.1.2, p. 438
U. Holzhauer	(b) **89**, 69 (1978)	*10.2*, p. *117*; 11, p. 128; 22.1.2, p. 436
B. Hönerlage	(b) **86**, 159 (1978)	*13.5.1*, p. *190*; 20.1, p. 321; 20.3, p. 342; 22.5, p. 474
	(b) **88**, 545 (1978)	*13.5.1*, p. *193*; *20.3*, p. *346*; 22.5, p. 475
Vo Hong Anh	see under letter "A"	
H. Hora	(b) **86**, 685 (1978)	*17*, p. *269*
W. Hörig	(b) **85**, K57 (1978)	*13.1*, p. *150*; *20.1*, p. *320*; 22.8, p. 503
S. Horiguchi	(b) **87**, 757 (1978)	*13.3*, p. *166*; 21.6, p. 402
A. Horst	(b) **85**, 239 (1978)	*13.5.1*, p. *189*; *20.1*, p. *319*; 22.4.1, p. 463
P. Höschl	(a) **48**, K43 (1978)	*13.4*, p. *178*; *16*, p. *265*; 22.4.3, p. 470
T. Hoshiai	(a) **46**, 671 (1978)	*2*, p. *44*; *12.1*, p. *135*; 21.1.1, p. 386
S. Hösl	(b) **85**, 87 (1978)	*5*, p. *60*; 22.5, p. 474
J. Th. M. de Hosson	(b) **87**, 151 (1978)	*10.1*, p. *102*; 21.1.1, p. 387
	(b) **90**, 225 (1978)	*4*, p. *59*; 6, p. 77; 10.1, p. 107; 21.2, p. 393
	(b) **90**, 643 (1978)	*10.1*, p. *107*; 21, p. 365
M. Hou	(a) **46**, K167 (1978)	*11*, p. *126*; 22, p. 412
M. E. Houghton	(a) **47**, 597 (1978)	*1.4*, p. *31*; *18.2*, p. *287*; 21.1, p. 374; 21.1.1, p. 387
	(a) **48**, 71 (1978)	*1.4*, p. *31*; *18.2*, p. *289*; 21.1, p. 375; 21.1.1, p. 388

B. HOULIER	(b) 89, K135 (1978)	*1.2*, p. *23*; *19*, p. *313*; 22.8.1, p. 516
W. HOYER	(a) 49, 111 (1978)	13.4, p. 180; *14.3.1*, p. *228*; *14.3.3*, p. *236*; 22.2.1, p. 448
	(a) 50, 771 (1978)	13.4, p. 180; *14.3.1*, p. *228*; *14.3.3*, p. *236*; 22.2.1, p. 448
L. HRIVNÁK	(a) 50, K97 (1978)	*13.4*, p. *184*; *14.3.4*, p. *246*; 22, p. 420
S. S. HSU	(a) 50, 393 (1978)	1.3, p. 28; *7*, p. *82*; 21.1, p. 382; 21.1.1, p. 390
W. F. HUANG	(a) 46, K107 (1978)	*10.1*, p. *102*; 21, p. 357
B. HUBER	(a) 47, K85 (1978)	*9*, p. *92*; 21, p. 359
D. HUBER	(a) 47, 489 (1978)	*14.3.3*, p. *235*; *20.3*, p. *345*; 22.2.1, p. 447
K. HÜBNER	(a) 46, 451 (1978)	2, p. 44; *4*, p. *57*; 22.6, p. 492
	(a) 48, 147 (1978)	2, p. 47; *4*, p. *59*; *13.1*, p. *157*; 14.4, p. 249; 22.6, p. 494
S. HÜFNER	(b) 90, 45 (1978)	*13.1*, p. *159*; *13.5.2*, p. *199*; 21.2, p. 393; 21.3, p. 395
B. HUGHES	(a) 46, 627 (1978)	*1.4*, p. *30*; *10.2*, p. *112*; 14.3, p. 220; 22.2.1, p. 446
M. E. VAN HULLE	(a) 50, K175 (1978)	*10.2*, p. *122*; *14.4*, p. *252*; 22.5.1, p. 477
R. HULTZSCH	(a) 47, 415 (1978)	10.2, p. 114; *20.2*, p. *339*; 22.5.3, p. 487
H. HUMBERG	(b) 89, 379 (1978)	*17*, p. *270*; 21.2, p. 393
J. HUMLÍČEK	(b) 86, 303 (1978)	13.1, p. 153; *14.3.4*, p. *241*; *20.1*, p. *322*; 22.1.1, p. 424
R. E. HUMMEL	(b) 88, 173 (1978)	*13.1*, p. *156*; *20.1*, p. *329*; 21.1, p. 375
K. HÜMMER	(b) 85, 271 (1978)	*13.5.1*, p. *189*; *20.1*, p. *320*; 22.4, p. 461
	(b) 86, 527 (1978)	*13.5.1*, p. *191*; *20.3*, p. *343*; 22.4, p. 461
O. HUNDERI	(b) 90, K69 (1978)	*13.1*, p. *160*; 21.1, p. 380
E. HUSTER	(a) 45, K23 (1978)	*18*, p. *273*; 21.1.1, p. 384
R. HUSTER	(a) 45, K23 (1978)	*18*, p. *273*; 21.1.1, p. 384
J. P. HUVENNE	(a) 48, 417 (1978)	*6*, p. *75*; *20.1*, p. *331*; 22.6, p. 495
P. IACCONI	(a) 50, 275 (1978)	10.2, p. 121; 13.4, p. 184; *20.3*, p. *350*; 22.6, p. 497
ST. IDZIAK	(a) 45, K29 (1978)	*4*, p. *57*; *19*, p. *303*; 22.8, p. 503
V. M. IEVLEV	(a) 48, 603 (1978)	*3*, p. *55*; *10.2*, p. *116*; 22.5.2, p. 483
	(a) 49, 413 (1978)	*1.4*, p. *33*; *1.5*, p. *38*; 3, p. 56; 21.6, p. 404
M. IGNAT	(a) 49, 675 (1978)	*10.1*, p. *106*; 21, p. 364; 21.1, p. 379
D. IHLE	(b) 86, 283 (1978)	13.1, p. 153; *14*, p. *203*
H. IHRIG	(a) 47, 437 (1978)	*14.4.1*, p. *255*; 22.8.1, p. 514
T. IKARI	(b) 86, 239 (1978)	13.1, p. 153; *14.3*, p. *218*; *15*, p. *259*; 22.3, p. 457
	(b) 90, 751 (1978)	13.1, p. 153; *14.3*, p. *218*; *15*, p. *259*; 22.3, p. 457
H. IKEDA	(a) 45, 107 (1978)	2, p. 42; *16*, p. *262*; 22, p. 409
E. N. ILICHEVA	(a) 50, K101 (1978)	*18.2.1*, p. *294*; 21.1, p. 381; 21.1.1, p. 390
M. N. ILIEV	(b) 86, 717 (1978)	*6*, p. *69*; 18.2, p. 285; *20.1*, p. *324*; 22.8.2, p. 518
V. ILKOVIČ	(b) 89, K103 (1978)	*18.2*, p. *290*
G. ILONCA	(a) 47, 357 (1978)	*14.2*, p. *214*
GH. ILONCA	(b) 88, 37 (1978)	*2*, p. *46*; *14.3*, p. *221*; 22, p. 414
IM-SAROEUN	(a) 47, K11 (1978)	*14.3.4*, p. *242*; 22, p. 413
V. L. INDENBOM	(a) 46, 379 (1978)	*10.2*, p. *111*; 22.1.2, p. 431
L. P. INGMAN	(a) 46, 213 (1978)	*19*, p. *306*; 22.9, p. 524
A. IOAN	(b) 89, 513 (1978)	*10.2*, p. *118*; 20.1, p. 334; 22.5.2, p. 485
D. E. IOANNOU	(a) 48, K1 (1978)	*10.2*, p. *115*; 22.1.2, p. 434
S. P. IONOV	(b) 85, 359 (1978)	*13*, p. *142*; *18*, p. *273*
	(b) 85, 683 (1978)	*13*, p. *143*; *18*, p. *273*

G. V. Ionova	(b) **85**, 359 (1978)	*13*, p. *142*; *18*, p. *273*
	(b) **85**, 683 (1978)	*13*, p. *143*; *18*, p. *273*
M. S. Iovu	(a) **50**, K59 (1978)	*2*, p. *52*; *14.3.1*, p. *229*; 22, p. 420
J. C. Irwin	(a) **46**, 81 (1978)	*20.1*, p. *323*; *20.3*, p. *343*; 22.8, p. 504
M. M. Islam	(b) **87**, 145 (1978)	*13*, p. *145*; *14.1*, p. *207*; 21.2, p. 392
N. Itoh	(b) **88**, 623 (1978)	*13.5.1*, p. *193*; 20.3, p. 346; 22.9, p. 526
V. Iuşan	(b) **88**, K181 (1978)	*18.1*, p. *278*; *18.4*, p. *301*; 21.1, p. 376; 21.1.1, p. 388
L. V. Ivaeva	(a) **50**, K153 (1978)	*18.2.1*, p. *294*; *20*, p. *318*; 21.1, p. 383; 21.1.1, p. 391
B. G. Ivanov	(a) **48**, 287 (1978)	*20.1*, p. *331*; 22.6, p. 495
M. A. Ivanov	(a) **45**, 403 (1978)	*10*, p. *96*
	(a) **47**, 731 (1978)	*10*, p. *96*
	(a) **49**, 517 (1978)	*10*, p. *97*
N. A. Ivanov	(a) **50**, 363 (1978)	*1.5*, p. *39*; *14.3.3*, p. *238*; 14.4, p. 252; 22.6, p. 497
N. R. Ivanov	(a) **46**, 533 (1978)	*14.4.1*, p. *254*; *20.1*, p. *325*; 22.9, p. 524
S. N. Ivanov	(b) **85**, 487 (1978)	6, p. 66; *7*, p. *80*; 22.2.3, p. 452
V. I. Ivanov-Omskii	(b) **85**, K39 (1978)	*16*, p. *262*; 18, p. 273; *20.1*, p. *320*; 22.4.3, p. 470
	(a) **46**, 77 (1978)	*13.4*, p. *173*; *20.3*, p. *343*; 22.4.4, p. 472
	(b) **87**, 381 (1978)	*13.5*, p. *187*; 14.4, p. 248; 22.4.3, p. 470
	(b) **90**, 11 (1978)	*6*, p. *64*; 13, p. 142; *19*, p. *303*; 22, p. 408; 22.2.3, p. 452; 22.4.1, p. 463
M. D. Ivanovitch	(a) **45**, 671 (1978)	*14.3.3*, p. *233*; 14.3.4, p. 240; 22.2.1, p. 445
E. L. Ivchenko	(b) **85**, 81 (1978)	*6*, p. *64*; *20.3*, p. *341*; 22, p. 408
A. Ivić	(b) **90**, K157 (1978)	*13*, p. *149*; *18.2*, p. *292*
M. Iwan	(b) **90**, 525 (1978)	*13.1*, p. *160*; *17*, p. *270*; 21, p. 365
T. Iwasaki	(b) **88**, 289 (1978)	*13.5.1*, p. *193*; 20.1, p. 329; 20.3, p. 345; 22.5, p. 475
V. A. Izvozchikov	(a) **49**, K173 (1978)	*14.4.1*, p. *256*; 22.6, p. 497
Yu. A. Izyumov	(b) **87**, 441 (1978)	*18.2*, p. *286*
A. Jabłonka	(b) **85**, K73 (1978)	*1*, p. *9*; 18.2, p. 282; 21.1.1, p. 385
	(b) **85**, K77 (1978)	*1*, p. *9*; 18.2, p. 282; 21.1.1, p. 385
C. Jaccard	(b) **87**, 497 (1978)	13.4, p. 175; *19*, p. *308*; *20.3*, p. *344*; 22.5.2, p. 481
	(b) **90**, K161 (1978)	*19*, p. *314*; 22.5.2, p. 486
M. Jachimowski	(a) **45**, 537 (1978)	*20.1*, p. *321*; 22.6.1, p. 499
B. Jacobs	(a) **49**, K139 (1978)	*14.3.3*, p. *237*; 22.2.2, p. 451
G. Jacobs	(b) **89**, 193 (1978)	*13.5.3*, p. *200*; *20.1*, p. *332*; 22.5.2, p. 483
	(b) **89**, K95 (1978)	*10.2*, p. *118*; 13.5.3, p. 200; *20.1*, p. *333*; 22.5.2, p. 484
G. Jacobsen	(a) **45**, 463 (1978)	*16*, p. *262*; 22.4.1, p. 463
J. L. Jacquemin	(a) **45**, 579 (1978)	*14*, p. *203*; 14.4, p. 248; 22.9, p. 523
M. Jaczyński	(b) **88**, 73 (1978)	*13.2*, p. *162*; *14.3*, p. *221*; 15, p. 260; 22.4.4, p. 473
K. Jagannadham	(a) **50**, 293 (1978)	*10*, p. *97*
	(a) **50**, 601 (1978)	*1.4*, p. *35*; *10*, p. *98*
N. K. Jaggi	(b) **87**, K81 (1978)	*18.2*, p. *287*; *18.4*, p. *300*
D. Jäger	(a) **49**, 431 (1978)	13.4, p. 182; *14.3.3*, p. *236*; *14.3.4*, p. *245*; 22.1.2, p. 437
T. Jagielinski	(a) **48**, 175 (1978)	2, p. 47; *18.2*, p. *289*; 21.1.1, p. 388
	(a) **50**, K25 (1978)	*2*, p. *52*; *18.2*, p. *292*; 21.1.1, p. 390
H. Jagodziński	(a) **49**, 119 (1978)	*1.6*, p. *41*
E. Jahne	(b) **86**, 517 (1978)	*6*, p. *68*; *20.1*, p. *324*; 22.2.2, p. 450; 22.2.4, p. 456

G. C. JAIN	(a) **46**, K151 (1978)	*9*, p. *92*; 22.1.2, p. 432
SUSHIL JAIN	(a) **50**, 11 (1978)	6, p. 64; *13.1*, p. *150*; 13.4, p. 169; *14.3*, p. *217*; 20.1, p. 319; 22.2, p. 443
S. C. JAIN	(a) **49**, 353 (1978)	*10.2*, p. *118*; *14.4*, p. *250*; 22.5.2, p. 484
V. C. JAIN	(b) **89**, 213 (1978)	*4*, p. *59*; 6, p. 75; 14.4, p. 250; 22.5.2, p. 483
V. K. JAIN	(b) **88**, K173 (1978)	*19*, p. *311*; 22.8.1, p. 515
	(a) **50**, 329 (1978)	*10.2*, p. *121*; 20.1, p. 336; *20.3*, p. *350*; 22.5.2, p. 485
R. JAKUBAS	(a) **50**, K137 (1978)	*14.4.1*, p. *257*; 22.8, p. 508
J. P. JAKUBOVICS	(a) **46**, 291 (1978)	*18.2*, p. *284*; 21.1.1, p. 386; 21.4, p. 397
W. JAKUBOWSKI	(a) **47**, K103 (1978)	*14.4*, p. *249*; 22.5.3, p. 487
G. J. JAN	(a) **48**, K181 (1978)	*1.4*, p. *32*; *2*, p. *48*; 21.1, p. 377; 21.6, p. 404
Z. JANÁČEK	(a) **45**, K55 (1978)	*1.1*, p. *11*; 22.1.2, p. 429
	(a) **50**, 285 (1978)	*1.1*, p. *14*; 10.2, p. 121; 22.1.2, p. 438
S. JANDL	(b) **88**, K135 (1978)	*14.3*, p. *223*; 22, p. 416
TH. JANNEMANN	(a) **49**, K21 (1978)	*18.2*, p. *290*; 21.1.1, p. 389
A. D. JANNUSSIS	(b) **88**, 341 (1978)	*13.1*, p. *156*; 18, p. 274
Š. JÁNOŠ	(b) **85**, K31 (1978)	*8*, p. *83*; 21.4, p. 396
	(b) **85**, 545 (1978)	*8*, p. *83*; 14.2, p. 213; 21, p. 354
Z. V. JANUŠKEVEČIUS	(a) **47**, 187 (1978)	*10.2*, p. *113*; *14.3*, p. *220*; 22.4.2, p. 468
E. JAPA	(a) **50**, K89 (1978)	1.3, p. 28; *5*, p. *62*; 21.1.1, p. 390
B. JAROSZ	(a) **47**, K103 (1978)	*14.4*, p. *249*; 22.5.3, p. 487
R. JAUCH	(b) **86**, 621 (1978)	*18.2*, p. *285*; *18.4*, p. *299*
P. R. JAY	(a) **46**, 199 (1978)	*9*, p. *92*; *14.3*, p. *219*; 22.2.2, p. 450
J. JEČNÝ	(a) **45**, 269 (1978)	*6*, p. *65*; 22.5, p. 474
M. JEDLIČKA	(a) **48**, K101 (1978)	*1.5*, p. *38*; 22.6, p. 495
E. JĘDRYKA	(a) **49**, K193 (1978)	*18.4*, p. *301*; *19*, p. *313*; 22.8.2, p. 520
R. J. JELITTO	(b) **85**, 247 (1978)	*18.2*, p. *281*
	(b) **85**, 535 (1978)	*13*, p. *143*; *18.2*, p. *282*
	(b) **90**, 135 (1978)	*18.2*, p. *291*
J. JELONEK	(a) **50**, 679 (1978)	*18.2*, p. *292*; 22.8.2, p. 521
L. C. JEN	(a) **50**, 469 (1978)	*10.2*, p. *121*; *20.1*, p. *338*; 22.5.2, p. 486
B. JENSEN	(b) **86**, 291 (1978)	*14.3*, p. *218*; *20.1*, p. *322*; 22.2.1, p. 445
A. JESMANOWICZ	(a) **48**, 191 (1978)	13.4, p. 177; *19*, p. *310*; 22.8, p. 506
H. JEX	(b) **85**, 511 (1978)	*6*, p. *66*; 8, p. 83; 22.9, p. 523
	(a) **50**, K117 (1978)	*1.2*, p. *24*; 22.8, p. 508
A. JEŻOWSKI	(a) **47**, 229 (1978)	*8*, p. *85*; 22.6, p. 493
Z. JIRÁK	(a) **50**, K21 (1978)	*1.1*, p. *14*; 22.8.2, p. 520
	(a) **50**, K131 (1978)	*18.3*, p. *297*; 22.8.1, p. 517
J. H. JISCHKE	(b) **87**, 581 (1978)	*10.2*, p. *113*; *20.1*, p. *327*; 22.5.2, p. 481
H. JO	(a) **47**, 631 (1978)	*1.2*, p. *19*; 1.3, p. 26; 21.1, p. 374
L. JOHN	(b) **89**, K183 (1978)	*13.5.1*, p. *195*; *20.1*, p. *334*; 22.4.4, p. 473
W. JOHN	(b) **86**, 569 (1978)	*13.1*, p. *154*; 14.2, p. 214; 21.1, p. 371; 21.1.1, p. 386
E. JOHNSON	(a) **49**, 585 (1978)	10.2, p. 119; *11*, p. *130*; 22.8, p. 507
M. L. JOHNSON	(a) **48**, 551 (1978)	*1.3*, p. *27*; 10.1, p. 105; 21, p. 362; 21.1, p. 376
WILLIAM C. JOHNSON	(a) **46**, 267 (1978)	*10.1*, p. *101*; *12.1*, p. *134*; 21, p. 356
G. B. JOHNSTON	(a) **49**, K177 (1978)	*18.3*, p. *296*; 18.4, p. 301; 21.1, p. 379
R. K. JOHRI	(b) **88**, 309 (1978)	*13.1*, p. *156*; *20*, p. *317*; 21, p. 360
I. P. JONES	(a) **47**, 163 (1978)	*1.1*, p. *12*; 21, p. 358
	(a) **47**, 385 (1978)	*1.1*, p. *13*; 21, p. 359
A. G. JORDAN	(a) **47**, 321 (1978)	*1.5*, p. *37*; 14.1, p. 207; 21.1, p. 373
B. JOSEPH	(a) **45**, K25 (1978)	*14.3.2*, p. *230*; 14.3.3, p. 233; 22.1.2, p. 429
C. JOUANIN	(a) **47**, 311 (1978)	*1.1*, p. *12*; 1.4, p. 30; 22.6, p. 493
J. C. JOUBERT	(a) **46**, 493 (1978)	*18.4*, p. *299*; 22.8.1, p. 512
	(a) **48**, 581 (1978)	*1.1*, p. *13*; *5*, p. *62*; 22.8.1, p. 515
	(a) **50**, 635 (1978)	*18.4*, p. *302*; 22.8.1, p. 516

C. Jourdan	(a) **49**, 529 (1978)	1.4, p. 33; *3*, p. *56*; 21, p. 363
J. C. Jousset	(a) **49**, 643 (1978)	*11*, p. *130*; 14.1, p. 211; 21, p. 363
P. Jung	(a) **46**, 527 (1978)	*10.1*, p. *102*; 11, p. 125; 14.1, p. 207; 21.6, p. 402
G. Jungk	(a) **49**, 629 (1978)	*1.5*, p. *39*; 13.4, p. 182; 14.3.1, p. 228; 20.1, p. 335; 20.3, p. 348; 22.2.4, p. 456
M. Jursich	(b) **88**, K39 (1978)	*18.2*, p. *288*
G. Juška	(a) **48**, K149 (1978)	*2*, p. *48*; *14.3*, p. *223*; 22.1.3, p. 441
W. Just	(a) **46**, 557 (1978)	1.3, p. 25; *10.1*, p. *102*; 21, p. 357
E. te Kaat	(a) **47**, 751 (1978)	*11*, p. *127*; 22.1.2, p. 433
O. Kabisch	(a) **49**, 299 (1978)	*10.1*, p. *106*; 21, p. 363; 21.1, p. 377
Z. S. Kachlishvili	(b) **88**, 283 (1978)	*13.4*, p. *177*; *14.3.4*, p. *243*; 18, p. 274; 22.1.1, p. 426
	(a) **49**, K135 (1978)	*13.4*, p. *182*; *14.3.4*, p. *245*; 22, p. 419
G. A. Kachurin	(a) **47**, K5 (1978)	1.5, p. 37; *11*, p. *126*; 22.2.3, p. 453
P. Kacman	(b) **89**, 521 (1978)	*13.1*, p. *159*; 14.4, p. 251; 18, p. 275; 22, p. 418
J. Kaczér	(a) **49**, K81 (1978)	2, p. 50; *18.2.1*, p. *294*; 21.1, p. 378; 21.4, p. 398
V. N. Kadchenko	(a) **46**, 315 (1978)	*13.4*, p. *173*; *20.3*, p. *343*; 22.5.2, p. 479
	(a) **46**, K61 (1978)	*17*, p. *269*; 22.5.2, p. 480
P. Kadura	(b) **85**, 225 (1978)	*13*, p. *142*
	(b) **88**, 537 (1978)	4, p. 59; *13*, p. *146*
E. I. Kafediiska	(a) **45**, 671 (1978)	*14.3.3*, p. *233*; 14.3.4, p. 240; 22.2.1, p. 445
V. M. Kaganer	(a) **46**, 379 (1978)	*10.2*, p. *111*; 22.1.2, p. 431
E. Kähler	(a) **49**, 431 (1978)	13.4, p. 182; *14.3.3*, p. *236*; *14.3.4*, p. *245*; 22.1.2, p. 437
O. Kahn	(b) **86**, 497 (1978)	*18.4*, p. *299*; 22.8, p. 504
O. A. Kaibyshev	(a) **45**, 77 (1978)	*10.1*, p. *99*; 21, p. 352; 21.3, p. 394
R. Kaiser	(b) **87**, 565 (1978)	*6*, p. *71*; 14.2, p. 214; *20.1*, p. *327*
D. Kakati	(a) **45**, K13 (1978)	*14.3.3*, p. *233*
H. Kalbfleisch	(a) **46**, K119 (1978)	*18.4*, p. *299*; 22.8.1, p. 512
V. N. Kalinenkov	(b) **89**, 503 (1978)	1, p. 10; *13*, p. *147*
	(b) **90**, 123 (1978)	1, p 10; *13*, p. *148*
V. V. Kalinin	(a) **45**, 377 (1978)	1.4, p. 29; *10.2*, p. *110*; *11*, p. *124*; 22.1.2, p. 429
G. Kalkoffen	(b) **88**, 135 (1978)	*13.1*, p. *156*; 17, p. 270; 20.1, p. 328; 22.1.2, p. 434
M. Kalm	(b) **87**, 575 (1978)	6, p. 71; *13.5.1*, p. *192*; *20.1*, p. *327*; 22.5, p. 475
A. E. Kalnin	(b) **88**, 445 (1978)	*13.4*, p. *178*; 20.1, p. 330; *20.3*, p. *346*; 22.5.4, p. 489
J. Kalus	(a) **48**, K179 (1978)	*9*, p. *93*; 22.5.1, p. 477
A. Kalz	(a) **50**, K247 (1978)	*1.6*, p. *41*; 11, p. 132; 21.1.1, p. 391
G. Kamarinos	(a) **46**, K13 (1978)	*13.3*, p. *165*; 14.3.3, p. 234; 22, p. 411
	(a) **48**, 137 (1978)	13.3, p. 166; *14.3.3*, p. *235*; *16*, p. *264*; 22.1.2, p. 434
R. Kamel	(a) **45**, 47 (1978)	*10.1*, p. *99*; 12.1, p. 134; 21, p. 352; 21.2, p. 392
	(a) **45**, 419 (1978)	*10.1*, p. *100*; 12.1, p. 134; 21, p. 354
	(a) **46**, 697 (1978)	*10.1*, p. *102*; 12.1, p. 135; 14.1, p. 207; 21, p. 357
G. Kamieniarz	(b) **87**, 455 (1978)	*13*, p. *145*; *18.1*, p. *278*
E. Kaminska	(a) **50**, 243 (1978)	*14.3.3*, p. *238*; 22.5.3, p. 488
E. P. Kaminskaya	(b) **88**, 359 (1978)	*13*, p. *146*; *14*, p. *203*

A. A. Kaminskii	(a) **49**, 305 (1978)	3, p. 56; 20.1, p. 333; *20.2*, p. *339*; *20.3*, p. *347*; 22.8.2, p. 520
	(a) **50**, 745 (1978)	13.4, p. 186; *20.1*, p. *338*; *20.2*, p. *340*; 22.8.1, p. 516
K. Kamiyoshi	(a) **45**, 513 (1978)	*1.2*, p. *16*; *7*, p. *80*; 14.4, p. 248; 22.5.4, p. 489
T. Kanadani	(a) **48**, K141 (1978)	*14.1*, p. *209*; 21, p. 362
N. G. Kanavina	(a) **50**, K101 (1978)	*18.2.1*, p. *294*; 21.1, p. 381; 21.1.1, p. 390
B. D. Kandilarov	(b) **86**, 425 (1978)	*13.3*, p. *164*; 14.3.3, p. 233
	(b) **87**, 361 (1978)	*13.3*, p. *165*; *14.3.3*, p. *235*
A. Kandušer	(b) **86**, K119 (1978)	*1.2*, p. *17*; 14.4.1, p. 254; *19*, p. *306*; 22.8.1, p. 512
A. R. Kangro	(a) **47**, 667 (1978)	*10.2*, p. *114*; *20.3*, p. *345*; 22.5.2, p. 481
A. M. Kapitonov	(a) **48**, K203 (1978)	*1.2*, p. *22*; 7, p. 82; *18*, p. *275*; 22.6, p. 496
	(a) **50**, 269 (1978)	*18.3*, p. *296*; *19*, p. *314*; 22.8.2, p. 520
V. V. Kaplin	(b) **86**, K35 (1978)	*11*, p. *124*; 13, p. 143; 22.1.2, p. 431
D. Kapor	(b) **90**, K157 (1978)	*13*, p. *149*; *18.2*, p. *292*
H. F. Kappert	(a) **47**, 751 (1978)	*11*, p. *127*; 22.1.2, p. 433
	(a) **50**, 221 (1978)	*10.2*, p. *120*; *14.3.3*, p. *238*; 22.1.2, p. 438
S. Kapphan	(a) **50**, K243 (1978)	10.2, p. 122; *14.4.1*, p. *257*; *20.1*, p. *338*; 22.8.1, p. 517
Th. Karakostas	(a) **50**, 579 (1978)	*1.2*, p. *24*; *1.3*, p. *28*; 10.2, p. 122; 22.3, p. 459
	(a) **50**, 703 (1978)	*1.1*, p. *14*; *1.4*, p. *35*
G. F. Karavaev	(b) **88**, 321 (1978)	*12.2*, p. *140*; *13.1*, p. *156*; 14.3, p. 222; 22.2.1, p. 447
R. N. Karekar	(a) **45**, K163 (1978)	14.3.1, p. 226; *16*, p. *263*; 22.9, p. 523
G. G. Kareva	(a) **46**, 343 (1978)	*1.1*, p. *11*; 1.6, p. 40; 17, p. 269; 22.1.2, p. 431
J. Karins	(a) **48**, K31 (1978)	*11*, p. *128*; *13.4*, p. *178*; 22.1.2, p. 435
A. E. Karkin	(a) **46**, K87 (1978)	11, p. 125; *14.2*, p. *214*; 21, p. 357
L. E. Karkina	(a) **47**, 731 (1978)	*10*, p. *96*
	(a) **49**, 517 (1978)	*10*, p. *97*
S.-E. Karlsson	(a) **47**, 85 (1978)	*1.6*, p. *41*; 22, p. 413
U. Karlsson	(b) **86**, 653 (1978)	*6*, p. *69*; *8*, p. *84*; 22.5.2, p. 480
A. G. Karoza	(b) **86**, K171 (1978)	*6*, p. *70*; *20.1*, p. *325*; 22.8, p. 504
W. Karthe	(a) **50**, 209 (1978)	11, p. 132; *19*, p. *314*; 22.1.2, p. 440
E. Kartheuser	(b) **90**, K73 (1978)	*6*, p. *78*; *14.3.4*, p. *246*; 22.2.3, p. 454
V. N. Kashcheev	(b) **86**, 359 (1978)	*18.2*, p. *284*; 19, p. 305
	(b) **89**, 45 (1978)	*6*, p. *75*; 14.4.1, p. 256
H. A. Kassim	(b) **86**, K107 (1978)	*6*, p. *69*; *8*, p. *84*; 22.1.1, p. 425
R. Kassing	(a) **49**, 153 (1978)	*13.4*, p. *180*; 14.3.3, p. 236; *14.3.4*, p. *244*; 22.1.2, p. 436
G. Kästner	(a) **49**, K1 (1978)	*10.2*, p. *118*; 22.4.3, p. 471
Atsuo Kasuya	(b) **89**, K145 (1978)	*17*, p. *270*; 22.2.1, p. 448; 22.5, p. 476
K.-H. Katerbau	(a) **47**, 479 (1978)	*1.4*, p. *31*; *10*, p. *96*
A. K. Kathuria	(a) **45**, K59 (1978)	*2*, p. *43*; *13.4*, p. *170*; 14.3.1, p. 226; 22.1.3, p. 441
S. P. Kathuria	(a) **50**, 329 (1978)	*10.2*, p. *121*; 20.1, p. 336; *20.3*, p. *350*; 22.5.2, p. 485
R. S. Katiyar	(b) **90**, 543 (1978)	*14.4.1*, p. *257*; 20.1, p. 337; 22.8.1, p. 516
J. Katriel	(b) **89**, 645 (1978)	*18.2*, p. *290*
A. A. Katsnelson	(b) **88**, 477 (1978)	*1.3*, p. *27*; *13*, p. *146*; 21.1, p. 375
L. P. Kaun	(b) **87**, 535 (1978)	*13.4*, p. *176*; 21.4, p. 397
P. V. Kavyuk	(b) **88**, 179 (1978)	*13.4*, p. *177*; 14.3, p. 222; *15*, p. *260*; 22.1.3, p. 441
S. Kawamura	(a) **50**, 669 (1978)	*13.4*, p. *186*; *14.3*, p. *225*; 22.6, p. 497
K. Kawasaki	(b) **85**, 707 (1978)	*18.1*, p. *277*; 18.2, p. 282; 18.4, p. 298
Balakrishna Rao Kaza	see under letter "R" (B. Rao Kaza)	
N. Kazama	(a) **49**, K163 (1978)	*2*, p. *51*; *18.2*, p. *291*; 21.1.1, p. 390

K. O. Kechechyan	(b) **87**, K123 (1978)	*10.2*, p. *113*; *13.4*, p. *176*; 22, p. 414
	(b) **88**, K15 (1978)	*10.2*, p. *115*; *13.4*, p. *177*
W. Keilig	(a) **49**, K67 (1978)	*2*, p. *50*; 11, p. 129; *18.3*, p. *296*; 21.1, p. 378; 21.4, p. 398
	(a) **50**, 537 (1978)	*18.3*, p. *297*; 21.1, p. 382; 21.4, p. 399
R. Keiper	(b) **85**, 155 (1978)	*16*, p. *262*; 22, p. 408
V. Kelhä	(a) **50**, 427 (1978)	*1.1*, p. *14*; 22.1.2, p. 439
V. M. Kenkre	(b) **89**, 651 (1978)	*20.3*, p. *348*; 22, p. 418
D. Kerkhoff	(b) **89**, 357 (1978)	*13.1*, p. *159*; 20, p. 317; 22.2, p. 444; 22.2.3, p. 454
F. Kersten	(a) **50**, K189 (1978)	*14.3.3*, p. *239*; 22.9, p. 529
F. R. Kessler	(b) **89**, 203 (1978)	13.1, p. 158; 13.5.1, p. 194; *18*, p. *275*; *20*, p. *317*; 22.1.1, p. 427
	(a) **49**, 255 (1978)	*1.5*, p. *38*; *2*, p. *49*; 22.1.1, p. 427
	(b) **89**, 389 (1978)	*13.5.1*, p. *194*; *16*, p. *266*; 22.1.1, p. 428
A. G. Khachaturyan	(a) **45**, 393 (1978)	*10.1*, p. *100*; 21.1, p. 370; 21.1.1, p. 385
Y. L. Khait	(b) **86**, 409 (1978)	*6*, p. *67*; 8, p. 84; 9, p. 92; 10, p. 96
	(b) **89**, 687 (1978)	*6*, p. *67*; 8, p. 84; 9, p. 92; 10, p. 96
	(b) **89**, K111 (1978)	*1.2*, p. *23*
Z. D. Khalafov	(b) **86**, K49 (1978)	*6*, p. *68*; *20.1*, p. *323*; 22.8, p. 504
J. A. A. Khan	(a) **50**, K161 (1978)	*10.2*, p. *122*; 11, p. 131; 22.5.2, p. 486
W. I. Khan	(a) **48**, 209 (1978)	*18.2*, p. *289*; 21.1, p. 375
L. G. Khandros	(a) **50**, K127 (1978)	*1.2*, p. *24*; 21.1, p. 383
K. N. Khanna	(b) **86**, K167 (1978)	*18.1*, p. *278*; 21.2, p. 392
	(b) **88**, 809 (1978)	*12.1*, p. *136*; 13.2, p. 162; 18.1, p. 278; 21.2, p. 393
Yu. I. Kharkats	(b) **85**, 371 (1978)	*1.2*, p. *15*
V. E. Khartsiev	(b) **85**, K53 (1978)	*13.5.1*, p. *189*; 22.1.2, p. 429
	(b) **86**, 505 (1978)	*13.5.1*, p. *190*; 22.1.2, p. 431
L. P. Khiznichenko	(a) **46**, K147 (1978)	*10.2*, p. *112*; 11, p. 126; *12.2*, p. *139*; 22.1.2, p. 432
A. A. Khomenko	(a) **50**, K213 (1978)	1.4, p. 35; *10.2*, p. *122*; 22.1, p. 423
A. A. Khoromenko	(a) **50**, 573 (1978)	*1.2*, p. *24*; *1.5*, p. *39*; 22, p. 421
V. A. Khramtsov	(b) **86**, K67 (1978)o	*19*, p. *305*; *20.1*, p. *323*; 22.5.2, p. 479
V. M. Khulugurov	(b) **89**, K91 (1978)	*10.2*, p. *117*; 20.1, p. 333; *20.3*, p. *347*; 22.5.2, p. 484
A. G. Khundzhua	(a) **49**, 803 (1978)	*1.2*, p. *23*; 1.3, p. 28; 21.1, p. 379
L. G. Khvostantsev	(a) **46**, 305 (1978)	*15*, p. *259*; 21.7, p. 406
Farid A. Khwaja	(b) **88**, 477 (1978)	*1.3*, p. *27*; *13*, p. *146*; 21.1, p. 375
N. P. Kibalina	(a) **47**, K91 (1978)	11, p. 127; *14.3.2*, p. *230*; 22.1.2, p. 433
H. Kiess	(a) **49**, 133 (1978)	*1*, p. *9*; *14.4*, p. *250*; 15, p. 261; 22.9, p. 527
B.-P. Kietis	(a) **50**, K63 (1978)	*7*, p. *82*; *16*, p. *267*; 22.4.1, p. 466; 22.4.2, p. 469; 22.4.3, p. 471
V. C. Kieu	(a) **45**, 571 (1978)	*13.4*, p. *171*; *14.3.4*, p. *240*; 22.2.1, p. 445
M. Kikuchi	(a) **48**, 175 (1978)	2, p. 47; *18.2*, p. *289*; 21.1.1, p. 388
K. Kilimann	(b) **88**, K59 (1978)	*13.5*, p. *187*; *13.5.1*, p. *193*; 22, p. 415
	(b) **90**, 175 (1978)	*13.5*, p. *188*; *13.5.1*, p. *196*; 22, p. 419
C. K. Kim	(b) **87**, 765 (1978)	*13.1*, p. *155*; 22.1.2, p. 433
Y. Kimura	(a) **47**, K111 (1978)	*1.2*, p. *20*; *14.1*, p. *208*; 21, p. 359
R. Kind	(a) **47**, 233 (1978)	*1.2*, p. *18*; 19, p. 307; 22.9, p. 525
J. Kinder	(b) **85**, 45 (1978)	*13.5.1*, p. *189*; 20.3, p. 341; 22.9, p. 522
P. J. King	(a) **45**, K33 (1978)	6, p. 65; *7*, p. *80*; 22.1.1, p. 424; 22.1.2, p. 429; 22.2.1, p. 445; 22.2.3, p. 452
N. Kinoshita	(a) **48**, 447 (1978)	*10*, p. *96*; 12, p. 133
A. A. Kirakossyan	(b) **87**, K123 (1978)	*10.2*, p. *113*; *13.4*, p. *176*; 22, p. 414
	(b) **88**, K15 (1978)	*10.2*, p. *115*; *13.4*, p. *177*
H. O. K. Kirchner	(a) **49**, 539 (1978)	*10.1*, p. *106*; 12.1, p. 136; 21.1.1, p. 389

D. L. Kirk	(a) **50**, K193 (1978)	*9*, p. *95*; 22.6, p. 498
K. I. Kirov	(a) **48**, 609 (1978)	*1.5*, p. *38*; *14.3.1*, p. *228*; 22.6, p. 495
	(a) **49**, 781 (1978)	*13.3*, p. *167*; *14.3.2*, p. *231*
	(a) **50**, 363 (1978)	*1.5*, p. *39*; *14.3.3*, p. *238*; 14.4, p. 252; 22.6, p. 497
S. A. Kirov	(a) **49**, 803 (1978)	*1.2*, p. *23*; 1.3, p. 28; 21.1, p. 379
R. Kirsch	(a) **46**, 459 (1978)	14.4, p. 248; *20.1*, p. *325*; 21.6, p. 402
J. Kirschner	(a) **45**, K117 (1978)	*1.6*, p. *40*; 21, p. 355
F.-G. Kirscht	(a) **47**, K17 (1978)	*10.2*, p. *113*; 22.1.2, p. 433
	(a) **48**, K11 (1978)	*10.2*, p. *115*; 22.1.2, p. 435
V. F. Kiselev	(a) **50**, 739 (1978)	*13.3*, p. *168*; 22.1.1, p. 428
R. Kishore	(b) **86**, K163 (1978)	*2*, p. *44*; *18.2*, p. *285*
V. F. Kitaeva	(a) **47**, 609 (1978)	*7*, p. *81*; *20.1*, p. *328*; 22.8.1, p. 514
M. Kitao	(a) **45**, 107 (1978)	2, p. 42; *16*, p. *262*; 22, p. 409
E. Klausmann	(a) **50**, 433 (1978)	11, p. 131; *14.3.2*, p. *232*; *14.3.3*, p. *238*; 22.1.2, p. 439
P. Kleinert	(b) **85**, 253 (1978)	*6*, p. *64*; *20.1*, p. *319*; 22, p. 409
	(b) **89**, 675 (1978)	*6*, p. *76*; *13*, p. *148*; 20.3, p. 348; 22, p. 418
K. Kleinstück	(a) **49**, 323 (1978)	*1.2*, p. *22*; *10.1*, p. *106*; 14.2, p. 215; 21.1, p. 377
W. Kley	(b) **89**, K29 (1978)	*9*, p. *93*; 21.6, p. 404
J. G. Kliava	(b) **89**, 531 (1978)	*10.2*, p. *119*; 19, p. 312; 22.5.2, p. 485
	(a) **49**, K43 (1978)	*19*, p. *312*; 22.8.1, p. 515
V. A. Klimov	(a) **46**, 259 (1978)	*10.2*, p. *111*; 20.1, p. 324; 22.5.2, p. 479
	(a) **46**, 421 (1978)	*1.4*, p. *30*; *10.2*, p. *112*; 22.5.2, p. 480
C. Klingshirn	(b) **89**, 431 (1978)	13.5.1, p. 195; *20.2*, p. *339*; *20.3*, p. *348*; 22.4, p. 462; 22.4.1, p. 465; 22.4.3, p. 471
	(b) **90**, 67 (1978)	13.4, p. 183; *13.5.1*, p. *195*; *20.3*, p. *349*; 22.4.1, p. 466
A. A. Klyukanov	(b) **87**, K159 (1978)	*19*, p. *308*; 22.4.1, p. 464
B. Knappe	(a) **47**, 523 (1978)	*1.5*, p. *37*; *2*, p. *45*; 18.3, p. 296; 21.1, p. 374; 21.4, p. 398
H. Knauer	(a) **47**, K161 (1978)	*14.2*, p. *214*
K. Kneipp	(b) **86**, 215 (1978)	*6*, p. *67*; *20.1*, p. *322*; 22.8.1, p. 511
J. Knof	(a) **45**, 433 (1978)	13.4, p. 171; *20.2*, p. *339*; 22.4.1, p. 463
O. Knop	(a) **50**, 55 (1978)	*1.2*, p. *24*; *20.1*, p. *337*; 23, p. 531
A. I. Knyazev	(a) **45**, 63 (1968)	*5*, p. *60*; 9, p. 91; 10.2, p. 109; 22.6, p. 491
A. Kobayashi	(a) **47**, K51 (1978)	*14.3.4*, p. *242*; *18.2*, p. *286*; 22.8, p. 505
M. Kobayashi	(a) **47**, 657 (1978)	*7*, p. *81*; 22.4.1, p. 464
S. Kobe	(b) **85**, 233 (1978)	*2*, p. *42*; 13, p. 142; *18.2*, p. *281*
N. P. Kobelev	(a) **50**, K185 (1978)	7, p. 82; *10.1*, p. *108*; *14.2*, p. *216*; 21, p. 367
V. Kobzarenko	(a) **49**, K5 (1978)	*1.5*, p. *39*; 22.8, p. 507
P. Kocevar	(b) **89**, 225 (1978)	*6*, p. *75*; 14.3, p. 224; 22.2.1, p. 448
E. E. Koch	(b) **90**, 525 (1978)	*13.1*, p. *160*; *17*, p. *270*; 21, p. 365
S. W. Koch	(b) **89**, 431 (1978)	13.5.1, p. 195; *20.2*, p. *339*; *20.3*, p. *348*; 22.4, p. 462; 22.4.1, p. 465; 22.4.3, p. 471
J. Kočka	(a) **45**, 559 (1978)	*2*, p. *43*; *14.3*, p. *218*; 22.3, p. 457
	(a) **48**, K59 (1978)	*2*, p. *47*; *14.3*, p. *222*; 14.3.4, p. 243; 22, p. 415
C. Koenig	(b) **88**, 569 (1978)	*1.3*, p. *27*; *10.1*, p. *105*; 21, p. 361
D. Kohl	(a) **49**, 27 (1978)	13.3, p. 167; 13.4, p. 180; *14.3.2*, p. *231*; 14.3.4, p. 244; *16*, p. *265*; 22.4, p. 462
H. Köhler	(b) **90**, 509 (1978)	*13.1*, p. *160*; 14.3, p. 225; *15*, p. *261*; 22.7, p. 502
H.-J. Köhler	(a) **49**, 569 (1978)	*14.2*, p. *215*
	(a) **50**, 417 (1978)	*14.2*, p. *216*
B. Koiller	(b) **87**, 351 (1978)	*13.4*, p. *175*; *20.1*, p. *326*; 22.5.2, p. 481
H. Kojima	(a) **47**, K119 (1978)	*2*, p. *46*; *18.2.1*, p. *293*; 21.1.1, p. 387

W. C. Kok	(b) **86**, K113 (1978)	*2*, p. *44*; *14*, p. *203*
T. I. Kolchenko	(a) **48**, 263 (1978)	*11*, p. *128*; 13.4, p. 178; 14.3, p. 222; 22.2.1, p. 447
E. Kolley	(b) **86**, 397 (1978)	*13*, p. *143*; *18.1*, p. *277*; 18.2, p. 284; 21, p. 355
	(b) **86**, K79 (1978)	*13*, p. *143*; 18.2, p. 284; 21, p. 356
	(b) **90**, K103 (1978)	*13.5.4*, p. *202*; 18.2, p. 291; 21, p. 365
W. Kolley	(b) **86**, 397 (1978)	*13*, p. *143*; *18.1*, p. *277*; 18.2, p. 284; 21, p. 355
	(b) **86**, K79 (1978)	*13*, p. *143*; 18.2, p. 284; 21, p. 356
	(b) **90**, K103 (1978)	*13.5.4*, p. *202*; 18.2, p. 291; 21, p. 365
E. P. Kolomeyko	(a) **50**, 195 (1978)	*2*, p. *52*; *14.3.3*, p. *237*; 14.3.4, p. 246; 22, p. 420
V. I. Kolomytsev	(a) **50**, K127 (1978)	*1.2*, p. *24*; 21.1, p. 383
V. A. Kolosanov	(a) **49**, 217 (1978)	*14.4*, p. *250*; 22, p. 417
M. N. Kolotov	(a) **49**, 211 (1978)	11, p. 129; *14.3.3*, p. *236*; *20.3*, p. *347*; 22.2.2, p. 451
O. S. Kolotov	(a) **50**, K101 (1978)	*18.2.1*, p. *294*; 21.1, p. 381; 21.1.1, p. 390
A. S. Komalov	(a) **50**, K153 (1978)	*18.2.1*, p. *294*; *20*, p. *318*; 21.1, p. 383; 21.1.1, p. 391
F. F. Komarov	(a) **45**, 343 (1978)	*1.2*, p. *15*; *1.5*, p. *36*; 11, p. 123; 21.1, p. 369
Y. Komem	(a) **48**, 555 (1978)	*1.5*, p. *38*; 21.6, p. 403
Č. Koňák	(a) **45**, K63 (1978)	*1.2*, p. *16*; 14, p. 203; 20, p. 315; 22.5, p. 474
R. V. Konakova	(a) **48**, K55 (1978)	*11*, p. *128*; 14.3.3, p. 235; 22.2.1, p. 447
A. A. Konova	(a) **50**, 627 (1978)	*1.5*, p. *39*; 22.6, p. 497
P. Konsin	(b) **86**, 57 (1978)	*1.2*, p. *16*; 6, p. 67; 14.4.1, p. 254; 22, p. 410
H. Konvent	(b) **88**, K43 (1978)	*13*, p. *146*; *18.2*, p. *288*
	(b) **88**, 747 (1978)	1.2, p. 21; *12.2*, p. *140*; *14.4.1*, p. *256*; 22.8.1, p. 515
M. Kopcewicz	(a) **46**, 675 (1978)	*5*, p. *61*; *18.2*, p. *285*; 21.1, p. 372; 21.1.1, p. 386
S. Kopta	(a) **46**, 285 (1978)	*11*, p. *125*; 22.1.2, p. 431
S. V. Koptsik	(a) **47**, 39 (1978)	*5*, p. *61*; 18.4, p. 300; 22.6, p. 493
U. Kh. Kopvillem	(b) **87**, K145 (1978)	*2*, p. *45*; *6*, p. *72*; 7, p. 81
J. Korec	(a) **48**, 225 (1978)	*3*, p. *55*; 22.1.2, p. 434
N. N. Koren	(a) **46**, K1 (1978)	*1.2*, p. *17*; *1.3*, p. *25*; 22.7, p. 501
A. P. Korenko	(a) **48**, K203 (1978)	*1.2*, p. *22*; 7, p. 82; *18*, p. *275*; 22.6, p. 496
V. A. Korneichuk	(b) **86**, K155 (1978)	13.5.1, p. 191; *16*, p. *263*; 22.4.1, p. 464
S. L. Korolyuk	(b) **86**, 765 (1978)	13, p. 144; *14.3*, p. *219*; 22.2.1, p. 446
L. I. Korovin	(b) **90**, 11 (1978)	*6*, p. *64*; 13, p. 142; *19*, p. *303*; 22, p. 408; 22.2.3, p. 452; 22.4.1, p. 463
F. P. Korshunov	(a) **48**, K127 (1978)	*11*, p. *128*; *14.3.2*, p. *231*; 22.1.2, p. 436; 22.6, p. 495
V. V. Korshunov	(b) **90**, K53 (1978)	*13.5.1*, p. *196*; *20.3*, p. *349*; 22.9, p. 528
N. E. Korsunskaya	(a) **50**, 767 (1978)	*13.4*, p. *186*; *14.3*, p. *225*; 16, p. 268; 22.4.1, p. 467
N. A. Korynevskii	(b) **88**, 51 (1978)	*1.2*, p. *20*; *14.4.1*, p. *255*; 22.8.1, p. 514
H.-J. Kos	(a) **50**, K165 (1978)	11, p. 131; *12.2*, p. *141*; *20.3*, p. *351*; 22.5.2, p. 486
A. Kosakov	(b) **85**, K11 (1978)	*13.1*, p. *150*; 17, p. 269; 22.7, p. 501
	(b) **89**, 619 (1978)	2, p. 50; *13*, p. *148*; *20*, p. *317*; 22.6, p. 496
J. Koshy	(a) **47**, K21 (1978)	*11*, p. *126*; 22.8.1, p. 513
R. Kosicik	(a) **49**, K67 (1978)	*2*, p. *50*; 11, p. 129; *18.3*, p. *296*; 21.1, p. 378; 21.4, p. 398
	(a) **50**, 537 (1978)	*18.3*, p. *297*; 21.1, p. 382; 21.4, p. 399

J. Kossut	(b) **86**, 593 (1978)	*13.1*, p. *154*; *13.4*, p. *174*; 14.3, p. 219; 22.4.4, p. 472
	(b) 88, 73 (1978)	*13.2*, p. *162*; *14.3*, p. *221*; 15, p. 260; 22.4.4, p. 473
M. E. Kost	(b) **87**, 369 (1978)	*1.2*, p. *18*; 12.1, p. 135; 21.4, p. 397
U. Köster	(a) **48**, 313 (1978)	*1.5*, p. *37*; 2, p. 48; *3*, p. *55*; 22.1.2, p. 435
H. Kostial	(b) **86**, K21 (1978)	*20.1*, p. *323*; 22, p. 410
	(b) 88, 675 (1978)	13.4, p. 178; 20.1, p. 330; *20.2*, p. *339*; 22, p. 415
A. K. Kostin	(b) **87**, 255 (1978)	*13.5.1*, p. *191*; 20.1, p. 326; 20.3, p. 344; 22.9, p. 524
A. A. Kosyachkov	(a) **50**, K113 (1978)	*17*, p. *270*; 22.3, p. 459
I. M. Kotina	(a) **46**, K21 (1978)	*13.4*, p. *173*; *14.3.3*, p. *234*; 22, p. 411
A. Kottwitz	(b) **87**, 239 (1978)	2, p. 44; *16*, p. *264*; 22.7, p. 501
V. I. Kovalechiv	(a) **45**, 249 (1978)	*14.4.1*, p. *253*; 22.8.1, p. 510
K. Koyama	(b) **87**, 759 (1978)	*13.3*, p. *166*; 21.6, p. 402
A. Kozanecki	(b) **89**, 313 (1978)	*13.5.1*, p. *194*; *20.3*, p. *347*; 22.4.4, p. 473
V. Kozhukharov	(a) **47**, 743 (1978)	*2*, p. *46*; 4, p. 58; 22.6, p. 494
M. J. Kozielski	(b) **87**, K25 (1978)	*6*, p. *70*; *20.1*, p. *326*; 22.4.1, p. 464
	(b) **90**, 649 (1978)	*6*, p. *78*; 10.2, p. 121; *20.1*, p. *337*; 22.4.1, p. 466
A. G. Kozorezov	(b) **85**, 487 (1978)	6, p. 66; *7*, p. *80*; 22.2.3, p. 452
W. Kraak	(b) 88, 105 (1978)	*13.4*, p. *177*; 14.1, p. 208; *14.3*, p. *221*; 22.7, p. 502
	(b) **89**, 547 (1978)	*14.1*, p. *211*; 21.7, p. 407
W. D. Kraeft	(b) 88, K59 (1978)	*13.5*, p. *187*; *13.5.1*, p. *193*; 22, p. 415
	(b) **90**, 175 (1978)	*13.5*, p. *188*; *13.5.1*, p. *196*; 22, p. 419
M. V. Kraevskii	(b) **85**, 111 (1978)	*13.5.1*, p. *189*; *20.3*, p. *341*; 22.4.1, p. 463; 22.4.2, p. 468
Ya. A. Kraftmakher	(a) **47**, K81 (1978)	1.2, p. 18; *14.1*, p. *208*; *15*, p. *260*; 21.4, p. 397
D. R. Krahn	(a) **46**, 209 (1978)	*14.2*, p. *214*; 21, p. 356
V. Kraševec	(a) **49**, 767 (1978)	*1.2*, p. *23*; *1.4*, p. *34*; 21.1, p. 379
A. N. Krasiko	(b) **87**, 613 (1978)	13.3, p. 165; *14.3.4*, p. *242*; *20.1*, p. *327*; 22.1.2, p. 433
V. V. Krasnova	(a) **46**, K1 (1978)	*1.2*, p. *17*; *1.3*, p. *25*; 22.7, p. 501
G. Krautheim	(a) **49**, K125 (1978)	*9*, p. *94*; 21.1, p. 379
S. P. Kravchuk	(b) **85**, 63 (1978)	*6*, p. *64*; *8*, p. *83*; 21.2, p. 392
	(b) **85**, 749 (1978)	*6*, p. *66*; *8*, p. *83*; 21.2, p. 392
A. R. Krebs	(a) **50**, K153 (1978)	*18.2.1*, p. *294*; *20*, p. *318*; 21.1, p. 383; 21.1.1, p. 391
K. Kreher	(a) **49**, 331 (1978)	*13.4*, p *181*; 13.5.1, p. 194; *20.3*, p. *347*; 22.2.2, p. 451
D. Kremp	(b) 88, K59 (1978)	*13.5*, p. *187*; *13.5.1*, p. *193*; 22, p. 415
	(b) **90**, 175 (1978)	*13.5*, p. *188*; *13.5.1*, p. *196*; 22, p. 419
H. Kresse	(a) **50**, K181 (1978)	*2*, p. *53*; *14.4*, p. *252*; 22.9, p. 529
K. Krezhov	(a) **47**, 743 (1978)	*2*, p. *46*; 4, p. 58; 22.6, p. 494
V. S. Krikorov	(a) **50**, K13 (1978)	3, p. 56; *14.4.1*, p. *257*; *20*, p. *318*; 22.8, p. 508
K. V. Krishna	(a) **47**, 375 (1978)	13.4, p. 176; *14.3.1*, p. *227*; 22.2.3, p. 453
V. G. Krishna Murty	see under letter "M"	
B. S. Krishnamurthy	(b) **86**, 725 (1978)	13.5.4, p. 201; *14.3*, p. *219*; 18.2, p. 285; 22.4.4, p. 472
	(b) 88, K117 (1978)	*13.5.4*, p. *202*; *15*, p. *260*; 18.2, p. 289; 22, p. 415
P. Krispin	(a) **47**, 91 (1978)	13.4, p. 175; *14.3.3*, p. *235*; *20.3*, p. *344*; 22.2.2, p. 450
I. K. Kristensen	(b) **89**, K155 (1978)	13.4, p. 182; *20.3*, p. *348*; 22.3, p. 459
J. Krištofik	(a) **45**, 559 (1978)	*2*, p. *43*; *14.3*, p. *218*; 22.3, p. 457

V. P. Krivitskii	(b) **87**, 129 (1978)	*13.1*, p. *154*; 20, p. 316; 21, p. 357; 21.1.1, p. 386
M. A. Krivoglaz	(b) **86**, 781 (1978)	*13*, p. *144*; 13.5.3, p. 200; 18.2, p. 285; 22, p. 412
N. P. Krivorotov	(b) **88**, 321 (1978)	*12.2*, p. *140*; *13.1*, p. *156*; 14.3, p. 222; 22.2.1, p. 447
I. V. Krivoshei	(a) **50**, K197 (1978)	*14*, p. *204*; *18*, p. *276*; 22.9, p. 529
M. A. Krivov	(a) **49**, 761 (1978)	*11*, p. *130*; 13.4, p. 182; *14.3*, p. *224*; 22.8, p. 507
	(a) **50**, 379 (1978)	11, p. 131; 13.4, p. 185; *14.3*, p. *225*; *20.1*, p. *337*; 22.8, p. 508
H. Krizek	(a) **46**, 39 (1978)	*14.1*, p. *206*; 21.4, p. 396
R. Kroggel	(a) **49**, 299 (1978)	*10.1*, p. *106*; 21, p. 363; 21.1, p. 377
F. Krok	(a) **47**, K103 (1978)	*14.4*, p. *249*; 22.5.3, p. 487
A. Król	(b) **87**, K25 (1978)	*6*, p. *70*; *20.1*, p. *326*; 22.4.1, p. 464
	(b) **90**, 649 (1978)	*6*, p. *78*; 10.2, p. 121; *20.1*, p. *337*; 22.4.1, p. 466
H. Kronmüller	(a) **47**, 451 (1978)	*18.2*, p. *287*; 21.4, p. 398
	(a) **50**, 187 (1978)	*14.2*, p. *215*; 18, p. 275; 20, p. 318; 21, p. 365
	(a) **50**, 475 (1978)	2, p. 52; *18.2*, p. *292*; *20*, p. *318*; 21.1.1, p. 390
N. Kroó	(a) **47**, 609 (1978)	*7*, p. *81*; *20.1*, p. *328*; 22.8.1, p. 514
K. Krop	(a) **50**, K89 (1978)	1.3, p. 28; *5*, p. *62*; 21.1.1, p. 390
A. Krotkus	(a) **50**, K73 (1978)	12.2, p. 141; *14.3.4*, p. *246*; 22.2.2, p. 454
H.-J. Krug	(b) **85**, 239 (1978)	*13.5.1*, p. *189*; *20.1*, p. *319*; 22.4.1, p. 463
E. Krüger	(b) **85**, 261 (1978)	13.1, p. 150; *14.2*, p. *213*; 21, p. 352
	(b) **85**, 493 (1978)	13.1, p. 151; *14.2*, p. *213*; 21, p. 354
	(b) **90**, 719 (1978)	13.1, p. 161; *14.2*, p. *216*; 21, p. 366
M. V. Kruglov	(a) **46**, 343 (1978)	*1.1*, p. *11*; 1.6, p. 40; 17, p. 269; 22.1.2, p. 431
M. Krupski	(a) **45**, 585 (1978)	*1.2*, p. *16*; 19, p. 304; 22.8, p. 503
	(b) **88**, K65 (1978)	*19*, p. *309*; 22.5.3, p. 488
J. Kružliak	(a) **47**, 287 (1978)	5, p. 61; *14.2*, p. *214*
V. I. Krylov	(b) **90**, 741 (1978)	*8*, p. *90*; *18.2*, p. *292*; 21.1, p. 382
J. Kubát	(a) **50**, 117 (1978)	*12*, p. *133*
J. Kuběna	(a) **45**, K55 (1978)	*1.1*, p. *11*; 22.1.2, p. 429
	(a) **50**, 285 (1978)	*1.1*, p. *14*; 10.2, p. 121; 22.1.2, p. 438
L. P. Kubin	(a) **45**, 453 (1978)	*10.2*, p. *110*; 22.8.1, p. 511
	(a) **45**, 493 (1978)	*10.1*, p. *100*; *18.2*, p. *283*; 21.1.1, p. 385
G. C. Kuczynski	(b) **87**, 581 (1978)	*10.2*, p. *113*; *20.1*, p. *327*; 22.5.2, p. 481
N. V. Kudrevathykh	(a) **45**, 71 (1978)	*18.2*, p. *281*; 21.1, p. 368; 21.4, p. 396
J. Kudrnovský	(b) **86**, K89 (1978)	*13.4*, p. *172*; 21, p. 356
T. A. Kudykina	(b) **87**, 507 (1978)	*13.5.3*, p. *200*; 22, p. 414
K. I. Kugel	(a) **48**, K131 (1978)	*1.2*, p. *22*; *14*, p. *204*; 22.3, p. 458
D. Kuhlmann-Wilsdorf	(a) **47**, 639 (1978)	*10.1*, p. *104*; *12.1*, p. *135*; 21, p. 359
G. Kühn	(b) **85**, K57 (1978)	*13.1*, p. *150*; *20.1*, p. *320*; 22.8, p. 503
W. Kuhn	(a) **49**, 51 (1978)	*1.2*, p. *22*; *14.4.1*, p. *256*; 20, p. 317; 22.8.1, p. 515
W. Kühn	(b) **90**, 167 (1978)	6, p. 77; 13.1, p. 159; 13.5.1, p. 195; *20.1*, p. *335*; 22.2.4, p. 456
R. Kühne	(b) **89**, 131 (1978)	*13.5.1*, p. *194*
A. Kühnel	(b) **85**, 319 (1978)	*14.4.1*, p. *253*; 19, p. 303; 22.9, p. 522
G. Kühnel	(a) **48**, K63 (1978)	13.4, p. 178; *14.3*, p. *222*; 22.8, p. 506
	(a) **49**, 705 (1978)	13.3, p. 167; *14.3.4*, p. *245*; 22.8, p. 507
	(a) **49**, K205 (1978)	13.4, p. 183; *14.3*, p. *224*; 22.8, p. 507
R. S. Kuindersma	(a) **46**, 687 (1978)	*18.1*, p. *278*; *18.4*, p. *299*; 22.8, p. 504
G. Kuka	(b) **89**, 95 (1978)	*13.1*, p. *158*; *19*, p. *311*; 21.7, p. 407
	(b) **89**, 547 (1978)	*14.1*, p. *211*; 21.7, p. 407

N. A. Kulagin	(b) **90**, K109 (1978)	*13.4*, p. *185*; 20.1, p. 337; 22, p. 421
B. I. Kuliev	(b) **86**, 103 (1978)	14.3.1, p. 226; *15*, p. *259*
N. I. Kulikov	(b) **86**, 83 (1978)	4, p. 57; *13.1*, p. *152*; 21.1, p. 371
S. K. Kulshreshtha	(a) **49**, K91 (1978)	*1.3*, p. *28*; *5*, p. *62*; 18.2, p. 290; 21.1.1, p. 389
N. Kumar	(b) **85**, 163 (1978)	*13*, p. *142*
R. Kumar	(b) **89**, K195 (1978)	*13.4*, p. *182*; 18.1, p. 279; 22.8, p. 507
V. Kumar	(b) **89**, K195 (1978)	*13.4*, p. *182*; 18.1, p. 279; 22.8, p. 507
R. Kümmel	(b) **87**, 99 (1978)	6, p. 70; *14.3.4*, p. *242*; 22.4.1, p. 464
W. Kümmerle	(a) **45**, 171 (1978)	*18.2.1*, p. *293*; 21.1.1, p. 384
A. G. Kundzich	(b) **88**, 797 (1978)	13.3, p. 166; *13.5.1*, p. *193*; *20.3*, p. *346*; 22.4.1, p. 465
L. Künne	(b) **88**, 537 (1978)	4, p. 59; *13*, p. *146*
A. B. Kunz	(b) **87**, 765 (1978)	*13.1*, p. *155*; 22.1.2, p. 433
C. Kunz	(b) **88**, 135 (1978)	*13.1*, p. *156*; 17, p. 270; 20.1, p. 328; 22.1.2, p. 434
Ch. S. Kuo	(b) **87**, 325 (1978)	*5*, p. *61*; 10.2, p. 113; 22.5.3, p. 487
T. B. Kurenkeev	(a) **48**, 281 (1978)	13.4, p. 179; *20.3*, p. *346*; 22.2.3, p. 453
	(a) **49**, K189 (1978)	*20.3*, p. *349*; 22.2.1, p. 448; 22.2.3, p. 454
	(a) **50**, K251 (1978)	13.4, p. 186; *20.3*, p. *351*; 22.2.1, p. 449; 22.2.3, p. 455
J. Kuriata	(b) **87**, K155 (1978)	*19*, p. *308*; 22.5.3, p. 487
K. Kuribayashi	(b) **90**, K31 (1978)	*10.1*, p. *107*; 21.1, p. 380
L. A. Kurochkin	(a) **47**, 677 (1978)	*11*, p. *127*; 19, p. 309; 22.1.1, p. 426
E. Kuroda	(a) **48**, 105 (1978)	*3*, p. *55*; 14.3, p. 222; 22.1.2, p. 434
I. A. Kurova	(a) **45**, 333 (1978)	*7*, p. *80*; *20.3*, p. *341*; 22.2.1, p. 445
A. Kurowska	(b) **87**, 69 (1978)	*13.1*, p. *154*; *20.1*, p. *325*; 22.4.4, p. 472
L. F. Kurtenok	(a) **48**, 323 (1978)	*14.3.4*, p. *243*; 22.1.2, p. 435
M. S. Kushwaha	(b) **87**, 247 (1978)	*6*, p. *70*; 12.2, p. 139; 23, p. 530
	(b) **89**, 583 (1978)	*6*, p. *76*; 8, p. 88; 23, p. 530
S. S. Kushwaha	(b) **87**, 247 (1978)	*6*, p. *70*; 12.2, p. 139; 23, p. 530
	(b) **89**, 583 (1978)	*6*, p. *76*; 8, p. 88; 23, p. 530
E. Kusior	(a) **45**, 537 (1978)	*20.1*, p. *321*; 22.6.1, p. 499
E. F. Kustov	(a) **50**, K13 (1978)	3, p. 56; *14.4.1*, p. *257*; 20, p. 318; 22.8, p. 508
R. Kutner	(b) **89**, K29 (1978)	*9*, p. *93*; 21.6, p. 404
H. Kuzmany	(b) **89**, K139 (1978)	11, p. 130; *20.1*, p. *334*; 22.9, p. 527
V. A. Kuzminikh	(a) **48**, K175 (1978)	11, p. 128; *20.3*, p. *346*; 22.5.2, p. 483
A. I. Kuznetsov	(a) **48**, 287 (1978)	*20.1*, p. *331*; 22.6, p. 495
V. N. Kuznetsov	(b) **85**, 95 (1978)	*13.1*, p. *150*
R. I. Kuznetsova	(a) **50**, 385 (1978)	1.4, p. 34; *10.1*, p. *108*; 21, p. 366
V. G. Kvachadze	(b) **85**, 769 (1978)	6, p. 66; 8, p. 83; *10.2*, p. *109*; 22.5.2, p. 479
G. F. Kventsel	(b) **89**, 645 (1978)	*18.2*, p. *290*
A. Lachter	(b) **90**, 147 (1978)	*20.1*, p. *335*; 22.5, p. 476
C. E. Laciana	(a) **45**, 315 (1978)	*10.1*, p. *100*; 21, p. 353; 21.1, p. 369
	(a) **47**, 425 (1978)	*10.1*, p. *103*; 21, p. 359; 21.1, p. 374
R. Lacroix	(a) **49**, K55 (1978)	*19*, p. *312*; 22, p. 417
S. Lahiry	(a) **46**, 153 (1978)	*18.1*, p. *277*; *19*, p. *305*; 22.9, p. 523
C. S. Lam	(b) **89**, 103 (1978)	*10.2*, p. *117*; *13.4*, p. *179*; 22.5.2, p. 483
D. Landgraf-Dietz	(a) **45**, K125 (1978)	*11*, p. *124*; *14.3.3*, p. *233*; 22.1.2, p. 430
P. T. Landsberg	(a) **50**, 423 (1978)	2, p. 52; 13.4, p. 185; *14.3.4*, p. *247*; 22.1, p. 423
G. Landwehr	(b) **85**, 311 (1978)	*6*, p. *65*; *20.1*, p. *320*; 22.7, p. 501
J. van Landuyt	(a) **45**, 553 (1978)	*1.2*, p. *16*; *1.4*, p. *29*; 21.1, p. 370
	(a) **46**, 479 (1978)	*1.1*, p. *12*; 1.2, p. 17; 22, p. 412
	(a) **47**, 241 (1978)	*1.2*, p. *18*; 1.3, p. 25; 1.4, p. 30; 21.1, p. 373; 21.6, p. 402
	(b) **87**, 419 (1978)	*1.1*, p. *13*; 22.1.1, p. 425; 22.6, p. 494

	(a) **47**, 539 (1978)	*1.2*, p. *19*; 1.3, p. 26; 1.4, p. 31; 21.1, p. 374; 21.6, p. 403
	(a) **48**, 39 (1978)	*1.2*, p. *20*; 1.3, p. 27; 1.4, p. 31; 21.1, p. 375; 21.6, p. 403
	(a) **48**, 383 (1978)	*1.3*, p. *27*; 1.4, p. 32; 21, p. 361; 21.1, p. 376
	(a) **48**, K5 (1978)	*1.4*, p. *31*; 10.2, p. 115; 22, p. 415
	(a) **49**, K217 (1978)	*1.4*, p. *31*; 10.2, p. 115; 22, p. 415
	(a) **50**, 587 (1978)	*1.3*, p. *28*; *1.4*, p. *34*; 10.1, p. 108; 21, p. 366; 21.1, p. 382
I. G. Lang	(b) **85**, 81 (1978)	*6*, p. *64*; *20.3*, p. *341*; 22, p. 408
	(b) **85**, 351 (1978)	*6*, p. *65*; *13.5.1*, p. *189*; 20.1, p. 320
	(b) **90**, K63 (1978)	*6*, p. *78*; *20.1*, p. *336*; 22, p. 420
H. Lange	(b) **90**, 167 (1978)	6, p. 77; 13.1, p. 159; 13.5.1, p. 195; *20.1*, p. *335*; 22.2.4, p. 456
G. Langouche	(b) **89**, K17 (1978)	*5*, p. *62*; 22.1.1, p. 427; 22.1.2, p. 436
V. Lantto	(a) **50**, 611 (1978)	*11*, p. *131*; 21.1, p. 382
D. Lapraz	(a) **50**, 275 (1978)	10.2, p. 121; 13.4, p. 184; *20.3*, p. *350*; 22.6, p. 497
V. M. Laptev	(b) **87**, 441 (1978)	*18.2*, p. *286*
V. Z. Latuta	(a) **50**, 573 (1978)	*1.2*, p. *24*; *1.5*, p. *39*; 22, p. 421
L. D. Laude	(b) **89**, 275 (1978)	11, p. 129; *13.1*, p. *158*; 20.2, p. 339; 22, p. 417
G. Lautenschläger	(b) **85**, K19 (1978)	*13.4*, p. *169*; 21, p. 352; 21.1, p. 368
G. Lautz	(a) **48**, 121 (1978)	*13*, p. *146*; *14.3.4*, p. *243*; 22.1.1, p. 426
	(a) **49**, 111 (1978)	13.4, p. 180; *14.3.1*, p. *228*; *14.3.3*, p. *236*; 22.2.1, p. 448
	(a) **50**, 771 (1978)	13.4, p. 180; *14.3.1*, p. *228*; *14.3.3*, p. *236*; 22.2.1, p. 448
A. G. Lavrushko	(b) **86**, 449 (1978)	*13.5.1*, p. *190*; *20.3*, p. *343*; 22.9, p. 524
C. Lawther	(a) **46**, 245 (1978)	*13.4*, p. *173*; *14.3.1*, p. *226*; 22.4.2, p. 468
	(a) **50**, 491 (1978)	*14.3.3*, p. *238*; *20.3*, p. *350*; 22.4.1, p. 467
D. P. Lazăr	(a) **45**, K145 (1978)	*18.1*, p. *277*; 18.3, p. 295; 21.1, p. 370; 21.1.1, p. 385; 21.4, p. 396
J. Lebech	(b) **87**, 527 (1978)	*13.2*, p. *162*; *14.1*, p. *208*; 21.6, p. 402
A. A. Lebedev	(a) **47**, 507 (1978)	*13.4*, p. *176*; *14.3.1*, p. *227*; 22.1.2, p. 433
L. Lebioda	(a) **49**, K27 (1978)	*1.2*, p. *22*; 22.9, p. 527
J. P. Leburton	(b) **90**, K137 (1978)	*13*, p. *149*; *14.3.4*, p. *246*; 22.2.1, p. 449; 22.2.3, p. 455
J. Leciejewicz	(a) **46**, K101 (1978)	*1.1*, p. *12*; *18.4*, p. *299*; 22.8, p. 504
	(a) **47**, 349 (1978)	*18.2*, p. *287*; *18.4*, p. *300*; 22.8, p. 505
E. W. Lee	(a) **50**, 101 (1978)	*18.3*, p. *296*; 21.1, p. 381; 21.4, p. 398
Jong K. Lee	(a) **46**, 267 (1978)	*10.1*, p. *101*; *12.1*, p. *134*; 21, p. 356
K. O. Lee	(a) **47**, K47 (1978)	*2*, p. *45*; *14.3.1*, p. *227*; 22.1.1, p. 425
Y. H. Lee	(a) **48**, K31 (1978)	*11*, p. *128*; *13.4*, p. *178*; 22.1.2, p. 435
	(b) **90**, 689 (1978)	*6*, p. *78*; 11, p. 131; *20.1*, p. *337*; 22.1.2, p. 439
	(a) **50**, 237 (1978)	10.2, p. 120; *11*, p. *130*; 19, p. 314; 22.1, p. 423
M. Lefeld-Sosnowska	(a) **48**, 565 (1978)	*1.4*, p. *32*; *10.2*, p. *116*; 22.1.2, p. 435
S. Lefrant	(b) **86**, 493 (1978)	*6*, p. *68*; *20.1*, p. *324*; 22.5.2, p. 480
P. Legrand	(a) **48**, 417 (1978)	*6*, p. *75*; *20.1*, p. *331*; 22.6, p. 495
A. Lehmann	(a) **46**, 451 (1978)	2, p. 44; *4*, p. *57*; 22.6, p. 492
G. Lehmann	(b) **85**, K85 (1978)	*13.4*, p. *171*; 21, p. 354
	(b) **86**, 249 (1978)	13.4, p. 172; *14.1*, p. *206*; 18, p. 273; 21, p. 355
	(b) **87**, K21 (1978)	*13.4*, p. *175*; 21, p. 358
S. Lehr	(a) **49**, 83 (1978)	*13.3*, p. *167*; 16, p. 265; 17, p. 270; 22.2.3, p. 453
K. Leibler	(a) **47**, K169 (1978)	*19*, p. *309*; 22.7, p. 502
L. Leibler	(b) **85**, 611 (1978)	13.1, p. 151; *19*, p. *304*; 22.2.4, p. 456; 22.4.4, p. 472

V. N. Leikin	(a) **46**, 365 (1978)	*1.2*, p. *17*; 22.2.1, p. 446; 22.2.2, p. 450; 22.2.3, p. 452
H. Leitz	(a) **49**, 73 (1978)	*1.5*, p. *38*; *2*, p. *49*; 21.1, p. 377; 22.1.1, p. 427
M. Lemaitre-Blaise	(b) **85**, 587 (1978)	*20.1*, p. *321*; *20.3*, p. *342*; 22.8, p. 503
V. Lemos	(b) **88**, 199 (1978)	*6*, p. *73*; *20.1*, p. *329*; 22.6, p. 494
R. Lenk	(b) **86**, 603 (1978)	*1*, p. *9*; 13, p. 144
	(b) **89**, 199 (1978)	*2*, p. *49*; 13, p. 147; 21.1, p. 377; 21.1.1, p. 389
G. Leonhardt	(b) **89**, 619 (1978)	2, p. 50; *13*, p. *148*; *20*, p. *317*; 22.6, p. 496
T. P. Lepetre	(a) **47**, K11 (1978)	*14.3.4*, p. *242*; 22, p. 413
Y. Lépine	(b) **89**, 53 (1978)	*13.5.3*, p. *200*; 22, p. 416
B. Lepley	(a) **46**, K55 (1978)	*14.3.3*, p. *234*; 14.3.4, p. 241; 22.2.2, p. 450
M. Leppihalme	(a) **47**, 245 (1978)	6, p. 71; 13.5.1, p. 191; *14.3.3*, p. *235*; *16*, p. *264*; 22.1.1, p. 425
V. I. Lesin	(b) **90**, K13 (1978)	*13.5.1*, p. *196*; 22.9, p. 528
B. Leśniewska	(a) **50**, K17 (1978)	*1.1*, p. *14*; *18.3*, p. *296*; 22.8.2, p. 520
D. Lesueur	(a) **48**, K123 (1978)	*10.1*, p. *105*; *14.1*, p. *209*; 21.7, p. 407
J. Leteurtre	(a) **50**, K161 (1978)	*10.2*, p. *122*; 11, p. 131; 22.5.2, p. 486
Y. C. Leung	(a) **46**, K107 (1978)	*10.1*, p. *102*; 21, p. 357
V. A. Levdik	(a) **48**, K97 (1978)	*11*, p. *128*; 21.1, p. 375
R. R. Levitskii	(b) **88**, 51 (1978)	*1.2*, p. *20*; *14.4.1*, p. *255*; 22.8.1, p. 514
A. Levstik	(b) **86**, K119 (1978)	*1.2*, p. *17*; 14.4.1, p. 254; *19*, p. *306*; 22.8.1, p. 512
	(a) **50**, 459 (1978)	*14.4.1*, p. *257*; 22.8, p. 508
F. Lévy	(b) **87**, K65 (1978)	*20.1*, p. *326*; 22.3, p. 458
	(b) **90**, 233 (1978)	*13.5.1*, p. *196*; *20.3*, p. *349*; 22.3, p. 459
	(a) **50**, 257 (1978)	*16*, p. *267*; 22.5, p. 476
K. Lewinski	(a) **49**, K27 (1978)	*1.2*, p. *22*; 22.9, p. 527
C. Lhermitte	(a) **50**, 651 (1978)	13.4, p. 186; *16*, p. *267*; 22.4.3, p. 471
Y. Li	(a) **46**, 603 (1978)	*12*, p. *133*
S. Ligenza	(b) **86**, 635 (1978)	13.5.4, p. 201; *18.3*, p. *295*; 22.8.2, p. 518
	(b) **90**, 319 (1978)	*13.4*, p. *183*; *18*, p. *275*; 22.8.2, p. 520
Li-Jen Cheng	see under letter "C"	
S. P. Likhtorovich	(b) **88**, K55 (1978)	*13.1*, p. *157*; 22.2.1, p. 447
Z. Lilienthal	(a) **49**, K1 (1978)	*10.2*, p. *118*; 22.4.3, p. 471
C. A. S. Lima	(b) **86**, 373 (1978)	*14.4.1*, p. *254*; 22.8.1, p. 511
P. J. Lin-Chung	(b) **85**, 743 (1978)	*13.1*, p. *151*; 22, p. 409
M. Lindén	(a) **45**, K167 (1978)	*11*, p. *124*; 22.2, p. 443
U. Lindner	(b) **86**, 621 (1978)	*18.2*, p. *285*; *18.4*, p. *299*
	(b) **89**, 123 (1978)	*13.5*, p. *188*; 18.4, p. 301
E. Linhart	(a) **46**, 511 (1978)	*14.4.1*, p. *254*; 22.8, p. 504
S. Linke	(a) **49**, 569 (1978)	*14.2*, p. *215*
A. Lipiński	(a) **49**, K41 (1978)	*13.4*, p. *181*; *14.3.4*, p. *245*; 22.9, p. 527
B. Lippold	(b) **87**, 535 (1978)	*13.4*, p. *176*; 21.4, p. 397
A. A. Litvin	(a) **49**, 211 (1978)	11, p. 129; *14.3.3*, p. *236*; *20.3*, p. *347*; 22.2.2, p. 451
D. B. Litvin	(b) **85**, 715 (1978)	*13*, p. *143*; *20*, p. *315*; 22, p. 409
V. I. Litvinov	(b) **90**, 423 (1978)	*1.2*, p. *23*; *13*, p. *148*; 14, p. 204; 18, p. 275
J. Livage	(a) **45**, 599 (1978)	*19*, p. *305*; 22.6.1, p. 499
F. M. Livingstone	(a) **46**, 595 (1978)	*14.3.3*, p. *234*; 16, p. 263; 22.1.2, p. 432; 22.4.1, p. 464
J. Llopis	(a) **49**, K9 (1978)	*10.2*, p. *118*; *20.3*, p. *348*; 22.2.2, p. 451; 22.6, p. 496
G. Löbl	(a) **45**, 529 (1978)	*1.1*, p. *11*; 11, p. 124; 21, p. 354
V. A. Lobodyuk	(a) **50**, K127 (1978)	*1.2*, p. *24*; 21.1, p. 383

W. Lochmann	(a) **45**, 423 (1978)	*6*, p. *66*; *13.1*, p. *152*; 20, p. 315; 22.1.1, p. 424; 22.1.2, p. 430
	(b) **89**, 357 (1978)	*13.1*, p. *159*; 20, p. 317; 22.2, p. 444; 22.2.3, p. 454
K. Lochner	(b) **88**, 653 (1978)	*13.1*, p. *157*; *16*, p. *265*; 22.9, p. 526
	(a) **50**, 95 (1978)	*1*, p. *10*; 22.9, p. 528
J. J. Loferski	(b) **89**, 313 (1978)	*13.5.1*, p. *194*; *20.3*, p. *347*; 22.4.4, p. 473
H. Löffler	(a) **48**, 369 (1978)	*1.3*, p. *27*; *10.1*, p. *105*; 14.1, p. 209; 21, p. 361; 21.1, p. 376; 21.3, p. 395
	(a) **48**, K67 (1978)	*1.2*, p. *21*; 10.1, p. 104; 21, p. 361
	(a) **49**, 299 (1978)	*10.1*, p. *106*; 21, p. 363; 21.1, p. 377
	(a) **50**, 123 (1978)	*10.1*, p. *107*; 14.1, p. 212; 21.1, p. 381
F. Loges	(b) **89**, K191 (1978)	*10*, p. *97*; 12, p. 133
F. Lohse	(b) **89**, 109 (1978)	*10.2*, p. *117*; 13.4, p. 180; *20.1*, p. *332*; 22.5.2, p. 483
W. Łojkowski	(a) **45**, 565 (1978)	*1.4*, p. *29*; 10.1, p. 100; 21, p. 354
S. Lokanathan	(b) **88**, 633 (1978)	*5*, p. *62*; 22.8, p. 506
V. M. Lomako	(a) **48**, 263 (1978)	*11*, p. *128*; 13.4, p. 178; 14.3, p. 222; 22.2.1, p. 447
A. Longoni	(b) **89**, 221 (1978)	*10.2*, p. *117*; 13.4, p. 180; *20.3*, p. *347*; 22.5.2, p. 484
J. Lorenc	(b) **88**, 747 (1978)	1.2, p. 21; *12.2*, p. *140*; *14.4.1*, p. *256*; 22.8.1, p. 515
B. Lorenz	(b) **89**, K99 (1978)	*18.1*, p. *279*; 22.6.1, p. 499
N. Lorenzelli	(a) **49**, 643 (1978)	*11*, p. *130*; 14.1, p. 211; 21, p. 363
P. P. Lottici	(b) **87**, 479 (1978)	*2*, p. *45*; *20.1*, p. *327*; 22, p. 413
R. N. Lovyagin	(a) **45**, 521 (1978)	9, p. 91; *14.3.3*, p. *233*; 22.1.2, p. 430
Yu. E. Lozovik	(b) **88**, 359 (1978)	*13*, p. *146*; *14*, p. *203*
W. Ludwig	(a) **49**, 667 (1978)	13.4, p. 182; *14.3.3*, p. *237*; *20.3*, p. *349*; 22.4.2, p. 469
V. I. Lugovoi	(b) **86**, 419 (1978)	*6*, p. *67*; *20.1*, p. *323*; 22.3, p. 457
V. I. Lukashova	(a) **47**, 677 (1978)	*11*, p. *127*; 19, p. 309; 22.1.1, p. 426
L. Lundgren	(a) **49**, K153 (1978)	*3*, p. *56*; 18.2, p. 291; 21.1, p. 379; 21.1.1, p. 389
A. G. Lundin	(b) **90**, K35 (1978)	*19*, p. *314*; 22.5, p. 476
E. Lüscher	(a) **49**, K17 (1978)	*11*, p. *129*; 14.3, p. 224; *19*, p. *312*; 22.1.2, p. 437
Ch. Lushchik	(a) **47**, 667 (1978)	*10.2*, p. *114*; *20.3*, p. *345*; 22.5.2, p. 481
H. Lüth	(b) **89**, 459 (1978)	6, p. 76; *14.3.1*, p. *228*; *20.1*, p. *334*; 22.9, p. 527
G. Luther	(a) **50**, 459 (1978)	*14.4.1*, p. *257*; 22.8, p. 508
F. Lüty	(b) **85**, 693 (1978)	10.2, p. 109; *14*, p. *203*; *20*, p. *315*; 22.5.2, p. 479
	(b) **86**, 127 (1978)	10.2, p. 110; *14*, p. *203*; *20*, p. *316*; 22.5.2, p. 479
	(b) **89**, K73 (1978)	12.2, p. 140; *13.4*, p. *180*; *20.3*, p. *347*; 22.5.2, p. 484
	(b) **89**, K149 (1978)	*6*, p. *76*; *20.1*, p. *334*; 22.8, p. 507
	(b) **90**, 277 (1978)	10.2, p. 120; *14.4*, p. *251*; 22.5.3, p. 488
V. S. Lysenko	(a) **49**, 405 (1978)	11, p. 129; 13.3, p. 167; *14.3.3*, p. *236*; 22.1.2, p. 437
H. Maaref	(b) **88**, 261 (1978)	*13.5.1*, p. *192*; 20.3, p. 345; 22.2.2, p. 450
M. Maček	(a) **49**, 767 (1978)	*1.2*, p. *23*; *1.4*, p. *34*; 21.1, p. 379
R. Mach	(a) **49**, 667 (1978)	13.4, p. 182; *14.3.3*, p. *237*; *20.3*, p. *349*; 22.4.2, p. 469
E. Macherauch	(a) **49**, 473 (1978)	*10.1*, p. *106*; 21.1, p. 378

W. MACIEJEWSKI	(b) **88**, 735 (1978)	*18.2.1*, p. *293*
P. MACKO	(a) **50**, K51 (1978)	2, p. 52; *8*, p. *89*; 11, p. 130; *14.3*, p. *225*; 22, p. 420
A. T. MACRANDER	(a) **46**, 397 (1978)	*1.2*, p. *17*; 23, p. 530
	(a) **48**, 571 (1978)	*1.2*, p. *22*; 8, p. 87; 23, p. 530
P. K. MADDEN	(a) **46**, 639 (1978)	*1.5*, p. *36*; 11, p. 125; 21.1, p. 372
A. MADUEMEZIA	(b) **88**, 805 (1978)	*13.1*, p. *157*; 13.2, p. 162; *20*, p. *317*; 21.2, p. 393
H. MAEDA	(a) **45**, 445 (1978)	*18.2*, p. *283*; 21.1, p. 370; 21.4, p. 396
	(a) **46**, 233 (1978)	*18.2*, p. *284*; 21.1, p. 371; 21.4, p. 396
K. MAEDA	(a) **48**, 587 (1978)	*10.2*, p. *116*; *12.2*, p. *140*; 22.4.3, p. 471
	(a) **49**, 685 (1978)	*2*, p. *50*; 10, p. 97; 12, p. 133
J. MAEGE	(a) **47**, 91 (1978)	13.4, p. 175; *14.3.3*, p. *235*; *20.3*, p. *344*; 22.2.2, p. 450
	(a) **48**, 407 (1978)	*1.4*, p. *32*; 14.3.2, p. 230; 20.3, p. 346; 22.2.2, p. 451
W. MAENHOUT-VAN DER VORST	(a) **45**, 463 (1978)	*16*, p. *262*; 22.4.1, p. 463
	(a) **50**, K175 (1978)	*10.2*, p. *122*; *14.4*, p. *252*; 22.5.1, p. 477
C. J. MAETZ	(b) **85**, 511 (1978)	*6*, p. *66*; 8, p. 83; 22.9, p. 523
	(a) **50**, K117 (1978)	*1.2*, p. *24*; 22.8, p. 508
B. MAFFEO	(b) **87**, 25 (1978)	*10.2*, p. *112*; *20.1*, p. *325*; 22.5.3, p. 487
	(b) **87**, 351 (1978)	*13.4*, p. *175*; *20.1*, p. *326*; 22.5.2, p. 481
L. I. MAGARILL	(b) **85**, K135 (1978)	*14.3.4*, p. *240*; 16, p. 262; 22.2, p. 443
T. J. MAGEE	(a) **49**, 313 (1978)	*1.3*, p. *28*; 1.5, p. 38; *14.3.2*, p. *231*; 21.6, p. 404; 22.1.2, p. 437
S. MAGER	(a) **47**, 157 (1978)	8, p. 85; *10.1*, p. *103*; 14.1, p. 207; 18.2, p. 286; 21.1, p. 373; 21.1.1, p. 387
G. MAGGIPINTO	(a) **46**, 589 (1978)	*14.3.4*, p. *241*; 22.8, p. 405
	(a) **50**, 257 (1978)	*16*, p. *267*; 22.5, p. 476
S. H. EL-MAHALAWY	see under letter "E"	
S. A. MAHMOUD	(a) **45**, K5 (1978)	*18.2*, p. *281*; 21.1, p. 369
H.-D. MAIER	(b) **89**, 587 (1978)	*14.4*, p. *251*; *19*, p. *312*; 22.9, p. 527
K. MAIER	(a) **45**, 253 (1978)	*9*, p. *91*; 10.1, p. 99; 21.6, p. 401
	(a) **50**, 171 (1978)	*9*, p. *94*; 21.6, p. 405
	(a) **50**, 543 (1978)	*9*, p. *95*; 21.6, p. 405
E. MAJERNÍKOVÁ	(b) **86**, 183 (1978)	2, p. 43; *13.1*, p. *152*; *13.5.1*, p. *190*; 18, p. 273; 22, p. 410
E. F. MAKAROV	(b) **85**, 359 (1978)	*13*, p. *142*; *18*, p. *273*
	(b) **85**, 683 (1978)	*13*, p. *143*; *18*, p. *273*
	(b) **87**, K101 (1978)	*5*, p. *61*; 18.3, p. 295; 22.8.2, p. 519
V. P. MAKAROV	(a) **50**, K5 (1978)	*18.2*, p. *292*; 21.1.1, p. 390
M. MAKI	(a) **48**, 105 (1978)	*3*, p. *55*; 14.3, p. 222; 22.1.2, p. 434
A. P. MAKOVETSKAYA	(b) **86**, 419 (1978)	*6*, p. *67*; *20.1*, p. *323*; 22.3, p. 457
G. I. MAKOVETSKII	(a) **47**, 219 (1978)	*18.4*, p. *300*; 22, p. 413
Z. MÁLEK	(b) **85**, 545 (1978)	*8*, p. *83*; 14.2, p. 213; 21, p. 354
J. MALINOWSKI	(a) **47**, 683 (1978)	*10.2*, p. *114*; 13.4, p. 176; 22.5.1, p. 477
R. SH. MALKOVICH	(a) **48**, 241 (1978)	*9*, p. *93*; 22, p. 415
YU. M. MALOZOVSKY	(a) **50**, 723 (1978)	*14.3*, p. *225*; 22.2.3, p. 455
I. A. MALTSEVA	(a) **50**, 139 (1978)	*10.2*, p. *120*; *20.3*, p. *350*; 22.8, p. 508
	(a) **50**, 445 (1978)	14.3, p. 225; *16*, p. *267*; 22.8, p. 508
V. A. MALTSEVA	(a) **46**, 77 (1978)	*13.4*, p. *173*; *20.3*, p. *343*; 22.4.4, p. 472
V. K. MALYUTENKO	(a) **50**, 723 (1978)	*14.3*, p. *225*; 22.2.3, p. 455
A. T. MAMADALIMOV	(a) **47**, 507 (1978)	*13.4*, p. *176*; *14.3.1*, p. *227*; 22.1.2, p. 433
A. MAMEDOV	(a) **50**, 139 (1978)	*10.2*, p. *120*; *20.3*, p. *350*; 22.8, p. 508
	(a) **50**, 445 (1978)	14.3, p. 225; *16*, p. *267*; 22.8, p. 508

A. Manabe	(b) 90, 157 (1978)	6, p. 77; 13.1, p. 159; 13.5.2, p. 199; *20.1*, p. *335*; 22.4.2, p. 469
A. M. Mancini	(a) 48, 293 (1978)	*13.4*, p. *179*; *14.3*, p. *223*; 22.3.1, p. 460
C. Mande	(b) 89, K21 (1978)	*13.3*, p. *167*; *13.5.2*, p. *198*; 20, p. 317; 21.6, p. 404
M. Manfredi	(b) 90, 657 (1978)	13.4, p. 185; *20.1*, p. *337*; *20.3*, p. *350*; 22.5.2, p. 486
C. Manfredotti	(a) 48, 293 (1978)	*13.4*, p. *179*; *14.3*, p. *223*; 22.3.1, p. 460
K. K. Mani	(b) 86, 759 (1978)	*6*, p. *59*; 12.2, p. 138; 22, p. 411
C. V. Manjunath	(b) 87, 395 (1978)	*14.4.1*, p. *254*; *19*, p. *307*; 22.8.1, p. 513
R. Mańka	(b) 87, 53 (1978)	*13.5.3*, p. *200*
C. Manolikas	(a) 47, 459 (1978)	*1.2*, p. *19*; 1.4, p. 31; 22.3, p. 458
R. Mansfield	(b) 87, 163 (1978)	*13.4*, p. *174*; *14.3.4*, p. *242*; 22.2.1, p. 446
N. V. Mansia	(a) 50, K197 (1978)	*14*, p. *204*; *18*, p. *276*; 22.9, p. 529
A. Mansingh	(a) 49, 773 (1978)	*14.4*, p. *251*; 22.6, p. 496
S. Mantl	(b) 89, 69 (1978)	*10.2*, p. *117*; 11, p. 128; 22.1.2, p. 436
S. Mantovani	(a) 50, K123 (1978)	*10.2*, p. *121*; *13.4*, p. *184*; 22.1.2, p. 438
M. Manzel	(a) 48, K161 (1978)	*14.3.2*, p. *231*; *14.3.4*, p. *244*; 22.1.2, p. 436
A. Marais	(a) 45, 503 (1978)	*18.3*, p. *295*; 22.8.2, p. 518
M. J. Marcinkowski	(a) 49, 725 (1978)	*1.4*, p. *34*; 10, p. 97
	(a) 50, 293 (1978)	*10*, p. *97*
	(a) 50, 601 (1978)	*1.4*, p. *35*; *10*, p. *98*
L. Mărculescu	(b) 85, 439 (1978)	*13.4*, p. *170*; *20.3*, p. *341*; 22.5.2, p. 478
	(b) 89, 335 (1978)	*13.4*, p. *170*; *20.3*, p. *341*; 22.5.2, p. 478
S. I. Marennikov	(a) 45, K17 (1978)	*13.4*, p. *170*; 14.4.1, p. 253; 16, p. 262; *20.1*, p. *321*; 22.8.1, p. 510
J. Mareš	(a) 50, K149 (1978)	*2*, p. *53*; *20.3*, p. *360*; 22.6, p. 498
M. Marezio	(a) 48, 581 (1978)	*1.1*, p. *13*; *5*, p. *62*; 22.8.1, p. 515
V. Marinković	(a) 49, 767 (1978)	*1.2*, p. *23*; *1.4*, p. *34*; 21.1, p. 379
K. Marinova	(b) 85, 111 (1978)	*13.5.1*, p. *189*; *20.3*, p. *341*; 22.4.1, p. 463; 22.4.2, p. 468
	(b) 88, K19 (1978)	*13.5.1*, p. *193*; 20.1, p. 329; 22.4.3, p. 470
I. V. Markevich	(a) 50, 767 (1978)	*13.4*, p. *186*; *14.3*, p. *225*; 16, p. 268; 22.4.1, p. 467
R. S. Markiewicz	(b) 90, 585 (1978)	*13.5.1*, p. *197*; 14.3, p. 225; 22.1.1, p. 428
S. Marklund	(b) 85, 673 (1978)	*13.4*, p. *170*; 22.1.1, p. 424; 22.1.2, p. 429
Y. K. Markolenko	(b) 87, K73 (1978)	*1.2*, p. *19*; *13.1*, p. *155*; 22.1, p. 422
I. A. Markova	(b) 86, K59 (1978)	*6*, p. *68*; *18*, p. *273*; 22.7, p. 501
E. Marschall	(b) 90, 241 (1978)	*13.1*, p. *160*; 18.1, p. 280; 18.2, p. 291; 21.1, p. 380
S. A. Marshall	(a) 48, 165 (1978)	*19*, p. *310*; 22.8.2, p. 519
	(b) 89, 267 (1978)	13.4, p. 180; *19*, p. *311*; 22.8.2, p. 520
T. Marshall	(a) 48, 165 (1978)	*19*, p. *310*; 22.8.2, p. 519
S. Martelli	(a) 49, K103 (1978)	*18.2.1*, p. *294*; 21.1, p. 379
J. L. Martin	(a) 49, 675 (1978)	*10.1*, p. *106*; 21, p. 364; 21.1, p. 379
M.-S. Martin	(b) 85, 597 (1978)	3, p. 54; 13.4, p. 170; *19*, p. *304*; 22.4.3, p. 470
A. Martín-Rodero	(b) 88, 591 (1978)	*13.3*, p. *166*; 22.2, p. 443
A. V. Martinaitis	(a) 46, K141 (1978)	10.2, p. 112; *14.3*, p. *220*; 22.4.2, p. 468
	(a) 47, 187 (1978)	*10.2*, p. *113*; *14.3*, p. *220*; 22.4.2, p. 468
	(a) 48, K43 (1978)	*13.4*, p. *178*; *16*, p. *265*; 22.4.3, p. 470
D. I. Marvakov	(b) 86, K7 (1978)	*4*, p. *57*; 12.2, p. 138; 22.1, p. 422; 22.1.1, p. 424; 22.1.2, p. 431
	(b) 88, K161 (1978)	*6*, p. *74*; 22.1, p. 422; 22.1.1, p. 427; 22.1.2, p. 435
	(b) 89, K5 (1978)	*4*, p. *59*; *13.1*, p. *158*; 22, p. 417
	(b) 89, K115 (1978)	*6*, p. *76*; 22.3, p. 459
	(b) 90, K5 (1978)	*4*, p. *59*; *13.1*, p. *160*; 22.3, p. 459

U. Marx	(b) **86**, K93 (1978)	*13*, p. *144*; 20, p. 316; 21.1, p. 372; 21.1.1, p. 386
	(b) **90**, 487 (1978)	*13.1*, p. *160*; *20*, p. *318*; 21.1, p. 381; 21.1.1, p. 390
M. Maryško	(a) **47**, 277 (1978)	*18.2*, p. *286*; 18.3, p. 295; *19*, p. *307*; 22.8.2, p. 519
	(a) **49**, K71 (1978)	18.3, p. 296; *19*, p. *312*; 22.8.2, p. 520
R. V. Masagutova	(a) **50**, 379 (1978)	11, p. 131; 13.4, p. 185; *14.3*, p. *225*; *20.1*, p. *337*; 22.8, p. 508
D. R. Mašović	(b) **89**, K57 (1978)	*13.1*, p. *158*; 21, p. 362
K. Masuda	(b) **87**, 739 (1978)	*13.3*, p. *165*; 21.1, p. 373
	(b) **90**, K179 (1978)	*13.3*, p. *168*; 21, p. 366
T. Masumoto	(a) **48**, 175 (1978)	2, p. 47; *18.2*, p. *289*; 21.1.1, p. 388
	(a) **49**, K163 (1978)	*2*, p. *51*; *18.2*, p. *291*; 21.1.1, p. 390
M. Maszkiewicz	(a) **47**, K77 (1978)	*8*, p. *86*; 21.1, p. 373
W. N. Mathews, Jr.	(b) **90**, 327 (1978)	*14.2*, p. *216*
P. C. Mathur	(a) **45**, K59 (1978)	*2*, p. *43*; *13.4*, p. *170*; 14.3.1, p. 226; 22.1.3, p. 441
	(a) **46**, 321 (1978)	*14.3.3*, p. *234*; 22.2.1, p. 445
	(a) **46**, 717 (1978)	*13.4*, p. *174*; *14.3*, p. *220*; 22.4.2, p. 468
	(a) **47**, 375 (1978)	13.4, p. 176; *14.3.1*, p. *227*; 22.2.3, p. 453
	(a) **47**, 699 (1978)	*13.4*, p. *176*; *14.3*, p. *221*; 22.4.2, p. 468
	(a) **50**, 11 (1978)	6, p. 64; *13.1*, p. *150*; 13.4, p. 169; *14.3*, p. *217*; 20.1, p. 319; 22.2, p. 443
M. Matlak	(b) **88**, 231 (1978)	*13.5.4*, p. *201*; 18.2, p. 288; 22, p. 414
M. Matsuda	(a) **48**, 105 (1978)	*3*, p. *55*; 14.3, p. 222; 22.1.2, p. 434
T. Matsuda	(b) **88**, 273 (1978)	13.2, p. 162; *14.1*, p. *209*; 21, p. 360
S. Matsuo	(b) **86**, 169 (1978)	*13.1*, p. *152*; 14.1, p. 206; 21.7, p. 406
H. Matsuoka	(a) **45**, 305 (1978)	*2*, p. *42*; 21.1.1, p. 384
	(a) **46**, 671 (1978)	*2*, p. *44*; *12.1*, p. *135*; 21.1.1, p. 386
E. Matsuura	(a) **45**, K101 (1978)	*19*, p. *305*; 22.1.2, p. 430
G. Matteucci	(a) **49**, K103 (1978)	*18.2.1*, p. *294*; 21.1, p. 379
R. Mattheis	(a) **45**, 219 (1978)	*14.1*, p. *205*; *14.2*, p. *213*; 21, p. 353
	(a) **49**, K67 (1978)	*2*, p. *50*; 11, p. 129; *18.3*, p. *296*; 21.1, p. 378; 21.4, p. 398
	(a) **49**, K97 (1978)	*14.1*, p. *210*; 21, p. 363
G. E. Matthews	(a) **47**, 617 (1978)	*8*, p. *86*; 10.2, p. 114; 22.5.3, p. 487
P. Matthys	(b) **85**, 639 (1978)	*19*, p. *304*; 22.5.2, p. 479
	(b) **89**, 151 (1978)	*19*, p. *311*; *20.1*, p. *332*; 22.5.2, p. 483
	(b) **89**, K25 (1978)	*19*, p. *311*; 22.5.2, p. 484
L. Mattioli	(b) **86**, K1 (1978)	*6*, p. *68*; *20.1*, p. *323*; 22.8.1, p. 511
V. A. Matveev	(a) **50**, 739 (1978)	*13.3*, p. *168*; 22.1.1, p. 428
E. E. Matyas	(a) **46**, K1 (1978)	*1.2*, p. *17*; *1.3*, p. *25*; 22.7, p. 501
D. Matz	(b) **89**, 53 (1978)	*13.5.3*, p. *200*; 22, p. 416
W. Matz	(b) **87**, 535 (1978)	*13.4*, p. *176*; 21.4, p. 397
B. N. Mavrin	(b) **85**, 381 (1978)	*6*, p. *65*; *20.1*, p. *320*; 22.3, p. 457
	(b) **86**, K49 (1978)	*6*, p. *68*; *20.1*, p. *323*; 22.8, p. 504
E. Maydell-Ondrusz	(a) **46**, 285 (1978)	*11*, p. *125*; 22.1.2, p. 431
M. Mayer	(a) **49**, 473 (1978)	*10.1*, p. *106*; 21.1, p. 378
R. M. Mayer	(a) **47**, 11 (1978)	*9*, p. *91*; *10.2*, p. *109*; 13.4, p. 169; 22.1.3, p. 441
D. Mazilu	(b) **87**, 61 (1978)	*18.2.1*, p. *293*
Ph. Mazot	(a) **48**, 99 (1978)	*10.1*, p. *104*; *12.1*, p. *135*; 21.1, p. 375
J. Mazzaschi	(b) **88**, 261 (1978)	*13.5.1*, p. *192*; 20.3, p. 345; 22.2.2, p. 450
E. Mazzega	(a) **50**, K123 (1978)	*10.2*, p. *121*; *13.4*, p. *184*; 22.1.2, p. 438

G. Della Mea	see under letter "D"	
H.-D. Mecke	(a) **45**, K29 (1978)	*4*, p. *57*; *19*, p. *303*; 22.8, p. 503
	(a) **46**, 547 (1978)	*1.2*, p. *17*; *4*, p. *57*; 19, p. 306; 22.8, p. 504
M. V. Medvedev	(b) **85**, 427 (1978)	*18.2*, p. *282*; *18.4*, p. *298*; 21.1, p. 369; 21.1.1, p. 384
	(b) **86**, 109 (1978)	18.1, p. 277; *18.2*, p. *283*; *18.4*, p. *298*
	(b) **88**, 117 (1978)	*18.2*, p. *288*
G. A. Medvedkin	(a) **45**, K95 (1978)	*16*, p. *262*; 22.8, p. 503
E. E. Meerson	(a) **48**, 31 (1978)	2, p. 47; *14.3.2*, p. *230*; *14.3.4*, p. *243*; 22, p. 415
P. Meertens	(b) **87**, 135 (1978)	*19*, p. *307*; 22.5, p. 475
R. M. Mehra	(a) **45**, K59 (1978)	*2*, p. *43*; *13.4*, p. *170*; 14.3.1, p. 226; 22.1.3, p. 441
H. Mehrer	(a) **45**, 253 (1978)	*9*, p. *91*; 10.1, p. 99; 21.6, p. 401
	(a) **50**, 171 (1978)	*9*, p. *94*; 21.6, p. 405
	(a) **50**, 543 (1978)	*9*, p. *95*; 21.6, p. 405
R. Mehrotra	(b) **88**, K131 (1978)	*6*, p. *74*; 20.1, p. 331; 22.5.2, p. 483
M. Meissner	(a) **49**, 59 (1978)	*8*, p. *88*; 22.1.3, p. 442
W. C. Meixner	(b) **86**, 339 (1978)	*13.3*, p. *164*
M. A. Mekhtiev	(b) **87**, 447 (1978)	*13.4*, p. *175*; *14.3*, p. *221*; 22.4.4, p. 473
K. Melchior	(a) **46**, K119 (1978)	*18.4*, p. *299*; 22.8.1, p. 512
K. Melidis	(a) **47**, K27 (1978)	*14.1*, p. *207*; 21, p. 358; 21.1, p. 373; 21.3, p. 394
G. H. Melkumyan	(a) **48**, 23 (1978)	*7*, p. *81*
H. Mell	(b) **88**, 531 (1978)	*2*, p. *47*; *16*, p. *265*; 22.1.2, p. 435
N. N. Melnik	(b) **85**, 381 (1978)	*6*, p. *65*; *20.1*, p. *320*; 22.3, p. 457
	(b) **87**, K115 (1978)	*6*, p. *72*; *20.1*, p. *328*; 22.3, p. 458
B. V. Melnikov	(b) **90**, K39 (1978)	*13.4*, p. *184*; *13.5.1*, p. *196*; 14.3.4, p. 246; 20.1, p. 336; 22.4.2, p. 469
V. I. Melnikov	(b) **88**, 377 (1978)	*14*, p. *204*; 22.9, p. 526
D. Melville	(a) **48**, 209 (1978)	*18.2*, p. *289*; 21.1, p. 375
A. N. Men	(b) **85**, 85 (1978)	*13.1*, p. *150*
G. Mende	(a) **45**, K121 (1978)	*11*, p. *124*; 22.1.2, p. 430
R. G. Mendiratta	(a) **46**, K127 (1978)	*14.4*, p. *248*; *16*, p. *264*; 22.4, p. 461; 22.5, p. 475
J. V. de Menezes	(a) **45**, K139 (1978)	*5*, p. *60*; 22.8.2, p. 518
H. Menniger	(a) **48**, 407 (1978)	*1.4*, p. *32*; 14.3.2, p. 230; 20.3, p. 346; 22.2.2, p. 451
F. Menzinger	(b) **86**, 77 (1978)	*1.2*, p. *16*; *6*, p. *67*; 21.1, p. 371
T. Merceron	(a) **45**, 503 (1978)	*18.3*, p. *295*; 22.8.2, p. 518
P. Meriel	(a) **46**, K35 (1978)	*18.4*, p. *299*; 21.3, p. 394; 21.4, p. 397
I. A. Meriloo	(a) **48**, 287 (1978)	*20.1*, p. *331*; 22.6, p. 495
J. F. Merklin	(a) **50**, 469 (1978)	*10.2*, p. *121*; *20.1*, p. *338*; 22.5.2, p. 486
A. I. Merkulov	(b) **87**, K141 (1978)	*14.3*, p. *221*; 22.8, p. 505
L. Merten	(b) **89**, 159 (1978)	*13.5*, p. *188*; 14.4, p. 250
J. Mertsching	(b) **87**, 599 (1978)	*13.1*, p. *155*; *18.4*, p. *300*
H. Merz	(b) **89**, 397 (1978)	*17*, p. *270*; 21.2, p. 393
F. Meseguer	(b) **89**, 281 (1978)	*20.1*, p. *332*; 22.5.2, p. 484
M. Yu. Messinev	(a) **47**, 39 (1978)	*5*, p. *61*; 18.4, p. 300; 22.6, p. 493
K. Mettler	(a) **49**, 163 (1978)	*16*, p. *266*; *20.3*, p. *347*; 22.2.1, p. 448
J. Metzdorf	(b) **89**, 203 (1978)	13.1, p. 158; 13.5.1, p. 194; *18*, p. *275*; *20*, p. *317*; 22.1.1, p. 427
H. Metzger	(a) **47**, 631 (1978)	*1.2*, p. *19*; 1.3, p. 26; 21.1, p. 374
V. M. Metzik	(b) **89**, K91 (1978)	*10.2*, p. *117*; 20.1, p. 333; *20.3*, p. *347*; 22.5.2 p. 484
H. Meuth	(a) **46**, 81 (1978)	*20.1*, p. *323*; *20.3*, p. *343*; 22.8, p. 504
A. Meyer	(a) **46**, K35 (1978)	*18.4*, p. *299*; 21.3, p. 394; 21.4, p. 397

B. Meyer	(a) **48**, K145 (1978)	*10.2*, p. *116*; 19, p. 311; 20.1, p. 331; 22.8, p. 506
C. Meyer	(a) **48**, 581 (1978)	*1.1*, p. *13*; *5*, p. *62*; 22.8.1, p. 515
E. N. Mgbenu	(a) **47**, K153 (1978)	*9*, p. *93*; 14.3.1, p. 227; 22.1.1, p. 426
	(b) **88**, 805 (1978)	*13.1*, p. *157*; 13.2, p. 162; *20*, p. *317*; 21.2, p. 393
M. G. Michailov	(a) **50**, 627 (1978)	*1.5*, p. *39*; 22.6, p. 497
K. J. Michalski	(b) **87**, K85 (1978)	*18.2*, p. *287*; 18.4, p. 300; 22.8, p. 505
	(b) **90**, K141 (1978)	*18.2*, p. *292*; 18.4, p. 302
B. Michel	(b) **85**, K49 (1978)	*1.6*, p. *40*; *10*, p. *96*
	(b) **89**, K179 (1978)	*1.6*, p. *41*; *10*, p. *97*
H. Michel	(b) **85**, 195 (1978)	*6*, p. *64*; *13.4*, p. *169*
M. E. Michel-Beyerle	(b) **85**, 45 (1978)	*13.5.1*, p. *189*; 20.3, p. 341; 22.9, p. 522
	(b) **85**, 473 (1978)	*13.5.1*, p. *189*; *20.3*, p. *342*; 22.9, p. 522
	(b) **89**, 261 (1978)	*13.5.1*, p. *194*; 20.1, p. 332; 22.9, p. 527
S. Mieke	(a) **50**, K165 (1978)	11, p. 131; *12.2*, p. *141*; *20.3*, p. *351*; 22.5.2, p. 486
A. Migdał-Mikuli	(a) **47**, 57 (1978)	*1.2*, p. *18*; 8, p. 85; 22.8, p. 505
D. Mihalache	(b) **87**, 61 (1978)	*18.2.1*, p. *293*
T. Mihara	(a) **50**, 165 (1978)	*2*, p. *52*; 21, p. 365
V. I. Mikhailov	(a) **48**, K169 (1978)	*2*, p. *48*; *10.2*, p. *117*; 22, p. 416
E. Mikuli	(a) **47**, 57 (1978)	*1.2*, p. *18*; 8, p. 85; 22.8, p. 505
A. Milchev	(b) **87**, 227 (1978)	*4*, p. *58*; 13, p. 145; 22.2, p. 443
	(b) **87**, 473 (1978)	*4*, p. *58*; 13, p. 145; 22.2, p. 443
	(b) **90**, 679 (1978)	*13.1*, p. *161*; *14.4*, p. *252*; 22.1.1, p. 428; 22.1.2, p. 439; 22.2.1, p. 449; 22.4.2, p. 469
F. Milia	(b) **86**, K119 (1978)	*1.2*, p. *17*; 14.4.1, p. 254; *19*, p. *306*; 22.8.1, p. 512
	(a) **46**, 239 (1978)	*10.2*, p. *111*; *19*, p. *306*; 22.9, p. 524
K. M. Miller	(b) **86**, 93 (1978)	*10.1*, p. *101*; 12.1, p. 134; 21.1, p. 371
W. A. Miller	(a) **49**, K31 (1978)	*2*, p. *49*; 21.6, p. 405
M. Milleville	(b) **89**, 495 (1978)	2, p. 50; *14.3.1*, p. *228*; *16*, p. *266*; 22.1.2, p. 437
B. Milsch	(b) **89**, 241 (1978)	13.4, p. 180; *14.4.1*, p. *256*; *19*, p. *311*; 22.8.1, p. 515
A. Minafra	(a) **46**, 589 (1978)	*14.3.4*, p. *241*; 22.8, p. 504
	(a) **50**, 257 (1978)	*16*, p. *267*; 22.5, p. 476
P. Minnhagen	(b) **85**, 135 (1978)	*13*, p. *142*; 17, p. 269; 20, p. 315; 21, p. 352
L. C. M. Miranda	(b) **86**, 373 (1978)	*14.4.1*, p. *254*; 22.8.1, p. 511
N. F. Miron	(a) **48**, K97 (1978)	*11*, p. *128*; 21.1, p. 375
A. G. Mironov	(b) **90**, 535 (1978)	*13.1*, p. *160*; *20.1*, p. *337*; 22, p. 420
L. V. Mironova	(a) **46**, 467 (1978)	*14.3.1*, p. *226*; *14.3.4*, p. *241*; 22.1.2, p. 432; 22.6, p. 492
N. Mirowska	(b) **88**, K153 (1978)	*13.1*, p. *158*; *20.1*, p. *331*; 22.3, p. 458
K. A. Mirza	(a) **47**, K63 (1978)	13.3, p. 165; 14.4.1, p. 255; *20.3*, p. *344*; 22.8.1, p. 513
K. Míšek	(b) **86**, K31 (1978)	*7*, p. *80*; 14.2, p. 213; 21, p. 355
J. Misiewicz	(b) **88**, K153 (1978)	*13.1*, p. *158*; *20.1*, p. *331*; 22.3, p. 458
H. Misiorek	(a) **47**, K137 (1978)	*8*, p. *87*; *10.1*, p. *104*; 21, p. 360; 21.1, p. 375
	(a) **50**, K261 (1978)	*8*, p. *87*; *10.1*, p. *104*; 21, p. 360; 21.1, p. 375
M. Mitera	(a) **49**, K163 (1978)	*2*, p. *51*; *18.2*, p. *291*; 21.1.1, p. 390
A. V. Mitin	(b) **87**, 433 (1978)	*7*, p. *80*; *18.2*, p. *286*
V. V. Mitin	(a) **48**, 323 (1978)	*14.3.4*, p. *243*; 22.1.2, p. 435
A. A. Mitrofanov	(b) **90**, 423 (1978)	*1.2*, p. *23*; *13*, p. *148*; 14, p. 204; 18, p. 275
K. P. Mitrofanov	(a) **48**, 183 (1978)	*1*, p. *9*; 5, p. 61; *14.4.1*, p. *255*; 18.2, p. 289; 18.4, p. 301; 19, p. 310; 22.8.1, p. 514
T. I. Mitrofanova	(b) **86**, K155 (1978)	13.5.1, p. 191; *16*, p. *263*; 22.4.1, p. 464

A. Mitsuishi	(b) **90**, 157 (1978)	6, p. 77; 13.1, p. 159; 13.5.2, p. 199; *20.1*, p. *335*; 22.4.2, p. 469
E. Mlitzke	(a) **47**, K181 (1978)	*13.4*, p. *177*; *20.3*, p. *345*; 22, p. 414
D. Möckel	(a) **49**, 103 (1978)	2, p. 49; *14.1*, p. *210*; *14.3.1*, p. *228*; 21.6, p. 404; 21.7, p. 407
M. Mocker	(b) **90**, 197 (1978)	*13.1*, p. *159*; 22, p. 419
S. Modesti	(b) **85**, 445 (1978)	*13.1*, p. *151*; 20.1, p. 321; 21, p. 354; 21.6, p. 401
F. Mohsen	(a) **45**, K5 (1978)	*18.2*, p. *281*; 21.1, p. 369
H. Möhwald	(a) **50**, 131 (1978)	*1.2*, p. *24*; 6, p. 78; 20.1, p. 336; 22.9, p. 528
U. V. Moish	(a) **45**, K159 (1978)	*10.1*, p. *101*; 21.1.1, p. 385
A. M. Moisin	(a) **47**, 305 (1978)	*18*, p. *274*; 21.1, p. 373; 21.1.1, p. 387
I. Mojzes	(a) **47**, K183 (1978)	*14.3.4*, p. *243*; 22.2.1, p. 447
L. P. Mokhracheva	(b) **87**, 49 (1978)	*4*, p. *58*; *13*, p. *145*; 22.3, p. 457
A. D. Mokrushin	(b) **90**, 99 (1978)	6, p. 77; *13*, p. *148*; 21, p. 364
W. Moldenhauer	(a) **47**, K99 (1978)	*8*, p. *86*; *14.3.4*, p. *242*; 22.6, p. 493
	(a) **47**, K129 (1978)	*1.2*, p. *20*; *14.3*, p. *221*; 22.6, p. 494
M. Moldovanova	(a) **48**, K193 (1978)	*14.3*, p. *223*; 15, p. 261; 22, p. 416
E. Mollwo	(b) **89**, 381 (1978)	13.4, p. 181; *20.1*, p. *333*; 22.4, p. 462
	(b) **90**, 33 (1978)	13.4, p. 183; *20.1*, p. *335*; 22.4, p. 462
B. V. Molotilov	(a) **50**, K5 (1978)	*18.2*, p. *292*; 21.1.1, p. 390
P. M. Mooney	(a) **48**, K31 (1978)	*11*, p. *128*; *13.4*, p. *178*; 22.1.2, p. 435
P. Moravec	(a) **48**, K43 (1978)	*13.4*, p. *178*; *16*, p. *265*; 22.4.3, p. 470
A. Moreau	(a) **50**, 41 (1978)	*1.3*, p. *28*; 14.1, p. 212; 21.1, p. 380
N. I. Moreva	(b) **87**, 535 (1978)	*13.4*, p. *176*; 21.4, p. 397
J.-M. Moret	(a) **49**, K55 (1978)	*19*, p. *312*; 22, p. 417
Z. Morlin	(b) **89**, 235 (1978)	1.2, p. 22; *14.4*, p. *250*; *20.1*, p. *332*; 22.5.2, p. 484; 22.5.4, p. 489
A. J. Morton	(a) **47**, 293 (1978)	*1.3*, p. *26*; *1.4*, p. *30*; 10.1, p. 103; 21.1, p. 373
P. Moser	(a) **46**, 89 (1978)	*10.1*, p. *101*; 11, p. 125; 21.1.1, p. 385
	(a) **50**, 617 (1978)	*10.1*, p. *108*; *12.1*, p. *137*; 21.1.1, p. 391
A. K. Moskalev	(a) **50**, K157 (1978)	*14.4.1*, p. *257*; *19*, p. *314*; 22.8, p. 509
V. N. Moskalev	(a) **45**, 71 (1978)	*18.2*, p. *281*; 21.1, p. 368; 21.4, p. 396
A. V. Moskalonov	(a) **49**, 231 (1978)	2, p. 49; *13.4*, p. *181*; 19, p. 312; *20.3*, p. *347*; 22, p. 417
N. A. Moskvina	(b) **90**, K39 (1978)	*13.4*, p. *184*; *13.5.1*, p. *196*; 14.3.4, p. 246; 20.1, p. 336; 22.4.2, p. 469
S. C. Moss	(a) **47**, 631 (1978)	*1.2*, p. *19*; 1.3, p. 26; 21.1, p. 374
Y. Motohashi	(a) **50**, 335 (1978)	*10.1*, p. *107*; 21.7, p. 407
	(a) **50**, 449 (1978)	*10.1*, p. *108*; *12.1*, p. *136*; 21, p. 366
V. Chandra Mouli	(b) **86**, K127 (1978)	*19*, p. *306*; 22.8.1, p. 512
B. Movaghar	(b) **88**, 531 (1978)	*2*, p. *47*; *16*, p. *265*; 22.1.2, p. 435
E. A. Movchan	(a) **48**, 323 (1978)	*14.3.4*, p. *243*; 22.1.2, p. 435
C. A. Moyer	(b) **87**, 287 (1978)	*14.1*, p. *207*; 21.1, p. 372; 21.6, p. 402
P. E. Mozol	(b) **86**, K155 (1978)	13.5.1, p. 191; *16*, p. *263*; 22.4.1, p. 464
E. Mrosan	(b) **85**, K19 (1978)	*13.4*, p. *169*; 21, p. 352; 21.1, p. 368
	(b) **85**, K85 (1978)	*13.4*, p. *171*; 21, p. 354
	(b) **86**, 249 (1978)	*13.4*, p. *172*; *141*, p. *206*; 18, p. 273; 21, p. 355
	(b) **87**, K21 (1978)	*13.4*, p. *175*; 21, p. 358
	(b) **89**, 553 (1978)	*13*, p. *148*; 21, p. 363
	(b) **89**, K175 (1978)	*13.1*, p. *159*; 21, p. 363
	(b) **90**, K95 (1978)	*13.4*, p. *184*; *14.1*, p. *212*; 21, p. 365
B. Mrygoń	(b) **85**, 387 (1978)	*18.2*, p. *281*
	(b) **90**, K99 (1978)	*1.2*, p. *24*; *8*, p. *89*; 21.1, p. 380
J. Mucha	(a) **48**, 221 (1978)	*8*, p. *87*; 21, p. 360
E. Mugeński	(b) **90**, 657 (1978)	13.4, p. 185; *20.1*, p. *337*; *20.3*, p. *350*; 22.5.2 p. 486

H. MUGHRABI	(a) **47**, 479 (1978)	*1.4*, p. *31*; *10*, p. *96*
	(a) **49**, 539 (1978)	*10.1*, p. *106*; 12.1, p. 136; 21.1.1, p. 389
K. YU. MUKHIN	(a) **50**, 213 (9978)	*10.2*, p. *120*; 22.1.2, p. 438
P. MUKHOPADHYAY	(a) **46**, K5 (1978)	*10.1*, p. *101*; 21, p. 356
B. G. MULIMANI	(b) **85**, K33 (1978)	*13.2*, p. *162*; 21.1, p. 368
B. MÜLLER	(a) **48**, 533 (1978)	*14.3.1*, p. *227*; *14.3.4*, p. *244*; 22, p. 416
CH. MÜLLER	(b) **87**, 129 (1978)	*13.1*, p. *154*; 20, p. 316; 21, p. 357; 21.1.1, p. 386
D. MÜLLER	(b) **88**, 99 (1978)	*14.4.1*, p. *255*; *19*, p. *309*; 22.8.1, p. 514
	(b) **89**, 587 (1978)	*14.4*, p. *251*; *19*, p. *312*; 22.9, p. 527
G. MÜLLER	(a) **49**, 279 (1978)	*9*, p. *94*; *11*, p. *129*; 22.2.4, p. 456
H. MÜLLER (Jena)	(b) **89**, 553 (1978)	*13*, p. *148*; 21, p. 363
H. MÜLLER (Braunschweig)	(b) **89**, K175 (1978)	*13.1*, p. *159*; 21, p. 363
	(a) **49**, 255 (1978)	*1.5*, p. *38*; *2*, p. *49*; 22.1.1, p. 427
H.-R. MÜLLER	(b) **87**, 203 (1978)	*2*, p. *44*; 3, p. 54; *18.2.1*, p. *293*; 21.1, p. 372; 21.4, p. 397
	(a) **47**, 523 (1978)	*1.5*, p. *37*; *2*, p. *45*; 18.3, p. 296; 21.1, p. 374; 21.4, p. 398
	(a) **49**, K63 (1978)	*2*, p. *50*; *18.3*, p. *296*; 21.1, p. 378; 21.4, p. 398
	(a) **50**, 537 (1978)	*18.3*, p. *297*; 21.1, p. 382; 21.4, p. 399
J. MÜLLER	(a) **47**, 489 (1978)	*14.3.3*, p. *235*; *20.3*, p. *345*; 22.2.1, p. 447
K. MÜLLER	(a) **49**, 39 (1978)	*1.6*, p. *41*; 21, p. 362; 21.6, p. 404; 22.1.2, p. 436
M. MÜLLER	(a) **48**, K51 (1978)	*1.2*, p. *21*; 5, p. 62; 21.1.1, p. 388
	(a) **50**, K189 (1978)	*14.3.3*, p. *239*; 22.9, p. 529
P. MÜLLER	(a) **49**, K75 (1978)	1.2, p. 22; *14.2*, p. *215*
R. MÜLLER	(b) **87**, K53 (1978)	*19*, p. *307*; 22.9, p. 525
W. MÜLLER	(b) **86**, 205 (1978)	13.1, p. 152; *14.3.4*, p. *241*; 20.1, p. 322; *20.3*, p. *343*; 22.2.3, p. 452
M. W. MULLER	(b) **86**, 345 (1978)	*1*, p. *9*; *18.3*, p. *295*; 22.8.2, p. 518
G. MÜLLER-VOGT	(a) **49**, 145 (1978)	1.3, p. 27; *3*, p. *55*; 22.8.1, p. 515
M. MÜLLNER	(a) **50**, K117 (1978)	*1.2*, p. *24*; 22.8, p. 508
G. MUNSCHY	(b) **88**, 713 (1978)	*13.4*, p. *178*; *13.5.1*, p. *193*; 14.3, p. 223; 22.2.1, p. 447
I. MUNTEANU	(a) **46**, K17 (1978)	*13.3*, p. *165*; 14.3.3, p. 234; 22, p. 411
F. M. MUNTYANU	(b) **86**, K71 (1978)	*15*, p. *259*; 21.7, p. 406
P. MUNZ	(a) **49**, 437 (1978)	*16*, p. *266*; 22, p. 418
T. MURA	(a) **48**, 447 (1978)	*10*, p. *96*; 12, p. 133
S. MURALIDHARA RAO		see under letter "R"
A. MURASIK	(a) **46**, K75 (1978)	*18.2*, p. *284*; *18.4*, p. *299*; 21, p. 356; 21.4, p. 397
P. MURILO OLIVEIRA		see under letter "O"
R. MURRI	(a) **48**, 293 (1978)	*13.4*, p. *179*; *14.3*, p. *223*; 22.3.1, p. 460
V. G. KRISHNA MURTY	(a) **45**, 83 (1978)	12.2, p. 138; *20*, p. *315*; 22.5.4, p. 489
M. I. MUSATOV	(a) **48**, 287 (1978)	*20.1*, p. *331*; 22.6, p. 495
H. E. MÜSER	(a) **49**, 51 (1978)	*1.2*, p. *22*; *14.4.1*, p. *256*; 20, p. 317; 22.8.1, p. 515
	(a) **50**, 459 (1978)	*14.4.1*, p. *257*; 22.8, p. 508
A. M. MUTAL	(b) **86**, K7 (1978)	*4*, p. *57*; 12.2, p. 138; 22.1, p. 422; 22.1.1, p. 424; 22.1.2, p. 431
	(b) **86**, K139 (1978)	*1.1*, p. *11*; 1.2, p. 17; 22.2.4, p. 456
	(b) **87**, K77 (1978)	*13.1*, p. *155*; 22.7, p. 502
W. MYCIELSKI	(a) **49**, K41 (1978)	*13.4*, p. *181*; *14.3.4*, p. *245*; 22.9, p. 527
S. MYHRA	(a) **49**, 285 (1978)	*10.2*, p. *118*; *14.3*, p. *224*; 15, p. 261; 22.2.3, p. 454
E. M. NADGORNYI	(a) **46**, 357 (1978)	*10.2*, p. *111*; 22.5.2, p. 480

S. Nadolski	(a) **46**, K47 (1978)	*18.4*, p. *299*; *19*, p. *306*; 22.8.1, p. 512
	(a) **49**, K193 (1978)	*18.4*, p. *301*; *19*, p. *313*; 22.8.2, p. 520
R. Nagarajan	(b) **89**, K79 (1978)	*5*, p. *62*; *18.1*, p. *279*; 22.6.1, p. 499
S. Nagel	(b) **87**, 117 (1978)	*13.1*, p. *154*; *20.1*, p. *326*; 22, p. 413
J. Nahum	(b) **85**, 629 (1978)	*13.5.1*, p. *190*; *20.3*, p. *342*; 22.4.3, p. 470
E. K. Naimi	(b) **85**, K43 (1978)	*14.2*, p. *213*
	(a) **46**, 351 (1978)	*10.2*, p. *111*; *12.2*, p. *138*; 22.5.2, p. 480
	(a) **46**, 411 (1978)	*10.1*, p. *101*; *12.1*, p. *135*; 21, p. 357; 21.1, p. 372; 21.3, p. 394
M. Naka	(a) **49**, K163 (1978)	*2*, p. *51*; *18.2*, p. *291*; 21.1.1, p. 390
K. Nakagawa	(a) **48**, 587 (1978)	*10.2*, p. *116*; *12.2*, p. *140*; 22.4.3, p. 471
Y. Nakao	(b) **89**, 663 (1978)	*13.4*, p. *181*; *20.3*, p. *348*; 22.4.1, p. 466; 22.4.2, p. 469
T. Nakayama	(b) **88**, 623 (1978)	*13.5.1*, p. *193*; 20.3, p. 346; 22.9, p. 526
R. S. Nakhmanson	(a) **46**, 59 (1978)	*14.3.2*, p. *230*; *14.3.3*, p. *233*; 16, p. 263; 22.1.1, p. 425; 22.1.2, p. 431
S. Nanao	(b) **90**, K31 (1978)	*10.1*, p. *107*; 21.1, p. 380
J. Narayan	(a) **45**, 625 (1978)	*10.2*, p. *110*; 22.6, p. 491
	(a) **46**, 281 (1978)	*1.4*, p. *29*; *10.1*, p. *101*; 22.6, p. 492
G. H. Narayanan	(a) **46**, 627 (1978)	*1.4*, p. *30*; *10.2*, p. *112*; 14.3, p. 220; 22.2.1, p. 446
K. Narayandas	(a) **48**, K71 (1978)	10.1, p. 104; *14.1*, p. *209*; 21.6, p. 403
	(a) **49**, K213 (1978)	1.2, p. 23; 2, p. 51; *14.1*, p. *211*; 21.6, p. 405; 22.1.3, p. 442
F. S. Nasredinov	(a) **45**, 11 (1978)	*5*, p. *60*; *11*, p. *123*; 22.5, p. 474; 22.8, p. 503; 22.9, p. 522
R. J. Nastaşi-Andrews	(b) **88**, 173 (1978)	*13.1*, p. *156*; *20.1*, p. *329*; 21.1, p. 375
P. Nath	(a) **48**, 257 (1978)	*1.5*, p. *37*; *14.3.1*, p. *227*; 22.1.1, p. 427
	(a) **48**, 345 (1978)	*1.5*, p. *37*; 10.2, p. 116; 22.6, p. 495
	(a) **49**, 379 (1978)	1.4, p. 33; *1.5*, p. *38*; 2, p. 49; 22.1.1, p. 427
Prem Nath	(a) **45**, 665 (1978)	*1.5*, p. *36*; 22.4.4, p. 472
	(a) **46**, 723 (1978)	*1.5*, p. *36*; *14.3.1*, p. *226*; 16, p. 263; 22.4.4, p. 472
R. Nath	(a) **46**, K127 (1978)	*14.4*, p. *248*; *16*, p. *264*; 22.4, p. 461; 22.5, p. 475
I. Natkaniec	(b) **85**, 331 (1978)	*6*, p. *65*; 22.9, p. 522
Th. Nattermann	(b) **85**, 291 (1978)	1.2, p. 15; *14.4.1*, p. *253*; 22.9, p. 522
	(b) **90**, 105 (1978)	*13*, p. *148*; *18.2*, p. *291*
A. Naurizbaev	(b) **87**, K1 (1978)	13.1, p. 155; *20.1*, p. *326*; 22.2.1, p. 446
	(a) **48**, K115 (1978)	*13.4*, p. *179*; 20.1, p. 331; 22.2.3, p. 453
	(a) **49**, 183 (1978)	*13.4*, p. *180*; *20.1*, p. *333*; 22.2.1, p. 448; 22.2.3, p. 454
W. Nawrocka	(b) **88**, 727 (1978)	*18.2*, p. *289*
W. Nazarewicz	(b) **86**, K135 (1978)	*6*, p. *69*; *20.1*, p. *325*; 22.4.4, p. 472
	(b) **87**, K25 (1978)	*6*, p. *70*; *20.1*, p. *326*; 22.4.1, p. 464
	(b) **90**, 649 (1978)	*6*, p. *78*; 10.2, p. 121; *20.1*, p. *337*; 22.4.1, p. 466
A. N. Nazarov	(a) **49**, 405 (1978)	11, p. 129; 13.3, p. 167; *14.3.3*, p. *236*; 22.1.2, p. 437
E. Nebauer	(a) **46**, K71 (1978)	*9*, p. *92*; 22.2.2, p. 450
	(a) **48**, K109 (1978)	*2*, p. *48*; *9*, p. *93*; 14.3.4, p. 244; 22, p. 416
	(a) **49**, 249 (1978)	*9*, p. *94*; 10.2, p. 118; 22.2.2, p. 451
P. Nédellec	(a) **90**, 207 (1978)	*6*, p. *77*; *14.2*, p. *215*; 21, p. 364
A. Neidhardt	(b) **49**, K125 (1978)	*9*, p. *94*; 21.1, p. 379
V. Nekvasil	(b) **87**, 317 (1978)	*13.4*, p. *175*; *20.1*, p. *326*; 22.8.2, p. 519

H. Nelkowski	(a) **49**, K13 (1978)	1.5, p. 39; 3, p. 56; *14.3.3*, p. *236*; *20.3*, p. *348*; 22.4.1, p. 465; 22.8, p. 507
V. V. Nemoshkalenko	(b) **87**, 129 (1978)	*13.1*, p. *154*; 20, p. 316; 21, p. 357; 21.1.1, p. 386
	(b) **88**, K83 (1978)	*13.1*, p. *157*; 21.4, p. 398
S. Neov	(a) **47**, 743 (1978)	*2*, p. *46*; 4, p. 58; 22.6, p. 494
S. Nešpůrek	(a) **46**, 273 (1978)	13.4, p. 173; *14.3.4*, p. *241*; 22.9, p. 524
	(a) **49**, K149 (1978)	13.4, p. 183; *14.3.4*, p. *245*; 22, p. 419
N. M. Nesterenko	(a) **46**, 577 (1978)	*1.2*, p. *18*; 22.8.1, p. 512
V. I. Nesterov	(b) **90**, 741 (1978)	*8*, p. *90*; *18.2*, p. *292*; 21.1, p. 382
H. Neubrand	(b) **86**, 269 (1978)	13.4, p. 172; *19*, p. *305*; 22.1.2, p. 430
	(b) **90**, 301 (1978)	13.4, p. 183; *19*, p. *313*; 22.1.2, p. 437
H. Neuhäuser	(a) **45**, 89 (1978)	*10.1*, p. *99*; 21.1, p. 369
	(a) **49**, 445 (1978)	*10.1*, p. *106*; 11, p. 130; 21.1, p. 378
G. Neumann	(b) **85**, 179 (1978)	*1.2*, p. *15*; 2, p. 42; *6*, p. *64*; 20.1, p. 319; 22.8.1, p. 510
	(b) **86**, 615 (1978)	1.2, p. 17; 6, p. 69; *20.1*, p. *324*; 22.8.1, p. 512
H. Neumann	(b) **85**, K11 (1978)	*13.1*, p. *150*; 17, p. 269; 22.7, p. 501
	(b) **85**, K57 (1978)	*13.1*, p. *150*; *20.1*, p. *320*; 22.8, p. 503
	(b) **87**, K65 (1978)	*20.1*, p. *326*; 22.3, p. 458
	(a) **47**, K35 (1978)	*20.1*, p. *327*; 22.2.1, p. 446
	(a) **49**, 629 (1978)	*1.5*, p. *39*; 13.4, p. 182; 14.3.1, p. 228; 20.1, p. 335; 20.3, p. 348; 22.2.4, p. 456
	(a) **49**, K139 (1978)	*14.3.3*, p. *237*; 22.2.2, p. 451
M. Neumann	(b) **90**, 525 (1978)	*13.1*, p. *160*; *17*, p. *270*; 21, p. 365
Tran Ngoc	(a) **50**, 745 (1978)	13.4, p. 186; *20.1*, p. *338*; *20.2*, p. *340*; 22.8.1, p. 516
Nguyen van Dong	see under letter "D"	
Nguyen Nhu Dat	see under letter "D"	
Nguyen Vinh Quang	see under letter "Q"	
Nguyen Van Trong	see under letter "T"	
P. H. Nguyen	(a) **46**, K55 (1978)	*14.3.3*, p. *234*; 14.3.4, p. 241; 22.2.2, p. 450
Hoang Xuan Nguyen	(b) **87**, K65 (1978)	*20.1*, p. *326*; 22.3, p. 458
J. R. Nicholls	(a) **49**, K85 (1978)	*10.1*, p. *106*; 21.1, p. 378; 21.1.1, p. 389
A. Nicula	(b) **85**, 75 (1978)	*19*, p. *303*; 22.8.1, p. 510
J. Nieber	(a) **46**, K131 (1978)	*1.1*, p. *12*; 22.1.2, p. 432
K. D. Nierzewski	(b) **86**, 697 (1978)	*10.2*, p. *112*; 20.1, p. 324; 22.5.2, p. 480
J. Nihoul	(a) **45**, 439 (1978)	*10.1*, p. *100*; *11*, p. *124*; 21.1, p. 370
	(b) **87**, K61 (1978)	6, p. 71; *10.1*, p. *102*; 21, p. 358
	(b) **88**, K93 (1978)	*4*, p. *58*; 6, p. 74; 21, p. 360
G. Nihoul-Boutang	(a) **45**, 639 (1978)	*1.1*, p. *11*; 1.4, p. 29; 22.6, p. 491
	(a) **47**, 311 (1978)	*1.1*, p. *12*; 1.4, p. 30; 22.6, p. 493
K. G. Nikiforov	(a) **49**, K197 (1978)	*14.3.4*, p. *246*; 22.8.2, p. 520
V. I. Nikitenko	(a) **46**, 379 (1978)	*10.2*, p. *111*; 22.1.2, p. 431
	(a) **50**, 213 (1978)	*10.2*, p. *120*; 22.1.2, p. 438
S. Nikitine	(b) **87**, 261 (1978)	*13.5.1*, p. *191*; *20.1*, p. *326*; 22.6, p. 493
J. S. Nikolov	(b) **86**, 431 (1978)	*6*, p. *68*; *20.1*, p. *323*; 22.9, p. 523
L. Nikolova	(a) **47**, 213 (1978)	*10.2*, p. *113*; *20.1*, p. *326*; 22.5.2, p. 481
H. T. G. Nilsson	(a) **48**, 345 (1978)	*1.5*, p. *37*; 10.2, p. 116; 22.6, p. 495
G. G. Nilsson	(b) **86**, 653 (1978)	*6*, p. *69*; *8*, p. *84*; 22.5.2, p. 480
N. G. Nilsson	(a) **50**, K43 (1978)	*14.3*, p. *224*; 22, p. 420
P. O. Nilsson	(b) **90**, K69 (1978)	*13.1*, p. *160*; 21.1, p. 380
M. Nimis	(b) **85**, K27 (1978)	*10.2*, p. *109*; *13.4*, p. *170*; 22.5.2, p. 478
	(b) **89**, 221 (1978)	*10.2*, p. *117*; 13.4, p. 180; *20.3*, p. *347*; 22.5.2, p. 484
	(b) **89**, K107 (1978)	*10.2*, p. *109*; *13.4*, p. *170*; 22.5.2, p. 478
M. Nippus	(b) **86**, 257 (1978)	*6*, p. *67*; *20.1*, p. *322*; 22.5, p. 474
V. N. Nishanov	(b) **88**, 359 (1978)	*13*, p. *146*; *14*, p. *203*

M. Nishida	(b) **85**, K69 (1978)	*13.3*, p. *164*; 22.1.1, p. 424
T. Nishimoto	(b) **86**, 169 (1978)	*13.1*, p. *152*; 14.1, p. 206; 21.7, p. 406
Y. Nishina	(b) **88**, 289 (1978)	*13.5.1*, p. *193*; 20.1, p. 329; 20.3, p. 345; 22.5, p. 475
	(b) **89**, K145 (1978)	*17*, p. *270*; 22.2.1, p. 448; 22.5, p. 476
N. Nisticò	(b) **85**, 445 (1978)	*13.1*, p. *151*; 20.1, p. 321; 21, p. 354; 21.6, p. 401
M. A. Nizametdinova	(b) **87**, K115 (1978)	*6*, p. *72*; *20.1*, p. *328*; 22.3, p. 458
S. Nizioł	(a) **45**, 591 (1978)	*18.2*, p. *283*; *18.4*, p. *298*; 22.8, p. 503
R. A. Noack	(b) **90**, 615 (1978)	12.2, p. 141; *13.1*, p. *161*; *20.3*, p. *350*; 22.2.3, p. 454
J. Noffke	(b) **89**, 367 (1978)	*13.3*, p. *167*; 21, p. 363
H. Noguchi	(b) **90**, 157 (1978)	6, p. 77; 13.1, p. 159; 13.5.2, p. 199; *20.1*, p. *335*; 22.4.2, p. 469
S. Noguchi	(b) **86**, 169 (1978)	*13.1*, p. *152*; 14.1, p. 206; 21.7, p. 406
W. Nolting	(b) **88**, 231 (1978)	*13.5.4*, p. *201*; 18.2, p. 288; 22, p. 414
P. Nordblad	(a) **49**, K153 (1978)	*3*, p. *56*; 18.2, p. 291; 21.1, p. 379; 21.1.1, p. 389
C. Norris	(b) **85**, 325 (1978)	*13.4*, p. *169*; *17*, p. *269*; 21.1, p. 368
L. M. Noskova	(a) **50**, K47 (1978)	*18.1*, p. *280*; *18.4*, p. *301*; 21.4, p. 399
E. Nossarzewska-Orłowska	(a) **48**, 225 (1978)	*3*, p. *55*; 22.1.2, p. 434
A. Nouailhat	(b) **89**, K85 (1978)	*10.2*, p. *117*; *20.1*, p. *333*; 22.5.2, p. 484
G. Nouet	(a) **50**, 703 (1978)	*1.1*, p. *14*; *1.4*, p. *35*
L. Novák	(a) **45**, K63 (1978)	*1.2*, p. *16*; 14, p. 203; 20, p. 315; 22.5, p. 474
P. Novák	(a) **50**, K21 (1978)	*1.1*, p. *14*; 22.8.2, p. 520
S. R. Novikov	(a) **46**, K21 (1978)	*13.4*, p. *173*; *14.3.3*, p. *234*; 22, p. 411
H.-J. Nowak	(a) **49**, 73 (1978)	*1.5*, p. *38*; *2*, p. *49*; 21.1, p. 377; 22.1.1, p. 427
Y. Z. Nozik	(a) **45**, 53 (1978)	1.2, p. 15; *14.4.1*, p. *253*; 19, p. 303; 20, p. 315; 22.8.1, p. 510
S. M. Oak	(b) **85**, 111 (1978)	*13.5.1*, p. 189; 20.3, p. 341; 22.4.1, p. 463; 22.4.2, p. 468
S. Öberg	(b) **89**, K127 (1978)	*13.4*, p. *181*; 22.2.1, p. 448
J. Obrda	(a) **46**, 273 (1978)	13.4, p. 173; *14.3.4*, p. *241*; 22.9, p. 524
Y. Ochiai	(a) **45**, K101 (1978)	*19*, p. *305*; 22.1.2, p. 430
A. A. O'Connel-Bronin	(b) **86**, K123 (1978)	*13.5*, p. *187*; 20.3, p. 343; 22, p. 412; 22.5.2, p. 480
	(a) **47**, 667 (1978)	*10.2*, p. *114*; *20.3*, p. *345*; 22.5.2, p. 481
T. H. O'Dell	(a) **48**, 59 (1978)	*18.2*, p. *288*
	(a) **50**, 179 (1978)	*18.2.1*, p. *294*
R. Oeder	(b) **87**, K35 (1978)	*10.2*, p. *113*; 11, p. 126; *19*, p. *307*; 22.8.1, p. 513
	(b) **89**, 375 (1978)	*19*, p. *312*; 20.3, p. 348; 22.8.1, p. 515
G. Oelgart	(b) **87**, 485 (1978)	*13.2*, p. *162*; *19*, p. *308*; 22.7, p. 502
	(b) **88**, 105 (1978)	*13.4*, p. *177*; 14.1, p. 208; *14.3*, p. *221*; 22.7, p. 502
	(b) **88**, 507 (1978)	*13.2*, p. *162*; 14.3, p. 223; 19, p. 310; 21.7, p. 407
Z. Ogorelec	(a) **47**, 703 (1978)	*2*, p. *46*; 22.1.1, p. 426
R. C. O'Handley	(a) **45**, 607 (1978)	*2*, p. *43*; *18.2*, p. *283*; 21.1, p. 370
S. M. Ohr	(a) **46**, 281 (1978)	*1.4*, p. *29*; *10.1*, p. *101*; 22.6, p. 492
M. Ohta	(a) **48**, K141 (1978)	*14.1*, p. *209*; 21, p. 362
S. Ohtake	(a) **50**, 335 (1978)	*10.1*, p. *107*; 21.7, p. 407
	(a) **50**, 449 (1978)	*10.1*, p. *108*; *12.1*, p. *136*; 21, p. 366
Y. H. Ohtsuki	(b) **87**, 757 (1978)	*13.3*, p. *166*; 21.6, p. 402

A. A. O'Konnel-Bronin	see under "O'Connel-Bronin"	
E. Okuno	(a) **47**, 617 (1978)	*8*, p. *86*; 10.2, p. 114; 22.5.3, p. 487
V. P. Oleinik	(b) **86**, 319 (1978)	*13.1*, p. *153*; 14.3, p. 218; 22, p. 410
	(b) **90**, 355 (1978)	*13.4*, p. *183*; *20.3*, p. *349*; 22.4.1, p. 466
S. Olejnik	(a) **50**, K137 (1978)	*14.4.1*, p. *257*; 22.8, p. 508
A. M. Oleś	(b) **87**, 625 (1978)	13.1, p. 155; *13.4*, p. *176*; *18.4*, p. *300*
	(a) **50**, K17 (1978)	*1.1*, p. *14*; *18.3*, p. *296*; 22.8.2, p. 520
A. L. de Oliveira	(b) **90**, K169 (1978)	*10.2*, p. *121*; *13*, p. *149*; 14.4, p. 252; 22.6.1, p. 500
J. de Oliveira	(b) **90**, K169 (1978)	*10.2*, p. *121*; *13*, p. *149*; 14.4, p. 252; 22.6.1, p. 500
L. E. Oliveira	(b) **87**, 25 (1978)	*10.2*, p. *112*; *20.1*, p. *325*; 22.5.3, p. 487
P. Murilo Oliveira	(b) **87**, 25 (1978)	*10.2*, p. *112*; *20.1*, p. *325*; 22.5.3, p. 487
M. T. Olm	(b) **88**, 705 (1978)	*10.2*, p. *116*; 16, p. 265; *19*, p. *310*; 22.5.1, p. 477
J. Staun Olsen	see under letter "S"	
W. Olsowski	(b) **89**, 389 (1978)	*13.5.1*, p. *194*; *16*, p. *266*; 22.1.1, p. 428
M. I. Onu	(b) **86**, K71 (1978)	*15*, p. *259*; 21.7, p. 406
Z. Onyszkiewicz	(b) **89**, K131 (1978)	*18*, p. *275*
C. P. van E. Oosthuizen	(a) **46**, K111 (1978)	*18.2*, p. *285*; 21.1.1, p. 386
H. Oppermann	(a) **45**, K77 (1978)	*14.1*, p. *205*; *14.3.1*, p. *226*; 22.6, p. 491
	(a) **47**, K129 (1978)	*1.2*, p. *20*; *14.3*, p. *221*; 22.6, p. 494
	(a) **48**, 377 (1978)	*1.2*, p. *21*; 6, p. 75; 14.3, p. 223; 22.6, p. 495
M. B. O'Reilly	(a) **48**, 489 (1978)	*10.2*, p. *116*; *14.4*, p. *250*; 22.5.1, p. 477
V. M. Orera	(b) **85**, 283 (1978)	*10.2*, p. *109*; *20.1*, p. *320*; 22.5.3, p. 487
A. Orlik	(a) **45**, 139 (1978)	*14.3.3*, p. *233*
N. N. Ormont	(a) **45**, 333 (1978)	*7*, p. *80*; *20.3*, p. *341*; 22.2.1, p. 445
S. E. Ormrod	(a) **50**, K193 (1978)	*9*, p. *95*; 22.6, p. 498
M. von Ortenberg	(b) **86**, 557 (1978)	*13.4*, p. *174*; 19, p. 306; 20.1, p. 324; 22.2.3, p. 452
V. V. Osiko	(a) **49**, 305 (1978)	3, p. 56; 20.1, p. 333; *20.2*, p. *339*; *20.3*, p. *347*; 22.8.2, p. 520
	(a) **50**, 745 (1978)	13.4, p. 186; *20.1*, p. *338*; *20.2*, p. *340*; 22.8.1, p. 516
W. Ossau	(b) **90**, 509 (1978)	*13.1*, p. *160*; 14.3, p. 225; *15*, p. *261*; 22.7, p. 502
E. Oteniyazov	(a) **46**, K147 (1978)	*10.2*, p. *112*; 11, p. 126; *12.2*, p. *139*; 22.1.2, p. 432
A. I. Otko	(a) **46**, 577 (1978)	*1.2*, p. *18*; 22.8.1, p. 512
P. Otschik	(b) **88**, 563 (1978)	*14.2*, p. *215*
G. J. den Otter	(a) **50**, 529 (1978)	*10.1*, p. *108*; *12.1*, p. *137*; 21.1, p. 382
A. A. Ovchinnikov	(b) **85**, 121 (1978)	*13*, p. *142*
M. Overhamm	(b) **86**, 205 (1978)	13.1, p. 152; *14.3.4*, p. *241*; 20.1, p. 322; *20.3*, p. *343*; 22.2.3, p. 452
I. A. Oxton	(a) **50**, K55 (1978)	*1.2*, p. *24*; *20.1*, p. *337*; 23, p. 531
C. Oytana	(a) **47**, 259 (1978)	*10.1*, p. *103*; 21, p. 358
K. Ozawa	(a) **46**, 217 (1978)	*1.3*, p. *25*; 6, p. 68; *11*, p. *125*; 21.6, p. 401
Yu. V. Ozerov	(b) **89**, 603 (1978)	*13.5.1*, p. *195*; *20.1*, p. *334*; 22.4.3, p. 471
G. Paasch	(b) **87**, 111 (1978)	*13.3*, p. *165*
	(b) **87**, 191 (1978)	*17*, p. *270*
S. K. Pabi	(a) **50**, K9 (1978)	*9*, p. *94*; 14.3.4, p. 246; 21, p. 365; 22.1.2, p. 438
O. M. Pachev	(b) **85**, 359 (1978)	*13*, p. *142*; *18*, p. *273*
	(b) **85**, 683 (1978)	*13*, p. *143*; *18*, p. *273*
H. Pagnia	(a) **49**, 83 (1978)	*13.3*, p. *167*; 16, p. 265; 17, p. 270; 22.2.3, p. 453

V. Paidar	(a) **50**, 483 (1978)	*9*, p. *94*; 10, p. 98
L. Palii	(a) **45**, K145 (1978)	*18.1*, p. *277*; 18.3, p. 295; 21.1, p. 370; 21.1.1 p. 385; 21.4, p. 396
S. Panchanadeeswaran	(a) **48**, K83 (1978)	*1.3*, p. *27*; *10.1*, p. *105*; 21, p. 361
A. N. Pandey	(a) **46**, K123 (1978)	*20.3*, p. *343*; 22.4, p. 461
	(a) **49**, K167 (1978)	*20.3*, p. *349*; 22.4, p. 462
S. D. Pandey	(b) **85**, K103 (1978)	*19*, p. *304*; 22.8.1, p. 511
M. M. Pant	(b) **88**, K131 (1978)	*6*, p. *74*; 20.1, p. 331; 22.5.2, p. 483
C. Paorici	(a) **50**, K235 (1978)	*8*, p. *90*; 22.3.1, p. 460
V. Papatheou	(b) **88**, 341 (1978)	*13.1*, p. *156*; 18, p. 274
C. Papatriantafillou	(b) **88**, 663 (1978)	*13*, p. *146*; 14.4, p. 249; 22.9, p. 526
K. I. Papunashvili	(b) **88**, K27 (1978)	*13*, p. *146*; *14.3.4*, p. *243*; 22.1.1, p. 426
B. Pardo	(a) **49**, 267 (1978)	*1.4*, p. *33*; 1.5, p. 38; 21.6, p. 404
I. A. Parfianovich	(b) **88**, K91 (1978)	*10.2*, p. *117*; 20.1, p. 333; *20.3*, p. *347*; 22.5.2, p. 484
G. Pastori Parravicini	(b) **88**, 773 (1978)	*20.1*, p. *331*; 22.5.2, p. 483
H. Pascher	(b) **86**, 557 (1978)	*13.4*, p. *174*; 19, p. 306; 20.1, p. 324; 22.2.3, p. 452
L. Pasemann	(b) **87**, 553 (1978)	*13.1*, p. *155*; 22.1.2, p. 433
R. Pässler	(b) **85**, 203 (1978)	*13.4*, p. *169*; 22, p. 408
	(b) **86**, K39 (1978)	*13.4*, p. *172*; 22, p. 410; 22.2.1, p. 445
	(b) **86**, K45 (1978)	*13.4*, p. *172*; 22, p. 411
A. R. Patel	(a) **47**, K1 (1978)	*1.5*, p. *37*; 21.7, p. 406
C. J. de Pater	(a) **48**, 503 (1978)	*14.4.1*, p. *256*; 22.5.4, p. 489
Anand P. Pathak	(b) **86**, 751 (1978)	*11*, p. *125*; 13, p. 144; 21.1, p. 372; 21.6, p. 402
	(b) **90**, 703 (1978)	*11*, p. *131*
C. G. Patil	(b) **86**, 725 (1978)	13.5.4, p. 201; *14.3*, p. *219*; 18.2, p. 285; 22.4.4, p. 472
	(b) **88**, K117 (1978)	*13.5.4*, p. *202*; *15*, p. *260*; 18.2, p. 289; 22, p. 415
J. D. Patterson	(b) **88**, K127 (1978)	*2*, p. *47*; *18.2*, p. *289*
	(b) **90**, K85 (1978)	*2*, p. *51*; *8*, p. *89*; 14.1, p. 212; 21.1.1, p. 390
G. Patzer	(a) **45**, 207 (1978)	4, p. 57; *10.2*, p. *109*; 22.2.1, p. 445; 22.2.2, p. 450; 22.2.3, p. 452
P. Paufler	(a) **49**, 323 (1978)	*1.2*, p. *22*; *10.1*, p. *106*; 14.2, p. 215; 21.1, p. 377
O. Pavlov	(a) **47**, K59 (1978)	*1.3*, p. *26*; 8, p. 86; *14.3*, p. *221*; 15, p. 260; 22, p. 413
S. T. Pavlov	(b) **85**, 81 (1978)	*6*, p. *64*; *20.3*, p. *341*; 22, p. 408
	(b) **85**, 351 (1978)	*6*, p. *65*; *13.5.1*, p. *189*; 20.1, p. 320
	(b) **90**, 429 (1978)	*13.1*, p. *160*; *20.1*, p. *336*; 22.3, p. 459
	(b) **90**, K63 (1978)	*6*, p. *78*; *20.1*, p. *336*; 22, p. 420
V. A. Pavlov	(a) **49**, 509 (1978)	*10.1*, p. *106*; 21.1.1, p. 389
V. N. Pavlovich	(b) **88**, 407 (1978)	*9*, p. *93*; *10*, p. *96*
P. Pawlicki	(b) **87**, K17 (1978)	*18.2*, p. *286*; 21, p. 357
J. M. Pawlikowski	(a) **46**, K9 (1978)	*8*, p. *84*; *13.1*, p. *153*; 22.4.4, p. 472
	(b) **88**, K153 (1978)	*13.1*, p. *158*; *20.1*, p. *331*; 22.3, p. 458
D. A. Payne	(a) **50**, 661 (1978)	*1.2*, p. *24*; 8, p. 90; 12.2, p. 141; 20.1, p. 338; 22.9, p. 528
A. J. Pedraza	(a) **45**, 315 (1978)	*10.1*, p. *100*; 21, p. 353; 21.1, p. 369
	(a) **47**, 425 (1978)	*10.1*, p. *103*; 21, p. 359; 21.1, p. 374
C. Pedrini	(b) **87**, 273 (1978)	*13.4*, p. *175*; *20.3*, p. *344*; 22.5.2, p. 481
H. Peisl	(b) **85**, 553 (1978)	*10.2*, p. *109*; *11*, p. *123*; 22.5.2, p. 478
G. S. Pekar	(b) **86**, K155 (1978)	13.5.1, p. 191; *16*, p. *263*; 22.4.1, p. 464
	(a) **48**, 249 (1978)	*13.4*, p. *178*; *16*, p. *264*; 22.4.1, p. 465
H. E. Peltner	(a) **47**, 497 (1978)	*9*, p. *93*; 14.1, p. 208; 21.6, p. 403
	(a) **48**, K79 (1978)	*3*, p. *55*; 21.6, p. 403

J. PELZL	(a) **48**, K35 (1978)	*1.2*, p. *21*; *7*, p. *81*; 22.8, p. 506
J. PENG	(a) **49**, 313 (1978)	*1.3*, p. *28*; 1.5, p. 38; *14.3.2*, p. *231*; 21.6, p. 404; 22.1.2, p. 437
CHANG PENG-HSIANG	(a) **47**, 397 (1978)	*10.1*, p. *103*; *18.2*, p. *287*; 21.1, p. 374
P. PENNING	(a) **47**, 577 (1978)	*10.1*, p. *104*; 21, p. 359
U. DEL PENNINO	(a) **50**, K123 (1978)	*10.2*, p. *121*; *13.4*, p. *184*; 22.1.2, p. 438
E. E. PENZINA	(b) **89**, K91 (1978)	*10.2*, p. *117*; 20.1, p. 333; *20.3*, p. *347*; 22.5.2, p. 484
R. PÉREZ ALVAREZ	see under letter "A"	
G. PERLUZZO	(b) **88**, K135 (1978)	*14.3*, p. *223*; 22, p. 416
R. PERRIN	(b) **90**, 667 (1978)	*18.4*, p. *302*; 21.1, p. 382
F. PERROT	(b) **87**, 707 (1978)	*10.1*, p. *103*; 14.1, p. 208; 21, p. 359
R. PERTHEL	(b) **87**, 203 (1978)	*2*, p. *44*; 3, p. 54; *18.2.1*, p. *293*; 21.1, p. 372; 21.4, p. 397
B. PERTZSCH	(b) **89**, 475 (1978)	12.2, p. 140; *13.4*, p. *181*; 22.2, p. 444
P. PESHEV	(a) **46**, K171 (1978)	*1.2*, p. *18*; 8, p. 85; 22.8.1, p. 513
P. PETELENZ	(b) **88**, 689 (1978)	*13.5.1*, p. *193*; 22.9, p. 526
	(b) **90**, 635 (1978)	*13.4*, p. *185*; *13.5.1*, p. *197*
K. PETERMANN	(a) **49**, K13 (1978)	1.5, p. 39; 3, p. 56; *14.3.3*, p. *236*; *20.3*, p. *348*; 22.4.1, p. 465; 22.8, p. 507
J. PETERSSON	(b) **88**, 99 (1978)	*14.4.1*, p. *255*; *19*, p. *309*; 22.8.1, p. 514
	(b) **89**, 587 (1978)	*14.4*, p. *251*; *19*, p. *312*; 22.9, p. 527
J. I. PETKOV	(b) **86**, 385 (1978)	*13.1*, p. *153*; 22.9, p. 523
V. E. PETRASHEN	(b) **88**, 439 (1978)	*13.4*, p. *178*; *19*, p. *310*; 22.8, p. 506
N. S. PETRENKO	(a) **47**, K145 (1978)	*8*, p. *87*; *14.2*, p. *214*; 21, p. 360
A. V. PETROV	(a) **48**, K153 (1978)	*8*, p. *88*; 22.6, p. 495
K. PETROV	(a) **46**, K171 (1978)	*1.2*, p. *18*; 8, p. 85; 22.8.1, p. 513
V. A. PETROV	(a) **48**, K131 (1978)	*1.2*, p. *22*; *14*, p. *204*; 22.3, p. 458
D. S. PETROVA	(a) **46**, 55 (1978)	*1.4* p. *29*; 20, p. 316; 22.8.2, p. 518
P. C. PETROVA	(b) **86**, 425 (1978)	*13.3*, p. *164*; 14.3.3, p. 233
	(b) **87**, 361 (1978)	*13.3*, p. *165*; *14.3.3*, p. *235*
A. G. PETUKHOV	(b) **88**, K161 (1978)	*6*, p. *74*; 22.1, p. 422; 22.1.1, p. 427; 22.1.2, p. 435
	(b) **89**, K5 (1978)	*4*, p. *59*; *13.1*, p. *158*; 22, p. 417
	(b) **90**, K5 (1978)	*4*, p. *59*; *13.1*, p. *160*; 22.3, p. 459
P. PFEFFER	(b) **88**, 247 (1978)	*6*, p. *73*; 13.1, p. 156; *14.3*, p. *222*; 22.2.3, p. 453
VU DUY PHACH	(b) **86**, 159 (1978)	*13.5.1*, p. *190*; 20.1, p. 321; 20.3, p. 342; 22.5, p. 474
	(b) **88**, 545 (1978)	*13.5.1*, p. *193*; *20.3*, p. *346*; 22.5, p. 475
S. D. PHADKE	(a) **45**, K163 (1978)	14.3.1, p. 226; *16*, p. *263*; 22.9, p. 523
M. T. PHAM	(a) **49**, 261 (1978)	*11*, p. *129*; *14.3.2*, p. *231*; 22.1.2, p. 436
A. PHILIPP	(b) **89**, 187 (1978)	*14.1*, p. *210*; 22, p. 417
Š. PICK	(b) **89**, 11 (1978)	13.1, p. 150; *13.3*, p. *164*; 21.1, p. 368; 21.1.1, p. 384
J. PIELASZEK	(a) **45**, K87 (1978)	*10.1*, p. *100*; 21.1, p. 370
H. PIEPER	(a) **49**, 137 (1978)	10.2, p. 118; *14.3*, p. *224*; 22.1.2, p. 436
B. PIETRASS	(a) **49**, 323 (1978)	*1.2*, p. *22*; *10.1*, p. *106*; 14.2, p. 215; 21.1, p. 377
C. K. PILLAI	(a) **46**, K127 (1978)	*14.4*, p. *248*; *16*, p. *264*; 22.4, p. 461; 22.5, p. 475
P. K. C. PILLAI	(a) **46**, K127 (1978)	*14.4*, p. *248*; *16*, p. *264*; 22.4, p. 461; 22.5, p. 475
R. G. PILLAY	(a) **45**, K109 (1978)	*5*, p. *60*; *18.2*, p. *283*; 21, p. 355
J.-J. PILLOUD	(b) **90**, K161 (1978)	*19*, p. *314*; 22.5.2, p. 486
A. PIMPALE	(b) **89**, K21 (1978)	*13.3*, p. *167*; *13.5.2*, p. *198*; 20, p. 317; 21.6, p. 404

P. Pinard	(a) **47**, 251 (1978)	*13.4*, p. *175*; *14.3.2*, p. *230*; 22.6, p. 493
T. Yu. Pinegina	(a) **47**, K81 (1978)	1.2, p. 18; *14.1*, p. *208*; *15*, p. *260*; 21.4, p. 397
H. Pink	(a) **46**, 81 (1978)	*20.1*, p. *323*; *20.3*, p. *343*; 22.8, p. 504
V. G. Pinko	(a) **50**, K153 (1978)	*18.2.1*, p. *294*; *20*. p. *318*; 21.1, p. 383; 21.1.1, p. 391
M. Pinnow	(a) **50**, K239 (1978)	*14.3.4*, p. *247*; 22.9, p. 529
A. Piotrowska	(a) **50**, 243 (1978)	*14.3.3*, p. *238*; 22.5.3, p. 488
K. Piotrowski	(a) **49**, K193 (1978)	*18.4*, p. *301*; *19*, p. *313*; 22.8.2, p. 520
R. Piotrzkowski	(a) **47**, K115 (1978)	13.4, p. 177; *16*, p. *264*; 22.1.1, p. 426
P. Pipinys	(a) **45**, K113 (1978)	*6*, p. *66*; 13.4, p. 171; *14.3.4*, p. *240*; 22.4.1, p. 463
J. Piqueras	(a) **49**, K9 (1978)	*10.2*, p. *118*; *20.3*, p. *348*; 22.2.2, p. 451; 22.6, p. 496
T. I. Pirozhkova	(a) **46**, K21 (1978)	*13.4*, p. *173*; *14.3.3*, p. *234*; 22, p. 411
Yu. G. Pismennyi	(a) **45**, 631 (1978)	13.4, p. 171; *14.3.4*, p. *240*; 22, p. 410
V. Platikanova	(a) **47**, 683 (1978)	*10.2*, p. *114*; 13.4, p. 176; 22.5.1, p. 477
V. G. Plekhanov	(b) **86**, K123 (1978)	*13.5*, p. *187*; 20.3, p. 343; 22, p. 412; 22.5.2, p. 480
P. Plenkiewicz	(b) **87**, 309 (1978)	*13.1*, p. *154*; 22.3, p. 458
S. Plesko	(a) **47**, 233 (1978)	*1.2*, p. *18*; 19, p. 307; 22.9, p. 525
P. de V. du Plessis	(a) **46**, K111 (1978)	*18.2*, p. *285*; 21.1.1, p. 386
F. Plicque	(a) **45**, 229 (1978)	*1.3*, p. *25*; 6, p. 65; 12.1, p. 134; 21.1, p. 369; 21.1.1, p. 384
K. Ploog	(a) **45**, K175 (1978)	*3*, p. *54*; *14.3.1*, p. *226*; 22.6, p. 491
A. F. Plotnikov	(a) **50**, 733 (1978)	*13.1*, p. *161*; *14.3.3*, p. *239*; 16, p. 268; 22.2.3, p. 455
S. V. Plotnikov	(a) **48**, K175 (1978)	11, p. 128; *20.3*, p. *346*; 22.5.2, p. 483
M. V. Plotnikova	(a) **48**, 183 (1978)	*1*, p. *9*; 5, p. 61; *14.4.1*, p. *255*; 18.2, p. 289; 18.4, p. 301; 19, p. 310; 22.8.1, p. 514
Ya. M. Pogosyan	(a) **49**, K3 (1978)	*18.2.1*, p. *294*; 21.1, p. 378; 21.1.1, p. 389
M. Pöhler	(b) **86**, 225 (1978)	*13.5.1*, p. *190*; *20.1*, p. *322*; 22.1.3, p. 441
B. Pohoryles	(a) **47**, K115 (1978)	13.4, p. 117; *16*, p. *264*; 22.1.1, p. 426
E. P. Pokatilov	(b) **85**, 403 (1978)	*14.3.4*, p. *240*; *20.2*, p. *339*; 22, p. 409
	(b) **87**, 725 (1978)	6, p. 72; 14.3.4, p. 242; *20*, p. *316*
	(b) **87**, K159 (1978)	*19*,p. *308*; 22.4.1, p. 464
N. A. Poklonskii	(b) **88**, K165 (1978)	*13.4*, p. *179*; *14.3*, p. *223*; 22, p. 416
V. A. Pokoeva	(a) **48**, 241 (1978)	*9*, p. *93*; 22, p. 415
K. Polák	(b) **89**, K187 (1978)	11, p. 130; *19*, p. *313*; 22.5.2, p. 485
M. Polcarová	(a) **46**, 567 (1978)	*1.1*, p. *12*
	(a) **47**, 179 (1978)	*1.1*, p. *12*
C. Politis	(b) **87**, 565 (1978)	*6*, p. *71*; 14.2, p. 214; *20.1*, p. *327*
I. K. Polushina	(a) **49**, 761 (1978)	*11*, p. *130*; 13.4, p. 182; *14.3*, p. *224*; 22.8, p. 507
V. P. Polyakova	(a) **49**, K117 (1978)	*8*, p. *88*; *14.1*, p. *211*; 21, p. 364
G. Pompe	(a) **50**, 147 (1978)	*8*, p. *89*; 14.1, p. 212; 21.1, p. 381
H. E. Ponath	(b) **86**, 215 (1978)	*6*, p. *67*; *20.1*, p. *322*; 22.8.1, p. 511
R. C. Pond	(a) **45**, 149 (1978)	*1.4*, p. *29*; 21, p. 353
B. K. Ponomarev	(b) **88**, K139 (1978)	*18.2*, p. *289*; 18.4, p. 301; 21.1, p. 376
	(a) **48**, K185 (1978)	*18.2*, p. *290*; 21.1, p. 377
E. G. Ponyatovskii	(b) **87**, 369 (1978)	*1.2*, p. *18*; 12.1, p. 135; 21.4, p. 397
	(a) **48**, K185 (1978)	*18.2*, p. *290*; 21.1, p. 377
E. Pop	(b) **88**, K181 (1978)	*18.1*, p. *278*; *18.4*, p. *301*; 21.1, p. 376; 21.1.1, p. 388
I. Pop	(b) 88, K181 (1978)	*18.1*, p. *278*; *18.4*, p. *301*; 21.1, p. 376; 21.1.1, p. 388
	(a) **49**, 207 (1978)	*18.1*, p. *279*; *19*, p. *311*; 21.1, p. 377

J. Popielawski	(b) 88, 241 (1978)	*13.5*, p. *187*; 20.1, p. 329
E. Popko	(a) 46, K9 (1978)	*8*, p. *84*; *13.1*, p. *153*; 22.4.4, p. 472
A. N. Popkov	(a) 48, K115 (1978)	*13.4*, p. *179*; 20.1, p. 331; 22.2.3, p. 453
B. N. Popov	(a) 46, 729 (1978)	*14.4.1*, p. *254*; *16*, p. *263*; 22.8.1, p. 512
	(a) 50, K255 (1978)	14.4.1, p. 258; *16*, p. *268*; *17*, p. *271*; 22.8.1, p. 517
D. E. Popov	(b) 86, K35 (1978)	*11*, p. *124*; 13, p. 143; 22.1.2, p. 431
L. K. Popov	(a) 46, 59 (1978)	*14.3.2*, p. *230*; *14.3.3*, p. *233*; 16, p. 263; 22.1.1, p. 425; 22.1.2, p. 431
S. V. Popova	(b) 90, K133 (1978)	*13.1*, p. *161*; 22, p. 421
P. Porcher	(b) 85, 587 (1978)	*20.1*, p. *321*; *20.3*, p. *342*; 22.8, p. 503
M. Porte	(a) 45, 503 (1978)	*18.3*, p. *295*; 22.8.2, p. 518
S. P. S. Porto	(b) 86, 373 (1978)	*14.4.1*, p. *254*; 22.8.1, p. 511
	(b) 90, 543 (1978)	*14.4.1*, p. *257*; 20.1, p. 337; 22.8.1, p. 516
D. Pöss	(a) 45, 263 (1978)	13.4, p. 170; *19*, p. *303*; 22.5.2, p. 478
V. S. Postinikov	(b) 85, K115 (1978)	*14.2*, p. *213*; 21.6, p. 401
	(b) 87, K11 (1978)	1.2, p. 15; *1.6*, p. *40*; *14.2*, p. *213*; 21, p. 352; 22, p. 408
	(a) 47, 65 (1978)	*7*, p. *80*; *12.2*, p. *139*; 14.4.1, p. 255; 22.8.1, p. 513
V. V. Postnikov	(b) 85, K115 (1978)	*14.2*, p. *213*; 21.6, p. 401
	(b) 87, 11 (1978)	1.2, p. 15; *1.6*, p. *40*; *14.2*, p. *213*; 21, p. 352; 22, p. 408
A. I. Potapov	(a) 49, 761 (1978)	*11*, p. *130*; 13.4, p. 182; *14.3*, p. *224*; 22.8, p. 507
H. W. Pötzl	(b) 86, 205 (1978)	13.1, p. 152; *14.3.4*, p. *241*; 20.1, p. 322; *20.3*, p. *343*; 22.2.3, p. 452
V. A. Povitskii	(b) 87, K101 (1978)	*5*, p. *61*; 18.3, p. 295; 22.8.2, p. 519
	(a) 47, 39 (1978)	*5*, p. *61*; 18.4, p. 300; 22.6, p. 493
L. V. Povstyanyi	(a) 46, 577 (1978)	*1.2*, p. *18*; 22.8.1, p. 512
B. M. Powell	(b) 85, 311 (1978)	*6*, p. *65*; *20.1*, p. *320*; 22.7, p. 501
J. Požela	(a) 45, 387 (1978)	*14.3.4*, p. *240*; 22.2.3, p. 452
	(a) 50, K73 (1978)	12.2, p. 141; *14.3.4*, p. *246*; 22.2.3, p. 454
J. Poźniak	(b) 89, 577 (1978)	*10.2*, p. *118*; 14.4, p. 251; 22.5.2, p. 485
	(b) 90, 261 (1978)	*10.2*, p. *118*; 14.4, p. 251; 22.5.2, p. 485
S. D. Pradhan	(a) 47, 691 (1978)	*10.2*, p. *114*; 11, p. 127; 14.4, p. 249; 19, p. 309; 22.9, p. 526
R. Prakash	(b) 85, 477 (1978)	*6*, p. *65*; 12.2, p. 138; 22.5.2, p. 478
	(b) 87, K127 (1978)	*4*, p. *58*; *12.2*, p. *139*; 22.6, p. 494
	(b) 89, 335 (1978)	*6*, p. *65*; 12.2, p. 138; 22.5.2, p. 478
A. Prasad	(a) 46, K151 (1978)	*9*, p. *92*; 22.1.2, p. 432
B. Prasad	(b) 85, 789 (1978)	*6*, p. *66*; *8*, p. *83*; 21.6, p. 401
	(b) 87, 771 (1978)	*6*, p. *72*; 8, p. 86; 21.2, p. 392
S. Prasad	(a) 46, 373 (1978)	*14.3*, p. *219*; 22.8.2, p. 518
B. N. Prasolov	(a) 47, 65 (1978)	*7*, p. *80*; *12.2*, p. *139*; 14.4.1, p. 255; 22.8.1, p. 513
R. Pratap	(a) 49, 637 (1978)	*14.3.3*, p. *237*; 14.4, p. 251; 22.6, p. 496
P. L. Pratt	(a) 49, 353 (1978)	*10.2*, p. *118*; *14.4*, p. *250*; 22.5.2, p. 484
M. Prejsa	(a) 47, 287 (1978)	5, p. 61; *14.2*, p. *214*
U. Preppernau	(b) 90, 517 (1978)	13.3, p. 168; *14.2*, p. *216*; 21, p. 365
H. Pretsch	(a) 47, 445 (1978)	*1.2*, p. *19*; 18.2, p. 287; 21.1, p. 374
P. Prevot	(b) 88, 327 (1978)	*6*, p. *73*; 7, p. 81; *20.1*, p. *329*; 22.5, p. 475
F. Prinz	(a) 49, 539 (1978)	*10.1*, p. *106*; 12.1, p. 136; 21.1.1, p. 389
A. I. Pristupa	(b) 87, 373 (1978)	13.5.1, p. 191; *16*, p. *264*; 20.3, p. 344; 22.9, p. 524
J. Přívratská	(b) 85, K1 (1978)	*13.5.3*, p. *200*; *18.2*, p. *281*; 22, p. 409

V. D. Prochukhan	(a) 49, 761 (1978)	*11*, p. *130*; 13.4, p. 182; *14.3*, p. *224*; 22.8, p. 507
	(a) 50, 379 (1978)	11, p. 131; 13.4, p. 185; *14.3*, p. *225*; *20.1*, p. *337*; 22.8, p. 508
A. Prodan	(a) 45, 469 (1978)	*1.2*, p. *16*; *1.3*, p. *25*; 22.8, p. 503
F. Prokert	(b) 87, 179 (1978)	*14.4.1*, p. *254*; 22.8.1, p. 513
A. M. Prokhorov	(a) 50, 745 (1978)	13.4, p. 186; *20.1*, p. *338*; *20.2*, p. *340*; 22.8.1, p. 516
A. V. Prokhorovich	(a) 45, K91 (1978)	*13.4*, p. *171*; *20.3*, p. *342*; 22.2.1, p. 445
	(a) 48, 593 (1978)	*13.4*, p. *179*; *20.3*, p. *346*; 22.2.1, p. 448
L. M. Protopopova	(a) 50, 269 (1978)	*18.3*, p. *296*; *19*, p. *314*; 22.8.2, p. 520
R. V. Prudnikov	(a) 50, 739 (1978)	*13.3*, p. *168*; 22.1.1, p. 428
J. Przedmojski	(a) 46, K115 (1978)	*1.2*, p. *18*; 22.8.1, p. 512
F. Przyborowski	(a) 50, K189 (1978)	*14.3.3*, p. *239*; 22.9, p. 529
M. Przybylski	(a) 50, K105 (1978)	*14*, p. *204*; 22.9, p. 528
H. Ptasiewicz-Bąk	(a) 46, K75 (1978)	*18.2*, p. *284*; *18.4*, p. *299*; 21, p. 356; 21.4, p. 397
	(a) 47, 349 (1978)	*18.2*, p. *287*; *18.4*, p. *300*; 22.8, p. 505
	(b) 90, 319 (1978)	*13.4*, p. *183*; *18*, p. *275*; 22.8.2, p. 520
G. A. Puchkovskaya	(b) 89, 27 (1978)	*13.5*, p. *188*; 20.1, p. 331; 22.6, p. 496
M. Pugnet	(b) 88, 261 (1978)	*13.5.1*, p. *192*; 20.3, p. 345; 22.2.2, p. 450
J. Puls	(b) 87, K111 (1978)	*13.5.1*, p. *192*; *20.1*, p. *328*; 22.4.1, p. 464
M. Punkkinen	(a) 46, 213 (1978)	*19*, p. *306*; 22.9, p. 524
B. Pura	(a) 46, K115 (1978)	*1.2*, p. *18*; 22.8.1, p. 512
J. J. Purāns	(b) 89, 531 (1978)	*10.2*, p. *119*; 19, p. 312; 22.5.2, p. 485
	(a) 49, K43 (1978)	*19*, p. *312*; 22.8.1, p. 515
D. S. Puri	(b) 87, 401 (1978)	*6*, p. *70*; *12.2*, p. *139*; 22.2, p. 443; 22.4, p. 461; 22.5, p. 475
T. Puritis	(a) 49, 225 (1978)	*13.5.2*, p. *199*; *14.3.4*, p. *244*; 22.1.2, p. 436
E. A. Pushkarev	(a) 47, K145 (1978)	*8*, p. *87*; *14.2*, p. *214*; 21, p. 360
D. I. Pushkarov	(b) 85, K89 (1978)	*13.5.4*, p. *201*; *18.2*, p. *283*
	(b) 88, 65 (1978)	*13.5.4*, p. *201*; *18.2*, p. *287*
	(b) 90, 361 (1978)	*13.5.4*, p. *202*; *18.2*, p. *291*
Kh. I. Pushkarov	(b) 85, K89 (1978)	*13.5.4*, p. *201*; *18.2*, p. *283*
	(b) 88, 65 (1978)	*13.5.4*, p. *201*; *18.2*, p. *287*
	(b) 90, 361 (1978)	*13.5.4*, p. *202*; *18.2*, p. *291*
H. Puslat	(b) 90, K121 (1978)	*14.1*, p. *212*; 21.1, p. 382
H. Puszkarski	(b) 88, K113 (1978)	*18.2.1*, p. *293*; *19*, p. *311*
H. Pykacz	(a) 47, 625 (1978)	*14.4.1*, p. *255*; 22.9, p. 525
Nguyen Vinh Quang	(b) 90, 597 (1978)	*13.1*, p. *160*; *20.2*, p. *340*
L. Quartapelle	(b) 86, 471 (1978)	*13.1*, p. *153*; 22, p. 411
S. L. A. de Queiroz	(b) 87, 351 (1978)	*13.4*, p. *175*; *20.1*, p. *326*; 22.5.2, p. 481
H. J. Queisser	(b) 89, 467 (1978)	*13.4*, p. *181*; *20.3*, p. *348*; 22.2.2, p. 451
X. Quillard	(a) 50, 41 (1978)	*1.3*, p. *28*; 14.1, p. 212; 21.1, p. 380
M. Quintard	(a) 46, K43 (1978)	*10.1*, p. *101*; *12.1*, p. *134*; 21, p. 356
	(a) 48, 99 (1978)	*10.1*, p. *104*; *12.1*, p. *135*; 21.1, p. 375
Tran Quoc Hai	see under letter "H"	
B. Rabin	(a) 49, 577 (1978)	*13.4*, p. *182*; *14.3.3*, p. *237*; 22.4.3, p. 471
M. Rachwalska	(a) 47, 57 (1978)	*1.2*, p. *18*; 8, p. 85; 22.8, p. 505
	(a) 47, 661 (1978)	*1.2*, p. *20*; 8, p. 86; 22.9, p. 525
	(a) 48, 297 (1978)	*1.2*, p. *21*; 8, p. 87; 22.8, p. 506
W. Raczyński	(a) 48, K27 (1978)	*9*, p. *93*; 21.1.1, p. 388
S. I. Radautsan	(b) 87, K141 (1978)	*14.3*, p. *221*; 22.8, p. 505
	(a) 49, K5 (1978)	*1.5*, p. *39*; 22.8, p. 507
	(a) 49, K197 (1978)	*14.3.4*, p. *246*; 22.8.2, p. 520

V. S. RADCHENKO	(b) **87**, K97 (1978)	*14.3.4*, p. *242*; 22.1.3, p. 441
S. RADHAKRISHNA	(a) **47**, 723 (1978)	13.4, p. 176; *19*, p. *309*; *20.1*, p. *328*; 22.8.1, p. 514
M. RADHAKRISHNAN	(a) **48**, K71 (1978)	10.1, p. 104; *14.1*, p. *209*; 21.6, p. 403
	(a) **49**, K213 (1978)	1.2, p. 23; 2, p. 51; *14.1*, p. *211*; 21.6, p. 405; 22.1.3, p. 442
A. P. RADLINSKI	(b) **86**, 41 (1978)	*13.4*, p. *172*; 20.1, p. 321; 20.3, p. 342; 22.4.2, p. 468
C. RADOMSKY	(a) **50**, 123 (1978)	*10.1*, p. *107*; 14.1, p. 212; 21.1, p. 381
J. RAFAŁOWICZ	(a) **47**, 229 (1978)	*8*, p. *85*; 22.6, p. 493
	(a) **47**, K137 (1978)	*8*, p. *87*; *10.1*, p. *104*; 21, p. 360; 21.1, p. 375
	(a) **48**, 215 (1978)	*14.1*, p. *209*; 21.7, p. 407
	(a) **48**, 221 (1978)	*8*, p. *87*; 21, p. 360
	(a) **50**, K261 (1978)	*8*, p. *87*; *10.1*, p. *104*; 21, p. 360; 21.1, p. 375
F. RAGA	(a) **50**, 643 (1978)	13.4, p. 186; *16*, p. *267*; 22.8, p. 508
H. RAIDT	(a) **48**, 407 (1978)	*1.4*, p. *32*; 14.3.2, p. 230; 20.3, p. 346; 22.2.2, p. 451
	(a) **49**, 629 (1978)	*1.5*, p. *39*; 13.4, p. 182; 14.3.1, p. 228; 20.1, p. 335; 20.3, p. 348; 22.2.4, p. 456
P. RAJ	(a) **49**, K91 (1978)	*1.3*, p. *28*; *5*, p. *62*; 18.2, p. 290; 21.1.1, p. 389
T. RAJASEKHARAN	(a) **45**, K51 (1978)	*19*, p. *303*; 22.7, p. 501
	(a) **50**, 303 (1978)	*18.2*, p. *291*; *19*, p. *314*; 22.8, p. 508
A. RAJCHEL	(a) **50**, K105 (1978)	*14*, p. *204*; 22.9, p. 528
B. RAJCHEL	(a) **46**, 285 (1978)	*11*, p. *125*; 22.1.2, p. 431
YU. V. RAKITIN	(b) **85**, 783 (1978)	1, p. 9; *13*, p. *143*; 22.5.4, p. 489
	(b) **89**, 503 (1978)	1, p. 10; *13*, p. *147*;
	(b) **90**, 123 (1978)	1, p. 10; *13*, p. *148*
V. RAMAMURTHY	(b) **85**, 761 (1978)	*6*, p. *66*; 12.1, p. 134; 21.2, p. 392
	(b) **90**, 117 (1978)	*6*, p. *77*; 12.1, p. 136; 21.2, p. 393
V. RAMAN	(a) **46**, K5 (1978)	*10.1*, p. *101*; 21, p. 356
A. RAMANAND	(b) **87**, 751 (1978)	*6*, p. *72*; 8, p. 86; 12.1, p. 135; 21.4, p. 397
R. RAMANI	(b) **86**, 759 (1978)	*6*, p. *69*; 12.2, p. 138; 22, p. 411
C. RAMASASTRY	(a) **45**, K51 (1978)	*19*, p. *303*; 22.7, p. 501
	(b) **90**, 465 (1978)	*10.2*, p. *121*; *20.1*, p. *337*; 22.5.2, p. 486
B. RAMBABU	(b) **90**, 465 (1978)	*10.2*, p. *121*; *20.1*, p. *337*; 22.5.2, p. 486
K. V. RAO	(b) **87**, 287 (1978)	*14.1*, p. *207*; 21.1, p. 372; 21.6, p. 402
	(a) **47**, K1 (1978)	*1.5*, p. *37*; 21.7, p. 406
T. BHASKAR RAO	(a) **47**, 723 (1978)	13.4, p. 176; *19*, p. *309*; *20.1*, p. *328*; 22.8.1, p. 514
S. MURALIDHARA RAO	(b) **90**, K175 (1978)	*5*, p. *63*; 22.6.1, p. 500
K. V. S. RAMA RAO	(a) **45**, K51 (1978)	*19*, p. *303*; 22.7, p. 501
	(a) **50**, 303 (1978)	*18.2*, p. *291*; *19*, p. *314*; 22.8, p. 508
R. RAMJI RAO	(b) **87**, 751 (1978)	*6*, p. *72*; 8, p. 86; 12.1, p. 135; 21.4, p. 397
A. RANGA RAO	(a) **45**, K131 (1978)	*10.1*, p. *100*; 12.1, p. 134; 21.3, p. 394
V. U. S. RAO	(a) **49**, 207 (1978)	*18.1*, p. *279*; *19*, p. *311*; 21.1, p. 377
B. RAOKAZA	(a) **46**, K123 (1978)	*20.3*, p. *343*; 22.4, p. 461
	(a) **49**, K167 (1978)	*20.3*, p. *349*; 22.4, p. 462
P. RAPPS	(a) **47**, 479 (1978)	*1.4*, p. *31*; *10*, p. *96*
M. RASOLT	(b) **86**, 37 (1978)	*13.2*, p. *162*; 21, p. 355
I. G. RATISHVILI	(b) **87**, 461 (1978)	*1.2*, p. *19*; 1.3, p. 26; 21, p. 359
K. N. RATNAKUMAR	(a) **45**, K13 (1978)	*14.3.3*, p. *233*
B. A. RATNAM	(a) **47**, K187 (1978)	*14.2*, p. *214*; 21, p. 360
V. V. RATNAM	(a) **48**, 235 (1978)	*10.2*, p. *115*; *20.3*, p. *345*; 22.5.2, p. 482
I. RATTAN	(a) **46**, 475 (1978)	*12.2*, p. *139*; *18.2*, p. *285*; 22.8.2, p. 519
H. RAUH	(a) **46**, 499 (1978)	*9*, p. *92*; 10, p. 96
	(b) **87**, 99 (1978)	6, p. 70; *14.3.4*, p. *242*; 22.4.1, p. 464
G. RAUNIO	(b) **86**, 653 (1978)	*6*, p. *69*; *8*, p. *84*; 22.5.2, p. 480

W. Raupach	(a) **50,**	K81 (1978)	*10.2,* p. *121*; 22.5.2, p. 485
B. Rauschenbach	(a) **47,**	79 (1978)	*2,* p. *45*; *11,* p. *126*
M. Rautenberg	(b) **86,**	205 (1978)	13.1, p. 152; *14.3.4,* p. *241*; 20.1, p. 322; *20.3,* p. *343*; 22.2.3, p. 452
	(b) **86,**	479 (1978)	*13.4,* p. *174*; *14.3,* p. *219*; 22, p. 411
B. Raveau	(a) **48,**	301 (1978)	*1.1,* p. *13*; *1.4,* p. *32*; 22.8, p. 506
	(a) **49,**	189 (1978)	*1.1,* p. *13*; *1.4,* p. *33*; 22.8, p. 507
S. Ravelet	(a) **46,**	K55 (1978)	*14.3.3,* p. *234*; 14.3.4, p. 241; 22.2.2, p. 450
R. D. Rawlings	(a) **47,**	407 (1978)	*1.4,* p. *31*; *14.1,* p. *208*; 21.1, p. 374; 21.1.1, p. 387
	(a) **49,**	K85 (1978)	*10.1,* p. *106*; 21.1, p. 378; 21.1.1, p. 389
M. Raychaudhury	(b) **90,**	609 (1978)	*13.1,* p. *161*; *14.2,* p. *216*
M. Razeghi	(b) **89,**	K135 (1978)	*1.2,* p. *23*; *19,* p. *313*; 22.8.1, p. 516
C. Razzetti	(b) **87,**	479 (1978)	*2,* p. *45*; *20.1,* p. *327*; 22, p. 413
M. N. Read	(a) **50,**	203 (1978)	*10.1,* p. *107*; *13.4,* p. *184*; 21.1, p. 381
I. Rechenberg	(b) **86,**	K21 (1978)	*20.1,* p. *323*; 22, p. 410
	(a) **47,**	K31 (1978)	*14.3.1,* p. *227*; 22, p. 413
	(b) **88,**	675 (1978)	13.4, p. 178; 20.1, p. 330; *20.2,* p. *339*; 22, p. 415
P. P. Reddy	(b) **89,**	K61 (1978)	*20.1,* p. *333*; 22.9, p. 527
Y. P. Reddy	(b) **89,**	K61 (1978)	*20.1,* p. *333*; 22.9, p. 527
W. Rehwald	(b) **90,**	61 (1978)	*1.2,* p. *23*; 7, p. 82; 12.2, p. 141; 22.9, p. 528
W. Reich	(b) **90,**	K141 (1978)	*18.2,* p. *292*; 18.4, p. 302
P. Reiche	(a) **49,**	305 (1978)	3, p. 56; 20.1, p. 333; *20.2,* p. *339*; *20.3,* p. *347*; 22.8.2, p. 520
J. Reichel	(a) **50,**	87 (1978)	9, p. 94; *10.2,* p. *120*; 22.1.2, p. 438
J. Reichelt	(b) **89,**	489 (1978)	*13.2,* p. *163*; 21, p. 363
W. Reichelt	(a) **45,**	K77 (1978)	*14.1,* p. *205*; *14.3.1,* p. *226*; 22.6, p. 491
	(a) **48,**	377 (1978)	*1.2,* p. *21*; 6, p. 75; 14.3, p. 223; 22.6, p. 495
H. Reichl	(a) **47,**	489 (1978)	*14.3.3,* p. *235*; *20.3,* p. *345*; 22.2.1, p. 447
M. B. O'Reilly	see under letter "O"		
B. Reimer	(b) **85,**	145 (1978)	*16,* p. *262*; 22.9, p. 522
G. Rein	(a) **45,**	253 (1978)	*9,* p. *91*; 10.1, p. 99; 21.6, p. 401
P. Reineker	(b) **89,**	131 (1978)	*13.5.1,* p. *194*
J. Reinhold	(b) **90,**	309 (1978)	*13.4,* p. *183*; *19,* p. *313*; 22.6, p. 497
U. Reinhold	(a) **49,**	323 (1978)	*1.2,* p. *22*; *10.1,* p. *106*; 14.2, p. 215; 21.1, p. 377
	(a) **49,**	K125 (1978)	*9,* p. *94*; 21.1, p. 379
F. J. Reiter	(b) **90,**	45 (1978)	*13.1,* p. *159*; *13.5.2,* p. *199*; 21.2, p. 393; 21.3, p. 395
M. Rekik	(a) **47,**	311 (1978)	*1.1,* p. *12*; 1.4, p. 30; 22.6, p. 493
R. Rėksnys	(a) **50,**	K63 (1978)	*7,* p. *82*; *16,* p. *267*; 22.4.1, p. 466; 22.4.2, p. 469; 22.4.3, p. 471
D. Renard	(a) **47,**	311 (1978)	*1.1,* p. *12*; 1.4, p. 30; 22.6, p. 493
	(a) **49,**	267 (1978)	*1.4,* p. *33*; 1.5, p. 38; 21.6, p. 404
P. Rennert	(b) **87,**	221 (1978)	*13.1,* p. *154*; 20, p. 316; 21.3, p. 394
B. I. Reser	(b) **87,**	41 (1978)	*13.1,* p. *154*; 22.1.2, p. 432
R. Resta	(b) **86,**	627 (1978)	*13.5.1,* p. *191*; 23, p. 530
G. Reuter	(b) **89,**	109 (1978)	*10.2,* p. *117*; 13.4, p. 180; *20.1,* p. *332*; 22.5.2, p. 483
T. Rewaj	(b) **87,**	K155 (1978)	*19,* p. *308*; 22.5.3, p. 487
	(b) **88,**	K65 (1978)	*19,* p. *309*; 22.5.3, p. 488
N. Rezlescu	(a) **50,**	K225 (1978)	*18.3,* p. *297*; 22.8.2, p. 521
M. Richard	(b) **90,**	77 (1978)	*13.5.1,* p. *195*; *20.1,* p. *335*; 22.4.1, p. 466
D. D. Richardson	(b) **86,**	K97 (1978)	*4,* p. *57*; *13,* p. *144*
A. Richter	(b) **88,**	189 (1978)	13, p. 146; *18,* p. *274*; 18.1, p. 278; 18.2, p. 288; 18.4, p. 300
F. W. Richter	(b) **89,**	69 (1978)	*10.2,* p. *117*; 11, p. 128; 22.1.2, p. 436

J. Richter	(b) **85**, 233 (1978)	*2*, p. *42*; 13, p. 142; *18.2*, p. *281*
	(a) **46**, K25 (1978)	*14.3.3*, p. *234*; 22, p. 411
	(b) **87**, K89 (1978)	*2*, p. *45*; *13*, p. *145*; 18.2, p. 287
	(a) **47**, K161 (1978)	*14.2*, p. *214*
	(b) **89**, 143 (1978)	*2*, p. *48*; *14.1*, p. *210*; 21, p. 362
	(b) **89**, 537 (1978)	13.1, p. 159; *14.1*, p. *211*; 21, p. 363
R. Richter	(b) **86**, 177 (1978)	*14.1*, p. *206*; 18.2, p. 283; 21, p. 355
	(b) **88**, 485 (1978)	*13.1*, p. *157*; 21.1, p. 376
	(b) **89**, 77 (1978)	13, p. 147; *14.1*, p. *209*; 18.4, p. 301; 21, p. 362
W. Richter	(a) **46**, K67 (1978)	*14.2*, p. *214*; 18.2.1, p. 293; 21.1.1, p. 386
	(a) **48**, 513 (1978)	*14.2*, p. *215*; 21, p. 362
	(a) **49**, K209 (1978)	*14.2*, p. *215*
R. de Ridder	(a) **48**, 383 (1978)	*1.3*, p. *27*; 1.4, p. 32; 21, p. 361; 21.1, p. 376
	(a) **49**, 655 (1978)	*1.1*, p. *14*; *1.4*, p. *33*; 21.1, p. 379; 21.6, p. 405
	(a) **50**, 349 (1978)	*1.1*, p. *14*; *1.4*, p. *34*; 22, p. 420
	(a) **50**, K67 (1978)	*1.6*, p. *41*; 22, p. 420
	(a) **50**, 579 (1978)	*1.2*, p. *24*; *1.3*, p. *28*; 10.2, p. 122; 22.3, p. 459
	(a) **50**, 587 (1978)	*1.3*, p. *28*; *1.4*, p. *34*; 10.1, p. 108; 21, p. 366; 21.1, p. 382
C. Ridou	(a) **48**, 67 (1978)	*1.2*, p. *21*; *8*, p. *87*; 22.8.1, p. 514
V. Riede	(b) **87**, K65 (1978)	*20.1*, p. *326*; 22.3, p. 458
C. Riekel	(a) **50**, K231 (1978)	*1.1*, p. *14*; 22, p. 421
H. Riemann	(a) **46**, 117 (1978)	13.4, p. 173; *14.3*, p. *219*; 22.1.2, p. 431
M. Rigdahl	(a) **50**, 117 (1978)	*12*, p. *133*
N. Rihon	(a) **49**, 697 (1978)	*17.1*, p. *272*; 21.1, p. 379
A. Rimeika	(a) **45**, K113 (1978)	*6*, p. *66*; 13.4, p. 171; *14.3.4*, p. *240*; 22.4.1, p. 463
E. Rimini	(a) **47**, 533 (1978)	*11*, p. *127*; 22.1.2, p. 433
U. Rindelhardt	(a) **47**, 157 (1978)	8, p. 85; *10.1*, p. *103*; 14.1, p. 207; 18.2, p. 286; 21.1, p. 373; 21.1.1, p. 387
L. Risch	(b) **88**, 111 (1978)	*13.3*, p. *166*; 13.5.1, p. 192; 20.1, p. 328; 22.4.1, p. 464
	(a) **49**, 177 (1978)	*13.3*, p. *167*; 16, p. 266; 20.1, p. 333; 22.4.1, p. 465
S. Ristić	(a) **50**, 153 (1978)	*14.3.3*, p. *237*; 22.1.2, p. 438
I. G. Ritchie	(a) **46**, 89 (1978)	*10.1*, p. *101*; 11, p. 125; 21.1.1, p. 385
	(a) **50**, 617 (1978)	*10.1*, p. *108*; *12.1*, p. *137*; 21.1.1, p. 391
J. P. Rivière	(a) **45**, 157 (1978)	10.1, p. 99; *14.1*, p. *205*; 21.1.1, p. 384
	(a) **47**, 469 (1978)	*10.1*, p. *104*; 14.1, p. 208; 21.1, p. 374; 21.1.1, p. 387
A. Rizzo	(a) **48**, 293 (1978)	*13.4*, p. *179*; *14.3*, p. *223*; 22.3.1, p. 460
D. J. Robbins	(a) **50**, 423 (1978)	2, p. 52; 13.4, p. 185; *14.3.4*, p. *247*; 22.1, p. 423
H. Robert	(b) **85**, 603 (1978)	*18*, p. *273*; 19, p. 304; 22.9, p. 523
W. H. Robinson	(a) **48**, 155 (1978)	*10.2*, p. *115*; 22.5.2, p. 482; 22.5.3, p. 488
E. M. Rodina	(b) **85**, 331 (1978)	*6*, p. *65*; 22.9, p. 522
V. E. Rodionov	(a) **48**, 593 (1978)	*13.4*, p. *179*; *20.3*, p. *346*; 22.2.1, p. 448
R. Rodloff	(a) **49**, 445 (1978)	*10.1*, p. *106*; 11, p. 130; 21.1, p. 378
J. A. Rodríguez	(b) **90**, 409 (1978)	*20.1*, p. *335*; 22.2.1, p. 448
J. Roggen	(a) **45**, 439 (1978)	*10.1*, p. *100*; *11*, p. *124*; 21.1, p. 370
A. Roizès	(a) **50**, 399 (1978)	13.4, p. 185; *14.3*, p. *225*; 22.1.2, p. 439
J. M. Roland	(a) **50**, 41 (1978)	*1.3*, p. *28*; 14.1, p. 212; 21.1, p. 380
U. Roll	(b) **89**, 459 (1978)	6, p. 76; *14.3.1*, p. *228*; *20.1*, p. *334*; 22.9, p. 527
M. Rollé	(b) **90**, 689 (1978)	*6*, p. *78*; 11, p. 131; *20.1*, p. *337*; 22.1.2, p. 439
N. Romeo	(a) **47**, 371 (1978)	*14.3.1*, p. *227*; 22.4.3, p. 470
J. Roos	(a) **47**, 233 (1978)	*1.2*, p. *18*; 19, p. 307; 22.9, p. 525
	(a) **48**, 53 (1978)	*1.2*, p. *21*; 22.8, p. 506

G. Röpke	(b) **88**, 189 (1978)	*13*, p. *146*; *18*, p. *274*; 18.1, p. 278; 18.2, p. 288; 18.4, p. 300
	(b) **88**, 457 (1978)	*13*, p. *146*; *18*, p. *274*
	(b) **88**, K59 (1978)	*13.5*, p. *187*; *13.5.1*, p. *193*; 22, p. 415
	(b) **90**, 175 (1978)	*13.5*, p. *188*; *13.5.1*, p. *196*; 22, p. 419
R. Rosei	(b) **85**, 445 (1978)	*13.1*, p. *151*; 20.1, p. 321; 21, p. 354; 21.6, p. 401
P. Rosemann	(a) **50**, 537 (1978)	*18.3*, p. *297*; 21.1, p. 382; 21.4, p. 399
V. M. Rosenbaum	(b) **88**, 385 (1978)	*13*, p. *146*; *14.3.4*, p. *243*; 16, p. 265; 22, p. 415
M. Rosenberg	(a) **48**, K35 (1978)	*1.2*, p. *21*; *7*, p. *81*; 22.8, p. 506
G. I. Rosenman	(a) **50**, K255 (1978)	14.4.1, p. 258; *16*, p. *268*; *17*, p. *271*; 22.8.1, p. 517
J. Rosenthal	(b) **89**, K167 (1978)	*13.5.1*, p. *195*; *19*, p. *313*; 22, p. 418
M. Rosenzweig	(b) **90**, 77 (1978)	*13.5.1*, p. *195*; *20.1*, p. *335*; 22.4.1, p. 466
P. L. Rossiter	(a) **47**, 597 (1978)	*1.4*, p. *31*; *18.2*, p. *287*; 21.1, p. 374; 21.1.1, p. 387
	(a) **48**, 71 (1978)	*1.4*, p. *31*; *18.2*, p. *289*; 21.1, p. 375; 21.1.1, p. 388
U. Rössler	(b) **89**, 475 (1978)	12.2, p. 140; *13.4*, p. *181*; 22.2, p. 444
W. Rössler	(a) **49**, 103 (1978)	2, p. 49; *14.1*, p. *210*; *14.3.1*, p. *228*; 21.6, p. 404; 21.7, p. 407
C. J. Rossouw	(a) **45**, 277 (1978)	*1.5*, p. *36*; 13.5.2, p. 198; 22, p. 409; 22.1, p. 422
M. van Rossum	(b) **89**, K17 (1978)	*5*, p. *62*; 22.1.1, p. 427; 22.1.2, p. 436
S. Roth	(b) **87**, K69 (1978)	*14.3*, p. *220*; *18.4*, p. *300*; 21, p. 358; 21.1, p. 372
L. Rothkirch	(b) **90**, 517 (1978)	13.3, p. 168; *14.2*, p. *216*; 21, p. 365
	(b) **90**, K115 (1978)	*13.3*, p. *168*; *14.2*, p. *216*; 21, p. 366
D. Rouby	(a) **48**, 439 (1978)	*7*, p. *81*; 10.1, p. 105; 21, p. 361
A. Rousset	(a) **50**, 109 (1978)	*14.3*, p. *224*; 18.3, p. 296; 22.8.2, p. 520
M. K. Roy	(a) **45**, 549 (1978)	*7*, p. *80*; *13.3*, p. *164*; 14.3.2, p. 230; 22.4.1, p. 463; 22.6.1, p. 499
U. Roy	(a) **46**, K171 (1978)	*1.2*, p. *18*; 8, p. 85; 22.8.1, p. 513
B. S. H. Royce	(a) **50**, 669 (1978)	*13.4*, p. *186*; *14.3*, p. *225*; 22.6, p. 497
E. W. van Royen	(a) **48**, 497 (1978)	*1.3*, p. *27*; *10.1*, p. *105*; 21.1, p. 376
B. Rozenfeld	(b) **89**, K159 (1978)	*13*, p. *148*; 21.5, p. 400
E. I. Rozum	(a) **48**, K175 (1978)	11, p. 128; *20.3*, p. *346*; 22.5.2, p. 483
I. Rückmann	(b) **87**, K111 (1978)	*13.5.1*, p. *192*; *20.1*, p. *328*; 22.4.1, p. 464
Yu. V. Rud	(a) **45**, K95 (1978)	*16*, p. *262*; 22.8, p. 503
	(a) **49**, 761 (1978)	*11*, p. *130*; 13.4, p. 182; *14.3*, p. *224*; 22.8, p. 507
	(a) **50**, 139 (1978)	*10.2*, p. *120*; *20.3*, p. *350*; 22.8, p. 508
	(a) **50**, 379 (1978)	11, p. 131; 13.4, p. 185; *14.3*, p. *225*; *20.1*, p. *337*; 22.8, p. 508
	(a) **50**, 445 (1978)	14.3, p. 225; *16*, p. *267*; 22.8, p. 508
V. N. Rudko	(b) **88**, 407 (1978)	*9*, p. *93*; *10*, p. *96*
	(b) **89**, 61 (1978)	*2*, p. *48*; *14*, p. *204*
G. Rudloff	(a) **49**, K121 (1978)	*13.4*, p. *182*; *17*, p. *270*; 22, p. 419
R. M. de La Rue	(a) **46**, 595 (1978)	*14.3.3*, p. *234*; 16, p. 263; 22.1.2, p. 432; 22.4.1, p. 464
E. L. Rumyantsev	(b) **85**, 427 (1978)	*18.2*, p. *282*; *18.4*, p. *298*; 21.1, p. 369; 21.1.1, p. 384
A. F. Rumynina	(a) **45**, 393 (1978)	*10.1*, p. *100*; 21.1, p. 370; 21.1.1, p. 385
W. A. Runciman	(a) **46**, K135 (1978)	*20.1*, p. *325*; 22.6, p. 492
S. Rundqvist	(a) **49**, K153 (1978)	*3*, p. *56*; 18.2, p. 291; 21.1, p. 379; 21.1.1, p. 389
H. Runge	(a) **45**, 509 (1978)	*9*, p. *91*; *11*, p. *124*; 14.3, p. 218; 22.1.2, p. 430
W. Ruppel	(a) **49**, 169 (1978)	*16*, p. *266*; 22.4.1, p. 465

R. Ruppin	(b) **87**, 619 (1978)	6, p. 71; *20.1*, p. *328*; 22.2.1, p. 446; 22.6, p. 494
G. J. Russell	(a) **46**, 433 (1978)	*1.6*, p. *41*; 22, p. 412; 22.4.1, p. 464
A. I. Russu	(b) **86**, K71 (1978)	*15*, p. *259*; 21.7, p. 406
Sh. A. Rustamov	(b) **88**, K55 (1978)	*13.1*, p. *157*; 22.2.1, p. 447
J. Rużyłło	(a) **48**, 199 (1978)	*1.2*, p. *21*; 1.6, p. 41; *3*, p. *55*; 22.1.2, p. 434; 22.6, p. 494
J. F. Ryan	(b) **88**, K5 (1978)	*6*, p. *73*; *20.1*, p. *329*; 22.8, p. 506
C. Rybka	(b) **89**, 253 (1978)	*10.2*, p. *117*; 22.9, p. 526
A. Ryska	(b) **85**, 545 (1978)	*8*, p. *83*; 14.2, p. 213; 21, p. 354
V. M. Ryzhkovskii	(a) **47**, K177 (1978)	1.2, p. 20; *18.3*, p. *296*; *18.4*, p. *300*; 22.8, p. 505
H. Rzewuski	(b) **89**, 313 (1978)	*13.5.1*, p. *194*; *20.3*, p. *347*; 22.4.4, p. 473
R. Kh. Sabirov	(b) **90**, 99 (1978)	6, p. 77; *13*, p. *148*; 21, p. 364
F. Sacchetti	(b) **86**, 77 (1978)	*1.2*, p. *16*; *6*, p. *67*; 21.1, p. 371
A. V. Sachenko	(b) **88**, 797 (1978)	13.3, p. 166; *13.5.1*, p. *193*; *20.3*, p. *346*; 22.4.1, p. 465
K. Sachse	(a) **48**, 121 (1978)	*13*, p. *146*; *14.3.4*, p. *243*; 22.1.1, p. 426
I. P. Sadikov	(b) **86**, K59 (1978)	*6*, p. *68*; *18*, p. *273*; 22.7, p. 501
K. Saermark	(b) **87**, 527 (1978)	*13.2*, p. *162*; *14.1*, p. *208*; 21.6, p. 402
	(b) **89**, 117 (1978)	*13.2*, p. *163*; 14.1, p. 210; 21, p. 362
V. P. Sapronov	(b) **85**, 63 (1978)	*6*, p. *64*; *8*, p. *83*; 21.2, p. 392
	(b) **85**, 749 (1978)	*6*, p. *66*; *8*, p. *83*; 21.2, p. 392
B. B. Sahay	(b) **90**, 93 (1978)	2, p. 51; *12.1*, p. *136*; 21, p. 364; 21.1, p. 379; 21.2, p. 393
V. C. Sahni	(b) **90**, 415 (1978)	*6*, p. *78*; *20.1*, p. *336*; 22.5.4, p. 490
K. Saiki	(b) **88**, 607 (1978)	13.1, p. 157; *13.5.1*, p. *193*; *20.1*, p. *330*; 22, p. 415
S. Saito	(b) **90**, 697 (1978)	*18.3*, p. *297*; 22.8.2, p. 521
A. Sakakibara	(a) **48**, K141 (1978)	*14.1*, p. *209*; 21, p. 362
A. P. Sakalas	(a) **46**, K141 (1978)	10.2, p. 112; 14.3, p. 220; 22.4.2, p. 468
	(a) **47**, 187 (1978)	*10.2*, p. *113*; *14.3*, p. *220*; 22.4.2, p. 468
	(a) **48**, 329 (1978)	*13.4*, p. *179*; *14.3*, p. *223*; 22.3, p. 458
	(a) **48**, K43 (1978)	*13.4*, p. *178*; *16*, p. *265*; 22.4.3, p. 470
	(a) **50**, K63 (1978)	*7*, p. *82*; *16*, p. *267*; 22.4.1, p. 466; 22.4.2, p. 469; 22.4.3, p. 471
E. Yu. Salaev	(b) **87**, K115 (1978)	*6*, p. *72*; *20.1*, p. *328*; 22.3, p. 458
J. Salardenne	(b) **90**, 147 (1978)	*20.1*, p. *335*; 22.5, p. 476
X. I. Saldaña	(a) **49**, 437 (1978)	*16*, p. *266*; 22, p. 418
N. Saleh	(a) **49**, 393 (1978)	*14.3.3*, p. *236*; *16*, p. *266*; 22.2.3, p. 454
E. A. Salkov	(b) **90**, 355 (1978)	*13.4*, p. *183*; *20.3*, p. *349*; 22.4.1, p. 466
V. M. Salmanov	(a) **47**, K157 (1978)	*20.3*, p. *345*; 22.3.1, p. 460
A. N. Salugin	(b) **87**, K101 (1978)	*5*, p. *61*; 18.3, p. 295; 22.8.2, p. 519
R. A. Salukvadze	(a) **50**, K213 (1978)	*1.4*, p. *35*; *10.2*, p. *122*; 22.1, p. 423
D. Samaras	(a) **46**, 493 (1978)	*18.4*, p. *299*; 22.8.1, p. 512
	(a) **50**, 635 (1978)	*18.4*, p. *302*; 22.8.1, p. 516
V. V. Samartsev	(a) **45**, 647 (1978)	*7*, p. *80*; *20*, p. *316*
	(a) **49**, 789 (1978)	*20*, p. *317*
M. Samoć	(a) **46**, 251 (1978)	*13.3*, p. *165*; 14.3, p. 219; 16, p. 263; 22.9, p. 524
A. G. Samoilovich	(b) **86**, 765 (1978)	13, p. 144; *14.3*, p. *219*; 22.2.1, p. 446
C. F. Sampson	(a) **49**, K177 (1978)	*18.3*, p. *296*; 18.4, p. 301; 21.1, p. 379
C. Sánchez	(b) **89**, 281 (1978)	*20.1*, p. *332*; 22.5.2, p. 484
M. van Sande	(a) **45**, 553 (1978)	*1.2*, p. *16*; *1.4*, p. *29*; 21.1, p. 370
	(a) **48**, 383 (1978)	*1.3*, p. *27*; 1.4, p. 32; 21, p. 361; 21.1, p. 376
	(a) **50**, 587 (1978)	*1.3*, p. *28*; *1.4*, p. *34*; 10.1, p. 108; 21, p. 366; 21.1, p. 382
W. Sander	(b) **89**, 459 (1978)	6, p. 76; *14.3.1*, p. *228*; 20.1, p. 334; 22.9, p. 527
S. Sapru	(a) **48**, K189 (1978)	*10.2*, p. *117*; *20.3*, p. *347*; 22.5.2, p. 483

J. Saraie	(a) 47, 123 (1978)	13.3, p. 165; *14.3.2*, p. *230*; 14.3.4, p. 242; 22.4.3, p. 470
R. S. Saraswat	(b) 87, K39 (1978)	*19*, p. *307*; 22.8, p. 505
O. G. Sarbey	(a) 48, 323 (1978)	*14.3.4*, p. *243*; 22.1.2, p. 435
R. M. Sardarly	(b) 87, K115 (1978)	*6*, p. *72*; *20.1*, p. *328*; 22.3, p. 458
	(b) 88, K5 (1978)	*6*, p. *73*; *20.1*, p. *329*; 22.8, p. 506
V. S. Sardaryan	(b) 85, 459 (1978)	*6*, p. *65*; *13*, p. *142*
C. K. Sarkar	(b) 88, K109 (1978)	13.1, p. 157; *15*, p. *260*; 22.2.3, p. 453
S. K. Sarkar	(b) 87, 517 (1978)	*4*, p. *58*; 12.2, p. 139; 22.5.2, p. 481
S. E. Sarkisov	(a) 49, 305 (1978)	3, p. 56; 20.1, p. 333; *20.2*, p. *339*; *20.3*, p. *347*; 22.8.2, p. 520
	(a) 50, 745 (1978)	13.4, p. 186; *20.1*, p. *338*; *20.2*, p. *340*; 22.8.1, p. 516
P. R. Sarode	(b) 88, K35 (1978)	*4*, p. *58*; *20*, p. *317*; 22.5.2. p. 482
G. Sarrabayrouse	(a) 46, 185 (1978)	*13.3*, p. *164*; *14.3.3*, p. *234*
	(a) 49, 603 (1978)	*14.3.3*, p. *237*; *16*, p. *267*
V. I. Sarrak	(a) 45, K159 (1978)	*10.1*, p. *101*; 21.1.1, p. 385
S. B. S. Sastry	(b) 85, K15 (1978)	*20.3*, p. *341*; 22.5.2, p. 478
	(b) 87, K163 (1978)	*20.3*, p. *341*; 22.5.2, p. 478
	(a) 47, 711 (1978)	*10.2*, p. *114*; 11, p. 127; *20.1*, p. *328*; 22.5.2, p. 481
	(a) 48, K189 (1978)	*10.2*, p. *117*; *20.3*, p. *347*; 22.5.2, p. 483
	(b) 90, 375 (1978)	*10.2*, p. *120*; *20.3*, p. *349*; 22.5.2, p. 485
G. Sivarama Sastry	(b) 86, K127 (1978)	*19*, p. *306*; 22.8.1, p. 512
S. Saterlie	(a) 48, 551 (1978)	*1.3*, p. *27*; *10.1*, p. *105*; 21, p. 362; 21.1, p. 376
M. Satishkumar	(b) 90, 117 (1978)	*6*, p. *77*; 12.1, p. 136; 21.2, p. 393
H. Sato	(b) 89, 571 (1978)	*14.1*, p. *211*; 21, p. 363
V. A. Savastenko	(a) 48, K135 (1978)	*12.2*, p. *140*; 22.2, p. 443
V. V. Savelev	(b) 87, 255 (1978)	*13.5.1*, p. *191*; 20.1, p. 326; 20.3, p. 344; 22.9, p. 524
M. M. Savin	(a) 47, 45 (1978)	*10.1*, p. *103*; 21, p. 358
E. J. Savino	(a) 45, 315 (1978)	*10.1*, p. *100*; 21, p. 353; 21.1, p. 369
	(a) 47, 425 (1978)	*10.1*, p. *103*; 21, p. 359; 21.1, p. 374
E. M. Savitskii	(a) 49, K117 (1978)	*8*, p. *88*; *14.1*, p. *211*; 21, p. 364
Sh. Sawada	(a) 50, K157 (1978)	*14.4.1*, p. *257*; *19*, p. *314*; 22.8, p. 509
B. D. Sawicka	(b) 86, K159 (1978)	*5*, p. *61*; 22.1.3, p. 441
J. A. Sawicki	(b) 86, K159 (1978)	*5*, p. *61*; 22.1.3, p. 441
Praveen Saxena	(a) 46, 321 (1978)	*14.3.3*, p. *234*; 22.2.1, p. 445
U. Saxena	(a) 49, 371 (1978)	*1.5*, p. *38*; 2, p. 49; 13.1, p. 158; *14.3.1*, p. *228*; 22.6, p. 496
V. K. Saxena	(b) 86, K163 (1978)	*2*, p. *44*; *18.2*, p. *285*
	(b) 88, K87 (1978)	*2*, p. *46*; *18.2*, p. *288*
M. Sayer	(a) 49, 773 (1978)	*14.4*, p. *251*; 22.6, p. 496
G. Sberveglieri	(a) 47, 371 (1978)	*14.3.1*, p. *227*; 22.4.3, p. 470
A. Scalabrin	(b) 86, 373 (1978)	*14.4.1*, p. *254*; 22.8.1, p. 511
M. A. F. Scarparo	(b) 90, 543 (1978)	*14.4.1*, p. *257*; 20.1, p. 337; 22.8.1, p. 516
L. C. Scavarda do Carmo	see under letter "C"	
P. Schäfer	(a) 47, K31 (1978)	*14.3.1*, p. *227*; 22, p. 413
	(b) 88, 675 (1978)	13.4, p. 178; 20.1, p. 330; *20.2*, p. *339*; 22, p. 415
R. Schäfer	(a) 50, 475 (1978)	2, p. 52; *18.2*, p. *292*; *20*, p. *318*; 21.1.1, p. 390
R. Schalge	(b) 87, 179 (1978)	*14.4.1*, p. *254*; 22.8.1, p. 513
F. W. Schapink	(a) 46, 653 (1978)	*10.1*, p. *102*; 21.6, p. 402
A. Scharmann	(b) 85, K15 (1978)	*20.3*, p. *341*; 22.5.2, p. 478
	(b) 87, K35 (1978)	*10.2*, p. *113*; 11, p. 126; *19*, p. *307*; 22.8.1, p. 513

	(b) **87**, K163 (1978)	*20.3*, p. *341*; 22.5.2, p. 478
	(b) **89**, 375 (1978)	*19*, p. *312*; 20.3, p. 348; 22.8.1, p. 515
	(b) **89**, 449 (1978)	*13.4*, p. *181*; *19*, p. *312*; 22.8.1, p. 516
F. Schauer	(a) **49**, K157 (1978)	*2*, p. *51*; 13.3, p. 167; *14.3.4*, p. *245*
G. Scheler	(b) **87**, K53 (1978)	*19*, p. *307*; 22.9, p. 525
S. Sckicktanz	(b) **87**, 565 (1978)	*6*, p. *71*; 14.2, p. 214; *20.1*, p. *327*
S. Schiefer	(a) **50**, 147 (1978)	*8*, p. *89*; 14.1, p. 212; 21.1, p. 381
W. Schiller	(b) **86**, 231 (1978)	*14.1*, p. *206*; 21, p. 355
	(b) **89**, 143 (1978)	*2*, p. *48*; *14.1*, p. *210*; 21, p. 362
R. Schimko	(a) **49**, 279 (1978)	*9*, p. *94*; *11*, p. *129*; 22.2.4, p. 456
K. Schindler	(a) **48**, K199 (1978)	*20.2*, p. *339*; 22.1.3, p. 441
C. J. Schinkel	(a) **45**, K39 (1978)	*18.2*, p. *282*; 22.8, p. 503
W. Schlapp	(a) **50**, K219 (1978)	*14.3.4*, p. *247*; 22.2.2, p. 451
P. Schlotter	(a) **48**, K15 (1978)	*13.3*, p. *166*; *16*, p. *264*; 22.9, p. 526
W. Schmatz	(a) **46**, 557 (1978)	1.3, p. 25; *10.1*, p. *102*; 21, p. 357
H. Schmid	(a) **49**, 539 (1978)	*10.1*, p. *106*; 12.1, p. 136; 21.1.1, p. 389
A. Schmidt	(a) **46**, 179 (1978)	*1.3*, p. *25*; 22.8.2, p. 518
G. Schmidt	(a) **46**, 533 (1978)	*14.4.1*, p. *254*; *20.1*, p. *325*; 22.9, p. 524
	(a) **50**, K109 (1978)	12.2, p. 141; *14.4.1*, p. *257*; 22.8.1, p. 516
G. Schmieder	(b) **89**, 431 (1978)	13.5.1, p. 195; *20.2*, p. *339*; *20.3*, p. *348*; 22.4, p. 462; 22.4.1, p. 465; 22.4.3, p. 471
B. Schnabel	(a) **50**, K209 (1978)	11, p. 132; *19*, p. *314*; 22.1.2, p. 440
G. Schneider	(b) **88**, 105 (1978)	*13.4*, p. *177*; 14.1, p. 208; *14.3*, p. *221*; 22.7, p. 502
J. Schneider	(a) **45**, K47 (1978)	*2*, p. *42*; *18.2*, p. *282*; 21.1, p. 369; 21.1.1, p. 384
M. Schneider	(a) **46**, K71 (1978)	*9*, p. *92*; 22.2.2, p. 450
Ch. Schnittler	(a) **45**, K179 (1978)	13.4, p. 171; *14.3.4*, p. *241*; 22, p. 410
	(a) **48**, 357 (1978)	13.4, p. 179; *14.3.4*, p. *244*; 22, p. 416
Ch. Schober	(b) **86**, 249 (1978)	13.4, p. 172; *14.1*, p. *206*; 18, p. 273; 21, p. 355
D. Schoemaker	(b) **88**, 615 (1978)	*10.2*, p. *116*; *20.1*, p. *330*; 22.5.2, p. 482
E. Schöll	(a) **50**, 423 (1978)	2, p. 52; 13.4, p. 185; *14.3.4*, p. *247*; 22.1, p. 423
R. Schöllhorn	(a) **50**, K231 (1978)	*1.1*, p. *14*; 22, p. 421
R. Schöpke	(b) **90**, K95 (1978)	*13.4*, p. *184*; *14.1*, p. *212*; 21, p. 365
J. Schreiber	(b) **88**, K43 (1978)	*13*, p. *146*; *18.2*, p. *288*
H. Schrey	(b) **90**, 67 (1978)	13.4, p. 183; *13.5.1*, p. *195*; *20.3*, p. *349*; 22.4.1, p. 466
P. Schrey	(b) **85**, 552 (1978)	*10.2*, p. *109*; *11*, p. *123*; 22.5.2, p. 478
U. Schröder	(b) **89**, 349 (1978)	*13.4*, p. *181*; *13.5.1*, p. *194*; 22, p. 418
G. Schroeder	(a) **47**, 145 (1978)	*10.1*, p. *103*; 18.2, p. 286; 21.1.1, p. 387
	(a) **47**, 397 (1978)	*10.1*, p. *103*; *18.2*, p. *287*; 21.1, p. 374
	(a) **50**, 475 (1978)	2, p. 52; *18.2*, p. *292*; *20*, p. *318*; 21.1.1, p. 390
D. Schroeer	(b) **87**, 325 (1978)	*5*, p. *61*; 10.2, p. 113; 22.5.3, p. 487
W. Schröter	(a) **49**, 489 (1978)	*5*, p. *62*; 9, p. 94; *10.2*, p. *119*; 22.1.2, p. 437
E. Schuberth	(b) **90**, 189 (1978)	*13.5.1*, p. *196*; *20.3*, p. *349*; 23, p. 531
R. Schuchardt	(b) **85**, 155 (1978)	*16*, p. *262*; 22, p. 408
E. M. Schulson	(a) **50**, 83 (1978)	*14.1*, p. *212*; 21.1, p. 380
D. Schultze	(a) **49**, 305 (1978)	3, p. 56, 20.1, p. 333; *20.2*, p. *339*; *20.3*, p. *347*; 22.8.2, p. 520
G. Schulz	(a) **45**, 89 (1978)	*10.1*, p. *99*; 21.1, p. 369
M. Schulz	(a) **49**, 483 (1978)	*14.3.4*, p. *245*; 22, p. 419
W. Schumacher	(b) **86**, 621 (1978)	*18.2*, p. *285*; *18.4*, p. *299*
B. Schumann	(b) **85**, K57 (1978)	*13.1*, p. *150*; *20.1*, p. *320*; 22.8, p. 503
P. Schuster	(b) **86**, 385 (1978)	*13.1*, p. *153*; 22.9, p. 523
R. Schuttler	(a) **50**, 399 (1978)	13.4, p. 185; *14.3*, p. *225*; 22.1.2, p. 439
J. U. von Schütz	(a) **49**, 563 (1978)	1.2, p. 23; 6, p. 77; *19*; p. *313*; 22.9, p. 528

C. Schwab	(a) **46**, K157 (1978)	*19*, p. *307*; 22.6, p. 493
	(b) **87**, 261 (1978)	*13.5.1*, p. *191*; *20.1*, p. *326*; 22.6, p. 493
	(b) **88**, 327 (1978)	*6*, p. *73*; 7, p. 81; *20.1*, p. *329*; 22.5, p. 475
	(a) **48**, K145 (1978)	*10.2*, p. *116*; 19, p. 311; 20.1, p. 331; 22.8, p. 506
M. Schwab	(b) **86**, 195 (1978)	*1.2*, p. *16*; 13.4, p. 172; 18.4, p. 298; 20.1, p. 322; 22.8.1, p. 511
D. Schwabe	(b) **85**, K15 (1978)	*20.3*, p. *341*; 22.5.2, p. 478
	(b) **87**, K35 (1978)	*10.2*, p. *113*; 11, p. 126; *19*, p. *307*; 22.8.1, p. 513
	(b) **87**, K163 (1978)	*20.3*, p. *341*; 22.5.2, p. 478
	(b) **89**, 375 (1978)	*19*, p. *312*; 20.3, p. 348; 22.8.1, p. 515
R. Schwabe	(b) **87**, 169 (1978)	*13.4*, p. *174*; *20.3*, p. *344*; 22.2.2, p. 450
	(b) **89**, 561 (1978)	*13.5.1*, p. *195*; *20.3*, p. *348*; 22.2.2, p. 451
	(b) **90**, 233 (1978)	*13.5.1*, p. *196*; *20.3*, p. *349*; 22.3, p. 459
G. Schwarz	(a) **49**, 279 (1978)	*9*, p. *94*; *11*, p. *129*; 22.2.4, p. 456
L. Schweitzer	(b) **88**, 531 (1978)	*2*, p. *47*; *16*, p. *265*; 22.1.2, p. 435
G. H. Schwuttke	(a) **48**, 335 (1978)	*1.5*, p. *37*; 3, p. 55; 22.3, p. 458
	(a) **50**, 221 (1978)	*10.2*, p. *120*; *14.3.3*, p. *238*; 22.1.2, p. 438
M. Scrocco	(b) **86**, K1 (1978)	*6*, p. *68*; *20.1*, p. *323*; 22.8.1, p. 511
V. L. Sedov	(a) **50**, 757 (1978)	*10.1*, p. *108*; 21.1, p. 382; 21.1.1, p. 391
R. Sedrakyan	(a) **45**, 387 (1978)	*14.3.4*, p. *240*; 22.2.3, p. 452
V. D. Sedykh	(b) **88**, K103 (1978)	*13.4*, p. *179*; 21.1.1, p. 388
K. Seeger	(b) **89**, 187 (1978)	*14.1*, p. *210*; 22, p. 417
D. Segers	(a) **48**, 133 (1978)	*10.1*, p. *104*; 21.7, p. 406
S. M. Seid-Rzaeva	(b) **87**, 447 (1978)	*13.4*, p. *175*; *14.3*, p. *221*; 22.4.4, p. 473
P. Seidel	(a) **46**, K25 (1978)	*14.3.3*, p. *234*; 22, p. 411
H. Seidlitz	(b) **85**, 45 (1978)	*13.5.1*, p. *189*; 20.3, p. 341; 22.9, p. 522
B. Seifert	(b) **85**, 569 (1978)	*13.1*, p. *151*; *13.5.1*, p. *190*; 20, p. 315; 22.1, p. 422
G. Seifert	(b) **89**, 553 (1978)	*13*, p. 148; 21, p. 363
	(b) **89**, K175 (1978)	*13.1*, p. *159*; 21, p. 363
T. Seisenbaev	(a) **50**, K251 (1978)	13.4, p. 186; *20.3*, p. *351*; 22.2.1, p. 449; 22.2.3, p. 455
R. P. Seisyan	(b) **88**, K177 (1978)	*13.1*, p. *158*; *13.5.1*, p. *194*; 22.2.1, p. 447
E. Seitz	(a) **46**, 557 (1978)	1.3, p. 25; *10.1*, p. *102*; 21, p. 357
R. Seldén	(a) **50**, 117 (1978)	*12*, p. *133*
J. Seliger	(a) **46**, 239 (1978)	*10.2*, p. *111*; *19*, p. *306*; 22.9, p. 524
D. J. Sellmyer	(b) **88**, K127 (1978)	*2*, p. *47*; *18.2*, p. *289*
G. Ya. Selyutin	(b) **88**, K79 (1978)	18.1, p. 278; 18.2, p. 288; *20.1*, p. *330*
M. A. Semary	(a) **45**, K5 (1978)	*18.2*, p. *281*; 21.1, p. 369
S. V. Semenovskaya	(b) **87**, 733 (1978)	*1.3*, p. *26*; 21.1.1, p. 387
A. K. Semenyuk	(a) **50**, K169 (1978)	11, p. 131; *14.3*, p. *225*; 22.1.2, p. 440
S. C. Sen	(b) **90**, 671 (1978)	*13.4*, p. *185*; 20.1, p. 337; 20.3, p. 350; 22.5, p. 476
	(a) **50**, 323 (1978)	*13.4*, p. *184*; *20.1*, p. *336*; 22.5.4, p. 490
S. K. Sen	(a) **45**, 657 (1978)	*9*, p. *91*; 21.1, p. 370; 21.1.1, p. 385; 21.6, p. 401
J. P. Senateur	(a) **45**, 591 (1978)	*18.2*, p. *283*; *18.4*, p. *298*; 22.8, p. 503
S. Sengupta	(b) **87**, 517 (1978)	*4*, p. *58*; 12.2, p. 139; 22.5.2, p. 481
W. Senske	(b) **89**, 467 (1978)	*13.4*, p. *181*; *20.3*, p. *348*; 22.2.2, p. 451
V. V. Serdyuk	(a) **50**, 717 (1978)	*13.4*, p. *186*; *14.3.4*, p. *247*; 16, p. 267; 22.4.2, p. 469
P. P. Seregin	(a) **45**, 11 (1978)	*5*, p. *60*; *11*, p. *123*; 22.5, p. 474; 22.8, p. 503; 22.9, p. 522

A. Serpi	(a) **50,** 643 (1978)	13.4, p. 186; *16*, p. *267*; 22.8, p. 508
J. Serughetti	(a) **45,** 453 (1978)	*10.2*, p. *110*; 22.8.1, p. 511
M. Servidori	(a) **49,** 347 (1978)	*11*, p. *129*; 22.1.2, p. 437
R. A. Serway	(a) **48,** 165 (1978)	*19*, p. *310*; 22.8.2, p. 519
	(b) **89,** 267 (1978)	13.4, p. 180; *19*, p. *311*; 22.8.2, p. 520
S. A. Seryshev	(b) **90,** K35 (1978)	*19*, p. *314*; 22.5, p. 476
V. Seshubai	(a) **45,** K51 (1978)	*19*, p. *303*; 22.7, p. 501
B. R. Sethi	(a) **46,** 717 (1978)	*13.4*, p. *174*; *14.3*, p. *220*; 22.4.2, p. 468
	(a) **47,** 699 (1978)	*13.4*, p. *176*; *14.3*, p. *221*; 22.4.2, p. 468
V. F. Shabanov	(a) **45,** 53 (1978)	1.2, p. 15; *14.4.1*, p. *253*; 19, p. 303; 20, p. 315; 22.8.1, p. 510
A. V. Shablovskii	(a) **50,** K205 (1978)	5, p. 63; *18.2*, p. *292*; *18.3*, p. *297*; 21, p. 367
Ya. L. Shaiovich	(a) **48,** 543 (1978)	*1.2*, p. *21*; 10.2, p. 116; 22.8.2, p. 519
G. N. Shakhlevich	(a) **47,** 219 (1978)	*18.4*, p. *300*; 22, p. 413
J. Shanker	(b) **89,** 213 (1978)	*4*, p. *59*; 6, p. 75; 14.4, p. 250; 22.5.2, p. 483
C. W. Shanley	(b) **88,** 173 (1978)	*13.1*, p. *156*; *20.1*, p. *329*; 21.1, p. 375
V. P. Shantarovich	(a) **45,** K159 (1978)	*10.1*, p. *101*; 21.1.1, p. 385
	(a) **48,** K97 (1978)	*11*, p. *128*; 21.1, p. 375
	(a) **48,** K169 (1978)	*2*, p. *48*; *10.2*, p. *117*; 22, p. 416
	(a) **50,** K213 (1978)	*1.4*, p. *35*; *10.2*, p. *122*; 22.1, p. 423
V. B. Sharan	(a) **50,** 323 (1978)	*13.4*, p. *184*; *20.1*, p. *336*; 22.5.4, p. 490
	(b) **90,** 671 (1978)	*13.4*, p. *185*; 20.1, p. 337; 20.3, p. 350; 22.5, p. 476
J. Sharma	(a) **46,** 153 (1978)	*18.1*, p. *277*; *19*, p. *305*; 22.9, p. 523
O. P. Sharma	(b) **86,** 483 (1978)	4, p. 57; *6*, p. *68*; 14.1, p. 207; 21.1, p. 371; 21.6, p. 401
	(a) **47,** 699 (1978)	*13.4*, p. *176*; *14.3*, p. *221*; 22.4.2, p. 468
P. K. Sharma	(b) **86,** K167 (1978)	*18.1*, p. *278*; 21.2, p. 392
	(b) **88,** 809 (1978)	*12.1*, p. *136*; 13.2, p. 162; 18.1, p. 278; 21.2, p. 393
Vijender Sharma	(a) **46,** 321 (1978)	*14.3.3*, p. *234*; 22.2.1, p. 445
V. M. Shatalov	(b) **87,** 79 (1978)	*13.1*, p. *154*; 13.5.1, p. 191; 20.1, p. 325; 22, p. 412
	(b) **88,** 129 (1978)	*13.1*, p. *155*; 13.5.1, p. 192; 20.1, p. 328; 22, p. 414
D. Shaw	(a) **49,** 593 (1978)	10.2, p. 119; *20.1*, p. *335*; 22.4.3, p. 471
V. N. Shchemelev	(a) **46,** 343 (1978)	*1.1*, p. *11*; 1.6, p. 40; 17, p. 269; 22.1.2, p. 431
N. F. Shchukin	(a) **47,** 39 (1978)	*5*, p. *61*; 18.4, p. 300; 22.6, p. 493
M. I. Sheboul	(b) **86,** 535 (1978)	*13.5.1*, p. *191*; 22.5, p. 474
S. S. Sheinin	(a) **45,** 289 (1978)	*1.4*, p. *29*; 21.1, p. 369
M. K. Sheinkman	(a) **48,** 249 (1978)	*13.4*, p. *178*; *16*, p. *264*; 22.4.1, p. 465
	(a) **50,** 767 (1978)	*13.4*, p. *186*; *14.3*, p. *225*; 16, p. 268; 22.4.1, p. 467
E. F. Sheka	(b) **85,** 331 (1978)	*6*, p. *65*; 22.9, p. 522
A. V. Shekoyan	(a) **48,** 23 (1978)	*7*, p. *81*
A. U. Sheleg	(a) **48,** K135 (1978)	*12.2*, p. *140*; 22.2, p. 443
V. A. Shenderovskii	(b) **88,** 179 (1978)	*13.4*, p. *177*; 14.3, p. 222; *15*, p. *260*; 22.1.3, p. 441
A. V. Shendrik	(b) **85,** 343 (1978)	2, p. 42; *13.4*, p. *169*; *19*, p. *303*; 22.8.1, p. 510
E. M. Shereghii	(b) **90,** 11 (1978)	*6*, p. *64*; 13, p. 142; *19*, p. *303*; 22, p. 408; 22.2.3, p. 452; 22.4.1, p. 463
A. V. Sherman	(b) **85,** 51 (1978)	*6*, p. *64*; *13.5.1*, p. *189*; 20.1, p. 319; 20.3, p. 341; 22, p. 408
Y. Shimada	(a) **47,** K119 (1978)	*2*, p. *46*; *18.2.1*, p. *293*; 21.1.1, p. 387
T. Shimizu	(b) **88,** K1 (1978)	*2*, p. *46*; *19*, p. *309*; 22, p. 414
Y. Shiroishi	(a) **50,** K157 (1978)	*14.4.1*, p. *257*; *19*, p. *314*; 22.8, p. 509
A. G. Shishkov	(a) **50,** K101 (1978)	*18.2.1*, p. *294*; 21.1, p. 381; 21.1.1, p. 390

G. K. Shivakumar	(a) **47**, K1 (1978)	*1.5*, p. *37*; 21.7, p. 406
B. I. Shklovskii	(b) **85**, K111 (1978)	*14.1*, p. *206*; *14.4*, p. *248*
S. P. Shrivastava	(b) **88**, 269 (1978)	*13.1*, p. *156*; 21.6, p. 403
V. Ya. Shteinshraiber	(b) **86**, 103 (1978)	14.3.1, p. 226; *15*, p. *259*
E. V. Shtrom	(a) **46**, 327 (1978)	*10.2*, p. *111*; 22.5.2, p. 479
E. I. Shtyrkov	(a) **45**, 647 (1978)	*7*, p. *80*; *20*, p. *316*
V. E. Shubin	(a) **50**, 733 (1978)	*13.1*, p. *161*; *14.3.3*, p. *239*; 16, p. 268; 22.2.3, p. 455
S. R. Shukla	(b) **85**, K103 (1978)	*19*, p. *304*; 22.8.1, p. 511
S. D. Shutov	(a) **50**, K59 (1978)	*2*, p. *52*; *14.3.1*, p. *229*; 22, p. 420
L. A. Shuvalov	(a) **45**, 249 (1978)	*14.4.1*, p. *253*; 22.8.1, p. 510
	(a) **46**, 533 (1978)	*14.4.1*, p. *254*; *20.1*, p. *325*; 22.9, p. 524
	(a) **46**, 547 (1978)	*1.2*, p. *17*; *4*, p. *57*; 19, p. 306; 22.8, p. 504
E. V. Shvedov	(a) **48**, 603 (1978)	*3*, p. *55*; *10.2*, p. *116*; 22.5.2, p. 483
L. S. Shvindlerman	(a) **45**, 359 (1978)	*1.4*, p. *29*; 9, p. 91; 21, p. 353
J. Shy-Yih Wang	see under letter "W"	
R. Shyam	(a) **50**, 11 (1978)	6, p. 64; *13.1*, p. *150*; 13.4, p. 169; *14.3*, p. *217*; 20.1, p. 319; 22.2, p. 443
Radhey Shyam	(a) **45**, K59 (1978)	*2*, p. *43*; *13.4*, p. *170*; 14.3.1, p. 226; 22.1.3, p. 441
G. Sicking	(a) **47**, K85 (1978)	*9*, p. *92*; 21, p. 359
S. K. Sidorov	(a) **46**, K87 (1978)	11, p. 125; *14.2*, p. *214*; 21, p. 357
V. A. Sidorov	(a) **46**, 305 (1978)	*15*, p. *259*; 21.7, p. 406
W. Siegel	(a) **48**, K63 (1978)	13.4, p. 178; *14.3*, p. *222*; 22.8, p. 506
	(a) **49**, 705 (1978)	13.3, p. 167; *14.3.4*, p. *245*; 22.8, p. 507
	(a) **49**, K205 (1978)	13.4, p. 183; *14.3*, p. *224*; 22.8, p. 507
S. Siek	(a) **46**, K101 (1978)	*1.1*, p. *12*; *18.4*, p. *299*; 22.8, p. 504
A. Siems	(a) **49**, 621 (1978)	*1.1*, p. *14*; 6, p. 77; 13.5.2, p. 199; 22.1.2, p. 437
A. Sienkiewicz	(a) **47**, K169 (1978)	*19*, p. *309*; 22.7, p. 502
H. Siethoff	(a) **45**, K135 (1978)	*9*, p. *92*; *10.2*, p. *110*; 22.1.1, p. 424; 22.1.2, p. 430
P. Siffert	(a) **49**, 577 (1978)	*13.4*, p. *182*; *14.3.3*, p. *237*; 22.4.3, p. 471
D. Sikorska	(b) **85**, K73 (1978)	*1*, p. *9*; 18.2, p. 282; 21.1.1, p. 385
A. R. Silin	(a) **50**, K149 (1978)	*2*, p. *53*; *20.3*, p. *350*; 22.6, p. 498
E. K. Silina	(b) **85**, 111 (1978)	*13.5.1*, p. *189*; *20.3*, p. *341*; 22.4.1, p. 463; 22.4.2, p. 468
V. M. Silonov	(b) **88**, 477 (1978)	*1.3*, p. *27*; *13*, p. *146*; 21.1, p. 375
A. Ferreira da Silva	see under letter "F"	
S. S. Simeonov	(a) **45**, 671 (1978)	*14.3.3*, p. *233*; 14.3.4, p. 240; 22.2.1, p. 445
O. Simmich	(a) **48**, K67 (1978)	*1.2*, p. *21*; 10.1, p. 104; 21, p. 361
D. Simon	(a) **46**, 499 (1978)	*9*, p. *92*; 10, p. 96
G. Simon	(b) **86**, K85 (1978)	*14.3*, p. *218*; 22.7, p. 501
P. A. Simonov	(a) **45**, 521 (1978)	9, p. 91; *14.3.3*, p. *233*; 22.1.2, p. 430
K. K. Singh	(b) **85**, 761 (1978)	*6*, p. *66*; 12.1, p. 134; 21.2, p. 392
M. Singh	(b) **86**, 691 (1978)	*6*, p. *69*; 8, p. 84; 10.2, p. 111; 22.1.1, p. 425
Mahiradhwai Singh	(b) **87**, 333 (1978)	*6*, p. *70*; 8, p. 85; 15, p. 260; 22.2.1, p. 446; 22.2.3, p. 453
	(b) **88**, 295 (1978)	*6*, p. *73*; 7, p. 81; 12.2, p. 140; 18, p. 274; 22.1.1, p. 426; 22.2.3, p. 453
	(b) **89**, 299 (1978)	*6*, p. *75*; 8, p. 88; 22.2, p. 444
	(b) **90**, 385 (1978)	*6*, p. *77*; 8, p. 89; 22.7, p. 502
	(b) **90**, K43 (1978)	*6*, p. *78*; *8*, p. *89*; 22.5.2, p. 485
R. Singh	(a) **49**, 733 (1978)	*14.4*, p. *251*; 22.6, p. 496
R. N. Singh	(b) **89**, K13 (1978)	*4*, p. *59*; *13*, p. *147*; 21.2, p. 393
R. P. Singh	(b) **86**, 759 (1978)	*6*, p. *69*; 12.2, p. 138; 22, p. 411
N. N. Sinha	(b) **90**, 369 (1978)	2, p. 51; *14.1*, p. *211*; *15*, p. *261*; 21, p. 364; 21.1, p. 380; 21.2, p. 393

S. K. Sinha	(b) 86, K55 (1978)	*1.2*, p. *17*; *14.2*, p. *214*; 21, p. 356
S. P. Sinitsa	(a) 48, 31 (1978)	2, p. 47; *14.3.2*, p. *230*; *14.3.4*, p. *243*; 22, p. 415
	(a) 49, 217 (1978)	*14.4*, p. *250*; 22, p. 417
V. A. Sinyak	(b) 86, 319 (1978)	*13.1*, p. *153*; 14.3, p. 218; 22, p. 410
M. Sinyukov	(b) 86, 563 (1978)	*6*, p. *68*; *20.1*, p. *324*; 22.2.2, p. 450
N. N. Sirota	(a) 47, 677 (1978)	*11*, p. *127*; 19, p. 309; 22.1.1, p. 426
J. Sitek	(a) 47, 287 (1978)	5, p. 61; *14.2*, p. *214*
S. D. Sivachenko	(a) 46, 77 (1978)	*13.4*, p. *173*, *20.3*, p. *343*; 22.4.4, p. 472
G. Sivarama Sastry	see under letters "Sa"	
R. Sizmann	(a) 45, 529 (1978)	*1.1*, p. *11*; 11, p. 124; 21, p. 354
J. Škácha	(a) 48, K93 (1978)	*2*, p. *47*; 16, p. 265; *20.1*, p. *330*; 22, p. 415
P. Skalicky	(a) 50, 73 (1978)	*1.4*, p. *34*; 10.1, p. 107; 21.1, p. 380; 21.4, p. 398
V. M. Skorikov	(a) 50, K141 (1978)	6, p. 79; 13.1, p. 161; *20.1*, p. *338*; 22.8, p. 508
T. Skośkiewicz	(a) 48, K165 (1978)	*14.2*, p. *215*; 21.1, p. 377; 21.6, p. 404
W. Skrotzki	(a) 46, 107 (1978)	*10.2*, p. *110*; 22.5.2, p. 479
Yu. N. Skryabin	(b) 87, 441 (1978)	*18.2*, p. *286*
L. N. Skuja	(a) 50, K149 (1978)	*2*, p. *53*; *20.3*, p. *350*; 22.6, p. 498
A. Ślebarski	(a) 46, K29 (1978)	*8*, p. *84*; *18.2*, p. *284*; 21, p. 356; 21.4, p. 397
P. S. Smertenko	(a) 45, 631 (1978)	13.4, p. 171; *14.3.4*, p. *240*; 22, p. 410
V. Šmíd	(a) 49, K157 (1978)	*2*, p. *51*; 13.3, p. 167; *14.3.4*, p. *245*
G. F. Smirnova	(b) 86, K171 (1978)	*6*, p. *70*; *20.1*, p. *325*; 22.8, p. 504
G. Smith	(a) 46, K135 (1978)	*20.1*, p. *325*; 22.6, p. 492
E. M. Smokotin	(a) 48, K203 (1978)	*1.2*, p. *22*; 7, p. 82; *18*, p. *275*; 22.6, p. 496
	(a) 50, 269 (1978)	*18.3*, p. *296*; *19*, p. *314*; 22.8.2, p. 520
F. Smutný	(a) 50, K201 (1978)	*8*, p. *90*; *14.4.1*, p. *257*; 22.8, p. 509
O. V. Snitko	(b) 87, 613 (1978)	13.3, p. 165; *14.3.4*, p. *242*; *20.1*, p. *327*; 22.1.2, p. 433
M. Šob	(b) 89, 159 (1978)	*13*, p. *148*; 21.5, p. 400
	(b) 90, K1 (1978)	*13.3*, p. *168*; 21.1, p. 380
A. Sobaszek	(b) 85, K73 (1978)	*1*, p. *9*; 18.2, p. 282; 21.1.1, p. 385
	(b) 85, K77 (1978)	*1*, p. *9*; 18.2, p. 282; 21.1.1, p. 385
L. Sobczyk	(a) 50, K137 (1978)	*14.4.1*, p. *257*; 22.8, p. 508
M. A. Sobeikh	(a) 47, K157 (1978)	*20.3*, p. *345*; 22.3.1, p. 460
L. M. Sobolev	(b) 89, K91 (1978)	*10.2*, p. *117*; 20.1, p. 333; *20.3*, p. *347*; 22.5.2, p. 484
N. Sobolev	(a) 50, K209 (1978)	11, p. 132; *19*, p. *314*; 22.1.2, p. 440
N. N. Sobolev	(a) 47, 609 (1978)	*7*, p. *81*; *20.1*, p. *328*; 22.8.1, p. 514
H. Sobotta	(b) 87, K65 (1978)	*20.1*, p. *326*; 22.3, p. 458
A. S. Sodeika	(a) 48, 329 (1978)	*13.4*, p. *179*; *14.3*, p. *223*; 22.3, p. 458
L. Sodomka	(a) 45, K171 (1978)	*20.3*, p. *342*; 22.3, p. 457
Ya. M. Soifer	(a) 50, K185 (1978)	7, p. 82; *10.1*, p. *108*; *14.2*, p. *216*; 21, p. 367
I. A. Sokolik	(b) 87, 373 (1978)	13.5.1, p. 191; *16*, p. *264*; 20.3, p. 344; 22.9, p. 524
A. L. Sokolov	(a) 49, K117 (1978)	*8*, p. *88*; *14.1*, p. *211*; 21, p. 364
O. B. Sokolov	(b) 85, 127 (1978)	*13*, p. *142*; 20, p. 315; 21, p. 352
V. I. Sokolov	(b) 90, K39 (1978)	*13.4*, p. *184*; *13.5.1*, p. *196*; 14.3.4, p. 246; 20.1, p. 336; 22.4.2, p. 469
V. V. Sokolov	(a) 48, K131 (1978)	*1.2*, p. *22*; *14*, p. *204*; 22.3, p. 458
V. I. Sokolova	(a) 45, K95 (1978)	*16*, p. *262*; 22.8, p. 503
H. Solbrig	(b) 86, K15 (1978)	*1.6*, p. *40*; 21.1, p. 371
A. Soliman	(a) 49, 393 (1978)	*14.3.3*, p. *236*; *16*, p. *266*; 22.2.3, p. 454
L. V. Solomatina	(a) 50, 757 (1978)	*10.1*, p. *108*; 21.1, p. 382; 21.1.1, p. 391
L. S. Solonchuk	(b) 87, K97 (1978)	*14.3.4*, p. *242*; 22.1.3, p. 441
	(b) 88, 179 (1978)	*13.4*, p. *177*; 14.3, p. 222; *15*, p. *260*; 22.1.3, p. 441

V. A. Solovev	(a) **49**, K145 (1978)	*10*, p. *97*; *12*, p. *133*
S. P. Soloviev	(a) **48**, 183 (1978)	*1*, p. *9*; 5, p. 61; *14.4.1*, p. *255*; 18.2, p. 289; 18.4, p. 301; 19, p. 310; 22.8.1, p. 514
T. Soma	(b) **86**, 263 (1978)	*4*, p. *57*; 13.1, p. 153; 22.1, p. 422; 22.1.1, p. 424; 22.1.2, p. 430
	(b) **87**, 345 (1978)	*6*, p. *70*; 12.2, p. 139; 21.7, p. 406; 22.1.1, p. 425; 22.1.2, p. 432
	(b) **88**, K69 (1978)	*4*, p. *58*; 22.1.1, p. 426; 22.1.2, p. 434
	(b) **90**, K129 (1978)	1.2, p. 24; *4*, p. *59*; *12.1*, p. *136*; 21, p. 366
G. Soncini	(a) **47**, 223 (1978)	*16*, p. *264*; 21, p. 358
M. Sondergeld	(b) **89**, 467 (1978)	*13.4*, p. *181*; *20.3*, p. *348*; 22.2.2, p. 451
J. Sonntag	(b) **87**, K69 (1978)	*14.3*, p. *220*; *18.4*, p. *300*; 21, p. 358; 21.1, p. 372
	(a) **50**, 147 (1978)	*8*, p. *89*; 14.1, p. 212; 21.1, p. 381
G. D. Sootha	(a) **46**, 153 (1978)	*18.1*, p. *277*; *19*, p. *305*; 22.9, p. 523
R. S. Sorbello	(b) **86**, 671 (1978)	*2*, p. *44*; *14.1*, p. *207*; 21.2, p. 392
I. Sosnowska	(b) **89**, K29 (1978)	*9*, p. *93*; 21.6, p. 404
M. de Souza	(b) **86**, 137 (1978)	*1.2*, p. *16*; *6*, p. *67*; 12.2, p. 138; 20.1, p. 321; 22.5, p. 474; 22.5.4, p. 489
J. M. Spaeth	(b) **88**, 581 (1978)	*13.4*, p. *178*; *19*, p. *310*; 22.5.2, p. 482
	(b) **89**, 109 (1978)	*10.2*, p. *117*; 13.4, p. 180; *20.1*, p. *332*; 22.5.2, p. 483
J. Spałek	(b) **87**, 625 (1978)	13.1, p. 155; *13.4*, p. *176*; *18.4*, p. *300*
W. Spengler	(b) **87**, 565 (1978)	*6*, p. *71*; 14.2, p. 214; *20.1*, p. *327*
V. I. Sperelup	(a) **49**, K117 (1978)	*8*, p. *88*; *14.1*, p. *211*; 21, p. 364
H. Spiering	(b) **85**, 87 (1978)	*5*, p. *60*; 22.5, p. 474
M. C. Spinelli	(b) **86**, 77 (1978)	*1.2*, p. *16*; *6*, p. *67*; 21.1, p. 371
H. W. G. Spraget	(b) **90**, 667 (1978)	*18.4*, p. *302*; 21.1, p. 382
B. Springmann	(a) **47**, 523 (1978)	*1.5*, p. *37*; *2*, p. *45*; 18.3, p. 296; 21.1, p. 374; 21.4, p. 398
B. Sprušil	(a) **48**, K113 (1978)	*10.1*, p. *105*; 21, p. 362
J. Spyridelis	(a) **47**, 459 (1978)	*1.2*, p. *19*; 1.4, p. 31; 22.3, p. 458
R. Srinivasan	(b) **87**, 395 (1978)	*14.4.1*, p. *254*; *19*, p. *307*; 22.8.1, p. 513
G. P. Srivastava	(b) **85**, K121 (1978)	2, p. 43; *13.1*, p. *152*; *20.1*, p. *321*; 22.1.2, p. 429
	(b) **90**, K125 (1978)	*6*, p. *78*; *8*, p. *90*; 22.1.1, p. 428; 22.1.2, p. 439; 22.2.1, p. 449; 22.2.3, p. 455
J. K. Srivastava	(b) **89**, K79 (1978)	*5*, p. *62*; *18.1*, p. *279*; 22.6.1, p. 499
	(b) **90**, 391 (1978)	*5*, p. *62*; 13.4, p. 184; 22.6.1, p. 499
	(b) **90**, 571 (1978)	*5*, p. *62*; 13.4, p. 185; 22.6.1, p. 499
	(b) **90**, K175 (1978)	*5*, p. *63*; 22.6.1, p. 500
M. P. Srivastava	(b) **90**, 703 (1978)	*11*, p. *131*
N. C. Srivastava	(a) **46**, 475 (1978)	*12.2*, p. *139*; *18.2*, p. *285*; 22.8.2, p. 519
O. N. Srivastava	(a) **49**, 371 (1978)	*1.5*, p. *38*; 2, p. 49; 13.1, p. 158; *14.3.1*, p. *228*; 22.6, p. 496
P. L. Srivastava	(b) **89**, K13 (1978)	*4*, p. *59*; *13*, p. *147*; 21.2, p. 393
	(b) **90**, 93 (1978)	2, p. 51; *12.1*, p. *136*; 21, p. 364; 21.1, p. 379; 21.2, p. 393
	(b) **90**, 369 (1978)	2, p. 51; *14.1*, p. *211*; *15*, p. *261*; 21, p. 364; 21.1, p. 380; 21.2, p. 393
R. Srivastava	(b) **90**, 543 (1978)	*14.4.1*, p. *257*; 20.1, p. 337; 22.8.1, p. 516
R. S. Srivastava	(b) **85**, 789 (1978)	*6*, p. *66*; *8*, p. *83*; 21.6, p. 401
	(b) **87**, 771 (1978)	*6*, p. *72*; 8, p. 86; 21.2, p. 392
D. N. S. Srivastwa	(b) **90**, 671 (1978)	*13.4*, p. *185*; 20.1, p. 337; 20.3, p. 350; 22.5, p. 476
	(a) **50**, 323 (1978)	*13.4*, p. *184*; 20.1, p. 336; 22.5.4, p. 490
J. L. Staehli	(b) **90**, 557 (1978)	*13.5.1*, p. *196*; 22.1.1, p. 428

L. Stals	(a) **45**, 439 (1978)	*10.1*, p. *100*; *11*, p. *124*; 21.1, p. 370
T. Stanek	(a) **48**, 297 (1978)	*1.2*, p. *21*; 8, p. 87; 22.8, p. 506
L. Stănescu	(a) **46**, K163 (1978)	*5*, p. *61*; 22.6, p. 493
J. Stankowski	(a) **45**, 585 (1978)	*1.2*, p. *16*; 19, p. 304; 22.8, p. 503
Yu. S. Stark	(b) **88**, K103 (1978)	*13.4*, p. *179*; 21.1.1, p. 388
M. Starke	(a) **50**, K189 (1978)	*14.3.3*, p. *239*; 22.9, p. 529
V. A. Starodub	(a) **50**, K197 (1978)	*14*, p. *204*; *18*, p. *276*; 22.9, p. 529
V. E. Startsev	(b) **89**, K53 (1978)	*14.1*, p. *210*; 21, p. 362
V. F. Stas	(a) **47**, K91 (1978)	11, p. 127; *14.3.2*, p. *230*; 22.1.2, p. 433
I. V. Stasyuk	(b) **88**, 51 (1978)	*1.2*, p. *20*; *14.4.1*, p. *255*; 22.8.1, p. 514
J. Stasz	(a) **47**, K23 (1978)	*18.2*, p. *286*; *19*, p. *308*; 22.8, p. 505
J. Staun Olsen	(a) **48**, 113 (1978)	*1.1*, p. *13*; 22.1.2, p. 434
G. Staupendahl	(b) **86**, 225 (1978)	*13.5.1*, p. *190*; *20.1*, p. *322*; 22.1.3, p. 441
	(a) **48**, K199 (1978)	*20.2*, p. *339*; 22.1.3, p. 441
B. Stébé	(b) **88**, 713 (1978)	*13.4*, p. *178*; *13.5.1*, p. *193*; 14.3, p. 223; 22.2.1, p. 447
W.-H. Steeb	(b) **86**, K131 (1978)	*13*, p. *145*
A. E. Stefanovich	(a) **46**, 45 (1978)	*13*, p. *144*; *14.3.4*, p. *241*; 22, p. 411
S. Yu. Stefanovich	(a) **50**, K13 (1978)	3, p. 56; *14.4.1*, p. *257*; *20*, p. *318*; 22.8, p. 508
R. Stegmann	(b) **87**, 485 (1978)	*13.2*, p. *162*; *19*, p. *308*; 22.7, p. 502
	(b) **88**, 507 (1978)	*13.2*, p. *162*; 14.3, p. 223; 19, p. 310; 21.7, p. 407
U. Steigenberger	(b) **86**, 557 (1978)	*13.4*, p. *174*; 19, p. 306; 20.1, p. 324; 22.2.3, p. 452
R. Steinbrech	(a) **46**, 107 (1978)	*10.2*, p. *110*; 22.5.2, p. 479
P. Steiner	(b) **90**, 45 (1978)	*13.1*, p. *159*; *13.5.2*, p. *199*; 21.2, p. 393; 21.3, p. 395
H. Steinhardt	(a) **49**, 93 (1978)	*10.2*, p. *118*; 22.2.1, p. 448
V. F. Stelmakh	(b) **88**, K165 (1978)	*13.4*, p. *179*; *14.3*, p. *223*; 22, p. 416
A. Stella	(a) **47**, 223 (1978)	*16*, p. *264*; 21, p. 358
S. I. Stenin	(a) **45**, 377 (1978)	1.4, p. 29; *10.2*, p. *110*; *11*, p. *124*; 22.1.2, p. 429
G. N. Stepanov	(a) **48**, K203 (1978)	*1.2*, p. *22*; 7, p. 82; *18*, p. *275*; 22.6, p. 496
J. D. Stephenson	(a) **50**, 427 (1978)	*1.1*, p. *14*; 22.1.2, p. 439
Kh. E. Sterin	(b) **86**, K49 (1978)	*6*, p. *68*; *20.1*, p. *323*; 22.8, p. 504
G. Stetter	(a) **49**, K17 (1978)	*11*, p. *129*; 14.3, p. 224; *19*, p. *312*; 22.1.2, p. 437
W. Streudle	(a) **49**, 563 (1978)	1.2, p. 23; 6, p. 77; *19*, p. *313*; 22.9, p. 528
G. C. Stevens	(a) **45**, 483 (1978)	13.4, p. 171; *19*, p. *304*; 22.9, p. 523
	(a) **46**, 141 (1978)	13.4, p. 173; *19*, p. *305*; 22.9, p. 523
	(a) **46**, 619 (1978)	13.4, p. 174; *19*, p. *306*; 22.9, p. 524
H. Stöhr	(b) **90**, 497 (1978)	*13.1*, p. *160*; 22.1.2, p. 439
T. Stoica	(b) **87**, 239 (1978)	2, p. 44; *16*, p. *264*; 22.7, p. 501
F. Storbeck	(a) **50**, K247 (1978)	*1.6*, p. *41*; 11, p. 132; 21.1.1, p. 391
W. Stössel	(b) **89**, 403 (1978)	*13.5.1*, p. *194*; *20.1*, p. *333*; 22.4.1, p. 465
H. Stötzel	(b) **87**, 239 (1978)	2, p. 44; *16*, p. *264*; 22.7, p. 501
L. Štourač	(a) **47**, 271 (1978)	*2*, p. *45*; *14.3*, p. *221*; 15, p. 260; 22, p. 413
S. Stoyanov	(b) **88**, K19 (1978)	*13.5.1*, p. *193*; 20.1, p. 329; 22.4.3, p. 470
E. Stracke	(a) **47**, 513 (1978)	*9*, p. *93*; 14.1, p. 208; 21, p. 359
R. Straubel	(a) **49**, K3 (1978)	*18.2.1*, p. *294*; 21.1, p. 378; 21.1.1, p. 389
R. Strehlow	(b) **86**, 311 (1978)	6, p. 67; 13.1, p. 153; 13.5.1, p. 190; *20.1*, p. *322*; 22.2.4, p. 456
	(b) **89**, 603 (1978)	*13.5.1*, p. *195*; *20.1*, p. *334*; 22.4.3, p. 471
	(b) **90**, 167 (1978)	6, p. 77; 13.1, p. 159; 13.5.1, p. 195; *20.1*, p. *335*; 22.2.4, p. 456

V. A. Streltsov	(a) **48**, K23 (1978)	*1.4*, p. *32*; *12.1*, p. *135*; 21, p. 360
	(a) **49**, K129 (1978)	10, p. 97; *12*, p. *133*
	(a) **49**, K145 (1978)	*10*, p. *97*; *12*, p. *133*
	(a) **50**, K33 (1978)	*10.1*, p. *107*; *12.1*, p. *136*; 21.1, p. 381
V. L. Strizhevskii	(b) **89**, 27 (1978)	*13.5*, p. *188*; 20.1. p. 331; 22.6, p. 496
B. Strocka	(a) **45**, K83 (1978)	*10.2*, p. *110*; 14.4.1, p. 253; *16*, p. *262*; 22.8.1, p. 511
F. Studer	(a) **48**, 301 (1978)	*1.1*, p. *13*; *1.4*, p. *32*; 22.8, p. 506
	(a) **49**, 189 (1978)	*1.1*, p. *13*; *1.4*, p. *33*; 22.8, p. 507
J. Stuke	(b) **89**, 495 (1978)	2, p. 50; *14.3.1*, p. *228*; *16*, p. *266*; 22.1.2, p. 437
J. Sturm	(b) **89**, 611 (1978)	*11*, p. *129*; 22.1, p. 422
R. Sturm	(b) **89**, 69 (1978)	*10.2*, p. *117*; 11, p. 128; 22.1.2, p. 436
D. Subašić	(a) **47**, 703 (1978)	*2*, p. *46*; 22.1.1, p. 426
V. Šubtra	(b) **87**, 647 (1978)	*13.5.1*, p. *192*; 22.9, p. 525
	(b) **90**, 295 (1978)	*13.5.1*, p. *196*; 22.9, p. 528
D. Suisky	(b) **85**, 217 (1978)	13.4, p. 169; *14.3.4*, p. *240*; *20.1*, p. *319*; 22.4.1, p. 463
N. Sumida	(a) **45**, 411 (1978)	*11*, p. *124*; 13, p. 143; 21, p. 354
K. Sumino	(a) **50**, 685 (1978)	*10.2*, p. *122*; 22.1.2, p. 439
C. S. Sunandana	(a) **48**, K19 (1978)	11, p. 128; *19*, p. *310*; 22.8.1, p. 514
J.-E. Sundgren	(a) **48**, 345 (1978)	*1.5*, p. *37*; 10.2, p. 116; 22.6, p. 495
P. Süptitz	(a) **48**, K109 (1978)	*2*, p. *48*; *9*, p. *93*; 14.3.4, p. 244; 22, p. 416
Z. Surowiak	(a) **50**, K145 (1978)	*14.4.1*, p. *257*; 22.8.1, p. 517
C. Suryanarayana	(a) **45**, K131 (1978)	*10.1*, p. *100*; 12.1, p. 134; 21.3, p. 394
R. Suryanarayanan	(b) **85**, 9 (1978)	13.1, p. 150; 14.1, p. 205; 14.3, p. 217; *20.1*, p. *319*; 22.7, p. 501
A. P. Sutton	(a) **45**, 149 (1978)	*1.4*, p. *29*; 21, p. 353
E. V. Suvorov	(a) **46**, 379 (1978)	*10.2*, p. *111*; 22.1.2, p. 431
	(a) **50**, 213 (1978)	*10.2*, p. *120*; 22.1.2, p. 438
K. Suzuki	(a) **47**, K51 (1978)	*14.3.4*, p. *242*; *18.2*, p. *286*; 22.8, p. 505
S. V. Svechnikov	(a) **45**, 631 (1978)	13.4, p. 171; *14.3.4*, p. *240*; 22, p. 410
P. Svoboda	(a) **45**, K43 (1978)	*14.1*, p. *205*; 21, p. 353; 21.6, p. 401
B. I. Swanson	(a) **47**, K95 (1978)	*1.2*, p. *19*; 6, p. 71; 20.1, p. 327; 22.8, p. 505
P. Świątek	(b) **89**, K1 (1978)	*13.1*, p. *158*; 14.4, p. 250; 20.1, p. 332; 22.4.3, p. 471
J. Sworakowski	(a) **46**, 273 (1978)	13.4, p. 173; *14.3.4*, p. *241*; 22.9, p. 524
	(a) **49**, K149 (1978)	13.4, p. 183; *14.3.4*, p. *245*; 22, p. 419
B. Sydzhimov	(a) **47**, 743 (1978)	*2*, p. *46*; 4, p. 58; 22.6, p. 494
M. Szczygielski	(a) **50**, 679 (1978)	*18.2*, p. *292*; 22.8.2, p. 521
J. Szczyrbowski	(a) **45**, 537 (1978)	*20.1*, p. *321*; 22.6.1, p. 499
W. Szkiełko	(b) **90**, K81 (1978)	*10.2*, p. *120*; *14.3.3*, p. *237*; 16, p. 267; 19, p. 314; 22.1.2, p. 438
J. Szňajd	(b) **86**, K11 (1978)	*1.2*, p. *16*; 18.1, p. 277; 18.4, p. 298; 22.6, p. 492
S. Szuba	(a) **49**, K201 (1978)	*20.3*, p. *349*; 22.4.1, p. 466
W. Szuszkiewicz	(b) **87**, 637 (1978)	*13.1*, p. *155*; 14.4, p. 249; 20.1, p. 328; 22.4.2, p. 468; 22.4.3, p. 470
H. Szymczak	(b) **88**, K97 (1978)	*2*, p. *47*; *18*, p. *275*; 21, p. 361
	(b) **90**, K59 (1978)	*19*, p. *314*; *20.1*, p. *336*; 22.8, p. 507
A. Szytuła	(a) **45**, 591 (1978)	*18.2*, p. *283*; *18.4*, p. *298*; 22.8, p. 503
	(a) **46**, K101 (1978)	*1.1*, p. *12*; *18.4*, p. *299*; 22.8, p. 504
H. Tabatabai	(a) **49**, 577 (1978)	*13.4*, p. *182*; *14.3.3*, p. *237*; 22.4.3, p. 471
V. I. Tagirov	(b) **86**, K49 (1978)	*6*, p. *68*; *20.1*, p. *323*; 22.8, p. 504
	(a) **47**, K157 (1978)	*20.3*, p. *345*; 22.3.1, p. 460
	(b) **89**, K43 (1978)	*6*, p. *75*; *20.1*, p. *332*; 22.3, p. 458
R. A. Tahir-Kheli	(b) **85**, 707 (1978)	*18.1*, p. *277*; 18.2, p. 282; 18.4, p. 298

K. Taira	(a) **46**, 671 (1978)	*2*, p. *44*; *12.1*, p. *135*; 21.1.1, p. 386
J. Takahashi	(a) **46**, 217 (1978)	*1.3*, p. *25*; 6, p. 68; *11*, p. *125*; 21.6, p. 401
K. Takahashi	(b) **89**, K39 (1978)	*18.2*, p. *290*
S. Takahashi	(a) **45**, 133 (1978)	*10.1*, p. *99*; *18.2*, p. *281*; 21.1, p. 369; 21.1.1, p. 384
K. Takano	(b) **88**, 273 (1978)	13.2, p. 162; *14.1*, p. *209*; 21, p. 360
T. Takebe	(a) **47**, 123 (1978)	13.3, p. 165; *14.3.2*, p. *230*; 14.3.4, p. 242; 22.4.3, p. 470
H. Takei	(a) **50**, 695 (1978)	*1*, p *10*; 10.2, p. 122; 14.4.1, p. 257; 22.8.1, p. 516
K. Takesawa	(a) **47**, 657 (1978)	*7*, p. *81*; 22.4.1, p. 464
N. Takeuchi	(b) **86**, K151 (1978)	*19*, p. *306*; 22.5.2, p. 480
S. Takeuchi	(a) **48**, 587 (1978)	*10.2*, p. *116*; *12.2*, p. *140*; 22.4.3, p. 471
	(a) **49**, 685 (1978)	*2*, p. *50*; 10, p. 97; 12, p. 133
T. Taki	(b) **86**, K147 (1978)	*19*, p. *306*; 22, p. 412
P. L. Talwar	(a) **47**, 699 (1978)	*13.4*, p. *176*; *14.3*, p. *221*; 22.4.2, p. 468
M. Tanaka	(b) **89**, K39 (1978)	*18.2*, p. *290*
S. Tanaka	(b) **88**, 607 (1978)	13.1, p. 157; *13.5.1*, p. *193*; *20.1*, p. *330*; 22, p. 415
T. Tanaka	(a) **47**, 123 (1978)	13.3, p. 165; *14.3.2*, p. *230*; 14.3.4, p. 242; 22.4.3, p. 470
P. N. Tandon	(a) **45**, K109 (1978)	*5*, p. *60*; *18.2*, p. *283*; 21, p. 355
O. P. Taneja	(a) **47**, 375 (1978)	13.4, p. 176; *14.3.1*, p. *227*; 22.2.3, p. 453
S. Tanigawa	(b) **90**, K31 (1978)	*10.1*, p. *107*; 21.1, p. 380
D. B. Tanner	(a) **46**, 209 (1978)	*14.2*, p. *214*; 21, p. 356
W. H. Tanttila	(b) **86**, K175 (1978)	*9*, p. *92*; 21.2, p. 392
	(b) **89**, K109 (1978)	*9*, p. *92*; 21.2, p. 392
V. A. Tarasov	(b) **87**, 433 (1978)	*7*, p. *80*; *18.2*, p. *286*
I. I. Tarasyuk	(b) **88**, 179 (1978)	*13.4*, p. *177*; 14.3, p. 222; *15*, p. *260*; 22.1.3, p. 441
A. Tari	(a) **46**, 173 (1978)	*18.2*, p. *284*; 19, p. 305; 21.1.1, p. 385; 21.4, p. 396
L. Tarricone	(a) **47**, 371 (1978)	*14.3.1*, p. *227*; 22.4.3, p. 470
M. G. Tashkova	(b) **86**, 425 (1978)	*13.3*, p. *164*; 14.3.3, p. 233
	(b) **87**, 361 (1978)	*13.3*, p. *165*; *14.3.3*, p. *235*
L. Taurel	(b) **86**, 493 (1978)	*6*, p. *68*; *20.1*, p. *324*; 22.5.2, p. 480
A. Tausend	(a) **49**, 59 (1978)	*8*, p. *88*; 22.1.3, p. 442
K. N. R. Taylor	(a) **46**, 39 (1978)	*14.1*, p. *206*; 21.4, p. 396
	(a) **50**, 203 (1978)	*10.1*, p. *107*; *13.4*, p. *184*; 21.1, p. 381
C. Tejedor	(b) **88**, 591 (1978)	*13.3*, p. *166*; 22.2, p. 443
R. V. Telesnin	(a) **50**, K101 (1978)	*18.2.1*, p. *294*; 21.1, p. 381; 21.1.1, p. 390
V. A. Telezhkin	(b) **87**, 79 (1978)	*13.1*, p. *154*; 13.5.1, p. 191; 20.1, p. 325; 22, p. 412
M. J. Tello	(a) **48**, 53 (1978)	*1.2*, p. *21*; 22.8, p. 506
G. van Tendeloo	(a) **45**, 553 (1978)	*1.2*, p. *16*; *1.4*, p. *29*; 21.1, p. 370
	(a) **46**, 399 (1978)	*1.3*, p. *25*; *1.4*, p. *30*; 21.1, p. 371; 21.6, p 401
	(b) **87**, 419 (1978)	*1.1*, p. *13*; 22.1.1, p. 425; 22.6, p. 494
	(a) **47**, 105 (1978)	*1.3*, p. *25*; *1.4*, p. *30*; 21.1, p. 373; 21.6, p. 402
	(a) **47**, 241 (1978)	*1.2*, p. *18*; 1.3, p. 25; 1.4, p. 30; 21.1, p. 373; 21.6, p. 402
	(a) **47**, 539 (1978)	*1.2*, p. *19*; 1.3, p. 26; 1.4, p. 31; 21.1, p. 374; 21.6, p. 403
	(a) **47**, 555 (1978)	*1.3*, p. *26*; *1.4*, p. *31*; 21.1, p. 374; 21.6, p. 403
	(a) **48**, 39 (1978)	*1.2*, p. *20*; 1.3, p. 27; 1.4, p. 31; 21.1, p. 375; 21.6, p. 403
	(a) **48**, 383 (1978)	*1.3*, p. *27*; 1.4, p. 32; 21, p. 361; 21.1, p. 376
	(a) **48**, K5 (1978)	*1.4*, p. *31*; 10.2, p. 115; 22, p. 415

	(a) **49**, 337 (1978)	*1.1*, p. *13*; *1.4*, p. *33*; 21.1, p. 377; 21.6, p. 404
	(a) **49**, 655 (1978)	*1.1*, p. *14*; *1.4*, p. *33*; 21.1, p. 379; 21.6, p. 405
	(a) **49**, K217 (1978)	*1.4*, p. *31*; 10.2, p. 115; 22, p. 415
	(a) **50**, 53 (1978)	*1.3*, p. *28*; *1.4*, p. *34*; 21.1, p. 380; 21.6, p. 405
N. Terao	(a) **47**, 565 (1978)	*2*, p. *45*; *12.2*, p. *140*; 22.8, p. 505
	(a) **49**, 237 (1978)	2, p. 49; *10.2*, p. *118*; *12.2*, p. *140*
	(a) **49**, K101 (1978)	*2*, p. *51*; *12.2*, p. *141*; 22.1, p. 422
S. F. Terekhova	(b) **90**, K49 (1978)	*13.5.1*, p. *196*; *20.1*, p. *336*; 22, p. 420
E. I. Terukov	(a) **45**, K77 (1978)	*14.1*, p. *205*; *14.3.1*, p. *226*; 22.6, p. 491
	(a) **48**, 377 (1978)	*1.2*, p. *21*; 6, p. 75; 14.3, p. 223; 22.6, p. 495
	(a) **48**, K153 (1978)	*8*, p. *88*; 22.6, p. 495
S. P. Teslenko	(a) **45**, 63 (1978)	*5*, p. *60*; 9, p. 91; 10.2, p. 109; 22.6, p. 491
W. Tews	(b) **86**, 329 (1978)	2, p. 43; 13.1, p. 153; *14.4*, p. *248*; *20.1*, p. *322*; 22.6, p. 492
V. E. Tezlevan	(b) **87**, K141 (1978)	*14.3*, p. *221*; 22.8, p. 505
	(a) **49**, K197 (1978)	*14.3.4*, p. *246*; 22.8.2, p. 520
A. Thaer	(a) **50**, 131 (1978)	*1.2*, p. *24*; 6, p. 78; 20.1, p. 336; 22.9, p. 528
W. A. Theiner	(b) **89**, 441 (1978)	*1.2*, p. *22*; *1.5*, p. *39*; 2, p. 50; 14.1, p. 210; 20.1, p. 334; 21.1.1, p. 389
J. Thery	(a) **45**, 599 (1978)	*19*, p. *305*; 22.6.1, p. 499
P. Thévenard	(a) **48**, 425 (1978)	*1.4*, p. *32*; 11, p. 128; 22.6, p. 495
W. Thielemann	(a) **47**, K35 (1978)	*20.1*, p. *327*; 22.2.1, p. 446
J. Thiemann	(b) **88**, 397 (1978)	10.1, p. 105; *19*, p. *310*; 21.1, p. 375
V. G. Thiessen	(b) **88**, K139 (1978)	*18.2*, p. *289*; 18.4, p. 301; 21.1, p. 376
	(a) **48**, K185 (1978)	*18.2*, p. *290*; 21.1, p. 377
D. B. Tran Thoai	(b) **85**, 561 (1978)	*13.5.1*, p. *190*; 13.5.2, p. 198; 22, p. 409
M. Thomas	(a) **50**, K231 (1978)	*1.1*, p. *14*; 22, p. 421
P. Thomas	(b) **86**, 541 (1978)	2, p. 43; *14.3*, p. *219*; 15, p. 260; 22.1.1, p. 425; 22.1.2, p. 432
	(b) **88**, K73 (1978)	*2*, p. *46*; *14.3*, p. *222*; 22.1.2, p. 434
Y. Thomas	(b) **87**, K13 (1978)	*6*, p. *70*; *8*, p. *85*; 21.6, p. 402
K. Thörmer	(b) **90**, 277 (1978)	10.2, p. 120; *14.4*, p. *251*; 22.5.3, p. 488
B. V. Thosar	(a) **47**, 691 (1978)	*10.2*, p. *114*; 11, p. 127; 14.4, p. 249; 19, p. 309; 22.9, p. 526
P. A. Thrower	(a) **47**, 11 (1978)	*9*, p. *91*; *10.2*, p. *109*; 13.4, p. 169; 22.1.3, p. 441
F. Thuselt	(b) **89**, 561 (1978)	*13.5.1*, p. *195*; *20.3*, p. *348*; 22.2.2, p. 451
	(b) **90**, 233 (1978)	*13.5.1*, p. *196*; *20.3*, p. *349*; 22.3, p. 459
B. Thuss	(a) **45**, 123 (1978)	*1.2*, p. *15*; *5*, p. *60*; 18.1, p. 277; 22.6.1, p. 499
I. I. Tichina	(b) **86**, 419 (1978)	*6*, p. *67*; *20.1*, p. *323*; 22.3, p. 457
B. Tieke	(b) **88**, 653 (1978)	*13.1*, p. *157*; *16*, p. *265*; 22.9, p. 526
M. Tilli	(a) **50**, 427 (1978)	*1.1*, p. *14*; 22.1.2, p. 439
C. W. M. Timmermans	(a) **47**, 589 (1978)	10.2, p. 114; *14.4*, p. *249*; *20.3*, p. *345*; 22.5.2, p. 481
V. V. Timofeenko	(b) **86**, K139 (1978)	*1.1*, p. *11*; 1.2, p. 17; 22.2.4, p. 456
	(b) **87**, K73 (1978)	*1.2*, p. *19*; *13.1*, p. *155*; 22.1, p. 422
	(b) **87**, K77 (1978)	*13.1*, p. *155*; 22.7, p. 502
M. I. Timoshechkin	(a) **49**, 305 (1978)	3, p. 56; 20.1, p. 333; *20.2*, p. *339*; *20.3*, p. *347*; 22.8.2, p. 520
P. P. Tirodkar	(a) **49**, 637 (1978)	*14.3.3*, p. *237*; 14.4, p. 251; 22.6, p. 496
V. S. Tishkov	(a) **45**, 343 (1978)	*1.2*, p. *15*; *1.5*, p. *36*; 11, p. 123; 21.1, p. 369
A. V. Titov	(a) **48**, 183 (1978)	*1*, p. *9*; 5, p. 61; *14.4.1*, p. *255*; 18.2, p. 289; 18.4, p. 301; 19, p. 310; 22.8.1, p. 514
V. V. Titov	(a) **48**, 13 (1978)	*3*, p. *54*; *11*, p. *123*; 22.1.2, p. 429
A. K. Tkachenko	(a) **48**, 329 (1978)	*13.4*, p. *179*; *14.3*, p. *223*; 22.3, p. 458
V. D. Tkachev	(b) **88**, K165 (1978)	*13.4*, p. *179*; *14.3*, p. *223*; 22, p. 416
Yu. A. Tkhorik	(a) **48**, K55 (1978)	*11*, p. *128*; 14.3.3, p. 235; 22.2.1, p. 447

T. Todorov	(b) **86**, 697 (1978)	*10.2*, p. *112*; 20.1, p. 324; 22.5.2, p. 480
	(a) **47**, 213 (1978)	*10.2*, p. *113*; *20.1*, p. *326*; 22.5.2, p. 481
J. Todorović	(a) **45**, 591 (1978)	*18.2*, p. *283*; *18.4*, p. *298*; 22.8, p. 503
S. Togashi	(b) **89**, K145 (1978)	*17*, p. *270*; 22.2.1, p. 448; 22.5, p. 476
V. V. Tokii	(a) **45**, 327 (1978)	*10.1*, p. *100*; 21, p. 353
V. A. Tolokonnikov	(a) **50**, 733 (1978)	*13.1*, p. *161*; *14.3.3*, p. *239*; 16, p. 268; 22.2.3, p. 455
K. B. Tolpygo	(b) **87**, 79 (1978)	*13.1*, p. *154*; 13.5.1, p. 191; 20.1, p. 325; 22, p. 412
	(b) **90**, 355 (1978)	*13.4*, p. *183*; *20.3*, p. *349*; 22.4.1, p. 466
I. Tomáš	(a) **49**, 747 (1978)	*18.2*, p. *291*
M. Tomášek	(b) **89**, 11 (1978)	13.1, p. 150; *13.3*, p. *164*; 21.1, p. 368; 21.1.1, p. 384
G. Tomasi	(a) **50**, 563 (1978)	*9*, p. *95*; 14.1, p. 212; 21.1.1, p. 391
M. Tomašich	(a) **47**, 287 (1978)	5, p. 61; *14.2*, p. *214*
P. M. Tomchuk	(b) **86**, 773 (1978)	*13*, p. *144*; *14.3*, p. *220*; 20, p. 316; 22, p. 412
A. Toneva	(b) **85**, K131 (1978)	*13.1*, p. *152*; 14.3, p. 218; 22.7, p. 501
L. Tongson	(a) **45**, K1 (1978)	*11*, p. *123*; 14.4.1, p. 253; 22.8.1, p. 510
V. Topa	(b) **89**, 513 (1978)	*10.2*, p. *118*; 20.1, p. 334; 22.5.2, p. 485
T. V. Torchinskaya	(a) **50**, 767 (1978)	*13.4*, p. *186*; *14.3*, p. *225*; 16, p. 268; 22.4.1, p. 467
E. Török	(a) **50**, 159 (1978)	*2*, p. *51*; 8, p. 89; 12.1, p. 136; *18.2*, p. *291*; 21.1, p. 381; 21.1.1, p. 390
P. D. Townsend	(a) **47**, K63 (1978)	13.3, p. 165; 14.4.1, p. 255; *20.3*, p. *344*; 22.8.1, p. 513
	(a) **47**, K67 (1978)	13.4, p. 175; *20.3*, p. *344*; 22.5.2, p. 481
C. Trallero Giner	(b) **85**, 351 (1978)	*6*, p. *65*; *13.5.1*, p. *189*; 20.1, p. 320
Tran Quoc Hai	see under letter "H"	
D. B. Tran Thoai	see under letters "Th"	
M. Trapp	(a) **49**, 279 (1978)	*9*, p. *94*; *11*, p. *129*; 22.2.4, p. 456
A. V. Trefilov	(b) **85**, 63 (1978)	*6*, p. *64*; *8*, p. *83*; 21.2, p. 392
	(b) **85**, 749 (1978)	*6*, p. *66*; *8*, p. *83*; 21.2, p. 392
M. Treilleux	(a) **48**, 425 (1978)	*1.4*, p. *32*; 11, p. 128; 22.6, p. 495
L. Trépied	(a) **49**, 713 (1978)	*1.4*, p. *33*; *10.2*, p. *119*; 22.6, p. 496
	(a) **50**, K37 (1978)	*1.4*, p. *34*; 22.6, p. 497
W. Treplin	(b) **85**, 247 (1978)	*18.2*, p. *281*
	(b) **90**, 135 (1978)	*18.2*, p. *291*
H. Treptow	(a) **49**, 667 (1978)	13.4, p. 182; *14.3.3*, p. *237*; *20.3*, p. *349*; 22.4.2, p. 469
M. M. Tribel	(b) **87**, 373 (1978)	13.5.1, p. 191; *16*, p. *264*; 20.3, p. 344; 22.9, p. 524
S. K. Trikha	(b) **87**, 513 (1978)	*6*, p. *71*; *8*, p. *86*; 23, p. 530
	(b) **88**, 815 (1978)	*6*, p. *74*; *8*, p. *87*; 23, p. 530
M. F. Trinkler	(b) **88**, 445 (1978)	*13.4*, p. *178*; 20.1, p. 330; *20.3*, p. *346*; 22.5.4, p. 489
R. P. Tripathi	(b) **88**, 633 (1978)	*5*, p. *62*; 22.8, p. 506
S. P. Tripathi	(b) **85**, 477 (1978)	*6*, p. *65*; 12.2, p. 138; 22.5.2, p. 478
	(b) **89**, 335 (1978)	*6*, p. *65*; 12.2, p. 138; 22.5.2, p. 478
R. Troć	(a) **45**, K39 (1978)	*18.2*, p. *282*; 22.8, p. 503
O. A. Troitskii	(a) **48**, 229 (1978)	*3*, p. *55*; 9, p. 93; 21, p. 360
R. Trojko	(a) **46**, K173 (1978)	*3*, p. *54*; 22.3.1, p. 460
R. Trommer	(b) **86**, 563 (1978)	*6*, p. *68*; *20.1*, p. *324*; 22.2.2, p. 450
Nguyen Van Trong	(b) **88**, 145 (1978)	*6*, p. *73*; 11, p. 127; 13.5.2, p. 198
A. N. Trukhin	(b) **86**, 67 (1978)	2, p. 43; *13.1*, p. *152*; 16, p. 263; 20.3, p. 342; 22.6, p. 491
C. Tsallis	(b) **86**, K101 (1978)	*13*, p. *145*; *18*, p. *274*
L. D. Tsendin	(a) **46**, 45 (1978)	*13*, p. *144*; *14.3.4*, p. *241*; 22, p. 411

P. K. Tseng	(a) 48, K181 (1978)	*1.4*, p. *32*; *2*, p. *48*; 21.1, p. 377; 21.6, p. 404
P. Tsilimigras	(b) 88, 341 (1978)	*13.1*, p. *156*; 18, p. 274
D. I. Tsiulyanu	(a) 50, 195 (1978)	*2*, p. *52*; *14.3.3*, p. *237*; 14.3.4, p. 246; 22, p. 420
V. A. Tskhai	(b) 87, 49 (1978)	*4*, p. *58*; *13*, p. *145*; 22.3, p. 457
A. A. Tsoi	(a) 47, K149 (1978)	10.2, p. 114; *11*, p. *127*; 22.1.2, p. 433
I. Tsolovski	(a) 46, K171 (1978)	*1.2*, p. *18*; 8, p. 85; 22.8.1, p. 513
S. Tsunekawa	(a) 50, 695 (1978)	*1*, p. *10*; 10.2, p. 122; 14.4.1, p. 257; 22.8.1, p. 516
N. Tsuya	(a) 48, 175 (1978)	2, p. 47; *18.2*, p. *289*; 21.1.1, p. 388
	(a) 50, K25 (1978)	*2*, p. *52*; *18.2*, p. *292*; 21.1.1, p. 390
E. A. Tsvetkov	(a) 46, 311 (1978)	*13.4*, p. *173*; 20.2, p. 339; 20.3, p. 343; 22.8, p. 504
L. T. Tsymbal	(b) 90, K17 (1978)	*13.3*, p. *168*; *14.1*, p. *211*; 19, p. 313; 21, p. 364
B. Tuck	(a) 46, 199 (1978)	*9*, p. *92*; *14.3*, p. *219*; 22.2.2, p. 450
N. S. Tulyaganova	(a) 46, K147 (1978)	*10.2*, p. *112*; 11, p. 126; *12.2*, p. *139*; 22.1.2, p. 432
T. Tuomi	(a) 47, 245 (1978)	6, p. 71; 13.5.1, p. 191; *14.3.3*, p. *235*; *16*, p. *264*; 22.1.1, p. 425
	(a) 50, 427 (1978)	*1.1*, p. *14*; 22.1.2, p. 439
P. Turchi	(a) 45, 229 (1978)	*1.3*, p. *25*; 6, p. 65; 12.1, p. 134; 21.1, p. 369; 21.1.1, p. 384
S. K. Turkov	(a) 47, 65 (1978)	*7*, p. *80*; *12.2*, p. *139*; 14.4.1, p. 255; 22.8.1, p. 513
R. B. Turner	(a) 50, 83 (1978)	*14.1*, p. *212*; 21.1, p. 380
E. A. Turov	(b) 85, 127 (1978)	*13*, p. *142*; 20, p. 315; 21, p. 352
V. I. Tutlis	(b) 90, K109 (1978)	*13.4*, p. *185*; 20.1, p. 337; 22, p. 421
V. Tvarožek	(a) 49, K81 (1978)	2, p. 50; *18.2.1*, p. *294*; 21.1, p. 378; 21.4, p. 398
V. A. Tyagai	(b) 87, 613 (1978)	13.3, p. 165; *14.3.4*, p. *242*; *20.1*, p. *327*; 22.1.2, p. 433
	(b) 88, 797 (1978)	13.3, p. 166; *13.5.1*, p. *193*; *20.3*, p. *346*; 22.4.1, p. 465
N. A. Tyapunina	(a) 46, 327 (1978)	*10.2*, p. *111*; 22.5.2, p. 479
	(a) 46, 351 (1978)	*10.2*, p. *111*; *12.2*, p. *138*; 22.5.2, p. 480
	(a) 46, 411 (1978)	*10.1*, p. *101*; *12.1*, p. *135*; 21, p. 357; 21.1, p. 372; 21.3, p. 394
T. Tyc	(a) 47, K125 (1978)	*8*, p. *86*; 21, p. 360
I. I. Tychina	(a) 48, 329 (1978)	*13.4*, p. *179*; *14.3*, p. *223*; 22.3, p. 458
N. N. Tyutyulkov	(b) 86, 385 (1978)	*13.1*, p. *153*; 22.9, p. 523
Y. Uchida	(a) 45, 411 (1978)	*11*, p. *124*; 13, p. 143; 21, p. 354
S. Ueda	(b) 88, K1 (1978)	*2*, p. *46*; *19*, p. *309*; 22, p. 414
L. A. Ugodnikova	(a) 50, K47 (1978)	*18.1*, p. *280*; *18.4*, p. *301*; 21.4, p. 399
Ch. Uihlein	(b) 87, 575 (1978)	6, p. 71; *13.5.1*, p. *192*; *20.1*, p. *327*; 22.5, p. 475
E. N. Ukraintsev	(b) 87, 381 (1978)	*13.5*, p. *187*; 14.4, p. 248; 22.4.3, p. 470
H.-J. Ullrich	(a) 49, 323 (1978)	*1.2*, p. *22*; *10.1*, p. *106*; 14.2, p. 215; 21.1, p. 377
B. Ulrici	(b) 86, 517 (1978)	*6*, p. *68*; *20.1*, p. *324*; 22.2.2, p. 450; 22.2.4, p. 456
G. R. Umarov	(b) 90, 733 (1978)	1, p. 10; *13.1*, p. *161*; 22, p. 421
M. Umeda	(a) 47, K111 (1978)	*1.2*, p. *20*; *14.1*, p. *208*; 21, p. 359
Yu. K. Undalov	(a) 50, 139 (1978)	*10.2*, p. *120*; *20.3*, p. *350*; 22.8, p. 508
	(a) 50, 445 (1978)	14.3, p. 225; *16*, p. *267*; 22.8, p. 508
K. Unger	(b) 87, 227 (1978)	*4*, p. *58*; 13, p. 145; 22.2, p. 443
	(b) 87, 473 (1978)	*4*, p. *58*; 13, p. 145; 22.2, p. 443

	(b) **89**, 561 (1978)	*13.5.1*, p. *195*; *20.3*, p. *348*; 22.2.2, p. 451
W. K. Unger	(a) **46**, 81 (1978)	*20.1*, p. *323*; *20.3*, p. *343*; 22.8, p. 504
	(a) **48**, K89 (1978)	*18.4*, p. *301*; 22.8.2, p. 519
S. Unterricker	(a) **46**, 125 (1978)	10.2, p. 110; *11*, p. *125*; 22.8, p. 504
G. C. Upreti	(b) **87**, K39 (1978)	*19*, p. *307*; 22.8, p. 505
K. Urade	(b) **90**, 697 (1978)	*18.3*, p. *297*; 22.8.2, p. 521
K. Urban	(a) **47**, 479 (1978)	*1.4*, p. *31*; *10*, p. *96*
A. Urbaniak	(b) **85**, 387 (1978)	*18.2*, p. *281*
R. W. Ure, Jr.	(a) **48**, K83 (1978)	*1.3*, p. *27*; *10.1*, p. *105*; 21, p. 361
M. Urner-Wille	(a) **50**, K29 (1978)	2, p. 52; *18.3*, p. *297*; *20*, p. *318*; 21.1.1, p. 390; 21.4, p. 398
P. Urwank	(b) **86**, 569 (1978)	*13.1*, p. *154*; 14.2, p. 214; 21.1, p. 371; 21.1.1, p. 386
R. G. Usmanov	(a) **49**, 789 (1978)	*20*, p. *317*
D. Vajda	(a) **47**, K123 (1978)	*7*, p. *81*; *14.3.4*, p. *243*; 22.2.3, p. 453
V. G. Vaks	(b) **85**, 63 (1978)	*6*, p. *64*; *8*, p. *83*; 21.2, p. 392
	(b) **85**, 749 (1978)	*6*, p. *66*; *8*, p. *83*; 21.2, p. 392
S. Valeri	(a) **50**, K123 (1978)	*10.2*, p. *121*; *13.4*, p. *184*; 22.1.2, p. 438
E. E. Valiev	(a) **46**, K87 (1978)	11, p. 125; *14.2*, p. *214*; 21, p. 357
R. Z. Valiev	(a) **45**, 77 (1978)	*10.1*, p. *99*; 21, p. 352; 21.3, p. 394
H. J. L. van der Valk	(b) **87**, 135 (1978)	*19*, p. *307*; 22.5, p. 475
Yu. A. Valov	(a) **45**, K95 (1978)	*16*, p. *262*; 22.8, p. 503
V. Valvoda	(a) **45**, 269 (1978)	*6*, p. *65*; 22.5, p. 474
	(a) **48**, K101 (1978)	*1.5*, p. *38*; 22.6, p. 495
E. R. Vance	(a) **46**, K135 (1978)	*20.1*, p. *325*; 22.6, p. 492
R. E. Vandenberghe	(a) **50**, K85 (1978)	*1*, p. *10*; 22.8.2, p. 521
M. Vanhaelst	(b) **85**, 639 (1978)	*19*, p. *304*; 22.5.2, p. 479
	(b) **87**, 719 (1978)	10.2, p. 113; *19*, p. *308*; 22.6, p. 494
	(b) **89**, 151 (1978)	*19*, p. *311*; *20.1*, p. *332*; 22.5.2, p. 483
	(b) **89**, K25 (1978)	*19*, p. *311*; 22.5.2, p. 484
V. D. Vankar	(a) **45**, 665 (1978)	*1.5*, p. *36*; 22.4.4, p. 472
	(a) **46**, 723 (1978)	*1.5*, p. *36*; *14.3.1*, p. *226*; 16, p. 263; 22.4.4, p. 472
	(a) **49**, 379 (1978)	1.4, p. 33; *1.5*, p. *38*; 2, p. 49; 22.1.1, p. 427
A. V. Vannikov	(b) **87**, 255 (1978)	*13.5.1*, p. *191*; 20.1, p. 326; 20.3, p. 344; 22.9, p. 524
D. Varchon	(a) **47**, 259 (1978)	*10.1*, p. *103*; 21, p. 358
M. D. Varentsov	(a) **50**, K169 (1978)	11, p. 131; *14.3*, p. *225*; 22.1.2, p. 440
H. Vargas	(b) **86**, 373 (1978)	*14.4.1*, p. *254*; 22.8.1, p. 511
	(a) **48**, K35 (1978)	*1.2*, p. *21*; *7*, p. *81*; 22.8, p. 506
R. A. Varin	(a) **45**, 565 (1978)	*1.4*, p. *29*; 10.1, p. 100; 21, p. 354
	(a) **46**, K79 (1978)	*10.1*, p. *102*; 21, p. 357
P. A. Varotsos	(a) **45**, K9 (1978)	*9*, p. *91*; 22.5.2, p. 478
	(a) **47**, K133 (1978)	*10.2*, p. *114*; 22.5.2, p. 482
	(a) **49**, K57 (1978)	*8*, p. *88*; *10*, p. *97*; 12, p. 133
	(b) **90**, 339 (1978)	4, p. 59; 6, p. 77; 12.2, p. 141; *14.4*, p. *251*; 22.5.1, p. 477; 22.5.2, p. 485
	(b) **90**, 629 (1978)	*8*, p. *89*; 9, p. 94; 10, p. 98
S. C. Varshney	(b) **85**, 733 (1978)	6, p. 66; *12.2*, p. *138*; *14.3*, p. *217*; 22, p. 409
Y. P. Varshni	(b) **89**, 103 (1978)	*10.2*, p. *117*; *13.4*, p. *179*; 22.5.2, p. 483
L. Vasanelli	(a) **48**, 293 (1978)	*13.4*, p. *179*; *14.3*, p. *223*; 22.3.1, p. 460
E. A. Vasilev	(a) **47**, K55 (1978)	*18.2*, p. *286*; *18.3*, p. *295*; 21, p. 359
	(a) **50**, K205 (1978)	5, p. 63; *18.2*, p. *292*; *18.3*, p. *297*; 21, p. 367
L. N. Vasilev	(a) **45**, 11 (1978)	*5*, p. *60*; *11*, p. *123*; 22.5, p. 474; 22.8, p. 503; 22.9, p. 522
M. A. Vasilev	(a) **50**, K113 (1978)	*17*, p. *270*; 22.3, p. 459

F. T. Vasko	(b) **88**, 385 (1978)	*13*, p. *146*; *14.3.4*, p. *243*; 16, p. 265; 22, p. 415
W. R. Vaughan	(a) **45**, 469 (1978)	*1.2*, p. *16*; *1.3*, p. *25*; 22.8, p. 503
C. Vautieř	(a) **50**, 651 (1978)	13.4, p. 186; *16*, p. *267*; 22.4.3, p. 471
A. van Veen	(a) **46**, 541 (1978)	*10.1*, p. *102*; 21, p. 357
	(a) **47**, 577 (1978)	*10.1*, p. *104*; 21, p. 359
	(b) **90**, 643 (1978)	*10.1*, p. *107*; 21, p. 365
	(b) **90**, K89 (1978)	6, p. 78; *10.1*, p. *107*; 21, p. 364
Yu. Kh. Vekilov	(b) **90**, 733 (1978)	1, p. 10; *13.1*, p. *161*; 22, p. 421
T. S. te Velde	(a) **50**, K29 (1978)	2, p. 52; *18.3*, p. *297*; *20*, p. *318*; 21.1.1, p. 390; 21.4, p. 398
B. Velický	(a) **47**, 271 (1978)	*2*, p. *45*; *14.3*, p. *221*; 15, p. 260; 22, p. 413
T. Velula	(b) **85**, 439 (1978)	*13.4*, p. *170*; *20.3*, p. *341*; 22.5.2, p. 478
	(b) **89**, 335 (1978)	*13.4*, p. *170*; *20.3*, p. *341*; 22.5.2, p. 478
Yu. N. Venevtsev	(a) **48**, 183 (1978)	*1*, p. *9*; 5, p. 61; *14.4.1*, p. *255*; 18.2, p. 289; 18.4, p. 301; 19, p. 310; 22.8.1, p. 514
S. Vengris	(a) **48**, K149 (1978)	*2*, p. *48*; *14.3*, p. *223*; 22.1.3, p. 441
P. Venkateswarlu	(b) **88**, K173 (1978)	*19*, p. *311*; 22.8.1, p. 511
G. Vergoten	(a) **48**, 417 (1978)	*6*, p. *75*; *20.1*, p. *331*; 22.6, p. 495
M. P. Verma	(b) **85**, 417 (1978)	*1.1*, p. *11*; 12.2, p. 138; 22.5.2, p. 478
	(b) **87**, 401 (1978)	*6*, p. *70*; *12.2*, p. *139*; 22.2, p. 443; 22.4, p. 461; 22.5, p. 475
M. S. Vershinin	(b) **90**, 429 (1978)	*13.1*, p. *160*; *20.1*, p. *336*; 22.3, p. 459
L. Vescan	(b) **87**, 239 (1978)	2, p. 44; *16*, p. *264*; 22.7, p. 501
R. Vetter	(a) **50**, 529 (1978)	*10.1*, p. *108*; *12.1*, p. *137*; 21.1, p. 382
G. C. Vezzoli	(a) **46**, K49 (1978)	*1.2*, p. *17*; 14.1, p. 206; 20.1, p. 324; 21, p. 356
	(a) **48**, K75 (1978)	*2*, p. *47*; *14.1*, p. *209*; 20.1, p. 330; 21.7, p. 407
J. Vidal	(a) **47**, 371 (1978)	*14.3.1*, p. *227*; 22.4.3, p. 470
N. A. Vidmont	(b) **90**, K53 (1978)	*13.5.1*, p. *196*; *20.3*, p. *349*; 22.9, p. 528
C. Viger	(a) **50**, 651 (1978)	13.4, p. 186; *16*, p. *267*; 22.4.3, p. 471
E. Vigil	(b) **90**, 409 (1978)	*20.1*, p. *335*; 22.2.1, p. 448
P. Viktorovitch	(a) **48**, 137 (1978)	13.3, p. 166; *14.3.3*, p. *235*; *16*, p. *264*; 22.1.2, p. 434
V. L. Vinetskii	(b) **87**, 507 (1978)	13.5.3, p. 200; 22, p. 414
Nguyen Vinh Quang	see under letter "Q"	
E. A. Vinogradov	(b) **87**, K115 (1978)	*6*, p. *72*; *20.1*, p. *328*; 22.3, p. 458
	(b) **89**, K43 (1978)	*6*, p. *75*; *20.1*, p. *332*; 22.3, p. 458
J. K. Viščakas	(a) **47**, 187 (1978)	*10.2*, p. *113*; *14.3*, p. *220*; 22.4.2, p. 468
I. K. Vitol	(a) **47**. 667 (1978)	*10.2*, p. *114*; *20.3*, p. *345*; 22.5.2, p. 481
B. Vitt	(b) **87**, K35 (1978)	*10.2*, p. *113*; 11, p. 126; *19*, p. *307*; 22.8.1, p. 513
	(b) **89**, 375 (1978)	*19*, p. *312*; 20.3, p. 348; 22.8.1, p. 515
D. Vivien	(a) **45**, 599 (1978)	*19*, p. *305*; 22.6.1, p. 499
E. Vlachá	(a) **48**, K113 (1978)	*10.1*, p. *105*; 21, p. 362
J. P. Vlahov	(b) **88**, 65 (1978)	*13.5.4*, p. *201*; *18.2*, p. *287*
	(b) **90**, 361 (1978)	*13.5.4*, p. *202*; *18.2*, p. *291*
N. Vlahovici	(b) **85**, 439 (1978)	*13.4*, p. *170*; *20.3*, p. *341*; 22.5.2, p. 478
	(b) **89**, 335 (1978)	*13.4*, p. *170*; *20.3*, p. *341*; 22.5.2, p. 478
Vo Hong Anh	see under letter "A"	
C. M. Vodenicharov	(a) **50**, 411 (1978)	*14.3.3*, p. *238*; *17*, p. *271*; 22.6, p. 497
M. Vodenicharova	(a) **49**, K49 (1978)	*14.3.3*, p. *236*; *14.3.4*, p. *245*; 22.9, p. 527
H. Vogel	(b) **85**, 87 (1978)	*5*, p. *60*; 22.5,p. 474
H. Vogt	(b) **85**, 179 (1978)	*1.2*, p. *15*; 2, p. 42; *6*, p. *64*; 20.1, p. 319; 22.8.1, p. 510
	(b) **86**, 615 (1978)	1.2, p. 17; 6, p. 69; *20.1*, p. *324*; 22.8.1, p. 512
O. Vöhringer	(a) **49**, 473 (1978)	*10.1*, p. *106*; 21.1, p. 378
G. Voigt	(a) **48**, 407 (1978)	*1.4*, p. *32*; 14.3.2, p. 230; 20.3, p. 346; 22.2.2, p. 451

J. Voigt	(b) **87**, K111 (1978)	*13.5.1*, p. *192*; *20.1*, p. *328*; 22.4.1, p. 464
S. V. Voitikov	(b) **88**, K165 (1978)	*13.4*, p. *179*; *14.3*, p. *223*; 22, p. 416
G. Vojta	(b) **85**, K81 (1978)	*14.1*, p. *206*; 18.2, p. 282; 21, p. 354; 21.1, p. 370
	(b) **86**, 177 (1978)	*14.1*, p. *206*; 18.2, p. 283; 21, p. 355
	(b) **89**, 77 (1978)	13, p. 147; *14.1*, p. *209*; 18.4, p. 301; 21, p. 362
T. R. Volk	(a) **45**, 249 (1978)	*14.4.1*, p. *253*; 22.8.1, p. 510
N. V. Volkenshtein	(b) **89**, K53 (1978)	*14.1*, p. *210*; 21, p. 362
V. L. Volkov	(b) **87**, K93 (1978)	*6*, p. *72*; *13.1*, p. *155*; 22, p. 414
H. D. Vollmer	(b) **85**, 301 (1978)	*13.3*, p. *164*; 13.5.2, p. 198; 20.1, p. 320; 21, p. 352; 22.4.1, p. 463
A. Vonlanthen	(b) **90**, 61 (1978)	*1.2*, p. *23*; 7, p. 82; 12.2, p. 141; 22.9, p. 528
S. A. Vorobev	(b) **86**, K35 (1978)	*11*, p. *124*; 13, p. 143; 22.1.2, p. 431
S. A. Vorobiev	(a) **47**, K149 (1978)	10.2, p. 114; *11*, p. *127*; 22.1.2, p. 433
	(a) **48**, K175 (1978)	11, p. 128; *20.3*, p. *346*; 22.5.2, p. 483
W. Maenhout-van der Vorst	see under letter "M"	
P. Vostrý	(a) **48**, K39 (1978)	*10.1*, p. *104*; *14.1*, p. *209*; 21, p. 361
	(a) **48**, K113 (1978)	*10.1*, p. *105*; 21, p. 362
R. Voszka	(b) **89**, 235 (1978)	1.2, p. 22; *14.4*, p. *250*; *20.1*, p. *332*; 22.5.2, p. 484; 22.5.4, p. 489
M. Voudouris	(b) **86**, K119 (1978)	*1.2*, p. *17*; 14.4.1, p. 254; *19*, p. *306*; 22.8.1, p. 512
V. I. Vovnenko	(a) **48**, 593 (1978)	*13.4*, p. *179*; *20.3*, p. *346*; 22.2.1, p. 448
S. Vratislav	(a) **50**, K21 (1978)	*1.1*, p. *14*; 22.8.2, p. 520
	(a) **50**, K131 (1978)	*18.3*, p. *297*; 22.8.1, p. 517
Vu Duy Phach	see under letter "P"	
Z. Vučić	(a) **47**, 703 (1978)	*2*, p. *46*; 22.1.1, p. 426
B. E. Vugmeister	(b) **90**, 711 (1978)	*18.1*, p. *280*; *19*, p. *314*
A. P. Vyatkin	(b) **88**, 321 (1978)	*12.2*, p. *140*; *13.1*, p. *156*; 14.3, p. 222; 22.2.1, p. 447
V. I. Vybornov	(b) **88**, 315 (1978)	*13.5.1*, p. *193*; 22.2.1, p. 447
Yu. M. Vysotskii	(a) **50**, K205 (1978)	5, p. 63; *18.2*, p. *292*; *18.3*, p. *297*; 21, p. 367
C. Wagner	(a) **47**, 131 (1978)	*11*, p. *126*; *14.3*, p. *220*; 22.1.2, p. 432
H. J. Wagner	(b) **89**, 403 (1978)	*13.5.1*, p. *194*; *20.1*, p. *333*; 22.4.1, p. 465
M. Wagner	(b) **85**, 195 (1978)	*6*, p. *64*; *13.4*, p. *169*
	(b) **88**, 517 (1978)	*6*, p. *74*; *10.2*, p. *115*; 22.5.2, p. 482
V. Wagner	(b) **85**, 311 (1978)	*6*, p. *65*; *20.1*, p. *320*; 22.7, p. 501
E. Walentynowicz	(a) **49**, K201 (1978)	*20.3*, p. *349*; 22.4.1, p. 466
Ch. Walkowiak	(a) **47**, K31 (1978)	*14.3.1*, p. *227*; 22, p. 413
K. Walther	(b) **87**, 535 (1978)	*13.4*, p. *176*; 21.4, p. 397
P. Walther	(a) **47**, K35 (1978)	*20.1*, p. *327*; 22.2.1, p. 446
W. Wanarski	(a) **47**, K45 (1978)	*11*, p. *126*; *14.4.1*, p. *255*; 22.9, p. 525
J. Shy-Yih Wang	(b) **86**, 37 (1978)	*13.2*, p. *162*; 21, p. 355
B. M. Wanklyn	(a) **49**, K193 (1978)	*18.4*, p. *301*; *19*, p. *313*; 22.8.2, p. 520
	(b) **90**, K59 (1978)	*19*, p. *314*; *20.1*, p. *336*; 22.8, p. 507
G. H. Wannier	(b) **88**, 757 (1978)	*13*, p. *147*
	(b) **88**, K147 (1978)	*13.1*, p. *158*
W. Wardzyński	(b) **90**, K59 (1978)	*19*, p. *314*; *20.1*, p. *336*; 22.8, p. 507
H. Warlimont	(a) **45**, 117 (1978)	*17.1*, p. *272*; 21, p. 353
Y. Waseda	(b) **87**, 777 (1978)	*2*, p. *45*; *14.1*, p. *208*; 21.1, p. 373
	(a) **49**, 387 (1978)	*2*, p. *49*; 21.1, p. 378
	(a) **49**, K31 (1978)	*2*, p. *49*; 21.6, p. 405
K. Watanabe	(a) **49**, K163 (1978)	*2*, p. *51*; *18.2*, p. *291*; 21.1.1, p. 390
Y. Watanabe	(b) **90**, 697 (1978)	*18.3*, p. *297*; 22.8.2, p. 521
E. Watson	(a) **49**, 593 (1978)	10.2, p. 119; *20.1*, p. *335*; 22.4.3, p. 471

A. Watterich	(b) 88, K51 (1978)	*10.2*, p. *115*; 20.1, p. 330; 22.5.2, p. 482
	(b) 89, K187 (1978)	11, p. 130; *19*, p. *313*; 22.5.2, p. 485
M. Wautelet	(b) 89, 275 (1978)	11, p. 129; *13.1*, p. *158*; 20.2, p. 339; 22, p. 417
	(b) 89, K199 (1978)	*1.2*, p. *23*
J. Wawryszczuk	(a) 47, K141 (1978)	*12.1*, p. *135*; 21.1.1, p. 388
	(b) 89, 253 (1978)	*10.2*, p. *117*; 22.9, p. 526
P. Weber	(a) 46, K67 (1978)	*14.2*, p. *214*; 18.2.1, p. 293; 21.1.1, p. 386
	(a) 49, 569 (1978)	*14.2*, p. *215*
H.-J. Weckner	(b) 90, 309 (1978)	*13.4*, p. *183*; *19*, p. *313*; 22.6, p. 497
R. Wedell	(b) 85, K67 (1978)	*11*, p. *123*
	(b) 87, 491 (1978)	*11*, p. *127*; 13, p. 145; 22.1.2, p. 433
G. Wegner	(b) 88, 653 (1978)	*13.1*, p. *157*; *16*, p. *265*; 22.9, p. 526
G. Weidner	(a) 48, K105 (1978)	*1.4*, p. *32*; 22.1.2, p. 435
G. Weimann	(a) 50, K219 (1978)	14.3.4, p. 247; 22.2.2, p. 451
P. Weinberger	(b) 89, 619 (1978)	2, p. 50; *13*, p. *148*; *20*, p. *317*; 22.6, p. 496
H. Weinert	(b) 89, 561 (1978)	*13.5.1*, p. *195*; *20.3*, p. *348*; 22.2.2, p. 451
G. Weinhausen	(a) 46, 517 (1978)	*14.3.3*, p. *234*; 14.3.4, p. 241; 22, p. 412
G. Weiser	(b) 87, 117 (1978)	*13.1*, p. *154*; *20.1*, p. *326*; 22, p. 413
W. Wendl	(a) 49, 145 (1978)	1.3, p. 27; *3*, p. *55*; 22.8.1, p. 515
M. Wendt	(a) 46, 179 (1978)	*1.3*, p. *25*; 22.8.2, p. 518
S. Wendt	(b) 85, 319 (1978)	*14.4.1*, p. *253*; 19, p. 303; 22.9, p. 522
Z. Werner	(b) 89, 313 (1978)	*13.5.1*, p. *194*; *20.3*, p. *347*; 22.4.4, p. 473
R. Wernhardt	(a) 48, K35 (1978)	*1.2*, p. *21*; *7*, p. *81*; 22.8, p. 506
D. Werneth	(a) 46, 99 (1978)	*10.1*, p. *101*; 12.1, p. 134; 18.2, p. 284; 21.1, p. 371
R. Wernicke	(a) 47, 139 (1978)	10.2, p. 113; *14.3*, p. *220*; 22.8.1, p. 513
R. M. Westervelt	(b) 90, 557 (1978)	*13.5.1*, p. *196*; 22.1.1, p. 428
D. G. Westlake	(a) 47, 631 (1978)	*1.2*, p. *19*; 1.3, p. 26; 21.1, p. 374
H. Wever	(a) 50, 563 (1978)	*9*, p. *95*; 14.1, p. 212; 21.1.1, p. 391
D. A. van Wezep	(a) 49, 651 (1978)	*16*, p. *267*; 22.8.1, p. 516
M. J. Whelan	(a) 45, 277 (1978)	*1.5*, p. *36*; 13.5.2, p. 198; 22, p. 409; 22.1, p. 422
M. A. B. Whitaker	(b) 87, 689 (1978)	6, p. 72; *19*, p. *308*; 22.8, p. 505
C. R. Whitehouse	(a) 47, K173 (1978)	*1.2*, p. *20*; 22, p. 414
C. E. Wickersham	(a) 47, 329 (1978)	13.4, p. 175; *14.3.1*, p. *226*; *20.1*, p. *327*; 22.6, p. 493
A. Wiegeleben	(a) 50, K181 (1978)	*2*, p. *53*; *14.4*, p. *252*; 22.9, p. 529
G. A. Wiegers	(a) 46, 479 (1978)	*1.1*, p. *12*; 1.2, p. 17; 22, p. 412
J. von Wienskowski	(b) 88, K169 (1978)	*13.3*, p. *166*; 22.1.1, p. 427; 22.1.2, p. 435
	(b) 90, K9 (1978)	*13.3*, p. *168*; 22.1.2, p. 438
E. Wieser	(a) 45, 123 (1978)	*1.2*, p. *15*; *5*, p. *60*; 18.1, p. 277; 22.6.1, p. 499
	(a) 45, K105 (1978)	*5*, p. *60*; *18.2*, p. *283*; 21.1, p. 370; 21.1.1, p. 385
	(a) 48, K51 (1978)	*1.2*, p. *21*; 5, p. 62; 21.1.1, p. 388
P. E. Wigen	(a) 46, 209 (1978)	*14.2*, p. *214*; 21, p. 356
M. Wilkens	(a) 47, 479 (1978)	*1.4*, p. *31*; *10*, p. *96*
	(a) 49, 555 (1978)	*1.4*, p. *33*; *10*, p. *97*
D. E. G. Williams	(b) 90, 667 (1978)	*18.4*, p. *302*; 21.1, p. 382
G. P. Williams	(b) 85, 325 (1978)	*13.4*, p. *169*; *17*, p. *269*; 21.1, p. 368
R. H. Williams	(b) 88, K157 (1978)	*13.1*, p. *158*; 17, p. 270; 22, p. 416
J. A. Wilson	(b) 86, 11 (1978)	6, p. 64; 13.1, p. 150; *14.1*, p. *205*; 22, p. 408
E. Wimmer	(b) 89, 619 (1978)	2, p. 50; *13*, p. *148*; *20*, p. *317*; 22.6, p. 496
W. Windsch	(b) 89, 241 (1978)	13.4, p. 180; *14.4.1*, p. *256*; *19*, p. *311*; 22.8.1, p. 515
S. Winter	(b) 85, K95 (1978)	10.2, p. 110; *13.4*, p. *171*; *20.1*, p. *321*; 22.1.1, p. 424
	(b) 90, 289 (1978)	*13.4*, p. *183*; 22.1.1, p. 428

M. C. Wintersgill	(a) **47**, K67 (1978)	13.4, p. 175; *20.3*, p. *344*; 22.5.2, p. 481
K. Winzer	(b) 89, 489 (1978)	*13.2*, p. *163*; 21, p. 363
E. Wiszt	(b) 87, 675 (1978)	*1.2*, p. *19*; 12.2, p. 139; 22.5, p. 475
A. M. Witowski	(b) 87, 637 (1978)	*13.1*, p. *155*; 14.4, p. 249; 20.1, p. 328; 22.4.2, p. 468; 22.4.3, p. 470
	(b) 89, K1 (1978)	*13.1*, p. *158*; 14.4, p. 250; 20.1, p. 332; 22.4.3, p. 471
D. Włosewicz	(b) **48**, 215 (1978)	14.1, p. 209; 21.7, p. 407
D. Wobig	(a) **49**, 59 (1978)	*8*, p. *88*; 22.1.3, p. 442
A. Wojtowicz	(a) **47**, 371 (1978)	*14.3.1*, p. *227*; 22.4.3, p. 470
H. Wolf	(b) 86, K15 (1978)	*1.6*, p. *40*; 21.1, p. 371
H. C. Wolf	(a) **49**, 563 (1978)	1.2, p. 23; 6, p. 77; *19*, p. *313*; 22.9, p. 528
M. Wolf	(a) **48**, 377 (1978)	*1.2*, p. *21*; 6, p. 75; 14.3, p. 223; 22.6, p. 495
R. Wolf	(a) **47**, 105 (1978)	*1.3*, p. *25*; *1.4*, p. *30*; 21.1, p. 373; 21.6, p. 402
	(a) **47**, 241 (1978)	*1.2*, p. *18*; 1.3, p. 25; 1.4, p. 30; 21.1, p. 373; 21.6, p. 402
	(a) **47**, 539 (1978)	*1.2*, p. *19*, 1.3, p. 26; 1.4, p. 31; 21.1, p. 374; 21.6, p. 403
	(a) **48**, 39 (1978)	*1.2*, p. *20*; 1.3, p. 27; 1.4, p. 31; 21.1, p. 375; 21.6, p. 403
A. Wolfenden	(a) **48**, 155 (1978)	*10.2*, p. *115*; 22.5.2, p. 482; 22.5.3, p. 488
J. C. Wolfrat	(b) 89, 85 (1978)	*13.4*, p. *179*; 22.1.2, p. 436
	(b) 89, 541 (1978)	*18.1*, p. *279*
A. Wolkenberg	(a) **48**, 203 (1978)	*14.3.2*, p. *230*; *14.3.3*, p. *235*; 22.1.2, p. 434
	(a) **50**, 249 (1978)	13.3, p. 168; *14.3.2*, p. *232*; *14.3.3*, p. *238*; 22.1.2, p. 438; 22.2.1, p. 449
F. Wondre	(b) 88, K5 (1978)	*6*, p. *73*; *20.1*, p. *329*; 22.8, p. 506
H. Wonn	(b) 87, 129 (1978)	*13.1*, p. *154*; 20, p. 316; 21, p. 357; 21.1.1, p. 386
	(b) 88, K83 (1978)	*13.1*, p. *157*; 21.4, p. 398
J. Woods	(a) **46**, 163 (1978)	*3*, p. *54*; 13.4, p. 173; 16, p. 263; 22.6, p. 492
	(a) **46**, 245 (1978)	*13.4*, p. *173*; *14.3.1*, p. *226*; 22.4.2, p. 468
	(a) **46**, 433 (1978)	*1.6*, p. *41*; 22, p. 412; 22.4.1, p. 464
	(a) **50**, 491 (1978)	*14.3.3*, p. *238*; *20.3*, p. *350*; 22.4.1, p. 467
H. Wörl	(a) **45**, 509 (1978)	*9*, p. *91*; *11*, p. *124*; 14.3, p. 218; 22.1.2, p. 430
J. Wortmann	(a) **46**, 69 (1978)	*13.3*, p. *164*; *17*, p. *269*; 21, p. 356
	(a) **50**, 509 (1978)	*13.3*, p. *168*; *17*, p. *271*; 21, p. 366
Z. Wróbel	(a) **45**, K67 (1978)	*14.4.1*, p. *253*; 22.8.1, p. 511
D. Wruck	(b) **89**, 177 (1978)	*6*, p. *75*; 20.1, p. 332; 22.1.2, p. 436
Tr. H. Wu	(b) 87, K159 (1978)	*19*, p. *308*; 22.4.1, p. 464
D. Wuertz	(b) **86**, 541 (1978)	*2*, p. *43*; *14.3*, p. *219*; 15, p. 260; 22.1.1, p. 425; 22.1.2, p. 432
	(b) 88, K73 (1978)	*2*, p. *46*; *14.3*, p. *222*; 22.1.2, p. 434
H.-J. Wünsche	(b) 85, K53 (1978)	*13.5.1*, p. *189*; 22.1.2, p. 429
	(b) **86**, 505 (1978)	*13.5.1*, p. *190*; 22.1.2, p. 431
T. Wydro	(b) 88, 787 (1978)	*18*, p. *275*
J. W. Wyrzykowski	(a) **45**, 565 (1978)	*1.4*, p. *29*; 10.1, p. 100; 21, p. 354
	(a) **46**, K79 (1978)	*10.1*, p. *102*; 21, p. 357
J. K. Wysocki	(a) **50**, 411 (1978)	*14.3.3*, p. *238*; *17*, p. *271*; 22.6, p. 497
Hoang Xuan Nguyen	see under letter "N"	
Yu. V. Yablokov	(b) 88, 439 (1978)	*13.4*, p. *178*; *19*, p. *310*; 22.8, p. 506
N. R. Yafaev	(a) **50**, 711 (1978)	2, p. 53; *13.4*, p. *186*; *20.1*, p. *338*; 22.8.1, p. 516
J. V. Yakhmi	(a) **49**, K91 (1978)	*1.3*, p. *28*; *5*, p. *62*; 18.2, p. 290; 21.1.1, p. 389
M. Yamada	(a) **48**, K141 (1978)	*14.1*, p. *209*; 21, p. 362

S. YAMADA	(a) **45**,	107 (1978)	2, p. 42; *16*, p. *262*; 22, p. 409
Y. YAMADA	(b) **85**,	723 (1978)	*13.1*, p. *151*; 22.3, p. 457
S. YAMAGUCHI	(a) **46**,	217 (1978)	*1.3*, p. *25*; 6, p. 68; *11*, p. *125*; 21.6, p. 401
R. YAMAMOTO	(a) **45**,	305 (1978)	*2*, p. *42*; 21.1.1, p. 384
	(a) **46**,	671 (1978)	*2*, p. *44*; *12.1*, p. *135*; 21.1.1, p. 386
	(a) **50**,	165 (1978)	*2*, p. *52*; 21, p. 365
N. YAMASHITA	(b) **89**,	663 (1978)	*13.4*, p. *181*; *20.3*, p. *348*; 22.4.1, p. 466; 22.4.2, p. 469
F. H. YANG	(b) **89**,	267 (1978)	13.4, p. 180; *19*, p. *311*; 22.8.2, p. 520
K. H. YANG	(a) **48**,	335 (1978)	*1.5*, p. *37*; 3, p. 55; 22.3, p. 458
	(a) **50**,	221 (1978)	*10.2*, p. *120*; *14.3.3*, p. *238*; 22.1.2, p. 438
B. YANGUI	(a) **45**,	639 (1978)	*1.1*, p. *11*; 1.4, p. 29; 22.6, p. 491
	(a) **47**,	311 (1978)	*1.1*, p. *12*; 1.4, p. 30; 22.6, p. 493
V. M. YANKOVSKII	(a) **45**,	343 (1978)	*1.2*, p. *15*; *1.5*, p. *36*; 11, p. 123; 21.1, p. 369
Y. D. YAO	(b) **89**,	K201 (1978)	*14.1*, p. *211*; 18.2, p. 290; 21.1, p. 378
V. K. YARMARKIN	(a) **45**,	63 (1978)	*5*, p. *60*; 9, p. 91; 10.2, p. 109; 22.6, p. 491
L. YARMUS	(b) **89**,	K167 (1978)	*13.5.1*, p. *195*; *19*, p. *313*; 22, p. 418
YU. N. YASHKIR	(b) **89**,	27 (1978)	*13.5*, p. *188*; 20.1, p. 331; 22.6, p. 496
B. M. YAVADOV	see under "B. M. DSHAVADOV"		
	(b) **89**,	K43 (1978)	*6*, p. *75*; *20.1*, p. *332*; 22.3, p. 458
I. YOKOTA	(a) **47**,	657 (1978)	*7*, p. *81*; 22.4.1, p. 464
K. YONEMITSU	(b) **88**,	273 (1978)	13.2, p. 162; *14.1*, p. *209*; 21, p. 360
	(b) **89**,	571 (1978)	*14.1*, p. *211*; 21, p. 363
I. YONENAGA	(a) **50**,	685 (1978)	*10.2*, p. *122*; 22.1.2, p. 439
M. YOSHIMI	(b) **88**,	607 (1978)	13.1, p. 157; *13.5.1*, p. *193*; *20.1*, p. *330*; 22, p. 415
O. YOSHINARI	(a) **46**,	217 (1978)	*1.3*, p. *25*; 6, p. 68; *11*, p. *125*; 21.6, p. 401
R. T. YOUNG	(a) **48**,	K31 (1978)	*11*, p. *128*; *13.4*, p. *178*; 22.1.2, p. 435
D. M. YUDIN	(b) **85**,	343 (1978)	2, p. 42; *13.4*, p. *169*; *19*, p. *303*; 22.8.1, p. 510
A. E. YUNOVICH	(b) **86**,	K21 (1978)	*20.1*, p. *323*; 22, p. 410
	(b) **88**,	675 (1978)	13.4, p. 178; 20.1, p. 330; *20.2*, p. *339*; 22, p. 415
N. B. YUNUSOV	(b) **88**,	87 (1978)	*13.4*, p. *177*; 19, p. 309; 22.5.3, p. 488
S. P. YURCHENKO	(a) **50**,	K1 (1978)	*1.5*, p. *39*; 21, p. 365; 21.6, p. 405
V. I. YUZVAK	(a) **45**,	53 (1978)	1.2, p. 15; *14.4.1*, p. *253*; 19, p. 303; 20, p. 315; 22.8.1, p. 510
Z. ZAGÓRSKI	(a) **47**,	K45 (1978)	*11*, p. *126*; *14.4.1*, p. *255*; 22.9, p. 525
W. ZAHN	(a) **49**,	573 (1978)	*14.2*, p. 215
B. N. ZAITSEV	(a) **50**,	573 (1978)	*1.2*, p. *24*; *1.5*, p. *39*; 22, p. 421
V. I. ZAITSEV	(a) **50**	K33 (1978)	*10.1*, p. *107*; *12.1*, p. *136*; 21.1, p. 381
J. ZAJÍČEK	(a) **50**,	K131 (1978)	*18.3*, p. *297*; 22.8.1, p. 517
N. A. ZAKHAROV	(a) **50**,	K13 (1978)	3, p. 56; *14.4.1*, p. *257*; *20*, p. *318*; 22.8, p. 508
M. I. ZAKHAROVA	(a) **49**,	803 (1978)	*1.2*, p. *23*; 1.3, p. 28; 21.1, p. 379
J. R. ZAKIS	(a) **49**,	231 (1978)	2, p. 49; *13.4*, p. *181*; 19, p. 312; *20.3*, p. *347*; 22, p. 417
T. ZAKRZEWSKI	(a) **47**,	K137 (1978)	*8*, p. *87*; *10.1*, p. *104*; 21, p. 360; 21.1, p. 375
	(a) **50**,	K261 (1978)	*8*, p. *87*; *10.1*, p. *104*; 21, p. 360; 21.1, p. 375
V. ŽALUDEK	(a) **49**,	K157 (1978)	*2*, p. *51*; 13.3, p. 167; *14.3.4*, p. *245*
N. A. ZAMYATINA	(a) **50**,	K153 (1978)	*18.2.1*, p. *294*; *20*, p. *318*; 21.1, p. 383; 21.1.1, p. 391
M. ZANINI	(b) **90**,	151 (1978)	20.1, p. 335; 22.1, p. 423
A. ZARĘBA	(a) **45**,	K183 (1978)	*14.3*, p. *218*; 22.1, p. 422
E. V. ZAROCHENTSEV	(b) **85**,	63 (1978)	*6*, p. *64*; *8*, p. *83*; 21.2, p. 392
	(b) **85**,	749 (1978)	*6*, p. *66*; *8*, p. *83*; 21.2, p. 392
W. ZAWADZKI	(b) **88**,	247 (1978)	6, p. 73; 13.1, p. 156; *14.3*, p. *222*; 22.2.3, p. 453

S. Zawisz	(b) **86**, 367 (1978)	*13*, p. *143*
S. Zazubovich	(b) **86**, 733 (1978)	*13.4*, p. *174*; *20.1*, p. *324*; 22.5.2, p. 480
Z. Zbǫiński	(a) **46**, 251 (1978)	*13.3*, p. *165*; 14.3, p. 219; 16, p. 263; 22.9, p. 524
	(b) **87**, 663 (1978)	*13.5.1*, p. *192*; 20.3, p. 344; 22.9, p. 525
L. Zecchina	(a) **50**, K235 (1978)	*8*, p. *90*; 22.3.1, p. 460
U. Zeimer	(a) **49**, K97 (1978)	*14.1*, p. *210*; 21, p. 363
A. Ch. Zeinally	(a) **50**, K141 (1978)	6, p. 79; 13.1, p. 161; *20.1*, p. *338*; 22.8, p. 508
S. Zeković	(b) **89**, K57 (1978)	*13.1*, p. *158*; 21, p. 362
V. P. Zentsov	(b) **88**, 87 (1978)	*13.4*, p. *177*; 19, p. 309; 22.5.3, p. 488
R. R. Zeyfang	(a) **46**, 511 (1978)	*14.4.1*, p. *254*; 22.8, p. 504
M. Z. Zgierski	(b) **87**, 653 (1978)	*6*, p. *72*; 13.5.1, p. 192; 22.9, p. 525
E. V. Zharikov	(a) **49**, 305 (1978)	3, p. 56; 20.1, p. 333; *20.2*, p. *339*; *20.3*, p. *347*; 22.8.2, p. 520
R. A. Zhitnikov	(b) **86**, K67 (1978)	*19*, p. *305*; *20.1*, p. *323*; 22.5.2, p. 479
F. A. Zhivopistsev	(b) **90**, 441 (1978)	*13*, p. *149*; 13.5.2, p. 199; 21.2, p. 393
N. N. Zhukov	(a) **50**, 385 (1978)	1.4, p. 34; *10.1*, p. *108*; 21, p. 366
E. Ziegler	(a) **48**, K63 (1978)	13.4, p. 178; *14.3*, p. *222*; 22.8, p. 506
	(a) **49**, 705 (1978)	13.3, p. 167; *14.3.4*, p. *245*; 22.8, p. 507
	(a) **49**, K205 (1978)	13.4, p. 183; *14.3*, p. *224*; 22.8, p. 507
O. Ziep	(b) **88**, 221 (1978)	*13.1*, p. *156*; *14.3.4*, p. *243*; 14.4, p. 249; 22.4.4, p. 473
	(b) **90**, 197 (1978)	*13.1*, p. *159*; 22, p. 419
	(b) **90**, 535 (1978)	*13.1*, p. *160*; *20.1*, p. *337*; 22, p. 420
P. Ziesche	(b) **87**, 129 (1978)	*13.1*, p. *154*; 20, p. 316; 21, p. 357; 21.1.1, p. 386
	(b) **88**, K83 (1978)	*13.1*, p. *157*; 21.4, p. 398
	(b) **89**, K175 (1978)	*13.1*, p. *159*; 21, p. 363
R. Zimmermann	(b) **86**, K63 (1978)	*13.5*, p. *187*; *20.1*, p. *323*; 22.2.1, p. 445
	(b) **88**, K59 (1978)	*13.5*, p. *187*; *13.5.1*, p. *193*; 22, p. 415
	(b) **90**, 175 (1978)	*13.5*, p. *188*; *13.5.1*, p. *196*; 22, p. 419
G. M. Zinenkova	(a) **46**, 327 (1978)	*10.2*, p. *111*; 22.5.2, p. 479
	(a) **46**, 411 (1978)	*10.1*, p. *101*; *12.1*, p. *135*; 21, p. 357; 21.1, p. 372; 21.3, p. 394
W. Zingg	(a) **45**, 117 (1978)	*17.1*, p. *272*; 21, p. 353
V. E. Zinovev	(a) **49**, K117 (1978)	*8*, p. *88*; *14.1*, p. *211*; 21, p. 364
V. E. Zirap	(a) **47**, 667 (1978)	*10.2*, p. *114*; *20.3*, p. *345*; 22.5.2, p. 481
Z. Żołnierek	(a) **45**, K39 (1978)	*18.2*, p. *282*; 22.8, p. 503
I. Zschokke-Gränacher	(b) **85**, 505 (1978)	*20.3*, p. *342*; 22.9, p. 523
R. Zossel	(b) **88**, 727 (1978)	*18.2*, p. *289*
V. I. Zubov	(b) **87**, 385 (1978)	*6*, p. *70*; *8*, p. *85*; 23, p. 530
	(b) **88**, 43 (1978)	*6*, p. *73*; *8*, p. *87*; 23, p. 530
W. Zurek	(b) **89**, 203 (1978)	13.1, p. 158; 13.5.1, p. 194; *18*, p. *275*; *20*, p. *317*; 22.1.1, p. 427
R. A. Zvinchuk	(a) **48**, 543 (1978)	*1.2*, p. *21*; 10.2, p. 116; 22.8.2, p. 519
A. D. Zvonkov	(b) **86**, 83 (1978)	4, p. 57; *13.1*, p. *152*; 21.1, p. 371
S. D. Zvonkov	(b) **88**, K103 (1978)	*13.4*, p. *179*; 21.1.1, p. 388
I. P. Zvyagin	(b) **85**, 105 (1978)	13.4, p. 169; *14.3*, p. *217*; 22, p. 408
	(b) **88**, 149 (1978)	*2*, p. *46*; 6, p. 73; *14.3*, p. *222*
M. Zwick	(a) **47**, 445 (1978)	*1.2*, p. *19*; 18.2, p. 287; 21.1, p. 374
A. Zygmunt	(a) **46**, K75 (1978)	*18.2*, p. *284*; *18.4*, p. *299*; 21, p. 356; 21.4 p. 397
	(a) **47**, 349 (1978)	*18.2*, p. *287*; *18.4*, p. *300*; 22.8, p. 505
A. N. Zyuganov	(a) **45**, 631 (1978)	13.4, p. 171; *14.3.4*, p. *240*; 22, p. 410

19. Magnetic Resonance

20. Optical Properties
    20.1 Spectra. Optical Constants (X-Ray Spectra See 20)
    20.2 Lasers
    20.3 Luminescence (X-Ray Spectra See 20)

21. Metals. Alloys
    21.1 Transition Metals (Group Sc to Zn) and Their Alloys
        21.1.1 Fe and Fe Alloys
    21.2 Alkali Metals
    21.3 Alkali-Earth Metals
    21.4 Rare-Earth Metals
    21.5 Actinides
    21.6 Noble Metals
    21.7 Semi-Metals (Semiconducting Alloys See 22.7)

22. Semiconductors and Ionic Crystals
    22.1 Elements
        22.1.1 Germanium
        22.1.2 Silicon
        22.1.3 Group VI Elements
    22.2 III–V Compounds
        22.2.1 Arsenides
        22.2.2 Phosphides
        22.2.3 Antimonides
        22.2.4 Mixed Crystals
    22.3 IV–IV, II–V, and III–VI Compounds
        22.3.1 Mixed Crystals
    22.4 II–VI Compounds
        22.4.1 Sulphides
        22.4.2 Selenides
        22.4.3 Tellurides
        22.4.4 Mixed Crystals
    22.5 Halides
        22.5.1 Silver Halides
        22.5.2 Alkali Halides
        22.5.3 Alkali-Earth Halides
        22.5.4 Mixed Crystals
    22.6 Simple Oxides (CdO, ZnO See 22.4)
        22.6.1 Mixed Crystals
    22.7 Semiconducting Intermetallic Compounds
    22.8 Tertiary and Higher Compounds
        22.8.1 Salts of Oxy-Acids (Carbonates, Phosphates, Silicates Including Mica etc.)
        22.8.2 Spinels, Garnets, Ferrites
    22.9 Organic Semiconductors

23. Solidified Gases

For comments on the use of the classification scheme see p. 6.

32 913/32 914

Manuscripts and letters for physica status solidi (b) — basic research and physica status solidi (a) — applied research

should be sent to

DDR-102 Berlin, Neue Schönhauser Str. 20

or

Internationaler Buchversand GmbH, 1 Berlin 19, Eschenallee 21.

The object of "physica status solidi (b)" and "physica status solidi (a)" is to publish rapidly important and topical results concerning fundamental research in the field of solid state physics and aspects of application and experimental techniques of fundamental solid state research, respectively. The issues of "physica status solidi (b)" and "physica status solidi (a)" will be published on the 1st and the 16th day of each month, respectively. Either series will contain the following categories:

1. Review articles written by arrangement with the editors. They will be remunerated.
2. Original papers selected from those which have not been submitted for publication elsewhere, after being reviewed. (The latest date for acceptance of contributions is the 8th and the 23rd day of each month for "physica status solidi (b)" and "physica status solidi (a)", respectively, 50 days before the delivery of the journal.)
3. Short notes published in the off-set part of the journal. They are, if necessary after being reviewed, selected by the editorial staff. (The latest date for acceptance of manuscripts for this part is the 12th and the 27th day of each month for "physica status solidi (b)" and "physica status solidi (a)", respectively, 16 days before the delivery of the journal.)
4. Pre-printed titles of papers to be published in the next issue of both series.

Editorial Information

Manuscripts should be submitted preferentially in English. Besides, papers in German, French, and Russian are admitted. The final decision on the incorporation of the papers in the series (b) or (a) is taken by the editorial staff after consultation of the referee. Authors are asked to observe the following rules:

1. Manuscripts have to be written as concisely as possible. As most, 20 and 4 pages (28 lines per page; abstracts, tables, references, figures, and captions of figures included, where 3 figures are counted as about one page) are admitted for original papers and short notes, respectively. All manuscripts should be type-written on one side only. Two copies are required. Badly arranged manuscripts are returned.
2. The author's name, initials, and institution should be given.
3. One abstract of about ten lines in English and one in German, French, or Russian are required for original papers.
4. Because of lack of space conclusions (if any) must not have the character of a summary (i. e. substantially a repetition of the contents of the abstract).
5. Formulae should not be written too small and with pencil. Separate lines for formulae are desirable. Indistinct writing will lead to a return of the manuscript.
6. Drawings should be enclosed on separate sheets of paper, not exceeding the maximum and minimum sizes of 20×30 and $10 \times 15\ cm^2$, respectively. Photographs should be submitted, on white paper, only for review articles and original papers. Duplicates are required.
7. Captions for the figures should be submitted on separate sheets of paper, and special instructions with regard to the size of illustrations should be given on these.
8. References should be numbered in order of appearance and should be given, on a separate sheet, in the form used in physica status solidi.
9. SI-units should be used.
10. Each paper should contain a proposal for the subject classification according to the classification scheme appearing on the second and third cover of each issue.

The authors will be sent galley proofs of review articles and original papers. Proof reading of short notes and pre-printed titles will be done by the editorial staff.
For each original paper and short note a total of 75 reprints is available free of charge. The number of reprints free of charge per review article is 25. Requests for additional reprints (against payment) should be given together with the submitted manuscript. **The deadline for orders of additional reprints is the 15th and the 30th day of each month** for "physica status solidi (b)" and "physica status solidi (a)", respectively, two weeks before the delivery of the journal.

Bei Fragen zur Produktsicherheit wenden Sie sich bitte an:
If you have any questions regarding product safety,
please contact:

Walter de Gruyter GmbH
Genthiner Straße 13
10785 Berlin
productsafety@degruyterbrill.com